NEBOSH Health and Safety Management for Construction (UK)

Technical Member (TechIOSH)

This category recognises the qualifications and competence needed by professionals working in a range of operational health and safety roles. To become a Technical member, you must have a suitable level of experience in a health and safety role, as well as an IOSH accredited qualification.

Learners on a course of study towards Tech IOSH will find this **Essential Health and Safety Guide** particularly useful in developing the knowledge requirements for an IOSH accredited qualification at this level.

iosh

RMS Publishing

Suite 3, Victoria House, Lower High Street, Stourbridge DY8 1TA

© RMS Publishing ™

Second Edition October 2007.

Third Edition January 2012.

Fourth Edition May 2015.

Fifth Edition April 2020.

Fifth Edition Reprint September 2022.

Any resemblance to NEBOSH examination questions is coincidental; the questions have been created to illustrate the type of question which may be asked at examination to check understanding of the relevant element learning outcomes.

Printed and bound in Great Britain by CPI Antony Rowe.

ISBN-13: 978-1-906674-96-0

CONTENTS

PREFACE

Publication users

This Essential Health and Safety Guide to the NEBOSH Health and Safety Management for Construction qualification provides an excellent reference for those who have responsibility for or involvement in construction activities. The guide provides an understanding of the health and safety risks affecting this important sector of work and assists those responsible to manage them effectively.

The practical and informative approach of the publication will provide managers, health and safety practitioners and others who work in the construction industry with a perspective of the significance of health and safety risk in construction activities and an understanding of the control measures that can be applied to manage risk. This publication would be particularly useful for clients, designers, principal contractors, principal designers, contractors and site managers. Those working in facilities and event management industries will also find the guide useful, as many of the risks they encounter are similar.

Successful completion of the NEBOSH NC qualification will enable applicants to apply for first level membership of many national professional bodies related to health and safety, for example, Associate Membership (AIOSH) of the Institution of Occupational Safety and Health (IOSH). The qualification also meets the academic requirements for Technical Membership (Tech IOSH) of IOSH.

Scope and contents

Scope

Topics covered by the guide include:

- Construction law and management.
- Improving health and safety culture and assessing risk.
- Managing change and procedures.
- Identification of workplace hazards and methods of risk control.

Reference is made to the harm that construction hazards may have and the prevalence of injuries they cause in the construction industry. The publication contains an emphasis on practical solutions to construction workplace health and safety issues.

Full colour photographs, tables and sample documents are provided to enable an understanding of how these risks can be managed.

Photographs and schematics

We have taken particular care to support the text with a significant number of full colour photographs and schematics. They are illustrative of both good and bad working practices and should always be considered in context with supporting text. Readers will find this a useful aid when trying to relate their background and experience to the broad based construction activities that may be encountered.

Where photographs, diagrams and text extracts are known to be drawn from other publications, a clear source reference is shown and RMS wish to emphasise that reproduction of such items within the Essential Health and Safety Guide is for educational purposes only and the original copyright has not been infringed.

All statistics shown throughout this publication are the latest available at time of going to press.

Supporting study features

The authors have taken particular care to reinforce learning and understanding, following many years' experience in producing such publications. Where relevant within each Element of study there are sections included entitled: Review; Consider and Case Study.

Review

After a review of a particular Element take an opportunity to answer a question to check your understanding.

Consider

Tasks to try or think about, at times relating to your own workplace or job role, in support of your learning process.

Case study

Applicable case studies to reflect on in context with your learning.

The additional sections at the back of the publication are designed to help the learner through study tips, revision and exam techniques and study question practice together with open book examination (OBE) guidance.

RMS
PUBLISHING

Legal requirements

This study book has at its heart the fact that health and safety should be managed as a risk. However, as one of the risks to a business is the risk of prosecution, care has been taken to relate topics to current legislation. Legislation is referred to in context in the various elements that comprise this study book, it reflects the syllabus of the NEBOSH Health and Safety Management for Construction and is targeted to meet the needs of learners studying to this level. In addition, this study book has a section 'Relevant Statutory Provisions' which is dedicated to legislation, and here the learner will find a useful summary of the legislation relevant for this qualification.

Assessment

The NEBOSH Health and Safety Management for Construction qualification is assessed by a single unit 'NC1: Managing construction safely' examination and a 'Pass' is required in this unit to be awarded the qualification.

The assessment is a 48-hour open book examination (OBE), which will be based on a realistic scenario, that will test both what you know and what you can do. You will be required to carry out a series of tasks using evidence from the scenario and the underpinning knowledge gained through your studies and revision.

In order that users may check their understanding of the core learning and assessment criteria for this qualification, a number of study questions have been included at the end of each element.

Revision and examination guidance

There are ways of making study time more effective to give a learner the greatest possible chance of success in their exam. This section is included to aid learners with useful study tips, time management and revision techniques.

Syllabus

This 5th Edition Essential Health and Safety Guide has been thoroughly revised and updated to embrace the December 2019 specification of the syllabus for the NEBOSH Health and Safety Management for Construction qualification and focuses on construction hazards and related losses, risk control measures and core management of health and safety principles as they relate to construction activities.

It has been designed to reflect the order and content of the NC syllabus which is structured in a very useful way, focusing on core management of health and safety principles (Elements 1 to 3) then hazards and their control (Elements 4 to 13); and in this way, the learner studying for this qualification can be confident that this guide reflects the themes of the syllabus and forms an excellent study book for that purpose.

Sub-elements	Learning outcome The learner will be able to:	Assessment criteria
1.1	Justify health and safety improvements using moral, financial and legal arguments	1.1 Discuss the moral, financial and legal reasons for managing health and safety in the workplace
1.2-1.7	Advise on the main roles, competencies and duties under construction legislation and on a range of general construction site issues.	1.2 Summarise the main health and safety duties under the Construction (Design and Management) Regulations 2015 and how contractors should be selected, monitored and managed. 1.3 Summarise the types of construction work and range of activities 1.4 Identify what to consider during a construction site assessment 1.5 Identify how to keep a site secure and in good order 1.6 Summarise how to manage temporary construction works 1.7 Outline what welfare arrangements should be on site and other particular construction issues
2.1-2.3	Positively influence health and safety culture and behaviour to improve performance in their organisation	2.1 Describe the concept of health and safety culture and how it influences performance 2.2 Summarise how health and safety culture at work can be improved 2.3 Summarise the human factors which positively or negatively influence behaviour at work in a way that can affect health and safety
3.1	Recognise workplace changes that have significant health and safety impacts and effective ways to minimise those impacts	3.1 Discuss typical workplace changes that have significant health and safety impacts and ways to minimise those impacts
3.2-3.4	Develop basic safe systems of work (including taking account of typical emergencies) and knowing when to use permit-to-work systems for special risks	3.2 Describe what to consider when developing and implementing a safe system of work for general activities 3.3 Explain the role, function and operation of a permit-to-work system 3.4 Discuss typical emergency procedures (including training and testing) and how to decide what level of first aid is needed in the workplace
3.5	Take part in incident investigations	3.5 Explain why and how incidents should be investigated, recorded and reported
2.4, 4-13	Assess risks, recognising a range of common hazards, evaluating risks (taking account of current controls), recommending further control measures, and planning actions	2.4 Explain the principles of the risk assessment process and 4-13 Produce a risk assessment that considers a wide range of identified construction hazards (drawn from elements 4-13)

Production of the publication

Managing Editor

Ian Coombes, Managing Director ACT, CMIOSH; former member of NEBOSH Council, former member of NEBOSH Board of Trustees, former member of NEBOSH Qualifications and Technical Council and former NEBOSH examiner. Former member of IOSH Professional Affairs Committee and chairman of the Initial Professional Development sub-committee. Member of the Safety Groups UK (SGUK) management committee.

Acknowledgements

RMS Publishing Ltd wishes to acknowledge the following contributors and thank them for their assistance in the preparation of the publication:

Kevin Coley, CMIOSH; a NEBOSH examiner with many years' health and safety experience in the private and public sector. Kevin has been a production manager in a large foundry, so has a heavy engineering background as well as a senior safety manager for large government bodies.

Robert Weeks, Dip NEBOSH, Cert Ed, Grad IOSH, MIIRSM; an experienced health and safety consultant in a range of sectors spanning more than 20 years. These include power generation; nuclear, thermal and wind turbines, construction, aerospace, petro-chemical including HAZOP implementation, water utility and transport industries. A NEBOSH examiner for the National Diploma Unit A and various Certificate Units.

Geoff Littley CMIOSH, NEBOSH Diploma, CSPA; experienced health and safety professional. Areas of work include manufacturing, NHS Trusts, Local Authorities, construction and power generation in the UK and Africa. A lead tutor for NEBOSH Diploma and General and Construction Certificate courses. Also an experienced examiner for NEBOSH Diploma Units A and C and both General and Construction Certificate courses.

Stephen Whitehouse; an experienced health and safety professional with more than 20 years' experience working with the energy sector.

Brian Christie CMIOSH, MIIRSM, M I Chem E; Experienced Health and Safety Advisor with over 20 years' experience in that role, including as a tutor for the NEBOSH General Certificate. Brian has been a senior manager in the Plastics and Petrochemicals industry with roles encompassing process engineering; maintenance; research and development and manufacturing on plant and equipment using and producing highly flammable, toxic and corrosive substances.

Julie Skett, Support and Development Manager. Rhiannon Davies, Robert Sannwald-Dunn and Rosie Hierons, layout and formatting.

![RMS PUBLISHING]

PUBLICATIONS ALSO AVAILABLE FROM RMS:

Publication	Edition	ISBN
A Study Book for the NEBOSH National General Certificate in Occupational Health and Safety	Eleventh	978-1-906674-93-9
A Study Book for the NEBOSH International General Certificate in Occupational Health and Safety	Sixth	978 1-906674-88-5
A Study Book for the NEBOSH Certificate in Fire Safety and Risk Management	Seventh	978-1-913439-03-3
The Management of Environmental Risks in the Workplace	Fifth	978-1-913439-18-7
A Guide to International Oil and Gas Operational Safety	Second	978-1-906674-65-6
Managing Safety the Systems Way - Implementing ISO 45001	First	978-1-906674-91-5
Study Books for the NEBOSH National Diploma in Occupational Safety and Health:		
Unit ND1: Know - Workplace health and safety principles (UK)	Seventh	978-1-913439-06-4
Unit ND2: Do - Controlling workplace health issues (UK)	Seventh	978-1-913439-07-1
Unit ND3: Do - Controlling workplace safety issues (UK)	Seventh	978-1-913439-08-8
Study Books for the NEBOSH International Diploma in Occupational Safety and Health:		
Unit ID1: Know - Workplace health and safety principles (international)	Fifth	978-1-913439-12-5
Unit ID2: Do - Controlling workplace health issues (international)	Fifth	978-1-913439-13-2
Unit ID3: Do - Controlling workplace safety issues (international)	Fifth	978-1-913439-14-9

LIST OF ABBREVIATIONS

Legislation

CAOR	Control of Artificial Optical Radiation at Work Regulations 2010
CAR	Control of Asbestos Regulations 2012
CDMR	Construction (Design and Management) Regulation 2015
CEMFAW	Control of Electromagnetic Fields at Work Regulations 2016
CER	Control of Explosives Regulations 1991
CLAW	Control of Lead at Work Regulations 2002
CLP	Classification, Labelling and Packaging of Substances and Mixtures Regulations 2009
CNWR	Control of Noise at Work Regulations 2005
COSHH	Control of Substances Hazardous to Health (Amendment) Regulations 2002
CSR	Confined Spaces Regulations 1997
CVWR	Control of Vibration at Work Regulations 2005
DSE	Health and Safety (Display Screen Equipment) Regulations 1992
DSEAR	Dangerous Substances and Explosive Atmospheres Regulations 2002
EESR	Electrical Equipment (Safety) Regulations 2016
EPA	Environmental Protection Act 1990
EPSIUPEA	Equipment and Protective Systems Intended for Use in Potentially Explosive Atmospheres Regulations 2016
ESQCR	Electricity Safety Quality and Continuity Regulations 2002
EWR	Electricity at Work Regulations 1989
FAR	Health and Safety (First-Aid) Regulations 1981
FSA	Fire (Scotland) Act 2005
FSECE	Fire Safety (England) Regulations 2010
FSECW	Fire Safety (Employees' Capabilities) (Wales) Regulations 2012
FSSR	Fire Safety (Scotland) Regulations 2006
HASAWA	Health and Safety at Work etc. Act 1974
HSCER	Health and Safety (Consultation with Employees) Regulations 1996
HSIER	Health and Safety Information for Employees Regulations 1989
IRR	Ionising Radiation Regulations 2017
LOLER	Lifting Operations and Lifting Equipment Regulations 1998
MAR	Health and Safety (Miscellaneous Amendments) Regulations 2002
MHOR	Manual Handling Operations Regulations 1992
MHSWR	Management of Health and Safety at Work Regulations 1999
MRRA	The Health and Safety (Miscellaneous Repeals, Revocations and Amendments) Regulations 2013
NRSWA	New Roads and Street Works Act 1991
PPER	Personal Protective Equipment Regulations 1992
PUWER	Provision and Use of Work Equipment Regulations 1998
REACH	Registration, Evaluation, Authorisation and Restriction of Chemicals EC Regulation No 1907/2006
RIDDOR	Reporting of Injuries, Diseases and Dangerous Occurrences Regulations 2013
RRFSO	Regulatory Reform (Fire Safety) Order 2005
RTA	Road Traffic Act 1988, 1991
SMSR	Supply of Machinery (Safety) Regulations 2008
SRSCR	Safety Representatives and Safety Committee Regulations 1977
SSCPR	Social Security (Claims and Payment) Regulations 1979
SSSR	Health and Safety (Safety Signs and Signals) Regulations 1996
WAH	Work at Height Regulations 2005
WHSWR	Workplace (Health, Safety and Welfare) Regulations 1992

General

2HC	Two-hand Controls
3D	Three dimensional
AA	Automobile Association
ABS	Anti-lock Braking System
AC	Alternating Current
ACAS	Advisory, Conciliation and Arbitration Service
ACMs	Asbestos-Containing Material
ACOP	Approved Code of Practice
ACSNI	Advisory Committee on the safety of Nuclear Installations
AETR	European Agreement concerning the work of Crews of Vehicles Engaged in International Road Transport
AIB	Asbestos Insulating Boards
AIDS	Acquired Immune Deficiency Syndrome
APF	Assigned Protection Factor
APS	Association of Project Supervisors
BBVs	Blood Borne Viruses
BIM	Building Information Modelling
BMGVs	Biological Monitoring Guidance Values
BSI	British Standards Institute
CAD	Computer Aided Design programmes
CAT	Cable Avoidance Tool
CBI	Confederation of British Industry
CBM	Condition-based Maintenance
CBT	Cognitive Behavioural Therapy
CCDO	Scheme for the Certification of Competence of Demolition Operatives
CCTV	Closed Circuit Television
CE	Conformité Européene
CECA	Civil Engineering Contractors Association
CEN	European Standards
CIOB	Chartered Institute of Building
CIRA	Construction Industry Research and Information Association
CITB	Construction Industry Training Board
CLP	Classification, labelling and packaging of chemical substances and mixtures
CO	Carbon Monoxide
CO_2	Carbon Dioxide
CONIAC	Construction Industry Advisory Committee
COPD	Chronic Obstructive Pulmonary Disease
CPR	Cardio Pulmonary Resuscitation
CTS	Carpal Tunnel Syndrome
dB	Decibel
DC	Direct Current
DNA	Deoxyribonucleic Acid
DoC	Declaration of Conformity
DSD	Dangerous Substances Directive
DSE	Display Screen Equipment
DVLA	Driver and Vehicle Licensing Agency
EC	European Commission
ECITB	Engineering Construction Industry Training Board
EMFs	Electromagnetic fields
EFAW	Emergency First-Aid At Work
ESA	European Safety Agency
ETA	Event Tree Analysis
EU	European Union

FAW	First Aid at Work
FLT	Fork-lift Truck
FMEA	Failure Mode and Effect Analysis
FRA	Fire Risk Assessment
FTA	Fault Tree Analysis
GAD	Generalised Anxiety Disorder
GB	Great Britain
GCS	Ground Control System
GDP	Gross Domestic Product
GHS	Globally Harmonised System of Classification and Labelling of Chemicals
H_2S	Hydrogen Sulphide
HAVS	Hand Arm Vibration Syndrome
HAZOP	Hazard and Operability Studies
HEPA	High Efficiency Particulate Air
HGV	Heavy Goods Vehicle
HIV	Human Immunodeficiency Virus
HML	High, Medium and Low Values
HSE	Health and Safety Executive
HSIER	Health and Safety Information for Employees Regulations (HSIER) 1989
HV	High Voltage
Hz	Hertz
IAEA	International Atomic Energy Agency
ICAO	International Civil Aviation Organisation
ICE	Institute of Civil Engineers
IEE	Institute of Electrical Engineers
IET	Institute of Engineering and Technology
IIRSM	International Institute for Risk and Safety Management
ILO	International Labour Organisation
IOSH	Institute of Occupational Safety and Health
IP	International Protection
IQ	Intelligence Quotient
ISO	International Organisation for Standardisation
IT	Information Technology
Kg	Kilograms
Km	Kilometres
kN	Kilonewton
kV	Kilovolts
LEV	Local Exhaust Ventilation
LFS	Labour Force Survey
LGV	Large Goods Vehicle
LOTO	'Lock Out' and 'Tag Out'
LPG	Liquefied Petroleum Gas
LTEL	Long-term Exposure Limit
LV	Low voltage
M	Meters
mA	MilliAmpere
MAC	Maximum allowable/acceptable concentration
MCWP	Mast Climbing Work Platform
MDI	Methylene Bisphenyl Di-isocyanate
MDMA	3,4-methylenedioxy-N-methylamphetamine
MEWP	Mobile Elevated Work Platform
MOT	Ministry of Transport test
MpH	Miles per hour
MRI	Magnetic resonance imaging
MS	Milliseconds

MSD	Musculoskeletal Disorder
MSF	Manufacturing, Science and Finance Union
N_2	Nitrogen
NASC	National Access and Scaffolding Confederation
NEBOSH	National Examination Board in Occupational Safety and Health
NEC	National Exhibition Centre
NHS	National Health Service
NICE	National Institute for Health and Care Excellence
NIHL	Noise-induced Hearing Loss
NNLW	Notifiable Non-Licensed Work
NVQ	National Vocational Qualification
OCD	Obsessive Compulsive Disorder
ORR	Office of Rail and Road
ORSA	Occupational Road Safety Alliance
Pa	Pascal
PASMA	Prefabricated Access Suppliers and Manufactures Association
PAT	Portable Appliance Testing
PCV	Passenger Carrying Vehicle
PTS	Permanent Threshold Shift
PTSD	Post-traumatic Stress Disorder
PHE	Public Health England
PNA	Predicted Noise Attenuation
PPR	Parts Per Billion
PPE	Personal Protective Equipment
PPM	Planned preventive maintenance
PTW	Permit-to-work
PVC	PolyVinyl Chloride
R	Resistance
RAC	Royal Automobile Club
RCD	Residual Current Device
RFID	Radio Frequency Identification
RMS	Root Mean Square
RoES	Representatives of Employee Safety
ROPS	Roll over protection structures
ROSPA	Royal Society for the Prevention of Accidents
RPA	Radiation Protection Adviser
RPE	Respiratory Protective Equipment
RTFLT	Rough Terrain Fork Lift Truck
SAMH	Scottish Association for Mental Health
SEM	Scanning Electron Micrograph
SDS	Safety Data Sheets
SLD	Service Locating Devices
SMART	Specific, Measurable, Achievable and Reasonable withTimescales
SNR	Single Number Rating
SPF	Sun Protection Factor
STEL	Short-term Exposure Limit
Sv	Sievert
SWL	Safe Working Load
TAU	Temporary Accommodation Unit
TDI	Toluene Di-isocyanate
THOR-GP	The health and occupation reporting network of general practitioners
TILE	Task, Individual, Load, and Environment
TNA	The National Archives
TTS	Temporary Threshold Shift
TUC	Trade Union Congress

TWA	Time Weighted Average
UAVs	Unmanned aerial vehicles
UK	United Kingdom
UKAS	United Kingdom Accreditation Service
UKCG	UK Contractors Group
UN	United Nations
UN GHS	United Nations' Globally Harmonised System
UNRTDG	UN Recommendations on the Transport of Dangerous Goods
USA	United States of America
μSv	Microsieverts
UV	Ultraviolet
V	Volt
VLOS	Visual Line of Sight
VOCs	Volatile organic compounds
VWF	Vibration White Finger
WBVS	Whole Body Vibration Syndrome
WEL	Workplace Exposure Limit
WHO	World Health Organisation
WRULD	Work-related Upper Limb Disorder

Element 1

The foundations of construction health and safety management

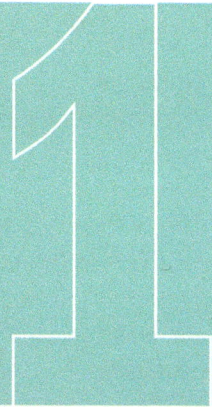

Contents

1.1 Morals and money

GENERAL ARGUMENT

The main reasons why organisations should manage workplace health and safety are:

1) **Moral**

Work-related injuries and ill-health result in a great deal of pain and suffering for those affected. Clearly, we must all do everything that we can to avoid this. One of the ways society expresses its moral expectations is in the form of legal expectations of good standards of health and safety, which aim to protect the health and safety of workers and others that might be affected by the organisation's work activities.

2) **Financial**

Accidents/incidents at work cost a great deal of money, particularly when the costs of fines and compensation are considered. There are also many other less obvious costs that can be very significant, including the costs relating to interruption to production, harm to the quality of our products or damage to the environment. Costs can be enormous – and perhaps already are many times larger than most individuals consider.

The moral and financial reasons provide a strong motivation to promote good health and safety standards and are discussed further in this section.

CASE STUDY

Effect of a serious accident/incident at work

A 32 year-old engineer fell 5 metres from a ladder while doing maintenance work on a building. He suffered major injuries to his head, arm and back when he hit the ground. He was taken to hospital where he remained unconscious for 24 hours and had surgery to repair the fractures he had suffered. He spent many days in hospital receiving treatment for his injuries.

The engineer's manager had to tell his family that he had been injured and was in hospital as a result of the fall at work. His manager said "It was really difficult to tell his family knowing that he was one of my workers. I knew his family well and felt that I had let them down by letting him get injured. It was hard to tell his wife and although his three young children did not understand I could see they were very upset to know he was in hospital."

His wife spent all her available time at the hospital with him. She shared the pain and emotional stress of his injuries. His condition affected everyone in the family. His wife said "One of his sons did not recognise him because of his injuries. I think it was very hard for them to see him like that."

The engineer used to bicycle to work every day. Although he had received treatment for his injuries and a long programme of exercises to improve his mobility the damage to his spine meant that he had difficulty in walking. He could no longer use a bicycle and would not be able to walk without a walking stick. This meant he could not return to his previous job.

The engineer said "I was once fit and healthy, now look at me, all because I didn't have the right equipment to work at height safely."

CONSIDER

Consider the case study above and identify three human effects or consequences that came from the incident.

MORAL EXPECTATIONS OF GOOD STANDARDS OF HEALTH AND SAFETY

Summary

Workplace injuries and ill-health can result in a wide number of effects, including a great deal of pain and suffering for those that are harmed. A worker should not have to expect that, by coming to work, their life is at risk. They should also not expect to be affected by hazardous substances that could shorten their life or cause long-term harm. Nor should others be adversely affected by work activities. Injuries and ill-health could impact on a worker's family life, cause them to have financial hardship due to not being able to work and have consequential effects on their financial position leading to pressure on their home life and socially. It is useful to consider that health relates to physical health (absence of disease) as well as psychological health (mental well-being).

It is recognised that in the construction industry workers experience a significant level of work-related mental ill-health. A report from the Office for National Statistics showed that male suicide deaths registered in England between 2011 and 2015, whose occupations was in the construction industry, were 3.7 times higher than the national suicide average. While this number may indicate a large problem, it does not reflect the suffering of the relatives, dependants, friends and work colleagues who may be impacted by the loss.

Ensuring good health and safety standards at work may therefore be seen as the right way for organisations to conduct themselves and harming people through work activities as the wrong way. The Chief Executive Officer/Top Manager and other Directors/Managers of an organisation are responsible for the governance of that organisation. Society expects that the organisation will be operated with due regard to health and safety and the leadership team of the organisation has a moral (and legal) duty to achieve health and safety in their workplaces.

Taking care not to harm people through work activities is a widely accepted custom of conduct and the right thing to do. This is reflected in many of the world's religions and cultures. This moral reason to prevent harm is usually further reinforced by societal expectations of behaviour, which requires the consideration of others that may be affected by interaction with them. In particular this includes work activities and how they may harm those involved or affected by the activity. This societal expectation is often expressed in both civil law and criminal law as, without the potential for litigation or regulatory action, many employers would not act upon their moral obligation to provide protection. In many countries, it is a specific legal requirement to safeguard the health and safety of workers and others that might be affected by an employer's work activities.

Discussion

The moral reason for achieving good standards of health and safety in work activities is founded on the desire to prevent harm to those that may be affected by the work activities.

Health and Safety Executive (HSE) statistics revealed that 123 workers were killed in work-related accidents during the year 2021/22 representing 0.38 deaths per 100,000 workers. In the same period, 80 members of the public were killed due to work-related activities. The manufacturing industry (22 deaths), construction industry (30 deaths) and agriculture (22 deaths) accounted for most of these. Falls from height, being struck by a vehicle and being struck by a falling object accounted for approximately 70 of these deaths.

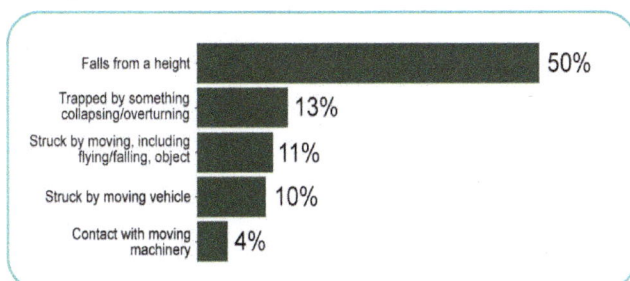

Figure 1-1: Top five causes of fatal injuries in construction.
Source: HSE, Construction statistics in Great Britain, 2021.

In the same period, in all industries, there were 51,211 non-fatal injuries reported. Slips, trips and falls on the level accounted for 33% of all non-fatal injuries. The second most common injuries were sustained whilst lifting, handling or carrying, representing 19% and struck by a moving object were 10% of the non-fatal injuries.

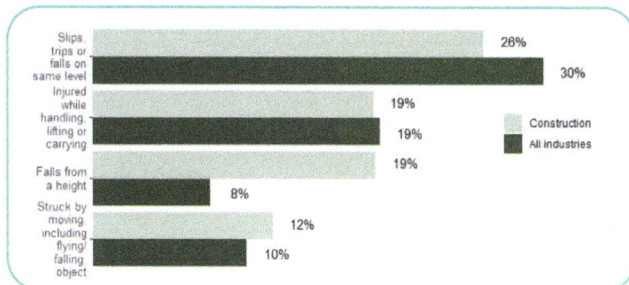

Figure 1-2: Non-fatal injuries in construction.
Source: HSE, Construction statistics in Great Britain, 2021.

In addition to work-related injuries, it was reported in 2020/21 that 1.7 million people were suffering from an illness (long-standing as well as new cases) they believed was caused or made worse by their current or past work. Of these illnesses 0.85 million of them were new conditions that started during the year. In 2020, 2,554 people died from mesothelioma, a type of cancer caused by exposure to asbestos fibres that usually starts in the membrane located between the lungs and chest wall. Thousands more people died from other occupational cancers and diseases such as chronic obstructive pulmonary disease (COPD), which is a work related lung disease caused by breathing in certain dusts, fumes, chemicals or gases.

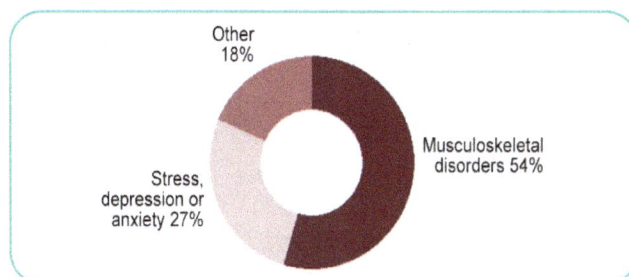

Figure 1-3: Self-reported work-related ill health in construction.
Source: HSE, Construction statistics in Great Britain, 2021.

In 2019/20, a total of 38.8 million days were lost due to work-related harm, 32.5 million days lost were due to work-related ill-health and 6.3 million due to workplace injury. The total cost of work-related ill-health and injury was estimated at £16.2 billion in 2018/19, illness represented £10.6 billion of these costs and injuries cost £5.6 billion. In 2018/19 it was calculated that there was an average cost of £1.7 million per fatality and £8,800 per non-fatal injury. *The coronavirus (COVID-19) pandemic has impacted health and safety statistics in 2020/21. Therefore, no new data on working days lost and economic costs is available.*

It can be seen by the statistics outlined that harm to people from work-related activities is very large and provides a strong moral reason to ensure good health and safety in the workplace.

In the UK, societal expectation of standards of health and safety are very strong, causing organisations to have good standards that prevent injury and ill-health. Societal expectations tend to influence health and safety standards in two ways:

1) Strategically, the general mass of the public influence societal expectations concerning tolerance or intolerance of specific workplace risks or situations, for example, intolerance of risks of work-related major disasters.

2) Locally, these influences tend to surround tolerance or intolerance of the practices of a specific organisation. This influence is often strongest following an accident/incident in an organisation and has led to the closure of some smaller organisations.

Therefore there is a broad agreement that society should expect and demand a safe and healthy work environment that does not harm workers, contractors, the self-employed and the general public.

This confirms that employers need to pay particular attention to the following five important health and safety factors in order to prevent work-related injury and ill-health:

1) A safe and healthy place of work.

2) Safe plant and equipment.

3) Safe systems of work.

4) Training and supervision.

5) Competent workers.

These health and safety factors are reflected in UK civil and criminal laws related to workplace health and safety, which further endorses the societal expectation to prevent harm from work-related activities.

The UK established these health and safety factors as requirements in the Health and Safety at Work etc Act (HASAWA) 1974, which sets out specific legal requirements to provide a safe place of work, safe plant and equipment, safe systems of work, training and supervision.

The five important health and safety factors have also been adopted within civil law to create a separate 'duty of care' towards those that might be affected by work activities. The employers' duty of care in civil law requires employers to take 'reasonable care of those that might foreseeably be affected by their acts or omissions'. The duty of care provides protection of workers and others (for example, visitors, members of the public, etc.) who might foreseeably be affected. It follows that workers have the right to work in a workplace and in a way that provides reasonable protection from harm. If an employer fails to

meet this duty the employer may be required to compensate the person for the harm done.

The civil and criminal responsibility to meet the five important health and safety requirements could be considered as society's formalisation of the moral responsibilities held by an individual or organisation towards another individual or organisation.

REVIEW

List two good reasons, with a description of each, for preventing accidents in the workplace.

THE FINANCIAL COST OF INCIDENTS

Summary

Employers can sustain costs in the event of accidents/incidents and ill-health at work. These can include legal fees, fines, compensation payments, investigation time and lost goodwill from the workforce, customers and wider community. Accidents/incidents and ill-health at work can cost organisations a great deal of money, especially when the costs arising from damage accidents/incidents are added and particularly as they may interrupt production, harm the quality of products or impair the environment. The Government realises that poor occupational health and safety performance also gives rise to additional costs to the State, for example, payments to the incapacitated worker and the costs of medical treatment.

Insured and uninsured costs

Direct costs

Direct costs are those costs that directly relate to the accident/incident or incidence of ill-health. Direct costs may be insured or uninsured depending on the cause of the loss. In addition, many smaller costs may not be covered by insurance policies because the insurance company expects the employer to meet some of the costs of losses. Sometimes this is referred to as a 'threshold' or 'excess' amount that the employer is required to pay before the insurance company pays their part of the compensation.

Insured direct costs often include:

- Claims for compensation made against the employer by workers and other people. This may be covered by employer's liability insurance or public liability insurance or similar insurance policies.

- Damage to buildings, tools, equipment, vehicles and materials, specifically covered by insurance policies, this may depend on the type of accident/incident, for example damage by a fire may be extensively covered by insurance. However, vehicles may only have insurance cover for accidental damage on public roads.
- Medical costs related to harm caused to a worker may be insured. This can include treatment, hospital and physician charges, occupational therapy, prescription medicines and medical equipment.
- Legal costs of defending a claim for compensation or prosecution are often insured.
- Long term loss of business may be insured, usually related to a major disaster like a fire.

Uninsured direct costs often include:

- Lost time of an injured worker or worker suffering ill-health.
- Continued payments to the worker or the worker's family as part of an employer's worker absence scheme.
- Additional payments to other workers or contractors to do the work the worker would have done.
- Wage costs due to decreased productivity once the injured worker returns to work. This is may be due to restricted movement or nervousness/cautious-ness on the part of the injured worker and time spent discussing the accident/incident with other workers etc.
- Damage to equipment, tools, property, plant or materials not covered by specific insurance.
- Most first-aid costs.

- Time and materials to clean up after the accident/ incident or incidence of ill-health.
- Increases in insurance premiums resulting from accidents/incidents and ill-health.
- Charges related to citations, notices and other forms of enforcement agency intervention.
- Fines imposed by a criminal court following prosecution.

Indirect costs

Indirect costs are those costs that indirectly relate to the accident/incident or incidence of ill-health. Some of these costs may be caused by a single accident/ incident or incidence of ill-health or may be the result of an accumulation of occurrences over time. Indirect costs can also be insured or uninsured, though not many can be insured.

Insured indirect costs include:

- Claims for consequential loss related to effects on the supply chain, from customers or suppliers.

Uninsured indirect costs include:

- Lost time by other workers who stop work or reduce performance.
  - Out of curiosity.
  - Out of sympathy.
- Workers and managers may feel that health and safety is a low priority, which could weaken morale and lead to reduced productivity.
- Loss of important experience that the worker had.
- Lost time by supervisors or other managers.
  - Assisting the injured worker.

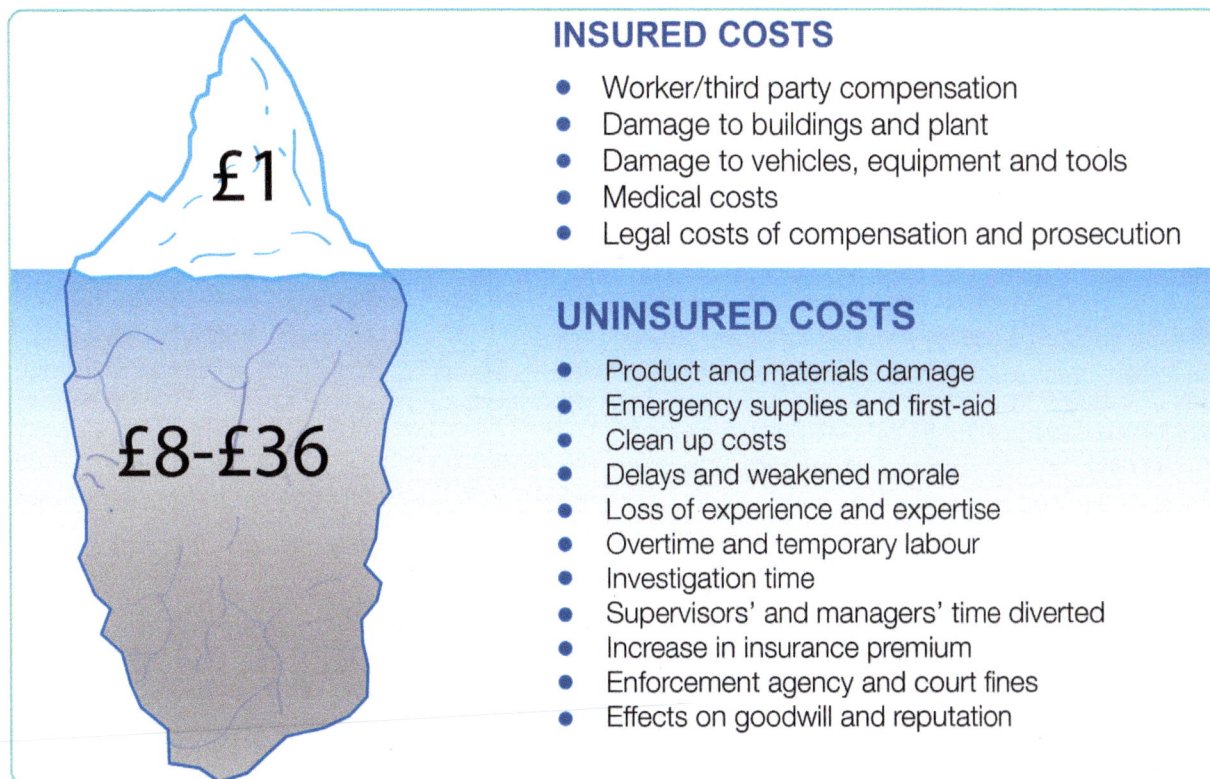

INSURED COSTS
- Worker/third party compensation
- Damage to buildings and plant
- Damage to vehicles, equipment and tools
- Medical costs
- Legal costs of compensation and prosecution

£1

UNINSURED COSTS
- Product and materials damage
- Emergency supplies and first-aid
- Clean up costs
- Delays and weakened morale
- Loss of experience and expertise
- Overtime and temporary labour
- Investigation time
- Supervisors' and managers' time diverted
- Increase in insurance premium
- Enforcement agency and court fines
- Effects on goodwill and reputation

£8-£36

Figure 1-4: Main insured and uninsured costs, typical ratio of costs. Source: RMS/HSE HSG96 (no longer available).

- Investigating the causes of the accident/incident or ill-health.
- Arranging for the injured workers' production to be continued by other workers.
- Organising activities to resume the operation of the organisation.
- Selecting or training a new worker to replace an injured worker.
- Preparing reports, attending enforcing agency meetings.
- Attending hearing or inquest courts.

- Interference with production leading to failure to fill orders on time, loss of bonuses, penalty payments and similar losses that do not benefit from insurance.
- Effects on goodwill and the reputation of the organisation.

Many case studies over recent years have illustrated the difference between insured costs and uninsured costs. It has been shown that uninsured costs were often between 8 and 36 times greater than the insured costs.

Studies carried out on behalf of the Health and Safety Executive found that uninsured costs outweighed insured costs by up to 36 times, see *Figure 1-4*, the 'Accident Costs Iceberg'.

CONSIDER

What sort of financial costs would a company incur as a result of having poor health and safety standards?

1.2 The Construction (Design and Management) Regulations 2015

ROLES AND DUTIES UNDER THE CONSTRUCTION (DESIGN AND MANAGEMENT) REGULATIONS 2015

The Construction (Design and Management) Regulations 2015 (CDMR 2015) establish requirements to establish management controls for all construction projects.

One of the main aims of CDMR 2015 is to integrate health and safety into the management of the project and to encourage everyone involved to work together to:

- Improve the planning and management of projects from the very start.

- Identify hazards early on, so they can be eliminated or reduced at the design or planning stage and the remaining risks can be properly managed.

- Target effort where it can do the most good in terms of health and safety and to discourage unnecessary bureaucracy.

In order to assist with this the CDMR 2015 contain specific duties for those who have particular roles in projects, in particular:

- Clients.
- Principal designers.
- Designers.
- Principal contractors.
- Contractors.
- Workers.
- Domestic Clients.

Client

The CDMR 2015 establishes that the client is an organisation or individual for whom a construction project is undertaken. This definition includes non-domestic (commercial) and domestic clients. A non-domestic client could be an organisation or individual, for example, a Local Authority, school governors, insurance companies and project originators on private finance initiative projects. Domestic clients are people who have work carried out on their own home or the home of a family member that does not relate to trade or business, whether for profit or not. CDMR 2015 makes the client accountable for any adverse effects their decisions and approaches have on health, safety and welfare.

Regulations 4 to 6 of CDMR 2015 establish a number of specific duties for clients. Where there is more than one client an agreement can be made that one or more client accepts the duties on behalf of the others. Non-domestic clients must meet these duties in full. However, Regulation 7 of CDMR 2015 clarifies how Regulations 4 to 6 apply to domestic clients. Regulation 7 of CDMR 2015 transfers most of the domestic client's specific duties to other duty holders.

The duties of clients (non-domestic clients) are to:

a) Make suitable arrangements for managing a project, including the allocation of sufficient time and other resources. Arrangements must be maintained and reviewed throughout the project.

b) Where there is more than one contractor, appoint in writing a principal contractor and a principal designer.

c) Take reasonable steps to ensure designers and contractors are competent.

d) Provide relevant pre-construction health and safety information to designers and contractors.

e) Provide information and instruction as soon as is practicable.

f) Co-operate with others in relation to the project.

g) Take reasonable steps to ensure the principal designer and principal contractor meet their duties.

h) Ensure that a construction phase plan is drawn up by the contractor/principal contractor before the construction phase begins.

i) Ensure that the principal designer prepares a health and safety file for the project and keeps it available for inspection.

j) If the project is classed as notifiable, notify the enforcing authority, which is usually the Health and Safety Executive.

k) Ensure a copy of the notification is displayed in a prominent position at the site for workers to read.

l) Ensure adequate welfare facilities are provided.

m) Ensure the health and safety file is passed on if the client no longer has an interest in the structure to which it relates.

Principal Designer

A principal designer is the designer with control over the pre-construction phase of the project. This may be combined with other roles in the project, such as project manager or architect. The principal designer is appointed in writing by the client for projects involving more than one contractor. The principal designer can be an organisation or an individual with sufficient knowledge, skills and experience to carry out the role.

The principal designer's role is to plan, manage and monitor and co-ordinate the pre-construction phase. For example, any designers appointed should not carry out any work beyond initial design unless the principal designer has confirmed that the client is aware of the duties a client has under CDMR 2015.

The principal designer will need to follow the general principles of prevention to identify and eliminate or control risks within the design that affect the health and safety of individuals. It is also their responsibility to ensure that the health and safety file is prepared and handed over to the client at the end of the project.

The principal designers' duties are to:

- Not accept an appointment to a project unless they have the skills, knowledge and experience, and, if they are an organisation, the organisational capability, necessary to fulfil their role in a manner that secures the health and safety of any person affected by the project.

- Plan, manage, monitor and co-ordinate the pre-construction phase so that it is carried out, so far as is reasonably practicable, without risks to health and safety. In particular when design, technical and organisational decisions are being made or the time to complete stages of work is being estimated.

- Take into account the general principles of prevention, any construction phase plan and any health and safety file.

- Identify, eliminate or reduce foreseeable risks related to the construction work, maintaining/cleaning of the structure or using the structure as a workplace.

- Ensure designers comply with their duties.

- Ensure co-ordination and co-operation between all parties working on the pre-construction phase.

- Co-operate with others in relation to the project.

- Assist the client in the provision of pre-construction information, in particular to the principal contractor.

- At the pre-construction phase, prepare a health and safety file and ensure it is reviewed, updated and revised as necessary. On completion of the project the principal designer must pass the health and safety file to the client.

- Liaise and share with the principal contractor information relevant to the planning, management and monitoring of the construction phase and the preparation of the construction phase plan.

- Ensure information or instruction is comprehensible and provided as soon as is practicable.

- When working on a project under the control of another, report to that person anything they are aware of in relation to the project which is likely to endanger their own health or safety or that of others.

Designer

With regard to CDMR 2015, a designer is an organisation or individual who, in the course of a business, prepares or modifies a design for a construction project (including the design of temporary works) or arranges for someone under their control to do so. Typically designers will prepare drawings, design details, specifications, bills of quantities, specification (or prohibition) of articles and substances, as well as the related analysis, calculations and preparatory work for their workers or others under their control to prepare designs relating to a structure or part of a structure.

Designers include architects, structural engineers, building surveyors, landscape architects, interior designers, quantity surveyors, consulting engineers and manufacturing design specialists contributing to or having overall responsibility for any part of a design.

The designers' duties are to:

- Not accept an appointment to a project unless they have the skills, knowledge and experience, and, if they are an organisation, the organisational capability, necessary to fulfil their role in a manner that secures the health and safety of any person affected by the project (general duties).

- Not commence work unless the client is aware of the duties a client has under CDMR 2015.

- Apply the principles of prevention and any pre-construction information when preparing or modifying a design.

- Provide information to the principal designer on residual risks that could not be eliminated by design.

- Ensure appropriate information related to residual risks that could not be eliminated is provided to the principal designer and included in the health and safety file.

- Provide information with the design to assist the client, other designers and contractors.

- Ensure information or instruction is comprehensible and provided as soon as is practicable (general duties).

- Co-operate with others in relation to the project.

- Ensure information or instruction is comprehensible and provided as soon as is practicable.

- To report to the person in control of the work anything they are aware of in relation to the project which is likely to endanger their own health or safety or that of others.

Principal Contractor

The principal contractor is the organisation or individual responsible for planning, managing, monitoring and co-ordinating the construction phase of a project to ensure, so far as is reasonably practicable, construction work is carried out without risks to health and safety. For projects involving more than one contractor the client appoints a principal contractor, in writing.

The principal contractors' duties are to:

- Not accept an appointment to a project unless they have the skills, knowledge and experience, and, if they are an organisation, the organisational capability, necessary to fulfil their role in a manner that secures the health and safety of any person affected by the project.

- Plan, manage, monitor and co-ordinate the construction phase to ensure that it is carried out, so far as is reasonably practicable, without risks to health and safety. In particular when design, technical and organisational decisions are being made or the time to complete stages of work is being estimate.

- Draw up a construction phase plan and ensure it is reviewed, updated and revised as construction work is carried out.

- Take into account the general principles of prevention.

- Liaise and share with the principal designer information relevant to the planning, management and monitoring of the construction phase and co-ordination of the pre-construction phase.

- Organise co-operation between contractors.

- Co-ordinate implementation by the contractors of applicable legal requirements.

- Ensure that employers and self-employed persons apply the principles of prevention and follow the construction phase plan.

- Where the project is notifiable, display the notification details in a prominent place so that workers can read them.

- Prevent unauthorised access to the construction site.

- Provide a suitable site induction.

- Ensure adequate welfare facilities are provided and maintained.

- Provide the principal designer with information relevant to the health and safety file.

- Ensure information or instruction is comprehensible and provided as soon as is practicable.

- Co-operate with others in relation to the project.

- Consult with workers or their representatives on matters which may affect their health, safety or welfare.

- To report to the person in control of the work anything they are aware of in relation to the project which is likely to endanger their own health or safety or that of others.

Contractors

Contractors are engaged in a project to do construction work, under the control of the principal contractor, where one is appointed. Contractors can be engaged to conduct a range of construction work including specialist construction work, for example, provision of utilities and special trades.

Figure 1-5: Contractors.
Source: Pexels.

In the context of CDMR 2015 anyone who directly employs or engages construction workers or manages construction is a contractor. A contractor may be an individual, a sole trader, a self-employed worker or a business that carries out, manages or controls construction work in connection with a business.

In some projects there will be only one contractor, who will liaise with the client and designer(s) and have responsibility for the completion of the construction phase of the project.

The contractors' duties are to:

- Not accept an appointment to a project unless they have the skills, knowledge and experience, and, if they are an organisation, the organisational capability, necessary to fulfil their role in a manner that secures the health and safety of any person affected by the project.

- Not start work unless they are satisfied that the client is aware of the duties a client has under CDMR 2015.

- Plan, manage and monitor the construction work under their control to ensure that it is carried out, so far as is reasonably practicable, without risks to health and safety.

- Where there is more than one contractor, co-operate with the principal designer or the principal contractor.

- Comply with the construction phase plan.

- Not employ a person who does not have the necessary skills, knowledge, training and experience to work safely and healthily.

- Provide each worker under their control with appropriate supervision, instruction and information so that work can be carried out safely and healthily. The information provided must include:

  - A suitable site induction.
  - Emergency procedures.
  - Information identified by risk assessments.
  - Any other information that the worker ought to be aware of.

- Not commence work unless unauthorised access to the site has been prevented.

- Provide adequate welfare facilities.

- Co-operate with others in relation to the project.

- To report to the person in control of the work anything they are aware of in relation to the project which is likely to endanger their own health or safety or that of others.

Where there is only one contractor, the contractor's duties include duties to:

- Take account of the general principles of prevention. In particular when design, technical and organisational decisions are being made or the time to complete stages of work is being estimate.

- Draw up a construction phase plan prior to setting up a construction site.

Figure 1-6: Worker.
Source: RMS.

Workers

A worker is anyone working for or under the control of a contractor on a construction site. Examples of workers include: plumbers, electricians, scaffolders, painters, decorators, steel erectors and labourers, as well as supervisors like foremen and charge hands.

Workers have an important role and should take an active part in helping to manage health and safety risks. In particular, workers must:

- Only carry out construction work if they have the relevant skills, knowledge, training and experience - or they are provided with the training and supervision that enables them to do it safely and without risk to health.

- Make themselves aware of the health and safety risks involved in work on every site and the way those risks are managed.

- Always follow site rules and procedures.

- Co-operate with other duty holders, such as the contractor in control of their work and the principal contractor (this may be who controls the overall project when there is more than one contractor).

- Report any risks they find to the person who controls work on site, whether the risks affect their own health and safety or anyone else, including other workers and members of the public.

Domestic clients

Domestic clients are people who have work carried out on their own home or the home of a family member, that does not relate to a trade or business, whether for profit or not. If the work is in connection with the furtherance of a business attached to domestic premises, such as a shop, the client is not a domestic client.

Local authorities, housing associations, charities, landlords and other businesses may own domestic property, but they are not domestic clients. Similarly, a company formed by independent leaseholders of flats to undertake maintenance of the common parts of a structure is not a domestic client, and will have duties under CDMR 2015.

Domestic clients are not required to carry out the duties of commercial clients. The effect of Regulation 7 of CDMR 2015 is to transfer most of the domestic clients' duties to other duty holders. Where there is only one contractor, most of the client's duties must be carried out by the contractor in addition to the duties that the contractor already has as a contractor. Similarly, if there is more than one contractor, most of the client duties must be undertaken by the principal contractor in addition to the principal contractor's own duties. As an alternative to this, the client can have a written agreement with the principal designer to carry out most of the client's duties, in addition to the principal designer's duties.

Most of the client's duties are passed on to the contractor, principal contractor or principal designer. However, where there is more than one contractor, domestic clients retain the following duties to:

- Appoint in writing a principal designer. If the domestic client fails to appoint a principal designer the designer in control of the pre-construction phase project will be the principal designer.

- Appoint in writing a principal contractor. If the domestic client fails to appoint a principal contractor the contractor in control of the construction phase of the project will be the principal contractor.

The role of designers, principal designers, principal contractors and contractors when working on a project for a domestic client, is normally no different to their role when working for a commercial client. They have the same duties and should carry them out in the same way as they would for a commercial client.

Designers and contractors working for domestic clients have to manage their own work and co-operate with and coordinate their work with others involved with the project in order to ensure the health and safety of all involved in the project. The requirements in Schedules 2 (welfare facilities) and Part 4 (Regulations 16-35) of CDMR 2015 still apply to the work they do for clients. The client has a specific duty to make arrangements to ensure welfare facilities are provided and, where they control the way in which construction work is carried out, must comply with the requirements of Part 4 as far as they are within the client's control.

APPOINTMENT AND COMPETENCE REQUIRED OF RELEVANT PARTIES

Regulation 5 of CDMR 2015 states:

"(1) Where there is more than one contractor, or if it is reasonably foreseeable that more than one contractor will be working on a project at any time, the client must appoint in writing:

(a) A designer with control over the pre-construction phase as principal designer.

(b) A contractor as principal contractor.

(2) The appointments must be made as soon as is practicable, and in any event, before the construction phase begins."

Figure 1-7: Appointment of the principal designer and the principal contractor. Source: CDMR 2015 Regulation 5.

A designer (including principal designer) or contractor (including principal contractor) appointed to work on a project must have the personal skills, knowledge and experience to identify significant risk and manage or control them to an acceptable level. Designers and contractors must not accept an appointment to a project unless they fulfil the conditions regarding skills, knowledge, experience and capability. This will require them to verify their own competence and, if they are an organisation, the resources and competence of their own personnel working on the project.

Clients must take reasonable steps to satisfy themselves that designers and contractors fulfil the conditions regarding skills, knowledge, experience and capabilities before appointing them. Reasonableness of the steps necessary to assure this will vary according to the level and complexity of risk, for higher risk more assurance is required.

Similarly, any designers or contractors seeking appointment must ensure they have the necessary qualities. If they are an organisation, they must display the organisational capability necessary to fulfil the role in a manner that secures the health and safety of any person affected by the project. Organisational capability is effectively the policies and systems that an organisation has in place to set acceptable health and safety standards which comply with the law, and the resources and people to ensure the standards are delivered.

The principal designer must be a designer and must be formally appointed, in writing, as principal designer by the client. The appointment of a principal designer to a project is only required where there is or is likely to be more than one contractor. As the principal designer is also a designer they must comply with the designer's duties established by CDMR 2015 in addition to their duties as a principal designers.

The principal contractor must be a contractor and must be formally appointed, in writing, as principal contractor by the client. The appointment of a principal contractor to a project is only required where there is or is likely to be more than one contractor. As the principal contractor is also a contractor they must comply with the contractor's duties established by CDMR 2015 in addition to their duties as a principal contractor.

When a contractor employs or appoints an individual to work on a construction site, they should make enquiries to confirm the individual:

a) has the skills, knowledge, training and experience to carry out the work they will be employed to do in a way that ensures health and safety for anyone working on the site; or

b) is in the process of obtaining them, for example they may be developing them through workplace learning and assessment.

"(7) A contractor must not employ or appoint a person to work on a construction site unless that person has, or is in the process of obtaining, the necessary skills, knowledge, training and experience to carry out the tasks allocated to that person in a manner that secures the health and safety of any person working on the construction site."

Figure 1-8: Contractor appointing competent people to carry out work on site. Source: CDMR 2015 Regulation 15.

WHEN THE HSE NEEDS TO BE NOTIFIED

CDMR 2015 applies to all construction projects. However, construction projects that have a construction phase longer than 30 days and more than 20 workers working on the project at any one time or involving more than 500 person days of construction work are notifiable to the enforcing authority, which is usually the Health and Safety Executive (HSE). Notification to the enforcing authority by the client must be in writing as soon as is practicable before the construction phase begins.

The client can make a written agreement that someone else carry out the notification on their behalf, but they remain accountable for the notification taking place correctly. The client can therefore establish a written agreement with the principal designer, designer, principal contractor or contractor to do this.

Where the client is a domestic client they are not required to carry out the duties placed on commercial clients for notification. This responsibility to notify the enforcing authority usually becomes the contractor's if there is only one contractor or if there is more than one contractor the principal contractor's responsibility. A written agreement can be made by the domestic client with their designer/principal designer that they manage the project, in which case the designer/principal designer will be responsible for notification of the project.

The notification must contain particulars specified in Schedule 1 of CDMR 2015 and notification can be carried out by using the form F10 on the HSE's website. The information to be notified is:

- Date of forwarding.

- Exact address of the construction site.

- The name of the Local Authority where the site is located.

- A brief description of the project and the construction work which it includes.

- Contact details of the client (name, address, telephone number and any e-mail address).

- Contact details of the principal designer (name, address, telephone number and any e-mail address).

- Contact details of the principal contractor (name, address, telephone number and any e-mail address).

- Date planned for the start of the construction phase.

- The time allowed by the client under Regulation 4(1) for the construction work.

- Planned duration of the construction phase.

- Estimated maximum number of people at work on the construction site.

- Planned number of contractors on the construction site.

- Name and address of any contractor already appointed.

- Name and address of any designer already engaged.

- A declaration signed by or on behalf of the client that the client is aware of the client duties under CDMR 2015.

The client must ensure that an up-to-date copy of the notification is displayed in the site office for workers to read.

PRE-SELECTION AND ONGOING MANAGEMENT OF CONTRACTORS

Pre-selection of contractors

The client's selection of contractors for contracted work usually involves balancing a number of issues, one of which is the financial aspects of the contract. Selection of a contractor based primarily on the price the contractor wants to charge the client for the work may not meet the client's health and safety standards or reflect legal responsibilities. The price a contractor charges is not usually a good indicator of their likely health and safety performance during the contract. Therefore an approach where health and safety is considered before the price is considered is a preferred way to select contractors.

There are five main elements to a health and safety focused selection procedure for contractors. The extent to which each element is relevant will depend upon the degree of risk and nature of work to be contracted.

The elements are:

1) Identification of the likely hazards and risks related to the work defined in the contract specification.

2) Identification of suitable bidders from those with an interest in bidding (sometimes called development of a preferred list of bidders). This is done by assessing those contractors with an interest in bidding against defined health and safety factors. The assessment might consider factors such as:

- The status and adequacy of the contractor's health and safety policy.

- Whether the contractor has a management system and if it is certified by a recognised organisation.

- The status of the competent person appointed to assist the contractor with health and safety.

- Previous accident/incident and ill-health records of the contractor.

- Details of any prosecutions or significant civil claims against the contractor.

- Details of any enforcing agency action against the contractor, for example, enforcement notices.

- The competence of the contractor's workers.

- The level and type of insurance the contractor holds.

- References from previous clients of the contractor.

Contractors that participate in 3rd party pre-qualification schemes typically have been assessed to determine the extent to which a range of factors similar to those set out earlier have been met. As they are usually audited periodically it could provide a client with a relatively easy means to confirm that the general requirements for pre-selection are met. Where the contract involves very specific or complex health and safety criteria further evidence is likely to be required.

3) Providing the contract specification to suitable bidders.

4) Assessing the health and safety aspects of bids, including how they intend to control hazards and manage health and safety risks.

5) Deciding on the contractors to proceed to final selection based on other factors than health and safety, in particular, quality, environmental and financial factors.

After final selection, the contractor should be managed and monitored during work activities. After completion of the contract the contractor's health and safety performance should be reviewed and their status on the list of preferred contractors updated.

Figure 1-9: Reviewing health and safety measures with contractors. Source: Pexels.

Ongoing management of contractors

One of the basic principles of the management of contractors is that, reflecting the level of risk, health and safety conditions should be set down in the contract.

For example, in situations where the contractor conducts work on the client's premises an important condition of the contract should be that the contractor agrees to abide by all the provisions of the client's health and safety policy that may affect the contracted work activities, including compliance with site health and safety rules and contractors' rules.

Before any contract work is commenced, a responsible person representing the main contractor should discuss with the client the health and safety control measures necessary to ensure the health and safety of contracted workers and others that may be affected by the contracted work.

Control measures for the ongoing management of contractors are likely to include: *Checklist.*

- Client's appointment/nomination of a person or team to co-ordinate all aspects of the contract, including health and safety matters.
- A pre-contract commencement meeting held with the contractor to review all health and safety aspects of the work.
- Provision by the contractor of risk assessments and/or written method statements in advance of undertaking particular work.
- Development of a contract plan and agreement of how the work will be co-ordinated.
- Control of access to the contractor's work area.
- Carrying out induction for contractor's workers and health and safety briefings for the client's workers.
- Control of any sub-contracting of work.
- Arrangements for storage of materials and equipment.
- Arrangements for welfare.
- Emergency arrangements.
- Arrangements for regular progress meetings, where health and safety is the first agenda item.
- Regular monitoring of the contractor's activities by the client's representative, at a frequency reflecting the risk.
- Participation in contractor's health and safety committees and consideration of contractor representation on the client's committees.
- Formal reporting to the client by the contractor of all lost-time injuries, work-related ill-health absences and dangerous occurrences, including those to sub-contractors.
- Arrangements for the contractor to leave the work area clean and tidy, removing all waste, materials, tools and equipment and for this to be formally checked.

The client and contractor share a responsibility to evaluate health and safety performance related to the contracted work activity. This should be planned in advance so that it can help to enable the successful management of the work and the evaluation of its effectiveness. As the client retains overall responsibility for the contracted work activity, if the level of health and safety performance is not acceptable the client should take action to ensure the contractor meets the health and safety requirements of the contract.

PLANNING AND CO-ORDINATION OF CONTRACTED WORK

The Management of Health and Safety at Work Regulations (MHSWR) 1999 require that employers make arrangements to plan, organise, control, monitor and review activities that it controls. This would, naturally, extend to contract work commissioned by a client. Regulation 11 of MHSWR 1999 specifically requires employers who work together in a common workplace, such as where contracted work is carried out in the workplace of a client, to co-operate in order to discharge their duties under relevant statutory provisions.

In addition, CDMR 2015 set out specific requirements for all construction projects and additional duties for those that are notifiable to the Health and Safety Executive

Clients and contractors therefore have a responsibility to build health and safety into the contract and work methods. The shared responsibilities for health and safety mean that the client and contractor need to collaborate throughout the contract period to ensure health and safety is achieved, including prior to the contract being signed, before commencement of work and during the work. To this end, it is essential that they co-operate with each other, seeking to plan and co-ordinate activities.

To meet their separate responsibilities, it is important that the client and contractor provide each other with health and safety information that might affect the contracted work activity, for example, the client might provide information on known hazards in the workplace, like asbestos, and the contractor should provide information on aspects of their work that affect health and safety of the client's workplace, for example, when the contracted work will require fire detection systems to be shut down.

The effective planning and co-ordination of contracts will include conducting appropriate risk assessments and foreseeing how the work activities and those involved in the work interact with each other.

The client and contractor should co-operate with each other by considering each other's activities when conducting risk assessments and share the information with each other to enable appropriate control measures to be put in place.

The information provided and risk assessments should be used to develop plans and statements of how the work is to be conducted and co-ordinated, for example, the contractor might produce method statements that describe the steps in the work activity and specific risk control measures to manage the risks. The client and contractor should also co-ordinate any control measures used to deal with any shared risks identified in the risk assessments. This includes in relation to emergencies, for example, when establishing fire evacuation procedures on a site occupied by the client and multiple contractors.

As part of the planning and co-ordination of contracted work the contractor needs to consult and engage with workers. This will help to ensure risk control measures are understood and effective.

> *"The principal contractor must:*
>
> *(a) make and maintain arrangements which will enable the principal contractor and workers engaged in construction work to cooperate effectively in developing, promoting and checking the effectiveness of measures to ensure the health, safety and welfare of the workers;"*

Figure 1-10: Worker consultation and engagement.
Source: CDMR 2015 Regulation 14.

It is important that plans identify situations where the client and contractor have to co-ordinate their work to ensure risks are minimised, including consideration of any interaction with existing staff. This might mean carrying out work activities at times that the client's workers do not work, planning to exclude the client's workers from the work area while the activity takes place or shutting down part of the client's activities that could harm the contractor.

Where possible, taking regard to the risks, plans should be formalised and documented in writing to enable progress and problems to be identified. It is essential that any time constraints of the contract and the associated plan have full regard for the need to ensure health and safety during the contract and the pressures that unrealistic time constraints can have.

INFORMATION, PLANS AND HEALTH AND SAFETY FILES

Pre-construction information

Purpose of pre-construction information
The purpose of pre-construction information is to provide information to those bidding for or planning work and for the development of the construction phase plan. Pre-construction information is essentially a collection of information about the significant health and safety risks of the construction project that the principal contractor will have to manage during the construction phase.

> *"pre-construction information" means information in the client's possession or which is reasonably obtainable by or on behalf of the client, which is relevant to the construction work and is of an appropriate level of detail and proportionate to the risks involved, including -*
>
> *(a) information about:*
>
> > *(i) the project;*
> >
> > *(ii) planning and management of the project;*
> >
> > *(iii) health and safety hazards, including design and construction hazards and how they will be addressed; and*
>
> *(b) information in any existing health and safety file;"*

Figure 1-11: Definition of pre-construction information.
Source: CDMR 2015 Regulation 2.

The pre-construction information serves two main purposes:

1) The pre-construction information plays a vital role in the tender and selection process. Pre-construction information provides the health and safety information needed by designers and contractors who are bidding for work on the project. It enables prospective designers and contractors to be fully aware of the project's health, safety and welfare requirements and take it into account in their tender submissions. The pre-construction information provides a template against which different tender submissions can be measured. The pre-construction information helps the principal designer to advise the client on the provision of resources for health and safety and to assess the competence of prospective principal contractors/contractors.

2) The pre-construction information provides the health and safety information needed by principal designers and principal contractors in planning, managing, monitoring and coordinating the project. In particular, it is used to guide the development of construction phase plan. It can provide a framework for the construction phase plan, allowing the detail of the plan to be added to the framework. The pre-construction information provides may require the plan to be developed in response to the information provided, for example, information provided on risks may require the plan to show how the risks will be managed.

Requirements for pre-construction information

The pre-construction information will mainly come from the client, designers or the principal designer.

> *"(4) A client must provide pre-construction information as soon as is practicable to every designer and contractor appointed, or being considered for appointment, to the project."*

Figure 1-12: Client duty regarding pre-construction information.
Source: CDMR 2015 Regulation 4.

Appendix 2 of CDMR 2015 gives guidance on the requirements for pre-construction information and the actions of each duty holder.

The client

The client has the main duty to provide information relevant to health and safety to the principal designer or designer and contractors (including the principal contractor) at the earliest opportunity, for example when they are being considered for appointment.

In cases where there is a single contractor the responsibility remains entirely with the client to provide the pre-construction information, but this does not prevent them liaising with the contractor and any designer involved in the project. In cases where there is more than one contractor, the client must appoint a principal contractor and the client can expect the principal designer to help collate the pre-construction information, who must assist the client in drawing up the information. The client must liaise with the principal designer and then:

- Ensure that the pre-construction information is adequate.

- Take steps to fill any gaps.

- Provide the pre-construction information to every designer and contractor.

The pre-construction information provided to those seeking appointment as principal contractors/contractors must be sufficient and in good time to allow them to put together a bid based on a clear understanding of the nature of the work involved.

The information should be in a convenient form, clear and concise. This will help other duty holders involved in the project to understand it so that they carry out their duties effectively.

Designers and principal designers

Principal designers must assist the client in gathering the pre-construction information assist the client to assess the adequacy of the information and take reasonable steps to fill any gaps.

The information must:

- Be relevant to the project.

- Have an appropriate level of detail.

- Be proportionate to the level of risk.

It is unlikely that information gathered at the initial stages of design will be sufficient for the needs of the whole duration of the project. Information must be collected and added to the pre-construction information as the design progresses and reflect new information about the health and safety risks identified through the design process and how they should be managed. Designers must take account of the pre-construction information when preparing or modifying designs.

Content of the pre-construction information

Pre-construction information is that information which is already in the possession of the client or which could be obtained on the client's behalf, such as an existing health and safety file and could include existing drawings, surveys of the site or premises, information on the location of services, etc. The contents of the pre-construction information will depend on the nature of the project itself. However, the following areas should be considered:

Description of the project
- Client brief.

- Key dates of the construction phase.

Planning and management of the project
- Arrangements for planning and managing the construction work.

- Allocation of resources and time.

- Arrangements to ensure coordination and co-operation between parties.

- Site security and welfare.

- Requirements for health and safety of client's workers and customers, including vehicles, permits to work, restrictions and emergency procedures.

- Activities on or adjacent to the site during the works.

- Site delineation and security arrangements.

Managing health and safety hazards on site
a) Safety hazards, including:

- Boundaries and access, including temporary access.

- Restrictions on delivery, storage or removal of materials.

- Adjacent land uses.

- Existing storage of hazardous materials.

- Location of existing services - water, electricity, gas, etc.

- Ground conditions.

- Existing structures - stability, or fragile materials.

- Difficulties and damage to structure, for example, height restrictions or fire damage.

b) Health hazards, including:

- Asbestos, including results of any surveys.

- Existing storage of hazardous materials.

- Contaminated land, including results of surveys.

- Existing structures, hazardous materials.

- Health risks arising from client's activities.

- Materials requiring particular precautions.

Information in an existing health and safety file
- Asbestos surveys.

- Drawings and plans.

- Safe working loads of floors.

- Key structural items.

- Information regarding maintenance and cleaning.

The construction phase health and safety plan

Purpose of the construction phase plan

The purpose of the construction phase plan is to set out how health and safety risks are to be managed during the construction phase. In particular, the purpose of the plan is to outline the health and safety arrangements for the construction phase, site rules and, if appropriate, how the risks identified in Schedule 3 - 'Work involving particular risks' of CDMR 2015 will be managed. For example, risks related to drowning or work near high voltages.

Requirements for the construction phase plan

"(5) A client must ensure that:

(a) before the construction phase begins, a construction phase plan is drawn up by the contractor if there is only one contractor, or by the principal contractor;"

Figure 1-13: Client duty regarding the construction phase plan.
Source: CDMR 2015 Regulation 4.

The construction phase plan is developed on behalf of the client by the principal contractor or, for single contractor projects, the contractor and must be drawn up **before** the construction phase commences. The plan is the foundation on which the health and safety management of the construction work is based.

"(1) During the pre-construction phase, and before setting up a construction site, the principal contractor must draw up a construction phase plan or make arrangements for a construction phase plan to be drawn up."

Figure 1-14: Principal contractor's duty regarding the construction phase plan.
Source: CDMR 2015 Regulation 12.

The client must ensure that the principal contractor/ contractor is provided with all available relevant information they need to draw up the plan, including the pre-construction information. The principal designer must help the principal contractor to prepare the construction phase plan by providing any relevant information they hold. This includes any information given to them by designers about the risks that have not been eliminated through the design process and the steps taken or required to reduce or control those risks.

"(3) The principal designer must assist the principal contractor in preparing the construction phase plan by providing to the principal contractor all information the principal designer holds that is relevant to the construction phase plan including:

(a) pre-construction information obtained from the client;

(b) any information obtained from designers under regulation 9(3)(b)."

Figure 1-15: Principal designer's duty regarding the construction phase plan.
Source: CDMR 2015 Regulation 12.

The construction phase plan must record the arrangements and rules for managing the significant health and safety risks associated with the construction phase of a project. The client must ensure that the plan adequately addresses the risks. The plan is also the basis for communicating these arrangements to all those involved in the construction phase, so it should be easy to understand and as simple as possible.

The client also must ensure that the principal contractor (or contractor) regularly reviews and revises the plan to ensure it takes account of any changes that occur as construction progresses and continues to be appropriate for its purpose.

"(4) Throughout the project the principal contractor must ensure that the construction phase plan is appropriately reviewed, updated and revised from time to time so that it continues to be sufficient to ensure that construction work is carried out, so far as is reasonably practicable, without risks to health or safety."

Figure 1-16: Principal contractor's duty to review, update and revise the construction phase plan.
Source: CDMR 2015 Regulation 12.

Appendix 3 of CDMR 2015 provides guidance on requirements for the construction phase plan.

Content of the construction phase plan

The contents of the construction phase health and safety plan will depend on the nature of the project itself and be proportionate to the risks involved in the project. Many of the items in the construction phase plan reflect the information considered at the pre-construction phase, but with further consideration for the construction phase.

When deciding what information should be included in the plan consideration should be given to:

- Its relevance to the project.

- The level of detail needed to clearly set out the arrangements, site rules and any special measures needed to manage the construction phase.

- The balance of the level of detail and the level and complexity of risk involved in the project.

The plan should not include documents that prevent the communication of a clear understanding of what is required to manage the construction phase, such as generic risk assessments, records of how decisions were reached or detailed health and safety method statements. However, the following areas should be considered for inclusion:

Description of the project

- A description of the project and programme, which will include details of important dates.

- Details of other parties involved in the project and the extent and location of existing records and plans.

Management of the work

- The management structure and responsibilities of the various parties involved in the project, including the client and other members of the project team, whether based at site or elsewhere.

- The health and safety standards to which the project will be carried out, including health and safety goals.

- These may be set in terms of statutory requirements or high standards that the client may require in particular circumstances.

- Means for informing contractors about risks to their health and safety arising from the environment in which the project is to be carried out and the construction work itself.

- Means to ensure that all contractors, the self-employed and designers to be appointed by the principal contractor are properly selected (i.e. they are competent and will make adequate provision for health and safety).

- Means for communicating and passing information between the project team (including the client and any client's representatives), the principal designer, designers, the principal contractor, other contractors, workers on site and others whose health and safety may be affected.

- Arrangements for reporting and identification of accidents, including passing information to the principal contractor about accidents, ill-health and dangerous occurrences that require to be notified by the principal contractor to the HSE under the Reporting of Injuries, Diseases and Dangerous Occurrences Regulations 2013 (RIDDOR 2013).

- Arrangements for the provision and maintenance of welfare facilities.

- Arrangements for health and safety induction and training.

- Arrangements that have been made for consulting and coordinating the views of workers or their representatives.

- Arrangements for site rules and for bringing them to the attention of those affected.

- Arrangements for risk assessments and written systems of work.

- Emergency arrangements for dealing with and minimising the effects of injuries, fire and other dangerous occurrences.

- Arrangements should be set out for the monitoring systems to achieve compliance with legal requirements; and the health and safety rules developed by the principal contractor.

Arrangements for specific site risks

- Arrangements of controls safety risks, including preventing falls, control of vehicles and safety when working with services.

- Arrangements for control of health risks, including removal of asbestos, manual handling, noise and vibration.

- In particular, attention should be given to the arrangements for managing the risks associated with any work that falls within one or more of the categories listed in Schedule 3 of CDMR 2015.

Preparation of the health and safety file

Purpose of the health and safety file

The purpose of the health and safety file is to provide a source of information to be taken into account and enable future construction work, alterations, refurbishment and demolition, including cleaning and maintenance, to be carried out in a safe and healthy manner.

Requirements for the health and safety file

The health and safety file is only required for projects involving more than one contractor. The duty to create a health and safety file is assigned to the principal designer and the appointment of a principal designer is only required where there is or is likely to be more than one contractor involved in the project.

Clients, principal designers, designers, principal contractors and other contractors all have legal duties in respect of the health and safety file:

- The principal designer must prepare, review, amend or add to the file as the project progresses, and give it to the client at the end of project.

- Clients, designers, principal contractors and other contractors must supply the information necessary for compiling or updating the file.

- Everyone providing information must make sure that it is accurate, comprehensible and provided promptly.

- Clients must keep the file to assist with future construction work.

"(5) A client must ensure that…

(b) the principal designer prepares a health and safety file for the project, which:

(i) complies with the requirements of regulation 12(5);

(ii) is revised from time to time as appropriate to incorporate any relevant new information; and

(iii) is kept available for inspection by any person who may need it to comply with the relevant legal requirements."

Figure 1-17: Client duties regarding the health and safety file. Source: CDMR 2015 Regulation 4.

Preparing the health and safety file

The principal designer is responsible for ensuring the health and safety file is prepared. Putting together the health and safety file is a task which should ideally be a continual process throughout the project and not left until the construction work is completed.

Early on in the construction project the principal designer may find it useful to discuss the health and safety file with the client. This will help determine what information the client requires and how the client wishes the information to be stored and recorded.

When the client's requirements are known, procedures may need to be drawn up by the principal designer so that all those who will be contributing to the health and safety file (for example, designers and contractors) are aware of:

- The information that is to be collected.

- How the information is to be collected, presented and stored.

The principal designer may find it useful to detail at the pretender stage requirements on how and when the information for the health and safety file is to be prepared and passed on. The principal contractor may also find it useful to include similar procedures in the health and safety plan for the construction phase.

Throughout the project those who carry out design work (including contractors) will need to ensure, so far as is reasonably practicable, that information about any feature of the structure which will involve significant risks to health and safety during the structure's lifetime are passed to either the principal designer or to the principal contractor. Providing this information on drawings will allow for amendments if any variations arise during construction. It will also allow health and safety information to be stored on one document, therefore reducing the amount of paperwork that needs to be retained.

"(5) During the pre-construction phase, the principal designer must prepare a health and safety file appropriate to the characteristics of the project which must contain information relating to the project which is likely to be needed during any subsequent project to ensure the health and safety of any person.

(6) The principal designer must ensure that the health and safety file is appropriately reviewed, updated and revised from time to time to take account of the work and any changes that have occurred."

Figure 1-18: Principal Designer's duties regarding the health and safety file. Source: CDMR 2015 Regulation 12.

The principal contractor may need to obtain details of services, plant and equipment which are part of the structure from specialist suppliers and installers, for example, mechanical and electrical contractors and pass this information on.

Contractors have a specific duty in CDMR 2015 to pass information for the health and safety file to the principal contractor, who in turn has to pass it to the principal designer. This information could include 'as built' and 'as installed' drawings as well as operation and maintenance manuals.

"(7) During the project, the principal contractor must provide the principal designer with any information in the principal contractor's possession relevant to the health and safety file, for inclusion in the health and safety file."

Figure 1-19: Principal Contractor's duty regarding the health and safety file. Source: CDMR 2015 Regulation 12.

At the end of the project the principal designer has to hand over the health and safety file to the client. In some cases it might not be possible for a fully developed file to be handed over on completion of the project. This may happen because the construction work was finished rapidly to meet a tight deadline and completion of the health and safety file was impossible. Clearly a common sense approach is needed so that the health and safety file is handed over as soon as practical after a completion certificate or similar document has been issued.

Appendix 4 of CDMR 2015 provides guidance on requirements for the health and safety file.

Contents of the health and safety file

The contents of the health and safety file will vary depending on the type of structure and the future health and safety risks that will have to be managed.

Typical information which may be put in the health and safety file includes:

- A brief description of the work carried out.

- Any hazards that have not been eliminated through the design and construction processes, and how they have been addressed (for example, surveys or other information concerning asbestos or contaminated land).

- Key structural principles (for example, bracing, sources of substantial stored energy - including pre or post-tensioned members) and safe working loads for floors and roofs.

- Hazardous materials used (for example, lead paints and special coatings).

- Information regarding the removal or dismantling of installed plant and equipment (for example, any special arrangements for lifting such equipment).

- Health and safety information about equipment provided for cleaning or maintaining the structure.

- The nature, location and markings of significant services, including underground cables; gas supply equipment; firefighting services etc.

- Information and as-built drawings of the building, its plant and equipment (for example, the means of safe access to and from service voids and fire doors).

CONSIDER

Consider maintenance or construction projects at home or work. List the main duties under the CDM Regulations.

1.3 Type, range and issues relating to construction activities

TYPES OF CONSTRUCTION WORK AND RANGE OF ACTIVITIES

> *"construction work" means the carrying out of any building, civil engineering or engineering construction work"*

Figure 1-20: Definition of construction work.
Source: CDMR 2015 Regulation 2.

Types of construction work
Building

Building involves most trades within the construction industry such as ground workers, steel erectors, brick layers, carpenters, plasterers, etc. all working closely together with the common goal of creating a new finished building or other structure.

Figure 1-21: Building.
Source: RMS.

Figure 1-22: Civil engineering.
Source: ICE.

Civil engineering and works of engineering construction

Civil engineering and engineering works normally relate to heavy construction activities requiring large items of plant and equipment such as cranes and excavators. It requires specialist knowledge and experience in order to undertake civil engineering activities such as highway construction, bridge construction, piling works, large foundations, large concrete structures, excavations and utility projects.

Engineering construction activities include the building of power stations, wind farms, chemical process plant and vehicle assembly plant.

Range of activities

Regulation 2 of CDMR 2015 establishes what construction work includes.

> *"(a) the construction, alteration, conversion, fitting out, commissioning, renovation, repair, upkeep, redecoration or other maintenance (including cleaning which involves the use of water or an abrasive at high pressure, or the use of corrosive or toxic substances), de-commissioning, demolition or dismantling of a structure;*
>
> *(b) the preparation for an intended structure, including site clearance, exploration, investigation (but not site survey) and excavation (but not pre-construction archaeological investigations), and the clearance or preparation of the site or structure for use or occupation at its conclusion;*
>
> *(c) the assembly on site of prefabricated elements to form a structure or the disassembly on site of the prefabricated elements which, immediately before such disassembly, formed a structure;*
>
> *(d) the removal of a structure, or of any product or waste resulting from demolition or dismantling of a structure, or from disassembly of prefabricated elements which immediately before such disassembly formed such a structure;*
>
> *(e) the installation, commissioning, maintenance, repair or removal of mechanical, electrical, gas, compressed air, hydraulic, telecommunications, computer or similar services which are normally fixed within or to a structure,*
>
> *but does not include the exploration for, or extraction of, mineral resources, or preparatory activities carried out at a place where such exploration or extraction is carried out;."*

Figure 1-23: What construction work includes.
Source: CDMR 2015 Regulation 2.

Examples of the activities included in the definition of construction work are outlined below.

Alteration

Alteration activities are required when the layout of an existing building, premises or other structure no longer suits the use for which it was originally intended. This can include elements of both new building works and renovation works.

Alterations can comprise an extension to an existing structure or demolition and removal of sections of the internal structure to make premises more spacious.

Alternatively, an alteration may involve dividing the existing structure into smaller, separate sections by the introduction of partition walls of various materials (block-work, brick-work or studding and plasterboard).

Renovation

Renovation activities involve restoring an existing building or structure to a condition that is representative of its original condition or improved by repair and modernisation using more up-to-date materials and practices. As with new building works, this also involves most of the common trades normally used within the construction industry.

Figure 1-24: Renovation.
Source: RMS.

Figure 1-25: Maintenance of existing structures.
Source: RMS.

Maintenance of existing structures

Maintenance work is an essential element to ensure that the condition of an existing building, premises or other structure does not deteriorate and that it remains in as good a condition as is possible. Work is normally carried out on a regular scheduled basis to deal with issues of wear and tear, but can also be necessary when a problem suddenly occurs that requires urgent attention, for example, loss of roofing materials following a storm.

Maintenance can be carried out on all components of existing structures, including the fabric of the structure, services, and any final finish or coating. Maintenance would include cleaning or applying protective coatings to a structure.

Cleaning can involve applying water, steam or various abrasive or chemical agents to the surfaces to be treated, for example, exterior walls using high pressure jets.

In some cases, the tools involved with applying protective coatings are handheld, but can also be powered. Various access systems, including mobile elevating work platforms, are usually required for this type of work.

Decommissioning

Decommissioning is concerned with the preparation necessary to take (permanently or temporarily) a piece of equipment or plant or other structure out of use. For example, the decommissioning of an air compressor that requires replacement or a building for a change of use (a warehouse into a manufacturing unit). Decommissioning also applies to taking a production site out of service, for example, a nuclear electrical power generation plant. The decommissioning process will include removal of mechanical, electrical, gas, compressed air, hydraulic, telecommunications, computer or similar services, which are normally fixed within or to a structure. In addition, it will include the safe disposal of any product or waste resulting from decommissioning.

Figure 1-26: Demolition equipment.
Source: RMS.

Figure 1-27: Demolition of a building.
Source: RMS.

Demolition

The term demolition refers to 'breaking down' or 'removing'. In construction activities, demolition is carried out on buildings and other structures that are no longer required or are possibly derelict and unsafe.

Account needs to be taken of potential hazards that may arise from demolition, for example, the risk of premature collapse, presence of live services, asbestos dust and falling debris. A pre-demolition survey is made to identify hazards and enable options for the demolition method to be considered.

Demolition must be carried out in a well-planned and controlled manner in compliance with a safe system of work. The arrangements for demolition should be written down before the work begins. This safe system of work may be in the form of a safety method statement identifying the sequence required to prevent accidental collapse of the structure.

A safe system of work may include:

- Establishing exclusion zones and hard-hat areas, clearly marked and with barriers or hoardings.

- Covered walkways, to protect pedestrians if materials should fall from above.

- Using high-reach machines for mechanical demolition.

- Reinforcing machine cabs so that drivers are not injured from falling debris.

- Training and supervising site workers, in site rules and incident reporting.

Dismantling

The term dismantling refers to taking things apart, often for reuse in another location. In construction, this is applied to parts of buildings and other structures that are no longer required where they are located, such as temporary structures used for outdoor events, for example, concert staging or temporary athletic tracks/ stadiums. In a similar way to demolition, arrangements for dismantling should be written down before the work begins. The safe system of work may be in the form of a safety method statement identifying the sequence to be followed to prevent accidental collapse of the structure and any temporary supports necessary to maintain stability.

Regulation 2 of CDMR 2015 refers to structures when establishing what construction work includes. Regulation 2 also defines what a structure is in relation to CDMR 2015.

"structure means:

(a) any building, timber, masonry, metal or reinforced concrete structure, railway line or siding, tramway line, dock, harbour, inland navigation, tunnel, shaft, bridge, viaduct, waterworks, reservoir, pipe or pipeline, cable, aqueduct, sewer, sewage works, gasholder, road, airfield, sea defence works, river works, drainage works, earthworks, lagoon, dam, wall, caisson, mast, tower, pylon, underground tank, earth retaining structure or structure designed to preserve or alter any natural feature, and fixed plant;

(b) any structure similar to anything specified in paragraph (a);

Figure 1-28: Definition of the term 'structure'.
Source: CDMR 2015 Regulation 2.

As can be seen by the definition, structure includes buildings (for commercial or private use), extends to other structure like bridges and transport systems, as well as water, gas, chemical and power transmission systems. The term structure includes supplementary structures used in the construction process.

Clearance and preparation of the site for use

Clearance consists of preparing the site prior to the works being undertaken. This may involve removal of hazardous waste, obstructive trees, unwanted scrub and landscaping. Where there are high levels of contaminants it may also be necessary for ground remediation.

Preparation of the site for use can include landscaping, which usually takes place during the final stages of the construction phase when there is little or no construction plant travelling around the site. Landscaping may include altering site levels and the introduction of trees, shrubs, grass turf/seed etc. The works generally consist of loading and unloading of materials, manual handling, and cleaning work (footways and roads).

Following completion of construction works clearance of the site will involve removal of all waste associated with the construction activities to a licensed waste disposal site, for example, brick and timber off-cuts, packaging and spoil. It will also include the removal of all plant and equipment used on the construction project, with the aim of leaving the site in a clean and tidy state ready for its intended use.

Figure 1-29: Excavation.
Source: RMS.

Excavation

Excavation consists of digging below ground level to various depths in order to create a cavity that can be used for exposure of buried utility services, trenching for installation of utility services, casting building foundations or ground investigations. Methods of digging used for excavation work include the use of hand tools (pick, fork, shovel), and also by using a mechanical excavator. Major excavations for basements, sub-structures etc., remain as a permanent part in the construction operation.

Figure 1-30: Structural work.
Source: RMS.

Structural work

Structural work can include steel erecting, welding and form-working. Quite often it can involve working at height (where specialist work platforms and fall arrest equipment should be used) and adequate control is essential to prevent the falling of people, materials and tools. A variety of specialist equipment may be required and should only be used by competent persons, for example, welding machine, bolt gun, or nail gun. Steel or concrete members have to be lifted into position, usually by means of a crane, which can present risks of being struck by materials and the crane overturning.

Site movements

Construction sites contain various types of heavy mobile plant and equipment and large numbers of site workers. This presents risks of collision with workers, as well as temporary or permanent structures. The operators of the plant and equipment may have limited view, which may increase risk. Construction sites use a wide variety of different materials in the construction process that will require loading, movement and unloading. Wherever possible this should be done by mechanical means, avoiding the need for manual handling. This will require construction site projects to be well planned so they take into account vehicles moving around the site. In particular, consideration to safe access/egress, adequacy of space for manoeuvring and ensuring operator visibility. Storage requirements should be identified and planned for the whole project. Routes for both vehicles and pedestrians should be provided and be suitably surfaced, clearly defined and separated.

Installation, removal and maintenance of services

Various utility services are required on both existing and new construction sites and are usually buried underground, for example, electricity, gas and water.

In new installations this involves a great deal of liaison with the planning authority, utility companies and the designer concerning the route the services will take and will involve excavation with heavy plant, loading and unloading of materials. New services are often connected to their source at a location that is situated outside the construction site boundary. Where this occurs, additional hazards to the general public may be created and suitable precautions will be required. For example, traffic control (by means of signing and lighting) as well as guarding and protection of the public by the use of barriers and warning signs. Removing or maintaining utilities can present other hazards. Utility services should be isolated correctly prior to any work commencing, as this may require involving the utility companies so proactive planning is essential.

Figure 1-31: Site movements.
Source: Unsplash.

Figure 1-32: Services.
Source: RMS.

WHY YOU NEED TO MAINTAIN THE STABILITY OF STRUCTURES

Construction activities carried out during construction work could threaten the stability of temporary or permanent structures. For example, structures can collapse unexpectedly during demolition if action is not taken to prevent instability. Scaffolds can collapse because ties are either forgotten or removed, or when the scaffold is overloaded or struck by moving vehicles/equipment. Structures under construction may also collapse, for example, steel frames that have not been adequately braced or formwork that is prematurely loaded could lose stability and collapse. Walls could fall because their

foundations are undermined by nearby excavations and whole buildings might become unstable and collapse during alteration due to becoming weakened and/or overloaded.

Workers could fall from an unstable structure or they could be struck or buried by falling materials when structures become unstable and collapse.

This could result in major injuries or death and could lead to severe damage to nearby structures or equipment and injury to members of the public. It is therefore important to maintain the stability of both temporary and permanent structures during construction work in order to prevent their premature or unplanned collapse.

Regulation 19 of CDMR 2015 emphasises requirements to maintain stability of structures.

"(1) All practicable steps must be taken, where necessary to prevent danger to any person, to ensure that any new or existing structure does not collapse if, due to the carrying out of construction work, it:

(a) may become unstable; or

(b) is in a temporary state of weakness or instability.

(2) Any buttress, temporary support or temporary structure must:

(a) be of such design and installed and maintained so as to withstand any foreseeable loads which may be imposed on it;

(b) only be used for the purposes for which it was designed, and installed and is maintained.

(3) A structure must not be so loaded as to render it unsafe to any person."

Figure 1-33: Stability of structures.
Source: CDMR 2015 Regulation 19.

1.4 Site assessment and control measures

FACTORS TO CONSIDER IN INITIAL SITE ASSESSMENTS

Historical and current use

The historical and/or current use of a site may present many hazards that need to be identified in an initial assessment before construction work starts. If, for example, it is a *'green field' (undeveloped) site*, it may provide public access/right of way to members of the public or a recreation area for children. The site could be private and used for agriculture or grazing of livestock.

If a **'brown field' (previously used, developed) site**, it could contain occupied or unoccupied buildings/ premises. Occupied premises will mean regular traffic on the site whilst unoccupied premises may be in a state of disrepair and dereliction. Any existing premises that are either occupied or unoccupied will, more than likely, be or have been connected to various below ground or overhead services that may require further investigation.

Figure 1-34: Site assessment - previous use, asbestos.
Source: RMS.

Figure 1-35: Site assessment - brown field site, access.
Source: RMS.

CONSIDER

Think of examples of 'structure' that could be subject to the CDM Regulations.

Unoccupied premises/buildings can pose other hazards, with unstable groundwork, footings, walls, beams supporting floors and roofing. In addition, unoccupied buildings are likely to be infested with vermin or suffer intrusion by pigeons in the rafters where roof integrity may have failed; this can increase the risks of **biological hazards.** Unoccupied premises may have been used by itinerant trespassers sleeping rough and the potential risk of needle stick injuries from discarded syringes will also need to be assessed. Again, these hazards should be highlighted, suitable risk assessments should be made and controls implemented.

It is important to take into account the full history of a site, as the site may present hazards in the form of **asbestos**

or **chemical contamination** which requires specialist waste removal and land reclamation services. In order to assess the risks from the possible presence of asbestos a specialist survey may be required.

There may be mineshafts present or other types of underground voids, such as abandoned cellars, manhole chambers or large diameter drains. Action should be taken to make site surveys and obtain current and/or historic plans that may identify any or all of the above circumstances.

Area of site, topography and features of the surrounding area

Area of site

The location or area of the site should be considered and any possible restrictions noted, for example, there may be trees that are protected and unable to be felled or other natural obstacles that could cause problems to the works, for example water courses. If a site is bounded by a main railway line or other premises, then the available space to store or operate any construction related plant and equipment may be restricted once work begins.

Topography

Topography relates to the physical surface conditions of the site and is an important factor to be considered along with the ground conditions below the surface.

The landscape may be flat and even or is it made up of banks, dips and hills therefore making any operations on site far more difficult to carry out. The ground conditions may be soft soil, clay or rock, each condition presenting its own individual problems and hazards. The site could have a high water table or be susceptible to becoming flooded or waterlogged.

Roads

Roads and highways that surround a site boundary can be a significant source of additional hazard and are a main area for consideration as access to any site is primarily via some form of roadway. Factors such as the type of road (dual carriageway, main road, one-way system, country lane), road speed encountered, volume of traffic (high all day, cyclical, rush hour), type of traffic using the road (agricultural plant, cars, heavy goods vehicles (HGVs), well-lit or unlit, and capacity of road (weight, height or width restrictions).

It may be that special requests or notifications are to be made with the local authority regarding access and site traffic proposals.

If the site is within a residential area or in close proximity to a school, permissions or restrictions may be enforced regarding when or if surrounding roadways may or may not be used. In any of the described situations, there is a potential for danger.

Figure 1-36: Roads.
Source: RMS.

Figure 1-37: Footpaths.
Source: RMS.

Footpaths

Footpaths are a means of providing pedestrians with a safe means of travelling by foot usually alongside a highway. It should be identified how the footpath is used (for example, for a school journey by children or by people queuing for a bus). The risk of injury may be increased at the entrance to a construction site where pedestrians cross the access opening and might encounter heavy site mobile plant or goods vehicles delivering materials to site. Footpaths may also skirt the construction site boundary, where pedestrians may also become at risk due to the activities within the site (for example, falling objects from a scaffold structure, flying objects from cutting, drilling or hammering operations, fumes, dust, chemicals).

Railways

Railways present the hazard of heavy, high speed trains (in excess of 125 mph) that do not have the ability to respond to dangerous circumstances that may arise as other transport modes are able to, i.e. quick emergency stopping, avoidance by changing direction.

In addition to these hazards, there may be overhead cables carrying 25,000 volts or rails carrying 750 volts. Work near to railways requires suitable planning, as any clash between a travelling train and site equipment, plant or vehicles could, and most likely would, lead to disastrous consequences. Rail authorities have laid down strict procedures that are to be followed and any

party working on railways should be in possession of a 'Personal Track Safety' certificate.

High-visibility clothing that is worn on or near to a railway line should be of the correct standard and colour (high-visibility orange). Restrictions on colours worn on or near to a railway should be strictly followed and nothing that is red or green be worn due to the fact that it may be mistaken as a signal by a train driver.

Communications should be maintained with the rail authority and notification given of any works being carried out on or near to the railway in order that all issues can be complied with correctly.

Figure 1-38: Railways.
Source: Welsh Highland Railway.

Figure 1-40: Waterways.
Source: RMS.

Waterways

Waterways located on or near construction sites present a risk of drowning; the water does not have to be fast flowing to cause a worker to get into difficulty. Other factors to consider if waterways are in close proximity to a construction site are: the likelihood of floods occurring, environmental pollution of the waterway, damage to associated wildlife by site activities. Conversely, it is necessary to consider exposure of site staff to hygiene hazards through contaminants or disease (chemical pollutants, Weil's disease) within the waterway. Waterways are used by a considerable number of boat operators that may be affected by the activities on site (for example, falling objects from a scaffold structure, flying objects from cutting, drilling or hammering operations, fumes, dust, chemicals). Boat operators may equally affect site safety, for example, collision of the boat with a scaffold structure.

Residential/commercial/industrial properties

Construction can take place in local or immediate proximity of residential, industrial or commercial property that may be either unoccupied or occupied and fully operational. Construction traffic will generally add significantly to the normal traffic loading in the area, which may result in congestion, for example, during the 'school run' period and lunch times. This may well be exacerbated by vehicles queuing near the site entrance to gain access. The increase in large goods vehicles (LGVs) in the area will lead to increased engine noise, vehicle exhaust fumes and the risk of collisions with other vehicles, pedestrians. Young children and the elderly may be particularly at risk, as they may choose to weave through semi-stationary traffic and may not be visible to large construction vehicle drivers when they decide to move their vehicles.

There are additional risks where work is carried out in commercial property (shops, offices) as the work may take place in busy areas where members of the public and staff are present. In the case of industrial properties (factories, workshops), people may be at work when construction activities take place and the premises can present additional hazards from machinery, plant, equipment, chemicals, etc. being used. Construction workers may not be familiar with the hazards presented, which could increase the risk of personal injury.

Figure 1-40: Commercial properties.
Source: RMS.

Figure 1-41: Industrial properties.
Source: RMS.

Schools

Schools are very busy areas and accommodate children of various ages that may have no or little perception of danger or risk and by their nature are often very inquisitive of their surroundings. In addition to children, parents or carers that deliver children to and collect children from the school create pedestrian and vehicular traffic hazards around the immediate area. Children are frequently tempted to try to gain access to construction sites and normally achieve this when site security is poor and consideration has not been given to access through small openings. Sites often underestimate the size of openings, which some small children are able to fit through, particularly when compared with that which is required to restrict adult access.

Means of access

Restrictions and hazards relating to safe access to a construction site may include the conditions of the highway from which access is being gained (size, speed, use). The traffic that will access and operate on the site needs to be considered (size, type, frequency, volume). Height restrictions should be assessed to determine how they could affect access, for example, service pipelines.

Figure 1-42: Means of access - overhead restrictions.
Source: RMS.

Figure 1-43: Overhead services.
Source: RMS

Presence of overhead and buried services

Overhead services in the form of electricity cables present the obvious risk of electrocution through either making direct contact with the electricity cable or where arcing (discharge by electrical energy 'jumping' to a near earth point) occurs. Overhead cables should therefore be identified prior to site works and the risk factors determined.

Buried services (electricity, gas, water, etc.) are not obvious and the likelihood of striking a service when excavating, drilling or piling is increased without a site survey to identify any service present. The results of striking an underground service are varied, and the potential to cause injury or fatality is high. Incidents can include shock, electrocution, explosion and burns from power cables, explosion, burns or unconsciousness from gas or power cables, and impact injury from dislodged stones/flooding from ruptured water mains.

SITE CONTROL MEASURES

Site planning

Following an initial site assessment that has identified hazards and risks associated with the works, control measures should be implemented to ensure the safety, health and well-being of all those who are affected by the construction site and its activities. A plan must be assembled to take into consideration the factors identified in the initial site assessment and reflecting the intended construction activities.

Regulation 13 of CDMR 2015 establishes requirements that the principal contractor plan construction work to prevent and reduce risks, which would include site planning. The principal contractor needs to take account of the general principles of prevention set out in Appendix 1 of CDMR 2015 when carrying out site planning, for example ensuring collective protective measures are given priority over individual protective measures.

> *"(1) The principal contractor must plan, manage and monitor the construction phase and coordinate matters relating to health and safety during the construction phase to ensure that, so far as is reasonably practicable, construction work is carried out without risks to health or safety;*
>
> *(2) In fulfilling the duties in paragraph (1), and in particular when:*
>
> *(a) design, technical and organisational aspects are being decided in order to plan the various items or stages of work which are to take place simultaneously or in succession; and*
>
> *(b) estimating the period of time required to complete the work or work stages;*
>
> *the principal contractor must take into account the general principles of prevention."*

Figure 1-44: Duty to plan construction work.
Source: CDMR 2015 Regulation 13.

Where possible planning should ensure that facilities are in place to minimise the risks before work activities commence. For example, in order to reduce the risks to people and equipment roads and walkways should

be installed at the earliest possible time. Similarly, where possible services for the final structure should be planned to be installed at an early stage in the development of the site so they are available for site activities and to prevent the need to rely on temporary arrangements for electricity, water and drainage. The planning process should enable workers to have safe access and to work from a permanent place of safety, where practicable, by establishing permanent access to workplaces at height as construction progresses.

Figure 1-45: Means of access - provided at early stage
Source: RMO.

Figure 1-46: Signs warning of buried services.
Source: Pixabay.

Signs

The provision of suitable signs is an important factor in making people aware of the hazards and precautions that relate to construction work. The entrance to a construction site should have an information board that provides details of the client, architect, supervisor, contractor and any other important parties. It will also include mandatory health and safety signs such as those requiring the use of hard hats or ear protection. Other information could include a basic site layout plan and details of where to report upon arrival at the site.

Visitors

Make sure site visitors report to the person in charge of the site and know where to go – notices may be required at the site entrance. Visitors should not be allowed to wander around the site alone. Therefore, a waiting area and booking-in system may be needed.

Offices

There should be an office available on site to keep the project notification to the enforcing authority and construction phase health and safety plan (if applicable), first-aid facilities and equipment, emergency procedures, health and safety information, site induction documentation, construction drawings and specifications. General site office equipment and records are also stored in the site office, for example telephone, computer, training records, visitor personal protective equipment and filing facility.

Provision of utility services

The planning of the provision of utility services to the site will involve consideration of the initial needs of the site, further needs as the site becomes established and the long term needs of the client who will take over the site. This may initially involve the use of portable, self-contained services, for example, water stored in a mobile tanker, electricity supplied by generators and portable welfare facilities with contained waste.

As the site develops, demands for utility service may increase and a more robust provision may be required. It may be possible to utilise the services that have to be put in place for the benefit of the client on completion of the construction work or it may be necessary to establish alternative sources. The identification of utility services required for the site should be determined in liaison with local supply companies before work commences, at the planning stage.

Certain utilities, such as gas or fast internet broadband, may not be available. Other services may be limited, such as water supply (particularly pressure), drainage or electrical capacity. Limitations in the water supply may need to be addressed by the provision of storage tanks, particularly if there is an identified need for a water sprinkler system or other firefighting provisions.

Drainage may need to be provided through the use of septic tanks or other means, such as organic reed bed installations. Electrical supply may require the organisation and suitable location of an additional electrical sub-station to increase the capacity of supply.

Figure 1-47: Poor storage
Source: RMS.

Figure 1-48: Electricity generator.
Source: RMS.

Storage

The storage requirements are dependent on the type of material to be stored. Different material types should be stored separately, to avoid cross contamination and the potential for a harmful adverse chemical reaction. If combined storage is permitted, it is advisable that different materials be kept separated for easy identification and retrieval. In addition to authorised access of site operatives, the storage area should be designed to allow safe access by fork-lift trucks and for the use of suitable mechanical lifting aids to eliminate the need for manual handling where possible. Housekeeping should be of a good standard and kept clean and tidy with suitable lighting and ventilation. There should be appropriate fire precautions and provision of correct fire extinguishers.

Storage areas should be restricted to authorised site staff only and have the correct safety information signs that warn of any dangers or mandatory signs enforcing the wearing of personal protective equipment (PPE).

They should be used for solely that purpose and must not be used for any other purposes such as mixing operations or as a rest or smoking area.

Specific arrangements should be made for the storage of flammable substances, see Element 11.2 Fire – Preventing fire and spread.

Loading/unloading

Construction sites are constantly changing with a variety of tasks all happening at once, including continuous worker, vehicles and material movements around the site. With this in mind, it is very important that any loading and unloading of materials used on the construction site is carried out in its own dedicated area/zone under competent supervision. The delivery of materials or equipment to a construction site should be scheduled and planned to minimise the volume of traffic at site at any one time, and drivers/operators of delivery vehicles given a site specific induction to ensure an awareness of hazards, site safety rules and procedures.

Loading and unloading of materials or equipment should be planned and organised. Materials should not be allowed to be placed too high or to lean over creating a potential risk of collapse with the possible result of major injury or a fatality. A number of people have been killed on construction sites due to carelessness in handling. Generally, loads should be kept small with goods on pallets that are in good repair. Lightweight boxes that offer no strength and can be crushed or bags of free-flowing solid (gravel) should always be palletised and never be stacked on top of each other. Different types of container should be stacked separately.

Loads should not block or be placed near emergency exits, fire points/extinguishers, vehicle routes or block light or visibility. They should be placed on firm, solid ground or floors that are strong enough to provide adequate support. Workers should be trained in the correct way to handle loads and also in the dangers associated with unsafe practices/or incorrect loading methods. Site staff should be instructed never to climb up or stand on top of racking or palletised material.

Lighting

Adequate lighting should be provided that allows safe access and egress into and around the construction site along roadways, vehicle and pedestrian routes. Special consideration should be made to areas such as scaffold structures, excavations, flammable stores or fuel bunkers/tanks. Signs warning of dangers at site should be clearly lit to allow approaching people advance notice.

Remediation works

Remediation works may need to be carried out prior to the development of a new site or the modification to an existing structure. Remediation can be necessary due to the effects on the site of its prior use, which may lead to contamination of land, water and structures. Former sites that may need remediation include fuel stations, gasworks, chemical storage facilities, industrial factories (for example, wire, rubber and chemical manufacture).

The substances forming the contamination can represent significant health hazards. They may include benzene, hydrocarbons, acids, cyanide and metals, for example, arsenic, nickel, cadmium and chromium.

Remediation work can vary from removal of contaminated soil, for example, with the redevelopment of a brown field site to the identification and establishment of controls for building contaminates, such as asbestos prior to an extension being made to an existing structure. The work may involve removal of hazardous materials left over from prior processes, excavation of contaminated soil, removal of contaminated surface water, cleaning of contaminated soil/water, stabilisation of land, treatment of invasive species of weeds, building temporary bunds

to control water runoff and the recovery, movement or protection of human or infected animal remains.

Site preparation/remediation work that involves the removal of topsoil that is contaminated with heavy metals or other hazardous substances will require specific arrangements in relation to personal protective equipment requirements and practices to be carried out. The work is likely to require the provision of gloves, overalls, eye protection and respirators. Additionally, the work will necessitate campaigns of awareness training to ensure good levels of personal hygiene are maintained, with emphasis on ensuring that open cuts are covered with waterproof dressings whenever required.

Consideration will need to be given to specific welfare facilities provision, for example, a decontamination unit may be needed, which has a dirty area where contaminated clothing can be removed after work, an area with a shower or other means of washing and a clean area where normal clothes could be kept while workers are working on site. There should be arrangements in place in order to prevent contamination when eating and smoking. First-aid and emergency decontamination facilities should be located near to the place of work.

Site preparation for specialist activities

Lifting

Lifting operations on construction sites can vary widely dependent upon the size and weight of the load required to be lifted and the lifting equipment required to carry out the lift. Specialist lifting equipment on construction sites is usually either mobile crane or tower crane type.

Prior to undertaking lifting operations with any piece of lifting equipment, vital preparation is required to ensure the safety of all those affected. This includes thorough planning of the lift to include a safe system of work, method statements, risk assessments, permits-to-work (if required) and competent authorised persons assigned to manage the lifting operation.

The ground on which the lifting equipment is to be situated should be firm and level and suitable load spreading decking provided to ensure stability. Depending on the nature and frequency of the lifting operations, it may be necessary and appropriate for the ground to be made up with hardcore or concrete to ensure strength and stability in all weather conditions.

Consideration should also be given to surrounding structures and overhead power cables, ensuring the minimum safe working distances from them. Access to the area below the lifting area should be controlled or restricted to protect people at site from items falling during the lift. The equipment used for the lift must be inspected prior to use and regularly as per the specified frequencies and all test certificates made available.

The Lifting Equipment and Lifting Operation Regulations (LOLER) 1998 are specifically related to mechanical lifting and associated equipment.

Piling

Piling is a method of creating long, straight underground cavities that are used for stabilising/strengthening the ground and providing foundations for a building or structure. This is performed by either using drilling or by compacting force.

The most common method of piling involves the use of a guide 'tube' in which a heavy solid 'driving piece' supported by lifting equipment is released down the guide tube causing the material it strikes to displace upon impact and thus create a cavity when withdrawn. This operation is repeated until the required depth of bore is achieved. Another method involves a steel liner being mechanically 'hammered' into the ground to the required depth.

Preparations for piling works may include identifying buried services, hidden voids, the density of the material being piled and ground stability for equipment, water table levels and any local buildings or structures that may be affected by vibrations. Piling is a very noisy operation and consideration should be given to surrounding areas.

Figure 1-49: Lifting.
Source: RMS.

Figure 1-50: Steelworks.
Source: RMS.

Steelworks

Steelwork is often used to form the basic 'skeleton' on which a structure is built. The majority of steelwork is designed and manufactured off site and is then assembled using the various pre-fabricated sections and steel beams during the build. Associated activities usually involve working at height with materials that are very heavy and difficult to manoeuvre.

Site preparation involves the creation of large, level concrete surfaces, on which the steelwork stands and is fixed. The erection of steelworks requires thorough planning to include safe systems of work, risk assessments, method statements and permits to work where required.

Where practicable, risks should be eliminated or reduced to the minimum possible level by combating hazards at source (for example, use of mobile elevated work platform for access, cranes for lifting). Any lifting equipment or access equipment should only be used on firm and level ground.

Security

General security arrangements

The principal contractor has the responsibility to ensure necessary steps are taken to prevent unauthorised access onto the construction site. Regulation 18 of CDMR 2015 requires that, depending on the level of the risk and what is reasonably practicable, the perimeter of the site must be identified by suitable signs and site boundaries must be readily identifiable and/or be fenced off. This would establish minimum requirements for site security. The principal contractor should also take steps to ensure that only those authorised to access the site do so, this may mean by provision of identification and physical or electronic checks.

> *"(4) The principal contractor must ensure that:*
>
> *(b) the necessary steps are taken to prevent access by unauthorised persons to the construction site;"*

Figure 1-51: Duty to prevent access to a construction site.
Source: CDMR 2015 Regulations 13.

Arrangements for site security should be planned at an early stage and will need to reflect the nature and level of risks that the construction site poses. For example, if houses are being built from timber construction it is important to have sufficient site security in place to act as a deterrent to possible arsonists, for example, perimeter fence, lighting, alarms and security officers. Similar arrangements may be required if the site is located in an area where it could be an attraction to young people, in order to deter them from entering the site. Similarly, special consideration will be required for sites that have rights of way through them or other work areas next to them, for example, when carrying out a shop refurbishment in a shopping centre.

It is a requirement of CDMR 2015 that contractors must not start work until reasonable steps to prevent access are in place. The Principal contractor should liaise with the contractors on site to physically define the site boundaries by using suitable barriers which take account of the nature of the site and its surrounding environment.

> *"(10) A contractor must not begin work on a construction site unless reasonable steps have been taken to prevent access by unauthorised persons to that site."*

Figure 1-52: Duty to not start work unless measures in place to prevent access. Source: CDMR 2015 Regulation 15.

Means of securing plant, chemicals, etc.

Plant, equipment, materials or chemicals should be suitably secured to prevent injury by unauthorised access. Plant should be locked up at all times when not in use and keys held in a secure location (site office, safe). It may be practical to keep plant in an additional internal site compound when it is not being used. In order to improve security certain items of heavy mobile plant are provided with steel sheets or shutters fitted around the cab, which are padlocked in position when the site is closed. Chemicals, for example flammable liquids, should be kept in secure storage areas.

Client/Occupier arrangements

Occupied premises

Construction related activities that are carried out at a client's premises, while the premises are occupied, must take into account the lack of knowledge of construction hazards of the client's workers and visitors using the premises. This lack of knowledge may put them at risk from the hazards. Hazards from paint, dust, falling masonry and excavations should be considered, suitable and sufficient risk assessments should be made and controls implemented. Particular attention will need to be given to assessing where the contractor - occupier interface presents a risk. Risks will increase when a high volume of the occupier's workers interact with the construction work, such as at the start of work, lunch period and close of work.

Careful planning of construction or maintenance work should be taken prior to work commencing on occupied premises. Factors to be considered include establishing controls to segregate contractor workers from the occupier's workers. This may require site rules to be amended and be supported by induction, personal protective equipment issue, signs, barriers, lighting or verbal instruction. Where there is a risk from the construction work, such as falls of materials from a height, walkways will need to be covered to protect pedestrians. This may include establishing separate parking, traffic and pedestrian routes.

Site rules

Site rules will vary at different sites or premises due to the wide range of activities that may be undertaken. Site rules provide instructions that must be followed by permanent site staff and visitors and also other important information relating to site/location specific hazards. Occupiers of premises or clients may have different standards of site rules and some may enforce them more stringently than others. Contractors should always enforce their own site rules in addition to client/occupier rules.

Co-operation

Co-operation between client, contractor and occupier is a very important factor. The occupier of a premises or site will have a detailed knowledge of any site specific hazards that may or may not be obvious to a contractor undertaking construction works and this may impact upon the works. In addition, the client or occupier has the authority to place controls and restrictions on the site. The contractor should be experienced in the activities that will be carried out and will have assessed any hazards related to the activities that are to be carried out at site. It is vital that all parties co-operate and communicate in order that this knowledge and information can be assessed to determine any new hazards that may arise and to enable appropriate information to be cascaded to other people at risk on the site. Co-operation will also be required where site activities need to be controlled or access restricted or where a shared knowledge is required to undertake a task, for example, decommissioning or removal of machinery.

Shared facilities

Occupied premises will quite often have various facilities available for existing workers/occupiers, for example, hot and cold water, toilets and rest facilities. It may be acceptable, with the agreement of the person in control of the premises/facilities, for these to be shared for mutual benefit. CDMR 2015 establishes a duty on the client, contractors and the principal contractor to ensure that welfare facilities are provided. Therefore, where the client occupies the premises where construction work is to take place, it is in the interests of both the client and contractors to agree what facilities are required and what can be made available to the contractors.

First-aid and accidents

It is the duty of the employer (this could be the main contractor), under the Health and Safety (First Aid) Regulations (FAR) 1981, to provide first-aid provisions and to inform the workers of these arrangements. These facilities should be in a clean environment, probably sited in a separate site office or portacabin.

The Reporting of Diseases and Dangerous Occurrences Regulations (RIDDOR) 2013 require a record of prescribed accidents to be kept on site. All reportable injury accidents and dangerous occurrences must be recorded, deaths being additionally notified by telephone directly to the Health and Safety Executive (HSE).

1.5 Site order and security

THE NEED FOR SAFE ENTRY AND EXIT FROM THE SITE

Regulation 17 of CDMR 2015 establishes requirements for safe access and egress, which includes safe entry and exit from the site.

"(1) There must, so far as is reasonably practicable, be suitable and sufficient safe access and egress from:

(a) every construction site to every other place provided for the use of any person whilst at work; and

(b) every place construction work is being carried out to every other place to which workers have access within a construction site.

(3) Action must be taken to ensure, so far as is reasonably practicable, that no person uses access to or egress from or gains access to any construction site which does not comply with the requirements of paragraph (1)".

Figure 1-53: Safe access and egress from a construction site.
Source: CDMR 2015 Regulation 17.

Arrangements for a site access

Access to a construction site must be planned to minimise any hazards identified during the initial assessment. When a suitable location for site access has been identified, this must be controlled at a point or points that are designated as authorised access point/s and the remainder of the construction site boundary must be secure against unauthorised access. Site safety information should be displayed at access points to inform all attendees of contact names, rules and emergency procedures. Barrier systems are normal on complex undertakings, accompanied by procedures for admittance and exit of people and plant. Special rules for site safety induction and issue of any site security identification are often used to maintain control.

There should be safe access onto and around the site for people and vehicles. Controls need to be in place to avoid vehicles entering the site queuing back onto the public highway; these will include the close monitoring of delivery schedule arrangements. A banksman may be needed to direct the vehicles and other traffic, partic-

ularly if the site entrance is restricted and a number of forward and backward movements of delivery vehicles are necessary to enter the site.

Figure 1-54: Site access.
Source: RMS.

Figure 1-55: Site access and roadways.
Source: Unsplash.

Public roadways often extend to or continue into construction sites and are subject to pedestrian and vehicular traffic. Construction site roads should be subject to site safety rules such as speed limits, safety restraints (seat belts), direction of flow (one-way systems where possible), and there should be segregation of pedestrian and vehicular traffic similar to that used on the public highway. Adequate space should be provided for parking and to allow vehicles and plant to manoeuvre safely and the area should be well lit as required. Vehicles and mobile plant should never be allowed to block roadways or access points as this may prevent emergency vehicles and crews gaining access in the event of a major incident. Site slurry/mud on the road surface that can create skid hazards for site plant and vehicles should not be allowed to accumulate and there should be a system for inspection and cleaning of the surface and vehicles' wheels.

Consideration should be made at the planning stage to ensure how vehicles will be kept clear of pedestrians, especially at site entrances, at vehicle loading/unloading areas, parking and manoeuvring places and areas where drivers' vision may be obstructed. It may be necessary to provide separate doors or gates for vehicles and pedestrians to achieve effective segregation at site entrances. Where practical, separate routes should be established for vehicles and pedestrians, involving raised pathways, barriers or markings. Doors that open onto traffic routes may need viewing panels or windows.

Means of controlling dangers on public highways

Areas that surround a construction site are subject to mud and debris from the tyres and chassis of vehicles that exit the site, which creates additional hazards to other road users as highway surfaces become slippery and create skid hazards. This can be controlled by the implementation of regular highway cleaning with road sweeper vehicles. Action can also be taken at site exits prior to vehicles leaving the site by routing site traffic through a tyre and undercarriage cleaning system, which assists in preventing mud and debris leaving the site.

If work continues in the hours of darkness, additional lighting may be necessary at site entrances and exits to improve the delivery drivers' general perception of hazards and pedestrian visibility.

SAFE AND SUITABLE ARRANGEMENTS OF THE WORKING SPACE

Regulation 17 of CDMR 2015 establishes requirements for the site to be kept safe, sufficient working space and how it is organised, including layout, storage and good housekeeping.

> *"(2) A construction site must be, so far as is reasonably practicable, made and kept safe for, and without risks to, the health of any person at work there.*
>
> *(3) Action must be taken to ensure, so far as is reasonably practicable, that no person uses access to or egress from or gains access to any construction site which does not comply with the requirements of paragraph …(2).*
>
> *(4) A construction site must, so far as is reasonably practicable, have sufficient working space and be arranged so that it is suitable for any person who is working or who is likely to work there, taking account of any necessary work equipment likely to be used there."*

Figure 1-56: Safe places of construction work.
Source: CDMR 2015 Regulation 17.

All workplaces must be laid out and workstations arranged so that there is no harmful effect on the health and safety workers. Specifically, each worker should have sufficient unobstructed working space to perform their work. This will mean giving consideration to the equipment and materials they need and any waste produced by their work. There should be sufficient space so that they can pass each other in the working space and manoeuvre materials and use the equipment

without the risk of being struck by other workers or that they are so close that they will be injured by the operation of the equipment.

Good housekeeping includes following the philosophy of 'a place for everything and everything in its place.'

Good housekeeping will benefit from the use of the following control measures:

- Identification of storage areas.
- Tidy storage of tools, equipment and materials.
- Marking areas to be kept clear.
- Providing areas for waste material.
- Paying particular attention to keeping emergency routes clear.

Introduce procedures for reporting defects and for dealing with obstacles on the floor that could cause workers to trip or spills that might reduce the slip resistance of surfaces. This will help to ensure high standards of housekeeping are maintained and keep floors clear of obstructions, debris and spills. Maintenance activities should include repair of such things as damaged tiles and carpets - particularly on routes, at the threshold of doorways and at changes of level. Suitable and sufficient lighting should be provided and maintained in order to preventing slips and trips, enabling people to avoid contamination that is on the floor and obstructions in the workplace.

Competent supervision of activities is essential to monitor and control any changes to the work, making sure any additional housekeeping is conducted at regular intervals to prevent slips and trips from debris created during the work.

Regulation 18 of CDMR 2015 specifically deals with good order and cleanliness, including good housekeeping, it requires that:

> *"(1) Each part of a construction site must, so far as is reasonably practicable, be kept in good order and those parts in which construction work is being carried out must be kept in a reasonable state of cleanliness."*

Figure 1-57: Good order.
Source: CDMR 2015 Regulation 18.

Employers should follow an effective cleaning regime for floors and walkways in order to maximise the surface roughness and slip resistance of floors. Consider available advice from the flooring manufacturer to ensure the regime is repeated often enough and is effective in removing layers of contamination and residues of cleaning agent. Frequent spot cleaning can supplement cleaning the whole of a floor, which can be useful where spills occur or where there are routes that are used higher than the average amount.

Indoor locations in the workplace may be subject to frequent wetting during normal use, for example, welfare facilities such as shower areas, or cleaning operations, for example, a car wash facility. Where the floor or pedestrian route in these areas is likely to frequently get wet, drainage should be provided.

Dust is a common hazard generated by many processes and could, if allowed, enter the atmosphere and spread throughout the workplace. Apart from the health issues that could affect workers, the dust may present a risk of slipping where it has built up on floors and walkways, particularly where the dust is large and granular. Dust that does settle on floors or walkways should be removed by means that do not cause dust to spread in the atmosphere, for example, vacuuming or wet scrubbing methods.

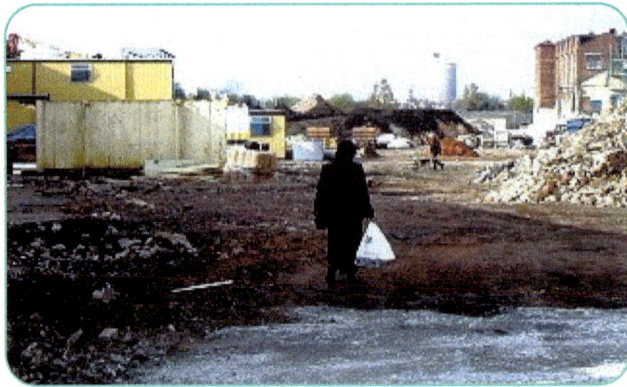

Figure 1-58: Unauthorised access.
Source: RMS.

THE REQUIREMENTS TO IDENTIFY THE SITE PERIMETER, EITHER WITH SUITABLE SIGNS OR FENCING

As discussed earlier when considering site security, regulation 13 of CDMR 2015 places a duty on the principal contractor to ensure that:

"The necessary steps are taken to prevent access by unauthorised persons to the construction site."

Figure 1-59: Duties of a principal contractor to prevent unauthorised access.
Source: CDMR 2015 Regulation 13.

This provides an obligation on the principal contractor to consider what steps are necessary, which will depend on the nature of the construction work and the risks related to it.

As illustrated in regulation 18 of CDMR 2015 this may, in low risk situations, be by means of suitable signs or a perimeter fence may be required.

"(2) Where necessary in the interests of health and safety, a construction site must, so far as is reasonably practicable, and in accordance with the level of risk posed, comply with either or both of the following:

(a) have its perimeter identified by suitable signs and be arranged so that its extent is readily identifiable; or

(b) be fenced off."

Figure 1-60: Site perimeter identification and fencing.
Source: CDMR 2015 Regulation 18.

Figure 1-61: Simple fence and sign demarcation.
Source: RMS.

Perimeter fencing

Where the level of risk requires, construction sites must be contained within a perimeter fence. The main purpose of fencing is to keep out unauthorised persons (for example, members of the public including children) and to prevent injury or death.

Perimeter fencing also provides security against theft of materials, plant or equipment from the site. Fences should be adequate and suitable and installed at a reasonable distance from the structure to allow unrestricted movement on site of people and mobile plant, and prevent any activities being undertaken affecting the environment outside the fence. The fence should be regularly inspected to ensure there is no damage, breaks or gaps to allow unauthorised entry. When one piece of fencing is fixed to another, the fixing should face into the protected site, where possible, so as to impede tampering and removal of the fences by unauthorised persons.

Figure 1-62: Perimeter fencing and signs.
Source: RMS.

Figure 1-63: Solid perimeter fencing.
Source: RMS.

Signs

Signs should be fixed at regular intervals on the perimeter fence to warn of the dangers within the site and instruct people to 'keep out'. Quite often, the name of the security company that is responsible for 'out of hours' security will also be displayed with a telephone number for emergency contact or to report any trespass.

Safe viewing points

Members of the public are quite often intrigued by construction sites and can be attracted to the perimeter fence to see for themselves what is going on.

This may result in injury from flying particles, dust, fumes or splashes, even though the person is outside the perimeter fence. This can be avoided by arranging for a pre-planned viewing point that consists of a wire mesh panel integrated into the fence that allows members of the public to view the site's activities. The viewing point will be planned and situated in an area that is not exposed to hazards.

OUT OF HOURS SECURITY ARRANGEMENTS

Arrangements for out of hours site security should be considered and planned at an early stage and will need to reflect the nature and level of risks that the construction site poses. For example, if houses are being built from timber construction it is important to have sufficient site security in place to act as a deterrent to possible arsonists when the normal workforce are not working, for example, movement or heat sensitive lighting/alarms and security officers. Similarly, equipment and materials kept on site may be vulnerable to theft when the site is not occupied.

Many children see construction sites as adventure playgrounds. Even though they may be entering the site without authority out of hours, they should still be protected from site dangers, many will be too young to appreciate the risks they are running.

At the end of the working day the following action should be taken:

- Isolate and immobilise vehicles and plant, if possible lock them in a compound.

- Lock away hazardous substances.
- Store building materials so that they cannot fall, for example, pipes, bricks and blocks.
- Remove or 'lock off' access ladders from scaffolds and excavations.
- Barrier off or cover over excavations.

Cameras may be used to enable local or remote monitoring of the site. In addition, it may be necessary for security officers to patrol the site to check security precautions are in place and that at risk items, for example, chemicals and vehicles, are satisfactorily secured.

1.6 Management of temporary works

Temporary works are defined under BS 5975 as "Parts of the works that allow or enable construction of, protect, support or provide access to, the permanent works and which might or might not remain in place at the completion of the works". Examples include, but are not limited to:

- Earthworks such as support to trenches, excavations and temporary slopes.
- Structures such as edge protection, falsework, shoring, edge protection, access and support scaffolding, bhoarding and signage and site fencing.
- Equipment and plant foundations such as tower crane bases, supports, anchors and ties for construction hoists and mast climbing work platforms; groundworks to provide suitable areas for plant erection, for example, mobile cranes, piling rigs and MEWPs, site cabin foundations, access roads and hard standing for plant and material storage.

The management of temporary works is an essential component of effective risk control.

"BS 5975:2019 - Code of practice for temporary works procedures and the permissible stress design of falsework" provides recommendations and guidance on the procedural controls to be applied to all aspects of temporary works in the construction industry, as well as specific guidance on the design, specification, construction, use and dismantling of falsework.

Contractors should be able to demonstrate that they have arrangements in place for controlling risks that arise from the use of temporary works. Industry best practice suggests that these arrangements will follow a procedure based on the following steps:

- Appointment of a Temporary Works Co-ordinator (TWC).
- Preparation of an adequate design brief.

- Completion and maintenance of a temporary works register.
- Production of a temporary works design (including a design risk assessment and a designer's method statement where appropriate).
- Independent checking of the temporary works design.
- Issue of a design/design check certificate, if appropriate.
- Pre-erection inspection of the temporary works materials and components.

There is no legal requirement to have a TWC however the legal requirement is that the party in control of a project (regardless of size) must ensure that work is allocated and carried out in a manner that does not create unacceptable risk of harm to workers or members of the public. On small projects it might not be necessary to appoint a TWC. However, the risks associated with temporary works still need to be properly managed to ensure safety. On medium or larger projects, a dedicated TWC will hold a vital role to ensure the safety of all involved.

The role of the TWC includes but is not restricted to:

- Coordinating the design, erection, inspection, use, monitoring, maintenance and dismantling ogf temporary works.
- Ensuring that adequate design briefs are created.
- Ensuring that those involved in the temporary works process are competent.
- Ensuring that a temporary works register is created.
- Ensure that responsibilities are allocated and accepted.
- Ensuring that residual risks are identified.
- Ensuring adequate design of temporary works is carried out.

1.7 Other construction issues including welfare arrangements

PROVISION OF WELFARE FACILITIES

CDMR 2015, Regulation 4, establishes an absolute duty on the client to make suitable arrangements to ensure that Schedule 2 of CDMR 2015, regarding welfare facilities, is complied with. The principal contractor also has an absolute duty under Regulation 13 to ensure that Schedule 2 is complied with, throughout the construction phase. A similar, but conditional duty, so far as is reasonably practicable, exists for contractors under Regulation 15, which requires them to ensure welfare facilities are provided for themselves and any other worker under their control.

Generally, on construction sites temporary welfare facilities are provided so that they may be relocated as the build phase of the work is carried out. If the work involves refurbishment or extension to an existing building, separate facilities may be required to segregate construction site workers from the occupiers. This may be a requirement of the contract with the client, but often is necessary to maintain the required number of facilities, such as sanitary facilities, for the extra number of people on site. Other reasons may be to meet specific legal requirements, such as the Control of Asbestos Regulations (CAR) 2012, where there will be a requirement to provide separate washing and clothing facilities for those working with asbestos or its removal. Construction activities may involve workplaces away from the construction site, for example, offices or depots. Where this is the case the health and welfare facilities for these workplaces will need to comply with the Workplace (Health, Safety and Welfare) Regulations (WHSWR) 1992.

For more information see the 'Relevant Statutory Provisions' section.

Sanitary conveniences

Schedule 2 of CDMR 2015 requires that readily accessible, suitable and sufficient sanitary conveniences (toilets) must be provided on construction sites. The conveniences must be adequate for the numbers and gender employed, lit, kept clean and maintained in an orderly condition. Separate conveniences for male and female workers must be provided, except where the convenience is in a separate room and the door is capable of being locked from the inside. Provision should be made for regular cleaning and replenishment of consumable toilet items.

Figure 1-64: Temporary sanitary conveniences.
Source: RMS.

Washing facilities

Schedule 2 of CDMR 2015 requires that suitable and sufficient washing facilities must be provided at readily accessible places on construction sites. This includes showers where they are necessary because of the nature of the work or for health reasons. Where work is particularly dirty or workers are exposed to toxic or corrosive substances, showers may also be necessary.

In addition, washing facilities must be provided:

- In the immediate vicinity of every sanitary convenience.
- In the vicinity of changing rooms required by Schedule 2.

On all sites, a suitable number of wash basins big enough to allow a person to wash their hands, face and forearms should be available. There must be a supply of clean, hot and cold or warm running water, so far as is practicable, soap or other means of cleaning, towels or other means of drying and the rooms that contain these facilities must be kept clean, ventilated, lit and maintained.

Other than facilities for washing hands, forearms and face separate washing facilities must be provided for men and women, unless provided in individual rooms that may be locked from the inside.

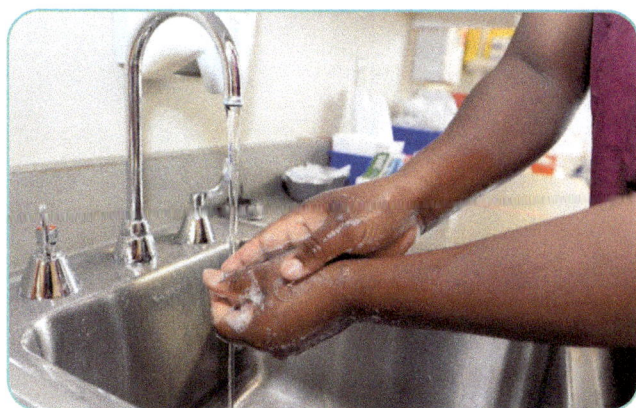

Figure 1-65: Washing facilities.
Source: Health.mil.

Changing rooms

Schedule 2 of CDMR 2015 requires that where special clothing must be worn at work or for reasons of health or propriety a person cannot change in another room then suitable and sufficient changing facilities must be provided on construction sites.

Separate facilities or separate use of facilities for male and female workers must be taken into account. The changing rooms must be provided with seating and, where necessary, include facilities to dry work clothing, personal clothing and personal effects.

The changing facilities must be readily accessible to workers, for example, when starting work/completing work and near eating facilities where there is a requirement to remove work clothing. The facilities provided should be sufficiently large to enable the maximum number of workers to use them comfortably and quickly at any one time.

Lockers

Schedule 2 of CDMR 2015 requires suitable and sufficient accommodation must be provided at readily accessible places to enable workers on construction sites to lock away their personal clothing not worn at work, clothing worn at work that is not taken home and personal effects.

Clothing accommodation must be suitably secure; it must be suitable for the work clothing and would typically be separate from personal clothing where it was necessary to avoid health risks due to contamination or damage. Clothing accommodation should be in a suitable location near to changing facilities and drying facilities for work clothing, personal clothing and personal effects.

Rest and eating facilities

Schedule 2 of CDMR 2015 requires readily accessible, suitable and sufficient rest rooms or rest areas to be provided or made available at readily accessible locations on construction sites. Schedule 2 of CDMR 2015 sets out specific requirements that the rest rooms or rest areas must conform to. Rest rooms and rest areas must:

> *"(b) Be equipped with an adequate number of tables and adequate seating with backs for the number of persons at work likely to use them at any one time.*
>
> *(c) Where necessary, include suitable facilities for any person at work who is a pregnant woman or nursing mother to rest lying down.*
>
> *(d) Include suitable arrangements to ensure that meals can be prepared and eaten.*
>
> *(e) Include the means for boiling water.*
>
> *(f) Be maintained at an appropriate temperature."*

Figure 1-66: Rest rooms and rest areas.
Source: CDM 2015 Schedule 2.

Canteens, restaurants etc. may be used as rest facilities providing there is no obligation to buy food.

Drinking water

Schedule 2 of CDMR 2015 requires that an adequate supply of wholesome drinking water must be provided on construction sites.

The supply needs to be accessible. If not provided in the form of a fountain, then drinking vessels must also be provided. The supply outlet from taps should be marked by an appropriate sign to indicate 'suitable for drinking' or 'unsuitable for drinking', as appropriate.

> **CONSIDER**
>
> Consider a construction site you may have recently visited. What range of welfare facilities were available on the site?

Figure 1-67: Rest and eating facilities.
Source: RMS.

Figure 1-68: Temporary accommodation unit.
Source: Wikimedia Commons.

Figure 1-69: Poor access and egress from TAU.
Source: RMS.

Figure 1-70: Means of access to TAU.
Source: RMS.

TYPES OF TEMPORARY ACCOMMODATION UNITS (TAU) REQUIRED FOR SITES

Temporary accommodation units include site offices, canteens, rest rooms, drying rooms and toilets. They are often made of timber, with the addition of insulation and fire resisting surface materials. They may be used as a single storey structure combining a number of functions, for example an office and facilities for changing and lockers or they may be sited as interlinking units on one level or over a number of floors.

REQUIREMENTS OF LOCATION OF TAU

TAUs should be sited in the open air and, in order to provide a fire break, separate from structures under construction, nearby buildings or flammable substance storage compounds. A fire risk assessment should be completed to consider the specific circumstances and needs for siting TAUs, however as a general principle separation of a minimum of 6m should be established, which should be increased to 20m on higher risk sites, such as those constructing timber frame buildings. Alternatively, the fire resistance of the TAU and/or adjacent structures should be assured and at least 30 minutes fire resistance provided.

Care should be taken to ensure good standards of health and safety related to TAUs, even though they are temporary. Aspects like access/egress and lighting must be to a good standard.

Where it is necessary to site the TAU inside a building it should be considered and assessed as part of the building and risk assessed accordingly.

Fire control measures would usually include means of fire detection, linked to an alarm.

TAUs for sleeping accommodation require special siting and fire arrangements.

PARTICULAR CONSTRUCTION ISSUES

Use of migrant workers

The construction industry is characterised by its intermittent, temporary, transitory nature. Generally, building and construction contractors hire a work force on a project basis. Thus, workers in the construction industry are accustomed to travelling from areas where work is not plentiful to fill short-term labour shortages created by expansion and contraction of local construction activity elsewhere. This includes the use of migrant workers, which has become a common part of the construction workforce.

Some of the issues related to the use of migrant workers is that they may be employed through an employment agency or employment business. This may extend and confuse the communication chain, which can present difficulties in establishing the identity and competence of the worker before they are put to work. If the agency or business operates from overseas the person using the worker will take primary responsibility for the health and safety of the worker. This will mean ensuring the worker has the competence required for the work and may mean establishing their previous experience or carrying out tests to confirm competence.

Another issue can be the difference in values of the migrant workers in regard to health and safety, they may be more willing and used to taking risks and working in a way that makes harm of themselves and others more likely.

One of the issues that can present difficulties is that they may have limited/no understanding of the English language or limited/no ability to communicate with others, including other migrant workers who speak a different language. This can mean providing site induction in different languages and ensuring instructions are communicated in a form that is understood by the worker. This may be achieved through an interpreter, who may be the supervisor of the worker. This is discussed further in the following paragraphs.

Migrant workers may be less reliable/stable than other workers, because their work permit may run out or issues in their home country may demand their time and attention. Migrant workers may also find access to primary care difficult due to a limited understanding of the local system of appointments and prescriptions. They will often have few ties to the UK and can quickly relocate to their home country or another country offering better prospects.

Temporary nature of activities and the constantly changing workplace

Construction sites constantly change through the build phase, as the trades that are associated with the construction vary greatly at each stage. Consideration needs to be given to site induction for new workers as appropriate.

The safe systems of work, risk assessments, site safety procedures and site inductions will need to be updated regularly to suit the most current situation. A construction site will always be an unfamiliar workplace with new hazards and dangers posed as each phase moves to the next.

Time pressures

Clients have one of the biggest influences on the health and safety of those working on a construction project. Their decisions have substantial influence on the time, money and other resources available for the project. Because of this, the Construction (Design and Management) Regulations (CDM) 2015 make them accountable for the impact their decisions have on health and safety. In the same way, the principal contractor may exert pressure on contractors to fulfil tasks within time limits. This is a natural part of the efficient organisation of a construction activity where a number of contractors work together to achieve the goals. Each activity may depend on the timely completion of another. However, if the time available to complete the overall project becomes limited, possibly due to delays or underestimation of the necessary time for completion of an activity, the principal contractor may put contractors under pressure to complete work on time or earlier than was planned or reasonably practicable. This can lead to less planning time and the potential for risks to be taken in order to save time.

Weather conditions

For those working outdoors, adverse weather can create numerous problems. Short term exposure to the sun can cause excessive sweating, dehydration and fatigue. There are fears that prolonged exposure can cause skin cancer. Strong wind increases risk when working at height and can cause unexpected movement of loads suspended on cranes. Heavy rain may cause soft ground conditions which can increase problems with site traffic and undermine the stability of scaffolds and excavations. Extreme cold leads to snow and ice which increases the likelihood of slips and falls. It may also increase the risk of brittle failure of equipment.

Levels of numeracy and literacy of workers

Poor levels of literacy and numeracy can significantly impede the progress of workers at a workplace. The level of understanding and critical information retention required, for example, when attending a health and safety induction, can be significantly reduced. Where numbers and the written word are utilised instead of a pictographic approach, then the meaning can be marred regarding written critical safety instructions or directions that may need to be followed could be unintentionally ignored. Weights and measures, dates and times critical to processes could also be misconstrued or overlooked if workers are innumerate.

Non-English speaking workers

As workers come to the United Kingdom from other countries, the possible lack of understanding of the English language presents significant communication problems. Effective steps must be taken so that workers who cannot speak English can work safely and without risks to their own health and safety or the health and safety of others who may be affected.

Regulation 10 of the Management of Health and Safety at Work Regulations (MHSWR) 1999 requires employers to provide information to employees which is comprehensible and relevant, i.e. capable of being understood by the person for whom it is intended.

For workers with little or no understanding of English, or who cannot read English, employers may need to make special arrangements. These could include providing translation, using interpreters, or replacing written notices with clearly understood symbols or diagrams.

Employers need to consider the following with reference to their non-English speaking workers:

- Check that prospective workers have sufficient command of English for their role.
- Check worker understanding of instructions and health and safety related notices.
- If other members of staff are available who can speak the worker's native language, then they may be able to help the worker understand the working culture and environment in the UK.

This approach may need to be extended to all workers on site who do not speak English or measures put in place to ensure those employing them communicate effectively.

Sources of reference

Reference the information provided, in particular web links, these are correct at time of publication, but may have changed.

Managing Health and Safety in Construction, Construction (Design and Management) Regulations 2015, Guidance on regulations, L153, HSE Books, ISBN: 978-0-7176-6623-3, http://www.hse.gov.uk/pubns/priced/l153.pdf

The Construction (Design and Management) Regulations 2015, Industry guidance for Clients, Produced by CONIAC, published by CITB, ISBN: 978-1-85751-389-9, https://www.citb.co.uk/media/bwzklrv5/cdm-2015-clients-interactive.pdf

The Construction (Design and Management) Regulations 2015, Industry guidance for Contractors Produced by CONIAC, published by CITB, ISBN: 978-1-85751-391-2, https://www.citb.co.uk/media/5dgbe5ol/cdm-2015-principal-contractors-interactive.pdf

The Construction (Design and Management) Regulations 2015, Industry guidance for Designers, produced by CONIAC, published by CITB, ISBN 978-1-85751-393-6, https://www.citb.co.uk/media/ndlbnb5v/cdm-2015-designers-interactive.pdf

The Construction (Design and Management) Regulations 2015, Industry guidance for Principal Designers, produced by CONIAC, published by CITB, ISBN 978-1-85751-390-5 https://www.citb.co.uk/media/nrnns1l1/cdm-2015-principal-designers-interactive.pdf

The Construction (Design and Management) Regulations 2015, Industry guidance for Workers, produced by CONIAC, published by CITB, ISBN 978-1-85751-394-3 https://www.citb.co.uk/media/oczbsxvn/cdm-2015-workers-interactive.pdf

Using contractors, a brief guide, INDG368, HSE Books, http://www.hse.gov.uk/pubns/indg368.pdf

Workplace health, safety and welfare, Workplace (Health, Safety and Welfare) Regulations 1992, ACOP, L24, HSE Books, ISBN 978-0-7176-0413-5 www.hse.gov.uk/pubns/priced/l24.pdf

Web links to these references are provided on the RMS Publishing website for ease of use – www.rmspublishing.co.uk

Statutory provisions

Construction (Design and Management) Regulations (CDMR) 2015 / Construction (Design and Management) Regulations (Northern Ireland) 2016

Management of Health and Safety at Work Regulations (MHSWR) 1999 / Management of Health and Safety at Work Regulations (Northern Ireland) 2000

Workplace (Health, Safety and Welfare) Regulations (WHSWR) 1992 / Workplace (Health, Safety and Welfare) Regulations (Northern Ireland) 1993

STUDY QUESTIONS

1. What information needs to be provided when completing an F10 in accordance with the Construction (Design and Management) Regulations 2015?

2. What topics should be included in an induction for site visitors?

3. What should be considered when assessing the adequacy of lighting on a construction site?

4. What features of a Temporary Accommodation Unit (TAU) should be addressed in order to meet the health, safety, and welfare requirements for workers?

For guidance on how to answer these questions please refer to the 'study question answer guidance' section located at the back of this guide.

Element 2

Improving health and safety culture and assessing risk

2

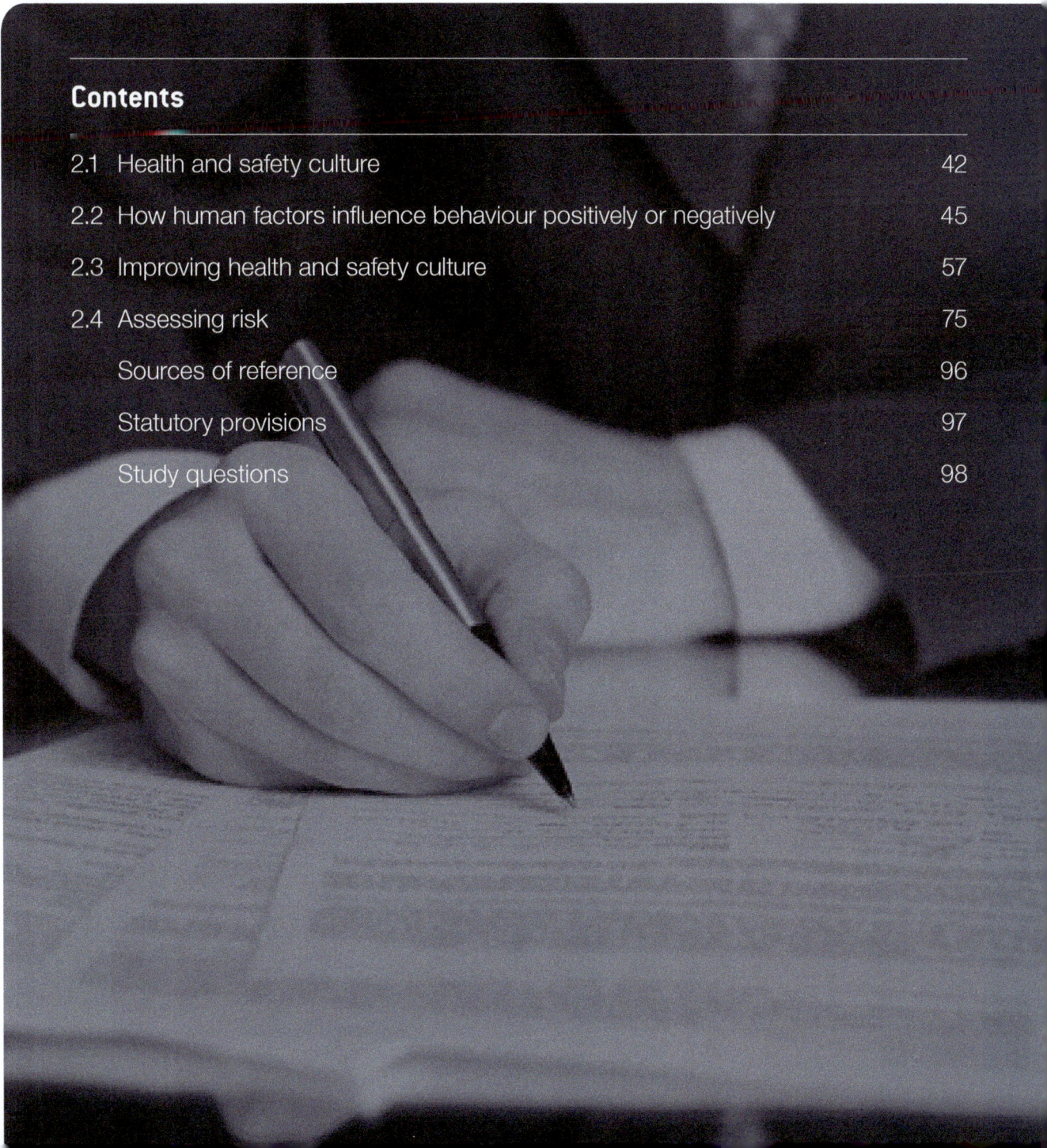

2.1 Health and safety culture

MEANING OF THE TERM 'HEALTH AND SAFETY CULTURE'

The term 'safety culture' was first used in the International Atomic Energy Agency's (IAEA) initial report following the Chernobyl disaster, in 1986. One definition of the term 'safety culture' was given in 1993 by the UK, Health and Safety Executive (HSE) Advisory Committee on the Safety of Nuclear Installations (ACSNI), not long after the IAEA report on Chernobyl. The definition was included in the ACSNI Human Factors Study Group's third report - 'Organising for safety', **see Figure 2-1.**

> "The safety culture of an organisation is the product of individual and group values, attitudes, perceptions, competencies, and patterns of behaviour that determine the commitment to, and the style and proficiency of, an organisation's health and safety management".

Figure 2-1: Safety culture of an organisation.
Source: ACSNI.

In more recent years the term 'safety culture' has been widened in its general use to 'health and safety culture' to reflect the increased interest in work-related ill-health.

To better understand the term health and safety culture a useful approach is to distinguish between three inter-related aspects of health and safety culture, specifically:

- 'How people feel' about health and safety.
- 'What people do' about health and safety in the organisation.
- 'What the organisation has' in place for health and safety.

'How people feel' about health and safety encompasses the values, beliefs, attitudes and perceptions of individuals and groups at all levels of the organisation, which are often referred to as the health and safety climate of the organisation.

'What people do' within the organisation includes the health and safety related activities, actions and behaviours of individuals and groups at all levels.

For example, individuals making time for health and safety and giving it due priority when making decisions.

'What the organisation has' is reflected in the organisation's policies, operating procedures, management systems, control systems, communication and workflow systems. For example, health and safety is integrated in planning work activities and design of the workplace.

The table in **Figure 2-2** provides examples of how these three aspects relate to a positive health and safety culture.

The HSE Advisory Committee on the Safety of Nuclear Installations provided a useful perspective on what a positive health and safety culture is, see **Figure 2-3.**

> "Organisations with a positive safety culture are characterised by communications founded on mutual trust, by shared perceptions of the importance of safety and by confidence in the efficacy of preventive measures."

Figure 2-3: Positive safety culture.
Source: ACSNI.

RELATIONSHIP BETWEEN HEALTH AND SAFETY CULTURE AND HEALTH AND SAFETY PERFORMANCE

Because the culture of an organisation has such a strong influence on the way it does things, there is a direct

Aspect of culture	A positive health and safety culture
'How people feel'	For example, values and beliefs: • People in the organisation value health and safety, regard it as mandatory and operate in a safe and healthy way. • People in the organisation believe that health and safety makes commercial and professional sense, that individuals are not the sole cause of accidents/incidents and ill-health and that the next accident/incident or ill-health event is waiting to happen.
'What people do'	For example, problem solving methods: • People in the organisation use risk assessment and accident/incident and ill-health investigation/analysis, they also use cost-benefit analysis. • People in the organisation actively search for problems in advance of accidents/incidents or ill-health events.
'What the organisation has'	For example, working practices: • Health and safety is integral to design and operations practices. • Health and safety is the first item on meeting agendas up to Board level.

Figure 2-2: Aspects of a positive health and safety culture.
Source: RMS.

relationship between the culture of an organisation and its health and safety performance. Each decision taken by an organisation is influenced by its culture. The way the decisions affect health and safety may be positive or negative, depending on the culture of the organisation. Organisations may have a culture that focuses only on the short-term or they may have a culture that values a longer term, more controlled, preventive approach to health and safety.

Organisations with a culture that supports a preventive approach have a greater chance of a good health and safety performance. For example, an organisation with a positive health and safety culture naturally includes health and safety in their decisions and gives it due priority. This means that the organisation is more likely to have good standards of health and safety and take early action to minimise risks. As organisations with a positive health and safety culture also tend to have managers and workers that are committed to health and safety they are more likely to behave in a way that leads to good health and safety performance.

The relationship between health and safety culture and health and safety performance can be illustrated by research that has been conducted. Research has shown that influences on the health and safety culture of an organisation made by the introduction of a health and safety programme significantly improves health and safety performance. For example, after the introduction of a health and safety programme in a range of forestry and logging organisations in Columbia, it was found by Painter and Smith (1986) that there were dramatic improvements in performance. The accident/incident frequency rate was reduced by 75% and the workers' compensation costs were reduced by 62%. In further research, Lauriski and Guyman (1989) found that after a health and safety management programme had been introduced at the Utah Power and Light Company, lost time injury rates were reduced by 60% over a period of five years. From 1980 to 1988, the accident/incident frequency rate was reduced from 40 to 8 per annum, while production more than doubled.

A number of performance indicators can be used to assess an organisation's health and safety culture. Some of these indicators relate to health and safety outcomes and include:

- The number of accidents/incidents occurring.
- The number of work-related illnesses occurring.
- Sickness and absenteeism rates.
- The number of complaints about working conditions.
- The number of workers leaving employment.

Where these outcomes are high, for example, if there are a lot of accidents/incidents, it usually indicates the presence of a negative health and safety culture.

The HSE recognises that health and safety performance in the construction industry has improved over the past decade reflecting a concerted effort to improve the health and safety culture. This has resulted in the number and rate of fatal injuries, other work-related injuries and ill health showing a general, long term downward trend. In the period 2000 to 2009 fatal injuries to workers were 739 and in the following period from 2010 to 2019 there were 403, which tended to support the conclusion that an improvement in health and safety culture can influence health and safety performance.

However, statistics also show that the levels of injuries and ill health remain high, with some recent signs of the numbers levelling off. The construction industry currently still accounts for around 25% of all fatal injuries to workers. In addition, the construction industry has a statistically significantly higher rate of occupational lung disease and musculoskeletal disorders (MSDs) than the average for all industries.

This indicates that the culture driving developments in the management of health risks has not kept pace with that for safety risks and that based on performance outcomes the health and safety culture in the construction industry requires further effort.

A reduction in accident/incident rates experienced by an organisation with a positive health and safety culture would be seen by it as a positive step forward and could lead to a further, positive influence on the health and safety culture and probable future improvement in performance.

Other indicators used to assess health and safety culture relate to what an organisation does to manage health and safety. Where there are a good range of indicators present and they are effective it usually indicates the presence of a positive health and safety culture.

These indicators include:

- Values and beliefs expressed in a health and safety policy.
- Visible leadership and commitment.
- Health and safety as the first item on the agenda of meetings.
- Health and safety performance reported in the annual report.
- Giving health and safety an equal priority to other matters.
- Directors/senior management visits to workplaces.
- The amount of risk assessments conducted.
- The amount of health and safety training conducted.

- Communication of health and safety matters.
- Manager and worker involvement.
- Compliance with health and safety rules and procedures.
- The amount of workplace and work equipment inspections.
- The use of investigation and analysis processes.
- Timely completion of corrective and improvement actions.

The accident/incident investigation process used by an organisation can provide a particularly useful indication of how positive the health and safety culture is. If investigation is limited to only identifying the immediate causes of accidents/incidents it would indicate that the culture is not strongly positive. If underlying and root causes are identified routinely by the accident/incident investigation process it would indicate a more positive health and safety culture.

In addition, the management of the various causes identified in the investigation process may indicate how well developed the health and safety culture is. If the procedure for accident/incident investigation requires the structured management of recommendations and these were completed appropriately it would indicate a positive health and safety culture.

However, if the same types of causes are repeatedly resulting in accidents/incidents it may indicate a lack of commitment to the prevention and improvement actions identified in the investigation recommendations. This would indicate the presence of a negative health and safety culture.

When assessing an organisation's health and safety culture, it is usual to consider both types of indicator, what an organisation does to manage health and safety, and the outcomes.

INFLUENCE OF PEERS ON HEALTH AND SAFETY CULTURE

The term 'peer' is usually used to refer to those people that work with an individual. Peers are usually at the same level in an organisation as the individual, they may also be members of the same group, for example, members of a health and safety committee or management team.

Many individuals are influenced by what people think of them and may act in a particular way to ensure that the people around them feel positive towards them. In group situations this can include the individual acting in the same way as the group to maintain a feeling of belonging in the group. This desire for belonging and to be liked can lead to an individual consciously or subconsciously behaving in a similar way to those they work

with, their peers. An individual's peers will also try to influence the individual to behave in a similar way to the way they behave. This influence of peers may lead to peer pressure to behave in a particular way and can be a very strong influence when peers act as a group. This can have a significant effect on an organisation's health and safety culture as the behaviour of people in an organisation is an important part of what influences the culture.

The influence of peers may have the effect of promoting good health and safety, for example, within a work team. The influence of peers means that if the members of a team believe that working safely is the right way to do the job, each team member will watch over the activities of the others in the team to ensure they conform to the way the team works. The group as a whole will ensure any new member follows their example to ensure they work safely. However, peer pressure may have a negative influence. The influence of peers may lead to managers and workers receiving direct or indirect pressure not to put effort into achieving good health and safety standards. This may particularly be the case where the peers do not value health and safety or consider it prevents them from achieving things that they feel are important. For example, a manager or worker may feel that working safely will limit their ability to achieve productivity rewards and they may put pressure on their peers to ignore health and safety requirements.

It is therefore important to anticipate the influence of peers and plan to use this influence to promote health and safety, helping to establish a positive health and safety culture. This will usually involve obtaining the commitment of the senior manager and using it to influence other managers, who in turn influence those they work with. In this way the influence of peers can have a significant positive effect on an organisation's health and safety culture.

REVIEW

Explain the following aspects of health and safety culture:

'What people think', 'What people do', What the organisation has'.

failure should be taken, including the use of risk assessment processes and the consideration of **work patterns** to ensure fatigue and boredom are minimised.

6) **Procedures and standards** for all aspects of work involving significant hazards and mechanisms for reviewing them. This should be supported **by effective monitoring** systems to check the implementation of the procedures and standards.

JOB FACTORS

Job factors are often caused by the presence of organisational factors and directly influence individual performance and the control of risks. There are a number of job factors that could influence health and safety behaviour in the workplace, including:

- The tasks individuals carry out.
- The effects of workload and the workplace environment.
- The design and maintenance of equipment and procedures.

The tasks that an individual is expected to do may be physically or mentally demanding, for example, a task may require a large amount of strength or have many complicated features to consider in order to come to a decision. The high demands of tasks compared with an individual's capabilities provide a potential for them to make errors.

Poor design of equipment can lead those using them to make errors, for example, information displays may be misread or the wrong controls operated in error. These errors can cause direct harm to the person using them or in some cases could lead to incorrect decisions that later result in harm. Equipment used by individuals should be designed in accordance with ergonomic principles to take into account any limitations in human performance of those using them, in particular, mental and physical ability. Ergonomics, in relation to equipment, is the study of the design and arrangement of equipment so that people will interact with the equipment in a safe, healthy, comfortable and efficient manner. One of the main principles of ergonomics is to match, as far as possible, the equipment and how it is used to the user, rather than match the user to the equipment. Following these ergonomic principles and, where possible, matching the job to the individual will help to minimise errors. For example, the use of ergonomic principles can greatly improve work equipment displays and the layout of controls. This can reduce the human errors made when reading displays and operating equipment controls.

Where equipment and tools are poorly maintained and they do not work correctly this can cause high levels of frustration and fatigue and could lead an individual to act in an unsafe or unhealthy way in order to get their work done, for example, if the drill bit fitted to a drilling machine frequently became loose and the worker had to keep removing the guard to re-fit the drill bit the worker may be tempted to leave the guard off.

The health and safety procedures and instructions an organisation uses are important job factors that should be considered. Where procedures or instructions communicated to an individual are unclear or have missing information, the individual receiving them may make an error, for example, they may mix two chemicals that react to each other or might administer the wrong amount of medicine to a patient.

In addition, there are a number of other job factors that could cause mental and physical fatigue to an individual and cause them to make errors. For example, fatigue may result where the workload is unacceptably high, where the individual is affected by interruptions or where the work environment is unpleasant, for example, noisy or very hot.

To manage the job aspects of human factors, organisations therefore need to ensure the:

1) Identification and comprehensive analysis of the **tasks** expected of individuals and the assessment of likely errors related to the tasks. This should include the evaluation of manager and worker **decision making.**

2) Application of **ergonomic principles** to the design of person-equipment interfaces, including **instrument displays** of process information and the suitable positioning and labelling of **control devices.** Care should be taken to ensure the provision and maintenance of the **correct tools and equipment** for the job.

3) Design and consistency of **presentation of procedures**, supported by the provision of efficient and suitable communication of information related to the job.

4) Scheduling of work patterns, including **shift organisation and workload**, to control fatigue and stress. This should include procedures to cover for absence and for emergencies that might arise when doing the job. Where concentration is required to prevent errors, arrangements should be made to **prevent interruptions**. This should be supported by the organisation and control of the working **environment** so that it reduces fatigue and stress, including the workspace, lighting, noise and thermal conditions.

INDIVIDUALS FACTORS

Although organisational and job factors are important, individual factors can also greatly influence the health and safety behaviour of individuals in the workplace.

These individual factors include the individual's personality, attitude, risk perception, skills and competence. All individuals have different physical and mental characteristics that can affect these factors and the way they behave. The individual differences related to mental characteristics come from a combination of the 'inherited characteristics' (passed on from the parents of the individual) and the various 'life experiences' through which the individual passes.

Typical life experiences that influence the mental characteristics of an individual include:

- Experiences in the womb.
- Birth trauma.
- Family influences.
- Geographical location.
- Pre-school influences.
- Education - opportunities, quality, support.
- Occupational factors - training and retraining.
- Hobbies and interests.
- Own family influences - marriage, children.
- Ageing.

These life experiences shape the mental characteristics (personality) of each person, such as their attitudes and perception. In addition, each individual develops physical characteristics, including their size, strength and any limitations arising from disability or illness. These characteristics combine to create a unique person, different from all other individuals. *Figure 2-6* shows some examples of individual factors and how they may influence the health and safety aspects of work.

It is important to know what a particular job involves so that the effects of individual factors can be minimised, especially for high hazard jobs. Jobs should be analysed to identify features of the job that may be affected by individual factors and the required characteristics of individuals to do the job safely and healthily.

Following the analysis of jobs, a review process should be used to assess if the job can be modified to enable

a wide range of individuals to conduct it without harm to their health and safety. The job may also have to be modified to take account of individuals with special needs, for example, disabilities. Where possible, the job an individual does should be adjusted by following good ergonomic principles to minimise the effects of individual factors on behaviour and performance.

However, the effects of individual factors may not always be controlled by good ergonomic principles, other methods may also be required. The information gained from analysis of jobs can also be used to establish person specifications for the jobs. Where necessary, appropriate selection techniques may need to be used to match individuals to jobs that suit their characteristics, particularly where these characteristics directly influence health and safety. Also, some individual characteristics can be modified by training and experience, such as skills, attitudes and risk perception.

Training should aim to give individuals the knowledge to allow them to understand their work processes, risks and control measures, and the skills that enable them to perform their job well and ensure health and safety. This training helps to improve human reliability because managers and workers become more aware of the process and task, their perception of risk increases to an appropriate level and an improvement in their attitude and motivation should take place. Further information on training and competence is provided later in this element.

Monitoring of individual performance should be carried out at all levels in an organisation. In particular, all workers should be monitored through direct supervision. This will ensure work activities are performed correctly and requirements for correct health and safety behaviour are met. An old maxim states: 'what gets measured, gets done.' This is particularly relevant for health and safety and if individuals are monitored they are more likely to behave safely and healthily when doing their work. Supervision also provides an opportunity to commend people and take action to improve behaviour.

Physical	Mental
Gender - for example, females of child bearing age should not be exposed to lead.	Attitude - for example, a person may have the attitude that all personal protective equipment is uncomfortable and refuse to wear it.
Size - for example, the large size of a person may restrict their movement in a confined space.	Motivation - for example, a person may feel that while doing their work in a particular way presents additional risks it is worth taking the risk to make the work easier or to complete the work quicker.
Health - for example, colour blindness may prevent the correct identification of wiring by an electrician.	Perception - for example, a person may not recognise the risk from moving machinery and remove a guard or they may not recognise an emergency alarm and fail to respond to it.
Capability/strength - for example, an individual may not have sufficient capability to manually handle a load.	Mental capability - for example, a person may have limited mental ability, which leads them not to follow complex or lengthy health and safety instructions.

Figure 2-6: Examples of individual factors and their possible effect on health and safety.
Source: RMS.

The degree of monitoring and supervision should be based on the level of health and safety risk related to the task and the level of competency of the worker carrying out the job.

For certain jobs there may be specified medical standards for which pre-employment/periodic health surveillance is necessary. These medical standards may relate to the practical requirements necessary to carry out the job and identify specific characteristics necessary to perform the job safely and healthily. For example, medical standards may be set for vehicle drivers and a medical examination required before an individual is allowed to become a driver. There may also be a need for routine health surveillance to determine the effects on individuals of their exposure to workplace hazards that influence behaviour, for example, fatigue due to an individual being exposed to high temperatures.

Similarly, it is useful to review the individual factors that could affect an individual's health and safety behaviour when they return to work following a significant period of sickness absence. This provides an opportunity to consider the reason for their absence, any arrangements that may enable their smooth return to normal duties and any underlying reasons for their absence. The underlying reasons for absence could include anxiety or stress related to their work and personal reasons like alcohol or drug abuse. Access to specialist assistance or counselling may be appropriate and temporary or permanent redeployment to other work may be necessary.

CASE STUDY

Managing individual factors, the health of a construction worker

Steven is a recently qualified civil engineer who has epilepsy. Because of the hazards presented while working near moving plant machinery or at heights, the recruitment standards for the post adopted by the company excludes employment of applicants with epilepsy. When this policy was challenged by the Human Resources manager, the company's health and safety officer confirmed that the role was likely to involve work at height and around hazardous machinery. The Human Resources manager requested a report on the practicalities and cost of eliminating the hazards from the post and a specialist medical assessment of Steven's prognosis and its relevance to the responsibilities of the post.

The management report concluded that while the work in hazardous situations could not be eliminated, it could be substantially reduced with comparatively little cost by reallocating responsibilities involving working at heights to other members of the team and rescheduling

duties in the vicinity of plant machinery to times when the machines would not be operating. Hazardous work could effectively be minimised to no more than a few hours annually.

The specialist medical report identified that Steven's epilepsy was well controlled and placed the annual risk of seizure recurrence between low and moderate. The report enabled the management team and the individual to agree that with reasonable adjustments, and a review if circumstances changed, the residual risks to the applicant of accident or injury would be well below that accepted on a daily basis in activities such as driving, and the risks to others would be negligible. Steven was recruited to the post.

In summary, to manage the individual aspects of human factors organisations should:

1) Write **job and person specifications**. Consider age (if this affects health and safety), physique, aptitude, personality, knowledge, skill, qualifications, experience.

2) Use **ergonomic principles**, where possible matching the job to the individual. Where necessary, **match the aptitudes and skills** of individuals to the requirements of jobs. Consider the needs of **special groups of individuals**, for example, those with disabilities.

3) Implement an effective **training** system.

4) **Monitor individual performance**, particularly supervision for individuals that do high risk jobs.

5) Provide pre-employment and periodic **health surveillance** where this is needed.

6) Review individual factors affecting an individual **after periods of absence**, including absence due to sickness or injury due to accidents/incidents.

7) Provide **counselling and support** for ill-health or stress.

Personality

The personality of an individual is a set of mental characteristics (pattern of thoughts, feelings, and behaviours) that influence an individual's values, attitudes, information processing (including perception), emotions and motivations in various situations. This makes the individual a unique person. Personality arises from within the individual and is influenced by their experiences in life and their interaction with other individuals. It is believed that an individual's personality usually remains fairly consistent throughout their life. The personality of an individual can therefore influence all aspects of their health and safety behaviour at work.

Attitude

Attitude may be defined as *'the tendency to respond in a particular way to a certain situation'*.

An individual's attitudes will influence the way in which situations are viewed by them and will create a response or pattern of behaviour.

Attitudes are one way in which individuals are different from each other. Attitudes guide how we react to other people, what things are important to us and our preferences. Attitudes are formed (not necessarily consciously) through a lifetime of experiences that shape the beliefs and values that establish our attitudes, for example, our attitude to working safely.

Attitudes are used to provide the individual with a structure to their world, organising an individual's standardised response to situations, which saves the effort of thinking of a response that is acceptable to their values and beliefs each time they meet a situation they need to respond to. Attitudes are not directly observable, but can be assessed by observing an individual's thoughts, feelings and behaviours (physical or verbal).

An individual's behaviour can therefore express their attitude towards something and reflect how they feel about that thing. Individuals often develop similar attitudes to those people they like and seek others with similar attitudes. This is very important when considering an individual's health and safety behaviour.

Individuals may develop attitudes about working safely, which can be formed over a long period of time and through interaction with other people and their attitudes. An individual may therefore develop a negative attitude to health and safety, which causes them to respond in a negative way to attempts to get them to work safely.

This attitude may develop through the individual working in situations where other individuals hold a negative attitude and copying their behaviour and attitude. For example, an individual may develop a negative attitude to health and safety when working in situations where there is a negative health and safety culture in an organisation. The individual may express their negative attitude in their behaviour, verbally rejecting the need to work safely and physically refusing to work in a safe way. Because attitudes are formed over time, sometimes many years, they are not easily changed. Attempts to change such a fundamental part of an individual's personality will usually be resisted.

The individual will often feel their values and beliefs are under threat. This is worth remembering in the context of health and safety propaganda campaigns that attempt to change attitudes. The difficulty in changing attitudes was summarised by Chairman Mao Tse Tung, see *Figure 2-7.*

> "People's attitudes and opinions that have been formed over decades of life cannot be changed by holding a few meetings or giving a few lectures."

Figure 2-7: Observation made by Chairman Mao Tse Tung.
Source: "Little Red Book".

Examples of verbal expression of negative attitudes affecting safe and healthy working include:

1) 'It will never happen to me'.

2) 'We have never had an accident/incident'.

3) 'I am not planning to injure myself'.

4) 'I know my limits'.

Everyone at work should attempt to change their own attitudes to health and safety, and those of other people, from - I work safely and healthily because:

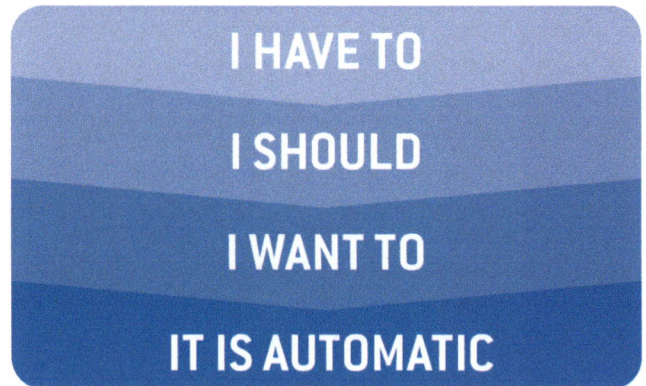

I HAVE TO
I SHOULD
I WANT TO
IT IS AUTOMATIC

Actions to influence attitudes might include:

- Providing strong leadership that shows a positive attitude to health and safety.

- Providing training and provision of retraining when there is a need for reinforcement.

- Providing positive experiences (involvement), for example, improving an individual's attitude to wearing personal protective equipment by involving them in the selection of the equipment.

- Discussing the basis of the attitudes that individuals hold. For example, asking individuals to consider whether their values and beliefs are correct, providing them with other knowledge that could adjust their values and beliefs and helping them to reform their attitude.

- Providing individuals with an interaction with other people who have the correct attitude. For example, arranging for individuals to work with a group of people with the correct attitude to health and safety. The individuals may change their behaviour to make working with the other people in the group easier and become more positive to health and safety to get the group to be positive towards them.

- Providing positive reinforcement of the correct attitude and the health and safety behaviour that accompanies it. For example, recognition of an individual who is using personal protective equipment correctly.

Motivation

Motivation may be defined as *'the driving force behind the way a person acts in order to achieve a goal'*. In the context of work situations there have been many attempts to identify what motivates people to work. The earliest theory (by F. W. Taylor) suggested that people worked for money and fear of losing their job. Financial reward was seen as the main motivator. This approach suggested that the more individuals were paid, the harder they worked. This led to a management philosophy that encouraged payment by results in the form of incentive schemes and motivation to do things that people did not like doing by provision of extra payments, for example, extra money to do dangerous or dirty jobs. This philosophy was found to be unsuitable for the management of health and safety as most incentive schemes encouraged people to work unsafely, for example, rushing to get the job done.

This often caused health and safety working practices to be ignored or compromised. Other motivational theories have been developed that establish a number of other themes related to motivation. Some of these themes emphasise the importance of involvement of managers and workers (team working and social interaction), setting goals/objectives, personal achievement and recognition of positive behaviour. Positive motivation (where safe and healthy working is encouraged and rewarded by recognition and praise) tends to be more effective than negative motivation (where workers fear disciplinary action for not ensuring health and safety) although both have a place and may be appropriate in different circumstances.

Actions to influence motivation might include:

- Establishing a positive health and safety culture, where risk taking is not accepted by managers or workers.

- Involvement of managers and workers in health and safety policy setting.

- Setting realistic goals/objectives with regard to accident/incident and work-related illness rates.

- Clarification of roles and responsibilities.

- Providing information and training to improving managers' and workers' knowledge of their responsibilities and the consequences of not working safely and healthily.

- Showing the commitment of the organisation to health and safety by providing resources and a safe and healthy working environment.

- Involving managers and workers in health and safety decisions, by consultation, team meetings and health and safety committees.

- Monitoring health and safety performance, including personal performance.

- Recognising and rewarding health and safety achievement, by developing a reward structure that recognises positive health and safety behaviour and performance.

Perception of risk

Perception may be defined as *'the way that a person views a situation'*. Perception is the process of getting, selecting, organising and interpreting sensory information.

The process of perception is a sequence of steps that begins with a stimulus in the environment and leads to our perception of the stimulus and an action in response to the stimulus. This process is continual and takes place without the individual thinking about the actual process at any given moment. The process of perception is illustrated in *Figure 2-8.*

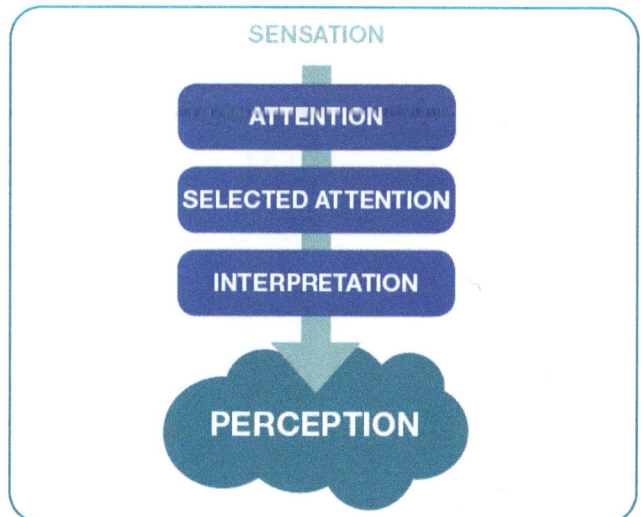

Figure 2-8: Model of perception.
Source: RMS

Within each individual, the attention and interpretation steps of the perception processes are closely interlinked and perception is influenced by a variety of factors, including:

- Attention factors, such as readiness of the individual to respond to a stimulus.
- The individual's past experience.
- The individual's motivation and emotional state.

The factors that influence the process and the way in which they operate are often referred to as the 'perceptual set' of the individual. This represents an accumulation of the individual's attention, experience, motivation and emotional state. These factors can cause an individual's perception to be distorted so that they perceive what they expect to perceive.

For example, in the case of the two images in *Figure 2-9* the individual sees what they expect to see, influenced by what is common in their experience. In the case of image *a)*, some people may see one box in the centre, sitting on a base with two sides or they may see two boxes to the side, hanging from a surface.

Similarly, in image *b)*, some people may see two people facing each other or a vase in the middle of the square. In work situations perception of health and safety hazards and risks is important. These examples illustrate that in regard to health and safety the subconscious process of perception takes place without the individual realising and may mean they genuinely may not see hazards in the workplace.

The influence an individual's perceptual set has on the perceptual process could be summarised by the phrases:

We do not see what is there.

We see what we expect to be there.

We do not see what we do not expect to be there.

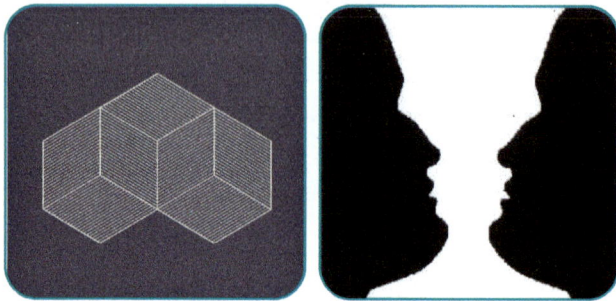

Figure 2-9: Examples of perception images.
Source: Ambiguous.

Factors that influence the effectiveness of an individual's perception of hazards and risks may include:

If an individual is doing a boring or repetitive job it may result in them thinking of other things than their work. This may result in a lowering of the impact of a stimulus and the stimulus may not get the individual's attention.

- Tiredness may reduce an individual's attention level and affect their perception of hazards. For example, they may not see a forklift truck that is approaching them and may step in front of it.

- Similarly, intense concentration on one task may make paying attention to another stimulus difficult or impossible. For example, an individual may fail to hear a warning of danger shouted by another person.

- An individual may not perceive a hazard they are aware of as a risk. For example, where an individual has no experience of the harm electricity can do they may not perceive it as a risk.

- Similarly, warnings of the hazard or risk may not be strong enough to get through the individual's perceptual set.

- Patterns of behaviour and habits that are learnt through experience can be carried from one situation to another where they are not appropriate. For example, a delivery driver may get used to driving at speed on a public highway and not perceive the need to change their speed when they enter a busy site.

- Some hazards may not be obvious or they could be hidden, making perception of their presence difficult, such as the hazard of electricity or certain gases like carbon monoxide.

- The presence of hazards may be masked by environmental issues such as poor lighting or ambient noise. Similarly, the use of personal protective equipment may interfere with the user's senses, limiting their perception.

- Individuals can get 'used to' a stimulus and, if it is not reinforced, it ceases to command their attention and is ignored. For example, they may get used to seeing obstructions in a walk route and after a time no longer perceive them.

Actions to influence perception might include:

- Making what we want people to perceive more obvious, for example, marking hazards so they are more visible.

- Providing information, instruction and training to influence an individual's interpretation of what should be perceived, for example, to improve recognition of hazards and risks.

- Providing experience to reinforce and organise the perception process so that the individual becomes experienced in what the correct perception is, for example, providing practice fire drills so that when an individual hears a fire alarm it is promptly perceived as requiring action to leave the building.

Competence

If managers and workers are to be effective in contributing to the correct health and safety behaviour at work it is essential that they are competent. Ensuring competence improves health and safety behaviour because managers and workers become more aware of the health and safety aspects of work activities, their perception of risk and danger increases to an appropriate level and there should be an improvement in their attitude and motivation. In addition, competent managers and workers will have a better understanding of the rules and procedures and are more likely to follow them. The decisions made by competent managers are more likely to give proper regard to health and safety, and lead to actions that support workers following rules and behaving appropriately.

Competence means more than just providing training and involves ensuring that the relevant **knowledge, skill and experience** are established. Studies have shown that human errors can occur through an absence of knowledge and skill. It is important that competencies held by managers and workers include the correct health and safety behaviour for their job. Where possible, correct health and safety behaviour should be learnt as part of the main job competencies, for example, when a manager is learning to manage contractors the correct health and safety behaviour expected of them and the contractors should be included. This will ensure that having the correct health and safety behaviour is seen as part of doing a job well and is more likely to become a normal part of doing the job.

Because individuals may not be capable of developing the competencies and behaving in the correct way needed for certain jobs, their suitability should be assessed. This will help to prevent situations where human errors and violations occur because a person is not capable of doing the job. For example, an individual may not have the aptitude, dexterity and physical ability needed to do certain jobs competently. Similarly, not everyone will be able to work at height or in confined spaces.

When new appointments of managers and workers are being made it should not be assumed that they have the competence that will provide the right health and safety behaviour, even if they are good at other aspects of their work, for example, quality. Managers and workers should be assessed for their range of competencies prior to appointment. Where competencies need to be developed it is important that managers and workers gain competence in a controlled way to ensure they learn the correct things and develop the right skills and behaviour. If this is not done they may not develop the correct health and safety behaviour, instead they may just become very skilled at their bad habits. In order to minimise human errors it may be necessary to arrange for individuals to gain experience under controlled conditions, for example, by close supervision. The Management of Health and Safety at Work Regulations (MHSWR) 1999 defines competence in Regulation 7, see **Figure 2-10.**

Competent people are defined as those who have:

"sufficient training and experience or knowledge and other qualities to enable them to perform their functions."

Figure 2-10: Requirements for competence.
Source: Management of Health and Safety at Work 1999.

Competence may be achieved and maintained by actions that include:

- Defining the competencies needed to carry out jobs and tasks safely and healthily.
- Assessing the competencies that managers and workers have and planning competence development.
- Providing the means to ensure that all managers and workers, including temporary workers, develop competencies they require in a controlled way, particularly for those involved in high risk work.
- Arranging for help, advice and supervision while they develop the correct competencies and health and safety behaviour.
- Ensuring individuals know the limits of their competency.
- Reviewing and refreshing competencies as needs change or where health and safety behaviour shows it is required.

Human failure - errors and violations

It is important to develop a good knowledge of how human factors influence health and safety behaviour at work, in particular with regard to understanding accident/incident causation. Human factors includes consideration of the individual and should include how human error and violations (human failures) are caused as this will help to develop a better understanding of accident/incident causation and provide wider opportunities for actions to reduce the likelihood of errors and violations.

It is estimated that approximately 80%–90% (Heinrich et al., 1980) of accidents/incidents are caused by human factors. Studies of major accidents/incidents have shown the importance of human factors as a cause in these events. They have also challenged the widely held belief that accidents/incidents are solely the result of 'human error' by the individual(s) directly involved in the activity related to the accident/incident. Attributing accidents/incidents to the 'human error' of these individuals alone has often been seen as sufficient explanation of the cause of accident/incidents and something that is beyond the control of managers. This ignores the failures of others, including managers, which can lead to accidents/incidents, for example, the failure of individuals involved in the organisation's management and decision-making processes. These failures are usually the result of 'human errors' by people other than those more immediately involved in the accidents/incidents and in many cases these failures can directly influence the behaviour of those individuals that are often accused of causing accidents/incidents. The table in **Figure 2-11** illustrates the influence of human failure in some major accidents/incidents. Similar human failures have been identified in other major accidents/incidents that have taken place worldwide since these events.

Accident/incident	Consequences	Human contribution
Three Mile Island, Nuclear industry, USA, 1979.	Serious damage to the core of the nuclear reactor.	Operators failed to recognise a valve that was stuck open due to poor design of the control panel. Maintenance failures had happened before, but no steps had been taken to prevent a recurrence.
Union Carbide Bhopal, Chemical processing, India,1984.	The plant released a cloud of toxic methyl isocyanate. The death toll was 2,500 and over one quarter of the city's population was affected by the gas.	The leak was caused by a discharge of water into a storage tank. This was the result of a combination of operator error, poor maintenance, failed safety systems and poor health and safety management.
Space Shuttle 'Challenger', Aerospace, USA, 1986.	Explosion killed all 7 astronauts on board.	There was an inadequate response to internal warnings about the faulty design of a seal. A decision was taken to go ahead with the launch in very cold temperatures despite the faulty seal. The decision made was a result of conflicting scheduling/safety goals, the mindset of individuals and the effects of fatigue.
Chernobyl, Nuclear industry, USSR, 1986.	A 1,000 MW nuclear reactor exploded releasing radioactivity over much of Europe. High environmental and human cost.	Causes were much debated, but the Soviet investigative team admitted "deliberate, systematic and numerous violations" of health and safety procedures by operators.
Herald of Free Enterprise, -Transport industry, UK, 1987.	Ferry sank killing 189 passengers and crew.	There was no system for checking that bow doors were shut. An inquiry reported that the company was "infected with the disease of sloppiness". The priority was to turn the ship around in record time.
Kings Cross fire, Transport industry, UK, 1987.	Major fire killed 31 people.	Organisational changes had led to poor escalator cleaning. The fire took hold because of inadequate firefighting equipment and poor staff training. There was a culture that viewed fires as inevitable.
Piper Alpha, Petrochemical industry, UK, 1988.	Major explosion on North Sea oil platform killed 167 workers.	The maintenance error that eventually led to the leak was the result of inexperience, poor procedures and poor learning. There was a breakdown in communications and the permit-to-work system at shift changeover and health and safety procedures were not properly practised.

Figure 2-11: Human failure in accidents/incidents.
Source: HSE, HSG48.

As can be seen from the examples in *Figure 2-11* the prevention of human failure should be recognised as an important part of an accident/incident and ill-health prevention programme. There are two different types of human failure that need to be assessed and managed effectively - errors and violations. A human error is an action or decision that was not intended, which involved a deviation from an accepted standard, and which led to an undesirable outcome. A violation is a deliberate deviation from a rule or procedure.

Human errors

Human error is an important part of human factors and a particularly significant part of the health and safety related human failures that take place in the workplace. They are not the deliberate human failures, but are the unintended actions or decisions of individuals, often when trying to do their job well. Human errors fall into two categories, skill-based errors and mistakes, see *Figure 2-12*. Skill based errors are then divided into two types, slips of action and lapses of memory. Mistakes are also divided into two types, rule-based mistakes and knowledge-based mistakes.

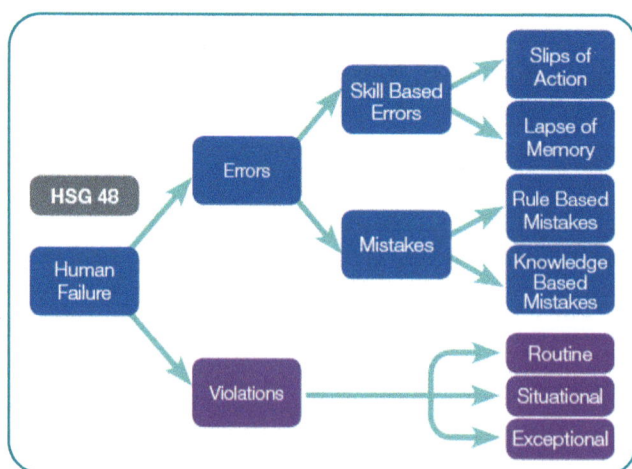

Figure 2-12: Human failures flow chart.
Source: HSE, HSG48.

Slips and lapses

Slips and lapses are the human errors that occur in the workplace during very familiar tasks, which can be carried out without much need for conscious attention. Once an individual has learned a skill, there is little need for conscious thought about what they are doing. They can carry out a task without having to think too much about the next step. An individual learns to ride a bicycle or drive a car in this way. They need to pay attention to the road and the traffic, but manipulate the pedals and change gear without thinking about it. If their attention is diverted by a distraction or interruption, they may fail to carry out the next action of a task, *a slip*, or forget the next action or lose their place, *a lapse*. Therefore, slips and lapses are the errors that are made by even the most experienced, well-trained and highly-motivated individuals whilst conducting routine tasks.

A slip is *'not doing what you are meant to do'*. Examples of slips include:

- Performing an action too soon in a procedure or leaving it too late, for example, raising a load on the forks of a forklift truck while the truck is travelling along.
- Carrying out an action with too much or too little strength, for example, over-tightening a bolt.
- Performing an action in the wrong direction, for example, an operator of a crane pushing the directional movement control to the left instead of the right.

A lapse is *'forgetting to do something, or losing your place part of the way through a task'*. Examples of lapses include:

- Forgetting to switch the local exhaust ventilation (LEV) system on when operating equipment that produces dust.
- Forgetting to nail down a wooden joist when building a floor.
- Taking a respiratory protection mask off to talk to someone and then forgetting to put it back on.
- Failing to secure a scaffold because the individual erecting it was interrupted during the task.
- Forgetting to empty the oil in a pump before moving it for maintenance, allowing it to leak on the floor.

Mistakes

Mistakes are also human errors, but are a little more complex than slips and lapses. Mistakes are decision making failures. Mistakes involve an individual doing the wrong thing, believing it to be right. The two main types of mistake are rule-based mistakes and knowledge-based mistakes.

Rule-based mistakes

There is a tendency for individuals at work to use familiar rules or procedures and sometimes they apply them in situations where they should not be applied. The wrong application of a rule to a situation can result in an error in the form of a rule-based mistake, for example, the use of a water-based fire extinguisher on a fire involving live electrical equipment. In this situation the individual applied the rule that required them to act promptly to extinguish the fire, but used the wrong type of fire extinguishing medium (water) to extinguish the fire. Using the wrong type of fire extinguisher could have caused them to have an electrical shock.

Knowledge-based mistakes

In some situations, such as unfamiliar circumstances, an individual may have to apply their knowledge and reason out what action is required. If the situation is misdiagnosed or the action is miscalculated, a knowledge-based mistake may occur.

For example, a driver may make a poor judgement when overtaking, leaving insufficient room to complete the manoeuvre before oncoming vehicles get dangerously close. Knowledge-based mistakes often occur with untrained and inexperienced people as they may base their decisions on misunderstandings and a lack of perception of risk. These errors also occur with trained, experienced people in situations where there are time pressures and where they are doing too many complex tasks at once.

An individual may be influenced by a number of job factors that can contribute to them making a mistake, for example, where there are too few people to do a task, distractions, confusing instructions, poor design of control panels or poor work environment. Individual factors like lack of sleep, fatigue, stress, medication, drugs and alcohol can also contribute to people making mistakes.

When considering the various types of human error as a whole there are various actions that may be taken to influence the reduction of errors including:

1) Reducing work environment factors that can lead to increased errors, for example, extremes of heat, humidity, noise, vibration, poor lighting, and restricted workspace.

2) Reducing the effects of extremely demanding tasks, for example, tasks where there is a high workload, tasks demanding high levels of concentration, tasks that are very monotonous or repetitive and tasks where there are many distractions and interruptions.

3) Reducing organisational factors, for example, insufficient numbers of managers or workers, inflexible or over-demanding work schedules, peer pressure and conflicting attitudes to health and safety.

4) Reducing individual factors, for example, inadequate training and experience, high levels of fatigue, reduced alertness, family problems, ill-health, misuse of alcohol and drugs.

5) Reducing equipment factors, for example, poorly designed displays and controls, inaccurate and confusing operating instructions.

6) Simplifying complex tasks and planning for emergencies and other non-standard situations.

7) Ensuring procedures and instructions are clear, concise, available and up-to-date.

8) Ensuring proper supervision, particularly for inexperienced managers and workers.

9) Considering the possibility of human error when undertaking risk assessments.

10) Identifying causes of human errors during accident/incident and ill-health investigations.

11) Ensuring arrangements for the evaluation of health and safety include monitoring the effectiveness of measures taken to reduce error.

Violations

Violations are intentional human failures and involve an individual deliberately doing the wrong thing. Although violations are a deliberate failure to follow a rule or procedure, they are rarely acts of vandalism or sabotage, but are often carried out in order to get the job done, for example, using a convenient ladder of insufficient length. Many injuries and cases of ill-health are caused by the violation of health and safety rules or procedures. There are three types of violation - routine, situational and exceptional violations.

Routine violations

Routine violations are where breaking the health and safety rules or procedure has become the normal way of working. They can develop in a number of ways. The incorrect method may be developed and used because it is a quicker way to work or because the rules are seen as too restrictive. In some cases, new workers can join an organisation and learn the incorrect method from other workers, not realising they are wrong. Examples of routine violations include:

- Not removing dirty work clothes when taking refreshment breaks in a facility provided for eating food.
- Removing the guard on a dangerous machine so that the user of the machine can see the operation easier.
- Driving a fork-lift truck too fast.
- Not wearing hearing protection in a noisy workplace.

Actions to minimise routine violations include:

1) Increasing the chances of violations being detected, by monitoring and supervision.

2) Identifying and removing unnecessary rules.

3) Making rules and procedures remain relevant and practical.

4) Explaining the reasons for and relevance of the rules and procedures.

5) Improving design factors that affect the likelihood of individuals not following rules or procedures. For example, equipment that is excessively awkward, tiring or slow to use, frequent false alarms from instrumentation and uncomfortable personal protective equipment.

6) Involving managers and workers in writing rules and procedures to help to ensure their acceptance.

Situational violations

Situational violations may occur due to pressures related to the work being done. This can include time pressures, extreme weather conditions, provision of an inadequate number of people and provision of the wrong equipment/materials to do a task. These pressures may make it very difficult to comply with rules in a particular situation or workers may think that the rules are inappropriate under the circumstances. For example, rules for safe working at height may be ignored in a situation where work on a roof may need to be completed and the correct work at height equipment had not been provided.

Actions to minimise situational violations include:

1) Providing the correct equipment.

2) Improving the working environment.

3) Providing appropriate supervision.

4) Improving job design and planning.

5) Establishing a positive health and safety culture.

Exceptional violations

Exceptional violations occur when something has gone wrong and a decision is made to solve the problem in a way that involves breaking a rule and taking a risk.

For example, an individual may enter a confined space without suitable breathing apparatus if another person has collapsed in the confined space from lack of oxygen and needs to be rescued. Though the individual carrying out the violation may perceive it to be necessary in the exceptional circumstances it is an incorrect belief that the benefits outweigh the risk.

Actions to minimise exceptional violations include:

1) Identify the possibility of violations in work activities as part of the risk assessment process.

2) Planning for exceptional situations.

3) Reducing the time pressure on managers and workers to act quickly in exceptional situations.

4) Providing safe and healthy procedures for exceptional situations.

5) Providing more training for abnormal and emergency situations.

LINK BETWEEN INDIVIDUAL, JOB AND ORGANISATIONAL FACTORS

The examination of human factors has shown that what may appear to be simple failings of individuals are often failings that have been shaped by the circumstances the individuals found themselves in. This will include the individual factors affecting them, the job factors and the organisational factors that cause these job factors to exist. The effects of a number of human factors may combine, causing individuals to fail to comply with health and safety rules or procedures and have poor health and safety behaviour.

An individual's attitude to health and safety (an individual factor) can be influenced by job factors that directly affect the individual. For example, if an individual is being affected by requirements to meet other priorities than health and safety (a job factor), this could lead to them developing an attitude that health and safety is not important. Further analysis might identify that this particular job factor was created by the presence of a poor health and safety culture (an organisational factor). This shows the link between individual, job and organisational factors.

This link is important to consider when analysing problems and devising solutions to improve health and safety behaviour. Attempts to improve the attitude of workers by the enforcement of health and safety rules are unlikely to be successful in a situation where organisational factors are creating job factors that strongly and negatively influence the attitude of workers. The enforcement of rules may change the behaviour of the individual for a short period, but it is unlikely to change their attitude or behaviour in the long term. Therefore, in this situation, it is important to ensure that the organisational factors that create the job factors are controlled in order for them to have a positive influence on the individual factors.

When the human factors, (organisational, job and individual), work together positively, they can help to ensure individuals adopt the correct behaviour. It is therefore important to ensure that all three factors are considered in order to adequately manage the human factors and ensure health and safety.

REVIEW

Explain the terms attitude, motivation and perception and give an example of each.

2.3 Improving health and safety culture

Health and safety culture at work can be improved by the use of a range of measures and by the coordinated effort of everyone in the organisation. Measures that can improve health and safety culture and contribute to establishing a positive health and safety culture include securing management commitment, leadership, promoting health and safety standards, holding people to account for their behaviour, ensuring the competence of individuals, effective communication, encouraging co-operation, arranging consultation and the provision of training.

GAINING MANAGEMENT COMMITMENT

Securing management commitment is one of the most important steps in establishing a positive health and safety culture, which is recognised as an effective way of ensuring appropriate health and safety behaviour. The absence of management commitment would indicate that health and safety was a low priority in the organisation and that it was not important. This could lead to managers and workers behaving in a way that is not supportive to health and safety. For example, when decisions are made, health and safety may not be given the resources required.

Securing a commitment from management would help to ensure health and safety was properly integrated in the processes of the organisation. This includes preparing and implementing a health and safety policy that reflects positive values and beliefs. The policy should be supported by effective management systems and procedures that include health and safety. Management should be committed to giving an equal priority to health and safety issues as other objectives, such as production and quality. This will involve managers actively ensuring that their work and management practices give appropriate priority to health and safety. For example, providing a pleasant working environment with good welfare facilities will indicate a commitment to health and safety and the good standards expected of managers and workers.

By securing the commitment of management, health and safety would be identified as one of the core values of the organisation and individuals would be more motivated to behave safely and healthily.

PROMOTING HEALTH AND SAFETY STANDARDS BY LEADERSHIP, EXAMPLE AND DISCIPLINARY PROCEDURES

Management actions at all levels should send clear signals to other managers and workers of the importance of the health and safety standards set by the organisation.

Managers should promote the health and safety standards by showing their commitment to them when they are carrying out their normal management activities and specific health and safety activities assigned to them. This will often include managers showing leadership by setting a good example. For example, managers should use the correct personal protective equipment, follow health and safety rules and show commitment by attending health and safety training. This will reinforce the need to conform to health and safety standards.

Managers should also show leadership by making health and safety important in meetings and decision making, for example, ensuring health and safety is dealt with as a priority item on the agenda of meetings. This will promote the importance of managers putting effort and priority into achieving good health and safety standards. Managers would then be more likely to give proper consideration to health and safety during meetings and in their other work. Managers would also have an opportunity to promote health and safety standards in these meetings by referring to them when operational issues are discussed and ensuring the standards are not compromised when making decisions. In situations where there is a pressure to compromise health and safety standards because of time constraints managers should emphasise the importance of maintaining standards and, where it is appropriate, promoting the acceptance that a delay will be necessary to complete the task safely. This strong leadership will set a good example to other managers and the workers affected by the decision, which will encourage positive health and safety behaviour in general.

There are many other opportunities for managers to show leadership and set an example, including during visits made to the workplace and when health and safety audits or reviews are conducted. Senior managers show good leadership when they promote health and safety standards during their opening presentation at health and safety training sessions that other managers and workers attend. This will also emphasise the importance of the training being provided and show a commitment to good health and safety standards.

It is important that managers show leadership and support for health and safety standards during problem solving processes. This will demonstrate that health and safety is valued and that there is a management desire for good health and safety standards. It is therefore useful for managers to be involved in the investigation of accidents/incidents and ill-health and is particularly useful for them to be involved in preventive measures like risk assessment processes. Both of these processes provide an opportunity to promote health and safety standards by identifying if the standards were in place, encouraging standards to be met and enabling measures to be put in place for individuals to work safely and healthily.

The manager's involvement in risk assessment processes would promote the importance of preventive measures and could encourage supervisors and workers to behave in a way that is more likely to reduce risks.

To have the maximum effect on health and safety behaviour the organisation's health and safety standards should also be promoted through the good example set by supervisors and influential individuals, such as worker representatives and committee members.

If regular checks are made by managers to ensure health and safety standards are being met it would also show positive leadership to supervisors and workers and reinforce the agreed standards. Where health and safety standards are not being met, the reasons for this should be established by identifying the underlying and root causes. A blame approach should be avoided where there are acceptable reasons for not meeting standards. However, *disciplinary procedures* should be used where they are appropriate, to reinforce the need to maintain health and safety standards.

COMPETENT WORKERS

If managers and workers are to be effective in contributing to the correct health and safety behaviour at work it is essential that they are competent. Ensuring competence improves health and safety behaviour because managers and workers become more aware of the health and safety aspects of work activities, their perception of risk and danger increases to an appropriate level and there should be an improvement in their attitude and motivation. In addition, competent managers and workers will have a better understanding of the rules and procedures and are more likely to follow them. The decisions made by competent managers are more likely to give proper regard to health and safety and lead to actions that support workers following rules and behaving appropriately.

Competence means more than just providing training and involves ensuring that the relevant *knowledge, skill and experience* are established. Studies have shown that human errors can occur through an absence of knowledge and skill. It is important that competencies held by managers and workers include the correct health and safety behaviour for their job. Where possible, correct health and safety behaviour should be learnt as part of the main job competencies, for example, when a manager is learning to manage contractors the correct health and safety behaviour expected of them and the contractors should be included. This will ensure that having the correct health and safety behaviour is seen as part of doing a job well and is more likely to become a normal part of doing the job.

Because individuals may not be capable of developing the competencies and behaving in the correct way needed for certain jobs, their suitability should be assessed. This will help to prevent situations where human errors and violations occur because a person is not capable of doing the job. For example, an individual may not have the aptitude, dexterity and physical ability needed to do certain jobs competently. Similarly, not everyone will be able to work at height or in confined spaces.

When new appointments of managers and workers are being made it should not be assumed that they have the competence that will provide the right health and safety behaviour, even if they are good at other aspects of their work, for example, quality. Managers and workers should be assessed for their range of competencies prior to appointment. Where competencies need to be developed it is important that managers and workers gain competence in a controlled way to ensure they learn the correct things and develop the right skills and behaviour. If this is not done, they may not develop the correct health and safety behaviour, instead they may just become very skilled at their bad habits. In order to minimise human errors, it may be necessary to arrange for individuals to gain experience under controlled conditions, for example, by close **supervision.** ISO 45001:2018 emphasises the need to ensure competence in Clause 7.2, see **Figure 2-13**.

> The organization shall:
>
> a) determine the necessary competence of workers that affects or can affect its OH&S performance;
>
> b) ensure that workers are competent (including the ability to identify hazards) on the basis of appropriate education, training or experience;
>
> c) where applicable, take actions to acquire and maintain the necessary competence, and evaluate the effectiveness of the actions taken;
>
> d) retain appropriate documented information as evidence of competence

Figure 2-13: Requirements for competence.
Source: ISO 45001:2018, Clause 7.2.

Competence may be achieved and maintained by actions that include:

- Defining the competencies needed to carry out jobs and tasks safely and healthily.
- Assessing the competencies that managers and workers have and planning competence development.

- Providing the means to ensure that all managers and workers, including temporary workers, develop competencies they require in a controlled way, particularly for those involved in high risk work.
- Arranging for help, advice and supervision while they develop the correct competencies and health and safety behaviour.
- Ensuring individuals know the limits of their competency.
- Reviewing and refreshing competencies as needs change or where health and safety behaviour shows it is required.

GOOD COMMUNICATION WITHIN THE ORGANISATION

Studies of major accidents/incidents have shown that without effective communication, managers and workers are less likely to make the right decisions and errors or violations may occur. It is therefore important to provide effective communication within an organisation in order to help to ensure the correct health and safety behaviour takes place. ISO 45001:2018, Clause 7.4, shows the type of communication arrangements and procedures required in an organisation, see **Figure 2-14.**

> The organization shall establish, implement and maintain the process(es) needed for the internal and external communications relevant to the OH&S management system, including determining:
>
> a) on what it will communicate;
>
> b) when to communicate;
>
> c) with whom to communicate:
>
> > 1) internally among the various levels and functions of the organization;
> >
> > 2) among contractors and visitors to the workplace;
> >
> > 3) among other interested parties;
>
> d) how to communicate.
>
> The organization shall take into account diversity aspects (e.g. gender, language, culture, literacy, disability) when considering its communication needs.

Figure 2-14: Communication arrangements and procedures.
Source: ISO 45001:2018, Clause 7.4.

Effective communication involves providing what is required in the right way and at the right time. This will mean using the best communication method for the information or instruction being provided and the circumstances. It is therefore necessary to consider the individual receiving the communication to ensure that the information is understood and the correct health and safety behaviour results.

For example, when managers communicate with individuals it is important that managers are aware of the individual's needs, desires, capabilities and expectations in order to communicate with them effectively.

In some cases, this will mean ensuring that the communication provides information on why action is to be taken as well as what is to be done. This is particularly the case with regard to health and safety. For example, a manager may communicate to maintenance workers that they must wear eye protection when doing a task, but without emphasising why they should wear eye protection they may not behave correctly and wear it. It is also important that managers at all levels communicate their commendation of positive health and safety behaviour when they see it, as this will make it more likely that the positive behaviour will continue.

Communication within an organisation should include workers informing managers on concerns they have regarding health and safety. It is important that there are means for workers to do this in a positive way, so that they are encouraged to report hazards and deficiencies that could be a danger to themselves and others. To encourage workers to continue to report concerns managers should provide workers with feedback on concerns they communicate.

The communication process

Communication may be defined as *'a process by which information is exchanged between individuals through a common system of symbols, signs or behaviour'.* Exchanging information may seem a simple task, but communicating in an effective way is often much more difficult than it appears. Poor communication, while common, can result in serious misunderstandings, leading to incorrect health and safety behaviour, accidents/incidents and work-related harm to health.

Barriers to effective communication

It is essential that communications are understood in order for them to be acted on and for individuals to have the correct health and safety behaviour. However, communication is not always effective as there are a number of barriers that may prevent it being understood, including:

- Noise and other similar distractions.
- Sensory impairment (poor hearing or eyesight).
- Complexity of the information.
- Language/dialect of the speaker.
- Illogically presented information
- Ambiguity of the information.
- Use of technical and local terms or abbreviations.

Information			
Purpose	To improve awareness about health and safety generally and in relation to specific hazards, their controls and management performance to bring about these controls. In itself passive, it relies on the recipient to interpret.		
Subjects	Legislation. Company policy statements.	Incident and ill health statistics. General hazards and controls.	Names of appointed first-aiders.
Means of communication	Bulletins and news sheets. Noticeboards, propaganda, films.	Team briefing. Written material for visitors.	Site signs and labels.

Instruction			
Purpose	To control the behaviour of workers, contractors and visitors with regard to general and specific health and safety arrangements. Typically, one-way communication, often no real check of understanding.		
Subjects	Health and safety rules. Policy, arrangements and plans.	Use of PPE. Specific hazards, for example, smoking.	Emergency procedures. Reporting incidents and ill health.
Means of communication	Carried out formally using verbal, written and visual material - noticeboards, induction and job training, direct issue of document, 'tool-box talks'.		

- The timeliness of the communication.
- Lengthy communication methods.
- Inattention of the person receiving the information.
- Lack of trust or respect.
- Capabilities of the person receiving the information to understand it, due to reduced health, fatigue, stress or limitations on their mental processing ability.

General principles of communication

Communication is a skill that people often take for granted. Like any other skill, some individuals are better at it than others. A one-way communication process involves sending information to an individual and not obtaining confirmation of effective communication. This type of process has the limitation that confirmation of receipt and understanding of the information is not achieved. Therefore, communication is generally a two-way process and is most effective in this form. In a two-way communication process the needs of the person receiving the information are equally as important as the needs of the person sending the information. The person sending the information should not assume that the person receiving it has understood what was sent. The two-way process requires the person sending the information to confirm that the other person has received and understood it. To ensure the success of two-way communication the person sending the information should:

- Communicate to the person receiving the information at a time that is suitable for their circumstances.

- Communicate in a form capable of being understood by the person receiving the information.

- Keep the content of the communication concise and relevant to the information being communicated.

- Provide the information in an order that is logical to the person receiving it.

- Use clear and unambiguous terms.

- Ensure that what is communicated matches the way it is communicated. If something is important, the communication of it should not be rushed. Non-verbal communication may communicate something different to what is said. Consider the need for respect and whether anger or humour may distract from what is communicated.

- Plan and encourage feedback to confirm understanding, including time, method and the form of feedback. The person receiving the communication may have to respond in a particular way, for example, repeating the information back to the person who gave the information.

- Use open ended questions to investigate understanding.

- Use closed questions, which tend to provide yes/no answers, to confirm understanding.

- Repeat the communication or use a different method of communication if understanding is unclear.

Communication methods

In ensuring effective communication, the correct method of communication should be selected and used for the particular information and circumstances in which it is being provided. This will mean considering the principles of good communication and the benefits and limitations of different methods. The different methods used to communicate health and safety typically include verbal, written and graphic methods.

Verbal communication

Verbal communication is widely used in relation to health and safety and involves direct speech between individuals or through another medium, for example, a telephone, radio, video link or electronic speaker system. Two-way forms of verbal communication provide the advantage of being able to communicate and obtain confirmation of understanding efficiently and promptly. This allows communication to be done in small amounts where necessary, ensuring understanding after each part of the information is communicated. This is particularly useful where information needs to be communicated accurately and understanding confirmed.

The barriers to communication outlined earlier are all relevant to verbal communication. In particular this method of communication can be influenced by the language spoken, dialects, accents and the unintended wrong use of words. These barriers can lead to misunderstandings or the wrong meaning being communicated. Where clarity of small amounts of important information is to be communicated the convention of spelling the letters of words and using a phonetic spelling alphabet is sometimes adopted. For example, the phonetic spelling alphabet of the International Civil Aviation Organisation (ICAO) assigns code words to letters of the English language, for example, A - Alpha, B - Bravo, C - Charlie, etc. This is particularly useful when communicating by radio or telephone and helps to avoid errors in verbal communication.

Where the information to be communicated is complex it may be preferable to use written communication to ensure clarity or a combination of written and verbal communication.

Written communication

Written communications are widely used for health and safety purposes, including policy documents, work instructions, procedures, site rules, permits-to-work, contractor contracts, meeting agendas and minutes, and reports.

The written method of communication enables the information to be retained by the person receiving it and referred to if they forget what the information was. It may also enable an individual to receive information prior to needing it and means they can read it at a time that suits them. Obtaining confirmation of understanding of written communications can be as important as for verbal communication. However, written communications are often used as a form of one-way communication, for example, notices or instructions posted on an information board.

One advantage of the written form of communication is that it can provide information to a large number and range of people at the same time. However, this can make it difficult to take account of the range of people reading it and the need to ensure effective communication to each of them. In particular, it may be hard to obtain or manage feedback from all of the people sufficiently to ensure they understood what was communicated. For example, this difficulty would affect a situation where all workers in an organisation were issued with a new health and safety handbook. It could be difficult to confirm whether they had all read and understood the handbook.

When communicating in the written form it is important to consider the barriers to and principles of communication, this will help to make the communication effective. Particular attention should be paid to the use of appropriate structure, language, writing style and terms used to ensure the person receiving the communication understands it. Clarity and understanding of written communications can be tested by getting a sample of those who will receive the written communication to read it and checking they understood it. For example, procedures may be read by a sample of workers who have to follow them.

Graphical communication

Graphical communications may be used to replace or support other forms of communication. Verbal presentations and written reports will often include graphical images to illustrate or clarify complex relationships, for example, changes in accident/incident frequency rates over a period of time may be illustrated by a graph.

Graphical communication is used to good effect with health and safety signs, where the colour and pictorial representation of hazards and control measures are used. This avoids problems where workers may have limited language abilities and would struggle to read words. Graphical communication is also used in posters and moving image communications, for example, electronic induction packages and message boards.

Graphical communication in the form of drawings, photographs, still or moving digital images may be used to clarify information on hazards or risk control measures.

They can provide an opportunity to make the communication of information more effective by containing examples of hazards and control measures from the actual workplace of those persons with whom they are communicating. The visual effect of graphical communication of this type may enable managers and workers to better understand practical aspects of health and safety. For example, a rule that requires a worker to wear a hard hat correctly could easily be demonstrated and understood by showing an image of a worker wearing a hard hat correctly.

CONSIDER

How effective is health and safety communication in your organisation?

What could be done to improve it?

Use and effectiveness of various communications measures

Notice boards

A traditional communication measure is to post health and safety information on notice boards. The advantage of this communication measure is that the communication is easily available to everyone in a particular work area where the notice board is located and to those that pass near it.

Figure 2-15: Health and safety notice board - obstructed, poor attitude. Source: RMS.

They can be a cheap and effective technique of communication when used to make general statements or to keep workers aware of current information and proposed developments. Notice board information relies on an individual's ability to read, understand and apply the information correctly. Therefore, particular care should be taken to ensure appropriate language(s) are used for information placed on a notice board and they should not be used as a substitute for more effective two-way communication that can provide confirmation of receipt and understanding of information. Also, the information posted on the notice board must be kept up to date and maintained in a legible condition if it is to be effective. This usually requires someone being given responsibility for the control of items placed on and removed from notice boards.

Health and safety media

Moving image media

Moving images of relevant health and safety topics are often used to provide and refresh awareness and understanding of health and safety. They can provide information in a way that improves motivation and leads to better health and safety behaviour. The visual impact in them is also a strong stimulus that maintains attention while learning and enables training to provide practical examples of hazards and measures to minimise risk. A common use for moving image media is at site induction of new workers and contractors, where it can provide information about the site, its rules and correct health and safety behaviour. The site induction may be supplemented by this media to enable the audience to relate better to the hazards and the control measures.

Shocking images of injuries and work-related ill-health are sometimes used to illustrate what might happen if procedures are not followed. It has been found that their effect may not change attitude in the long term, but can gain the attention of those seeing the images and allow them to accept the importance of health and safety measures when this is explained after seeing the images. However, for some people the shocking images cause them to put up a barrier to receiving further information and can prevent the desired health and safety message being received.

Digital media

Digital media can be a particularly effective tool for communicating health and safety messages as it can be viewed by a single person or used for larger audiences. Digital media is often made available through the internet, enabling it to be provided from a main source in an organisation to individuals that are difficult to communicate with by traditional means because of their location or work patterns. A form of digital media that is growing in use is electronic products that take a student through structure learning and provide them with an on-line assessment for them to confirm their understanding. These e-learning products are a form of digital media that lends itself to induction of workers/contractors and providing awareness.

Poster campaigns

Posters displaying health and safety information are sometimes seen as an inexpensive and visible way of showing commitment to health and safety. This approach can be ineffective if management place too much reliance on them and use them to inform workers to take care to avoid hazards that should have been managed by better control measures. To be effective, messages communicated by posters should be positive, aimed at the correct audience and be believable.

a) **Positive** - posters warning of the consequences if a particular action is not taken can be ineffective. To be effective, messages should emphasise the positive health and safety benefits of working safely and healthily to those reading the posters.

b) **Aimed at the correct audience** - posters quickly lose their impact and blend into the background, therefore posters should be changed regularly to maintain their impact as a means of communication. There can also be a reduction in impact when the message being communicated is perceived to be irrelevant by those reading the poster. Therefore, campaigns should be carefully targeted and posters positioned so they are seen by those the campaign is aimed at. For example, a poster reminding workers to wear the correct clothing when working with machinery should be sited close to the relevant machines.

c) **Believable** - messages should be convincing and realistic. Care must be taken to avoid offending or distracting the person reading the poster from the message by the use of inappropriate images of people. Similarly, pictures of horrific injuries can lead to a rejection of the intended message as the person reading it may focus on the image rather than the message. Some images may be so distressing that individuals refuse to look at the poster at all.

The advantages of using posters for communicating health and safety include:

- They have a relatively low cost.
- They provide visual impact to health and safety messages, allowing brief messages to be easily understood.
- They are flexible, enabling them to be displayed in the most appropriate positions and moved easily.
- They can be used to reinforce verbal instructions and provide a constant reminder of important health and safety issues.
- They can enable workers to be involved in health and safety locally, by encouraging them to suggest ideas for posters, selecting or designing posters and locating the posters where they will have most effect.

The disadvantages of using posters for communicating health and safety include:

- There is a need to change posters on a regular basis to maintain attention on them.
- They may become soiled, defaced and out of date.
- There is a possibility that they might be seen to trivialise serious matters.
- They might offend people if inappropriate stereotypes are used.
- It is not easy to assess whether the message has been understood.
- They may be used by some employers as an easy, if not particularly effective, way of discharging their responsibility for health and safety by moving the responsibility onto workers.

Toolbox talks

'Toolbox talks' are short communication sessions provided by supervisors and team leaders to explain or remind workers of important health and safety risks and control measures. The term toolbox talk comes from a scheme where short talks on health and safety were provided by supervisors in the workplace, sometimes near where the workers kept their tools. They were designed to take place in the workplace so that workers were not distracted from their work for too long. The term is now often used to describe any form of communication where workers are gathered together for a short talk in or near to the worker's workplace.

Because toolbox talks are planned and organised in advance, the technique provides fast and reasonably consistent communication of specific subjects. The toolbox talks often cover only one subject to ensure the attention of workers is focused and two-way communication is encouraged. Subjects covered in talks may provide a reminder of a particular risk related to work activities or discuss recent good or poor health and safety experience. The talks may be used to tell workers about proposed changes, for example, changes in personal protective equipment, work practices or procedures. Toolbox talks enable health and safety issues to be discussed by supervisors and workers, providing increased awareness/understanding of risks and control measures and reinforcement of rules. They can be regularly used to encourage workers to have appropriate health and safety behaviour.

Electronic messages and e-mails

Electronic messages and e-mails are a useful way to provide health and safety communication on a timely basis. For example, they can give information on proposed changes to procedures before the amendment and distribution of the whole procedure. Electronic messages are often used as a means of one-way communication. One of the difficulties in using electronic messages is that the person sending the message might assume that everyone to whom they sent the message received it, however this might not be correct as loss of power of battery operated equipment and poor signal can influence receipt of the message. Where the message is important to health and safety there should be a process that confirms receipt. Similar problems can exist with the use of e-mails, although software is available to check whether a person has received and opened an e-mail. However, confirmation of receipt of a memo or e-mail does not confirm that it has been read, understood or actioned.

Many people complain about the number of emails they receive. Where individuals receive a high volume of e-mails there is a risk that important health and safety e-mails may be missed. This situation is reported to be particularly difficult in situations where individuals are copied in to e-mails in bulk, causing people to receive e-mails that are of very low or no relevance to them. This could reduce the perceived value of e-mails received from these people and lead to important health and safety messages they send being missed.

Worker handbooks

Workers need to know the current rules and procedures so they can follow them and have the correct health and safety behaviour. Issuing a worker handbook will help to provide this information, but if it is not kept up to date it can lead to workers making errors.

Handbooks are often issued to new workers at induction. They are useful in communicating site rules and information such as accident/incident and ill-health reporting procedures. Similarly, they will often contain information relating to on-site emergency arrangements such as fire and first-aid. To be effective, the organisation should establish a mechanism to recall and reissue the handbooks when changes occur. Some organisations operate a loose-leaf folder design to enable amendments to be made. If paper-based handbooks are provided, consideration needs to be given to where each person keeps the handbook so that it is available for use and updating. Some organisations have computer systems available and therefore choose to provide handbooks in an electronic format. This can enable changes to the electronic copy of the handbook to be made quickly. However, unless workers can be equally quickly informed of the changes, they are not likely to follow the new rules and procedures because they will not be aware of them.

Toolbox talks or other means of communication may have to be used to inform workers affected by the changes. If toolbox talks are used, they would provide an opportunity to discuss the changes and ensure people knew why the changes were taking place and that they understood what the changes were.

This could provide additional confidence that everyone would follow the new rules and procedures.

When choosing an appropriate communication process that uses a particular communication method and communication measure, it is important to consider the risk that might result from poor communication. For example, it may not be appropriate to rely on a one-way communication process, like notice boards, for important communications related to high risk work. In most communication situations, confirmation that the information communicated was received is important and experience shows that confirmation of understanding of the information is essential if the correct health and safety behaviour is to result from it. Therefore, the communication methods and measures that provide a two-way communication process are usually preferred.

Company intranet

Intranets use Internet Protocol technology to share information, operational systems, or computing services within an organisation. A company intranet is a powerful internal communications tool, which can unify the whole team, provide access to essential health and safety information and improve health and safety performance. An intranet can help managers and workers find information more easily and ultimately perform their health and safety duties more effectively. They are particularly useful in enabling the provision and maintenance of the most up to date health and safety policy documents and procedures to be made available. The intranet assists with document control as it enables a single controlled copy to be made available on demand to multiple users.

The same system can be used to communicate health and safety issues and solutions on a timely basis by posting alerts, newsletters and health and safety instructions to the intranet. An intranet can also be used to provide web-based health and safety tools that can assist in the management of health and safety, for example, tools to assist in the risk assessment process or produce reports on health and safety. The intranet enables remote reporting and therefore communication of pro-active and reactive health and safety matters, including incidents that may take place in the workplace.

Specific legal requirements for health and safety information

The Health and Safety Information for Employees Regulations (HSIER) 1989 require that the employer is required by law to either display the HSE-approved poster (see *Figure 2-16*) or to provide each of their workers with the equivalent 'leaflet'. The poster tells workers what their employers need to do with regard to health and safety in the workplace, what workers must do and what to do if there are concerns with health and safety in the workplace. Details may be added of any worker safety

representatives or other health and safety contacts, but this is not compulsory. If a poster is used, the information must be legible and up to date and the poster must be prominently located in an area to which all workers have access. If a leaflet is used, new leaflets must be issued to workers when any similar changes occur.

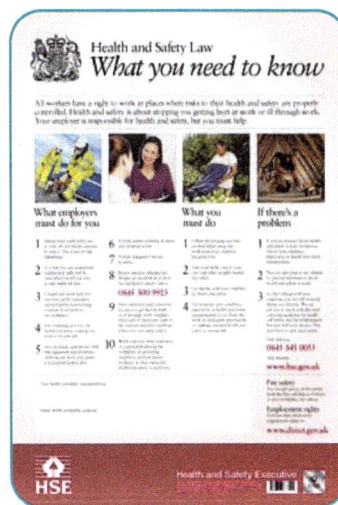

Figure 2-16: Health and safety law poster.
Source: HSE Books.

The MHSWR 1999, Regulation 10 requires that workers are provided with relevant information about hazards to their health and safety arising from risks identified by risk assessments, including those arising from contact with other employer's activities.

Clear instruction concerning any preventative or protective control measures must also be provided, including those relating to serious and imminent danger and fire risk assessments. Details of any competent persons nominated in relation to MHSWR 1999 must also be communicated.

Before employing a child (a person who is not over compulsory school age), the employer must provide those with parental responsibility for the child with information on the risks that have been identified and preventative and protective measures to be taken.

CO-OPERATION AND CONSULTATION WITH THE WORKFORCE AND CONTRACTORS

Those organisations that recognise the importance of establishing a positive health and safety culture have arrangements in place for good management and worker (including contractors, where applicable) co-operation and consultation. It is important to establish that there is a difference between informing people and consulting them. Informing is a one-way process involving the provision of relevant information, for example, by management to workers, whereas consulting is a two-way process where account is taken of the views of those being consulted before a decision is taken.

> "The principal contractor must:
>
> (a) make and maintain arrangements which will enable the principal contractor and workers engaged in construction work to cooperate effectively in developing, promoting and checking the effectiveness of measures to ensure the health, safety and welfare of the workers;"

Figure 2-17: Principal contractor's - arrangements for worker co-operation.
Source: CDMR 2015, Regulation 14.

Consultation with workers can be undertaken in various ways. It does not need to be a formal process and can be as simple as talking to them regularly and considering their views when making health and safety decisions. This may be achieved by consulting the workforce (and contractors, where applicable) on a one-to-one basis, consulting with worker health and safety representatives or by using health and safety committees.

In some organisations co-operation and consultation would extend to the involvement of contractors, this is particularly important where contractors provide services in a workplace on a long-term contract, for example maintenance and security contractors. Co-operation and consultation with contractors can be difficult to achieve, sometimes because of the nature of the work contractors are involved in or because the extended lines of communication through the contractor can make it difficult to organise. However, as both the organisation and the contractor must work together and effectively co-ordinate their activities, it is worth making an effort to ensure arrangements for co-operation and consultation are made.

One of the ways of enabling co-operation and consultation to take place is to have regular meetings with contractors throughout the period of contracted activities. Alternatively, a representative of contractors could attend the organisation's health and safety committee. The level of co-operation and consultation needed will depend on the activities the contractors do, the risks involved and the number of contractors involved.

Worker participation

The main *role of worker participation* in health and safety is to provide the employer with a wider view of how risks affect workers, obtain feedback on the effectiveness of current control measures and gain their view on proposed control measures. Worker participation in health and safety through *consultation has the benefit* that it will draw on the worker's experience and knowledge. This can be used to improve the effectiveness of health and safety measures. This improved effectiveness and participation of workers in the development of new methods can provide better worker acceptance of health and safety measures and lead to long term improvements in worker health and safety behaviour.

In addition, to providing an opportunity for workers to contribute to health and safety, the role of worker participation is to show management commitment and motivate workers to work safely and healthily. The benefit of worker participation of this type is that it has been identified as one of the most significant factors to influence health and safety behaviour and promote a positive health and safety culture in organisations. Workers feel valued and involved in decision making. This can lead to shared health and safety values and the motivation of those involved to work together to improve health and safety. Worker participation provides an effective way of ensuring worker feedback on health and safety, for example proposed changes to the layout of a workplace, options for content and delivery of training, hazard identification and procedures to control risks. Workers often have a good understanding of the risks in the workplace and involving them shows them that health and safety is taken seriously and their views on problems and solutions are valued.

Legislation in the UK establishes requirements that employers consult their workers or the workers' representatives. This is specified in the Safety Representatives and Safety Committees Regulations (SRSCR) 1977 (as amended), which relates to union appointed representatives, and in the Health and Safety (Consultation with Employees) Regulations (HSCER) 1996 (as amended), which relates to situations where workers are not represented by a union.

> "Regulation 3 - Duty of employer to consult
>
> Where there are employees who are not represented by safety representatives under the 1977 Regulations, the employer shall consult those employees in good time on matters relating to their health and safety at work"

Figure 2-18: Duty to consult employees.
Source: Health and Safety (Consultation with Employees) Regulations 1996 (as amended).

In addition, CDMR 2015 places a specific duty on the principal contractor to make arrangements for worker co-operation, which must specifically include consulting workers or their representatives on health and safety matters connected with the project, in good time.

> "The principal contractor must:
>
> (b) consult those workers or their representatives in good time on matters connected with the project which may affect their health, safety or welfare, in so far as they or their representatives have not been similarly consulted by their employer;"

Figure 2-19: Principal contractor's – duty to consult.
Source: CDMR 2015, Regulation 14.

The employer should therefore consult workers on matters of health and safety that affect them, in particular before any changes in arrangements made. The two consultation Regulations also establish a duty for the employer to consult on the following specific health and safety matters:

(i)   Introduction of any measures at the workplace which may substantially affect worker health and safety - for example, new or different procedures, types of work, equipment, premises, ways of working (shift patterns, hours of work).

(ii)   Arrangements for appointing/nominating competent health and safety people to help the organisation meet its legal obligations, for example, a health and safety advisor.

(iii) Information that must be given to workers – for example, on likely risks in their workplace and precautions they should take or related to substances workers are exposed to or the provision of an worker handbook/rule book on health and safety.

(iv) Planning and organisation of health and safety training for workers – for example, induction training or how to operate equipment safely.

(v)   Health and safety consequences of introducing new technology - for example, the introduction of drones or driverless vehicles into the workplace.

This may be done by consulting individuals directly or indirectly to obtain the participation of workers in the consultation process. There are a range of ways to consult with workers directly, including: one-to-one discussions, regular tours/walkabouts carried out by managers, inclusion of health and safety on meeting agendas, toolbox talks, special worker meetings, working groups. Ways to consult workers indirectly could include company internet sites/apps that provide information and ask for views, worker surveys, notice boards and suggestion schemes.

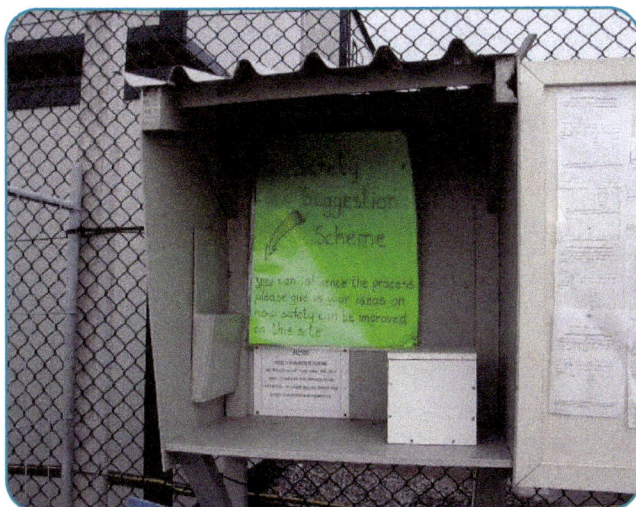

Figure 2-20: Health and safety suggestion scheme.
Source: RMS.

Alternatively, consultation may be by the participation of worker representatives, where groups of workers are represented by a specific individual. The representative would participate in consultation with managers on behalf of the group of workers. This could be as single worker representatives or in groups of representatives at a health and safety committee.

Workplaces where workers or their representatives are consulted and engaged in decisions about health and safety measures are generally safer and healthier. Consultation about health and safety is two way. It involves giving information to workers, listening to them and taking account of what they say before decisions are made.

One of the **benefits of using worker representatives** is that it is easier to provide them with time to be consulted, rather than consulting each individual worker or a number of small groups of workers. Consultation with worker representatives can also save time, as explaining what they are being consulted on will be easier to do for a smaller number of people, enabling them to quickly understand what the organisation's aims are.

The employer may not have a duty to adopt any suggestions made by workers or their representatives, but there is a value in obtaining their views. Because the consultation process can lead to different views and disagreement on health and safety matters, a grievance procedure should be in place to deal with any disputes or misunderstandings that may occur in the consultation process.

REVIEW

What actions could an organisation take that would help to establish a positive health and safety culture?

Appointment, functions and rights of worker representatives

The HSE has produced an approved code of practice related to the two consultation Regulations - 'L146 Safety Representatives and Safety Committees Regulations (SRSCR) 1977 (as amended) and Health and Safety (Consultation with Employees) Regulations (HSCER) 1996 (as amended); Approved Codes of Practice and Guidance'. In addition, a guide called 'HSG263 Involving your workforce in health and safety' is available from www.hse.gov.uk.

The guide is mainly aimed at medium to large employers. It is designed to help employers in their duty to consult and involve their workers on health and safety matters. The guide provides practical help, supported by case studies.

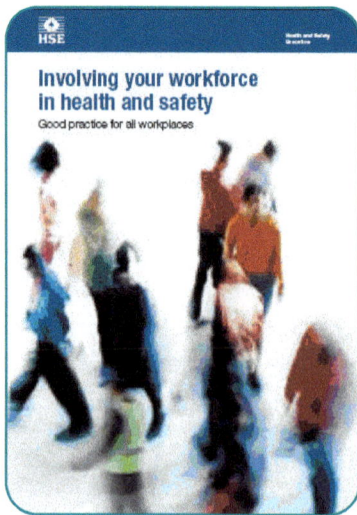

Figure 2-21: Involving your workforce in health and safety. Source: HSE, HSG263.

Trade union appointed - Safety Representatives (SR)

The HASAWA 1974 made provision for the appointment of Safety Representatives by recognised trade unions and the formation of Safety Committees. The Safety Representatives and Safety Committees Regulations (SRSCR) 1977 establish their appointment, functions, and rights.

Appointment

A trade union safety representative must be elected by the workers they represent and appointed by a recognised trade union in writing to the employer. This should set out the group of workers they represent.

The safety representative should have experience in the workplace in question or a similar one for a period of 2 years and would cease to be a representative on leaving the union, the employer's workplace or on removal notified to the employer in writing by the trade union.

Functions

The SRSC Regulations grant safety representatives the opportunity to carry out certain functions as outlined in the following list.

- To carry out investigations into potential hazards and dangerous occurrences.
- Examine the causes of accidents in the workplace.
- To carry out investigations into complaints by any worker they represent relating to health, safety or welfare.
- To make representations to the employer on the above and general matters of health, safety and welfare affecting those they represent.
- To carry out inspections of the workplace:
- Provided it has not been inspected in the last 3 months.
- When reasonable notice is provided in writing to the employer.
- At more frequent intervals if the employer agrees.
- Or when a substantial change has occurred.

- Or when new information is published by the Health and Safety Executive.
- Or following an over-three-day injury, notifiable accident, dangerous occurrence or notifiable disease.
- To inspect/copy documents relevant to those they represent.
- To represent the workers they were appointed to represent in consultations with enforcing authority inspectors.
- To receive information from an enforcing authority inspector on behalf of workers they represent.
- To attend meetings of safety committees related to matters affecting workers they represent.
- In order to fulfil the functions the safety representative should:
- Take all reasonably practical steps to keep themselves informed.
- Encourage co-operation between their employer and his workers.
- Bring to the employer's notice, normally in writing, any conditions that come to their attention.

Rights

The SRSC 1977 set out duties on employers towards safety representatives - these in effect provide rights to the Safety Representative that make the provision of their function easier.

If two or more trade union appointed safety representatives request in writing the formation of a safety committee, the employer must implement this request within three months. Consultation must take place with the representatives making the request and recognised trade unions the members of which work in the workplace that the committee relates to.

Information

The SRSC 1977 require employers to make available to safety representatives any information, within their knowledge, necessary to enable them to fulfil their functions.

This should include:

- Information about the plans and performances and any changes that may affect the health and safety at work of their workers - for example, a plan to re-organise the layout of a manufacturing site.
- Information of a technical nature about hazards to health and safety and precautions deemed necessary to eliminate or minimise them - for example, about substances that workers are exposed to.
- Information which the employer keeps relating to the occurrence of any accidents, dangerous occurrences or notifiable industrial disease and any statistical records relating to such accidents, dangerous occurrences or cases of notifiable industrial disease - for example, a copy of a report of an accident to an enforcing authority.

- Any other information relating to matters affecting the representative's workers, including any measurements to check the effectiveness of health and safety arrangements.
- Information on articles or substances which the employer issues to home workers.

However, an employer is not obliged to disclose the following information:

- Information that would be against the interests of national security or prohibited by law.
- Information relating specifically to an individual without the consent of the individual concerned – for example, personal medical information.
- Where its disclosure would, for reasons other than health and safety, substantially damage the employer's business interests.
- Where the information was obtained by the employer in connection with legal proceedings.
- Information not relating to health and safety.

Time off work to carry out functions and to attend training

The union appointing the safety representative may wish them to be trained on a Trades Union Congress (TUC) approved course. However, there is much to be gained by employers approaching the trade unions active in their workplace with the objective of holding courses that representatives from different trade unions can attend together. This has a particular advantage in that management may also be involved and enable the training to focus on the specific issues affecting the employer's organisation. In any event it is prudent for the employer to carry out company/industry orientated training to supplement the broad-based TUC course. The functions and training of the safety representatives may be carried out during normal working hours. The representative must receive normal earnings whilst carrying out their functions or training and this must take account of any bonuses that would have been earned if carrying out their normal work activities.

A safety representative may make a complaint to an Employment Tribunal if:

- The employer has failed to permit the safety representative to take reasonable time off to perform their functions.
- The employer has failed to allow the safety representative to attend reasonable training.
- The employer has failed to pay him for the time taken to perform these functions or attend training.

The complaint to the Employment Tribunal must be presented within three months of the date when the failure occurred.

Facilities and assistance

An employer must provide reasonable facilities and assistance to enable safety representatives to carry out their functions. Depending on the particular circumstances, such facilities might include a private room, telephone, email and internet access, storage facilities (physical/electronic), photocopier and relevant reference material.

Elected - Representatives of Employee Safety (RoES)

The Health and Safety (Consultation with Employees) Regulations (HSCER) 1996 create a duty for employers to consult workers who are not represented by trade union appointed safety representatives. Employers can consult either directly with workers or, in respect of any group of workers, with one or more elected representatives of that group. These are referred to as 'Representatives of Employee Safety (RoES)'. If the latter option is chosen, then employers must tell the workers the name of the representative and the group they represent. In a situation where an employer has been consulting a representative they may alternatively or in addition choose to consult the whole workforce. However, the employer must inform the workers and the representatives of that fact.

Appointment

It is the employer's decision as to the manner of consultation; if it is deemed that appointment of RoES is appropriate it is necessary inform the workers and allow them to decide who they would like to appoint. This would usually be carried out by election, which need not be a complicated or overly formal approach. Time off and suitable facilities are to be given where any person is standing as a candidate for election as a representative. It would be reasonable that RoES be appointed for local workplace issues and some of those be appointed for consultation on organisation-wide issues. When electing RoES, factors to be considered are: the size and location of the workplace, the number of workers, any workers not represented by trade unions, shift patterns and the nature of the work being carried - types and level of risk, the variety of jobs workers have and activities carried out.

Functions

Representatives of employee safety (RoES) have the following functions:

- To make representations to the employer on potential hazards and dangerous occurrences at the workplace which affect, or could affect, the represented workers.
- Make representations to the employer on general matters of health and safety, in particular matters the employer consults them on.
- To represent the workers in workplace consultations with enforcing authority inspectors.

Rights

Information

If employers consult workers directly then they must make available such information, within the employers' knowledge, as is necessary to enable them to participate fully and effectively in the consultation. If a representative is consulted, then the employer must make available all necessary information to enable them to carry out their functions. In addition, the employer must make available any record made under the RIDDOR 2013 which relates to the represented group of workers. This does not provide a right to inspect or copy any document which is not related to health and safety and the limitation on what information does not need to be provided by the employer is the same as for a union appointed Safety Representative.

Time off work to carry out their function or for training

Representatives of employee safety must be given reasonable training in order to carry out their functions. Employers must meet the costs of the training and any travel and subsistence. They must also permit the representatives to take time off with pay during working hours in order for them to carry out their functions and for any person standing as a candidate for election as a representative.

Facilities

Employers must provide reasonable facilities and assistance to enable RoES to carry out their functions. Depending on the particular circumstances, such facilities might include a private room, telephone, email and internet access, storage facilities (physical/electronic), photocopier and relevant reference material. Assistance might include access to lists of people and the workplaces they represent. This could also include access to these people to obtain their views, access to health and safety professional advice and to management to discuss health and safety issues.

Benefits of worker participation

The benefit of worker participation is that, by encouraging everyone in the organisation to participate in health and safety it enables people with different views and experience to contribute to solving problems.

Figure 2-22: Worker participation.
Source: Shutterstock.

This provides an opportunity for workers to provide feedback on proposals to solving problems and can often lead to workers providing good ideas that have not been considered or improvements to the proposals. The participation of workers allows managers to implement the organisation's health and safety policy giving full consideration to those affected by it.

If a health and safety problem is discussed with the workers affected, the fact that a number of people have considered it and agreed actions to solve the problem can mean that the actions are more likely to be taken and be effective. If a sample of workers affected by the problem are involved in the problem solving process this can provide a persuasive argument to help convince other workers to accept the solutions to the problem.

In addition to helping to solve problems, experience has shown that regular involvement of the workforce can stimulate a more active (proactive) approach to health and safety, by providing an opportunity to consider and discuss ideas for health and safety improvements. Worker participation in this way can therefore promote co-operation on health and safety matters and support the normal worker/employer systems for the reporting and control of workplace health and safety risks. Ways that the participation of workers can be improved can be summarised as follows:

- Define formal requirements and arrangements to consult and participate.
- Provide consultation training to both management and workers.
- Plan direct involvement of workers as part of the standard operation of the organisation, for example, at departmental meetings, team meetings, toolbox talks.
- Consult with worker health and safety representatives when logistical barriers make it difficult to consult directly with workers.
- Establish health and safety committees to enable planned involvement of workers and managers in a way that encourages discussion and recommendations to be made.
- Use questionnaires, surveys and suggestion schemes.
- Consult as part of the development process for safe systems of work and procedures.
- Consult as part of accident/incident or ill-health investigations or as part of the risk assessment process.
- Enable informal consultation to take place between supervisors and their team.

WHEN TRAINING IS NEEDED

General points

Without training, managers and workers would develop their own way of working. Their way of working could also be dependent on knowledge gained informally from unreliable sources. This lack of training may lead individuals to unknowingly be in breach of rules and procedures causing errors and violations to take place. For example, fork-lift truck drivers who have not been trained might drive with the forks of the truck raised too high because they are following the bad habits of other drivers.

Systematic training ensures individuals develop the correct health and safety competence (knowledge, skills and experience) and appropriate health and safety behaviour for their job. Where individuals have received appropriate training they are more likely to reliably identify hazards and apply the correct controls. They are also more likely to make the right decisions when risks are encountered in unusual circumstances or new situations.

Training provides individuals with a framework of expected behaviour that leads them to give health and safety the correct priority, ensuring actions are taken in the correct order and on a timely basis. Training, therefore, directly improves human reliability by reducing the likelihood of errors and violations.

Providing training contributes to establishing a positive health and safety culture that reinforces correct behaviour and helps to identify the many human factors that can cause human failure in work activities. Through good training, opportunities to improve the organisational, job and individual factors can be identified and measures can be put in to place to reduce failures.

Section 2 of the HASAWA 1974 establishes a general duty that employers provide such training as is necessary to ensure, so far as is reasonably practicable, the health and safety at work of their workers and Regulation 13 of the MHSWR 1999 expands this requirement with more specific duties.

1) Every employer shall, in entrusting tasks to his employees, take into account their capabilities as regards health and safety.

(2) Every employer shall ensure that his employees are provided with adequate health and safety training—

(a) on their being recruited into the employer's undertaking; and

(b) on their being exposed to new or increased risks because of -

(i) their being transferred or given a change of responsibilities within the employer's undertaking,

(ii) the introduction of new work equipment into or a change respecting work equipment already in use within the employer's undertaking,

(iii) the introduction of new technology into the employer's undertaking, or

(iv) the introduction of a new system of work into or a change respecting a system of work already in use within the employer's undertaking.

(3) The training referred to in paragraph (2) shall -

(a) be repeated periodically where appropriate;

(b) be adapted to take account of any new or changed risks to the health and safety of the employees concerned; and

(c) take place during working hours."

Figure 2-23: Requirements for training.
Source: Management of Health and Safety at Work Regulations 1996 (as amended).

This emphasises that if training is to be effective and influence health and safety behaviour it is important that it takes account of the needs of different people. Matching training to the needs of different people will require an analysis of training needs to consider the capabilities of individuals, the requirements of their work and what training is needed. This will enable training to be organised and provided that meets individual needs, is effective and leads to competence.

Training should be carried out from the start of an individual's employment with an organisation, during induction, and throughout their career. Training may also be required when transferring between jobs or work departments or promotion to a management role. The training provided should not only consider routine aspects of work, but also work that takes place infrequently, such as maintenance and emergencies. A suitable training programme should be defined and implemented for all individuals, from worker to director level.

To ensure individuals retain knowledge, skill, and the right experience, refresher training should be given at regular intervals, where appropriate. When considering the need for refresher training, particular attention should be paid to infrequent tasks, complex tasks or tasks that are critical for ensuring health and safety. When risks are assessed and accidents/incidents and ill-health are investigated, they may identify a need for training or refresher training. Training will also be required when situations change that can affect health and safety, for example, the introduction of new technology, new systems of work or new work equipment.

Induction training for new managers and workers

Induction training is generally defined as the training given when an individual starts working for an organisation for the first time or works at a new location of the same organisation. The purpose of a health and safety induction is to orientate individuals to the organisation's approach to health and safety and provide them with essential information that will enable them to be safe and healthy in their work and contribute to the health and safety of the organisation. It provides an opportunity to explain the organisation's values and commitment to health and safety. The induction process should motivate individuals and make them aware of the health and safety behaviour the organisation expects of them. It provides confidence to the workforce that health and safety is an important part of doing work in the right way. It also provides clear communication on the line of reporting for supervision and explains where workers report to if they identify a conflict of interest between what is being expected and getting the work done.

Induction training is an important integral part of the prevention process. It is essential that systematic arrangements are in place to ensure a consistent standard of induction for health and safety. Induction training for new managers and workers can be done in two stages, by providing a general induction to the organisation/location in which they work and a specific induction to their job. When a job induction is carried out as well as a general induction it enables the information provided to be more specific to the individual's workplace. In smaller organisations the content of a general induction may be essentially the same as the content of a job induction, making a job induction not necessary. However, in larger organisations both types of induction may be needed. For example, there may be general things to know about site rules and emergency arrangements, which can be provided in a general induction, but there may also be very specific additional rules and emergency arrangements related to the individual's job. These important additional items may not be relevant to all the people attending the general induction and may therefore best be covered in a job induction.

General induction training for new managers and workers should include:

- Review and discussion of the health and safety policy.
- Hazards of the workplace and control measures.
- Health and safety signs.
- Specific training requirements.
- Welfare facilities.
- Personal protective equipment (PPE) provisions - limitations, use and maintenance etc.
- Fire and emergency procedures.
- First-aid procedures and facilities.
- Accident/incident and ill-health reporting and investigation.

It is necessary for workers to know the relevant hazards and controls for each location in which they are working. For example, this will mean knowing areas they can and cannot go to, the use of walkways, where rest facilities are, who the first-aiders are and to whom they report hazards/accidents.

For job induction training to be meaningful, it has to be relevant to the trainee and focus on the controls that should be used as well as the hazards that may be encountered.

A job induction for a driver who makes both internal and external deliveries could include:

- Information on internal traffic routes.
- Speed limits.
- Parking/loading areas.
- Any hazardous substances on site.
- The need to carry out regular inspections of the vehicle and how to report defects.
- Company rules regarding the carrying of passengers, mobile phones, etc.
- How to report accidents and incidents.

Specific health and safety training

The specific health and safety training required by an individual will greatly depend on the individual's job, for example, whether they are a worker or manager. The need for specific training for a particular job should be defined through an analysis of training needs. Some examples of specific health and safety training that may be required include training related to:

- Safe systems of work. The actual training provided would depend on the individual's role. The training a worker needs may be more practically based than that of a manager, who needs to understand the health and safety requirements and how to ensure the work takes place safely and healthily.

- Equipment training, for example, specific training to develop competence in driving a fork-lift truck.

- Personal protective equipment training. For example, involving the practical aspects of selection, use and maintenance of respiratory protection equipment (RPE).

- Fire training, for example, how to act as a fire marshal for a specific part of a building.

- First-aid training, for example, how to give first-aid, this may be very specific training relating to the types of injury that might be experienced in a particular workplace.

- Workplace inspections and risk assessments. This type of training may involve learning the theory of the process and then the practical application of the technique in the workplace under controlled conditions.

When job specific health and safety training is provided it is common for competence to be formally assessed and recorded. For example, an individual may work under a higher than usual level of supervision for an initial period following the training to confirm that the important aspects of training have been transferred to the individual's work. This systematic approach helps to ensure the individual can apply what they learnt and that they have the correct health and safety behaviour.

Supervisor and manager training

General health and safety training should be provided to supervisors and managers to ensure responsibilities are known and to enable them to carry out the organisation's health and safety policy. This will enable them to fully understand and participate in the management of health and safety in their organisation. It will also help them to consistently and reliably respond to health and safety risks, which will ensure the organisation's standards for health and safety are maintained.

General health and safety training for supervisors and managers should include:

- The health and safety policy of the organisation.

- Causes and consequences of accidents/incidents and ill-health.

- The range of hazards and risks related to the organisation's activities.

- The legal framework the organisation works in and the legal responsibilities of the organisation, managers, workers and others.

- The structure and operation of the health and safety management system.

- How health and safety is organised and the responsibilities everyone has been given. Including methods available to guide health and safety behaviour, disciplinary procedures and commendation procedures.

- How health and safety is planned and implemented, including risk assessment, risk prevention and risk reduction techniques. This should be supported by examples of measures to control the hazards and risks related to the organisation's activities and an explanation of specific arrangements/processes/procedures for the management of risks, for example, permits-to-work.

- How health and safety performance is evaluated, including reactive and active monitoring, for example, accident/incident investigation, inspection procedures, auditing, reporting to more senior managers and review processes.

- How corrective and improvement actions are managed.

Refresher training

As time passes, a manager's or worker's approach to the health and safety aspects of their job can change from the way they were originally trained. This may be because they have forgotten the way they were trained, sometimes because of infrequent use, or because they have developed a different way of working that they prefer. This could mean that they work in unexpected ways when they work with other people and the practices they have developed could put themselves and others at unnecessary risk. It is therefore important that regular refresher training is used to reinforce the correct way of working and maintain an awareness of risks.

Many countries require annual refresher training for first-aiders and a similar approach may be taken for work activities where the consequences of the individual not working in the way they were initially trained could result in high risks.

Training			
Purpose	To develop the knowledge and skill of people, their attitudes, perception and motivation with regard to health and safety to ensure acceptable actions. Training should use two-way communication - information/instruction given and understanding/skill checked. This may be by observation of a person's practical skill, for example, by driving a fork-lift truck or using a simulator, and/or by written or verbal assessment of understanding.		
Subjects	Accident investigation. Conducting risk assessments. Conducting inspections/audits. How to comply with instructions.	How to set up your display screen workstation. How to use work equipment, for example, rough terrain fork-lift truck.	Use of personal protective equipment. Manual handling techniques. Emergency procedures.
Means of communicating	On/off the job. Internal/external trainers.	Explanation, demonstration, discussion and practice.	One-to-one or group. Written, oral and visual material.

However, two or three years may be an acceptable interval between training and refresher training periods for other work activities. Refresher training may also be required where the causes of accidents/incidents or ill-health are linked to competence.

Similarly, when an individual returns to work after a period of sustained absence, perhaps for ill-health reasons, it is often useful to provide them with refresher training to remind them of important health and safety aspects of their job. Some organisations have well established refresher training programmes for workers who need to retain skills, it is also important to ensure refresher training is provided to managers as well.

Changes that create a need for training

As an individual's *job or the work process* changes, training is required to update techniques and ensure awareness of the correct methods. Without effective training the individual will not be fully prepared for their new job or may try to apply their old knowledge to new processes. The training should be designed to prevent the individual transferring old methods to the new job/process. This may require a significant amount of practice under controlled conditions to ensure the new methods have been learnt. Whether the individual is a manager or worker the absence of training could lead to incorrect behaviour and risk of harm.

Where *new legislation/standards* are introduced it can have the effect of introducing new requirements that the organisation has to meet. This will make it necessary to train managers and workers in these requirements and the action taken by the organisation to meet them. For example, a change in legislation could involve a reduction in occupational exposure limits and may require managers and workers to use more stringent exposure controls. This could mean managers and workers would need training to understand the importance of the new exposure limit and what specific actions they need to take to ensure the limit is not exceeded. As the introduction of new health and safety measures can take some time to complete, training may need to be provided in stages as the work to introduce the measures is completed and the effect on work activities is known.

The introduction of *new technology* will often require the adoption of new work practices, for example, the introduction of mechanical aids to improve manual handling risks will require new methods of working. The use of new technology will often include a need to develop skills to interpret equipment control layouts and data display. This would require careful and systematic training to ensure the individual using the new technology was able to use it as intended.

Preparing for a training session

Effective training is an important part of managing health and safety and careful preparation prior to delivering a training session is necessary to ensure its success. One of the first considerations is to identify the particular aspect of health and safety that the training relates to, so that the objectives and the content (breadth and depth) of the training session can be established. Other factors to be considered when preparing a training session include:

- The training style and methods to be used. For example, lecture, visual presentation, role play, group work, use of equipment or site visit.

- The target audience. For example, their existing knowledge and skills, relevance of the subject being taught to them, their motivation.

Figure 2-24: Health and safety training.
Source: Shutterstock.

- The number of trainees. Considering the number of trainees the tutor can train at one time when related to the training method.

- The time available for the training. It may be necessary to break the training down into a number of small parts and provide the training over a period of days.

- The knowledge, skills and experience required of the trainer; this will be particularly influenced by the training method to be used. For example, if a trainer is required to train a number of people in the safe use of a fork-lift truck, they will need to have sufficient skill in using a fork-lift truck to demonstrate the safe use. Assistance from trainers outside the organisation may be required to ensure the right expertise.

- Audio-visual and other training aids required. Effective training is often supported by training aids. It is important when preparing for training to ensure they work and that the trainer knows how to use them.

- The facilities needed to conduct the training. For example, their location (near work activities or off-site), room layout, size and lighting.

- Provision of refreshments, if necessary.

- How the effectiveness of the training is going to be evaluated. This can be done at the time of training (for example, by using course evaluation forms) and after the training (for example, by measuring the level of compliance with procedures or number of accidents/incident).

2.4 Assessing risk

MEANING OF HAZARD, RISK AND RISK ASSESSMENT

Hazard

A hazard is something that has the potential to cause harm (or loss).

"The potential to cause harm, including ill-health and injury, damage to property, plant, products or the environment, production losses or increased liabilities."

Figure 2-25: Definition of hazard.
Source: Successful health and Safety Management, HSG65, HSE.

Hazards can include:

- Articles, for example, tools such as chisels.
- Substances and chemicals such as pesticides or cement.

- Equipment, for example, mobile cranes or a fixed grinder for sharpening tools.
- Methods of work, for example, production line workers or workers at height.
- The working environment, for example, cold environments such as a frozen food storage warehouse, or hot and humid ones such as an industrial laundry.
- Other aspects of work organisation such as shift or lone working.

A practical example would be that of a substance, such as cement, which has a particular inherent hazard, i.e. chemical (alkaline) burn, which may occur if the cement or concrete comes into contact with the skin or eyes of a worker.

Risk

'Is a measure of the likelihood of a hazardous event occurring combined with the severity of the harm from a hazard being realised'.

"The likelihood that a specified undesired event will occur due to the realisation of a hazard by, or during, work activities or by the products and services created by work activities."

Figure 2-26: Definition of risk.
Source: Successful Health and Safety Management, HSG65, HSE.

For example, the risk from a substance is the likelihood that it will harm a person in the actual circumstances of use. This will depend upon:

- The hazard presented by the substance.
- How it is controlled.
- Who is exposed, to how much of the substance, for how long, and what they are doing.

Consider another example to show how risk is affected by the circumstances - when a worker drives a vehicle at work, such as a fork-lift truck, it has the potential to cause harm. The vehicle is therefore a hazard. When the vehicle is being driven, the potential exists for the vehicle to strike another worker. If there are a lot of workers in the area, it is more likely that a worker will be struck by the vehicle. The impact may be very severe and could break a bone or kill, or it may have less impact and push them over, depending on the speed of the vehicle and how the worker encounters the vehicle. If the vehicle driver checks by looking to see that no one is nearby when the vehicle is moving then the vehicle is less likely to strike a worker. Therefore, the likelihood of hitting a worker is reduced by adopting control measures.

Risk assessment

Risk assessment relates to the process of evaluating the risks arising from the hazards, identifying preventive and protective measures, taking into account the adequacy of any existing controls, and deciding whether or not the risk is acceptable. There are a number of definitions for the term risk assessment, including one used in the NEBOSH glossaries.

> "The process of evaluating the risks to safety and health arising from hazards at work."

Figure 2-27: Definition of risk assessment.
Source: NEBOSH, Technical Glossary.

Risk assessment is an analytical process that identifies hazards, who may be harmed and in what way they may be harmed, and also takes into account factors that make the risk more likely and those that make it less likely.

PURPOSE OF RISK ASSESSMENT

The purpose of health and safety risk assessment is to provide information to enable organisations to manage risks and prevent workplace incidents, in particular accidents and ill-health. Risk assessment processes provide organisations with the means to identify hazards, sometimes before they result in harm, and to evaluate the priority and resources needed to control them. This enables preventive and precautionary measures to be applied on a timely basis enabling risk to be reduced to an acceptable level or in some cases eliminated from work activities.

This is achieved through the:

- Identification of all the factors that may cause harm to workers and others (the **hazards**).

- The consideration of the chance of that harm being done to someone in the circumstances of a particular case, and the possible consequences that could come from it (the **risks**).

- Deciding what are suitable preventive and precautionary measures and the priority to take action to implement them.

Risk assessment involves:

1) The identification of hazards.

2) Identification of the population at risk, who might be harmed and how.

3) The evaluation of the risks from the hazards (likelihood and severity (consequence)) and deciding on precautions (adequacy of current controls and need for additional controls).

4) Recording significant findings and the implementation of them.

5) Reviewing the risk assessment and updating it if necessary (periodically or when there is a significant change, for example, process or legislative change).

Most people undertake risk assessment as a normal part of their everyday lives. Routine activities, such as crossing the road and driving to work, call for a complex analysis of the hazards and risks involved in order to avoid injury. Therefore, most people are able to recognise hazards as they develop, and take corrective action. People do, however, have widely different perceptions regarding risk and may find it difficult to apply their experience to the formal workplace risk assessments required.

The assessment of risks must be **suitable and sufficient** and should cover the whole of the organisation's activities. It should also be broad enough so that it remains valid for a reasonable period. Risks arising from routine activities associated with life in general can usually be ignored unless the work activity compounds these risks.

Legal requirements to conduct risk assessments

Management of Health and Safety at Work Regulations (MHSWR) 1999

Regulation 3 of MHSWR 1999 requires employers (and the self-employed) to assess the risk to the health and safety of their workers and to anyone else who may be affected by their work activity. This is necessary to ensure that the preventive and protective steps can be identified to control hazards in the workplace. Where five or more workers are employed, the significant findings of risk assessments must be recorded in writing (the same threshold that is used in respect of having a written health and safety policy). This record must include details of any workers being identified as being especially at risk. The duty under Regulation 3 of MHSWR 1999 is a general duty, as it relates to assessments for general risks. Other legislation sets out legal requirements for risk assessment for specific risks.

Other legislation

Requirements for the assessment of specific risks includes:

- Substances harmful to health, including biological substances - Control of Substances Hazardous to Health Regulations (COSHH) 2002 (as amended).
- Manual handling - Manual Handling Operations Regulations (MHOR) 1992.
- Display screen equipment - Health and Safety (Display Screen Equipment) Regulations (DSE) 1992.
- Fire - Regulatory Reform (Fire Safety) Order (RRFSO) 2005.
- Noise - Control of Noise at Work Regulations (CNWR) 2005.
- Vibration - Control of Vibration at Work Regulations (CVWR) 2005.
- Ultraviolet, visible light, and infrared radiation - Control of Artificial Optical Radiation at Work Regulations (CAOR) 2010.

The general duty to conduct risk assessments that is set out in the MHSWR 1999 exist in conjunction with these more specific ones, however, this does not mean that risk assessments have to be carried out twice. For example, if you have carried out a risk assessment to comply with COSHH 2002 (as amended) you will not have to do it again for the same hazardous substances to comply with the MHSWR 1999. As a rule, a specific duty will take the place of a general one that duplicates it.

A 'SUITABLE AND SUFFICIENT' RISK ASSESSMENT

The term 'suitable and sufficient' when used in relation to conducting risk assessments relates to how well they are done and the standard it needs to reach. The Health and Safety Executive's Managing for Health and Safety HSG65 considers that a risk assessment must be suitable and sufficient and says "it should show that:

- A proper check was made.
- You asked who might be affected.
- You dealt with all the obvious significant risks, taking into account the number of people who could be involved.
- The precautions were reasonable, and the remaining risk is low.
- You involved your workers or their representatives in the process."

Suitable is generally taken to relate to the type of process used to assess risk, and **sufficient** is usually taken to relate to the amount of process done. Where risk assessments are required to be suitable and sufficient, it means that the type of assessment and amount of assessment may vary in different circumstances. The level of detail in a risk assessment should be proportionate to the risk and appropriate to the nature of the work. This means that where risks are not complex, a relatively simple approach could be taken, but where risks are more complex, a more detailed approach would be necessary. Sufficient means just enough and is designed to avoid organisations having to waste resources.

General considerations may include the following:

- For small organisations presenting few or simple hazards, a simple approach based on informed judgement and reference to appropriate guidance will be enough.

- In many intermediate cases the risk assessment will need to be more sophisticated. Some areas of the risk assessment may require specialist advice, for example, those relating to complex processes or where specialist analytical techniques are required to measure worker exposure to the hazard and to assess its impact.

- Large and hazardous sites will require the most developed and sophisticated risk assessments, particularly where there are complex or novel processes.

- Risk assessments must consider all those who might be affected, for example, contractors and members of the public.

- Employers are expected to take reasonable steps to help themselves identify risks, for example, by looking at appropriate sources of information or seeking advice from competent sources.

- The risk assessment should be appropriate to the nature of the work and should identify the period of time for which it is likely to remain valid.

The risk assessment should be conducted in a systematic way. Adopting a step-by-step approach, as described in the HSE guidance – 'Risk assessment A brief guide to controlling risk in the workplace' (INDG 163 rev4), will assist in assuring that a suitable and sufficient risk assessment is conducted. This approach is also reflected in the International Labour Organization (ILO) method, which is illustrated in *Figure 2-28.*

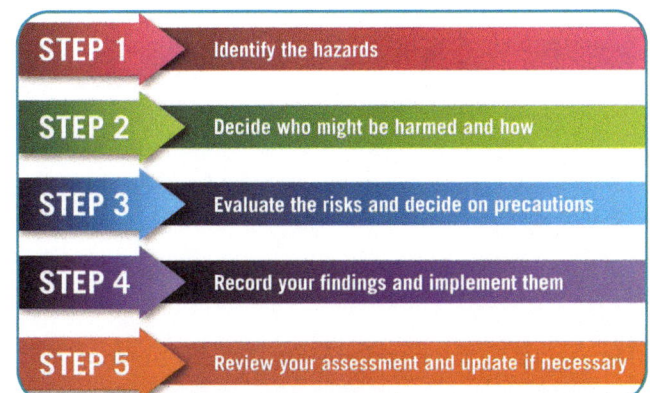

STEP 1 — Identify the hazards

STEP 2 — Decide who might be harmed and how

STEP 3 — Evaluate the risks and decide on precautions

STEP 4 — Record your findings and implement them

STEP 5 — Review your assessment and update if necessary

Figure 2-28: 5 steps to risk assessment.
Source: ILO OSH management system: a tool for continuous improvement.

Risk assessments should be good enough to fulfil legal requirements and prevent foreseeable injuries and ill-health. The risk assessment should be appropriate to the nature of the work, so the approach proposed in the five steps would not be suitable and sufficient for all of the activities carried out in hazardous high-risk establishments, for example, oil refineries or nuclear establishments. For some activities it may be necessary to use more advanced hazard identification techniques, for example, hazard and operability studies (HAZOP) and methods to calculate the probability of failure of control measures.

CONSIDER

For each of the stages in a risk assessment process, identify from your experience one difficulty (barrier) that managers might experience when carrying out the stage and one tip (remedy) that can help when carrying it out.

A GENERAL APPROACH TO RISK ASSESSMENT

Identifying hazards

Sources and forms of harm

The first step in the risk assessment process is to identify the work tasks and associated hazards.

Sources of hazard may be related to:

- People - they may carry infections or be violent.
- Equipment - which may have mechanical or electrical hazards associated with it.
- Materials - which may be sharp, heavy or toxic.
- Environment - which could have characteristics like height or slipperiness.
- Systems/situations - the way work is carried out may put undue pressure on a worker.

Forms of hazards can be grouped under two main headings: health hazards and safety hazards.

The health hazard categories are:

Categories	Examples
Chemical	Paint, solvents, transport exhaust fumes
Biological	Bacteria, pathogens
Physical	Noise, vibration
Psychological	Occupational stress
Ergonomic	Repetitive strain injuries, manual handling

Safety hazard categories include:

Machinery.
Flying or falling objects.
Moving vehicle.
Struck against something fixed or stationary.
Mechanical handling, lifting or carrying.
Slips, trips and falls on the same level.
Working from a height.
Pressure equipment.
Dangerous substance.
Fire.
Confined space.
Electricity.
Animals.
Workplace violence.

The identification of hazards can be done in many ways. For complex activities it may be necessary to break the activity down into its component parts by using task analysis, for example. *Task analysis* identifies hazards before work starts. This is achieved by breaking the task down into its component steps and identifying the hazards associated with each step. Having established the hazards, safe and healthy working methods can be established that deal with them. For a large machine, task analysis could mean looking at:

- Installation.
- Normal operation.
- Breakdown.
- Cleaning.
- Adjustment.
- Dismantling.

The hazards associated with each part of the job or task could then be identified more easily and thoroughly. It is necessary to identify contingent hazards that could arise from failures of a system, component, or checking and maintenance, as well as continuing hazards, i.e. those that are present continuously. Examples may include:

- Mechanical hazards.
- Electrical hazards.
- Thermal hazards.
- Noise and vibration.
- Radiation.
- Toxic materials.
- Ergonomic design.

Once the area/activity has been selected, the method(s) of hazard identification can be chosen. If a hazard is defined as being something with the potential to cause harm, then hazard identification can be carried out by observing the activity and noting the hazards as they occur in the actual work setting, for example by using task analysis. This has advantages over carrying out hazard identification as a desktop exercise using the health and safety manual, because the manual may not reflect how work is actually done and workers may have developed their own method of working, contrary to instructions and training.

In more complex organisations, it may be appropriate to use other, more advanced, hazard identification techniques:

- Hazard and operability studies (HAZOP), which are much used by the chemical industry at the design stage of processes and equipment.
- Reliability analysis and failure mode and effect analysis (FMEA), which are inductive techniques.
- Fault tree analysis (FTA) and event tree analysis (ETA), which are deductive techniques.

CONSIDER

Step 1: Identify the hazards

Walk around your workplace and look at what could reasonably be expected to cause harm.

Check manufacturer's instructions or data sheets for chemicals and equipment as they can be very helpful in spelling out the hazards and putting them in their true perspective.

Remember to think about long-term hazards to health, for example, high levels of noise or exposure to harmful substances.

SOURCES OF INFORMATION TO BE CONSULTED

Internal to the organisation

Incident and absence records
Incident, for example accident and ill-health, and absence records can provide information on the organisation's workers actual experience of harm from hazards, this can be useful in identifying hazards that exist in the workplace, indicating how likely it is that harm can result from them, and the probable type and extent of harm that can result from exposure to the hazard.

Investigation reports
Investigation reports can provide information on the causes of accidents and ill health, which may identify hazards and the effectiveness of control measures. Immediate, underlying and root causes should be analysed and care taken to take this into account in the risk assessment process.

Results of audits/inspections
Audits and inspections can provide information on hazards in the workplace, including those that have not

resulted in harm. They can also provide information on how effective the current control measures are, indicating where they are effective or not. The results of audits and inspections are particularly useful in identifying actual circumstances in the workplace, for example, whether procedures are followed, training takes place or guards sufficiently enclose hazards. In this way the circumstances can be taken into account when deciding the adequacy of current control measures and the need for further control measures.

Maintenance records
If a system of planned maintenance is implemented then, by definition, any breakdown of equipment is likely to be unintended. Therefore, breakdowns should be reported and investigated to determine the causes. Trends and patterns of failures can be analysed and taken into account during the risk assessment. The absence of failure of equipment used to control risks can support the effectiveness of the equipment and confirm its continued use as a control measure identified in the risk assessment documentation.

External to the organisation

Manufacturers' data
The HASAWA 1974, Section 6, sets out a requirement that manufacturers make adequate information available to those they supply with machinery, articles or substances. The information should be sufficient to enable the person to use the machine, article or substance safely and healthily. The information should cover reasonably foreseeable risks, including those arising from disposal and dismantling. This information can be used in the risk assessment process to compare how the organisation uses the machinery, article or substance. This may show an exposure to a hazard that was not expected, confirm the adequacy of current control measures or indicate opportunities to improve the control measures.

If, as a result of testing or research, new hazards come to light, manufacturers may provide new information to those supplied. This should then be used to review the current risk assessment to see if the hazard had already been identified by other means and the adequacy of control measures compare with any advice provide by the manufacturer.

Legislation
Legislation, for example, Acts and Regulations, are prime sources of information that set out requirements to improve health and safety at international and national levels. Legislation can be difficult to read without some legal understanding, and it is also easy to miss changes and amendments unless an updating service is used. However, it is helpful to workplace organisations in setting out requirements that can encourage an organisation to put systems in place that ensure the identification and management of risks.

Legislation may set specific technical limits to the exposure of workers to hazards, for example levels of noise. It may also specify minimum requirements for control measures. This information is important to take into account when conducting risk assessments as the organisation may have standards that are below the legal requirements. UK Legislation is available through the website www.legislation.gov.uk or in publications produced by the HSE.

European

The European Safety Agency (ESA) was set up by the European Union (EU) to bring together the vast reservoir of knowledge and information on OSH-related issues and preventive measures. This information is informative and can widen the perspective of those in the organisation conducting risk assessments and help them to understand the significance of hazards and what good practice control measures should be used.

The ESA makes good-practice information available covering a range of topics and sectors. These include good prevention practices, campaigns and an active publications programme producing everything from specialist information reports on hazards to fact sheets covering a wide variety of occupational health and safety problems. In addition, the ESA:

- Collects and disseminates technical, scientific and economic information.
- Promotes and supports co-operation and exchange of information and experience amongst the Member States.
- Organises conferences and seminars and exchanges of experts.
- Supplies the Community bodies and the Member States with the objective available technical, scientific and economic information they require to formulate and implement judicious and effective policies designed to protect the safety and health of workers.
- Provides the Commission in particular with the technical, scientific and economic information it requires to fulfil its tasks of identifying, preparing and evaluating legislation and measures in the area of the protection of the safety and health of workers, notably as regards the impact of legislation on enterprises, with particular reference to small and medium-sized enterprises.

HSE Publications

The UK Health and Safety Executive (HSE) publish both approved codes of practice (ACOPs) and guidance notes. The HSE provides a range of information in different forms. The range includes information related to different industries, including air transport, construction, food and drink, oil and gas, railways and textiles. Topics covered by this information include asbestos, confined spaces,

falls from height, nanotechnology, stress and welding. Information is available directly from the website, in leaflets, practical guides, detailed codes, and DVDs and as electronic tools to aid health and safety risk assessment. Electronic versions of HSE publications are available free via their website, http://www.hse.gov.uk.

Trade associations

Good practices are technical and/or practical documents prepared by a variety of interested parties, including insurance companies, trade/industry associations, and practitioners' associations. They often demonstrate solutions to occupational health and safety hazard control. As these solutions can be transferred to other employers, or other countries, information on good practice provides a useful source of reference. They contribute to the risk assessment process and the decisions regarding the suitability and effectiveness of control measures.

Unlike regulatory standards, these good practice documents are not drafted according to a specified format or procedure, and are generally not submitted to an approval procedure by officially recognised standards organisations. Because of this, good practice solutions and guidance are not considered legally binding, but are often used in judicial processes to demonstrate the expected level of health and safety performance required so that a preventive action can be achieved.

British, European and international standards

The British Standards Institute (BSI) provides some high quality advice, which is often above the legal minimum standard. However, there is a distinct trend towards linking some British Standards with legislation, for example, to comply with the Safety Signs Regulations safety signs need to meet BS 5378 requirements.

Of particular interest to health and safety is the ISO 45001:2018 'Occupational health and safety management systems – requirements with guidance'. ISO 45001 is intended to help organisations develop a framework for managing occupational health and safety so workers and others who may be affected by the organisation's activities are adequately protected. As part of the standard it includes requirements for risk assessment and defines a hierarchy of control measures, the guidance that is part of the document provides further advice on this.

There has also been progressive harmonisation to European Standards (CEN) and the use of CE marking. Some British and European standards are being used as a basis for international standards and are being adopted as ISO standards, such as that for machinery health and safety BS EN ISO 12100-1. The British Standards Institute (BSI) provides some high-quality advice which is usually above the legal minimum standard. Some of these standards provide detailed technical information that can inform more complex risk assessment process, providing good practice standards for control measures for a variety of hazards.

ILO and other authoritative sources

The International Labour Organization (ILO) deals with labour standards, fundamental principles and rights at work. They have a website and a database listing all types of work-related information. The ILO produces an Encyclopaedia of Occupational Health and Safety, a widely respected comprehensive publication which is now available free of charge on the ILO website. This provides comprehensive information on hazards, their risks and control measures, which is useful and supports the risk assessment process.

The World Health Organization is the United Nation's specialised agency for health and was established in 1948. It has information on health, where good health is defined as *"a state of complete physical, mental and social well-being and not merely the absence of disease or infirmity"*.

Professional bodies

There are a number of health and safety related professional bodies, such as the Institution of Occupational Safety and Health (IOSH). These provide local newsletters, regular meetings and a subscription to their magazine as part of the membership package. People who are interested in health and safety can join as affiliate members with no formal qualifications.

IT sources

There is an increasing amount of information available online that can be used when conducting risk assessments. The internet provides access to many health and safety organisations, including the ILO (http://www.ilo.org).

Care should be taken to verify information from an unknown web source through an alternative source; also, some material found may be out of date, applicable to another territory or incorrect.

There is also a range of electronic subscription services available that aim to keep the subscriber informed of relevant health and safety issues, providing access to information and an updating service.

REVIEW

Where might information about health and safety come from?

Provide two external and two internal sources.

IDENTIFYING PEOPLE AT RISK

In the risk assessment process it is necessary to identify the **people at risk**. In practice, this may mean noting groups or types of people at risk, for example, all nursing staff in the operating theatre or members of the public or patients. It is not necessary to list all individual workers in a category as this would be too much detail for most risk assessments.

A suitable and sufficient risk assessment will identify all groups of people at risk. When considering the people who might be affected, it is important to remember certain groups of workers who may work unusual hours, for example, security staff and cleaners. Similarly, maintenance workers need to be considered and, where relevant, the fact that they may be workers not employed by the organisation should be identified. Also, people who do not work at a fixed workplace, such as welfare workers, street cleaners, delivery workers, consultants and sales people, should be considered. For some workers in a shared workplace, it will be necessary to consult with another employer to determine any risks associated with the shared workplace. Members of the public, visitors, students, work experience people and even trespassers should be considered in the risk assessment process.

Specific groups at risk

Vulnerable people

Vulnerable people might be at risk from a particular hazard or may be affected by all hazards because of their lack of knowledge or understanding. For example, women of childbearing age (or more particularly any unborn foetus they may be carrying) may be deemed to be at risk from exposure to the hazards presented by lead.

The range of people that may be considered to be vulnerable includes new workers, those with disabilities, workers that have ill-health or are on medication, young people, pregnant women or nursing mothers, workers that have come from other countries where language or work practices may be different.

Operators

Typically, operators are individuals engaged in production-type activities where they have little control over their environment or work routine. Issues include repetitive strain, slips, trips and falls, together with a variety of equipment hazards. Consideration of the task and issues of fatigue and loss of concentration are usually significant.

Maintenance staff

Maintenance issues that may result in injury are usually associated with poor access and egress.

Some maintenance work is carried out infrequently, resulting in a lack of familiarity, which can lead to serious mistakes.

The fact that a maintenance worker is working alone may make the risk of harm and possible injuries worse.

Cleaners

Cleaners may be at risk from the materials that they use or that which they clean. Often the turnover of cleaners is high and their competency, in terms of such issues as to correct use and health effects of the materials they use or remove, is low.

Figure 2-29: Car washing.
Source: BOSH Service.

Contractors

Contractors may be involved in work that has a particularly high risk due to its unusual nature or complexity. They are a particular group to identify as they may not be as familiar with the workplace as other workers. They may not understand the hazards of the workplace and may not be as equipped as other workers to deal with the hazards.

CONSIDER

Step 2: Decide who might be harmed

This does not mean listing everyone by name, but by general group, for example, welders, people working at height or the general public.

In each case, identify how they might be harmed and what type of injury or ill-health might occur, for example, falls of objects onto workers or the general public.

Some workers have particular requirements. New or young workers and people with disabilities may be at a higher risk, as may cleaners, visitors, contractors and maintenance workers etc., i.e. those who may not be in the workplace all the time.

If you share your workplace, you will need to think how your work affects others, as well as how their work affects your workers.

Visitors/public

Visitors and the public are particular groups of people that need to be identified because they may not perceive or understand hazards and may behave differently from workers. They are seen as a vulnerable group because of their lack of awareness and ability to protect themselves from hazards.

EVALUATING RISK AND ADEQUACY OF CURRENT CONTROLS

Likelihood of harm and probable severity

After the hazards have been identified, it is necessary to evaluate the risk. The risk assessment process requires a judgement for each hazard to decide, realistically, what the most likely outcome is and how likely this is to occur. It may be a matter of simple subjective judgement or it may require a more complex technique depending on the complexity of the situation. In order to do this, at least two factors must be considered - the likelihood and the probable severity of harm.

Likelihood - when conducting a risk assessment, we take account of the circumstances in which the hazard may be encountered and the current controls in place, as these can greatly influence the likelihood of a person being harmed by a hazard. The circumstances may relate to environmental factors that can mask or make a hazard more obvious, for example, poor or good lighting; where the hazard is located, away from normal workers or in a busy walkway. The person encountering the hazard is another factor affecting likelihood. Someone who does not perceive the hazard because of lack of knowledge or reduced visual ability or other sensory impairment is more likely to come into contact with the hazard. Hazard controls in use may have only a limited effect, may fail, be defeated or become inactive at various times and these factors also influence the likelihood. Reliance on a control such as personal protective equipment would normally increase the likelihood compared to a control that put the hazard behind a protective barrier.

Other factors that might be considered include:

- Competence of workers.
- Levels and quality of supervision.
- Attitudes of workers and supervisors.
- Environmental conditions, for example, adverse weather.
- Frequency and duration of exposure.
- Work pressures.

Probable severity - this considers the probable outcome of contact with the hazard, which may include death, major injury, minor injury, damage to plant/equipment/product, or damage to the environment. It is important that this is the most probable outcome, and not a possible outcome, as it may be possible to think of some extreme circumstances in which all hazards may have the outcome of death. Again, it is important to take into account the nature of the hazard and the circumstances in which the hazard is encountered. The severity of contact with electricity is greatly reduced if the nature of the hazard is that it is operating at low voltage. Also, the circumstances may be that electricity is encountered in wet conditions, which could increase the probable severity due to the increased conductivity. Similarly, the person coming into contact with the hazard may have an influence on severity; for a large healthy adult the severity may be different than it would be for an unhealthy child.

Possible acute and chronic health effects

Risk assessment should consider both the acute and chronic effects. A risk, for example from exposure to a chemical, may present a harmful acute effect and if exposure to the chemical continues over time a worker may develop chronic effects. An acute effect may be irritation to the lungs when the chemical is breathed in, this may only sustain for a short period while being exposed. However, the repeated exposure over time may mean that sufficient dose of the chemical is absorbed

into the lungs and was likely to lead to a chronic effect on the workers organs, for example, a form of cancer or the lungs may be affected causing chronic obstructive pulmonary disease (COPD).

Risk rating and prioritisation

In order to manage risks, it is important to understand their scale: are they significant or not? One way to evaluate this is to describe the two components of risk, likelihood and probable severity, in terms of how large they are as a factor of the risk.

Different people may have a different perception of risk, as a result of training, life experiences and background. A method is therefore required in order to have a common approach and attempt to overcome individual differences. Risk can be rated according to the likelihood and probable severity of harm resulting from a hazard by using a guided approach to assigning a value to likelihood and severity. Words commonly associated with the scale of likelihood and severity are provided to guide the decision process. These words have a numeric value attached to them. The risk rating is a combination of the likelihood and severity value.

Risk rating = likelihood × severity

Where the:	Likelihood is how likely a loss will occur as a result of contact with the hazard.
	Severity is the probable amount of harm from contact with the hazard.
	Risk rating is the level of risk after current controls have been taken into account.

Likelihood categories

The likelihood can be assessed on a scale of 1 to 5.

5.	Almost certain	Absence of any management controls. If conditions remain unchanged there is almost a 100% certainty that an accident/incident will happen (for example, broken rung on a ladder, live exposed electrical conductor, and untrained personnel).
4.	High	Serious failures in management controls. The effects of human behaviour or other factors could cause an accident/incident, but it is unlikely without this additional factor (for example, ladder not secured properly, oil spilled on floor, poorly trained personnel).
3.	Medium	Insufficient or sub-standard controls in place. Loss is unlikely during normal operation; however, it may occur in emergencies or non-routine conditions (for example, keys left in fork-lift trucks, obstructed gangways, refresher training required).
2.	Low	The situation is generally well managed; however, occasional lapses could occur. This also applies to situations where people are required to behave safely in order to protect themselves but are well trained.
1.	Improbable	Loss, accident/incident or illness could only occur under exceptional conditions. The situation is well managed and all reasonable precautions have been taken.

Severity categories

The severity can be assessed on a scale of 1 to 5.

5.	Major	Causing death to one or more people or loss or damage is such that it could cause serious business disruption (for example, major fire, explosion or structural damage).
4.	High	Causing permanent disability (for example, loss of limb, sight or hearing, cancer, chronic obstructive pulmonary disease (COPD)).
3.	Medium	Causing temporary disability (for example, fractures, electrical burns, chemical burns, heat exhaustion).
2.	Low	Causing significant injuries (for example, sprains, bruises and lacerations, lung irritation, irritant dermatitis).
1.	Minor	Causing minor injuries (for example, cuts, scratches). No lost time likely other than for first-aid treatment, or repair of superficial damage, for example to interior decorations.

Using the formula stated (Risk rating = Likelihood × Severity), the risk rating can be calculated. It will fall into the range of 1-25. This risk rating is used to prioritise the observed risks. This approach is known as a 'semi-quantitative' risk evaluation, and different organisations will use different numbers and descriptions to assign a value to risk. It is the general principle of evaluating and managing risk that is important, not the exact words used on a form. The risk rating is then classified.

Principles to consider when controlling risk

Regulation 4 and Schedule 1 of the Management of Health and Safety at Work Regulations (MHSWR) 1999 require an employer to implement preventive and protective measures on the basis of general principles of prevention defined in the Regulations.

The general principles of risk prevention are broad principles that are applied to risks in order to prevent them.

Severity (consequence)		Probability (likelihood)				
		Certain	Probable	Possible	Remote	Improbable
		5	4	3	2	1
Fatal	5	25	20	15	10	5
Major	4	20	16	12	8	4
Lost Time	3	15	12	9	6	3
Minor	2	10	8	6	4	2
Negligible	1	5	4	3	2	1

Risk Rating Low	1-8	Green
Risk Rating Medium	9-15	Amber
Risk Rating High	16-25	Red

Figure 2-30: Risk rating 5x5 matrix. Source: RMS.

There are many variations on this method and other variables can be used other than those proposed in the 5x5 matrix used in this example. The size of the matrix can be increased to 10x10 and many more subdivisions for severity and consequence can be added. The more complex and hazardous the process, the more sophisticated the approach to assessing risk must be. Ultimately, the numbers used can be scientifically developed, such as the hours before a pump fails in use, or the static electricity charge crude oil will gain when travelling down a pipeline. Assessing risk on this level would be known as a 'quantitative' risk assessment.

The principles of prevention encourage that risks be avoided or combated at source, but they also establish further risk-prevention principles, such as adapting work to the individual. They are not hierarchical, but illustrate a good-practice approach to the prevention of risk. Therefore, for any given risk, all the principles of prevention should be considered and applied.

(a)   Avoiding risks.

(b)   Evaluating the risks which cannot be avoided.

(c)   Combating the risks at source.

(d)   Adapting the work to the individual, especially as regards the design of workplaces, the choice of work equipment and the choice of working and production methods, with a view, in particular, to alleviating monotonous work and work at a predetermined work rate and to reducing their effect on health.

(e)   Adapting to technical progress.

(f)   Replacing the dangerous by the non-dangerous or the less dangerous.

(g)   Developing a coherent overall prevention policy which covers technology, organization of work, working conditions, social relationships and the influence of factors related to the working environment.

(h)   Giving collective protective measures priority over individual protective measures.

(i)   Giving appropriate instructions to employees.

Figure 2-31: Principles of prevention.
Source: Schedule 1, Management of Health and Safety at Work Regulations (MHSWR) 1999.

Avoiding risks

If risks are avoided completely, they do not have to be either controlled or monitored. For example, not using pesticides or not working at height.

Evaluating unavoidable risks

Carry out a suitable and sufficient assessment of risks.

Controlling (combating) hazards at source

Repairing a hole in the floor is safer than displaying a warning sign. Other examples are the use of local exhaust ventilation to remove a substance at source, the design of equipment so that mechanical movement is enclosed and does not create a hazard, and the replacement of a defective bearing to control the hazard of noise at source.

Adapting work to the individual

This emphasises the importance of human factors in modern control methods. If the well-being of the person is dealt with, there is less chance of the job causing ill-health and less chance of the person making mistakes which lead to accidents. Consideration should be given to the design of any equipment used. Frequently used controls should be close to the operator, start buttons should be positioned to avoid inadvertent use and stop buttons should be close to the operator and easy to operate in an emergency.

All equipment should be clearly labelled. Consideration should also be given to minimisation of fatigue. Alleviating monotonous work by breaks or task rotation can help the individual to remain alert and pay attention to the task.

Adapting to technical progress

Technical progress can lead to improved, safer and healthier working conditions, for example, the provision of new non-slip floor surfaces or the bringing into use of less hazardous equipment such as new sound-proofed equipment to replace old noisy equipment. In recent years, this has included the use of waste chutes for removal of materials from scaffolds and the use of equipment which produces lower levels of vibration.

Replacing the dangerous by the non/less dangerous

For example, using a battery-operated drill rather than a mains-powered tool, providing compressed air at a lower pressure or providing water-based chemicals instead of solvent-based chemicals.

Developing an overall coherent prevention policy

Taking a holistic stance to the control of risk involves consideration of the organisation through the establishment of risk/control identification systems, consideration of the job, the use of task analysis and the selection of people, which includes consideration of human factors that affect an individual, such as mental and physical requirements.

Giving priority to collective protective measures over individual protective measures

Organisations with a less developed approach to health and safety may mistakenly see the solution to risks is to provide individuals with protective measures, such as warning people of hazards and provision of personal protective equipment - using the *'safe person'* (and healthy person) strategy. A more developed and effective approach is to give priority, where possible, to collective measures that provide protection to all workers, such as provision of barriers around street works or cleaning up a slippery substance spill rather than putting signs to warn workers - using the *'safe place'* (and healthy place) strategy.

These two strategies are supported by a third strategy, often called the *'safe system'* (and healthy system) strategy. This strategy establishes the correct way to do things in the form of rules and procedures, for example, the rule that says 'clean up after you do work' in order that the place is left in a safe and healthy condition.

Reliance on only a safe/healthy person strategy is the weakest of controls. The preferred strategy is the safe/healthy place and priority must be given to using it where possible. By establishing a safe/healthy place, an organisation will ensure that all people that find themselves in it will gain protection. This approach is reflected in the hierarchy used for safeguarding dangerous parts of equipment - our first priority is to provide guards around the dangerous parts (making it a safe place), and for the remainder that cannot be enclosed in this way we provide information, instruction and training to those that use the equipment (making the safe person).

In practice, the most successful organisations use a combination of the three strategies, with the emphasis on making the place safe/healthy, supporting this with systems of work (procedures) and paying attention to the human element so that they support rather than undermine the other strategies.

Many organisations that feel they have invested effort in getting the place and procedures right are taking a fresh look at the actions necessary to ensure the person is right also. Studies have shown that this too is a critical aspect of effective management of health and safety.

Providing appropriate instructions to workers

In order to meet the principle of providing appropriate instructions to workers it is essential to support the other principles by ensuring that workers know what is expected of the in relation to action taken to prevent risk.

General hierarchy of control

A hierarchy of control is a list of measures designed to control risks that are considered in their order of importance and effectiveness. The approach used in applying the hierarchy is to address each risk in the order of the priority expressed in the control hierarchy. The general hierarchy of control for the elimination of hazards and reduction of health and safety risks referred to in Clause 8.2 of ISO 45001:2018 is set out below:

a) Eliminate the hazard.

b) Substitute with less hazardous processes, operations, materials or equipment.

c) Use engineering controls and reorganization of work.

d) Use administrative controls, including training.

e) Use adequate personal protective equipment.

This general hierarchy of controls begins with the preferred measure to be used, elimination of the hazard, and goes through to the lowest level of protection, the use of personal protective equipment. The hierarchy of risk controls encourages the use of the highest level of control first. Only when we have found that it is not possible to use this control should we move to the next highest control. Often a combination of measures is used to control a risk adequately.

This means that when the current control measures are considered during the risk assessment process, the extent they relate to the hierarchy of controls should be determined and whether the appropriate control measure has been used or if other, better, control measures should be used.

It is important that the risk assessment evaluate the adequacy of the current control measures in place.

Eliminate the hazard/risk

Risks can be eliminated by avoiding situations that create them. This is the best option for controlling risk as it means that everyone is protected and there is no residual risk to manage. If hazards/risks are avoided completely they do not have to be either controlled or monitored. For example, not using pesticides or not working at height.

In many cases this may be a difficult option to achieve. However, it is something that should be considered by designers at the conception stage of any work activity. For example, in a construction project, in order to avoid the risk of falling it is critical to design out the need to work at height or when a new workplace is being designed good ergonomic practices should be used to eliminate ergonomic hazards.

Substitution of the hazard/risk

Replacing the dangerous substance or practice with something less dangerous can be a very effective preventive measure and provides a means to control the hazard/risk at source. Risks can be reduced to an acceptable level by substituting what is causing the risk with something less hazardous, for example substituting a water-based chemical for a solvent-based chemicals or by using a hazardous substance in a dilute form. Similarly, risks can be controlled by substituting the energy causing the risk for one of a lower level, for example, using a battery-operated drill rather than a mains-powered tool or providing compressed air at a lower pressure or replacing equipment that has high vibration characteristics with equipment that has low vibration characteristics.

Engineering controls or organisational control measures

It is possible to combat many risks at source by the use of engineering controls or organisational control measures, particularly as they are collective protective measures rather than personal protective measures. Controlling the hazard/risk at source is preferable to the use of other control measures, for example, repairing a hole in the floor is better than displaying a warning sign.

A common engineering control involves the *isolation* of the hazard from people. For example, enclosing the hazard, so there is a controlled barrier between people and the hazard. This could include fitting fixed guards around dangerous parts of a machine, installing sound absorbing enclosures around noise sources, totally enclosing dusty processes or those involving toxic chemicals, providing guard rails on scaffolds or barriers round temporary work activities. Other measures could include providing machinery that is remotely operated and to which materials are fed automatically, therefore

Elimination of hazards	-	Stopping using hazardous chemicals, applying ergonomic techniques when planning new workplaces, eliminating monotonous work or work that causes negative stress, removing mobile equipment like fork lift trucks from a work area.
Substitution with less hazardous processes, operations, materials or equipment	-	Replacing the hazardous with less hazardous, (for example, solvent based paint with water based paint), or change the physical form of materials used (for example, dust to pellets) or replace workplace materials (for example, slippery floor material with material that provides better grip for walking), changing processes (for example, requiring clients that present a risk of violence to visit controlled offices instead of seeing them in their home).
Engineering controls and reorganisation of work to limit exposure to the hazard	-	Providing collective preventive measures (for example, machine guarding, fume/dust extraction or isolation of noise with insulation, guard rails for workplaces at height, security glass screen between worker and visitors), isolating people from hazards (for example a 'glove box' for handling hazardous biological agents), segregation of workers from a hazard by distance (for example, locating observation points away from hot processes). Reorganisation of work to limit unhealthy work hours or workload or victimisation (for example, limiting hours worked without a break, job rotation, reassignment to a different team).
Administrative controls to remain alert to the hazard	-	Conducting periodic inspections (for example, of safety equipment), conducting training (for example, to prevent bullying and harassment, induction training), administrating competence (for example, forklift driving licences), coordination of activities (for example, managing contractors), control of work patterns (for example shifts or who does work at night), managing a health/medical surveillance programme (for example, related to hearing, hand-arm vibration, respiratory disorders, skin disorders), giving instructions to workers (for example, entry control processes like permit-to-work) and supervision.
Personal protective equipment to limit the effects of exposure to the hazard	-	Personal protective equipment, a physical barrier between the worker and the hazard (for example, protective glasses, hearing protection, gloves, protective helmet, anti-slip shoes).

Figure 2-32: Hierarchy of controls.
Source: Adapted from ISO 45001:2018.

separating the machine operator from danger areas or using robots for tasks that have high risk, such as spraying of isocyanate paints.

Some engineering controls limit the chance or amount of exposure to hazards. For example, control of the amount of dust/fumes released into the atmosphere by an activity may be achieved by using local exhaust ventilation. Other examples of engineering controls include those that respond to work conditions that can present an increased risk. For example, provision of an overload or over-run device on a crane/hoist, pressure relief valves fitted to boilers or other pressure vessels, automatic water sprinklers linked to heat detectors and devices that prevent equipment running if the operator is not in the correct position to avoid hazards it creates or that stop the equipment if the operator approaches the hazards.

Organisation controls could include:

- Organising high hazard tasks so they take place when workers are not nearby.

- Organising work to minimise the exposure of any one worker to the hazard, for example, by reducing the frequency and duration of exposure to radiation, noise, vibration, heat or cold.
- Controlling the hours worked to ensure rest and avoid fatigue.

Admininstrative controls

Administrative controls are those that rely on the administration of a procedure as the control measure. A *safe system of work* is an example of an administrative control; it is a formal procedure which results from systematic examination of a task in order to identify all the hazards and the controls necessary for health and safety. It defines safe and healthy methods of working, to ensure that hazards are eliminated or risks minimised.

In some cases, the system of work is controlled by the use of structured instructions that help to ensure the system of work is carried out as intended, this is often called a method statement or procedure.

Where work involves a significant risk, structured checklists are often used to control that each step in the system of work is carried out, this is often in the form of what is often called a *permit-to-work*.

For further information see section '3.2 - Safe systems of work for general work activities' and '3.3 - Permit-to-work systems' in Element 3 Managing change and procedures.

Other examples of administrative controls include provision of licences to control who can do higher hazard tasks, managing health surveillance programmes, information, instruction, training, co-ordination and supervision of activities.

Supervision is used to reinforce the use of systems of work and work performance standards in order to ensure health and safety. It includes monitoring that agreed systems of work are followed and the use of positive reinforcement to encourage continued good behaviour and, where necessary, disciplinary action to correct poor behaviour.

The supervisor has a crucial part to play in ensuring administrative controls and other control measures are effective.

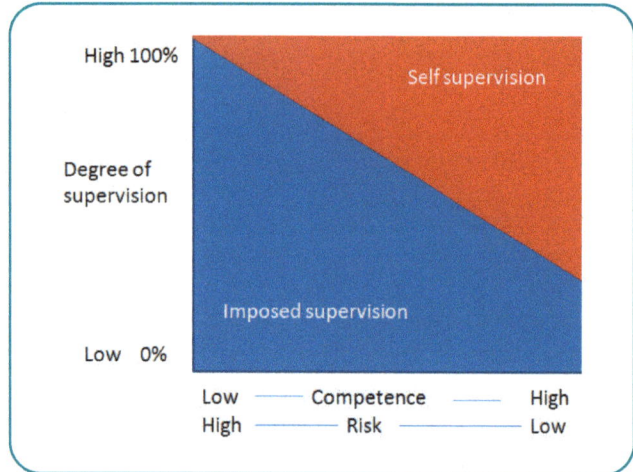

Figure 2-33: Competence vs supervision.
Source: RMS.

It is important to balance the amount and timing of supervision against the nature of the work being done. It is generally appropriate that the level of supervision necessary increases with the level of risk involved in the work. When considering supervision of individuals, it is important to also take account of their level of competence. In cases where the individual has qualifications, but no experience, and is therefore low in competence (for example, a young person straight from full-time education), supervision must increase accordingly.

Signs and warnings may accompany administrative controls to help to ensure people are aware of the hazards and respond as intended.

For further information on the application of signs to manage different hazards see the appropriate sections, for example '4.2 - Control measures for excavation work.

The Health and Safety (Safety Signs and Signals) Regulations (SSSR) 1996 require employers to provide specific health and safety signs whenever there is a significant risk which has not been avoided or controlled by other means, for example, by engineering controls and safe systems of work. Where a safety sign would not help to reduce that risk there is no need to provide a sign. Signs which no longer apply should be removed to ensure the credibility of the signage elsewhere.

In addition, employers must:

- Maintain the signs that are provided by them.
- Explain unfamiliar signs to their workers.
- Tell people what they need to do when they see a safety sign.
- They require, where necessary, the use of road traffic signs within workplaces to regulate road traffic.

In order to avoid possible confusion due to language differences and reading skills, signs combine shape, colour and a pictorial symbol to provide specific health and safety information and instruction, see *Figure 3-34.* General health and safety signs manufactured in the UK since 2012 comply with BS EN ISO 7010:2012. Any new signs purchased and used will be to this standard and the pictorial designs will be the same as those in the international standard ISO 7010. The introduction of signs that meet an internationally agreed standard helps to ensure standardisation globally and that people coming to the UK are familiar with the signs.

The SSSR 1996 require that fire safety signs comply with BS 5499. The SSSR 1996 do not cover the colour coding of pipework containing hazardous substances, a system of marking for pipework is contained in BS 1710.

Supplementary signs provide additional information. For example, where a noise hazard is identified by a warning sign - 'hearing protection available on request' may be added as supplementary information. Supplementary safety signs can also be used to mark hazards, for example, the edge of a raised platform, dangerous locations where objects may fall, or an area where work is going on that the public should not access. These are in distinct colours to attract the attention of workers and others that may encounter them. The colours used include yellow and black or red and white - in each case they consist of alternate colour stripes, often set at 45° angles.

1.	*Prohibition*	Circular signs. Red and white.	For example: ● No smoking. ● No pedestrian access. ● No children. ● No unauthorised access.	
2.	*Warning*	Triangular signs. Black on yellow.	For example: ● Caution - slippery floor surface. ● Toxic substance. ● Site traffic. ● Electrical hazard. ● Deep excavation.	
3.	*Safe condition*	Oblong/square signs. Green and white.	For example: ● Emergency assembly point. ● Fire exit. ● First-aid. ● Eye-wash station.	
4.	*Mandatory*	Circular signs. Blue and white.	For example: ● Hearing protection must be worn. ● Safety helmets must be worn. ● Safety boots must be worn. ● All visitors must report to site office.	

Figure 2-34: Categories and examples of typical health and safety signs, ISO 7010:2011. Source: RMS.

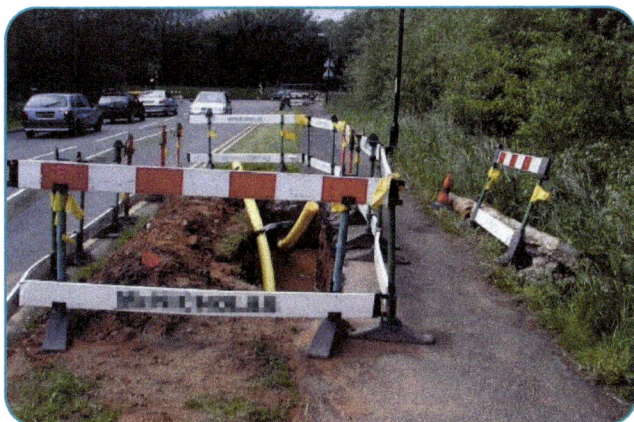

Figure 2-35: Hazard identification - obstruction. Source: RMS.

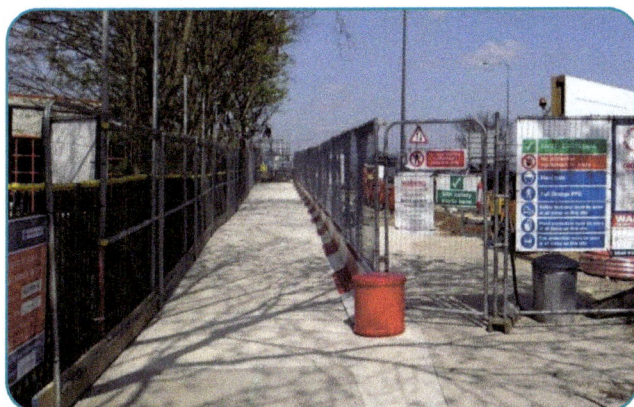

Figure 2-36: Hazard identification - red and white markings. Source: Bay Publishing.

The role of *signals* is to provide information, instruction and *warning* to workers. The SSSR 1996 establish principles for acoustic, verbal and hand signals.

Acoustic signals may be used to indicate the presence of a hazard, for example, the horn sounded by a driver of a fork-lift truck to indicate the truck's presence in a workplace, or that a hazard is about to be present, for example, a machine is about to start operating.

Verbal signals can be used to direct hazardous operations and may be made by either human, for example, when directing lifting operations, or artificial voices, such as those used to indicate that a vehicle is reversing. Spoken messages must be clear, concise and understood by the listener.

Code word	Meaning
Start	Start an operation
Stop	Interrupt or end an operation

Figure 2-37: Sample of codes for verbal signals.
Source: Health and Safety (Safety Signs and Signals) Regulations (SSSR) 1996.

Lifting equipment often works on congested construction sites where the signaller/slinger is out of the sight of the lifting equipment operator and the standard hand signals described in *Figure 2-39* cannot be used.

Therefore, verbal signals are often used to direct lifting operations, for example by a signaller/slinger using a two-way radio to communicate with a tower crane and direct the movement of a load, particularly when landing a load.

It is important that the verbal signals used have a common meaning to both the signaller/slinger and crane operator, for this reason the standard signals depicted in the SSSR 1996 should be used where possible. It is also important that the signals have the same contextual meaning, therefore a signal to direct movement of a load to the left or right of its current position must have the same contextual meaning to the crane operator.

A signaller/slinger has an important role on a construction site, which requires good observation and concentration skills to keep workers safe and prevent collisions with structures or loss of a load. A slinger needs to have excellent skills to work alongside lifting teams providing signals and controlling lifting operations. A signaller/slinger has the authority to halt operations if they deem the conditions too dangerous to continue working.

Figure 2-38: Verbal signals using 2 way radio.
Source: Pexels.

Hand signals are used to direct hazardous construction site operations, such as lifting materials using small mobile cranes or vehicle manoeuvres. It is essential that they be precise, simple, easy to make, easy to understand and clearly distinct from other such signals. A standard set of signalling codes, proposed by the SSSR 1996, a sample is shown below.

When lifting materials, the signaler should stand in a safe location, where they can see the load and can be seen by the lifting equipment operator. They should face the operator if possible and ensure each signals is distinct and clear.

Figure 2-40: Hand signals to a mobile crane.
Source: FCT Training.

REVIEW

What are the safety colours/shapes for the following safety signs: Prohibition, Warning, Safe Condition and Mandatory?

Give examples of where they may be used.

Personal protective equipment

Where residual hazards/risks cannot be controlled by collective measures, the employer should provide for appropriate personal protective equipment (PPE), including clothing, at no cost, and should implement measures to ensure its use and maintenance.

Personal protective equipment is equipment that will protect the user against health or safety risks at work.

Meaning	Description	Illustration
General signals:		
START Attention Start of Command	Both arms are extended horizontally with the palms facing forwards	
STOP Interruption End of movement	The right arm points upwards with the palm facing forwards	

Figure 2-39: Sample of hand signals.
Source: Health and Safety (Safety Signs and Signals) Regulations (SSSR) 1996.

PPE includes the following when worn for health and safety reasons at work:

- Aprons, for example, to protect from spills.
- Adverse weather gear, for example, thermal protection.
- High-visibility clothing, for example, work alongside a road.
- Gloves, for example, gauntlet hand-forearm protection.
- Safety footwear and helmets, for example, on a construction site.
- Eye protection, for example, goggles when dispensing acids.
- Life-jackets, for example, when working over or near water.
- Respiratory protection equipment (RPE), for example, dust mask.
- Safety harness, for example, protect from falls from a height.
- Breathing apparatus, for example, work in confined space.

PPE should only be used for low risk activities when all other controls have been considered or to provide additional protection when other control measure are used. Even when substitution or engineering controls and safe systems of work have been applied, some hazards might remain that may have to be controlled by the use of PPE. For example, a respiratory protector may be used to protect against limitations/failure of local exhaust ventilation used to control volatile toxic substances. Similarly, hearing protection may be used to further reduce noise exposure of workers where enclosure of a noise source has not reduced noise levels sufficiently. Other examples of residual hazards that may be controlled by the use of PPE could include:

- Contaminated air, which may be a hazard to lungs.
- Falling materials, which may be a hazard to a worker's head and feet.
- Flying particles or splashes of corrosive liquids, which may be a hazard to eyes.
- Corrosive materials, which may be a hazard to skin.
- Extremes of heat or cold, which may be a hazard to a worker's body temperature.

In general, PPE should only be used as a last option, after all other options have been considered. The reasons for this include:

- It is a legal requirement that other controls which provide collective protection of workers are considered first (for example, within the Control of Substances Hazardous to Health Regulations (COSHH) 2002 (as amended) and Management of Health and Safety at Work Regulations (MHSWR) 1999).

- PPE may not provide adequate protection because of such factors as poor selection, poor fit, incompatibility with other types of PPE, contamination, and misuse or non-use by workers.

- PPE is likely to be uncomfortable and relies for its effectiveness on a conscious action by the user.

- In certain circumstances, its use can actually create additional risks (for instance, warning sounds masked by hearing protection).

- PPE only protects the wearer and not others who may be in the area and also at risk.

- The introduction of PPE may bring another hazard such as impaired vision, impaired movement or fatigue.

- PPE may only be capable of minimising injury rather than preventing it.

- If PPE fails, it fails to danger and does not provide 100% protection, it may not be possible to measure actual exposure to hazards such as dust, vapours and noise.

"Whatever PPE is chosen, it should be remembered that, although some types of equipment do provide very high levels of protection, none provides 100%."

Figure 2-41: PPE quote.
Source: Guidance on the UK Personal Protection Equipment Regulations 1992.

In many cases, the effectiveness of protection provided by PPE will depend upon how well it fits, for example, with respirators, breathing apparatus and noise protection devices. Factors that can influence fit are whether adequate training and instruction have been given, and certain other individual conditions such as:

- Long hair.
- Wearing of spectacles.
- Facial hair. Where a person has shaved prior to work, hair growth during a work period may necessitate the progressive adjustment of respiratory protection. Beards may prohibit the wearing of respiratory protection.

The Personal Protective Equipment at Work Regulations (PPER) 1992 establishes that PPE will not be suitable unless:

- It is appropriate to the risk involved and the prevailing conditions of the workplace.
- It is ergonomic (user-friendly) in design and takes account of the state of health of the user.
- It fits the wearer correctly, perhaps after adjustment, and is comfortable to wear.
- It is effective in preventing or controlling the risk(s).
- It does not increase the overall risk to the user to an unacceptable level.

In addition:

- It must comply with relevant UK requirements for manufacture.
- It must be compatible with other equipment where there is more than one risk to protect against.

The main benefits of using PPE as a control measure include their relatively low cost and portability when compared with engineering strategies, which can be difficult to implement or be very costly.

Good-practice principles encourage employers to:

- Ensure PPE is suitable for hazard and person.
- Provide adequate training/instruction.
- Issue, obtain signature and record.
- Set up monitoring systems.
- Organise routine exchange systems.
- Implement cleaning/sterilisation.
- Issue written/verbal instructions, define when and where to use.
- Provide suitable storage.

Training and instruction in the use of PPE includes both:

- Theoretical - provision of a clear understanding of the reasons for wearing the PPE, factors which may affect the performance, the cleaning, maintenance and identification of defects.
- Practical - practising the wearing, adjusting, removing, cleaning, maintenance and testing of PPE.

From 6 April 2022, existing PPE regulations were amended, and the Personal Protective Equipment at Work (Amendment) Regulations 2022 came into force. They extend employers' and employees' duties regarding PPE to a wider group of workers. These are defined as 'limb (b) workers' to include those who have more casual employment relationships than employees, for example, gig, agency and temporary workers.

Application of control measures based on prioritisation of risk

Further or different control measures should be applied to risks that are determined to be inadequately dealt with by current controls. It is essential that the application of control measures be made based on the priority of the risk. The higher the risk, the greater the priority to apply control measures.

"It is essential that the work to be done to eliminate or prevent risk is prioritised. The prioritisation should take account of the severity of the risk, the likely outcome of an incident, the numbers who might be affected and the time necessary for prevention."

Figure 2-42: Prioritisation of control measures.
Source: European Commission Guidance on Risk Assessment at Work.

Where risk control measures will take a period of time to set up, it may be necessary to put in place temporary control measures that may not be so effective but give sufficient short-term risk control while better controls are being established.

Where risks are particularly high and only low-level temporary control measures are available, it may be necessary to suspend the work activity until the necessary effective controls are in place.

Use of guidance

When making a judgement about risks and whether controls are adequate, care has to be taken to consider relevant guidance.

This may be in the form of guidance to legislation, HSE guidance documents, industry-standard guidance and relevant international/local standards. If this is not done, then the risk assessment may be determined to be not suitable and sufficient.

Legislation requiring controls for specified hazards

Duties to apply specific control measures are found in the schedules to the Control of Substances Hazardous to Health Regulations (COSHH) 2002 (as amended) including a duty to comply with exposure standards for substances set out in EH40. Similar requirements are provided in the Control of Lead at Work Regulations (CLAW) 2002 and Control of Asbestos Regulations (CAR) 2012. The guidance to the Manual Handling Operations Regulations (MHOR) 1992 provides comprehensive information on the evaluation of risk and the application of controls to manual handling operations.

Where specific duties set out control measures that must be taken, the risk assessment must consider whether these are in place or whether additional control measures are required to meet the specific duties.

Applying controls to specified hazards

Controls need to be identified and applied to specific hazards, for example, electrical, chemical, manual handling. Methods for the reduction in the levels of risk for a particular hazard will vary in accordance with the hazard. For example, reduction for a chemical substance may involve changing to a less harmful substance or using it in a dilute form, whereas reduction of hazards created by manual handling may involve redesign of containers so each one is lighter to carry or a reduction in the distance that they are carried.

Many workplace hazards are supported by guidance and codes of practice, such as that provided to support COSHH 2002 (as amended), HSE guidance notes, such as permit-to-work (HSG250) and work sector standards or British Standards, such as BS EN ISO 12100-1 on machinery health and safety. These will aid the application of controls to specific hazards.

The controls applied should follow the general hierarchy of controls, see earlier in the element.

Residual risk

Residual risk is concerned with the risk which remains when controls have been decided. For example, whilst a slip which could result in a fall from height may be prevented by a guard rail, the potential to slip on the level would remain, and this would be the residual risk after the guard rail was provided. Similarly, if local exhaust ventilation equipment removes a substance from a workplace it will only work to a degree of efficiency. It may reduce the level of substance to just below the workplace exposure limit (WEL) set by HSE Guidance, but the amount of substance remaining in the atmosphere will have a residual risk.

If the residual risk is low after the control measures have been applied then it may be considered an *acceptable risk level.* The degree of acceptance will be influenced by what society considers as acceptable or tolerable. This may be influenced by how much knowledge there is on a given hazard or the level of risk to which people expect to be exposed. Societal standards change and risk acceptability reduces each year. For example, the noise action levels applicable in European Union (EU) countries have been reduced in over time, illustrating this change in acceptability.

If the residual risk is high a decision has to be made about whether the risk is tolerable or unacceptable. *Tolerable* implies that the risk is acceptable for the short term and further control measures should be developed to reduce the risk further. Unacceptable risk clearly implies that the level of risk is too high for the work to be allowed to continue. Much of the health and safety legislation created by countries places a general duty to reduce the level of risk so far as is reasonably practicable (assessment of the balance between the consequence of the risk against the time, trouble and cost to mitigate it to an acceptable level). When applying and meeting the term 'practicable', employers need to use any new improvements in technology. This will also assist in keeping pace with acceptability, as society may change its view on acceptability based on what level of risk reduction can easily be achieved.

Distinction between priorities and timescales

Reducing a high risk may have a high priority; however, the action needed might take a very long time to carry out. It is therefore not always helpful to class all high risks as requiring a solution within an immediate timescale. It is important not to ignore the risk and often it is possible to carry out some aspects of lesser control on an interim basis while the longer-term solution is being established. A low risk may be low priority, but it may also be a very simple low-cost solution to risk reduction. A law court may find it difficult to understand why a low-cost simple solution was not speedily implemented. Interim and simple precautions may reduce risk from 'tolerable' to 'acceptable'. They may not be long-term solutions to the nature of the problem but help manage short-term risk while longer-term strategies are being developed.

Record significant findings

Format

Employers should record the data considered in the risk assessment process and the significant findings of the risk assessments. This may be in writing or electronically, so long as it can be retrieved. The record should confirm when the assessment took place, the scope of the assessment (workplace, activity or specific risk) and who was involved. It should be noted that there are many forms and systems designed for recording assessments; while these may differ in design, the methodology broadly remains the same. The record will assist in providing proof that a suitable and sufficient risk assessment has been made.

No.	Hazard identi-fication	Associated risks	Persons at risk	Existing controls	Likelihood	Severity	Current risk rating	Comments and actions

Figure 2-43: Risk assessment form. Source: RMS.

Information to be recorded

The record should represent an effective statement of hazards and risks and should lead management to take action to eliminate and reduce risk.

In its simplest form, the record will include details of the task/plant/process/activity, together with the hazards involved, the associated risks, persons affected by the hazard and the existing control measures.

The necessary actions required to further reduce the risk are also recorded, sometimes separately. Some hazards, particularly those with a high-risk rating, may require a more detailed explanation or there may be a series of alternative actions.

Information on risk assessments and any controls must be brought to the attention of those assigned the task of work. Risk assessment information should be included in lesson plans to ensure items are not missed when staff are trained or retrained.

> **CONSIDER**
>
> Step 4: Record your findings and implement them
>
> Putting the results of your risk assessment into practice will make a difference.
>
> Writing down the results and sharing them with your workers will encourage you all to make a difference.
>
> Remember, prioritise and tackle the most important things first. As you complete each action, tick it off your plan.

Reasons for review

The risk assessment should be periodically reviewed and updated. In addition, a review of risk assessments should be carried out following any significant changes to a workplace.

Examples of circumstances that would require the review of the validity of a risk assessment are:

Internal to the organisation

- When the results of monitoring, for example, routine analysis of accident and ill-health reports, damage accident reports, and 'near-miss' reports are adverse and not as expected.
- Following the findings of an accident investigation.
- Following consultation with workers.
- Following internal health and safety audits.

Internal changes

- A change in process, work methods (introduction of shifts) or materials.
- Changes in personnel.
- The introduction of new plant or technology.

External to the organisation

- Changes in legislation, for example, work exposure limits.
- New information becoming available from research, for example, identification of new asthma sensitiers.
- Following enforcement action that called into question the health and safety of a process.
- Third party health and safety audits.

Periodic review

As time passes - the risk assessment should be periodically reviewed and updated. The validity of risk assessment should be monitored through a combination of monitoring techniques such as:

- Preventive maintenance inspections.
- Health and safety representative/committee inspections.
- Statutory and maintenance scheme inspections, tests and examinations.
- Health and safety tours and inspections.
- Occupational health surveys.
- Air monitoring.

> **CONSIDER**
>
> Step 5: Review and update your assessment
>
> Few workplaces stay the same. New equipment, substances and procedures could lead to new hazards. Review should be carried out at least annually.

APPLICATION OF RISK ASSESSMENT FOR SPECIFIC TYPES OF RISK AND SPECIAL CASES

Examples of when specific risk assessment methodologies are required

The general approach to risk assessment described previously is likely to be adequate for the identification and assessment of general risk in a workplace. However, some specific types of risk and special cases may require an enhanced approach where by specific factors are considered. For example, it is generally accepted that specific risks arising from fire, noise, vibration and chemical substances require specific factors to be considered during the risk assessment process. Similarly risks arising from display screen and manual handling activates benefit from an approach where specific risk factors are considered in the assessment process.

Why specific risk assessment methods are required

Specific risk methods that take account of factors specific to the particular risk are required to enable the proper, systematic consideration of all relevant issues that contribute to the risk.

Special case applications of risk assessment

Young people

Another specific group of people who must be considered in risk assessments is young people (for most countries this is a person under 18 years of age) and children (generally taken to be under compulsory school-leaving age). Regulation 19 of the MHSWR 1999 requires the employer to carry out the risk assessment before young workers start work.

Young person risk assessments should consider the following risks:

- Work which is beyond their physical or psychological capacity.
- Work involving harmful exposure to agents that are toxic, carcinogenic or cause heritable genetic damage, harm to the unborn child or other chronic health effects.
- Work involving harmful exposure to radiation.

- Work involving risk of accidents that it may reasonably be assumed cannot be recognised or avoided by young people owing to their insufficient attention to safety or lack of experience or training.
- Work in which there is a risk to health from extreme cold or heat, noise or vibration.

Where children are involved, the significance of assessments should be communicated to the people who have parental responsibility. Regulation 10 of the MHSWR 1999 requires that the significance of assessments must be communicated to the people who have parental responsibility for the children.

Young people tend to be more vulnerable because of their lack of knowledge, experience or training, and the tendency of young persons to take risks because of over-enthusiasm/poor perception and to respond more readily to peer pressure.

Young people may also demonstrate an eagerness or willingness for work and agree to undertaking tasks they are not skilled enough to undertake. Young people may also have less well developed communication skills and fail to understand warnings they are given about hazards. All these factors should be considered when carrying out risk assessments. It may be necessary, therefore, to restrict young persons from carrying out certain high-risk activities (for example, machine operation). It may also be necessary to restrict hours of work and to conduct a thorough induction into the workplace.

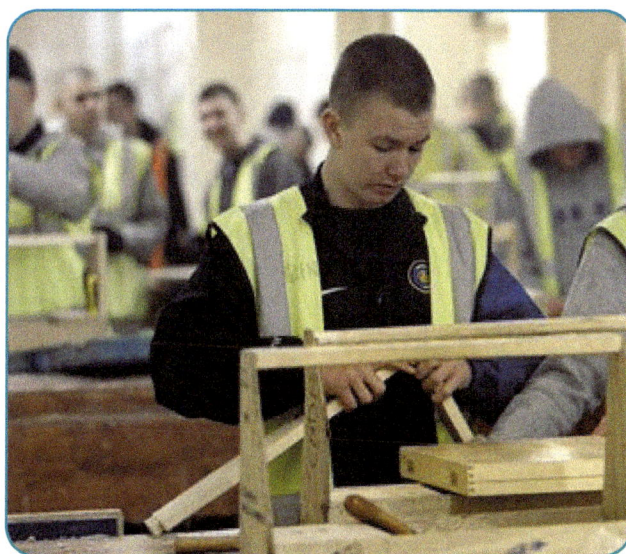

Figure 2-44: Young people involved in craft training.
Source: Don McPhee.

Expectant and nursing mothers

Regulation 16 of MHSWR 1999 requires risk assessments to consider specifically the risks to women of childbearing age and any new or expectant mother.

There are many factors that may increase risks to pregnant workers in the workplace. These include: exposure to chemicals such as pesticides, lead and those causing intracellular changes (mutagens) or affecting the embryo (teratogenic); biological exposures (for example, rubella or hepatitis); exposure to physical agents such as ionising radiation and extremes of temperature; manual handling (especially later in pregnancy); ergonomic issues relating to prolonged standing or the adoption of awkward body movements; stress; whole body vibration and issues associated with the use and wearing of personal protective equipment.

Regulation 18 of MHSWR 1999 clarifies that employers do not need take action to control risks to a specific individual until they are notified in writing by the woman that she is pregnant, has given birth in the last six months or is breastfeeding.

Disabled workers

When a general risk assessment is conducted, it is common that the assessor's main focus is the majority of the working population being considered. For most situations, this will be people who are not disabled or otherwise vulnerable. It is essential that if disabled workers are, or are going to be, involved in the work being considered, the risk assessment take full account of them. This may mean conducting a separate risk assessment for the disabled workers. Because disabilities vary greatly, it may be necessary to conduct separate risk assessments related to the specific disabilities that the workers have. In this way they are more likely to be both suitable and sufficient. Common worker impairments include difficulty with hearing, vision, mobility, and learning. Care should be taken to consider the particular disabilities in relation to the job when it is functioning normally, when problems arise from it and where emergencies may exist.

Lone workers

Risk assessments that relate to people working alone are a special case in that special risks can arise that are not present in other work. It should be remembered that work that is risk assessed on the basis of it normally being conducted in the main part of the day, whilst other people are around, can quickly become lone working when people work late, come in early or work weekends.

Essentially, the hazards from the work are often the same as they would be if the person was working with others, but the additional risk control of having others at hand to provide routine assistance, for example, to move something or hold something, are not available. Furthermore, lone working leads to the removal of another risk control - the ability of a co-worker to observe if another worker is suffering harm from the risks to which they are exposed or to respond or obtain assistance. When assessing the risks, it is therefore essential to identify how the risks are controlled to take account of the lone working and whether this is adequate.

> ### CONSIDER
>
> Hierarchy of controls. Think of one practical example for each control measure in the hierarchy, preferably from your own workplace.

> ### CASE STUDY
>
> A UK care worker notified her employer of her pregnancy. The employer looked back at the outcome of the initial risk assessment, which had identified that a possible risk for pregnant women was exposure to acts of violence (for example, difficult patients). The employer then conducted a specific risk assessment for the pregnant worker, who dealt with patients who were difficult and on occasion violent. As a result, the employer offered the care worker suitable alternative work at the same salary and reviewed the assessment at regular intervals. The employee accepted the alternative work and had a risk-free pregnancy. Following her maternity leave, the employee returned to work.

Sources of reference

Reference the information provided, in particular web links, these are correct at time of publication, but may have changed.

Common topic 4: Safety culture, HSE, http://www.hse.gov.uk/humanfactors/topics/common4.pdf

Consulting workers on health and safety, Safety Representatives and Safety Committee Regulations 1977 (as amended) and Health and Safety (Consultation with Employees) Regulations 1996 (as amended), ACoP, L146, HSE Books, ISBN: 978-0-717664-61-0 http://www.hse.gov.uk/pubns/priced/l146.pdf

Ergonomics and human factors at work, INDG90, HSE Books, http://www.hse.gov.uk/pubns/indg90.pdf

Example risk assessments, HSE, http://www.hse.gov.uk/risk/casestudies/index.htm

Human factors and ergonomics http://www.hse.gov.uk/humanfactors/

Human factors: Organisational change, HSE, http://www.hse.gov.uk/humanfactors/topics/orgchange.htm

Involving your workforce in health and safety: Good practice for all workplaces (HSG263), HSE Books, ISBN 978-0-717662-27-2 http://www.hse.gov.uk/pubns/priced/hsg263.pdf

Leading health and safety at work (INDG417 rev 1 2013) HSE Books www.hse.gov.uk/pubns/indg417.htm

Managing Health and Safety in Construction, Construction (Design and Management) Regulations 2015, Guidance on regulations, L153, HSE Books, ISBN: 978-0-7176-6623-3, http://www.hse.gov.uk/pubns/priced/l153.pdf

New and expectant mothers http://www.hse.gov.uk/mothers/

Organisational culture, HSE, http://www.hse.gov.uk/humanfactors/topics/culture.htm

Personal protective equipment (PPE) at work regulations from 6 April 2022, https://www.hse.gov.uk/ppe/ppe-regulations-2022.htm

Reducing Error and Influencing Behaviour (HSG48), HSE Books, ISBN: 978-0-717624-52-2 http://www.hse.gov.uk/pubns/priced/hsg48.pdf

Risk Assessment (INDG163, rev 4), HSE Books, ISBN: 978-0-717664-40-5 http://www.hse.gov.uk/pubns/indg163.pdf

Risk assessment, A brief guide to controlling risks in the workplace, INDG163, HSE Books http://www.hse.gov.uk/pubns/indg163.pdf

The health and safety toolbox, How to control risks at work, HSG268, HSE Books, ISBN: 978-0-7176-6587-7, http://www.hse.gov.uk/pUbns/priced/hsg268.pdf

Young people and work experience, A brief guide to health and safety for employers, INDG364 (rev 1), HSE books, http://www.hse.gov.uk/pubns/indg364.pdf

Web links to these references are provided on the RMS Publishing website for ease of use - www.rmspublishing.co.uk

Statutory provisions

Construction (Design and Management) Regulations (CDMR) 2015 / Construction (Design and Management) Regulations (Northern Ireland) 2016

Health and Safety (Consultation with Employees) Regulations (HSCER) 1996 / Health and Safety (Consultation with Employees) Regulations (Northern Ireland) 1996

Health and Safety at Work etc. Act (HASAWA) 1974

Health and Safety Information for Employees Regulations (HSIER) 1989

Management of Health and Safety at Work Regulations (MHSWR) 1999 (as amended)/Management of Health and Safety at Work Regulations (Northern Ireland) 2000

Personal Protective Equipment at Work Regulations (PPER) 1992 (as amended)

Personal Protective Equipment at Work (Amendment) Regulations 2022 (PPER 2022)

Safety Representatives and Safety Committee Regulations (SRSCR) 1977 / Safety Representatives and Safety Committee Regulations (Northern Ireland) 1979

STUDY QUESTIONS

1. What are ways in which organisations can positively influence the health and safety behaviour of their workers?

2. a) Explain the role of worker participation in health and safety.

   b) What are the benefits of participation for the employer?

3. What should be considered when preparing and presenting a training session on health and safety?

4. What is the criteria which must be met for a risk assessment to be 'suitable and sufficient'?

5. What are the sources of information that might be consulted to help determine the severity of the outcome of an incident?

For guidance on how to answer these questions please refer to the 'study question answer guidance' section located at the back of this guide.

Element 3

Managing change and procedures

3

Contents

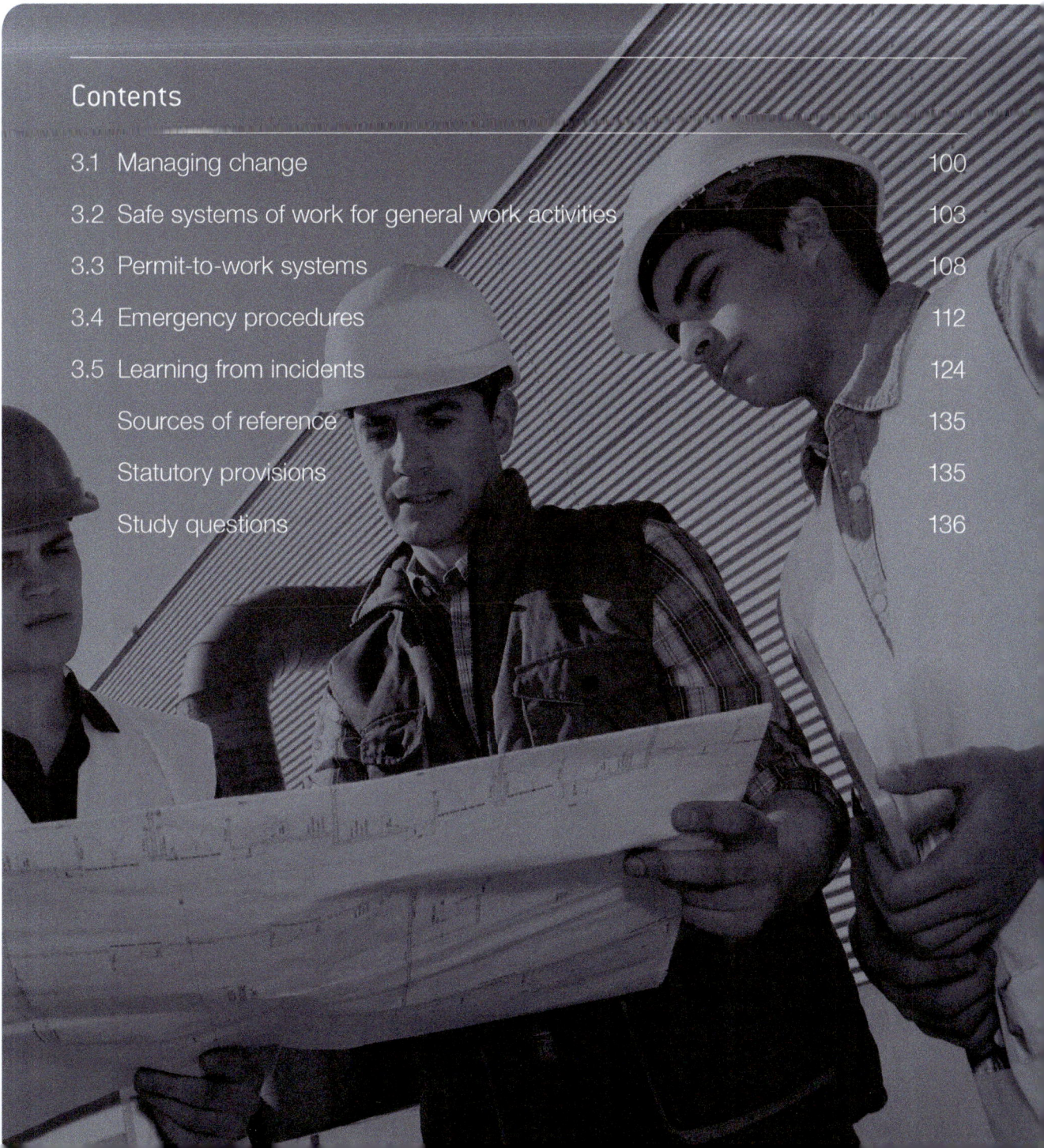

3.1 Managing change

The objective of a management of change process is to minimise the introduction of new hazards and risks into the workplace and to identify opportunities for improvements as changes occur. The efforts and resources put into the process should be proportionate to the complexity of the change, the scale of the hazards concerned and the degree to which the change may impact on the management of major hazards.

TYPICAL CHANGES

Typical types of change likely to be encountered in the workplace include:

- Organisational changes - for example, new workers, managers promoted, temporary contracted seasonal workers introduced or even a complete restructuring of the organisation.

- Activity changes – for example, changes to processes, equipment, infrastructure (construction work) and work practices.

- Material changes – for example, new chemicals, raw material changed from liquid form to powder, packaging changed from solid containers to plastic sacks, buying in bulk making the size of items larger.

POSSIBLE IMPACT OF CHANGE

Ineffective management of change is one of the leading causes of serious incidents. Some of the possible impacts of ineffective management of change could include:

- Changing a chemical to speed up a chemical reaction could lead to increased heat that the equipment is not designed to withstand and a fire may result.

- Changing equipment that works at a faster rate may lead to ergonomic hazards and fatigue of workers.

- Changing working practices from manual handling of goods to the use of a battery powered fork lift truck would introduce the hazard of moving vehicles in the workplace, electrical hazards and explosive gas hazards from the battery charging process.

- Where construction activities are to take place in a workplace this represents a change to what is normal in that workplace. This could introduce hazards that would not normally be there, for example flammable liquids or sources of ignition, closure of access routes and obstruction of emergency equipment/exits.

- In some cases construction changes to a building could reveal hidden hazards, for example the presence of asbestos in a building, which could lead to uncontrolled exposure of workers.

OVERALL APPROACH TO THE MANAGEMENT OF CHANGE

The overall approach to management of change includes consideration of health and safety hazards and risks prior to the introduction of the "change". This involves the identification of the hazards associated with the "change", assessment of the risks and implementation of the controls needed to mitigate the hazards and reduce the risks. In addition, the approach should identify and implement opportunities for improving risk by elimination of hazards or provision of control measures that can be applied as part of the change.

The main steps involved in management of change are as follows:

1) Identification of the change.

2) Preliminary screening of the health and safety impacts, for example design review.

3) Preliminary plans for the implementation of the change.

4) Risk assessment.

5) Development / approval of final plans, including control measures.

6) Implementation of the change.

7) Final assessment and close out.

MANAGING THE IMPACT OF CHANGE

The main control measures to mitigate the impact of change include:

- Communication and co-operation.

- Risk assessment.

- Appointment of competent people.

- Segregation of work areas.

- Amendment of emergency procedures.

- Welfare provision.

Communication and co-operation

One of the most important issues to consider when mitigating the impact of a proposed change is ensuring good communication and co-operation between those initiating the change and those affected by it. This is to ensure those affected by the change understand the need for the change, what it is, how it will affect them and their role in the change. This will help to maintain their support and involvement in the change process and provide an opportunity to communicate and discuss how health and safety risks related to the change are going to be managed.

This communication and co-operation will generally be best achieved through the usual communication and meeting structure established to ensure the effective management of health and safety of the construction project and the site, for example client liaison and planning meetings, Principal Contractor meetings with contractors and toolbox talks.

Where appropriate, communication and co-operation may have to be secured at the time the change is needed and require those in control of the change to make direct contact with those involved to ensure a clear understanding of what is needed and gain their personal co-operation. This should include checking if those involved have any concerns about how the proposed change will affect their health and safety or that of other workers. This approach will often provide useful information that can be used in the risk assessment process.

As workers in construction activities come from all types of backgrounds and different ethnic origins significant challenges to effective communication may exist, which could affect the management of any proposed change. Information and instructions must be communicated in a way they understand. This may mean engaging a bilingual person who can communicate in the appropriate language.

Risk assessment

Effective steps must be taken when managing the impact of change so that workers can work with minimal risks to their own health and safety and that of other people who may be affected. This must include carrying out a risk assessment of the health and safety impact of a proposed change.

A proposed change may involve a wide range of changes, for example, to a design, the order in which something is done, how work is carried out, where a work activity takes place or who is going to do it. These types of proposed change may have major implications to health and safety, which should be identified before the change is initiated. In some cases this may inform a decision to revise a proposed change because of a lack of capability to achieve it or an unactable level of health and safety risk arising from it.

Therefore, it is essential that the health and safety risks arising from the change are assessed at an early time

in order to determine what they are and the appropriate control measures to minimise them.

The risk assessment should consider how the proposed change deviates from the planned way of working, which may be set out formally in method statements, previous risk assessments or site scheduling. This will help to determine the significance of the change and whether the change can be achieved by utilising the current control measures or if additional control measures are required.

In particular, the risk assessment should determine who will be affected and how they would be affected. Depending on the proposed change the risks could impact on a wide range of people, including the principal contractor's workers, contractors, client's workers, those who are going to use what is being constructed or members of the public.

The assessment of risks may highlight the need for additional specific assessments to be undertaken, for example, manual handling, hazardous substances or noise.

Appointment of competent people

Everyone involved in a construction project is expected to take steps to satisfy themselves that the people who do the work are competent, which should be established before the change is made and work starts.

People who develop proposed changes or consider the risks arising from them must be sufficiently competent in health and safety. This may require the appointment of specific people with specialist skills in health and safety related to the proposed change. A little time and effort to involve competent people could prevent major problems. It is therefore important that those involved in the change process are aware of their own limitations and avoid being driven by time pressures to make the change even though they are not competent to determine the risks and appropriate control measures.

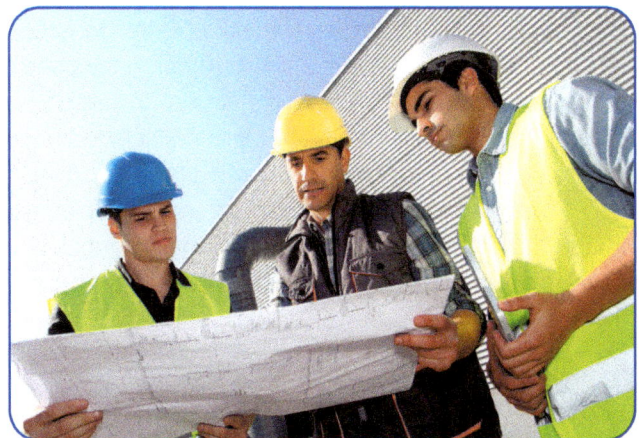

Figure 3-1: Competent people.
Source: Shutterstock.

It should be remembered that those most affected by the proposed change are often competent to consider the likely impacts and how to mitigate them, they may quickly identify issues that would need to be addressed in order to make the change.

The proposed change may require competent specialist health and safety monitoring and supervision in order to ensure the change is effective and risks are managed as the change takes place.

Segregation of work areas

When considering the management of the impacts of change care needs to be taken to ensure the maintenance of safe access onto and around the site for pedestrians and vehicles. Proposed changes need to be planned to show how vehicles will be kept clear of pedestrians, especially at site entrances, where it may be necessary to provide doors or gates to achieve sufficient segregation.

Proposed changes may bring groups of workers together that would not usually share the same work area to carry out their work activities. This should be considered when planning the change as it may be necessary to provide some level of segregation to ensure they do not affect each other's health and safety. This may involve segregation of work areas by the use of movable barriers to ensure each group of workers has enough space for the work they are doing, for example using machinery. It could also mean providing sheeting to segregate work areas where dust is generated or designating areas for mixing/preparation of wet construction materials.

Emergency procedures

It is important to manage the impact of changes that affect emergency procedures on site. Because of the transient nature of construction sites, emergency procedures will need constant review and amending as the site changes. The construction phase plan should anticipate the changes and put measures in place to maintain effective emergency procedures.

Changes to the main structural aspects of the construction site could have an effect on fire emergency procedures, for example the fire alarm system installed in a building being worked on. This could result in the need to provide a temporary means of raising the alarm if there was a fire. In a new build situation, it might be planned that as the project progresses the changes might allow a new fire alarm system installed in the building to be made available at an early stage in construction, to remove reliance on temporary measures.

Changes could lead to fire evacuation routes becoming blocked by new parts of the structure, by materials or while construction activities take place.

In addition, changes could affect the availability of electricity for lighting, which would be necessary to safely evacuate.

As work progresses, changes to work activities may require additional emergency procedures. This could include emergency procedures for when work at height activities commence and workers use personal fall arrest equipment or when a tower crane is brought on site and escape from the crane would need to be considered in emergency procedures.

Welfare provision

Good clean welfare facilities can play an important part in protecting the health of everyone involved in the construction work. When the principal contractor develops the health and safety plan for the construction phase of the project, it must take account of the changing welfare needs as the various stages of work and work activities of the project progresses.

Depending on the project, at the early stage there may be a need for specialist welfare facilitates to deal with decontamination of a site that contains chemicals or asbestos. As the site is prepared for development a relatively small number of people may be involved, but their work may be very dirty, requiring welfare facilities that can support this.

As the project progresses the welfare needs may be affected by changes in the weather, requiring facilities for wet clothing and warm areas to rest. As the type of work being carried out changes more people of various trades may be required on site. If this change is not taken into account there is a risk that facilities get overwhelmed to the extent that they become ineffective. The provision of welfare facilities for the site needs to keep pace with the changes that affect them.

REVIEW OF CHANGE

As the change progresses and when the change has been introduced, health and safety performance should be reviewed to ensure there are no adverse impacts due to hazards not identified or control measures not being effective. Adverse impacts identified at this stage may require further analysis and development of additional actions to manage health and safety. The review should also identify what went well in the change process in order to learn from this to benefit future changes.

3.2 Safe systems of work for general work activities

In criminal law the *Health and Safety at Work etc. Act (HASAWA) 1974* requires the provision and maintenance of systems of work that are, so far as is reasonably practicable, safe and without risks to health. Employers should also provide safe systems of work in order to fulfil their common law duty of care to their workers.

WHY WORKERS SHOULD BE INVOLVED WHEN DEVELOPING SAFE SYSTEMS OF WORK

Worker involvement in the development of safe systems of work is essential to ensure their safe behaviour and encourage them to work to the system. Workers often know the practical difficulties in working to theoretical systems of work, such as the absence of equipment specified or requirements that mean more workers are needed to do the task than are available. They may have encountered circumstances that make the work difficult and usually know how they get around them, such as poor access/egress.

By discussing proposed systems of work with workers, it is possible to learn what these practicalities are and ensure the system of work takes account of them. Involvement in this way often ensures a deeper understanding of hazards and risks and how safe systems of work will reduce those risks. This helps with ownership of the safe system of work and makes it more likely that the agreed system of work will be followed.

WHY PROCEDURES SHOULD BE RECORDED AND WRITTEN DOWN

Where the safe system of work is complex or the risks of not working to it are high, it is important to define what must be done in written procedures. This does not mean that written procedures have to be complicated or wordy; some of the most effective are those that are clear, simple and easily understood. The importance of creating written procedures is that they serve as a clear setting of standards of work, including how the risks of the work are to be combated. This is particularly relevant when communicating to a client how a contractor is going to conduct work safely. Written procedures can contribute to tender documents used to win work and will enable supervisors and others that monitor health and safety to do so against clear standards of performance.

Written procedures may relate to specific work to be done such as carrying out a cutting operation, or to general health and safety matters such as reporting accidents. The construction industry makes particular use of written procedures in the form of method statements. As the name suggests, these are documents that express how work is to be conducted, the order of events, the health and safety considerations and the measures that will make it safe and healthy for those that might be affected.

It is important that procedures express how the work is done, not just how it should be done in theory. If there is a poor match between the procedure and what can be done in practice it will devalue that procedure and others and cause confusion. If a good match is achieved, in a manner that is clear to workers and supervisors, there is a higher chance that it will be complied with and the supervisor will be more able to enforce it. Job/task analysis is not only necessary to identify the hazards and controls, but is essential for preparing written procedures. These can then be translated into comprehensive training plans to ensure the correct skill and knowledge content for the work to be carried out safely.

TECHNICAL, PROCEDURAL AND BEHAVIOURAL CONTROLS

In the workplace we use a variety of controls as part of safe systems of work. In order to give emphasis to them and to help identify how complete our approach is to health and safety we group controls into a number of categories. The categories tend to reflect the main strategies used to improve health and safety and the general principles of prevention referred to in *Element 2.4 'Assessing risk'*, but use slightly different words.

It is important to note that, in order to deal effectively with a hazard, we will need to use all three forms of control, to a greater or lesser degree, for example, the use of a hard hat for the control of a hazard of falling materials.

A hard hat is a technical control that will afford a degree of protection - the technical aspects of this come into play when we consider how long a hard hat lasts and what type is suitable for which work/person. Purchasing hard hats will not be effective if procedures are not in place to enable issue and replacement. These controls will be immediately undermined if we do not provide education and training to those that wear and enforce the wearing of hard hats in order to ensure the right behaviour of the wearer.

Technical	(place)	(job)
Procedural	(system)	(organisation)
Behavioural	(person)	(person)

Risk control measures:

Technical controls include:

- Equipment - design (for example, guarding) and maintenance.

- Access/egress - provision of wide aisles, access kept clear of storage items.
- Materials (substances and articles) - choice of packaging to make handling easier.
- Environment (temperature, light, dust, noise) - local exhaust ventilation (LEV).
- Issue of the correct personal protective equipment (PPE) for the task.

Procedural controls include:

- Policy and standards.
- Rules.
- Procedures.
- Permit-to-work.
- Authorisation and coordination of actions.
- Purchasing controls.
- Accident investigation and analysis.
- Emergency preparedness.
- Procedures in the issue, use and maintenance of PPE.

Behavioural controls include:

- Awareness, knowledge, skill, competence.
- Attitude, perception, motivation, communication.
- Supervision.
- Health surveillance.
- Training in the issue and use of PPE.

In practice, the most successful organisations use a combination of the three strategies, with the emphasis on making the place safe/healthy (providing technical controls) and supporting this with procedures and paying attention to the person. Many organisations that feel they have done all that is necessary to get the place and procedures correct are taking a fresh look at the actions necessary to ensure the person's behaviour is also correct. Studies have shown that this too is a critical aspect of effective management of health and safety.

DEVELOPING A SAFE SYSTEM OF WORK

Developing a safe system of work requires a systematic approach and generally requires the involvement of a number of people. The development process requires a number of stages:

- Analysing the task.
- Identification of hazards and risk assessment.
- Introducing controls and formulating procedures - including the definition of the safe and healthy method and the implementation of the system.
- Instruction and training in the operation of the system.
- Monitoring the system.

It is important that the development of a safe system of work involves relevant people; this could include managers, workers who will work to the system,

maintenance workers, competent health and safety practitioners and specialists.

Analysing the task

Analysis of the task must consider not only the work to be done, but the environment where it is to be done, to allow a full consideration of the hazards to be made and the establishment of an effective safe system of work. Simple work activities that involve low risk may be analysed in a structured way, but without the complexity of a detailed review using documentation. Though a simpler approach may be taken, the principles of analysis are similar to those where a detailed review is involved. More complex or higher risk work activities may involve a detailed review known as *job* or *task analysis*. Job/task analysis consists of a formal step-by-step examination of the work to be carried out.

Analysis can be:

1) Job based - considering all the work of a specific group of people, for example, all the work of operators of a process plant or fork-lift truck drivers.

2) Task based - considering specific work, for example, a manual handling activity that is part of the operation of process plant or the fork-lift truck task of stacking loads on the trailer of a vehicle.

The main difference between job and task analysis is that the steps of the work activity are smaller in task analysis. The steps in job analysis are made up of the tasks that comprise the job and the steps in task analysis are smaller parts of the task. There is a tendency for people to use the terms job and task to mean the same thing when discussing this process.

Job/task analysis for health and safety reasons is the identification of all hazards relating to the job/task, the accident/incident and ill-health prevention measures appropriate to a particular job/task and the behavioural factors that most significantly influence whether or not these measures are taken. The approach is formal, structured and is diagnostic as well as descriptive.

The process of job/task analysis involves a number of stages, typically this will include:

1) Select the job/task to be analysed.

2) Break the job/task down into a sequence of steps.

3) Identify potential hazards.

4) Identify current control measures.

5) Consider the effectiveness of the control measure and the factors that could affect this, for example, worker behaviour.

6) Determine what additional control measures are required.

The process is usually conducted by direct observation of the job/task, though a generic analysis may be conducted without direct observation by using the experience of a number of people to consider a job/task before it is introduced into the workplace. This can be useful in establishing initial safe systems of work that are adjusted as the job/task is actually introduced into the workplace. When conducting a job/task analysis, all aspects of the job/task should be considered and recorded in writing to ensure that nothing is overlooked.

Typical considerations in any analysis for development of a safe system of work are:

- The work to be done.
- Where the work is done.
- Equipment, materials and people used.
- The current controls.
- Adequacy of controls.
- Correct use by the workers of the controls.
- Behavioural factors relating to workers/supervisors/managers - consideration of errors and violations.

The objective of the analysis is therefore to establish the hazards and controls at each step of the work to ensure a safe and healthy result. How the progress of work, in particular health and safety conditions, will be monitored should also be considered, for example, gas testing, temperature or pressure levels, and measurement of emissions.

Select the job/task to be analysed

Ideally, all jobs/tasks should be systematically analysed, but it is usually necessary to decide which ones are done first. Factors to consider could include the frequency and severity of accidents/incidents and ill-health related to the job/task. Consideration should also be given to the existence of newly established or infrequent jobs/tasks, as these could present higher risks due to the lack of experience of workers.

Break the job/task down into a sequence of steps

When breaking the job/task down into steps, it is important to ensure that it enables sufficient depth of analysis for the purpose of the process. The steps should be recorded in the sequence that they occur. If the job/task is easily repeated it is often useful to concentrate on breaking down and recording the steps only and recording the other factors separately when the job/task is repeated. If this is not easy, it will be necessary to consider and record the other factors as they occur, as well. The steps should be recorded by starting each step with an action verb, for example, 'remove...' 'insert...'. The job/task should be observed in normal work conditions, which will mean that if the job/task is usually done at night the analysis should be done at night.

Identify potential hazards and current control measures

The hazards that may occur and the control measures in place are identified and recorded for each step. Hazards may be identified by observation, supported by asking the workers and others questions.

Consider such things as:

- Slips, trips and falls.
- Falling objects.
- Hazards presented by tools, machines or other equipment.
- Body parts getting caught in or between objects.
- Making harmful contact with moving objects.
- Strain from lifting, pushing, or pulling.
- Hot, toxic, or caustic substances.
- Dusts, fumes, mists or vapours in the air.
- Excessive noise or vibration.
- Exposure to extreme heat or cold.
- Lighting problems.
- Weather conditions that affect health and safety.

A range of control measures may be in place and include technical control measures like temperature limitation, procedural control measures like control of access to keys or behavioural control measures in the form of supervision. The control measures will also include the actions the worker takes to prevent harm from hazards, such as checking a lid is firmly in place on a drum of chemicals after use or holding a tool firmly.

Consider the effectiveness of the control measures

It is important to consider the effectiveness of the control measures and the factors that could mean the controls might not be used. This could include worker behaviour. The job/task may be observed 'as the work should be done', because the worker is being observed and taking care to do the work as expected. However, this may not be the way that workers usually do the job/task. Workers may do the work in a way they find easier, or external factors may force the worker's behaviour to change from the correct way of working. The job/task analysis should try to determine these influences on behaviour. The analysis may determine that the current control measures are not adequate and that improvements are necessary. This should be identified and recorded.

The results of a job/task analysis can be used to establish safe systems of work that set out the correct sequence for the job/task and the control measures that are to be used at each step to control the hazards. They may also make a specific contribution to the risk assessment process and improve such things as training, emergency procedures, reporting of information or the layout of work areas.

Example of part of a task analysis - changing a wheel of a vehicle

	Step	Hazards	Control measures	Additional control measures
1	Park vehicle.	Vehicle too close to passing traffic. Vehicle on uneven, soft ground. Vehicle may roll.	Drive to area well clear of traffic. Turn on hazard warning lights. Choose a firm level parking area. Apply the parking brake.	Place blocks in front and back of the wheel diagonally opposite the wheel to be removed.
2	Loosen wheel nuts of wheel to be removed.	Injury to hands. Muscle strain.	Correct design of spanner that contacts all sides of the wheel nut, to reduce slipping. Gloves for grip and to protect. Long-handled spanner to enable suitable force.	Apply steady pressure, slowly.
3	Raise vehicle with jacking device.	Difficulty in access under vehicle. Jacking device could slip. Vehicle could overbalance.	Suitable clothing or sheet to enable you to get down to level to fit jacking device. Locate jacking device in jacking point. Raise vehicle only high enough so wheel to be removed clears the ground and can be removed. If tyre of wheel being removed is flat raise sufficiently high to enable spare wheel to be fitted.	Use hand light to help see position of jacking point.
4	Continued…	Continued…	Continued…	Continued…

Figure 3-2: Example of part of a task analysis.
Source: RMS.

Hazard identification and risk assessment

The identification of hazards and the assessment of risks are important factors arising from the analysis stage of developing safe systems of work. The analysis will provide data on the steps that comprise the work and will also usually provide information on the hazards related to the work and current control measures and indicate the need for further controls. This naturally leads into the stage in system of work development that considers these hazards and assesses risks related to them. Where a significant risk is identified, this may require further, more detailed analysis of the work activity, involving job/task analysis if this has not already been done. This in turn will lead to a more detailed risk assessment.

This is an important reflective stage of the development of the safe system of work. The analysis stage may only confirm what hazards exist and the current controls, it may not strongly challenge whether the risk remaining is acceptable. This stage reflects on the current situation and proposed additional controls determined in the analysis stage and considers whether the controls relate appropriately to the risk.

For example, exposure to a hazardous substance may present a very high risk of death. This may have been identified in the analysis stage and the current control of workers using a nuisance dust mask determined as acceptable as it follows current practice. This risk assessment stage has to consider the scale of the risk and determine if the current or proposed control measures are adequate to manage the risk to an acceptable level.

In the example given, this would mean challenging the reliance on a simple nuisance dust mask for such a high risk. From the reflection and decision making conducted at this risk assessment stage, appropriate control measures can be introduced into the system of work and workers can be trained how to minimise risks through use of the system of work.

For further information on hazard identification and risk assessment see Element 2.4 – 'Assessing risk'.

INTRODUCING CONTROLS AND FORMULATING PROCEDURES

Define the safe and healthy methods

The first phase in introducing controls and formulating procedures is to define the safe and healthy methods. The earlier stages in the development of a safe system of work required the analysis of the job/task, identification of hazards and assessment of risk. These stages will have considered the control measures that are in place or are required to deal with hazards related to the work. This will enable a safe and healthy method of working to be developed and defined that incorporates these controls. In the example given in *Figure 3-2*, one of the controls would be to park the vehicle where there is minimal or no traffic. A further control is to ensure the parking brake has been applied.

The method can include setting out the correct order of the steps in the work, if this is applicable. This safe and healthy method should be defined to establish it as a formal decision by the organisation on the way work must be done. In situations where the risk related to the work is high, this must be done in writing, often in a procedure. In situations where risks are low, this defined method may not be written down but agreed between the people involved in the work.

The following should be considered when deciding and defining a safe and healthy work method, it should:

- Take account of the analysis, identification of hazards and risk assessment conducted.

- Adequately control hazards associated with the work. Where possible, hazards should be eliminated at source. The need for protective or special equipment should be identified, as should the need for the provision of temporary protection, guards or barriers.

- Comply with the organisation's standards.

- Comply with relevant legal or international best-practice standards.

- Take account of manufacturer and supplier instructions.

- Be as simple as possible and defined in a way that is understandable to workers.

- Identify specific responsibilities at various stages in the work, and the person in control of work should be clearly identified.

- Confirm who does what, where, when and how.

- Take account of likely emergencies, for example, fire or spillage. If there is a possibility that harm could result during the work, needs for emergency and rescue methods should be identified.

Where the defined safe and healthy method is complicated or is a significant change from current practice, it should be tested using a simulated work situation and/or introduced with a pilot group of workers to validate that it is effective.

Implementation

Once the system of work has been developed and agreed, preparation for implementation can proceed. Implementation is far harder than developing the safe system of work because managers and workers need to accept and adopt the new working practice and use the control measures that have been developed. It is important that care be taken to ensure that controls referred to when the system of work was being developed and defined are available to the workers when the system of

work is implemented. If they are not, this will devalue the system of work and cause the workers to have to work outside the defined method, possibly in an unsafe way. In the example given in *Figure 3-2*, if the correct spanner is not available then workers may use an oversize spanner that might slip when being tightened. The process of implementation will involve:

- The formal acceptance of the defined safe and healthy method, by the simple agreement to work in the proposed way by the people affected by it or by formal acceptance of the agreed written procedure by managers and workers. This will involve agreement on the date and, in some cases, the time the system of work becomes effective.

- Instruction and training in the operation of the system.

- A period of evaluation of the effectiveness of the system.

- If problems arise that necessitate modification to the system, formal approval should be gained, documentation of the modification should be made, and the modification communicated to those involved with the system of work.

Instruction and training in how to use the system

Provision for the communication of relevant information to all involved or affected by the system of work is an essential element of the development of a system of work. Many organisations may be tempted to issue procedures relating to safe systems of work without instruction and training. This can lead to a great deal of misunderstanding about what the system of work is and will often lead to people not being motivated to work to the system. This is particularly the case when the system is a change from the usual way that workers may have worked. Workers, and supervisors, can have a naturally high resistance to change.

Simply providing information will not usually be sufficient to change behaviour. A more assured way to introduce the system would be to utilise such things as 'tool box' talks, where team leaders brief their group of workers at the start of shift to explain any relevant day-to-day issues or the introduction of new methods, technology or equipment. Tool box talks provide a practical opportunity to develop the workers' understanding of how to use the system, and if these are done as part of a cascade training method the supervisor or similar person will have received training from someone committed to the new system and will be better placed to minimise any resistance of workers to the change.

In addition to worker training, it is important to ensure of those that manage workers receive training on systems of work. There may be specific legal requirements for this training, for example, the Provision and Use of Work Equipment Regulations (PUWER) 1998 require that supervisors and managers are adequately trained to enable them to identify the risks, control measures and methods of use associated with work equipment under their control. They particularly need to have sufficient training to identify that the defined system of working is being followed, when workers deviate from it and the consequences.

Job/task analysis is not only useful to identify the hazards and controls, but is essential for preparing written procedures and specifying the skill and knowledge content of the work to be carried out. The analysis will not only identify the sequence of work but often the work rate expected - sometimes timeliness is an important consideration, particularly in relation to certain chemical manufacturing processes.

The analysis can then be incorporated into job training programmes. For certain high-risk tasks, this will often involve a training course to develop knowledge and understanding. This is then followed by practical application, either utilising a work simulator (for example, train driver / aircraft pilot) or close one-to-one supervision while on the job (for example, fork-lift truck driver). Where training requirements have been identified, a record of those affected should be made. The required training should be conducted, confirmed as successful and recorded.

MONITORING THE SYSTEM

All safe systems of work should be formally monitored and records kept of compliance and effectiveness. This can be done by direct observation or by discussion at team meetings or safety committee meetings. The safe system of work should not simply be imposed upon the people responsible for their operation. A system of monitoring and feedback should be implemented to ensure it is effective. Audits and investigation of accidents/incidents and ill-health that may occur can provide a valuable insight into whether systems of work are effective. Any permanent record of monitoring of the system of work must be kept and regularly reviewed by a member of the management team responsible for the system of work in order to identify the need for improvements or modification. If improvements or modification to the system are required formal approval should be gained, documentation of the modification should be made and the modification communicated to those involved with the system of work. If the change to the system of work is significant, workers and managers may require retraining.

3.3 Permit-to-work systems

MEANING OF A PERMIT-TO-WORK SYSTEM

A permit-to-work system is a formal written (administrative) system used to control certain types of activities that have high hazard potential.

A permit-to-work system (PTW) should not be applied to every task (except in certain permanent high-risk work, for example, processing highly flammable liquids) as this may detract from its importance and lead to it being treated as nothing more than a bureaucratic, form-completion exercise.

The term 'permit-to-work' refers to the paper or electronic 'permit' document that is used as part of the overall system of work which has been devised to identify and control the risks involved in the work.

WHY PERMIT-TO-WORK SYSTEMS ARE USED

A permit-to-work system is an integral part of a safe system of work and can help to properly manage work activities. The purpose of a permit-to-work system is to ensure that full and proper consideration is given to the risks of the particular work that it relates to and to control the way of working in order to ensure a safe system of work is followed.

A permit-to-work system will:

- Ensure the proper authorisation of specified work.

- Confirm the identity, nature, timing, extent and limitations of the work.

- Establish criteria to be considered when identifying hazards and what they are.

- Confirm that hazards have been removed, where possible.

- Confirm that control measures are in place to deal with residual hazards.

- Confirm work is started, suspended, conducted, and finished safely.

- Control and confirm who has control of the location and equipment relating to the work when it passes between parties.

- Control change and consider other work activities that might interact with specified work.

- Provide a record of the steps in the process.

HOW PERMIT-TO-WORK SYSTEMS WORK AND ARE USED

Requirements of the system

The requirements of an effective permit-to-work system are that it must:

- Be formal and documented.
- Simple to operate.
- Have the commitment of those who operate and are affected by it.
- Provide concise and accurate information.
- Ensure liaison with controllers of other plant or work areas whose activities may be affected by the permit-to-work.
- Require that boundary or limits of work area are clearly marked or otherwise defined.
- Include contractors undertaking specific tasks in the permit-to-work system, including any briefing prior to commencement.
- Be supported by training in the system for all working under it and affected by it.

Permit-to-work document

The important features of a permit-to-work document are:

- A description of the task to be performed.

- An indication of the duration of the validity of the permit.

- Details and signature of the person authorising the work.

- Identification of the hazards affecting the work and the precautions to be taken.

- The isolations that have been made and the additional precautions required.

- A clear record that:

- All foreseeable hazards have been considered.

- All precautions are defined and taken in the correct sequence.

- Acceptor assumes responsibility for safe conduct of work.

- Establishes a safe system of work.

- An acknowledgement of acceptance by the workers carrying out the task, who would then need to indicate on the permit that the work had been completed and the area made safe in order for the permit to be cancelled.

- Overrides any other instructions until cancelled.

- Excludes work not described within the permit.

- In the event of a change to the programme of work, it must be amended or cancelled and a new permit issued.

- Only the originator may amend or cancel.

WHEN TO USE A PERMIT-TO-WORK SYSTEM

A permit-to-work system is a formal health and safety control system designed to prevent accidental injury/illness of workers, damage to plant, premises and product, particularly when work with a foreseeable high hazard content is undertaken and the precautions required are numerous and complex.

Good practice suggests that permit-to-work systems are normally considered most appropriate to:

- Non-production work (for example, maintenance, repair, inspection, testing, alteration, construction, dismantling, adaptation, modification, cleaning).

- Non-routine operations.

- Jobs where two or more individuals or groups need to coordinate activities to complete the job safely.

- Jobs where there is a transfer of work and responsibilities from one group to another.

Hot work

This typically involves welding operations, such as joining pipework where the risk of sparks and heat may ignite combustible materials. Elimination or protection of such combustible materials will need to be considered. The provision of firefighting equipment close at hand and trained personnel to deal with inadvertent ignition is also important.

The Dangerous Substances and Explosive Atmospheres Regulations (DSEAR) 2002 specify the application of permits-to-work, to be used in hazardous places or involving hazardous activities. *See Figures 3-3 and 3-4* for a sample hot work permit.

Work on non-live (isolated) electrical systems

The high risks associated with work on or near electrical systems often justifies the use of a permit-to-work system. Work on electrical equipment, such as a transformer, requires safe isolation, access and egress, work at a height and heavy lifting to be considered. The permit will help to ensure that adequate and effective isolation is in place, the electrical system is confirmed to be 'dead' and that all precautions are in place, including the competency of the workers involved.

Hot work permit - applies only to area specified below

Part 1

Site: .... Floor: ....

Nature of the job (including exact location) ....

....

The above location has been examined and the precautions listed on the reverse side have been taken.

Date:..

Time of issue:.. Time of expiry: ....

NB. This permit is only valid on the day of issue.

Signature of person issuing permit: ....

Part 2

Signature of person receiving permit: ....

Time work started: ....

Time work finished and cleared up: ....

Part 3 | Final check up

Work areas and all adjacent areas to which sparks and heat might spread (such as floors above and below and opposite side of walls) were inspected one hour after the work finished and were found fire safe.

Signature of person carrying out final check: ....

After signing return permit to person who issued it.

Figure 3-3: Hot work permit - front of form.
Source: Lincsafe.

Hot work permit – precautions

Hot Work Area

- Loose combustible material cleared.
- Non-moveable combustible material covered.
- Suitable extinguishers to hand.
- Gas cylinders fitted with a regulator and flashback arrester.
- Other personnel who may be affected by the work removed from the area.

Work on walls, ceilings or partitions

- Opposite side checked and combustibles moved away.

Welding, cutting or grinding work

- Work area screened to contain sparks.

Bitumen boilers, lead heaters, etc.

- Gas cylinders at least 3m from burner.
- If sited on roof, heat-insulating base provided.

Figure 3-4: Hot work permit - reverse of form. Source: Lincsafe.

Machinery/plant maintenance

Machinery/plant maintenance sometimes requires workers with different disciplines to work on large complex plant at the same time. This is aggravated by the fact that the plant may be spread over a number of floors in a building, such as power-generation plant, flour mills or lift systems in an office block.

This sort of work can involve many risks - related to a variety of services and energy sources, dangerous parts of the equipment, problems with access, risk of falling or being trapped inside the plant. These risks are best controlled by a well-established system of work, supported by a permit-to-work. The permit can ensure that the work is carefully planned and that power sources (electric or hydraulic) are appropriately isolated and locked off. The permit can also take into account the release of any stored energy, for example, electrical, hydraulic or kinetic energy.

Confined spaces

A confined space is a place which is substantially enclosed (though not always entirely), and where serious injury can occur from hazardous substances or conditions within the space or nearby (for example, lack of oxygen). Failure to appreciate the dangers associated with confined spaces has led not only to the deaths of many workers but also to the demise of some of those who have attempted to rescue them.

A confined space, as defined by Confined Space Regulations (CSR) 1997, may be any place, such as a chamber, tank, vat, silo, pit, well, pipe, sewer, flue, or similar, in which, by virtue of the enclosed nature, there is a foreseeable risk.

Work in a confined space may present a risk of:

- Injury arising from fire or explosion.
- Loss of consciousness from increased body temperature.
- Loss of consciousness from asphyxiation from gas, fume, vapour or lack of oxygen.
- Drowning of any person from an increase in the level of liquid.
- Asphyxiation of a person as a result of being trapped by a free-flowing solid.

It is worth noting that although this definition in CSR 1997 refers to an 'enclosed space', this does not mean that the space must be totally enclosed. Such examples as a 'pit', a 'trench' and a 'well' are included, all of which might be open to the sky. Nevertheless, the high risks associated with confined spaces are well documented and must be controlled by a permit-to-work system.

See Figure 3-5 for a sample confined space permit. See also Element '4.3 - Safe working in confined spaces' for more information on confined spaces.

Example Entry into confined spaces

Possible Lay-Out for A Permit-to-work Certificate

Permit-to-work Certificate

PLANT DETAILS (Location, identifying number, etc.)			ACCEPTANCE OF CERTIFICATE Accepts all conditions of certificate	
WORK TO BE DONE				Signed Date Time
WITHDRAWAL FROM SERVICE	Signed Date Time		COMPLETION OF WORK All work completed equipment returned for use	
ISOLATION Dangerous fumes Electrical supply Sources of heat	Signed Date Time			Signed Date Time
CLEANING AND PURGING of all dangerous materials	Signed Date Time		EXTENSION	Signed Date Time
TESTING for contamination	Contaminations tested Results Signed Date Time			
I CERTIFY THAT I HAVE PERSONALLY EXAMINED THE PLANT DETAILED ABOVE AND SATISFIED MYSELF THAT THE ABOVE PARTICULARS ARE CORRECT (1) THE PLANT IS SAFE FOR ENTRY WITHOUT BREATHING APPARATUS (2) BREATHING APPARATUS MUST BE WORN Other precautions necessary: Time of expiry of certificate: Signed Delete (1) or (2) Date Time			THIS PERMIT-TO-WORK IS NOW CANCELLED. A NEW PERMIT WILL BE REQUIRED IF WORK IS TO CONTINUE Signed Date Time	
			RETURN TO SERVICE	I accept the above plant back into service Signed Date Time

Figure 3-5: Example of a permit-to-work for entry into confined spaces. Source: HSE Guidance note on permit-to-work.

Work at height

Work at height (which may mean working near an excavation or cellar where a fall below ground level can cause serious injury) has been identified as presenting a significant risk to workers who may fall. The permit system used will ensure that avoiding work at height is considered and, if this does not prove possible, that fall-prevention strategies (for example, provision of a safe place) are used. If a safe place cannot be provided, then the permit must be used to consider minimisation of the distance of the fall by the use of safety devices (lanyards, nets, etc.). The permit is also likely to consider the influence of weather conditions on the safety of the task being undertaken, for example, high winds or ice.

Working at height includes:

- Ladders.
- Stepladders.
- MEWP (mobile elevated working platforms).
- Roof working.

Figure 3-6: Working at height.
Source: Shutterstock.

CONSIDER

A maintenance worker is required to enter a chemical process tank containing a residual amount of a substance that gives off significant quantities of an inert gas that would replace oxygen in the tank when disturbed. The inert gas is heavier than air. The maintenance worker will have to disturb the substance in order to remove it. As much of the substance as possible has been drained away.

Entry is via an opening at the top of the tank. The worker will have to gain access by a ladder and will need to take temporary lighting. The maintenance worker has been provided with a respirator mask for use while doing the cleaning. The cleaning task is defined as a one-person activity on the job card.

Using the risk rating matrix, determine:

The severity of risk related to this work	
Likelihood of harm	
Overall risk rating	

3.4 Emergency procedures

WHY EMERGENCY PROCEDURES NEED TO BE DEVELOPED

It is important to develop, implement and test emergency procedures for potentially major loss-causing events in order to bring the event under control promptly, reduce the effects of the event (on premises, equipment, materials, environment and people that might be affected) and enable a fast return to normal operations. In the absence of emergency procedures, there may not be a timely or suitable response to the emergency, allowing it to get out of control and cause more serious effects than if procedures were in place. Depending on the emergency, this could result in major loss of life, long term ill-health, environmental effects, significant damage to building, and equipment, and long delays in the organisation returning to normal operations. Some organisations or locations where emergencies take place that are not effectively controlled never recover from the effects.

Besides the major benefit of providing guidance during an emergency, developing emergency procedures has other advantages. Organisations might discover unrecognised hazardous conditions that would aggravate an emergency situation, allowing work can be carried out to eliminate them. The process of developing emergency procedures may bring to light deficiencies, such as the lack of resources (equipment, trained personnel, and supplies) or defective equipment that can be rectified before an emergency occurs.

Adequate emergency procedures should be in place, or developed, to control likely emergencies, for example, a fire, terrorist threat, spillage or release of hazardous substances, explosion, vehicle collision, severe weather, major injury, poisoning or exposure to pathogens, or release of radioactivity.

Requirements for an emergency procedure

The development of emergency procedures is required by Regulation 30 of the Construction (Design and Management) Regulations 2015 (CDMR 2015).

> "(1) Where necessary in the interests of the health or safety of a person on a construction site, suitable and sufficient arrangements for dealing with any foreseeable emergency must be made and, where necessary, implemented, and those arrangements must include procedures for any necessary evacuation of the site or any part of it."

Figure 3-7: Requirements for emergency procedures.
Source: CDMR 2015, Regulation 30.

In addition, ISO 45001:2018 'Occupational health and safety management systems' also has a specific clause, Clause 8.2, which sets out requirements.

> "8.2 Emergency preparedness and response
>
> The organization shall establish, implement and maintain a process(es) needed to prepare for and respond to potential emergency situations, as identified in 6.1.2.1, including:
>
> (a) establishing a planned response to emergency situations, including the provision of first aid;
>
> (b) providing training for the planned response;
>
> (c) periodically testing and exercising the planned response capability;
>
> (d) evaluating performance and, as necessary, revising the planned response, including after testing and, in particular, after the occurrence of emergency situations;
>
> (e) communicating and providing relevant information to all workers on their duties and responsibilities;
>
> (f) communicating relevant information to contractors, visitors, emergency response services, government authorities and, as appropriate, the local community;
>
> (g) taking into account the needs and capabilities of all relevant interested parties and ensuring their involvement, as appropriate, in the development of the planned response.
>
> The organization shall maintain and retain documented information on the process(es) and on the plans for responding to potential emergency situations."

Figure 3-8: Emergency preparedness and response.
Source: ISO 45001, Clause 8.2.

Therefore, developing emergency procedures is a legal requirement and emergency procedures are also a common part of health and safety management systems.

WHAT ARRANGEMENTS MUST BE MADE WHEN PLANNING EMERGENCY PROCEDURES AND FIRST AID PROVISION

When making arrangements for the planning of emergency procedures, under Regulation 30 of CDMR 2015 account must be taken of the type of work, the characteristics and size of the construction site, the work equipment being used, the number of persons likely to be present on the site at any one time and the physical and chemical properties of any substances or materials present.

When planning emergency procedures the principal contractor should identify and assess the nature of any likely emergency and the arrangements needed to minimise the effects of the emergency. As the primary duty lies with the principal contractor, it is important that they determine what is required to manage these risks and liaise with the emergency services to agree what support they can provide.

The Management of Health and Safety at Work Regulations (MHSWR) 1999 require that, where necessary, contact with external service should be made, see *Figure 3-9.*

> "Every employer shall ensure that any necessary contacts with external services are arranged, particularly as regards first-aid, emergency medical care and rescue work."

Figure 3-9: Requirements for contact with external emergency services.
Source: Management of Health and Safety at Work Regulations 1999.

Where emergencies procedures would usually rely on external emergency services for assistance with a fire, rescue and medical care it is important that the principal contractor consider response times and the capability of the external emergency services.

For example, if response times needed for the rescue of a person from suspension in a harness after a fall cannot be met by the external emergency services, then an on-site emergency rescue team would be required. Similarly if rescue from a tower crane is required, the local fire and rescue service may not have the capability and equipment to respond to this type of emergency.

Figure 3-10: Emergency station.
Source: Armorgard.

If the construction activities could lead to a major fire it is important to formalise arrangements with the local fire and rescue service to establish that an adequate and timely response can be made by them. This will usually mean agreeing the amount of support the fire and rescue services can provide and what must be provided by those managing the construction project.

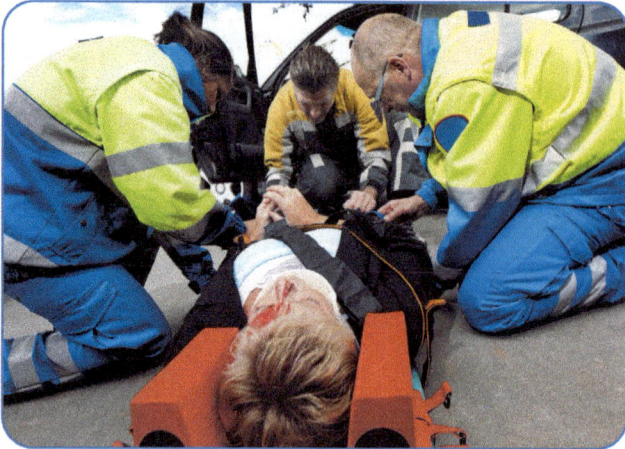

Figure 3-11: Emergency services.
Source: Shutterstock.

Similarly, a decision would need to be made on the provision of first-aid by the site and the appointment of specific people who are competent in emergency medical techniques, for example, people competent in resuscitation techniques. It may be necessary for the principal contractor to contract for specialist emergency medical support services or make arrangements with external medical emergency services for a mobile service to be available near the site at times of highest risk. If the activities being undertaken require the availability of special anti-toxins or isolation facilities formal arrangements should be made with the local medical emergency services.

Consideration should also be given to identifying and training those people who have a specific role and responsibility on site for contacting the external emergency services and who will be on site to receive the emergency services when they arrive, so that they can be informed of the nature of the emergency and directed to its precise location.

Emergency instructions should be posted throughout the site to guide managers and workers in the correct response to an emergency, for example, it is common to post instructions on how to respond to a fire. Contact telephone numbers for site emergencies should be well known and posted on emergency stations in a sufficient number of locations to ensure they are available when needed. The means of contacting the external emergency and rescue services should be agreed.

It may be necessary for the site to contact the external emergency and rescue services before complex hazardous work is carried out, to alert them that they may be needed.

Figure 3-12: Emergency telephone symbol.
Source: www.sksigns.co.uk.

This could include work in a confined space or other work where there is a significant risk that people may need to be rescued. This precautionary contact would enable the emergency and rescue services to be ready for the specific type of emergency and in some cases they may position a suitable team nearer to where the work is taking place.

Suitable and sufficient steps must be taken to ensure that everyone is familiar with the emergency procedures and the procedures are tested, for example, by carrying out tests of fire alarms and fire drills.

Typical scope of emergency procedures

Emergency procedures should cover:

- What might happen and how the alarm will be raised.

- How emergencies are communicated within the organisation, i.e. night and shift working, weekends, presence of temporary workers, contractors, members of the public and times when the premises are closed, for example, holidays.

- The planned responses to the emergency - for example, obtaining assistance from external emergency and rescue services and the first action of internal emergency response teams to limit the effects of the emergency.

- The people who will be involved in implementing the procedure and the training that they will need. This will mean identifying, where necessary, first-aiders, fire marshals, confined-space rescue teams, etc.

- Any equipment needed to deal with the emergency, for example, spill kits, firefighting equipment or rescue equipment.

Element 3

- Nomination of competent people to take control (a competent person is someone with the necessary skills, knowledge and experience to manage health and safety).

- What other key people are needed, such as a nominated incident controller, someone who is able to provide technical and other site-specific information if necessary, or first-aiders.

- Plan of essential actions, such as emergency plant shutdown, isolation or making processes safe. Clearly identify important items such as shut-off valves and electrical isolators, fire dry and wet risers and hydrant points.

- Evacuation of people that might be affected - including methods to evacuate disabled and other vulnerable people to agreed safe assembly points. Consider distance to reach a place of safety or to get rescue equipment, and the need for suitable forms of emergency lighting.

- Ensuring that there are enough emergency exits for everyone to escape quickly, and that emergency doors and escape routes are kept unobstructed and clearly marked.

- Actions to reduce the effects of the emergency - for example, the provision of first-aid, prevention of environmental contamination and shutting down plant/equipment that could make the emergency worse.

- Actions to clean up after the emergency and return to normal operations - for example, checking the atmosphere of the area affected by the emergency to see if people can be allowed to return to work.

FIRST-AID PROVISION

The purpose of first-aid is to:

- Preserve life.
- Prevent the condition requiring first-aid getting worse/minimise its consequences until medical help arrives.
- Promote recovery of the person requiring first-aid.
- Provide treatment where medical attention is not required.

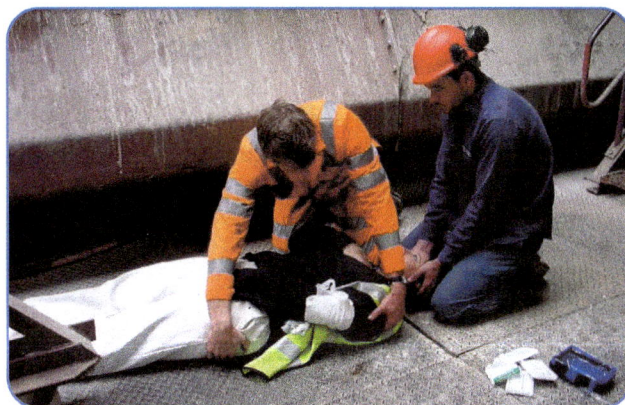

Figure 3-13: Providing first-aid on site.
Source: High Peak First Aid.

To ensure compliance with Regulation 3, an employer should make an assessment to determine the first-aid needs of the organisation. This will determine the organisation's requirements for first-aiders (quantity and competence) and first-aid equipment/facilities (quantity and type). When making the assessment, consideration should be given to a number of factors, including:

- The size of the organisation - in particular, the number and distribution of workers that need to be covered by the first-aid provision.

- The type of workforce - including the special needs of trainees, younger and older workers, those with known medical conditions and the disabled.

The Health and Safety (First-Aid) Regulations (FAR) 1981 - Main requirements

Reg 2	Regulation 2 defines first-aid as: '...treatment for the purpose of preserving life and minimising the consequences of injury or illness until medical (doctor or nurse) help can be obtained. Also, it provides treatment of minor injuries which would otherwise receive no treatment, or which do not need the help of a medical practitioner or nurse'.
Reg 3	Requires that every employer must provide equipment and facilities which are adequate and appropriate in the circumstances for administering first-aid to his employees.
Reg 4	An employer must inform his employees about the first-aid arrangements, including the location of equipment, facilities and identification of trained personnel.
Reg 5	Self-employed people must ensure that adequate and suitable provision is made for administering first-aid while at work.

Element 3

- The need to provide first-aid at all times when workers are on site - considering shift patterns and shift change-over periods, any work done outside normal working hours and cover for first-aiders due to their sickness or other absences.

- Different work activities - first-aid needs will depend on the type of work being done. Some activities, such as office work, have relatively few hazards and low levels of risk and will need less first-aid provisions than other activities that have more hazards or more specific hazards (construction or chemical sites).

- Past experience - previous accidents and ill-health should be considered, including their type, location and the type of harm caused.

- Ease of access to medical treatment - where external emergency medical services are not easily available, more first-aid provision may be required, for example, a first-aid room.

- Workers working away from the employer's premises - for some work activities first-aid arrangements may have to be made for workers working in locations away from the employer's fixed work places. For example, mobile teams of workers, particularly those working in remote locations.

- Workers of more than one employer working together - first-aid facilities may be shared, for example, when a number of contractors work together on a construction site they could share first-aid facilities or when individual workers go to an employer's site to do maintenance work they could use the employer's first-aid facilities.

- Provisions for non-employees – under FAR 1981 employers do not usually make first-aid provisions for any person other than their own workers. Liability issues and interpretation placed on employers by the HASAWA 1974 may alter the situation, as, for example, in the case of a theatre, hospital or other place where the public enter.

The first-aid assessment should be used to guide the employer in providing first-aid arrangements. A sufficient number of competent first-aiders should be readily available. Sometimes first-aiders will need specific competencies related to the risks of the work activities, for example, competencies in the use of a defibrillator or the treatment of chemical burns. There should be a sufficient number and size of first-aid boxes/containers to provide ready access to facilities.

Additional equipment and facilities may also be appropriate, for example, a stretcher, defibrillator, or first-aid room. When first-aid arrangements have been established, the employer should inform workers of where first-aid equipment is and who is responsible for performing first-aid duties.

The employer should ensure that first-aid treatment remains available *throughout the time the employer's work activities take place.* Additional first-aiders may be necessary to cover out of hours, shift or overtime working. In some organisations, security workers are available throughout the day and night, including at weekends, and it may be possible to train and appoint them as first-aiders to ensure first-aid can be provided when work is taking place out of normal working hours. When high hazard work is being done outside normal working hours it may be necessary to arrange for specific first-aiders to be available near to the high hazard work activity, rather than relying on general first-aid provision.

Where *workers work away* from their employer's premises or if the *geographical area* in which the work takes place is large, all workers may need to be trained and provided with first-aid equipment. For example, this may be appropriate for gas, oil, electrical or telecommunication workers, particularly where they are working in remote locations. A full first-aid box/container may be provided in vehicles and/or each individual could be provided with a personal first-aid pouch. A first-aid pouch is a first-aid container that has fewer contents than a full first-aid box/container and is usually provided for individuals to administer first-aid to themselves or for small teams to assist each other in case of minor injuries.

Where a *site is shared by a number of employers* it may be useful for them to make an agreement so that each employer does not have to make separate first-aid arrangements and to enable cover for the absence of each other's first-aiders. For example, if main contractors on construction projects make their first-aid facilities available to sub-contractors this would avoid the need for them to provide their own first-aid facilities. Similarly, where a contractor's workers work in another employer's premises, for example, doing maintenance on the employer's equipment, the contractor could make an agreement to access the employer's first-aid facilities. It is important that these agreements are made formally, preferably as a written agreement, as this will enable those making the agreement to be confident about the arrangements. The employer should inform workers of the arrangements made to provide first-aid.

First-aiders and first aid equipment

FAR 1981 Regulation 3(2) states that in order to provide first-aid to injured or ill workers:

> "an employer shall provide, or ensure that there is provided, such number of suitable persons as is adequate and appropriate in the circumstances for rendering first-aid to his employees if they are injured or become ill at work; and for this purpose a person shall not be suitable unless he has undergone such training and has such qualifications as may be appropriate in the circumstances of that case."

Figure 3-14: Appointment of suitable persons.
Source: Health and Safety (First-Aid) Regulations (FAR) 1981.

A first-aid needs assessment should help to employers decide how many people are required and what first-aid training they should have. For example, the employer should consider whether first-aiders should be trained in emergency first-aid at work (EFAW) only or receive standard first-aid at work (FAW) training.

In appropriate circumstances an employer can provide an 'appointed person' instead of a first-aider. The 'appointed person' is someone appointed by the employer to take charge of the situation (for example, to call an ambulance) if a serious injury occurs in the absence of a first-aider.

The employer must also ensure that an adequate quantity of suitable first-aid equipment is provided. Employers are usually expected to meet minimum requirements and take account of the particular circumstances related to their work activities. The number of first-aid kits provided and their contents will be influenced by the level of hazard related to the workplace and the number of workers the first-aid box/container is to cover.

British Standard BS 8599-1:2011 'Workplace first-aid kits; specification for the contents of workplace first-aid kits' (BS 8599) outlines a guide for employers to decide the most suitable size and number of first-aid kits for their workplace, presuming there are no special risks in the workplace, see *Figure 3-17.*

Category	Number of workers at location	Suggested number and type of suitable person
Low hazard For example, offices, shops, libraries	Less than 25	At least one appointed person
	25-50	At least one first-aider trained in EFAW
	More than 50	At least one first-aider trained in FAW for every 100 employed (or part thereof)
Higher hazard For example, light engineering and assembly work, food processing, warehousing, extensive work with dangerous machinery or sharp instruments, construction, chemical manufacture	Less than 5	At least one appointed person
	5-50	At least one first-aider trained in EFAW or FAW depending on the type of injuries that might occur
	More than 50	At least one first-aider trained in FAW for every 50 employed (or part thereof)

Figure 3-15 (Above): Suggested number and type of suitable first-aid person.
Source: HSE.

Figure 3-16: First-aid kit.
Source: RMS.

Size of first-aid kit	Small	Medium	Large
Low hazard workplaces (offices, shops, etc.)	Less than 25 workers	25-100 workers	Over 100 workers, 1 large per 100 workers
Higher hazard workplaces (light engineering and assembly work, food processing, warehousing, extensive work with dangerous machinery or sharp instruments, construction, chemical processing, etc.)	Less than 5 workers	5-25 workers	Over 25 workers, 1 large per 25 workers

Figure 3-17: Suggested size of first-aid kit for different workplaces and numbers of workers.
Source: BS 8599.

PRINCIPLES OF FIRE EVACUATION

When construction activities are not adequately controlled there is often a risk of a fire. Every year construction workers are badly injured as a result of exposure to fire, in the period 2018/2019 nine workers were seriously injured and 17 sustained injuries that required them to be off work for more than seven days. Children and other members of the public could also be killed or injured when they are in adjacent properties to a construction site where a fire takes place. It is therefore vital to have adequate means of dealing with this type of emergency.

Information on the principles of fire, preventing fire and its spread, fire alarms and fire-fighting is provided in Element 11 - 'Fire'. This section covers good practice for fire evacuation.

The Regulatory Reform (Fire Safety) Order (RRFSO) 2005 requires people responsible for fire safety to make arrangements to prevent fires and provide protective measures to minimise the effects when they occur. This is a general duty and applies to all people who have a controlling responsibility, including those involved in construction activities, which could include a landlord, client, designer, principal contractor or contractor.

"11(1) The responsible person must make and give effect to such arrangements as are appropriate, having regard to the size of his undertaking and the nature of its activities, for the effective planning, organisation, control, monitoring and review of the preventive and protective measures."

Figure 3-18: Responsible person's duty to make arrangements.
Source Regulatory Reform (Fire Safety) Order (RRFSO) 2005.

CDMR 2015 supports the general duty in clause 11 of the RRFSO 2005 in regard to establishing some specific requirements for escape routes and exits for construction sites and the RRFSO 2005 provides more detailed legal requirements that all emergency routes and exits in workplaces must meet.

Means of escape

Buildings are often at their most susceptible during the construction phase of a project. Exposed and untreated timbers, the presence of flammable substances and large quantities of combustible materials significantly increase the risk from fire.

In the event of a fire, people must be able to escape from it and the principles of fire evacuation require a responsible person to consider escape routes and fire exits in order to provide a means of escape. This is supported by requirements set out in CDMR 2015.

"(1) Where necessary in the interests of the health or safety of a person on a construction site, a sufficient number of suitable emergency routes and exits must be provided to enable any person to reach a place of safety quickly in the event of danger.

(2) The matters in regulation 30(2) must be taken into account when making provision under paragraph (1).

(3) An emergency route or exit must lead as directly as possible to an identified safe area.

(4) An emergency route or exit and any traffic route giving access to it must be kept clear and free from obstruction and, where necessary, provided with emergency lighting so that it may be used at any time.

(5) Each emergency route or exit must be indicated by suitable signs."

Figure 3-19: Requirements for emergency routes and exits.
Source: CDMR 2015, Regulation 31.

Proper provision for means of escape is therefore needed for all workers and visitors on site, however transient the activity and wherever they are on site, for example workers may be on a roof or in a confined space or on a scaffold. CDM 2015 emphasises that a sufficient number of suitable escape routes and exits must be provided. This will require careful planning and should form part of the construction phase plan. During construction, escape routes are likely to change and possibly become unavailable as work progresses. It is therefore important that replacement routes are identified and provided early. The RRFSO 2005 is helpful in setting out some additional requirements that emergency routes and exits must fulfil.

"14(1) Where necessary in order to safeguard the safety of relevant persons, the responsible person must ensure that routes to emergency exits from premises and the exits themselves are kept clear at all times.

(2) The following requirements must be complied with in respect of premises where necessary (whether due to the features of the premises, the activity carried on there, any hazard present or any other relevant circumstances) in order to safeguard the safety of relevant persons:

(a) emergency routes and exits must lead as directly as possible to a place of safety;

(b) in the event of danger, it must be possible for persons to evacuate the premises as quickly and as safely as possible;

(c) the number, distribution and dimensions of emergency routes and exits must be adequate having regard to the use, equipment and dimensions of the premises and the maximum number of persons who may be present there at any one time;

(d) emergency doors must open in the direction of escape;

(e) sliding or revolving doors must not be used for exits specifically intended as emergency exits;

(f) emergency doors must not be so locked or fastened that they cannot be easily and immediately opened by any person who may require to use them in an emergency;

(g) emergency routes and exits must be indicated by signs; and

(h) emergency routes and exits requiring illumination must be provided with emergency lighting of adequate intensity in the case of failure of their normal lighting."

Figure 3-20: Duties for emergency routes and exits.
Source: Regulatory Reform (Fire Safety) Order (RRFSO) 2005.

A basic principle of escape routes is that any person confronted by an outbreak of fire, or the effects of it, can turn away from it or pass it safely to reach a place of safety. The risk of being trapped in a 'dead-end' of a building should be minimised and escape routes should avoid workers having to go through areas where flammable substances are stored.

Escape routes on a construction site should lead as directly as possible to a place of safety, which, depending on the nature of the construction site, could present some challenges. Materials and features of the structure being constructed may obstruct or extend the route that must be taken to escape. In a situation where a passenger hoist is used to reach various workplaces consideration must be given to how workers will make their way down to a place of safety during a fire, without using the hoist.

The requirement that escape routes should lead as directly as possible to a place of safety makes it important to consider travel distance.

Travel distances

Travel distance is the distance from where someone is working to a place of safety – a place in the open air, at ground level and away from the building (for example, at an assembly point). The distance includes travel around obstructions in the workplace and may be greatly affected by any work in progress on a construction site.

If a worker is outside on a scaffold it is unlikely to be considered as a place of safety and the travel distance would be taken as that distance to reach the ground away from the building.

The travel distance should be kept to a minimum and there should normally be alternative routes leading in different directions from a workplace to assist with keeping travel distance to a minimum.

Figure 3-21: Fire escape - hazard of falling on exit due to height from ground. Source: RMS.

	Fire hazard		
	Lower	Normal	Higher
Enclosed structures			
Alternative	60m	45m	25m
Dead-end	18m	18m	12m
Semi-open structures			
Alternative	200m	100m	60m
Dead-end	25m	18m	12m

Figure 3-22: Maximum travel distances.
Source: HSE, HSG168.

Notes: Semi-open structures are completed or partially constructed structures in which there are substantial openings in the roof or external walls, which would allow smoke and heat from any fire to readily disperse, and which are not at risk of exposure from radiation or direct impingement from a fire on the site.

Stairs

In most cases it is preferable to use escape routes that are part of the structure, for example permanent stairways that are constructed from fire resisting materials and protected by fire doors. Where this is not possible, external temporary escape stairways may be required. Stairways can be constructed from scaffolding or system stairway equipment. Where possible, ladders should not be used as an escape route, stairway systems are more preferable. Escape routes must not be obstructed. Doors must not be wedged open and exit doors should need to open outwards, in the direction of travel.

Emergency lighting

Emergency lighting should be considered if escape is likely to be required in dark conditions. This could mean late afternoon in winter time, not just at night time.

Exit and directional signs

Fire escape signs are provided to guide escape from wherever people are, via a place of relative safety (the escape route) to the place of safety (the assembly point).

It should be possible for someone to see the signs that indicate the fire escape route from where they work on site, if this is not obvious. Care needs to be taken to provide signs for complicated escape routes and alternative escape routes. As the layout of the site changes it may be necessary to revise where signs are positioned and the direction to exits. In some cases, it may be necessary to provide signs that confirm routes that are not suitable as an escape route to avoid confusion.

See also 'Element 2.4 - Assessing risk' - Safety signs.

Figure 3-23: Fire escape sign.
Source: RMS.

Assembly points

The assembly point is a place of safety where people wait whilst any incident is investigated, and where confirmation can be made that everyone has evacuated. The main factors to consider are:

- Safe distance from buildings.
- Sited in a safe position.
- Not sited so that workers will be in the way of the fire and rescue service.
- Must be able to walk away from assembly point and back to a public road.
- Clearly signed.
- More than one provided to suit numbers and groups of people.
- Communications should be provided between assembly points.
- Measures provided to decide if evacuation was successful.
- Person must be in charge of assembly point and identified.
- Person to meet/brief the fire and rescue service.

Figure 3-24: Assembly point.
Source: RMS.

Emergency evacuation procedures

Article 15 of the RRFSO 2005 sets out requirements for the responsible person to establish emergency evacuation procedures.

> "Establish and, where necessary, give effect to appropriate procedures, including safety drills, to be followed in the event of serious and imminent danger to relevant persons."

Figure 3-25: Article 15 of RRFSO 2005.
Source: The Regulatory Reform (Fire Safety) Order (RRFSO) 2005.

Furthermore, it sets out a requirement to nominate a sufficient number of competent persons to implement evacuation procedures. The danger that may threaten people if a fire emergency occurs at work depends on many different factors. Consequently it is not possible to construct one model emergency evacuation procedure for action in the event of a fire. However, fire evacuation procedures need to reflect the work activities, the people affected and the premises involved. Many of the different issues to consider for different situations have common factors, for example, evacuation needs to be done in an efficient/effective manner, an agreed assembly location (which may be different for different workers) provided and checks made to ensure people are safe.

Emergency evacuation procedures should be in writing and regularly tested through drills and exercises. Quick and effective action may help to ease an emergency situation and reduce the consequences.

Fire instruction notices should be located in conspicuous positions in all parts of the location and adjacent to all fire alarm actuating (call) points. The notices should state, in concise terms, the essentials of the action to be taken upon discovering a fire, including the fire emergency evacuation procedure, and on hearing the fire alarm.

The emergency evacuation procedures should also be subject to regular review to determine if any new factors are affecting them. For example, fire emergency procedures may need regular review throughout a major construction project as construction work activities can create changes that could affect exit routes and assembly points, requiring them to be redefined and those individuals affected by the changes to be re-trained.

Role and appointment of fire marshals/ wardens

A person should be nominated to be responsible for co-ordinating the fire evacuation plan on each construction site. This may be the same person that organises fire instruction/training and drills and co-ordinates the evacuation at the time of a fire. They could appoint other people to assist them in fulfilling the role, if required.

For example, on larger sites or higher fire risk sites, the appointment of fire marshal/wardens may be appropriate to assist with the safe evacuation of the site.

The appointment of fire marshals/wardens should be made known to workers and they should be clearly identifiable at the time of an emergency so that those that are asked to evacuate understand the authority of the person requiring them to do so. It is important that when fire marshals/wardens are appointed, they are trained and given the necessary authority to carry out their tasks.

The way in which fire marshals/wardens assist will vary depending on the construction project and the preferences of those managing the site. For example, in the event of a fire some fire marshals might check assigned areas to ensure everyone has evacuated while other fire marshals lead the evacuation to show where to go.

Fire marshals/wardens will also liaise with the nominated person on the site with overall responsibility for fire evacuation to confirm the status of their part of the evacuation. In turn, the nominated person will liaise with the fire and rescue service and provide information on access, people trapped and any special hazards.

Fire drills

Workers and managers should receive training on what the emergency evacuation procedure requires. Training will provide them with an understanding of what the alarm means to them and the action they must take. It also increases the chance that they will take the alarm seriously and take action without delay. This will help them to act in a calm, orderly and efficient way if a fire occurs.

Those people designated with specific duties would need specific training in what these duties are and how to carry them out. In most cases this will require practical training, for example as a fire marshal/warden checking an area or a person using equipment to minimise the effects of the emergency.

In an emergency people are more likely to respond reliably if they:

- Have clearly agreed responsibilities.
- Are well trained and competent.
- Take part in regular and realistic practice (drills and exercises) to rehearse actions.

The results of the drills and exercises should be recorded and the procedures amended as necessary to make them effective.

Fire drills reinforce the knowledge provided in training, check the current understanding of all site workers/ visitors and how they act in the event of a fire.

They establish the evacuation actions as a practiced routine, making this familiar and more likely to be repeated without hesitation or doubt in the event of a real fire.

They will also identify practical problems with escape routes, such as 'bottlenecks' etc. Fire drills can also identify what works well in the evacuation procedure.

Fire drills, where the entire workforce evacuates the site, are a useful means of checking that the fire evacuation procedure works effective. However, due to the changing nature of constructions sites and the workforces on them it is recognised these can often be impracticable and of limited use.

However, as the risks of, and from, fire increase and the number of people on site rises, the need for a fire drill increases, especially when the main structure of the building is complete. The interval at which fire drills take place on a construction site will depend on the complexity of the site, the hazards and the evolving nature of the site. Records should be kept of all fire evacuations, including fire drills.

Provision for people with disabilities

When planning fire evacuation procedures responsible persons need to consider who may be in the workplace, their abilities and capabilities. Any disability for example, hearing, vision, mental or mobility impairment must be catered for. Some of the arrangements may be to provide the person with a nominated assistant(s) to support their escape, for example, with the use of a specially designed evacuation chair to enable them to make their way down emergency exit stairs and out of a building.

Figure 3-26: Fire refuge sign for disabled people.
Source: Safety Selector.

Part of the provisions for people with disabilities is to make sure they are capable of knowing that an emergency exists. This may mean providing them with special alarm arrangements that cater for their disability, for example, a visual and or vibrating alert for the hearing impaired.

In some cases, disabled people may need to use a refuge area, a relatively safe waiting area for short periods. A refuge area is separated from the fire by a fire-resisting structure and has access via a safe route to a fire exit. It provides a temporary space for disabled people to wait for others who will help them to evacuate. Some buildings may be equipped with an evacuation lift, which has been specifically designed within a fire-resisting enclosure and having a separate power supply.

To ensure the safety of all persons on site, fire marshals/wardens and others need to be kept up to date with changes to the site, including the location and number of people on site. This includes any vulnerable workers, including those who may have difficulty escaping unaided or may not hear/see an emergency alarm. This includes tower crane operators and workers in loan working situations. Adequate provision must be made to ensure everyone can escape safely in the event of a fire.

SUITABLE EMERGENCY ARRANGEMENTS WHEN WORKING NEAR WATER

CDMR 2015, Part 4: Duties Relating to Health and Safety on Construction Sites, Regulation 26, states that:

"(1) Where, in the course of construction work, a person is at risk of falling into water or other liquid with a risk of drowning, suitable and sufficient steps must be taken to -

(a)   prevent, so far as is reasonably practicable, the person falling;

(b)   minimise the risk of drowning in the event of a fall; and

(c)   ensure that suitable rescue equipment is provided, maintained and, when necessary, used so that a person may be promptly rescued in the event of a fall.

(2) Suitable and sufficient steps must be taken to ensure the safe transport of any person conveyed by water to or from a place of work.

(3) Any vessel used to convey any person by water to or from a place of work must not be overcrowded or overloaded."

Figure 3-27: Prevention of drowning.
Source: CDM 2015 Regulation 26.

This means that prior to work above water being undertaken, a full risk assessment must be made that takes into account the conditions of the water being worked over or near (tidal, depth, temperature, fast flowing) and suitable control measures are identified and implemented, including emergency rescue plans and procedures.

It is important that emergency arrangements and procedures ensure that if a worker falls into water that:

- The worker is kept afloat.
- Their location is continuously tracked.
- Rescue is achieved as quickly as possible.

In order to assist emergency and rescue personal buoyancy equipment must be worn by anyone working on or near water and at risk of falling in. Personal buoyancy equipment includes personal buoyancy aids and lifejackets. A lifejacket is preferred as it will support an unconscious person in a 'face upright flotation position', therefore reducing the likelihood of the wearer drowning. A buoyancy aid will only provide sufficient buoyancy to keep a conscious person afloat in a reasonable flotation position.

In fast flowing water channels that have no water-borne traffic, 'safety nets' or 'safety lines' may be stretched across the width of the channel to allow the person in the water to hang on and wait for rescue. This is only effective if the person in the water remains conscious.

Rescue teams (including first aiders) of suitable capability should be organised to deal with emergency situations. An effective communication system should be established between workers, supervisors observing the workers and the rescue team in case of emergency. Members of the rescue team should receive training, which needs to cover keeping the person in the water afloat, how to make their location known (especially in flowing water) and how to safely effect a rescue.

Methods of rescuing a worker who has fallen in water include using a rescue boat, reaching out from the bank, throwing out a means of personal buoyancy with a line (lifebuoy or personal buoyancy aid) and wading or swimming out. In addition, there remains the option to contact emergency services to rescue the worker.

The most appropriate method to rescue a worker will depend on the specific circumstances of the situation, how deep the water is, how close they are to the side, the temperature of the water and whether the water is moving and how fast. Whichever method is used it should not put the rescuer at unnecessary risk, such that they also need to be rescued.

Figure 3-28: Working over water.
Source: Blueprint Water Safety.

The majority of drownings occur close to the bank or water's edge, and this should be taken into account at the planning stage. It may be possible to reach out to the person from the bank or edge, but care needs to be taken to ensure the rescuer has a secure foothold and is sufficiently balanced to not overbalance and fall in the water when attempting to pull the person to safety.

Similarly, a rescuer may be able to get sufficiently close to a person by wading out, but this is only possible if there is a firm shallow water's edge. This method could easily result in the rescuer becoming out of their depth if the depth of water suddenly changes, particularly if this is not visible to them. Where possible, it is preferable to use proprietary equipment designed to be thrown from a safe position to the person in the water. This equipment would have a line attached to help with the rescue process without the need to enter the water.

Proprietary equipment of this type that can be used to assist someone who has fallen in water and remains near the edge are life buoys and personal buoyancy aids, with rescue lines. Life buoys are normally attached to approximately 30 metres of rescue line, but can only be thrown a short distance of 6 8 metres. Personal buoyancy aids are available in various forms with rescue lines ranging from 25-40 metres in length. They work by throwing a bag or capsule to the person in the water and the line deploying as the bag/capsule is thrown out. The line, bag and capsule all stay afloat and allow the person in the water to grab the line and be pulled to safety. Care must be taken that the person deploying the line is secure at the bank and unable to be pulled into the water during the rescue.

Where construction work is being carried out on fast flowing, or tidal waters then a rescue boat should be made available. This must have a reliable engine, carry oars, and be fitted with grab lines for people who have fallen into the water. If any work is carried out in the hours of darkness, then the boat will be required to be fitted with high efficiency lighting. It may be a requirement to have two-way radio installed on the boat for communication between boat and shore. The person that is operating the boat must be competent and experienced at handling small boats on flowing water and in most cases two people will be required to lift a person into the boat and operate it safely.

CONTINUAL REVIEW OF EMERGENCY PROCEDURES AS A BUILD CONTINUES

It is a requirement of Regulation 5 of the Management of Health and Safety at Work Regulations (MHSAWR) 1999 to review arrangements made for health and safety. For any emergency procedures to work well, all workers and managers must be aware of the procedures and be able to test them. The emergency procedures should be evaluated and modified taking into account the changing nature of the site, the activities being undertaken and the size of the site.

Specifically, there is a need to actively review and revise the fire evacuation arrangements that relate to means of escape. This is vital on construction sites as the layout may change as construction activities progress. Depending on the rate of construction, fire means of escape may need to be reviewed on a weekly or even daily basis. Checks and assessment should be made to ensure that the means of escape are still appropriate and that they have not being compromised:

- Travel distances remain within limits.

- Egress via stairs and passageways is achievable without restriction.

- Doors have not been filled in, moved or become unusable.

- Exit and directional signs still relate to the actual escape route.

- Emergency lighting has not become obscured by processes that could contaminate the surface or screened by new structures.

- Assembly points remain accessible and in place, for example, not built over.

INCLUSION OF TAU WITHIN THE EMERGENCY PLAN

A fire risk assessment should be completed to consider the specific circumstances and needs of temporary accommodation units (TAUs) used on site and what arrangements are required with regard to emergencies. It is essential that they are then included in the emergency plan for the site.

The emergency plan should consider where the TAUs are located relative to the construction work. TAUs should be located at a suitable distance away from the construction work, proportionate to the size of the site and risk of fire. As a general principle separation of a minimum of 6m should be established, which should be increased to 20m on higher risk sites, such as those constructing timber frame buildings. If TAUs must be located closer, the risk of a fire spreading is increased therefore the fire resistance of the TAU should be assured and at least 30 minutes fire resistance provided. In no circumstances should TAUs be sited within high fire risk structures such as timber frame.

TAU complexes can be assembled in many different combinations to perform a range of functions and as the complex increase in size and complexity, careful consideration needs to be given to ensuring that emergency arrangements reflect this. TAUs should have at least two means of escape, if escape is only possible in one direction, the escape route must be adequately fire protected and a short travel distance assured. Sufficient escape stairs should be provided and protected for multiple level TAUs and travel distances to a place of safety must also be considered.

A TAU complex may have a separate fire alarm system from the areas where construction work is taking place, this must be taken into account in the emergency plan for the site. For simple TAU complexes of one or two units, a simple fire alarm in the form of a hand-operated bell might be sufficient to provide an alarm throughout the complex. In larger, more complicated TAU complexes and those that include arrangements for cooking an electrical break-glass system and multiple call points may be required.

The emergency plan for the site must take account of the specific circumstance of the TAU complex. In some situations the location of the TAUs away from construction activities may mean that they do not need to be evacuated if a fire is only affecting the construction activities. The complex may also be on a different fire alarm system and have a different assembly point. It will then be necessary for those in control of evacuation of the construction area to liaise with the person in control of the TAU to decide on appropriate action when the fire starts and if circumstances change. Where the TAU is located within a building under construction the fire alarm for the TAU should be integrated into the rest of the building.

Where the TAU complex provides sleeping accommodation additional alarm and evacuation measures should be made appropriate to its use. These TAUs would usually be sited in an area of lower risk of fire, away from construction activities and have their own fire arrangements.

CONSIDER

What types of emergency could affect your organisation?

Who would need to be contacted if these emergencies occurred?

3.5 Learning from incidents

Employers have a responsibility to prevent harm to workers and learn from incidents that could have caused harm. Therefore, where an incident causes harm there is a need to investigate the causes and prevent a recurrence. In addition, the study by Frank Bird confirmed that in general organisations have more incidents that do not result in harm to people than those that do, this is illustrated by the 'incident ratio study' shown in *Figure 3-29*.

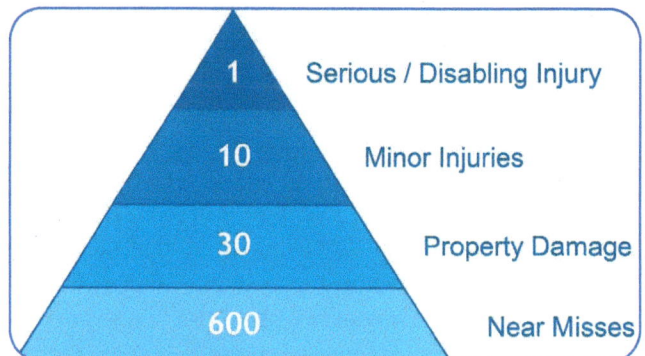

Figure 3-29: Incident ratio study.
Source: Frank Bird.

Many of the incidents that do not result in harm have causes that in slightly different circumstances could result in harm to people, which emphasises the importance of investigation of all incidents.

The investigation of 'near-miss' incidents and the identification of their underlying causes might allow preventive action to be taken before something more serious occurs. It also gives the right message that all failures are taken seriously by the employer and not just those that lead to injury.

Figure 3-30 demonstrates the difference between an accident, near-miss and unsafe conditions.

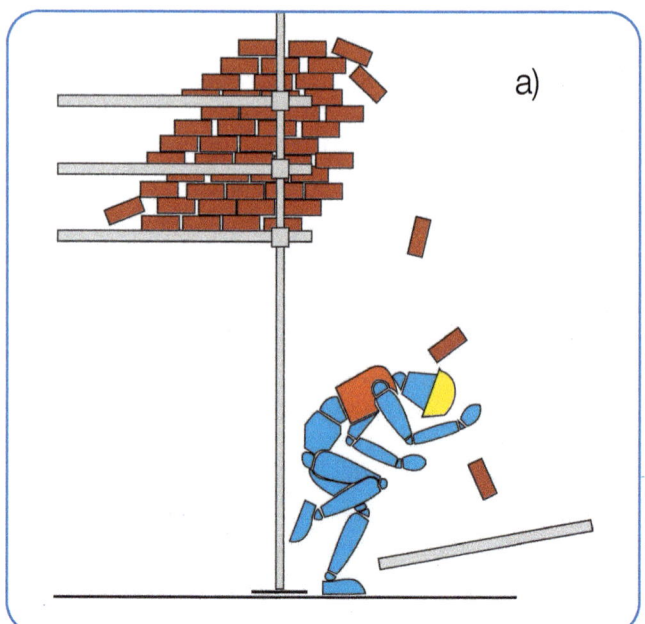

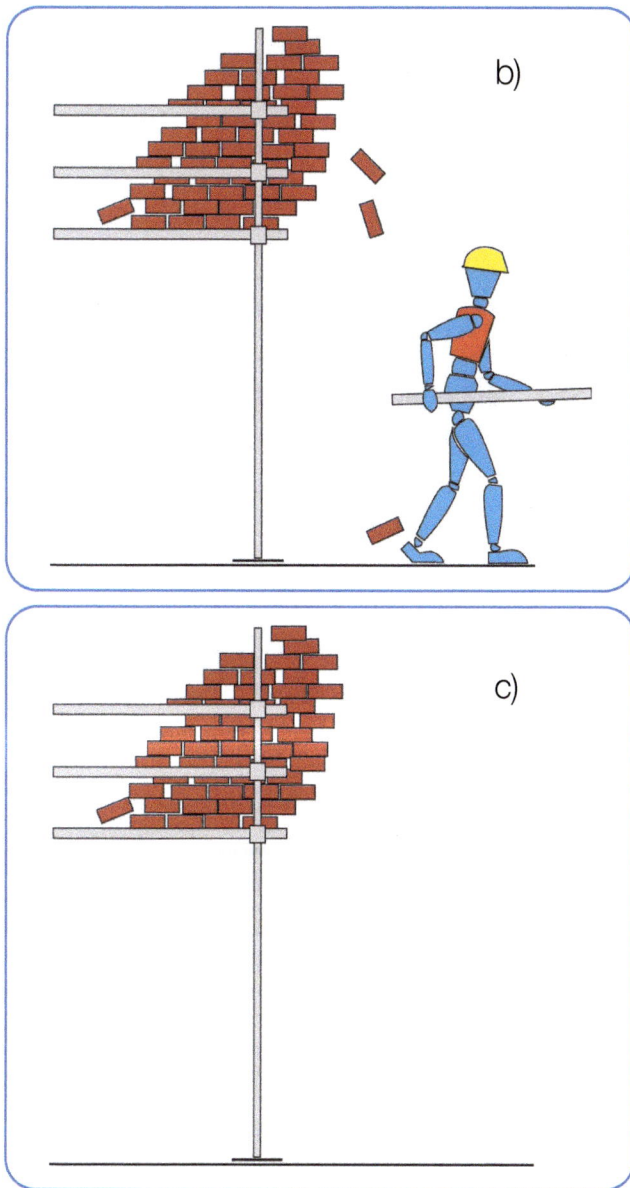

Figure 3-30: (a) Accident (b) Near-miss (c) Unsafe conditions.
Source: UK HSE, HSG45.

It is important that lessons learned from incidents are shared with as many people who would benefit from it as is possible. As a minimum this must include different departments within an organisation and with different contractors that are onsite. It would be regrettable if an incident happening in one part of the construction site was seen as of no relevance to others, without considering the causes. The more that the root causes are examined the more likely the lessons are to be relevant to other contractors. For example, examination of the causes of an accident/incident involving one contractor may reveal a need to improve the site induction processes, which may affect all contractors on site.

The role of incident investigation includes to:

- Learn from what happened.
- Discover underlying and root causes.
- Prevent a recurrence.
- Establish legal liability, prepare for legal action and ensure legal obligations are complied with.

- Provide enforcement agency/insurance/worker compensation data.
- Determine compliance with the organisation's policy processes and procedures.
- Identify weaknesses in health and safety systems, arrangements and procedures so that standards can be improved.
- Determine the economic loss caused by the incident.
- Gather data to produce statistics.
- Identify trends.
- Update risk assessments.
- Demonstrate the commitment of management to provide a safe and healthy place of work.
- Establish if internal disciplinary procedures are necessary.
- Support worker morale, not carrying out an incident investigation will have a negative effect on the health and safety culture as workers will assume the organisation does not value their health and safety.

In addition, there is a specific legal requirement to report accidents, dangerous occurrences and ill-health to enforcing authorities and contractual obligations to report details to insurance organisations. The information needed for these reports will cause the employer to investigate the circumstances leading up to an incident.

The role and directive of any investigation of this nature should never seek to blame any individual or group of individuals. If human error is believed to be a significant cause, the reasons for this must be investigated. Lack of knowledge, training or unsuitability for the job may be the causes of this error. These are management and not worker failings. Only when these have been evaluated can the conclusion of wilful and intentional acts or omissions be considered.

THE DIFFERENT LEVELS OF INVESTIGATION

Ideally all incidents should be investigated, but the level of investigation may vary depending on the circumstances of the incident. The level of investigation should depend on the severity of actual or potential loss resulting from the incident, whichever is the greater. For example, a scaffold collapse may not have caused any injuries, but had the potential to cause major or fatal injuries. Therefore the investigation should, in part, be based on the worst possible case of harm which is reasonably foreseeable as a result of the incident in question. It should also consider how likely a similar incident is to occur. In this way the risk related to the incident is considered.

The following table, detailed in the HSE guidance document HSG 245: 'Investigating accidents and incidents', will assist you in determining the level of investigation appropriate for the incident.

Risk	Level of investigation	
Minimal	Minimal level	In a minimal-level investigation, the relevant supervisor will look into the circumstances of the incident and try to learn any lessons which will prevent future occurrences.
Low	Low level	A low-level investigation will involve a short investigation by the relevant supervisor or line manager into the circumstances and immediate, underlying and root causes of the incident, to try to prevent a recurrence and to learn any general lessons.
Medium	Medium level	A medium-level investigation will involve a more detailed investigation by the relevant line manager, with the support of the health and safety practitioner and, where appropriate, worker representative and will look for the immediate, underlying and root causes.
High	High level	A high-level investigation will involve a team-based investigation, typically involving a senior manager, line manager, health and safety practitioner, technical specialist and, where appropriate, worker representative. It will be carried out under the supervision of top management/ directors and will look for the immediate, underlying, root causes and lessons that can be learned that can be applied across the organisation and other locations, if applicable.

Figure 3-31: Level of incident investigation.
Source: HSE HSG245: Investigating accidents and incidents.

Minimal and low-level investigations by the Supervisor

The supervisor as the person in immediate operating control of an area or activity is a logical person to gather information on all accidents that happen in their area of responsibility. It often will result in swift identification of the causes and remedial actions being implemented, and underlines the supervisor's responsibility for health and safety on a day to day basis. This investigation is normally all that is necessary for the majority of incidents.

Medium-level investigations by the Manager

Organisations that encourage management involvement identify incidents that warrant further investigation by an appropriate manager. Where there is a medium level risk a manager, independent of the line managers where the incident occurred, may be required to conduct a more detailed investigation. They would take advice, including from the health and safety practitioner, as appropriate.

Managers who receive reports of the investigation of minimal and low risk incidents conducted by supervisors may decide that a more detailed investigation is required. This may be in order to gain a better understanding of the root causes of the incident, something that may be outside the perspective of the supervisor.

High-level investigation

Where an incident has caused serious harm, or there has been the potential for serious harm to have occurred, for example, a near miss with high potential for harm, it will

be necessary to carry out a high-level investigation using a team. Any person whose responsibilities or actions may have been involved in the incident being investigated, for example, a supervisor, should be excluded from the investigation team, but would be valuable as a witness. A typical high-level investigation team would usually include the following people:

- A senior manager from another department to act as an independent chairperson.
- A health and safety practitioner to advise on specific health and safety issues.
- A health and safety practitioner to provide management experience and perspective.
- An engineer or technical expert to provide any technical information required.
- A worker health and safety representative to provide worker experience and perspective.

Health and safety practitioner monitoring and investigation of incidents

The organisation's health and safety practitioner should monitor and evaluate the incidents reported and help to decide the level of investigation required. They will support medium and high-level investigations and monitor lower level investigations to determine if the investigation has been effective. Where necessary lower level incidents may be investigated by the health and safety practitioner to a greater depth than the initial investigation. Ill-health events may require involvement of other specialists with a medical or environmental workplace assessment background.

BASIC INCIDENT INVESTIGATION

Basic incident investigation steps

When an incident occurs prompt emergency action should be taken, for example provision of first aid, making the area where the incident happened safe and preserving the scene of the incident to enable it to be investigated. If it is necessary to disturb the site, the employer should arrange for a competent person to make a record of the site of the incident, including photographs, plans and the identification of witnesses, before intervention.

Throughout the process of investigation, it must be clearly borne in mind that the objective is to prevent a recurrence of the incident, not to apportion blame. It is important to identify the true causes of the incident, not just the superficial ones. This cannot be achieved without the full commitment and assistance of witnesses and other persons who work in the area in which the incident happened. It follows that recommendations must be put into action, even though they may take a considerable amount of time, trouble and money.

Guidance on the approach to be taken when investigating incidents is provided by the HSE in the document HSG245 'Investigating Accidents and Incidents'.

It suggests a step-by-step approach to investigations:

Step 1: Gathering the information - the where, when, who, what and how of the incident. The information gathered will include results of interviews, photographs of the equipment involved and the area in which it was positioned at the time, sketches of the workplace layout, weather conditions, etc.

Step 2: Analysing the information - determine what has happened and analyse the information to identify the immediate, underlying and root causes. At this stage it should be considered if human error is a contributory factor. Job factors, human factors and organisational factors can all influence human behaviour and will all need to be considered in the analysis.

Step 3: Identifying suitable risk control measures. Possible solutions can be identified. This will involve looking at the technical, procedural and behavioural controls, with the technical or engineering risk control measures being more reliable than those that rely on human behaviour.

Step 4: The action plan and its implementation to identify which risk control measures should be implemented in the short and long-term. The risk control action plan should have SMART objectives, i.e. Specific, Measurable, Achievable and Reasonable, with Timescales. This will also state which risk assessments need to be reviewed and which procedures need to be updated; any trends that need further investigation; and the cost of the incident.

HSG245 contains useful investigation forms that could be used or developed further before use by organisations. It also has examples of simple, but effective investigation tools such as the accident/incident investigation tree.

Step 1 Gathering information

The step to gather information should explore all reasonable lines of enquiry, be timely and structured – determining clearly what is known, what is not known and starting a record of the investigation process. This step focuses on determining where and when the incident took place, what the result of the incident was (for example, injury or ill health), who was involved (including witnesses) and how it occurred (the events that led up to the incident).

It is important to gather information as soon as possible after the incident. This helps to maintain its integrity and prevents confusion, for example, items at the scene of the incident might get moved, control measures not used might be put in place (guards replaced, personal protective equipment worn) and witnesses may discuss what they believe they saw and cause confusion. If necessary, work must stop and unauthorised access to the site, witnesses and other potential evidence prevented.

Gathering information includes:

- Physical information - from the scene of the incident, for example photographs, closed circuit television (CCTV) footage, making a plan of the area and noting details of any equipment involved.

- Verbal information – from witness statements, available recordings of instructions and conversations.

- Documented information – for example, risk assessments, policies, procedures, work instructions permits-to-work, and training records.

The amount of time and effort spent on information gathering should be proportionate to the level of investigation. Collect all available and relevant information. That includes opinions, experiences, observations, sketches, measurements, photographs, check sheets, permits-to-work and details of the environmental conditions at the time etc. This information can be recorded initially in note form, with a formal report being completed later. These notes should be kept at least until the investigation is complete.

Interviews

Interviews are an important part of the investigation process and can quickly provide information on the incident and what lead up to it, providing further investigation trails. However, the witnesses may be distressed, very defensive and could feel that they might be blamed, so it is important to put the person being interviewed at ease, emphasising that the purpose of the interview is to help determine the facts to prevent a recurrence.

Good interview techniques will include:

- Interview of witnesses promptly after the incident, to avoid lapse of memory or distortion through witnesses debating what occurred.

- Conducting the interview in private one person at a time. Free from distractions and interruptions.

- Explanation of the purpose of the interview i.e. to prevent recurrence. The interviewer should explain what is to happen during and after the interview process.

- An explanation that notes will be taken of their responses during the interview.

- The use of open questions, such as, 'what were you doing before the incident occurred?'. Open, broad questions aid interviewee reflection, generally put them at ease and make them more responsive to the interview process.

- The use of closed questions, ones that can be answered with a single word or short phrase can be used to confirm a specific point, such as 'did the incident happen as soon as soon as you let go of the ladder?'

- The use of a sketch plan and photographs by the interviewer to assist in developing a good understanding of what happened and assist in determining root causes of the incident. Sketch plans of the workplace and the site of the incident can be used during interviews to provide a clear indication of the incident scene including the position of any injured person, witnesses, plant and equipment. Cameras can be used to record and preserve images of incident scenes or resultant injuries. This can be especially useful if the situation changes, through such things as corrective actions being put in place or changes in environmental conditions, while the investigation is continuing.

- Care to keep the person being interviewed at ease. Avoid complex language or jargon. It is especially important to observe the witnesses' body language to ensure that they are listening, giving full attention to the questions and remain at ease with the interview. The interviewer should maintain an even voice tone and manage their responses to answers in a calm and neutral manner.

- Recording the details: names of the interviewers, interviewee and anyone accompanying interviewees; place, data and time of the interview; and any significant comments or actions during the interview.

- Summarising the interviewer's understanding of the information provided by the witness to ensure there is a mutual understanding gained from the interview. If necessary and appropriate, ask the witness to write and sign a statement of truth to create a record of their testimony.

- Expressing appreciation for the information given by the witnesses. Thank them for their help, explain may need to discuss further at a later stage.

🔍 REVIEW

List five good practice techniques for conducting an effective interview of witnesses following an accident.

What are the main reasons for investigating accidents?

List the categories of staff who may be team members in an accident investigation team for a major incident.

Step 2 Analysing information

The step of analysing information should be objective, unbiased and evidence based. The analysis should identify the sequence of events and adverse conditions leading to the incident, identify immediate, underlying and root causes, determine common themes from interviews, record findings.

An analysis of the information gathered involves examining all the facts, determining what happened and why. All the detailed information gathered should be assembled and analysed to identify what information is relevant and what information is missing. The information gathering and analysis are usually carried out together, as a progressive process when further lines of enquiry requiring additional information develop.

To be thorough and free from bias, the analysis must be carried out in a systematic way, so that all the possible causes and consequences of the incident are fully considered.

Identifying immediate causes

Immediate cause: the most obvious reason why an incident happened, this is usually due to workplace conditions and the actions of people, for example:

Workplace condition – for example, an electrical cable on the floor (trip hazard), a guard missing from a machine, poor light level or a slippery floor.

Actions – for example, overreaching on a ladder, driving too fast, not looking in the direction of walking or not lifting a load properly.

There may be several immediate causes identified in any one incident. These workplace conditions and actions result in workers being exposed to uncontrolled hazards that present a risk of injury or ill health.

Underlying causes

Underlying cause: these are the causes that led to the immediate causes and relate to the 'organisational', 'job' and 'individual' reason for an incident happening, for example; the hazard has not been adequately considered via a suitable and sufficient risk assessment, inadequate time allocated for a task, poor maintenance of equipment, wrong or no equipment/materials, absence of training, unclear procedures/instructions, unclear responsibilities, absence of supervision, poor co-ordination of work activities, no inspection of equipment, poor motivation of worker.

For example, the underlying causes of a worker overreaching on a ladder could be that the ladder was placed too far away from where it was needed, the ladder was too short, an obstruction at the base of the ladder prevented it being placed nearer, the worker did not understand the risk of overreaching, the work activity forced the worker to overreach, the worker did not have time to reposition the ladder, the worker was rushing to complete a job, the worker decided it was easier to overreach than reposition the ladder.

Root causes

Root cause: these are the causes that enabled or led to the underlying causes and relate to management failings for example, failure to identify training needs and assess competence, low priority given to risk assessment, poor selection of equipment, inadequate provision of equipment, poor planning of work activities, pressure to complete work quickly, other priorities put above health and safety, poor leadership, poor prioritisation, inadequate resources, failure to maintain the workplace and equipment, poor management of change, ill-defined or inadequate standards, poor communication, inadequate worker engagement, failure to implement risk control measures, failure to monitor health and safety.

These *management system failings* occur when the organisation does not establish an adequate health and safety policy, organisation, planning, support, operation, performance evaluation, improvement where an appropriate approach to health and safety risk identification and control for the organisation's activities is established.

The analysis of incident information often show that an incident can be the result of many causes.

Examples:

A worker slipped on a patch of oil on a warehouse floor, was admitted to hospital with an injury and remained there for several days. The oil was found close to a stack of pallets that had been left abandoned on the pedestrian walkway.

Immediate cause:
- Oil that had leaked onto the floor from equipment.
- The floor remaining in a slippery condition because the spillage was not cleaned up.
- The abandoned pallets blocking the walkway causing the injured worker to make a detour.
- Inadequate lighting at the scene of the incident.
- The worker who slipped was wearing unsuitable footwear and not looking where they were walking.

Underlying cause:
- No materials available to clean up oil leak.
- Leak left for someone else to clean up.
- Oil on floor not considered to be a hazard by workers in the area.
- No time available for workers to clean up oil.
- No clear storage area for pallets.
- Lighting units not working.
- Worker who slipped was late and rushing.
- Lack of supervision.
- Walkway poorly identified.

Root cause:
- The absence of adequate risk assessments and safe systems of work.
- Failure to introduce procedures for routine maintenance of equipment and cleaning up leaks/spillages.
- Poor warehouse design with inadequate walkways.
- Failure by management to monitor working conditions in the warehouse to ensure workers are not exposed to risks to their health and safety.
- Failure to define responsibilities.
- Failure to make workers aware of hazards and consequences of not taking action to control them.
- Inadequate training or instruction of workers in those procedures that might have been introduced.
- A range of personal factors, for example, stress, fatigue, influence of drugs and alcohol.

Step 3 Identifying risk control measures

After identifying the causes, risk control measures that effectively remove the root causes need to be identified. The information analysis stage may have identified a number of risk control measures that either failed or could have prevented the incident, if they had been in place. This should indicate possible solutions that should prevent a recurrence of the incident. This could include the implementation of additional risk control measures to prevent risk controls failing and measures that were identified to be missing.

These possible solutions need to be systematically evaluated and best solutions should be considered for implementation. In deciding which risk control measures to recommend and their priority, you should choose measures based on the risk control hierarchy. This means, where possible, measures that eliminate risk should be given priority followed by measures that reduce the risk at source. Generally, measures that rely on human intervention are less effective than engineering controls that require no human input. Each risk control measure has to be evaluated in its own right to assess its effectiveness in preventing a recurrence and if they can be successfully implemented. The fact that some measures might be more difficult and/or expensive to implement should not be a barrier to them being considered.

When identifying if the risk control measures in place to prevent incidents are appropriate ensure:

- Missing/inadequate/unused controls are identified.
- Compare conditions/practices as they were with that required by current legal requirements, codes of practice and guidance.
- Risk assessments are reviewed.
- Identify additional control measures (application of hierarchy of control).
- Provide meaningful and realistic recommendations based on the outcomes of the investigation.

Recommended risk control measures should include both short and long-term controls measures. Short-term control measures may be necessary to help to prevent a recurrence until longer-term control measures can be implemented to remove the root causes.

Ensure recommendations are meaningful and reflect the causes identified. If the only identified risk control measure recommended is 'workers need to take more care' this will indicate that the investigation has not been thorough enough and failed to identify root causes.

Reports

The report should include a summarised version of the facts and recommendations for risk control measures, together with discussion of controversial points and if necessary, appendices containing specialist reports (medical and technical), photographs and diagrams. This virtually finishes the work of the investigator, but management is still responsible for seeing that the necessary corrective actions are implemented and monitored to ensure that the causes are satisfactorily controlled. The line manager, health and safety practitioner and health and safety committee/members should monitor the actions regarding risk control measures to ensure they are completed and effective. The information included in the investigation report should cover:

- Date, time and location of the incident.
- Personal details of the injured parties (name, role, work history, injuries sustained).
- Description of the activity being carried out at the time.
- Immediate, underlying and root causes.
- Drawings and photographs used to aid understanding of the scene.
- Identification of any breaches of law.
- Details of witnesses and witness statements.
- Recommended risk control measures that are prioritised and with estimated costs.
- The cost of the incident to the organisation.

Step 4 Action plan and its implementation

If several risk control measures are required, they should be carefully prioritised and put in the form of a risk control action plan, which sets out what needs to be done, when and by whom. In deciding priorities, the level of risk and the ability of each risk control measure to limit the risk should be considered. It may be necessary to implement some less effective short-term risk control measures while better longer term risk control measures are being developed and implemented.

It is important to ensure that the action plan deals effectively with the immediate, underlying and root causes of the incident investigated. It should also include actions related to any more general lessons that may be applied to prevent other incidents, for example assessment of competencies may be needed for other activities.

Senior managers who have the authority to influence performance must be responsible for creating an action plan based on the SMART (specific, measurable, achievable, reasonable, and time bound) principles. Responsibility for specific actions should be assigned to ensure the timetable for implementation is carried out. A senior manager should be appointed to be overall responsible for implementation of the action plan.

The results of the investigation, the action plan and arrangements to ensure the action plan is implemented and progress monitored should be communicate to everyone who needs to know.

Progress on the action plan should be regularly reviewed. Any significant departures from the plan should be explained and risk control measure rescheduled, if appropriate. Workers and their representatives should be kept fully informed of the contents of the risk control action plan and progress with its implementation.

> ### 🔍 REVIEW
>
> What are the stages of the accident investigation process?
>
> What documentation should be reviewed when investigating an accident?
>
> What information should be included in an accident report?

RECORDING AND REPORTING INCIDENTS

Organisational requirements

Organisational requirements for recording and reporting incidents will usually be wider than the statutory minimum, as statutory requirements tend to focus on the more serious incidents, those that clearly had or could have had serious consequences. Organisations will have an interest in a wider range of incidents, some of which may not be reportable to an enforcing authority because they involve minor equipment damage or a near miss. The organisation should have a system to record and report these incidents, as well as the more serious ones, as they provide an opportunity to identify immediate and root causes that may lead to more serious incidents.

In addition, although there may not be statutory obligations to report incidents that can only have a minor potential, they still represent a loss to the organisation and if they occur frequently will warrant corrective and preventive action to avoid a reoccurrence.

Organisations may have separate recording and reporting forms for such things as near misses in order to ensure they are focused on or they may have a single form on which the type of incident is identified.

There are many important reasons why employers need to ensure that a wide range of incidents are recorded and reported, among them are the following:

- It is an implied requirement of the Management of Health and Safety at Work Regulations (MHSWR) 1999 (Regulation 5).
- Reporting provides an opportunity for an investigation.
- The investigation, in turn, should help to identify flaws with existing controls and therefore assist in the implementation of improved controls.
- It enables analysis of reports which may identify trends or patterns that may emerge.
- It enables the gathering of statistical evidence to enable the employer to compare health and safety performance with industry and other standards (benchmarking).
- It provides evidence for use in legal actions that may follow.
- It helps identify an increase in the number of incidents and should prompt a review of risk assessments.
- It will help the employer to comply with reporting requirements of RIDDOR 2013.

It must be borne in mind that many incidents go unreported and there are many reasons for this. For example, the worker may fear disciplinary action; they might be embarrassed because the incident was caused by something they did wrong; there could be peer pressure; the worker might not want to break a long standing record for the period between incidents; they might not know what to report or the importance of reporting or how to report and who to report to. Managers should consider the possible reasons and put actions in place to improve reporting.

Accident 'book'

The Social Security (Claims and Payment) Regulations (SSCPR) 1979 requires that all employers keep an accident book or an equivalent recording method available to workers to enable them to record accidents they have at work, for which (industrial injury) benefit may be payable.

It is not necessary that the employer use the book designed specifically for this purpose, the BL 510, provided the record has the same headings. The employer has a duty to investigate all accidents recorded by workers to determine if they were accidents at work (for which benefit may be payable).

Following this, it is the duty of the employer to provide information to government organisations investigating a worker's claim for benefit.

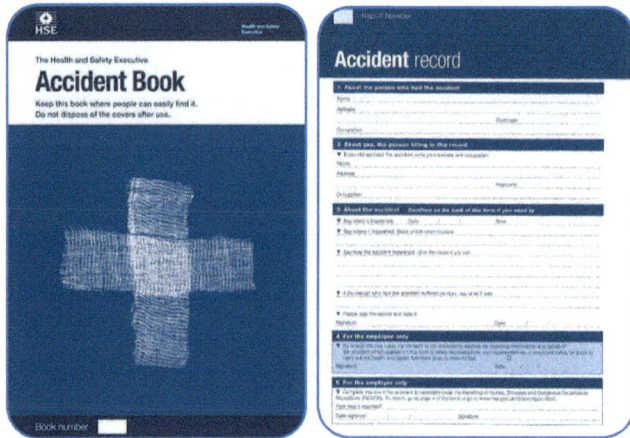

Figure 3-32: Examples of an accident book.
Source: HSE.

Accident record books (for example, BL 510) need to comply with the requirements of the Data Protection Act (DPA) 2018 with regard to protection of personal data and must be kept available for inspection for three years after the last entry.

Statutory requirements for recording and reporting to enforcing authorities

Employers, the self-employed or those in control of work premises, have duties under the Reporting of Injuries Diseases and Dangerous Occurrences Regulations (RIDDOR) 2013 to record and report certain injuries, dangerous occurrences and ill-health at work to the appropriate enforcing authority. The information enables the enforcing authority to identify where and how risks arise and to investigate serious injury and ill-health. The reports are used to compile statistics to show trends and to highlight problem areas, in particular industries or organisations. The enforcing authority can then provide advice on preventive action to reduce injury and ill-health.

Incidents defined in RIDDOR 2013

Work related accident

Includes an act of non-consensual physical violence done to a person at work.

Unspecified non-fatal injuries to workers

Injury to a worker arising from that work which results in incapacity for routine work for more than 7 consecutive days (excluding the day of the accident).

Non-fatal injuries to non-workers

Injury to a person not at work, as a result of a work-related accident:

a) That results in the person being taken to a hospital for treatment in respect of the injury; or

b) a specified injury on hospital premises.

Specified injuries

The list of specified non-fatal injuries in Regulation 4.1 is:

- Any bone fracture diagnosed by a registered medical practitioner, other than to a finger, thumb or toe.
- Amputation of an arm, hand, finger, thumb, leg, foot or toe.
- Any injury diagnosed by a registered medical practitioner as being likely to cause permanent blinding or reduction in sight in one or both eyes.
- Any crush injury to the head or torso causing damage to the brain or internal organs in the chest or abdomen.
- Any burn injury (including scalding) which:
  - Covers more than 10% of the whole body's total surface area.
  - Causes significant damage to the eyes, respiratory system or other vital organs.
- Any degree of scalping requiring hospital treatment.
- Loss of consciousness caused by head injury or asphyxia.
- Any other injury arising from working in an enclosed space which:
  - Leads to hypothermia or heat-induced illness.
  - Requires resuscitation or admittance to hospital for more than 24 hours.

Fatal injury

Anyone who dies as a result of a work-related accident or occupational exposure to a biological agent or where a worker dies within one year as a result of a non-fatal injury defined in Regulation 4.

Diseases

The list of occupational diseases in Regulation 8 is:

- Carpal Tunnel Syndrome (CTS), where the person's work involves regular use of percussive or vibrating tools.
- Cramp in the hand or forearm, where the person's work involves prolonged periods of repetitive movement of the fingers, hand or arm.
- Occupational dermatitis, where the person's work involves significant or regular exposure to a known skin sensitizer or irritant.
- Hand Arm Vibration Syndrome (HAVS), where the person's work involves regular use of percussive or vibrating tools, or the holding of materials which are subject to percussive processes, or processes causing vibration.
- Occupational asthma, where the person's work involves significant or regular exposure to a known respiratory sensitizer.
- Tendonitis or tenosynovitis in the hand or forearm, where the person's work is physically demanding and involves frequent, repetitive movements.

There is an additional list of specific occupational diseases in Regulation 9 is:

- Any cancer attributed to an occupational exposure to a known human carcinogen or mutagen (including ionising radiation).
- Any disease attributed to an occupational exposure to a biological agent.

Dangerous occurrences
Dangerous occurrences are events listed in schedule 2 of RIDDOR 2013 that have the potential to cause death or serious injury and so must be reported whether anyone is injured or not.

Examples of dangerous occurrences listed in schedule 2 include:

- The failure of any load-bearing part of any lifting equipment, other than an accessory for lifting.
- The failure of any pressurised closed vessel or any associated pipework.
- Any unintentional incident in which plant or equipment either:
  - Comes into contact with an uninsulated overhead electric line in which the voltage exceeds 200 volts.
  - Causes an electrical discharge from such an electric line by coming into close proximity to it.
- Electrical short circuit or overload attended by fire or explosion which results in the stoppage of the plant involved for more than 24 hours.
- The malfunction of breathing apparatus whether for training or operational purposes.
- The complete or partial collapse (including falling, buckling or overturning) of any scaffold more than 5 metres in height

Procedures for reporting under RIDDOR
Notify and report within 10 days
When a **person dies or a worker suffers a specified injury** (listed in Regulation 4.1 - non-fatal injuries to workers) or **a non-worker suffers a defined non-fatal injury** (listed in Regulation 5 - non-fatal injuries to non-workers) as a result of a work-related accident or **a person dies as a result of occupational exposure to a biological agent** or a **dangerous occurrences** occurs the responsible person must **notify** the relevant enforcing authority by the **quickest practicable means** (usually by telephone) **without delay** and must send them a **report** in an approved manner (online) within **10 days, see Figure 3-33.** This therefore includes work-related accidents where:

- A worker or a self-employed person at work is killed or suffers a specified injury.
- A member of the public is killed or taken to hospital (there is no need to report incidents where people are taken to hospital purely as a precaution when no injury is apparent).

Report without delay
In cases of **work-related diseases** listed in Regulation 8 the responsible person must send a **report** of the diagnosis in an approved manner (online) to the relevant enforcing authority **without delay**.

Notify
In cases of diseases related to carcinogens, mutagens and biological agents listed in Regulation 9 the responsible person must **notify** the relevant enforcing authority **in an approved manner.**

Report within 15 days
If injury to a worker results in more than **7 days incapacity for routine work**, but is not one of the specified non-fatal injuries, the responsible person must send a **report** to the relevant enforcing authority in an approved manner (online) as soon as is practicable and in any event **within 15 days** of the accident. The day of the accident is not counted, but any days which would not have been working days are included.

Responsible person
In relation to an injury, death or most dangerous occurrences the responsible person is the worker's employer or if it involves a person not at work or a self-employed person working in someone else's premises the responsible person is the person in control of the premises at the time. Where a self-employed person is working in premises they control the responsible person is the self-employed person or someone acting for them. In relation to diseases the responsible person is the worker's employer or the self-employed person diagnosed with the disease.

The enforcing authority for most workplaces is either the Health and Safety Executive or the Local Authority, for railway operations it is the Office of Rail and Road (ORR).

Procedures for recording under RIDDOR
In the case of a reportable injury at work, the following details must be recorded:

- Date and time.
- Name.
- Occupation or status (for example, customer, visitor).
- Nature of injury.
- Place of accident.
- Brief description of the circumstances in which the injury happened.
- The date on which the injury was first notified or reported to the relevant enforcing authority and the method used.

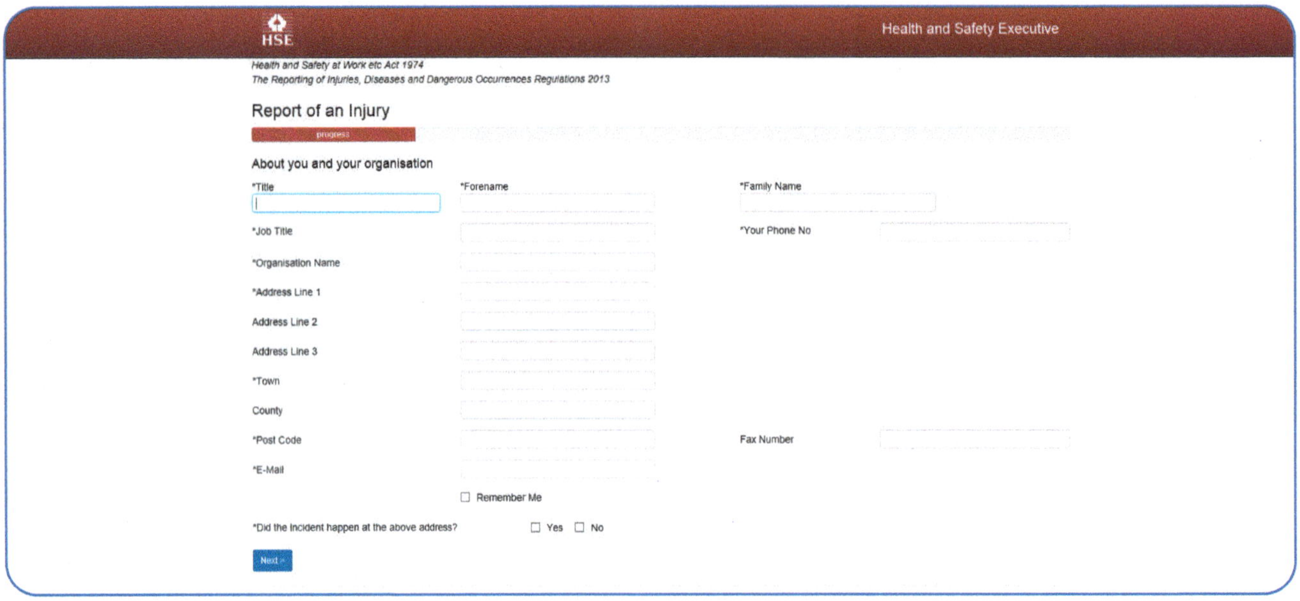

Figure 3-33: Reporting an injury, online form.
Source: HSE.

Similar information should be recorded for dangerous occurrences, except that details of injured persons will not be relevant. In the case of a diagnosis of a reportable disease, similar details to those relating to an injury must be recorded.

For non-reportable injuries that *incapacitate for more than 3 days* a record of notification or reporting is not relevant, but a similar *record* to that shown above must be kept, a record in the Accident Book would be enough to meet this requirement.

Records must be kept for at least 3 years and kept at the place where the work it relates to is carried out or at the usual place of business of the responsible person.

REVIEW

Consider how accidents/incidents are recorded in your workplace.

Who is responsible for initially reporting accidents?

Sources of reference

Reference the information provided, in particular web links, these are correct at time of publication, but may have changed.

Emergency Procedures, HSE website guidance, http://www.hse.gov.uk/toolbox/managing/emergency.htm

Example risk assessments, HSE, http://www.hse.gov.uk/risk/casestudies/index.htm

First aid at work, L74, third edition (amended 2018), HSE Books, ISBN: 978-0-7176-6560-0 http://www.hse.gov.uk/pubns/priced/l74.pdf

First aid in work, HSE, https://www.hse.gov.uk/simple-health-safety/firstaid/index.htm

Guidance on permit-to-work systems. A guide for the petroleum, chemical and allied industries (HSG250) HSE Books, ISBN: 978-0-717629-43-5 http://www.hse.gov.uk/pubns/priced/hsg250.pdf

Human factors: Organisational change, HSE, http://www.hse.gov.uk/humanfactors/topics/orgchange.htm

Human factors: Permit to work systems, HSE, http://www.hse.gov.uk/humanfactors/topics/ptw.htm

Managing Health and Safety in Construction, Construction (Design and Management) Regulations 2015, Guidance on regulations, L153, HSE Books, ISBN: 978-0-7176-6623-3, http://www.hse.gov.uk/pubns/priced/l153.pdf

Risk Assessment (INDG163, rev 4), HSE Books, ISBN: 978-0-717664-40-5 http://www.hse.gov.uk/pubns/indg163.pdf

Risk assessment, A brief guide to controlling risks in the workplace, INDG163, HSE Books http://www.hse.gov.uk/pubns/indg163.pdf

Specific topic 3: Organisational change and transition management, HSE, http://www.hse.gov.uk/humanfactors/topics/specific3.pdf

The health and safety toolbox, How to control risks at work, HSG268, HSE Books, ISBN: 978-0-7176-6587-7, http://www.hse.gov.uk/pUbns/priced/hsg268.pdf

Web links to these references are provided on the RMS Publishing website for ease of use - www.rmspublishing.co.uk

Statutory provisions

Confined Space Regulations (CSR) 1997 / Confined Space Regulations (Northern Ireland) 1999

Construction (Design and Management) Regulations (CDMR) 2015 / Construction (Design and Management) Regulations (Northern Ireland) 2016

Health and Safety (First Aid) Regulations (HSFAR) 1981

Management of Health and Safety at Work Regulations (MHSWR) 1999 (as amended) / Management of Health and Safety at Work Regulations (Northern Ireland) 2000

Reporting of Injuries Diseases and Dangerous Occurrences Regulations (RIDDOR) 2013

The Regulatory Reform (Fire Safety) Order (RRFSO) 2005

The Social Security (Claims and Payment) Regulations (SSCPR) 1979

STUDY QUESTIONS

1. An organisation is introducing a new work activity that requires a safe system of work.

   a) Why is it important to involve workers in the development of a safe system of work?

   b) Why is it important for safe systems of work to have written procedures?

2. What are the general details that should be included in permit-to-work?

3. What are the factors to consider when making an assessment of first-aid provision in a workplace?

4. An engineer was involved in a near-miss incident when he dropped a component whilst working on an overhead crane. The component narrowly missed an employee who was passing below. What are four reasons why the near-miss incident should be investigated?

For guidance on how to answer these questions, please refer to the assessment section located at the back of this guide.

Element 4

Excavation

4.1 Excavation work hazards and assessment

THE HAZARDS OF WORK IN AND AROUND EXCAVATIONS

Work in excavations and trenches, basements, underground tanks, sewers, manholes etc., can involve high risks and each year construction workers are killed with some buried alive or asphyxiated.

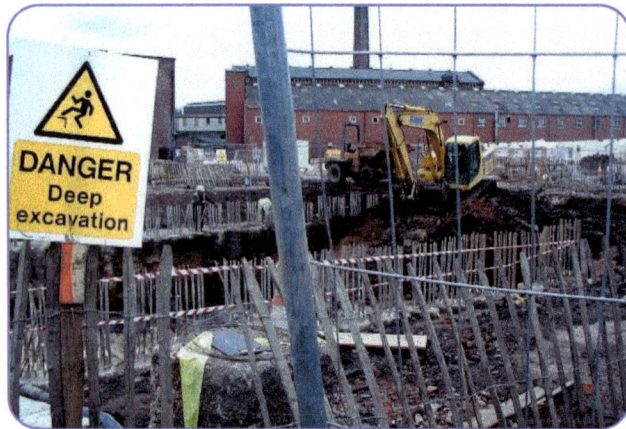

Figure 4-1: Excavation hazards.
Source: RMS.

Buried services

Although electricity cables provide the most obvious risk, gas pipes, water mains, drains and sewers can all release dangerous substances. Gas is particularly dangerous if there is a potential ignition source close by. Fibre-optic cables may carry laser light, which could be damaging to the eyes if severed accidentally and are very expensive to repair.

Buried services (electricity, gas, water, etc.) may not be obvious at the time of conducting a site survey. This increases the likelihood of striking a service when excavating, drilling or piling. The results of striking an underground service are varied, and the potential to cause injury or a fatality is high. As with overhead power lines, any underground service should be treated as live until confirmed dead by an authority (utility provider). Incidents can include shock, electrocution, explosion and burns from power cables, explosion, burns or unconsciousness from gas, impact injury from dislodged stones or flooding from ruptured water mains.

Before groundwork is due to commence it is common and good practice to check for the presence of any of the previously mentioned services or hazards by using a detection device. A common device used regularly in the construction industry is a cable avoidance tool (CAT), more commonly known as a CAT scanner when utilised as a detection device in a particular area.

Figure 4-2: Buried services.
Source: RMS.

Falls of people/equipment/material into the excavation

When people are working below ground in excavations, the problems are very similar to those faced when people are working at a height - falls and falling objects. Particular problems arise when:

- Materials, including spoil, are stored too close to the edge of the excavation.

- The excavation is close to another building and the foundations may be undermined.

- The edge of the excavation is not clear, especially if the excavation is in a public area.

- Absence of barriers or lighting.

- Poor positioning or the absence of access ladders allowing people to fall.

- Absence of organised crossing points.

- Badly constructed ramps for vehicle access which can cause the vehicle to topple.

- No stop blocks for back filling.

- Routing of vehicles too close to the excavation.

Collapse of sides

Often, the soil and earth that make up the sides of the excavation cannot be relied upon to support their own weight, leading to the possibility of collapse. The risk can be made worse if:

- The soil structure is loose or made unstable by water logging.

- Heavy plant or materials are too close to the edge of the excavation.

- Machinery or vehicles cause vibration.

- There is inadequate support for the sides.

The consequences of even a minor collapse can be very serious. A minor fall of earth can happen at high speed and bring with it anything (plant and machinery) that may be at the edge. Even if the arms and head of a person are not trapped in the soil, the material pressing on the person can lead to severe crush injuries to the lower body and asphyxiation due to restriction of movement of the chest.

Collapse of adjacent structures

Excavations that are carried out within close proximity to existing buildings or structures may result in their foundations becoming undermined and create the potential for significant settling damage to occur or worse still, collapse. Consideration should be given to the effects that excavation work might have on foundations of neighbouring buildings or structures, and control measures implemented to ensure that foundations are not disturbed or **undermined**. Building foundations that are at a distance of less than twice the excavation depth from the face of the excavation are more likely to be affected by ground movement; underpinning or shoring of such structures may be required to prevent structural damage.

Figure 4-3: Risk of collapse of sides and water ingress.
Source: RMS.

Water ingress

Ingress of water may occur through rainfall, flood (river, sea) or when an excavation extends below the natural groundwater level. In deep excavations, where access is not readily available, the combined effect of water and mud could lead to difficulty in escape and risk of drowning. In addition, this can lead to the sides of the trench becoming soft and the integrity of the excavation supports can be undermined.

Use of cofferdams

A cofferdam is a retaining wall that is built into a river, lake, or other body of water and is commonly made of steel sheet piling, or concrete. The retaining wall is intended to provide a dry workplace in which work can be carried out.

Figure 4-4: Cofferdam.
Source: USA Army Corps Library.

Cofferdams are typically used in the construction of the foundations of bridges, docks, and piers. The principle hazard involved in the use of a cofferdam is the risk of failure of the retaining wall, allowing water ingress. If this happens gradually it can make the ground very wet and difficult to work in or travel around. Without the use of pumps to control water ingress slips and falls may become a major problem. If the failure of the wall is sudden, perhaps caused by severe tidal or storm water activity, there is a high risk of workers drowning.

The nature of the soil, depth of water, amount of water movement, depth of excavation and type of structure can all influence the stability and strength of the retaining wall and therefore its effectiveness in keeping water out.

Use of caissons

A caisson is a retaining, water tight, structure that is constructed, floated or lowered into position where work needs to be conducted in a defined area, creating a type of 'box' to work in. Caissons are often made of concrete sections joined together to create the structure. When set in position, the water within the caisson is pumped out, providing a chamber as a workspace.

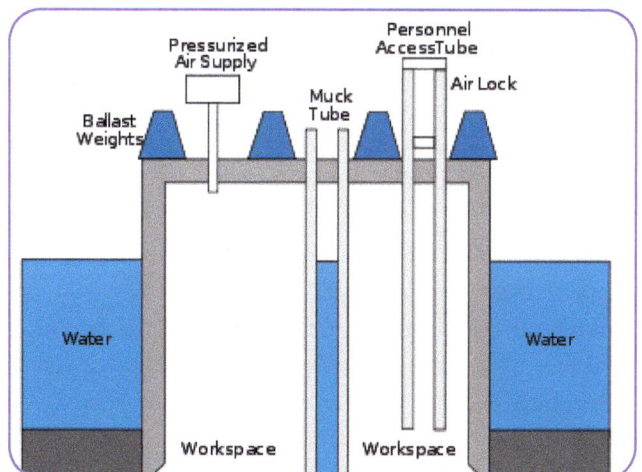
Figure 4-5: Pressurised caisson.
Source: YK Times.

In the case of hollow caissons, if additional depth is required the earth can be mechanically excavated from the centre of the caisson and the weight of the caisson causes it to sink lower. This can be continued by adding additional caisson sections until a firm foundation is reached. Caissons are used for a wide variety of work, for example, to create or work on the foundations of a bridge pier that may be submerged in water.

Pressurised (Pneumatic) caissons are shaped like an open box, turned upside down. They are placed where the work is needed and the bottom edges of the caisson sink into the earth, creating a seal. The inside is pressurised by compressed air to remove the water and prevent water ingress. Workers enter through an air lock. They remove earth from the bottom of the caisson by hand and place it in a (muck) tube connected to the surface, where it is removed by a crane and bucket. As with the hollow caisson, removing the earth at the bottom allows the caisson to sink deeper. When great strain or pressure is likely to be encountered in the construction of foundations or piers deep in a body of water, the pressurised caisson is generally used.

The principal hazard of work within caissons is the risk of water or slurry ingress and drowning. With hollow caissons, this may be because of failure of the sealing of joints between the sections that comprise the caisson. This failure could be due to damage at the time of assembly, the effects of water over time or the pressure due to the depth it is operating at overcoming the seal. Because caissons are frequently used in rivers and the sea there is the further hazard of service vessels colliding with the caisson, causing leaks or catastrophic failure.

In the case of pressurised caissons, they have the additional possibility of failure of the system maintaining pressure within the structure; this could lead to failure of seals or in the case of a homogenous structure the collapse of the sides. Work in pressurised conditions can be strenuous and workers leaving a pressurised caisson after hours of working under high pressure may suffer from a form of decompression sickness unless they are given special decompression treatment to accustom them to the lower atmospheric pressure.

Contaminated ground

Digging may uncover buried materials that have the potential to be hazardous to health. Sites that once were used as steel works may contain arsenic and cyanide dating back many years; farmyards may have been used as graves for animals and to dispose of pesticides and organophosphates. Excavation may also involve removing old drums of chemicals and other buried debris that might be contaminated. There is always the presence of vermin to consider - this can increase the risk of diseases such as leptospirosis.

Toxic and asphyxiating atmospheres

Excavations can under different circumstances be subject to toxic, asphyxiating or explosive atmospheres. Chalk or limestone deposits when in contact with acidic groundwater can release carbon dioxide, and gases such as methane or hydrogen sulphide can seep into excavations from contaminated ground or damaged services in built-up areas. These atmospheres can accumulate at the bottom of an excavation and result in potential fatalities from asphyxiation, poisoning, or explosion.

Mechanical hazards

Mechanical hazards of excavation work relate mainly to the equipment used to create the excavation and to lay equipment/materials in place. The principal mechanical hazard is the risk of being struck by this equipment as it moves its excavator arm or jib. The risk is greatly increased as people move close to the equipment, for example, workers or pedestrians passing by. People particularly at risk are those that place themselves in close proximity to the equipment when it is operating.

This would include people directing the equipment movement or those supervising the work, particularly if they stand in or close to the excavation at a position near to where work is going on. The closing movements of parts of this equipment, such as an excavator arm or bucket, present a significant crushing risk.

In addition, as the plant is usually capable of mobility there is a risk of people working in close proximity receiving crush injuries to their feet.

Figure 4-6: Mechanical hazard from excavator.
Source: RMS.

Overhead hazards

The overhead hazards influencing excavation activities will depend upon where the excavation work is carried out. If the work is on a 'green field' site the only overhead hazards present are likely to be from overhead telephone and power lines or trees.

If the work is on an existing site, then the access and egress route, as well as the working area will need to be assessed for overhead hazards. Hazards may include overhead pipes and pipe racks/bridges, pedestrian walkways between buildings and power/telephone lines.

Work on or adjacent to the highway will present a number of overhead hazards, such as street lights, traffic signs/signal systems and bridges. The bridges may additionally carry utilities such as gas, water and sewage below them. Work near to water or railways will have similar overhead hazards to the highway.

Wherever the excavation is taking place, if the site is well developed there is the additional overhead hazard of scaffolds, MEWPS or parts of the structure.

Hazards of overhead power lines

Every year people are seriously hurt by coming into contact with overhead power lines. A high proportion, about one third, proves to be fatal. These fatalities occur at voltages ranging from 230V to 400 kV. Any lines found on a site should always be treated as live until they are proved to be otherwise.

Overhead power lines usually consist of bare (uninsulated) conductors. These are often referred to as cables. They are supported overhead in a number of ways, the most common being wooden posts or metal towers. A common problem with the type supported by wooden posts is that they are mistaken for telephone cables, which may mean they are treated with less care than is deserved; an accidental contact with an excavator's bucket could be fatal to the operator.

If the excavator comes into contact with an overhead line it may not discharge immediately to earth through the vehicle; this can be because the tyres offer some resistance. Operators of excavators have received fatal shocks in these circumstances when, on getting out of the vehicle, they make contact with the ground and provide a passage for the current to earth.

One of the other risks related to overhead power lines is that of arcing or, as it is sometimes called, a flashover. This occurs where the excavator approaches the conductor, but does not touch it. Instead, the electricity in the conductor bridges the air gap between the conductor and excavator as an arc; current then flows through the excavator. The risk of flashover increases significantly as the voltage applied to the conductor increases.

If excavators are brought closer than the following distances there is a significant risk of flashover:

- 15m of overhead lines supported by steel towers.
- 9m of overhead lines supported by wooden poles.

See also - Element 10.3 - Electricity – 'Work near overhead power lines'.

RISK ASSESSMENT OF EXCAVATION WORK

Factors to consider

The depth of the excavation

Risks increase with the depth of excavation, risks from materials falling increase with the height they may fall and there is an increased risk of collapse at greater depths as the amount of material to support (comprising the excavation wall) increases. A collapse of an excavation deeper than the head height of a worker carries a high risk of suffocation.

It should be remembered that the work being carried out may mean a worker has to carry out task with their head low down in an excavation, so even a shallow excavation can present serious risks.

The type of soil being excavated

Different types of soil have different cohesive properties that encourage the particles of soil to bind together, therefore giving different strength. Some types of soil can be self-supporting, for example, moist clay, others like sand, gravel and silt are less cohesive and as they are more free flowing they are less strong. However, clay presents specific risks of collapse when it dries out or as it becomes more fluid when wet.

Soils generally change their strength significantly when they become wet as the water affects the cohesion of soil particles. Soil containing high levels of water can present an additional risk because water may make its way into the excavation and cause flooding.

Soil that has previously been excavated or recently laid down and not compacted will have reduced strength compared with soil that has had chance to settle and become compacted.

If the soil is chalky, carbon dioxide may be liberated as part of the excavation process, similarly if it is in contaminated land there may be toxic substances and risks from methane.

The type of work being undertaken

If the work to create and work in the excavation is to be carried out manually rather than mechanically the risks vary significantly. The tasks being conducted can increase the risks from the type of work, for example, pipe jointing operations carried out manually will usually bring the head of the worker below the top of the excavation for a significant time. Additional risks may also be created by hot working or work in a confined space in the excavation.

The use of mechanical equipment

Though this can reduce the risk to workers by not requiring them to enter the excavation to dig it, they carry risks of their own.

Where people work alongside this equipment there is a risk of contact or collision and as it carries a considerable amount of momentum major injuries or death would not be uncommon if contact was made. The mechanical equipment may also put an unacceptable amount of weight on the edge of the excavation and increase the risk of the sides collapsing.

Proximity of roadways and structures etc.

If vehicles pass on roadways nearby or materials are placed too close to excavations there is an increased risk of the excavation collapsing due to the additional weight acting on the side of the excavation, as well as a risk of materials/vehicles falling into the excavation.

Therefore, where excavations are within close proximity of roadways and structures, vehicles should be segregated and excluded from excavation areas wherever possible. Brightly painted baulks or barriers and or fencing should be used where necessary to protect and clearly identify the excavation area.

If vehicles have to unload materials into excavations, the use of stop blocks is recommended to prevent the vehicles falling into the excavation as a result of over-running the edge of the excavation. Where vehicles have to approach or pass the edge of an excavation special consideration should be given to the reinforcement of the sides of the excavation, which may need extra support.

Where excavations are made near to roadways and structures they could undermine and weaken their foundations, increasing the risk that the roadway or structure could collapse.

Excavations should not be allowed to affect the footings of scaffolds or the foundations of nearby structures. Walls, especially in older properties, may have very shallow foundations, which could be undermined by small excavations, therefore some structures may need temporary support even before digging can commence. It is always best practice to conduct surveys of the foundations and gain the advice of a structural engineer in such cases.

Presence of public

If the public are present in the area where an excavation is to take place they increase the risk of people falling into the excavation, particularly as they may not be familiar with the hazard. The level of risk increases with the number of people present and their vulnerability, for example, if they are old or young. In addition, the risk also increases if the area around the excavation is dark or if public vehicles pass near it.

Excavations in public areas should therefore be fenced off, fitted with toe boards and stop block type barriers to prevent pedestrians and vehicles falling into them.

Where children or trespassers might inadvertently get access to a site out of hours then certain precautions should be considered such as backfilling or securely covering excavations to reduce the potential of such people being injured.

The excavation should have suitable and sufficient identification and signage, with any pedestrian routes or detours clearly marked, fenced off and lit (if appropriate).

Weather

Heavy rain or fast melting snow may undermine the stability of the ground and therefore of the supports or sides of the excavation. Hence, careful consideration should be given to the type of shoring used, depending on the site and sub-soils.

Services

The presence or likelihood of services in the area of an excavation carries high risks of a different type. For example, contact with an electrical service may be immediately fatal to a worker or damage to a gas service may cause a risk of major explosion.

4.2 Control measures for excavation work

CONTROLS

Identification/detection and marking of buried services

The Health and Safety Executive (HSE) guidance, HSG47; 'Avoiding danger from underground services', outlines the dangers that can arise from work near underground services and gives advice on how to reduce the risk. The guidance defines 'service(s)' as all underground pipes, cables and equipment associated with the electricity, gas, water (including piped sewage) and telecommunications industries.

It also includes other pipelines that transport a range of petrochemical and other fluids. It does not include underground structures such as railway tunnels. The term 'service connection(s)' is used for pipes or cables between distribution mains and individual premises.

Identification

Excavation operations should not begin until all available service location drawings have been obtained and thoroughly examined to identify the location of services. Location drawings should not be considered as totally accurate, but do serve as an indication of the likelihood of the presence of services, their location and depth.

It is possible for the position of an electricity supply cable to alter if previous works have been carried out in the location. This can be due to the flexibility of the cable and movement of surrounding features since the original installation of the cable. In addition, location drawings often show a proposed position for the services that does not translate to the ground, such that services are placed in position only approximately where the plan says.

Detection

The main types of detection and locating devices are:

Hum detectors, which detect the magnetic field radiated by electricity cables that have a current flowing through. They do not detect where there is little or no current flowing, for example, service connection cables to unoccupied premises or street lighting cables in the daytime.

Radio frequency detectors, which generate a low frequency radio signal into a cable or pipe placed very close to it. The receiver can then detect this signal and locate the cable or pipe run.

Metal detectors, which usually locate flat metal covers, joint boxes etc., but may not detect round cables or pipes.

Ground probing radar, which is a method that is capable of detecting anomalies in the ground. When plotted into a continuous line they may indicate a cable, duct or pipe.

Commercially available instruments use more than one of the techniques listed and may also include a depth-measuring facility.

It is important that 'service location devices', such as a cable avoidance tool (CAT) are used by competent, trained operatives to assist in the identification and marking of the actual location and position of buried services.

Figure 4-7: Using a cable locator.
Source: Allpipe Stoppers.

Marking

Markers may have been used to indicate the presence of services when they were laid, these include:

- Marker tiles, for example, to mark gas pipes when they have been laid at shallow depths.
- Coloured plastic, for example, tape laid to over an electric cable.
- Marker posts/plates, for example, to show the position and size of valves on gas mains or water mains.

Markers are useful in alerting caution with regard to excavation activities. However, these markers cannot be relied upon to accurately establish the position of the buried service, as they may have been disturbed in later work. Also, the absence of markers does not indicate that services are absent from the location.

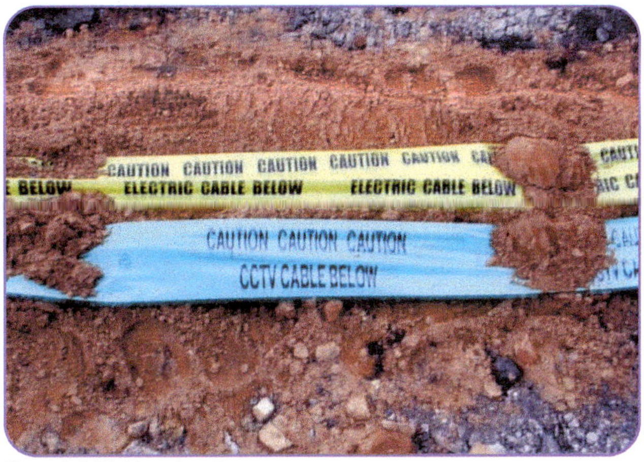

Figure 4-8: Marking of services.
Source: RMS.

When it is uncovered, the type of service may be identified by markings in the form of a colour coding, for example, blue for water and yellow for gas.

When services are identified it is essential that physical markings be placed on the ground to show where the services are located. This can include waterproof, biodegradable paint, applied to solid surfaces, or wooden (not metal) pegs, for un-surfaced areas.

The colour coding of paint/pegs and addition of tape indicators can help to identify the type of service present.

Safe digging methods

Safe digging methods should be implemented within 0.5 metres of a buried service. This involves the use of insulated hand tools such as a spade or shovel with curved edges (to be used with light force and not sudden blows).

Mechanical, probing or piercing tools and equipment such as excavators, forks, picks or drills should *not* be used when in the vicinity of buried services, as these may cause damage to any service if they strike it.

Careful hand digging using a spade or shovel should be used instead.

Example of a permit to dig

Contract:..	Contract No..
Principal Contractor:...	Sub Contractor:..

Permit No:	Date

1. Location:	
2. Size, detail and depth of excavation:	
3. Are Service Plans on site?	YES/NO
Comments:..	
4. Has cable locating equipment been used to identify services?	YES/NO
Comments:..	
5. Are all known services marked out? (site inspection by relevant statutory bodies)	YES/NO
Comments:..	
6. Are trial holes required?	YES/NO
Comments:..	
7. Have precautions been taken to prevent contact if overhead lines are in the vicinity of the operation or near approach to the operation?	YES/NO
8. Additional precautions, i.e. Shoring/Fencing/Access/Storage/Fumes/Record of setting out points to re-establish services routes	YES/NO
Comments...	
9. Sketch details or attach copy of plans 	
10. Date and time of excavation:..	

Signed... For and on behalf of issuing party	Accepted by:... (Work shall not commence unless all persons involved are aware of the safe systems of work.)

A new permit will be required for any further excavations. This permit is for guidance only. Persons carrying out work must take reasonable precautions when working around services.

Figure 4-9: Permit to dig. Source: Reproduced by kind permission of Lincsafe.

Methods of supporting excavations

"(1) All practicable steps must be taken to prevent danger to any person, including, where necessary, the provision of supports or battering, to ensure that:

(a) Any excavation or part of an excavation does not collapse.

(b) No material forming the walls or roof of, or adjacent to, any excavation is dislodged or falls.

(c) No person is buried or trapped in an excavation by material which is dislodged or falls."

Figure 4-10: Methods of shoring and battering.
Source: CDMR 2015 Regulation 22(1).

Precautions must be taken to prevent collapse. An alternative method to the use of supports to prevent collapse is where soil and material is removed from the sides of the excavation so that a shallow slope is created in the sides, generally called **battering**, or steps are created in the sides of the excavation, generally called **benching**, see **Figure 4-11.**

Battering relies on the material in the steps or slopes of the sides of the excavation creating minimal downward/sideways pressure, reducing the likelihood and effect of collapse.

The options available for providing support by shoring include closed boarding/sheeting see **Figures 4-13** and **4-15,** open sheeting see **Figures 4-14 and 4-16** and where the shoring needs to be repositioned frequently, such as where services are being laid, a trench box see **Figure 4-12.**

The application of a particular method of prevention of collapse depends on the:

- Nature of the subsoil - for example, when wet it may require close shoring with sheets.
- Projected life of the excavation - a trench box may give ready-made access where it is only needed for short duration work.

- Work to be undertaken, including equipment used - for example, the use of a trench box for shoring where pipe joints are made.
- Possibility of flooding from ground water and heavy rain - close shoring would be required.
- Depth of the excavation - a shallow excavation may use battering instead of shoring, particularly where shoring may impede access.
- Number of people using the excavation at any one time - a lot of space may be required so cantilever sheet piling may be preferred.

Figure 4-11: Benching.
Source: RMS.

Figure 4-12: Trench box - for shoring.
Source: RMS.

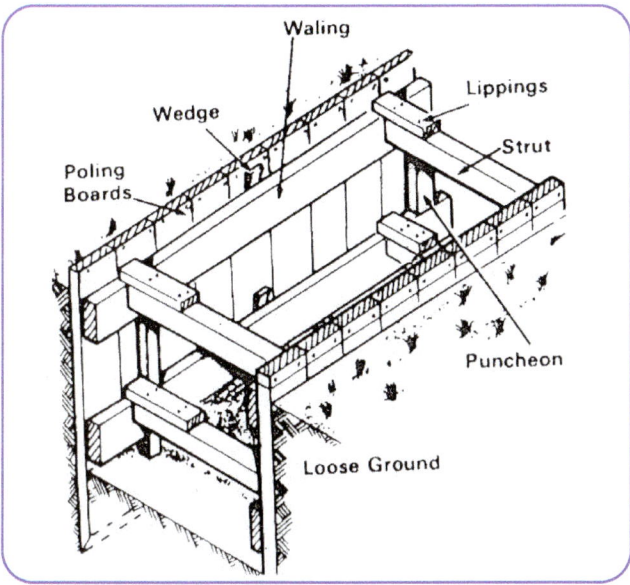

Figure 4-13: Close boarded excavation.
Source: BS6031.

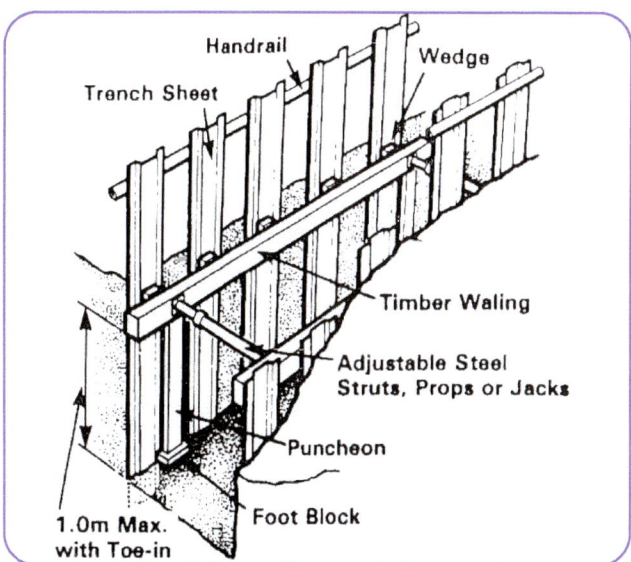

Figure 4-14: Open sheeting.
Source: BS6031.

Figure 4-15: Close sheeting.
Source: RMS.

Figure 4-16: Open sheeting.
Source: RMS.

Means of access

Ladders are the usual means of access and egress to excavations. They must be properly secured, in good condition and inspected regularly. The ladder should extend about one metre or three rungs above ground level to give a good handhold. To allow for emergency egress it is recommended that a minimum requirement of one ladder every 15 metres be provided.

Crossing points

Crossing excavations should only be allowed at predetermined points. The crossing point should be able to withstand the maximum foreseeable load and be provided with guardrails and toe boards. The spacing or location of crossing points should be such that workers and others are encouraged to use them, rather than to attempt other means of crossing the excavation.

Barriers, lighting and signs

Where people or materials can fall from height a form of edge protection must be provided. In some cases the shoring method used can provide this barrier by ensuring the top of the shoring extends sufficiently above the edge of the excavation, see *Figure 4-17.*

Figure 4-17: Access and guardrails.
Source: HSE, HSG150.

Guardrails that are provided must meet the same standards as those for working platforms used for work at height. It is also good practice to cover shallow trenches when they are left unattended.

"(2) Suitable and sufficient steps must be taken to prevent any person, work equipment, or any accumulation of material from falling into any excavation."

Figure 4-18: Prevention of falls into excavation.
Source: CDMR 2015 Regulation 22(2).

The large pieces of mobile plant and equipment that are commonly used for excavation work have the potential to cause serious harm to site workers and members of the public. In order to keep people and vehicles apart, exclusion zones identified by barriers, fencing, warning signs (see *Figures 4-19 and 4-20*) and lights should be provided.

Concrete or wooden blocks (usually old railway sleepers) may be placed some distance from the edge of an excavation to prevent vehicles from getting too close, particularly when the excavation is being 'back filled'. In this case, the wooden blocks provide a 'stop block'.

Figure 4-19: Barriers, fencing and sign.
Source: RMS.

Figure 4-20: Fencing.
Source: RMS.

Signs that comply with The Health and Safety (Signs and Signals) Regulations (SSSR) 1996 should be displayed to warn people of the excavation and any special measures to be taken. If working on a public highway, the police or the local authority must be consulted over the positioning of traffic lights. Appropriate lighting should be provided; it must provide sufficient illumination for those at work but should not create glare or other distractions for passers-by, especially motorists. Battery operated headlamps (to avoid trailing cables) may be considered for individual use. If excavations are present in dark conditions they must be suitably lit to prevent vehicles or people colliding or falling into them.

Safe storage of spoil

Excavated material (spoil), other materials, plant and vehicles should never be stored or parked close to the sides of any excavation. The additional pressure distributed on the ground from spoil, vehicles, etc. significantly increases the likelihood of collapse occurring at the sides of the excavation. Though it will depend on the weight of the material, it would normally be kept to a minimum of 1 metre from the edge of the excavation. See *Figure 4-21* which shows heavy pipes set away from the excavation and spoil set to the side.

Figure 4-21: Materials storage.
Source: RMS.

In addition, spoil heaps consist of loose materials that present a risk of falling into the excavation, setting spoil away from the excavation reduces this risk. A means of preventing spillage of spoil into an excavation is by positioning scaffold boards as toe boards, fixed along the outside of trench sheets. This provides additional protection to combat the risk of loose materials spilling into the excavation. An alternative to this is to allow boards or sheeting to protrude above the top of the excavation sufficiently to act as toe boards and prevent materials falling from ground level into the excavation. Head protection in the form of a hard hat must be worn by workers working in excavations to provide protection from small pieces of material falling either from above or from the sides of the excavation and possible impact with obstructions.

If stored at a suitable distance away from an excavation and at a suitable height, spoil heaps can form an effective barrier against vehicles travelling around the construction site and assist in preventing falls of vehicles and plant into an excavation and onto workers.

Control of water and dewatering

Consideration must be given to the likelihood of water entering the excavation and the measures to be implemented in order to control water entering the excavation and water levels within it. Ingress of water may occur through rainfall, flood or when excavating below the natural groundwater level. When an excavation is liable to water ingress the stability of the excavation walls can be undermined, increasing the risk of collapse and possible drowning.

The first priority is to control the amount of water entering the excavation; this will influence the choice of shoring, for example, close sheeting may be used rather than open sheeting. In addition, it may be necessary to position flood or rain water barriers to direct water away from the excavation; this may include the use of sandbags or similar methods that provide a portable barrier, see *Figure 4-22.*

Figure 4-22: Preventing water ingress.
Source: RMS.

Well points

Well point dewatering is often used to reduce the hydrostatic pressure from the back of a sheet pile wall, in conjunction with lowering the water table to the maximum excavation depth required for the construction site.

Well point dewatering systems are installed around the perimeter of the site against excavation shoring and are coupled with a perimeter discharge line via a pump set. The system can effectively draw the water table down to between three and eight metres depending on the type of pump set used and the type of soil conditions. For a well point dewatering system to be effective, the ground sub-strata generally needs to be permeable, to allow the well points to draw the groundwater.

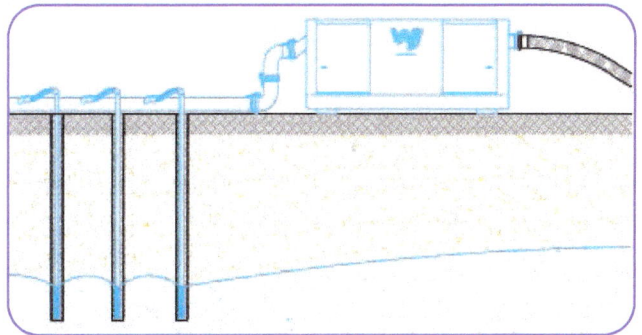
Figure 4-23: Well point schematic.
Source: W J Groundwater Ltd.

Figure 4-24: Well point trench installation.
Source: www.khansahebsykes.com.

Sump points

Usually water in excavations is drained to sumps, from where it can be pumped out for disposal. The sump should be at the lowest point where the water collects.

Typically a sump will be a minimum of 0.6 metres deep, so that the pump is covered by the water. This allows the pump to cycle on and off with rising and falling water levels within the sump.

It is normally permissible to distribute pumped groundwater from within an excavation over grassy areas where any silt deposits can be absorbed and not have a detrimental effect on the environment. However, when works are within close proximity to a watercourse, i.e. within 10 metres, then advice should be sought on the disposal of groundwater and a 'Consent for Works Affecting Watercourses' should be obtained from the Environment Agency.

Substances

Excavations should be treated with similar caution to that applied to confined spaces, and an assessment should be carried out prior to work commencing in excavations to identify the risk of toxic gas, oxygen deficiency and fire or explosion. It should also identify the appropriate risk control measures required, such as:

- Type of gas monitoring equipment to be provided.
- Testing of the atmosphere before entry into the excavation.
- Provision of suitable ventilation equipment.
- Training of workers.
- Use of a sufficient number of people, including one at ground level.
- Procedures and equipment needed for an emergency rescue.

Figure 4-25: Substances in excavation.
Source: RMS.

Positioning and routing of vehicles, plant and equipment

"(3) Suitable and sufficient steps must be taken, where necessary, to prevent any part of an excavation or ground adjacent to it from being overloaded by work equipment or material."

Figure 4-26: Overloading of excavation by work equipment or material.
Source: CDMR 2015 Regulation 22(3).

To prevent objects falling into excavations, the following precautions should be taken:

- Spoil and construction materials must not be stacked near to the edge.
- The weight of stacks should not be enough to cause the sides to collapse.
- Designated operating areas for vehicles and equipment must be established.
- Unnecessary vehicles and equipment should be routed away from the excavation.
- Where vehicles have to approach the excavation, stop barriers must be provided to prevent overrunning.

Personal Protective Equipment

As well as the need for hard hats to limit the risks from falling materials, other personal protective equipment (PPE) may be necessary, such as:

- Breathing apparatus.
- Safety harnesses.
- Hearing protection.
- Clothing to protect from the rays of the sun.
- Masks and respirators.
- Face masks and gloves for welding and grinding.
- Gloves for contaminated materials.
- Footwear.

Filling in

On completion of use of the excavation, experienced people should remove support materials. A competent person should inspect the site to ensure that all people and materials have been removed. The excavation may need water pumping from it before filling. Where vehicles approach the excavation to add materials it is essential that stop barriers are used at the edge of the excavation to avoid vehicles falling into it.

Materials added to the excavation must be sufficiently compacted to allow for the next use of the area, for example, as part of a roadway or to allow a scaffold to be placed on top of it. Only appropriate in-fill materials must be used. Uncontrolled tipping/burial is an offence.

REVIEW

What factors must be considered when deciding how to support a trench?

PARTICULAR REQUIREMENTS FOR CONTAMINATED GROUND

Contaminated ground can give rise to various health hazards, including biological and chemical hazards.

Exposure to the hazards can vary depending on the work to be carried out and assessments should be made to determine which controls are required.

Site preparation work which involves the removal of topsoil from land known to be contaminated with heavy metals will require specific arrangements in relation to personal protective equipment required and practices to be carried out. The work is likely to require the provision of gloves, overalls, eye protection and respirators. Additionally, the work will require campaigns of awareness training to ensure good levels of personal hygiene are maintained, with emphasis on ensuring that open cuts are covered with waterproof dressings whenever required. Consideration will need to be given to specific welfare facilities provision, for example, a decontamination unit. There should be arrangements in place in order to prevent contamination when eating and smoking. First-aid and emergency decontamination facilities should be made available located near to the place of work.

Soil testing

Surveys should be undertaken, and a soil-testing programme implemented *prior* to works beginning. As each site is different and may involve various contaminant types, the methods of testing and analysis used should be suitable to the particular local needs of each site.

Usually, specialist contractors are used to carry out this type of work. Historical information such as land deeds, maps, building plans and any other information relating to the site provide useful assistance. Contaminants will usually be found in the top 0.5 - 1m layer (strata) of ground and soil samples should be at least to the greatest depth of any excavation, remembering that contamination may seep from one part of a site to another. As a minimum precaution, any proposed control measures chosen should adequately protect against the highest concentrations of contaminant found on site.

Welfare facilities

Ensuring good hygiene standards is one of the more important elements of protective measures to be taken when working on construction sites and in particular with contaminated ground. Whilst the level of risk that is identified will establish the range of measures that are to be provided, the principles outlined should always be followed when working on a contaminated site.

Installation of a sufficiently ventilated and lit hygiene facility must be provided. This may be purpose built or a smaller standard unit may be sufficient depending upon the number of personnel involved at the site. The installation of the hygiene/decontamination unit should be situated at the most convenient point to or from the dirty zone.

The hygiene/decontamination facility should generally be in three stages, but prior to entering the unit a boot wash point should be provided that includes running water and fixed or hand brushes for removing contaminated matter.

The hygiene facility should then be arranged in its three stages as follows:

Stage 1

Should provide storage for ordinary clothing not used whilst at work.

Stage 2

A high standard washing facility. This includes the provision of hot and cold water with taps preferably operated by the elbow or foot, detergent, nail brushes and disposable towels. In certain circumstances (toxic or corrosive contamination) showers may be the necessary means of cleansing. Both men and women may use showers (at separate times) provided that it is in a separate room with a lock on the inside of the door.

Stage 3

Should provide storage for contaminated work wear such as overalls, boots, etc.

In addition to the hygiene facility, toilets should be provided. The location of toilet units on contaminated sites must be situated so workers are directed through the hygiene facility prior to reaching the toilet for use. Daily cleaning and decontamination must be carried out on toilet facilities. Everyone who works on construction sites must have access to adequate toilet and washing facilities, a place for warming up and a clean room set aside for taking meals and refreshment. Secure storage for clothing and facilities for drying wet clothing. First-aid facilities should also be provided.

Health surveillance

The employer should decide about the need for health surveillance as part of the Control of Substances Hazardous to Health Regulations (COSHH) 2002 assessment. Health surveillance is appropriate where:

- A disease or adverse effect may be related to exposure.
- It is likely that this could arise in the circumstances of the work.
- There are valid techniques for detecting the disease or effect.

For example, work with cadmium, phenol, or arsenic, will normally need health surveillance including biological monitoring of workers.

INSPECTION REQUIREMENTS FOR EXCAVATION SUPPORT SYSTEMS

CDMR 2015, Regulation 22, requires inspections and reports to be carried out for excavations.

A competent person must inspect excavations:

- At the start of each shift in which the work is to be carried out.
- After any event likely to have affected the strength or stability of the excavation.
- After any material unintentionally falls or is dislodged.

The competent person must:

- Inform the person, on whose behalf the inspection was carried out, before the end of the shift, of any matters that could give rise to a risk to the safety of any person. Work must not be carried out in the excavation until the matters have been remedied and the person who carried out the inspection is satisfied that construction work in the excavation can be safely carried out.

- Prepare a report and must, within 24 hours of completing the inspection to which the report relates, provide the report or a copy of it to the person on whose behalf the inspection was carried out.

CDMR 2015, Regulation 24, requires that the inspection report must include the following information:

1) Name and address of person on whose behalf the inspection was carried out.

2) Location of the place of construction work inspected.

3) Description of the place of construction work or part of that place inspected (including any work equipment and materials).

4) Date and time of the inspection.

5) Details of any matter identified that could give rise to a risk to the safety of any person.

6) Details of any action taken as a result of any matter identified in the last point.

7) Details of any further action considered necessary.

8) The name and position of the person making the report.

The person receiving the inspection report must keep a copy of the report at the site where the inspection was carried out until the construction work is completed and after that for three months.

REVIEW

The Construction (Design and Management) Regulations 2015 (CDM) require that inspections and reports for excavations are carried out by a competent person.

When should inspections be carried out under CDM? What information must the inspection report include?

4.3 Safe working in confined spaces

TYPES OF CONFINED SPACES AND WHY THEY ARE DANGEROUS

Types of confined spaces

A confined space is a substantially enclosed (though not always entirely) place which, by virtue of its enclosed nature, there is a foreseeable risk of serious injury from hazards within the space or nearby (for example, lack of oxygen). The Confined Spaces Regulations (CSR) 1997 define a confined space as any chamber, tank, vat, silo, pit, pipe, sewer, flue, well or similar space, in which, by virtue of its enclosed nature, there arises a reasonably foreseeable specified risk.

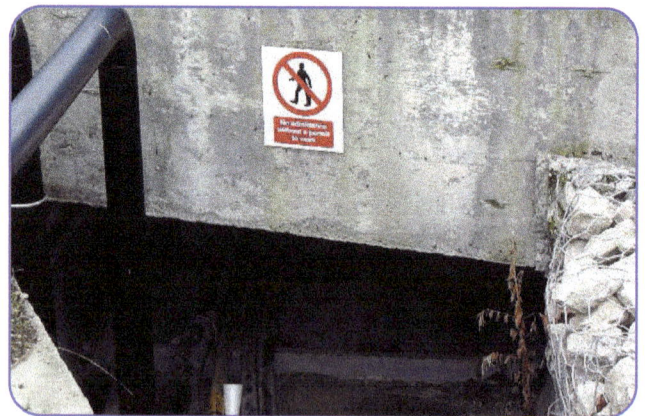
Figure 4-27: Confined space - chamber.
Source: RMS.

Figure 4-28: Confined space - sewer.
Soure: RMS.

Chamber	Caisson, cofferdam, or interception chamber for water.
Tank	Storage tanks for solid or liquid chemicals.
Vat	Process vessel which may be open, but by its depth confines a person.
Silo	May be an above-the-ground structure for storing cereal, crops.
Pit	Below ground, such as a chamber for a pump.
Pipe	Concrete, plastic, steel, etc. fabrication used to carry liquids or gases.
Sewer	Brick or concrete structure for the carrying of liquid waste.
Flue	Exhaust chimney for disposal of waste gases.
Well	Deep source of water.

"Specified risk", as defined under CSR 1997, means a risk of:

- Serious injury to any person at work arising from a fire or explosion.

- The loss of consciousness of any person at work arising from an increase in body temperature.

- The loss of consciousness or asphyxiation of any person at work arising from gas, fume, vapour or the lack of oxygen.

- The drowning of any person at work arising from an increase in the level of a liquid.

- The asphyxiation of any person at work arising from a free flowing solid or the inability to reach a respirable environment due to entrapment by a free flowing solid.

Figure 4-29: Confined space - worker in a pipe.
Source: Shutterstock.

Why they are dangerous

Confined spaces are dangerous because the hazards may not be obvious to workers until they enter the confined space and experience the effects of the hazard. Failure to appreciate the dangers associated with confined spaces has led to the deaths of many workers and often the death of some of those who have attempted to rescue them. This is because in order to rescue a worker, the rescuers urgently enter the confined space, often without establishing what the hazards affecting the worker were and without taking sufficient precautions to protect themselves, and they also become affected by the hazard. In addition, because of the confined nature of the space it can make rescuing a worker very difficult and slow, which means that a worker affected by hazards suffers serious harm or dies before the rescue is completed.

Figure 4-30: Confined space - cement silo.
Source: RMS.

Figure 4-31: Confined space - Tank entry.
Source: www.picstopin.com.

CONSIDER

Try and think of examples of confined spaces which you may have encountered whilst carrying out your work.

Hazards associated with working within confined spaces

The main hazards associated with working within a confined space include:

- Flammable vapours/dusts or excess oxygen that could cause fire or explosion.
- Heat that could cause loss of consciousness from increased body temperature.
- Exposure to a toxic gas/fume/vapour or lack of oxygen that could cause loss of consciousness or asphyxiation.
- Increase in the level of liquid that could cause drowning.
- Free-flowing solid that if trapped by it could cause asphyxiation.
- High concentrations of dust that could cause respiratory difficulties, for example dust found in a flour silo.
- Moving parts of machinery that could cause impact, crush, entanglement, etc.
- Steam that could cause burns.
- Restrictions in the space that could cause a worker to be trapped or difficulties in escaping.

Assessing risks from a confined space

Before work in a confined space can be carried out it must be shown that it not practicable to carry out the work without entering the confined space. If it is considered necessary to enter a confined space then the employer must carry out a risk assessment to identify risks and the precautions that will be necessary to ensure a safe system of work.

An assessment of risks from a confined space should consider the risks present in the confined space, those introduced by the work done in the confined space and those created by work carried out near to the confined space that can affect those working in the confined space.

The risk assessment should, in particular, consider the hazards and how they arise, including from the following typical situations:

- Previous contents of the confined space - whether they have been removed effectively or whether some material remains as a vapour or gas.

- Residues left in the confined space or equipment attached to the confined space - for example, sediment or material trapped in seals that may release toxic or flammable material when disturbed.

- Substances and other materials allowed to enter the confined space while working in it – for example, flood water entering a sewer or process substances entering a mixing tank.

- Oxygen deficiency and oxygen enrichment – normal oxygen content of air is approximately 21%.

- Shape and dimensions of the confined space - the atmosphere may vary in parts of the confined space, for example, low lying compartments.

- Temperature changes may also cause toxic materials to be released – for example, changes may be due to introducing heat in the form of a steam lance for cleaning, or welding tasks.

- Cleaning chemicals or chemicals introduced as part of the task being done in the confined space could affect the atmosphere - for example, vapours from adhesives used to glue a replacement tank lining or the products of combustion from a welding task.

- Substances or dangerous fumes entering the confined space - for example, from adjacent work or contamination being released from the ground (methane can leach into groundwater and be released or limestone may give off carbon dioxide in the presence of acid groundwater).

- The temperature in the confined space - ambient temperature may be too hot or too cold to allow ease of breathing or control a worker's body core temperature (welding activities will produce a lot of heat and with storage vessels that are located outside the temperature inside the vessels will rise significantly when exposed to sunlight).

- Mechanical equipment operating while workers are in the confined space – for example, a stirrer in a mixing vessel.

- Services connected to the confined space – for example, steam, electricity, water.

- Sources of ignition for flammable materials – in the confined space or brought in by workers, for example their equipment or portable lighting.

- Ventilation of the confined space – adequacy of natural ventilation or need to provide forced ventilation (consider monitoring the atmosphere, a source of fresh air for forced ventilation, the effects of failure of forced ventilation).

- Lighting limitations – adequacy of natural lighting or need to provide lighting (consider risks introduced by electrical lighting equipment).

It is important that the risk assessment also take account of:

- The task, as well as the materials and equipment involved.

- The people - including the competency of those involved and the suitability of those who will enter the confined space.

- What could influence safe access and egress from the confined space – for example, limitations of access routes and whether a hoist is required.

- What could influence the health and safety of the working environment.

- Reliability of control measures – need for a permit-to-work.

- Emergency arrangements – consider if a harness and lifeline are required, factors that will affect a rescue, the resources available to a rescue team and the location of external emergency services.

Precautions to include in a safe system of work for confined spaces

Similar to other risks, the priority would be to determine if entry into the confined space can be avoided, for example, by using cleaning and other equipment operated from outside the confined space (for example, water jets, long-handled tools or vibrating blockage clearing equipment) and use of viewing panels or cameras for inspection. If it is necessary to enter the confined space it is important to take into account the risks and put into place precautions to minimise the risks.

Some of the precautions to consider in a safe system of work for confined spaces are:

- Level of supervision - Supervisors should be given responsibility to make sure that the necessary precautions are taken, by checking precautions at each stage and they may need to remain present while work is carried out.

- Suitability of the workers – their competence (training and experience), size, fitness (ability to wear respiratory protective equipment) and claustrophobia.

- Isolation and lock-off of any service or material entry and exit points (such as water stop valves).

- Isolation of electrical systems and moving parts of the equipment.

- Removal of residues before entry into the space, by cleaning from outside the confined space.

- Atmosphere testing, before entry into the space and at intervals, personal monitors may also be required.

- Ventilation before and during entry into the space, forced ventilation may be needed.

- PPE requirements, including respiratory protection if required.

- Access (the way in) and egress (the way out), ladders or hoists may be required and it is important to ensure unobstructed routes for escape.

- Communication methods between those in the confined space and those outside.

- Fire precautions, deluge fire systems would usually need to be isolated so portable fire extinguishing equipment would need to be provided.

- Lighting levels, provision of additional temporary lighting. Consider flammability of the atmosphere when choosing the type of equipment.

- Equipment suitable for the confined space, extra low voltage (less than 25V) for work inside metal confined spaces, use of residual current devices, non-sparking tools.

- Control of the duration of a working period to manage fatigue and heat/cold exposure.

- Emergency rescue arrangements, for example a lifeline attached to a harness should run back to a point outside the confined space, someone to keep watch, communicate and raise the alarm.

The risk assessment will, in particular, help to identify the need for a formal administrative system such as a permit-to-work (**see Element 3.3 - Permit-to-work systems**). The permit-to-work-system will typically include consideration of precautions described above and, in particular, procedures for:

- Testing the atmosphere.

- Respiratory protective equipment and other personal protective equipment for those risks that cannot be controlled by other means.

- Equipment for safe access and egress.

- Suitable and sufficient emergency rescue arrangements.

Testing the atmosphere

Testing the air will need to be carried out to check that it is free from both toxic and flammable gases/vapours and that it is fit to breathe. Testing should be carried out by a competent person using suitable detectors, which are correctly calibrated. Where the risk assessment indicates that conditions may change, or as a further precaution, continuous monitoring of the air may be needed. The appropriate choice of testing equipment will depend on the particular circumstances of the confined space and the work to be done. Tests should be carried out by competent and experienced persons and records should be kept of the results of tests.

Personal air monitoring equipment should be worn whenever appropriate to minimise the risks from hazards present as local pockets of contaminant in the confined space.

Personal protective equipment

Where respiratory protective equipment is provided or used in connection with confined space entry (including emergency rescue), it must be suitable. Other equipment - ropes, harnesses, lifelines, resuscitating apparatus, first-aid equipment, protective clothing and other special equipment will usually need to be provided.

Figure 4-32: Typical entrance to a confined space.
Source: RMS.

Safe access to and egress from confined spaces

- Openings need to be sufficiently large and free from obstruction - to allow the passage of workers wearing the necessary protective clothing and equipment and to allow access for rescue purposes.

- Practice drills - to check that the size of openings and entry procedures are satisfactory.

- Identification of the confined space warning signs placed – to ensure the correct confined space is being worked on and to warn other that confined space work is in progress so they do not do anything that could harm those workers inside.

- An observer present at all times when work is in progress - the observer should stay in contact with the workers and be equipped to raise the alarm immediately if an emergency occurs.

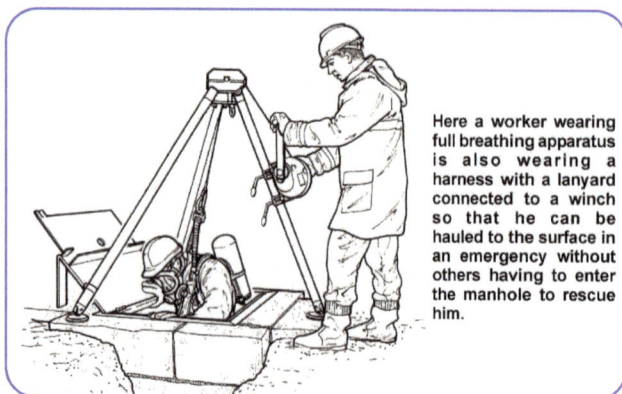

Here a worker wearing full breathing apparatus is also wearing a harness with a lanyard connected to a winch so that he can be hauled to the surface in an emergency without others having to enter the manhole to rescue him.

Figure 4-33: Access to a confined space.
Source: HSE, HSG150.

Emergency arrangements

Rescuers need to be properly trained, sufficiently fit to carry out their task, on standby, and capable of using any equipment provided for rescue, for example, breathing apparatus, lifelines and firefighting equipment. Rescuers also need to be protected against the cause of the emergency.

CSR 1997 prohibits any person to enter or carry out work in a confined space unless there are suitable and sufficient rescue arrangements in place.

The arrangements for emergency rescue will depend on the nature of the confined space, the risks identified and the likelihood of an emergency rescue being needed.

The arrangements might need to cover:

- Raising the alarm and rescue.

- Safeguarding the rescuers.

- First-aid, including resuscitation equipment.

- Special arrangements with local hospitals (for example, for foreseeable poisoning).

- Firefighting equipment.

- Public emergency services contact methods.

When a permit-to-work would not be required for a confined space

The decision not to adopt a permit-to-work system should be taken by a competent person, where necessary following consultation with specialists.

A permit-to-work would not be required for a confined space that did not present significant risks. This would typically be where a confined space was large enough for a worker to enter, but did not contain the hazards that could cause serious harm/death described earlier in this element and the work being done could also not introduce these hazards. This would be identified and confirmed through a risk assessment process.

REVIEW

List four possible risks when working in a confined space.

What should the competent person consider when developing a safe system of work for a confined space?

CASE STUDY

Prosecution of a company following a triple fatality which resulted from confined-space working.

A company transported waste water/sludge, using a tanker lorry, from food factories and abattoirs to various farms for spreading on agricultural land. This involved transferring the waste from the tanker into a holding tank (at the side of the field), from which it was pumped to a tractor for spreading on the agricultural site.

On occasions, a thick sludge built up in the bottom of the holding tank. This required an worker to climb onto the top of the tank and to use a high-pressure hose to dislodge and dilute the sludge. The person undertaking the task fell into the tank through an open hatch, became unconscious (the tank atmosphere was incapable of supporting life) and drowned in the liquid at the bottom of the tank.

Two men who attempted a rescue also suffered the same fate. A fourth man also tried a rescue but managed to get out.

In summary, there were no measures to prevent access to the tank, and training and risk assessments were said to be inadequate. The company and their directors were successfully prosecuted and fined.

Sources of reference

Reference information provided, in particular web links, is correct at time of publication, but may have changed.

Avoidance of danger from overhead electric power lines, Guidance Note GS6 (Fourth Edition), HSE Books, http://www.hse.gov.uk/pubns/gs6.pdf

Avoiding danger from underground services, HSG47, HSE Books, ISBN 978-0-7176-6584-6 http://www.hse.gov.uk/pubns/priced/hsg47.pdf

Managing Health and Safety in Construction, Construction (Design and Management) Regulations 2015, Guidance on Regulations, L153, HSE Books, ISBN: 978-0-7176-6626-3 http://www.hse.gov.uk/pubns/priced/L153.pdf

Personal Protective Equipment at Work (Guidance), L25, HSE Books, ISBN 978-07176-6139-3 http://www.hse.gov.uk/pubns/priced/l25.pdf

Protecting the public: Your next move, HSG151, HSE Books, ISBN 978 0 7176 6294 4 http://www.hse.gov.uk/pubns/priced/hsg151.pdf

Safe work in confined spaces. ACoP, regulations and guidance, L101, HSE Books, ISBN 978-0-7176-6233-3 http://www.hse.gov.uk/pubns/priced/l101.pdf

Web links to these references are provided on the RMS Publishing website for ease of use – www.rmspublishing.co.uk

Statutory provisions

Confined Space Regulations (CSR) 1997 / Confined Space Regulations (Northern Ireland) 1999

Construction (Design and Management) Regulations (CDMR) 2015 / Construction (Design and Management) Regulations (Northern Ireland) 2016

Personal Protective Equipment at Work Regulations (PPER) 1992 (as amended) / Personal Protective Equipment at Work Regulations (Northern Ireland) 1993

Work at Height Regulations (WAHR) 2005, as amended 2007 / Work at Height Regulations (Northern Ireland) 2005

STUDY QUESTIONS

1. What are the main hazards associated with excavation work adjacent to a busy dual carriageway with a speed limit of 40mph?

2. What are the control measures that could be considered for excavation work on a construction site?

3. (a)   Give FOUR examples of a confined space that may be encountered in construction activities?

   (b)   What are FOUR hazards associated with work in a confined space encountered in construction activities?

4. A leaking underground petrol tank has been emptied so that it can be visually inspected internally prior to repair. What are the features of a safe system of work for the inspection team in order to satisfy the requirements of the Confined Spaces Regulations 1997?

5. A trench is to be excavated to a depth of 1.5 metres through a park in order to repair the water mains supply to the on-site café.

(a) What are the possible hazards associated with this work?

(b) What control measures should be taken to minimise the risks to the workers and others who may be affected by the work?

6. A tank measuring 3 metres long, 2 metres wide and 2 metres deep is to be placed into the ground with the soil then back filled around and over it. What factors need to be considered in the risk assessment before excavation starts?

7. What control measures are needed to reduce the risks of excavation work on a construction site?

For guidance on how to answer these questions please refer to the 'study question answer guidance' section located at the back of this guide.

Contents

THE MEANING OF TERMS

Deconstruction

The term deconstruction is used to describe the taking apart and removal of a structure in such a way that it may be reused again. Reuse may be as before or in a different configuration. The term deconstruction may be related to any structure that is taken apart in an orderly way that enables reuse in part or whole, whether the structure was permanent or temporary.

The growth in environmental considerations has led to more permanent structures being deconstructed when redevelopment is needed, enabling the reuse of their component parts. Some industrial units are built in such a way that lends them to be deconstructed and used again elsewhere if they have not reached the end of their useful life, for example, when an industrial site grows, units may have to be changed or relocated.

Temporary structures are one particular group of structures that are designed for ease of construction and deconstruction. Temporary structures include those used at events, such as concerts and sporting events. This will include structures for spectator seating, hospitality gatherings, communication (large screens) and accommodation facilities for performers or athletes. The events may take place indoors in arenas, for example, the National Exhibition Centre (NEC), Birmingham, or outdoors under temporary cover, such as in marquees, and in the open air on show grounds. In addition, temporary film/television sets and exhibitions are constructed and later deconstructed.

The risks of deconstruction include major risks related to managing and controlling an orderly deconstruction that does not cause overloading of a small number of load-bearing parts and the structure's sudden collapse. Particular difficulties may also be experienced in deconstructing parts that have been strained in use and resist being taken apart. Procedures need to be in place to deal with these anticipated difficulties without putting workers at risk.

The timing and order of deconstruction can have a significant bearing on the hazards and their control, for example, where more than one aspect of the structure is being deconstructed at the same time they can conflict with each other leading to increased risks. Similarly, it is important to organise the isolation of power sources in such a way that hazards workers are exposed to be controlled. It may be necessary to isolate all power sources from the structure and provide independent supplies to enable the safe deconstruction. Other than these issues, the general hazards of deconstruction are essentially the same as those that might be experienced in constructing structures.

Planning should also involve the recognition that some of the control measures within the structure may not be available for the protection of workers as deconstruction progresses, for example, smoke detection and fire alarm facilities may be disabled at an early point in deconstruction. This would mean that protection of workers would have to be provided by different means.

Demolition

The term demolition is used to describe the tearing down, knocking down or blowing up of a structure that is no longer required. There are a number of techniques used in demolition processes, including piecemeal demolition, deliberate controlled collapse and explosive collapse.

Piecemeal demolition

Piecemeal demolition involves systematically removing pieces of the structure, usually in a similar order to which it was built. It is carried out by using hand held tools and/or machines. For example, when demolishing a tall chimney with occupied buildings in close proximity, the job may commence with the painstaking task of dismantling by hand - brick by brick. When the structure has been reduced to about 10 metres, then conventional heavy equipment can be used.

Deliberate controlled collapse

Deliberate controlled collapse involves the pre-weakening of the structure or building. This involves removing key structural members so the remaining structure collapses under its own weight.

There are several problems associated with this method of demolition:

- The structure may collapse to a greater extent than was anticipated, affecting neighbouring buildings.

- The planned collapse may only be a partial collapse and could leave the structure hazardous and insecure.

- The resultant debris may be projected over a wider area than anticipated, particularly dust.

- The pile of debris that is left after the collapse of the structure may be in a dangerous condition, presenting a serious risk of injury to those who remove it, for example, contaminated with biological hazards or asbestos.

One method of removing the key structural members is overturning by wire rope pulling. Wires are attached to the main supports, which are pulled away using a heavy tracked vehicle or a winch to provide the motive power.

The area must be cleared of workers for this operation and there must be enough clearance for the vehicle to move the distance required to pull out the structural supports. Problems associated with this method arise when the wire becomes overstressed. If it breaks, then whiplash can occur which can have sufficient force to slice through the human body.

The forces applied may be enough to overturn the winch or the tracked vehicle. Another problem can occur when the action of pulling has begun and there is inadequate power to complete it.

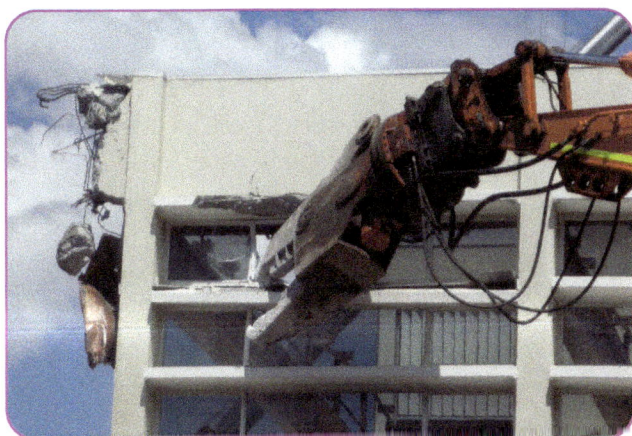

Figure 5-1: Hydraulic demolition jaws.
Source: Wikimedia Commons.

Figure 5-2: Hydraulic demolition grapple.
Source: Pixabay.

Explosive collapse

Explosives may be used to bring about the collapse of structures, for example, where other means are not practicable. The technique uses explosive charges to weaken load-bearing parts of the structure, causing it to collapse under its own weight.

Refurbishment

The term 'refurbishment' is a term used to describe a process of improvements by cleaning, redecorating, renovation or re-equipping and is usually carried out within or to the fabric of thean existing building or other structure.

THE SELECTION OF THE APPROPRIATE METHOD

A systematic approach to demolition and deconstruction projects is a team effort between many people, who all have responsibilities. Clients must appoint duty holders who are competent and adequately resourced. These will include a principal designer to survey the site and a principal contractor to plan effective site management that keeps people (site workers and the public) as far as possible from the risks, and provides principal contractors with as much information as possible.

The selection of an appropriate method will need to take into account the method of construction used for the original building and its proximity to other buildings, structures and the general public. These factors, together with location, the cost and availability of waste disposal and the desirability and economics of reuse, must also be considered.

'As-built' drawings and structural details may often be unavailable or unreliable, consequently a site investigation will be necessary. This will ascertain the way in which the building was originally constructed, and identify the stresses and strains which exist within it.

Failure to establish these facts may result in risks to the safety of both those involved in the demolition and those in close proximity to the site.

HAZARDS RELATING TO DECONSTRUCTION AND DISMANTLING

Premature collapse

Premature collapse can occur because the structure is in a poor and dilapidated condition or the method used to deconstruct or demolish the structure has been done ineffectively leading to premature collapse. A common reason for this type of structural failure is due to the lack of effective planning before demolition commences. Collapses may be:

- Localised collapse, for example, a collapse confined to a small part of the structure, without damage elsewhere.
- Progressive collapse, for example, the isolated failure of a critical member or section initiates a sequence of events, causing failure of the entire structure.

Collapse due to dilapidation

Dilapidation is the state of disrepair of a building. Dilapidation of structures and building may occur for a number of reasons, for example, a coal fired power station may no longer be required and abandoned, the owner of a house may no longer be in occupancy or a building may not be maintained for economic reasons, such as a non-profitable Victorian pleasure pier at a sea resort.

Dilapidation develops over time through lack of maintenance, resulting in decay of the outer fabric of the structure leading to water ingress, wind damage and ultimate weakening of the whole structure. Dilapidation presents particularly high risks to those conducting deconstruction/demolition work and may influence the decision for the structure to be demolished, rather than deconstructed, as it may be possible to demolish it from a safe distance.

Typical problems related to dilapidation include:

- Presence of weak/deteriorating materials and structural elements.
- Complex load paths, partial collapse may have caused additional load and stress on the remaining load-bearing parts.
- High levels of structural instability.

Premature structural collapses can occur with dilapidated structures for a variety of reasons. It can result from a major structural fault arising from dilapidation or smaller faults can contribute to a chain of events that leads to a collapse while work is being conducted.

Figure 5-3 shows the result of a sudden collapse of a three storey building in Berkhamsted High Street, Hemel Hempstead in February 2011. It is believed a structural fault in the building caused the collapse.

Figure 5-3: Building collapse.
Source: The Gazette.

Collapse due to the method used

In addition, even if the building is not in a dilapidated condition the process of deconstruction or demolition can create a hazard of premature collapse or increase the risk if it is already there. Removal of part of a supporting wall could be sufficient to release cast concrete floor slabs to drop from one level to another, and the resultant kinetic energy can lead to the further collapse of floors below.

The likelihood of premature collapse is a particular consideration when using the wrecking ball method or carrying out partial demolition. In some cases the façade of a building may be left intact for renovation and the rest of the structure removed. In these circumstances great care must be taken to avoid freestanding walls and in order to provide adequate support to the structure. When carrying out wrecking ball demolition the impact of the ball is designed to weaken the structure; if it is not used in a planned and careful manner the wrong parts of the structure, or too much of it, may be weakened leading to premature collapse.

Premature collapse can also be caused when carrying out systematic deconstruction or demolition by hand, particularly when load-bearing parts of the structure, such as beams, lintels, load-bearing walls and roof supports are removed. Premature collapse can occur in a similar way when carrying out building/structure undermining or dismantling.

Existence of services such as electricity, gas and water

Services that may be encountered during deconstruction/demolition present different hazards. Electricity may be still present in a building's mains supply because it had not been isolated, and presents a risk of electric shock. Sometimes there is a belief that mains circuits have been isolated, but because it is fed from more than one route only part of the supply has been dealt with. Some buildings have backup systems in the form of batteries and/or capacitors which can present an unexpected electrical hazard when encountered. High voltage power supplies may pass underground or overhead, and these will present a significant hazard if not isolated before work commences.

Gas supplies routed into or under a building, if not isolated, may carry a risk of fire or explosion if they are disrupted. Residual quantities of gas left in isolated services within a building can present a lesser hazard but should not be ignored.

Water supplies that are damaged can present an indirect hazard in that they can lead to flooding of lower areas of a building where workers may be present. In addition, it may undermine part of the structure leading to premature collapse, increase the risk of slips trips and falls or aggravate a risk of electric shock.

Falls and falling material

Depending on the type of building and method of deconstruction or demolition, workers may be expected to work at a significant height, for example, when demolishing a brick built structure by hand, and this can present a fall hazard. In the case of hand deconstruction/demolition it would be usual to remove parts of the structure in essentially the reverse order that it was erected. As such the same hazard of falling would exist as would be the case when constructing such a building. Fall hazards can often occur in the deconstruction/demolition process by the action of removing material, such as floors, walls and stairs, or when access doors are not secured when part of the structure is removed. Part of the process may mean the removal of equipment built into the structure to prevent falls when in the structure was in normal use, thus exposing hazards, such as a rail and barrier around an atrium walkway.

Figure 5-4: Risk of falls and materials falling.
Source: RMS.

As with other construction work there is a risk of a worker falling outside the framework of a scaffold erected as part of the deconstruction/demolition process, particularly where workers over-reach. If the site is not well controlled there is a possibility of unauthorised people gaining access to the area, and in so doing they could be exposed to the risk of falling.

The process of deconstruction/demolition involves the movement of large quantities of material to ground level. If this is not carried out in an organised way and material is allowed to fall it may expose workers and passers-by to the hazard. This is particularly the case where workers are carrying out demolition above other workers or near to access routes.

In addition, mechanical or explosive demolition may eject material, which could fall significant distances from the demolition process. Even small items of material represent a significant hazard when they are allowed to fall from a great height. In demolition processes, depending on how well they are conducted, material may fall in significant quantities and unexpectedly.

Figure 5-5: Risk of falling or premature collapse.
Source: RMS.

Figure 5-6: Falling materials.
Source: RMS.

Plant, vehicles and other equipment overturning

Demolition processes create large quantities of material that accumulate at low level. As this builds up, demolition equipment may be working on top of it to gain access to part of a structure. This can cause the centre of gravity of the equipment to be moved towards the edge of, or outside its stability base. Sudden movement or an unexpected heavy load can cause it to overbalance and overturn.

Other ways in which the hazard of overturning may present itself are:

- Weak ground support, for example, cellars, sump covers, soft ground, causing one side of the equipment to be lower than another.
- Operating outside the machine's capability, for example, too large a load at too great a distance.
- Striking or getting caught up in fixed obstructions.

Manual handling

In demolition processes there are large quantities of a wide variety of materials to be moved. This presents a potential manual handling hazard, though, in many cases, much of the manual handling hazards have been removed by the use of mechanical handling equipment. In situations where deconstruction/demolition is carried out by hand there is usually a lot of debris to move which may also be carried out by hand as the structure is reduced. In some cases the distance debris is moved is limited by the provision of chutes. Where material is being salvaged this can mean an increase in manual handling is necessary in order to safeguard the condition of the material.

Hazardous substances

The process of deconstruction/demolition can liberate large quantities of a range of materials relating to the structure being deconstructed or demolished. A lot of these materials may become present in the air as dust in areas where people are working or the public is passing by. If the process is conducted in dry conditions and windy conditions the dust may be more mobile and therefore presents more of a risk. Some of the dusts, such as **brick dust**, may be considered a nuisance in the open air, but when encountered in a confined area where workers are carrying out activities may be a hazard because of the quantity of dust in the air. Dust may be created from various construction materials, including concrete based materials that contain silica, which if breathed in over a period of time could cause **silicosis.**

Dust may also contain **fungal and bacterial matter** or other hazardous substances. As structures fall out of use it is not uncommon for pigeons to take up residence in them; their droppings contain a biological hazard that can lead to a lung disease called **psittacosis** if the dust from the droppings are breathed in.

During deconstruction/demolition sewers may become exposed and present a risk of biological contamination from substances within them or diseases carried by vermin, for example, **hepatitis** or **leptospirosis.**

In addition, the prior use of the structure may have led to the presence of chemical deposits in the form of contaminated dusts; this may include toxic substances such as **arsenic**. Hazardous substances may also be present on the deconstruction/demolition site because they are left over from prior use. In industrial buildings this may mean such things as **cadmium, arsenic, mercury, acids, alkalis and fuel oils and solvents** (some of which may be prohibited by current good practices). In laboratory or hospital buildings that are to be demolished biological hazards may be present.

Structures that are being demolished may contain certain construction materials used in the construction of the structure which present a hazard as a dust or fume, for example, **asbestos** as fire insulation, and **lead** as a protective paint for metalwork.

Figure 5-7: Noise and dust.
Source: Pixabay.

Figure 5-8: Noisy and dusty plant.
Source: RMS.

Asbestos has been commonly used in buildings for many years and when the substance is broken up and fibres get into the air it can present a significant health risk. Asbestos fibres may get into the lungs and it can take many years (sometimes up to 40 years) before the symptoms of illnesses, like mesothelioma (cancer of the membrane lining the lungs), become apparent. The asbestos fibres may also be released into the air and contaminate clothing, which may pose an additional risk of harm to other workers and the public in the vicinity.

Asbestos cement drain pipe and cladding

Decorative coating containing asbestos

Asbestos cement flue in a ceiling void

Figure 5-9: Asbestos use in buildings.
Source: HSE, HSG213.

The type of asbestos fibre used in the manufacture of different products and present in buildings today has varied over time, for example, white asbestos was the main form used in the manufacture of asbestos cement by the end of 1999; blue asbestos was used from 1950–69 and brown asbestos between 1945 to 1985. Asbestos cement containing blue or brown asbestos is more hazardous than that which contains white asbestos alone. Asbestos may be present in lagging of boilers and pipework, fire proofing of structures, decorative ceiling materials and roofing materials.

Where vehicles are being used to transport the demolition waste material to a disposal site, they may inadvertently carry hazardous substances that could come off their wheels or drop off when travelling. Deconstruction/demolition work also possesses the potential to generate waste substances, such as oils, which can leak away into the soil. In addition, hazardous substances from the site could contaminate sewers.

Carbon monoxide presents a chemical asphyxiant hazard and may be given off by equipment used to support the demolition process, for example, demolition plant or generators used to provide temporary power supply.

Noise and vibration

Much of the equipment used in the deconstruction/demolition process can present a noise hazard. Not surprisingly explosive charges present a noise hazard, as does impact equipment, such as wrecking ball plant. Crusher and manipulation equipment can also present a noise hazard as the hydraulic system draws its power from the equipment's engine. Debris crusher plant can emit a large amount of noise as debris such as bricks is reduced.

Hand deconstruction/demolition may be supplemented by the use of electric or pneumatic chisels which emit high noise levels. On a busy site this cumulative noise can have an effect on workers and people nearby. Neighbours or other residents of an adjoining property may be affected by deconstruction/demolition work in the form of ground borne vibration, surface shock wave, or air overpressure from demolition work. The use of hammer drills, pneumatic chisels, hammers and other similar equipment presents a vibration hazard that could lead to hand arm vibration syndrome (HAVS). The operation of demolition equipment such as manipulation and bucket equipment can expose the operator to whole body vibration syndrome (WBVS).

Fire and explosion

As with any other similar operation or undertaking, fire and explosion hazards are nearly always present in deconstruction/demolition activities. These can result from hot work (welding or cutting with oxy-acetylene torches), deliberate burning of waste materials on site or arson. Explosive gases could be present following accidental damage to underground or over ground piped services and flammable liquids may still be present on site in fuel tanks.

CONTROL MEASURES FOR DECONSTRUCTION AND DISMANTLING WORK

The Construction (Design and Management) Regulations (CDMR) 2015 apply to all construction projects, of which demolition is a part. CDMR 2015 focuses on the responsibilities that make the management of construction projects effective in managing health and safety risks, for example, competence, co-operation, co-ordination, principles of prevention and the roles of those involved in the project.

In addition, CDMR 2015, Part 4, sets out duties related to health and safety on construction sites, including - good order and site security, stability of structures, demolition and dismantling, explosives, excavations, energy distribution installations, traffic routes and vehicles.

Regulation 20, of Part 4, in CDMR 2015 requires that:

"(1) The demolition or dismantling of a structure must be planned and carried out in such a manner as to prevent danger or, where it is not practicable to prevent it, to reduce danger to as low a level as is reasonably practicable.

(2) The arrangements for carrying out such demolition or dismantling must be recorded in writing before the demolition or dismantling work begins."

Figure 5-10: Demolition or dismantling.
Source: CDMR 2015 Regulation 20.

Avoidance of premature collapse

Irrespective of which deconstruction/demolition process is used, it is important that the deconstruction/demolition be conducted in a systematic way that prevents premature collapse of the structure. This will mean working with a good understanding of the building's strong and weak points. It is important that the deconstruction/demolition process does not leave unsupported walls or parts of a structure. Those controlling the deconstruction/demolition process should have a good understanding of building structure and mechanics, to avoid removing parts of the structure that can place an excessive load on other fragile parts. Modern machines can be fitted with scissor jaws that can break down a wall in small pieces. The use of this process reduces the likelihood of major structural damage or premature collapse.

When the wrecking ball method or partial demolition is conducted, as these processes often leave free standing structures, it is important that a careful deconstruction/demolition plan is adhered to. As part of this plan it is important to leave the structure in its most stable condition when not working on it. This may mean not starting a phase of deconstruction/demolition because it cannot be completed before the end of the day or providing shoring to the structure that is left. If a part of a building, such as a façade is to be left intact for renovation it is important to provide adequate longer term support.

Services

Details and location of all the public utility services must be identified in order to identify the hazards they present and to take action to isolate, divert or otherwise protect them. This will involve the examination of service location drawings held by the utility service organisations.

Most of these will be buried services, including gas pipes, electricity cables, water pipes and sewerage. This will be supported by visual examination, test equipment and, in some cases, test digging to establish the specific location and status of the services.

Isolations

It is important to isolate services at the furthest point practicable from the structure to be demolished, for example, at the site boundary.

This may include gas or water services isolated in the pavement outside the demolition site. This will help to remove the hazard at source and is preferable to local isolation of the services within the building. It should be remembered that some systems will store energy from services and should not be presumed safe merely by isolating the service. This includes electricity stored in capacitors and batteries, stored water in tanks, and compressed oils/air in pressure systems.

Temporary services

It may be necessary to install temporary services such as electricity, lighting and fresh air. These services must be installed by competent people and, in the case of fresh air supply; there must be an audible and visual warning of failure.

Protection from falls and falling material

The technical advances in mechanical plant have considerably reduced the risk to deconstruction/demolition workers of working at height. For example, modern equipment enables a lot of demolition to be conducted by agile high reach equipment.

When deconstruction/demolition work is carried out by hand it involves the workers moving about at height on materials that may have degraded and are not able to sustain their weight. In this type of operation full scaffolding and other platforms should be provided and the building deconstructed/demolished in the reverse order of erection.

Where it is necessary for people to work at height the same standards of protection from falls must be provided to deconstruction/demolition workers as are provided to other construction workers. This will include provision of safe work platforms, use of mobile elevating work platforms (MEWPs), nets and fall arrest equipment.

In order to prevent debris falling on to passers-by, protective screens, and where appropriate fans, should be set up around scaffolds erected around structures undergoing deconstruction/demolition. In addition, brick guards, debris nets and sheeting will help to contain falling materials and prevent them falling outside the framework of a scaffold erected as part of the deconstruction/demolition process. Areas of a structure under deconstruction/demolition that present a risk from falling material must be clearly demarcated from other areas. Where buildings are being deconstructed/demolished by hand a safe access route to the workplace must be maintained, which may include demarcated routes and temporary erection of covered areas. The use of audible and visual warnings will assist in removing people from hazard areas at times that they are at risk from a particular demolition task. Head protection is an essential last line of defence for workers in this environment.

Access and egress from the site

Access and egress for site vehicles is an essential part of site planning. Access to the site should be planned and established at an early stage to enable the large plant and appliances to be taken or driven to the site safely. It may also be necessary to get permission to close roads or pathways for the duration of the work.

Consideration should be given to:

- The appropriateness of public roads in the proximity of the site. Select routes of suitable width and schedule access to avoid times when large numbers of domestic vehicles may be parked in the road whenever possible.

- The nature of the surrounding buildings and their function, for example, the presence of schools or hospitals for which traffic and noise would be a problem.

- The number and size of vehicles expected to enter and leave the site should be determined, and their arrival and departures scheduled to avoid back-up of waiting vehicles onto public highways.

- Speed limits should be imposed on site vehicles, limit signs should be displayed and the limit enforced.

- Reversing from the site must be avoided or if reversing is necessary a banksman should be used to control the process.

If necessary, vehicle movements should be planned to avoid rush hour traffic and steps taken to minimise contamination of public roads, such as vehicle wheel washing.

Siting and use of plant, vehicles and other equipment

The key to safe siting and use of plant is planning and control. The appointment of a competent person to organise the operation of the plant, vehicles and other equipment is essential.

Care with siting of plant is necessary in order to prevent accidental contact with nearby structures, the structure being deconstructed/demolished, power lines or other overhead obstructions. In restricted areas tight control of the use of plant and vehicles is essential; this may include physical barriers to restrict movement, where practicable.

Due to the dangerous nature of deconstruction/demolition work all plant, vehicles and equipment should only be used by trained, competent and authorised workers.

In order for safe operation of plant, vehicles and other equipment care should be taken to:

- Identify weak ground conditions.

- Avoid operating outside the equipment's capability.

- Create space sufficient to manoeuvre to avoid striking people and obstructions.

- Maintain safe access to work areas by the prompt removal of demolition debris.

Consideration should be given to the movement of vehicles and people on and around the site. Arrangements for access and egress for loading and unloading materials should be considered. Segregation of vehicle and pedestrian traffic should be established.

Access and egress routes should be kept clear of any obstructions, with traffic and pedestrian routes clearly marked and identified by signs. Emergency routes may have unique signage. All routes should be inspected on a regular basis for compliance. Where possible, it is good practice to operate a one-way traffic management system.

It is important to restrict out of hours access to plant, vehicles and other machinery to prevent unauthorised use, by, for example, children. This will include securing plant in compounds, locking doors and securing keys effectively. Taking steps to prevent unauthorised access to the deconstruction/demolition site by using barriers, fencing, and increased security is the responsibility of the principal contractor.

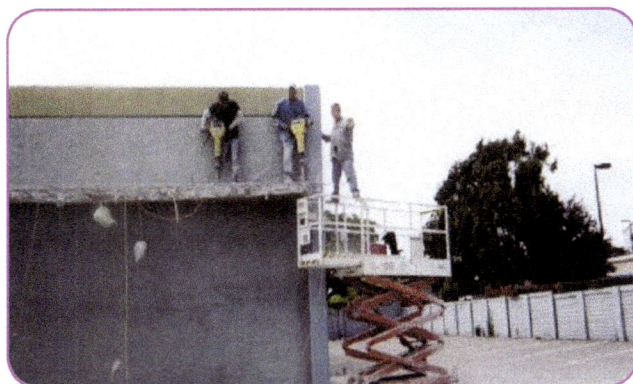

Figure 5-11: Doubtful competence of workers.
Source: Lincsafe.

Figure 5-12: Use of demolition equipment.
Source: RMS.

Competence of workforce

It is the client's responsibility to check the competence of the principal contractor. In turn, the principal contractor must ensure the competence of contractors appointed to conduct demolition and dismantling. Competence checks can include taking up references on past work and reviewing contractor's health and safety policies and procedures. It is important to recognise that deconstruction/demolition contractors may not be competent to act as principal contractors on larger or more complex projects.

It is important that deconstruction/demolition work is controlled by someone with an understanding of buildings/structures and in turn that workers understand the effects on the building or structure that their work will make. In this way premature collapse may be prevented. Operators of plant and equipment must be competent to operate the specific items they use, in demolition circumstances. Operation in deconstruction/demolition work can be quite different to operation in other building work and is particularly affected by the terrain of the demolition work area.

Proximity and condition of other structures and roadways

If a structure is close to the building/structure to be demolished it may be necessary to shore it up and route demolition plant and equipment away from it.

The proximity and condition of roads to a site is an essential item to consider. If roads are good, and wide enough to take plant transport vehicles and debris vehicles, this will have a positive influence on the safety of the deconstruction/demolition activity.

If restrictions exist, it may be necessary to obtain police co-operation to clear roads and to time the delivery of plant when least disruption will occur.

A new, dedicated access may be necessary to provide access to a larger road nearby.

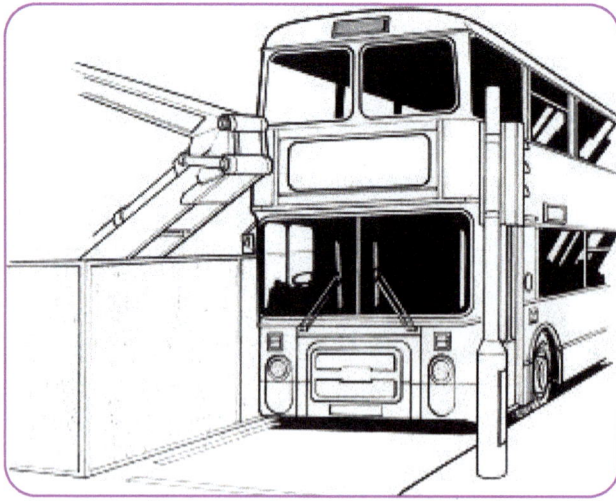

Figure 5-13: Risk of collision due to close proximity with public road. Source: HSE, HSG151.

This may influence the order of work to be conducted as part of the building/structure may need to be demolished to create better access. It is important to ensure that equipment and parts of equipment such as elbows of cranes, excavators, loaders etc. do not swing into the path of vehicles or pedestrians. Early identification of risk and nuisance to adjoining land, buildings or road users will enable early notification to those affected.

Use of explosives

The use of explosives requires the expertise of an experienced explosives engineer. Also, the Health and Safety Executive (HSE) should be consulted to ensure compliance with the Control of Explosives Regulations (CER) 1991 and HSE Construction information sheet No 45.

When designing for demolition using explosives it is important to plan for the risk of possible ejection of projectiles. An exclusion zone should be established at a distance from and surrounding any structure that is being demolished using explosives. Everyone, with the possible exception of the shot-firer, should be outside the exclusion zone at the time of the blast. If the shot-firer needs to remain within the zone, a position of safety must be created.

An exclusion zone is built up from four areas:

1) Plan area - the plan area of the structure that is to be demolished.

2) Designed drop area - the area covered by the main debris pile.

3) Predicted debris area - the maximum area in which fragments can fall outside the designed debris area.

4) Buffer area - the area between the predicted debris area and the exclusion zone perimeter.

Consideration should also be given to services, for example, ground vibration which could damage services.

Regulation 21, of Part 4, in CDMR 2015 requires that:

"(1) So far as is reasonably practicable, explosives must be stored, transported and used safely and securely.

(2) An explosive charge may be used or fired only if suitable and sufficient steps have been taken to ensure that no person is exposed to risk of injury from the explosion or from projected or flying material caused by the explosion."

Figure 5-15: Explosives.
Source: CDMR 2015 Regulation 21.

There are a number of factors to be considered for safe demolition with the use of explosives, which include:

- The local and structural conditions must be considered when fixing the size of the charges. The structure to be blasted can be divided into a number of sections and suitable charges applied to each section.

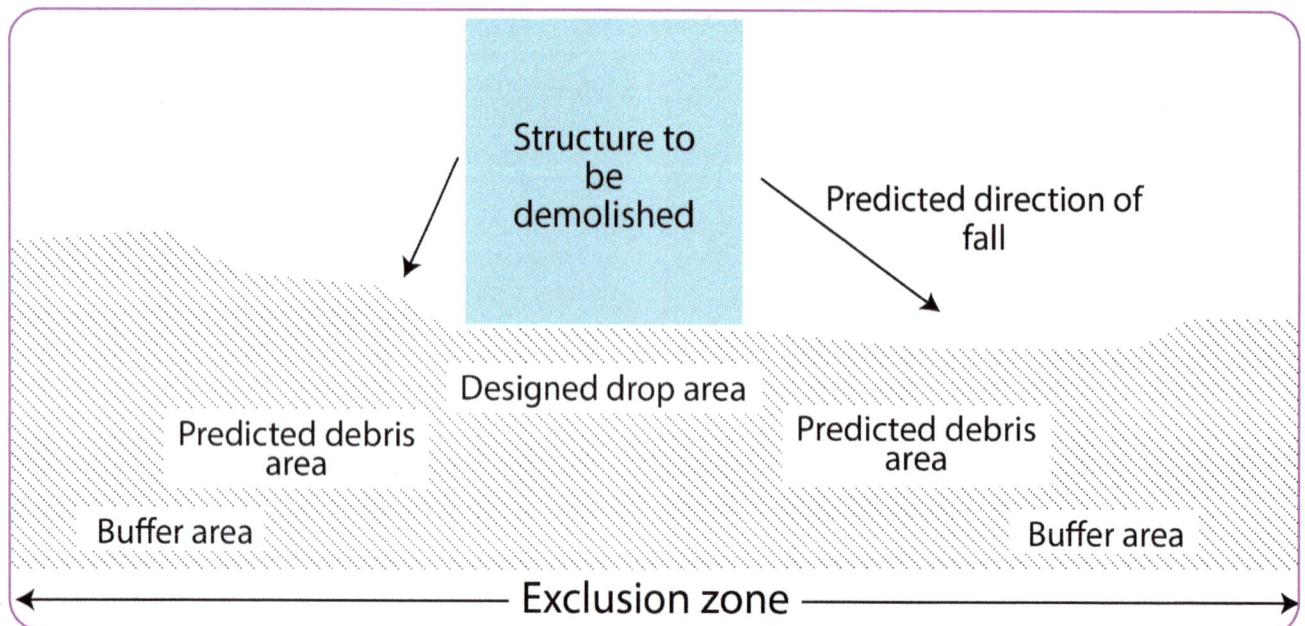

Figure 5-14: Demolition exclusion zone.
Source: Adapted from HSE construction sheet No 45.

- Shot holes should be drilled electrically. Drilling pneumatically could cause vibration, which could result in premature collapse.

- Charges should not be placed near cast iron as it easily shatters into shrapnel.

- The area around the structure being demolished and the firing point should be barricaded and unauthorised persons not allowed to enter.

- The demolition engineer must be satisfied that no dangerous situation or condition has been left or created. The danger zone must be barricaded until rendered safe.

Protection of public, third parties and surrounding structures

The protection of the public, third parties and surrounding structures is an essential pre-deconstruction/demolition control measure. Control measures should be put in place to remove and prevent people accessing the building/structure, such as children, persons sleeping rough and those who wish to salvage materials, as well as people who are just passing by. The boundaries of the demolition site need to provide an effective barrier to persons not involved with the work, typically 2.5 metres high fencing with warning signs posted around it.

Fencing must be maintained to a good standard and checks should be made at the end of each day to ensure it is in place, and continues to be effective. Where the work is close to where the public may pass they should be protected from falling materials by the use of sheeting, brick guards, nets, fans and covered areas.

Figure 5-17: Protection of public.
Source: RMS.

Figure 5-18: Disposal of waste.
Source: RMS.

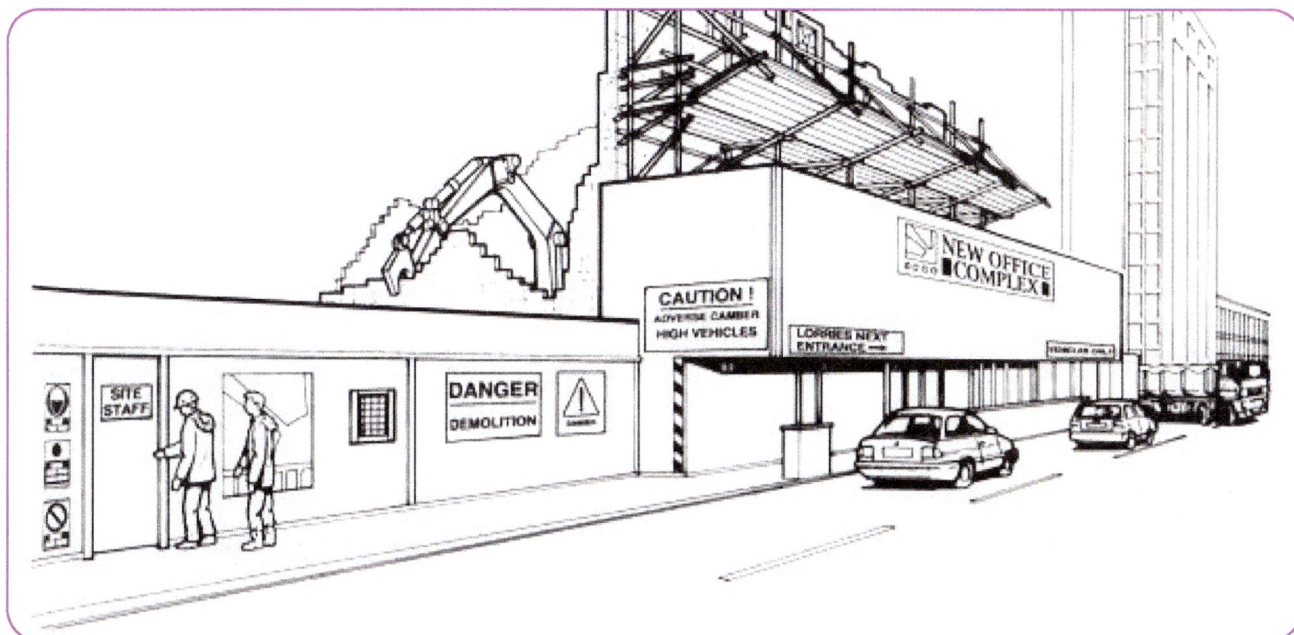

Figure 5-16: Demolition - seperate entrance for workers and protected pedestrian route.
Source: HSE, HSG151.

Manual handling

Most of the deconstruction/demolition process should, where possible, be carried out by using mechanical equipment to prevent or reduce any potential manual handling hazards. Chutes and conveyors should be used to transport debris from the upper floors of the building into skips, thereby eliminating any manual handling between floors.

Dust and hazardous substances

Large volumes of dust can be dissipated when deconstruction/demolition work is conducted. This can be in the form of brick dust, asbestos or cement based materials. Asbestos is a special case, and the substance should be identified and removed by specialists who are authorised and registered by the Health and Safety Executive (HSE) to remove asbestos material. All asbestos should be removed before any deconstruction/demolition work takes place. The liberation of other dusts can be limited and restricted by the amount of deconstruction/demolition activities conducted at any one time.

Consideration of when to demolish and dismantle is also a key factor as wind direction can play a significant role, affecting the direction of travel of dust and materials. Exclusion of workers while a dust producing activity is conducted and subsequent clean up processes can also control the exposure of workers to dust. Consider water suppression and extraction for processes that generate dust.

The build-up of fume emissions, typically from diesel powered equipment, can be reduced by limiting the amount of plant and equipment activities conducted at any one time. This can be combined with control measures or restrictions imposed to segregate plant and equipment from any person who is likely to come into contact with such fume emissions. The use of modern plant and equipment significantly reduces the level of emissions as they possess more efficient engines with cleaner exhaust systems.

Where fumes may be emitted from the actual deconstruction/demolition process, for example, when demolishing some form of storage vessel, such as in a chemical plant or hospital, potential emission should be restricted by draining any system or vessel holding fume producing substances in a controlled and planned manner.

Noise and vibration

Noise and vibration can be a major cause of harm to workers; therefore it is important that suitable and sufficient risk assessments are produced. Exposure to noise and vibration should be minimised by the careful selection and maintenance of equipment such that levels emitted are kept to a minimum. For example, vibration damped seats for cab operated equipment, vibration mats for areas where workers stand to operate crusher plant, and vibration damped handles for power tools.

Exposure levels can be controlled by the careful use of task rotation such that those exposed to high noise and vibration levels are periodically moved to tasks that have low exposure levels. Where levels of noise are above the action values hearing protection must be made available or provided, as appropriate and it is the duty of the employer to enforce and ensure that such protection is worn. A noise protected refuge may be required in order to allow workers to move out of noisy areas to limit their exposure during rest breaks.

Waste management - on site segregation and off site disposal

The construction industry is one of the largest producers of waste in the UK. Construction and deconstruction/demolition waste can be broken down into the following:

- Concrete, bricks and tiles.
- Asphalt, tar and tar by-products.
- Metals.
- Soil and rubble.
- Wood.
- Soil from ground works (which may be contaminated, through the sites previous use).

It is important that arrangements are established for waste management. This means that waste products should be identified before disturbance and an action plan developed to regulate how these products should be handled.

Taking into account the size of the site, waste material must not be allowed to accumulate. This means controlling build-up of waste at local points around the site, particularly where they impede access and egress, as well as any major accumulation points.

Waste should be segregated in terms of type, identifying waste that may be re-used, recycled and recovered and segregating it from waste for disposal.

All waste products and residues in containers must be identified and clearly labelled and steps taken to prevent the combined storage of incompatible materials prior to disposal. Offsite disposal must be carried out by a licensed contractor and taken to a licensed waste disposal site. If necessary, waste removal should be planned to avoid rush hour traffic and steps taken such as wheel washing to prevent contamination on public roads.

Protection of the environment

Where vehicles are being used to transport the demolition waste material to a disposal site, then provision should be made for wheel washes and the cleaning of the highway by a road sweeper. Waste substances, such as oils, must not be allowed to leak away into the soil and must be disposed of in a controlled manner. During demolition sewers may become exposed and be at risk of contamination from substances, so care should be taken to identify and protect them from entry of such materials. In the same way, it may be necessary to establish interceptor pits to control water run-off into water courses and/or fit temporary pressure drain covers.

Emergency arrangements

Emergency procedures and arrangements should be established for dealing with foreseeable emergencies, for example, fire, explosions, chemical leaks and premature collapse of the structure. Fire detection, fire alarms, fire-fighting equipment, first-aid provisions, site evacuation, and rescue plans should form part of these arrangements. Procedures should specify necessary training in the use of emergency techniques and equipment.

> **REVIEW**
>
> What hazards may be present during the demolition of a building?

5.2 Pre-demolition, deconstruction or refurbishment surveys

PURPOSE AND SCOPE OF SURVEYS

The purpose of a pre-demolition, deconstruction or refurbishment survey is to ensure that before any work begins, the structure to be worked on is assessed to determine inherent hazards related to the structure and those hazards that are likely to emanate from the proposed work on the structure.

The scope of the survey will include the following hazards:

- Premature collapse.
- Falls and falling materials.
- Plant, vehicles and other equipment overturning.
- Access and egress restrictions.
- Manual handling.
- Dust and fumes.
- Noise and vibration.
- Existence of services.
- Hazardous substances.
- Dilapidation.
- Presence of fragile and sharp materials.
- Weak surfaces, for example, cellars.
- Proximity of other structures, vehicles and people that might be affected.
- Likely waste issues.

Figure 5-19: Dilapidation of a structure.
Source: RMS.

The information determined by the survey will be used to determine a method of working that ensures satisfactory control measures and an orderly work process are in place.

DUTIES OF THE PROPERTY OWNER FOR CARRYING OUT A PRE-DEMOLITION SURVEY

Clients of projects that fall within the scope of CDMR 2015 must provide all designers and contractors with suitable pre-construction (demolition is a form of construction work under CDMR 2015) information that will allow hazards to be identified and work to be planned. It is not sufficient to state that "asbestos might be present".

The relevant information might be in an existing survey (for example, an asbestos survey) or in an existing health and safety file. If clients do not have sufficient information available, then they must make arrangements to obtain the information, for example, by arranging the necessary surveys to be carried out by a competent person.

When there is more than one contractor working on the project, the client must appoint a principal designer. The principal designer will assist the client in obtaining information to fill in any gaps and the provision of the pre-construction information to other designers and contractors (including the principal contractor).

It is the client's responsibility to ensure that any building/ structure to be demolished is adequately surveyed. Before any demolition work begins a pre-demolition survey should be carried out to ensure that the demolition can go ahead safely and that property surrounding the site is protected from the demolition work as it progresses.

The survey should give special consideration to the location of the intended demolition work, for example, if it is near a place of worship, school, hospital, shopping area, leisure centre, or housing estate then the survey may identify the need for extra provisions to control risks to members of the public. It is important to know the structural features and previous use of the building/ structure.

It may be made of pre-tensioned material and still contain toxic or flammable materials, biological hazards or even radioactive materials or sources.

IDENTIFICATION OF FEATURES OF THE STRUCTURE

Type of structure

The survey should clearly identify the type of structure to be worked on, for example, high rise office blocks, chimneys or old buildings, as this will have a considerable impact on the method of deconstruction/demolition, the hazards present and the precautions to be taken.

Figure 5-20: Building structure.
Source: RMS.

Figure 5-21: Proximity of other structures.
Source: Pixabay.

Key structural elements

It is important that the survey identifies the key structural elements in order to evaluate the effects on the structure of the proposed work. The location and condition of lintels, beams and other load-bearing members will need to be taken into account when planning how the work is to be carried out. The structural elements may be made of steel beams or concrete, which may be reinforced, pre-tensioned or post-tensioned.

Reinforced concrete is poured around rods of steel that are formed to the shape of the structure being created.

Pre-tensioned concrete is a method of overcoming the concrete's natural weakness in tension. It is used to produce beams, floors or bridges with a longer span than is practical with ordinary reinforced concrete.

Pre-tensioned tendons (generally of high tensile steel cable or rods) are used to provide a clamping force that produces a compressive stress to offset the tensile stress that the concrete compression member would otherwise experience due to bending load. Concrete is cast around already the pre-tensioned tendons.

Post-tensioned concrete is the descriptive term for a method of applying compression after pouring and curing concrete and is conducted in situ. Tendons are passed through the duct that provides the proposed shape of the member and the concrete is poured. Once the concrete has hardened, the tendons are tensioned by hydraulic jacks; when the tendons have stretched sufficiently they are locked in position and continue to transfer pressure to the concrete.

Demolishing reinforced concrete does not present any unusual hazards. However, demolishing pre-tensioned and post tensioned structures can be difficult because of the energy stored in the stressing tendons, which are under constant tensile stress. The uncontrolled release of this stored energy can cause a concrete component, or parts of it, to suddenly fly through the air in an uncontrolled way, which could cause serious injury. Planning and care needs to be taken when the tensioned tendons are cut. The tendons in tensioned structures are generally made of high tensile steel so there will be fewer to cut compared to standard reinforced concrete.

Presence of cellars

Basements, cellars, wells or storage tanks need to be identified in order to apply control measures to the risks they create, such as falls from a height and drowning. Workers and equipment may fall through into underground areas, while tanks pose a special risk if they have held or still hold toxic or flammable materials.

Location and type of services

The location and type of services connected to the structure being worked on should be identified. Consideration should be given to both over-ground, including, for example, pipe racks, power lines and telephone wires; and underground, such as pipes, cables and equipment associated with the electricity, gas, water (including piped sewage) and telecommunications industries. This will also include the identification of other pipelines that may transport a range of petrochemical and other fluids through the site.

See also 'Element 4.2 - Control measures for excavation work' - Identification and detection of buried services.

Significance and extent of dilapidation of the structure

Structural condition

The structural condition of the building/structure should be surveyed in order to identify any degradation of materials that might influence the method and order of demolition. The structural condition might present such risks that it is not safe to carry out demolition by going inside or working on the roof of the building/structure and other means, such as the use of mechanical demolition equipment, may be required.

The survey should also determine the extent of the dilapidation of the structure and its significance to the health and safety of those that will be affected by the work to be conducted. The dilapidation may mean that asbestos materials have become damaged and are exposed.

Floors may have collapsed and stairs that could have been able to be used for access may have become rotten as they became affected by water ingress. If part of the roof has collapsed this may add stress to other members holding up the remainder of the roof.

Fixtures and fittings may become damaged or parts stolen, leaving electrical services in a poor condition. Though power should have been removed from them this is not always the case.

Figure 5-23: Black mould.
Source: Premier preservation Ltd.

Figure 5-24: Ivy.
Source: Daily Mail.

Figure 5-22: Dilapidation of buildings.
Source: RMS.

Fungal and other harmful growth

Fungal growth occurs when there is moisture content in the walls of a structure. It flourishes in an environment of high humidity with lack of ventilation. Fungal growth may indicate the level of decline in the integrity of the structure.

Harmful growth includes creeping ivy and plants that can grow either on walls, roofs or gutters. Roots can go deep into the existing holes causing cracks and water penetration that lead to the erosion of mortar joints and dampness, which weaken the structure.

Cracking of walls and leaning walls

Vertical or diagonal cracks in the wall are common symptoms of structural instability. The survey should identify these cracks as symptoms and try to diagnose the cause.

This may be poor foundations, weak materials and joints; or any shrinkage or thermal movements such as those of timber window frames.

Diagonal cracks, usually widest at the foundations and terminating at the corner of a building, often occur when shallow foundations are laid on shrinkable sub-soil that is drier than normal. They may also be caused by a physical uplifting action of a large tree's main roots close to the walls.

Common causes of leaning walls include a spreading roof, which forces the weight of a roof down towards the walls, sagging due to soil movement, weak foundations due to the presence of dampness, shrinkable clay soil or decayed building materials; and disturbance of nearby mature trees with roots expanding to the local settlement.

Figure 5-25: Settlement diagonal wall cracking.
Source: diynot.com.

Decayed floorboards

Timber floorboards were widely used in many older buildings including churches, schools, and railway stations. Some timber floorboards will have been subjected to surface abuse and subsequently deteriorated to the extent that they can no longer hold the weight of workers walking on them.

The main natural causes of decay are pest attacks, such as dry rot or hidden cavities caused by death watch beetle.

Figure 5-26: Dry rot - characteristic cuboidal cracking.
Source: Safeguard Europe Ltd.

Figure 5-27: Timber damaged by death watch beetle.
Source: Rudders and Paynes Ltd.

Weakened or split boards may be caused by damage by occupants, loss of natural preservatives and the effects of corroded nails.

Roof defects

Common defects of roof tiles include corrosion of nails that fix the tiles to battens and rafters, the decay of battens, and the cracking of tiles caused by harmful growth. Another aspect to be considered is the mortar applied for ridge tiles, which tends to decay or flake off over the years. Roof light materials often deteriorate on exposure to sunlight and become brittle, lose structural strength and are easily fractured. Often they are covered with felt to prevent leaks or become obscured from the rest of the roof by the growth of moss. Hidden roof lights present a risk to workers of falling through.

Unstable foundations

Most of the common problems associated with foundations depend on the geology of the ground upon which a building stands. For example, settlement or subsidence may result if it is built prematurely on made up ground or the water table height is causing the ground to expand or contract.

The leaning Tower of Pisa is perhaps the most famous example of building subsidence. The tower began to sink after construction had progressed to the second floor in 1178. This was due to a shallow three-metre foundation, set in weak, unstable sandy subsoil, a design that was flawed from the beginning.

Unstable foundations may occur due to traffic vibrations, deterioration of building materials and through change of use, for example, increased floor loading.

Problems with unstable foundations may lead to an unstable building structure, which increases the risks of premature collapse during deconstruction or demolition.

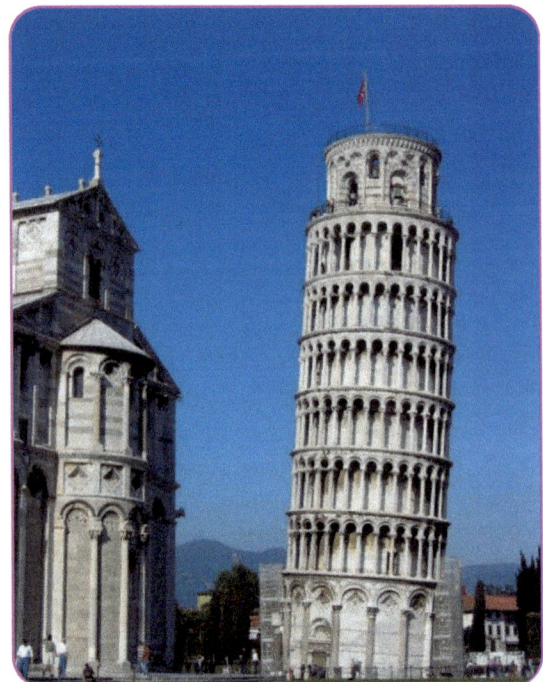

Figure 5-28: The leaning tower of Pisa, Italy.
Source: Marshaü.

Presence of hazardous substances

A particular health hazard that should be identified at time of survey is possible exposure to asbestos, which can present a significant risk, especially if workers are unaware of its presence and no precautions have been taken. It is important to obtain details of any prior survey conducted on the building/structure by the owner/occupier, for consideration as part of the pre-demolition survey.

The Control of Asbestos Regulations (CAR) 2012 prohibits the importation, supply and use of all forms of asbestos. They also prohibit the second-hand use of asbestos products such as asbestos cement sheets and asbestos tiles. If existing asbestos containing materials (ACMs) are in good condition they may be left in place and their condition monitored and managed.

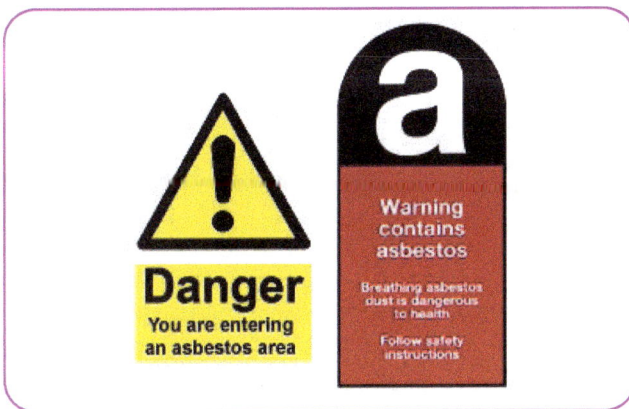

Figure 5-29: Hazardous substances.
Source: Scaftag.

CAR 2012 sets out the following general requirements:

- The presence of asbestos must be identified and must be labelled.
- An assessment must be carried out of work which exposes employees to asbestos.
- Training is mandatory for those that may be exposed to asbestos fibres at work, including deconstruction/demolition workers or others that may come into contact with or disturb asbestos.
- A written plan of work is required for work with asbestos.

The CAR 2012, regulation 4, requires that duty holder, for example, the owner of a building, to provide information on the location and condition of ACMs to anyone who is liable to work on or disturb them. Demolition or deconstruction work would often mean the removal of ACMs or at the very least disturbance of them. The information on ACMs may be provided as part of information obtained during the pre-demolition survey. In the absence of this information it may be necessary for the duty holder or those involved in the demolition or deconstruction activity to conduct their own asbestos survey.

The CAR 2012, regulation 3, requires that workers who work with asbestos must be licensed unless the circumstances are that:

a) The exposure of employees to asbestos is sporadic and of low intensity.

b) It is clear from the risk assessment that the exposure of any employee to asbestos will not exceed the control limit.

c) The work involves:

   i) Short, non-continuous maintenance activities.

   ii) Removal of materials in which the asbestos fibres are firmly linked in a matrix.

   iii) Encapsulation or sealing of asbestos-containing materials which are in good condition.

   iv) Air monitoring and control, and the collection and analysis of samples to ascertain whether a specific material contains asbestos.

Other controls required by CAR 2012 related to asbestos include:

- Work with asbestos that requires a licence must be notified to the employer's enforcing authority.
- Exposure must be prevented or reduced by controls.
- Controls must be used and maintained.
- The employer is responsible for the cleaning of personal protective clothing.

CAR 2012 introduced a category of work called Notifiable Non-Licensed Work (NNLW). This required the appropriate enforcing authority to be notified if this category of work was to be carried out. Records of work carried out must be kept and workers must have a periodic medical examination.

For further information see 12.4 'Duty to manage asbestos'.

There may be other health hazards present that could affect the deconstruction/demolition worker or others; for example, Weil's disease from rats, breathing difficulties from the dust of disturbed pigeon droppings or chemicals from the previous use of the building.

REVIEW OF FACTORS AFFECTING THE STRUCTURE

Drawings, structural calculations, health and safety

Before work commences every effort should be made to find the original design drawings, together with the calculations made by the architect. If the building is of recent construction these will be found in the health and safety file related to the structure.

This will provide information, including that drawn from the pre-construction phase plan, on how the structure was built, for example, reinforced concrete frame or steel frame. If it is a large structure, it will be important to identify any pre-tensioned or post-tensioned concrete beams present within the structure and whether any floor slabs or piles were involved in the build. It is also important to determine the age of the structure and its previous use.

It is the client's responsibility to provide such information to the contractor. Prior to deconstruction/demolition, any structural alterations that might affect the load-bearing capacity of walls and floors should be reviewed to avoid premature collapse.

Structural alterations carried out on the structure in the past

The pre-demolition survey should consider past alterations. This is particularly important in order to identify any changes, for example, the addition of doors or windows or the removal of walls, which may have compromised the building's original structural stability.

For older buildings that pre-date CDMR 2015, similar requirements to maintain a health and safety file, the alterations may not have been recorded or records maintained. If the alterations had been carried out by the owner's own workers this may have been done without structural drawings, calculations and records. This absence of records may require a detailed physical survey of the structure to determine any significant alterations and the implications to the demolition process intended.

> **REVIEW**
>
> What should be included in a pre-demolition survey?

Sources of reference

Reference information provided, in particular web links, was correct at time of publication, but may have changed.

Asbestos: The survey guide, HSG264 (Second Edition), HSE Books, ISBN 978-07176-6502-0 http://www.hse.gov.uk/pubns/priced/hsg264.pdf

Code of practice for demolition, BS 6187, British Standards Institution, ISBN 978-0-5803-3206-7

Establishing exclusion zones when using explosives in demolition, HSE Construction information sheet No 45. http://www.hse.gov.uk/pubns/cis45.pdf

Health and Safety in Construction, HSG150, HSE Books, ISBN 978-0-7176-6182-2

http://www.hse.gov.uk/pubns/priced/hsg150.pdf

Managing Health and Safety in Construction, Construction (Design and Management) Regulations 2015, Guidance on Regulations, L153, HSE Books, ISBN: 978-0-7176-6626-3 http://www.hse.gov.uk/pubns/priced/l153.pdf

Personal Protective Equipment at Work (Guidance), L25, HSE Books, ISBN: 978-0-7176-6139-3 http://www.hse.gov.uk/pubns/priced/l25.pdf

Protecting the public: Your next move, HSG151, HSE Books, ISBN 978-0-7176-6294-4 http://www.hse.gov.uk/pubns/priced/hsg151.pdf

Web links to these references are provided on the RMS Publishing website for ease of use – www.rmspublishing.co.uk

Statutory provisions

Construction (Design and Management) Regulations (CDMR) 2015

Control of Explosives Regulations (CER) 1991

Environmental Protection Act (EPA) 1990

Personal Protective Equipment at Work Regulations (PPER) 1992

Provision and Use of Work Equipment Regulations (PUWER) 1998

STUDY QUESTIONS

1) A multi-storey car park in a city centre is to be demolished.

   (a) What are FOUR possible hazards to the environment that could be caused by the demolition?

   (b) What control measures should be considered to reduce the risk to the environment during the demolition work?

2) What are the control measures that should be considered when demolishing a group of ten terraced houses in a series of twenty terraced houses within an occupied street?

3) What are the main issues to be addressed in a pre-demolition survey of a three storey office block in a city centre?

4) A steel framed warehouse that was part of a large factory is to be demolished. What are the main control measures to be included in a demolition method statement?

5) A single-storey farm building is to be demolished using an excavator. What control measures should the method statement contain for this activity?

6) What hazards may affect workers or the public during the demolition of a building?

For guidance on how to answer these questions please refer to the 'study question answer guidance' section located at the back of this guide.

This page is intentionally blank

Element 6

Mobile plant and vehicles

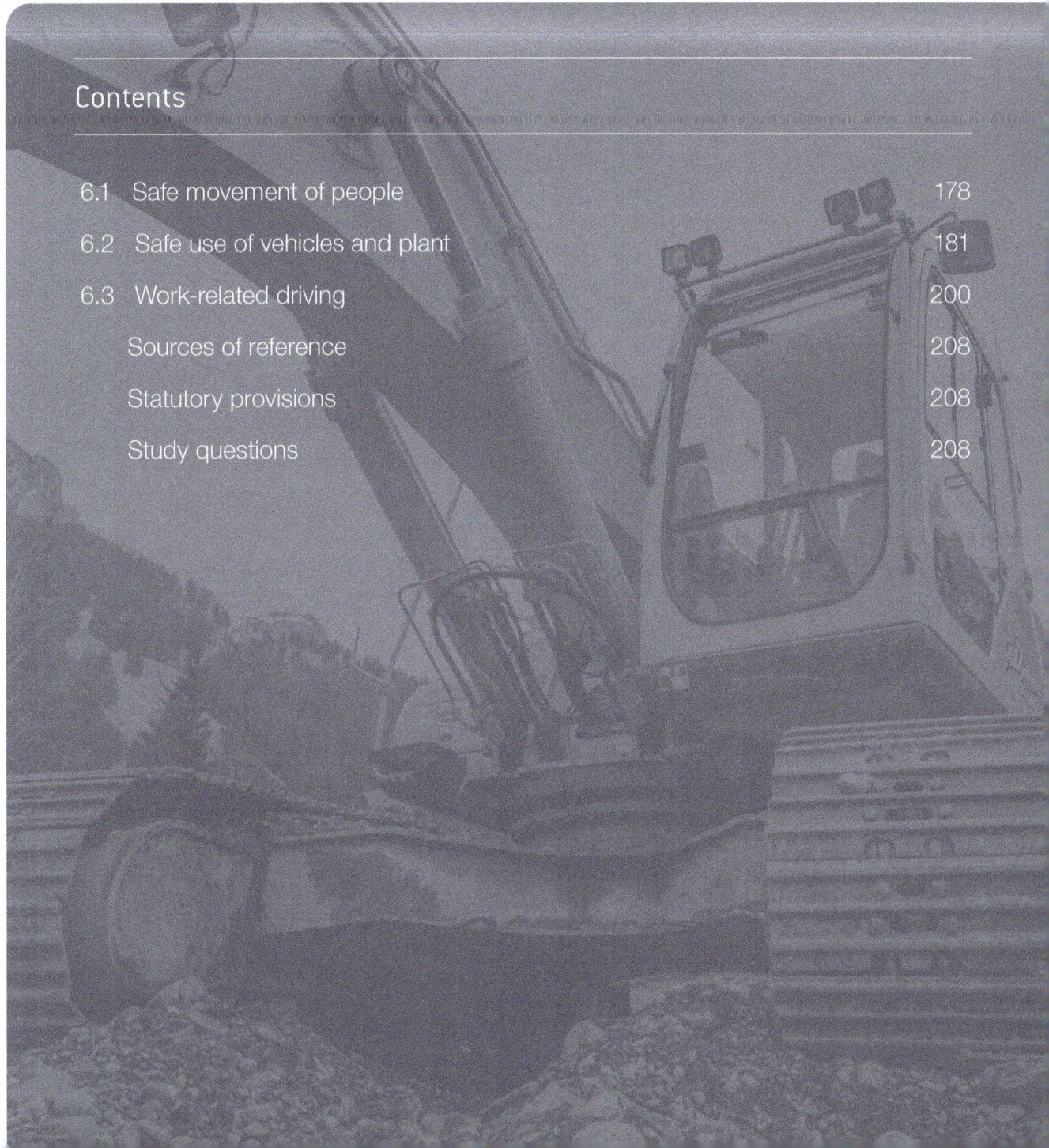

6.1 Safe movement of people

HAZARD TO PEDESTRIANS

Being struck by moving, flying or falling objects

Being struck by moving, flying or falling objects are hazards to pedestrians in the workplace and may also affect people in areas adjoining the workplace if precautions are not taken. Areas adjoining the workplace could include a public footpath, road, yard of a dwelling or other building beside a workplace.

The Health and Safety Executive (HSE) reported in 2020/21 that, in the UK, workers who suffered serious injuries by being struck by a moving object (including flying and falling objects) in the workplace made up 10% of non-fatal injuries to employees, and this type of hazard accounted for 14% of fatal injuries .

Examples of moving, flying or falling objects that could strike a pedestrian include:

- An object falling from a structure – for example scaffolding components, tools, roofing materials, bricks, ceiling panels that have not been securely fixed and materials that fall from over-stacked shelving.

- An object falling from lifting equipment, a vehicle or other moving equipment - including loads being lifted that are not well secured or are unstable.

- An object or material ejected while using machinery or hand tools – including parts of the machine/tool and waste material.

- The collapse of an unstable structure - including stacks of material, shelves, benches, racking and mezzanine floors not strong enough to bear the weight of objects on them.

- An object that has been deliberately thrown when carrying out a task - including throwing a scaffold coupling.

- Unstable objects falling - for example, an unsecured ladder.

Some of the situations that increase the likelihood of these hazards and the risk of injury to pedestrians include:

- Stacking materials too high – for example, bricks, blocks, sacks, pallets of materials and drums.

- Use of damaged or unsuitable pallets.

- Overloading of materials on work places at height – for example scaffolds or roofs.

- Lack of use of barriers or mesh screens to prevent materials falling from height or being blown off.

- Unstable workplace structure, for example, insufficient number of scaffold tie points or uneven ground.

- Loose materials stacked at too steep an angle - for example, excavated soil.

- Faulty or inappropriate means of lifting or moving materials – for example, poor slinging methods, driving a fork-lift truck so fast that part of the load fall from the forks.

- Insufficient clearance between people and load when materials are being moved by lift trucks and cranes.

- Unstable loads on vehicles - for example, not correctly supported, tied or shrink wrapped.

- Insecure components or workpiece in moving machinery.

- Products of machining processes not contained within the machine - for example, waste material being ejected.

- Free dust from work processes or outside areas blown into the eyes of workers.

> "(1) Each part of a construction site must, so far as is reasonably practicable, be kept in good order and those parts in which construction work is being carried out must be kept in a reasonable state of cleanliness."

Figure 6-2: Duty to keep good order on site.
Source: CDMR 2015 Regulation 18.

Figure 6-1: Risk from falling materials.
Source: RMS.

Figure 6-3: Risk from falling materials - broken pallet.
Source: HSE.

A father of four was killed while working for an Essex based granite manufacturer. The worker was unloading a vehicle which had delivered stone slabs to the warehouse. He was moving the slabs over to A-frames to be stored when the accident happened. He was lowering a load which weighed around 3 tonnes when the pile of slabs fell and crushed him up against a wall. The worker was pronounced dead at the scene aged just 57.

The investigation had shown that the system of work which was being followed by the worker was unsafe, and therefore he was put at risk by his employers.

During the investigation it was discovered that the A-frames were not in a suitable position in the warehouse. Simple changes made to the way the storage had been designed and used could have prevented the death of the worker and saved his family from so much suffering. The employer pleaded guilty in court and was fined £20,000 with additional costs of £40,000.

Source: HSE.

Collisions with moving vehicles

According to the HSE, on average, each year, about 7 workers die as a result of accidents involving vehicles or mobile plant on construction sites and a further 93 are seriously injured

The Health and Safety Executive (HSE) reported in 2021/22 that in the UK the number of workers killed in the workplace by a moving vehicle was 23 and 1,640 others suffered serious injuries. The figures reported by the UK HSE are for injuries that took place in workplaces and do not include injuries involving workers on the public highway.

Figure 6-4: Collision with moving vehicle.
Source: DT Specialized Services.

Situations that increase the likelihood of a pedestrian being involved in a collision with moving vehicles include:

- Restricted space that does not allow for vehicle manoeuvring and the pedestrians in an area.

- Obstructed designated pedestrians routes.

- Undefined routes that do not segregate vehicles from pedestrian - for example, people walking where there are heavy goods vehicles, excavators, dumper trucks and rough terrain vehicles move in loading bays.

- Large vehicles reversing without anyone providing assistance.

- Disregard for rules concerning vehicles - for example, speed restrictions, competent drivers.

- Insufficient devices fitted to vehicles to warn pedestrians of their presence.

- Failure to provide recognised pedestrian crossing points.

- Lack of lighting, particularly at pedestrian crossing points and pedestrian exits.

- Lack of barriers to segregate pedestrian from vehicles.

- Absence of separate exits from buildings or the site for pedestrians and vehicles.

- Pedestrian exits from work areas that open out onto vehicle routes.

"(1) A construction site must be organised in such a way that, so far as is reasonably practicable, pedestrians and vehicles can move without risks to health or safety."

Figure 6-5: Duty to ensure safe movement of vehicles on site.
Source: CDMR 2015 Regulation 27.

A building contractor was working at a nuclear site when one of their workers was run over by a cherry picker (mobile elevating work platform). The injured worker was guiding the vehicle on foot when he was hit, and as a result of the accident he lost his leg.

The Health and Safety Executive prosecuted his employer as an investigation into the accident revealed a number of problems. The injured employee had been put at risk because there wasn't a safe system of work created or followed for the task he was performing.

Striking against fixed or stationary objects

The Health and Safety Executive (HSE) reported in 2018/19 that in the UK the number of workers killed in the workplace by striking against a fixed or stationery object was 3 (none in construction work) and 2,451 (121 in construction work) suffered serious injuries.

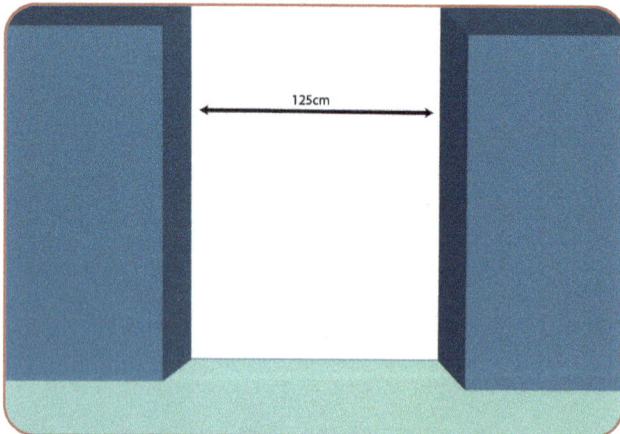

Figure 6-6: Wide enough walkways.
Source: ILO, Ergonomic checkpoints.

Situations that increase the likelihood of striking against fixed or stationary objects include:

- Poorly positioned machinery, equipment and materials - avoid sharp corners protruding out.

- Insufficient space for storing tools and materials causing poor access and egress.

- Inadequate width of walkways.

- Inadequate lighting.

- Work in enclosed areas - such as restricted height in a confined spaces.

- Crane lifting accessories left in a position that workers can walk into them - such as hanging hooks/slings.

- Low overhead fixtures - such as pipework fixed at head height.

- Workers rushing, distracted or not looking where they are going.

- Nails or similar sharp objects protruding from timber or other material.

"(4) A construction site must, so far as is reasonably practicable, have sufficient working space and be arranged so that it is suitable for any person who is working or who is likely to work there, taking account of any necessary work equipment likely to be used there."

Figure 6-7: Duty to ensure sufficient working space on site.
Source: CDMR 2015 Regulation 17.

Figure 6-8: Obstructions.
Source: Lincsafe.

Treading on or contacting sharp objects

Sharp objects represent a foot hazard to pedestrians. Construction sites have a large amount of materials moved onto and around them. Sometimes materials, such as sections of trunking (such as may be used to carry electrical wiring) and copper pipe are cut to length and sharp edged off cuts may fall to the ground and be left there. Material can often arrive on site secured by wire binding and when the binding is removed it is often left on the ground, presenting a sharp trip hazard.

In addition, nailed together timber is sometimes used for improvised tasks, and when it is finished with it may be broken apart and discarded, leaving nails protruding. Abandoned broken pallets will present a similar sharps hazard.

"(3) No timber or other material with projecting nails (or similar sharp object) must:

(a) be used in any construction work; or

(b) be allowed to remain in any place,

if the nails (or similar sharp object) may be a source of danger to any person."

Figure 6-9: Duty to protect from projecting nails.
Source: CDMR 2015 Regulation 18.

6.2 Safe use of vehicles and plant

HAZARDS FROM WORKPLACE TRANSPORT OPERATIONS AND PLANT

Many workers die each year because of workplace transport and plant accidents/incidents. The HSE has reported that from 2016 to 2021, 10% of all fatal injuries on constructions sites were a result of accidents involving vehicles or mobile plant.

Very few accidents involving movement of vehicles and plant result in minor injury, many result in serious injury, for example, spinal damage, amputation and crush injuries. In addition, accidents relating to the movement of vehicles and plant on site cause damage to structures, plant, equipment, infrastructure and vehicles.

Typical hazards relating to vehicle and plant movement

Please note - unless specified, a reference to vehicles should also be taken to be a reference to mobile plant.

Vehicle and plant movement can present a number of hazards, some of which may be caused by driver error, for example, driving too fast. Features of the workplace can also cause vehicle and plant movement hazards; for example, lack of signs may cause a driver to carry out an unexpected manoeuvre or to take a wrong route and have to reverse their vehicle or plant.

Other hazards may relate to the vehicle or plant itself, including mechanical failure (for example, the brakes may fail) or the 'silent' operation of some vehicles and poor visibility to the rear from the operating position or around loads that are carried. Environmental conditions (for example, mud on the road, ice or snow) will also present a hazard that can influence safe vehicle and plant movement.

Driving too fast

The hazard of driving too fast may occur because the workload is too great for one driver to complete in the time available or there are incentives to complete the work as fast as possible, for example, the incentive created when fork-lift truck or delivery vehicle drivers are allowed to go home when the work is finished, even though their shift may have several hours left.

Often accidents/incidents occur when the driver who is driving too fast changes direction, for example, driving around a bend or steering to avoid an obstacle in the vehicle's path.

A driver's inexperience may lead them to drive too fast for the characteristics of the vehicle or the terrain being driven over, causing them to have to brake too late or make too sharp a turn.

Reversing

The potential for collision with pedestrians/fixed items is greatly increased when vehicles and plant are required to reverse. This is partly because the field of vision of the driver to the rear of the vehicle or plant may be limited by the fixed position of the vehicle mirrors, or because drivers have to adopt an awkward posture in order to be able to turn to see where they are going. In addition, large vehicles may have a significant area at their rear that the driver cannot see.

The hazard of reversing is also influenced by the driver having to operate the steering of the vehicle in the opposite way to that used when travelling forwards.

Close proximity of vehicles and pedestrians

The layout and nature of the construction activities may lead workers to be in close proximity to vehicles and plant, which presents a risk of the vehicle/plant colliding with pedestrians or pedestrians being crushed against fixed or stationery objects.

Silent operation of some vehicles and plant

Vehicles and plant that operate 'silently', typically those that are battery-powered, present a particular hazard to pedestrians as their silent operation may mean that they are not heard in time to avoid injury.

This is a significant hazard where segregation of pedestrians and vehicles and plant is not practicable, for example, where mobile elevating work platforms and pedestrians work together or where a fork-lift truck brings materials to a worker inside a building.

Workplace background noise or workers wearing ear muffs or ear plugs may also reduce the pedestrian's perception of such vehicles and plant approaching.

Unstable loads

Materials may not be sufficiently secured to the vehicle or plant and part or all of the load may become dislodged or out of balance as it moves, particularly in areas of the site that involve rough terrain.

Large and heavy loads

Some materials moved by vehicles and plant are large and could make the vehicle/plant unstable when the load is moved or obstruct the view of the driver. These loads may make maneuvering the vehicle or plant very difficult, particularly in confined areas of the construction site.

Mechanical failure of a vehicle or plant

Load bearing or safety critical parts of a vehicle or plant could fail. For example, the hoist of a rough terrain fork-lift truck could fail or a braking system on a vehicle could fail allowing it to move or fail to stop when moving.

Poor visibility

The load on a vehicle or plant may obstruct the driver's vision and cause them to look to alternate sides of the load in order to see past it. This difficulty could mean the driver may not see pedestrians, vehicles or fixed objects sufficiently to avoid impact with them. Parts of the vehicle or plant, for example, the lowered jib of a mobile crane, may prevent good visibility for the driver. The height of the driver above the road and the position of windows in a vehicle may also create poor visibility, in particular to the rear of the vehicle.

Similar difficulties can occur where the driver sits in an enclosed cab and environmental conditions like dust/sand, high humidity, fog, rain or ice make visibility through the windows of the cab poor. When vehicles move through areas with changing light levels, for example from high light levels to low light levels, it can lead to poor visibility while the driver's eyes adjust to the change.

This can occur in a number of situations, for example, when moving from indoor areas to outdoor areas at night or moving from an area in bright sunlight to an area of shade during daylight hours. In addition, reliance on vehicle lights when driving outdoors at night may not be adequate to enable good visibility.

Poor visibility also relates to being able to see the vehicle. Poor light levels may make it hard to see a vehicle approaching, particularly if it does not have lights, its colour is not readily distinguishable from the background and it is moving quickly. Poor visibility can exist in many workplaces where doors, corners of buildings, and stacked materials prevent pedestrians and vehicle drivers having a clear view of vehicles approaching.

Site conditions

Site conditions, for example, mud, ice or snow on traffic routes will also present a hazard that can influence safe vehicle and plant movement. This can lead to reduced control of the vehicle and risk of collision and/or overturning of the vehicle.

Construction sites will often have slopes, edges and gradients, particularly when the ground is being cleared or prepared. These present a hazard to vehicles travelling across or near to them as the edge may collapse or the gradient could be too steep for the vehicle to drive across safely, causing it to overturn.

Typical risks related to vehicle and plant movement

The main areas of risks associated with workplace transport operations and plant are:

- Vehicle or plant collision with other vehicles, plant, pedestrians or fixed objects.
- Crushed or struck by a vehicle or plant overturning.
- Injury caused due to mechanical failure of a vehicle or plant component.
- A person being struck by a load or materials falling from a vehicle or plant.
- Coming into contact with the inside of the vehicle or plant while travelling in it.
- A person striking a fixed or moving object whilst travelling in a moving vehicle or plant.

Collisions with other vehicles, plant, pedestrians or fixed objects

People may unexpectedly appear from a part of a building/structure, or workers intent on the work they are doing may step away from where they are working to collect materials or tools. Through these actions they may step in front of moving vehicles or plant and cause the vehicle/plant to collide with them or the driver to take emergency action.

Figure 6-10: Potential collision with fixed objects.
Source: RMS.

"(5) No vehicle is to be driven on a traffic route unless, so far as is reasonably practicable, that traffic route is free from obstruction and permits sufficient clearance."

Figure 6-11: Traffic routes - free from obstruction and sufficient clearance.
Source: CDMR 2015 Regulation 27.

Often, the space in workplaces, such as construction sites, are restricted. Stacks of construction materials may be increased in height to maximise the floor space. This can then lead to poor visibility and can cause a risk of collisions with other vehicles, plant, pedestrians and fixed objects, such as roof and racking supports.

Factors that affect the risk of collisions include:

- Inadequate lighting and direction signs.

- Inadequate signs or signals to identify the presence of vehicles and plant.

- Drivers unfamiliar with the site.

- Needing to reverse a vehicle or plant.

- Poor visibility - for example, sharp bends, mirror/windscreen misted up, design of plant.

- Poor identification of fixed objects - for example, overhead pipes, doorways, storage tanks, corners of buildings.

- Lack of safe crossing points on roads and other vehicle or plant operating routes.

- Lack of separate entrance/exit for vehicles/plant and pedestrians.

- Lack of segregation of pedestrians and vehicles/plant in the workplace.

- Pedestrians using doors provided for vehicle only use.

- Lack of barriers to prevent pedestrians suddenly stepping from an exit/entrance into a vehicle's path.

- Poor maintenance of vehicles or plant - for example, tyres or brakes.

- Excessive speed of vehicles.

- Lack of vehicle and plant management - for example, use of traffic control, signaller/banksman/reversing assistant (appointed individuals who control or direct vehicles/plant).

- Environmental conditions - for example, poor lighting, rain, mud, strong winds, dust/sand storms, snow or ice.

Figure 6-12: Restricted view to rear of an excavator
Source: RMS.

Figure 6-13: Restricted view to rear of a crane.
Source: RMS.

Overturning

There are various circumstances that may cause a vehicle or plant to overturn. They generally relate to factors that cause the centre of gravity of the vehicle or plant to be acting so that the force created is mainly to one side of the vehicle or plant instead of down to the ground. These uneven forces can contribute to causing vehicles and plant to overturn. Factors that could cause vehicles or plant to overturn include:

- Overloading or uneven loading, for example overloading of lifting equipment.

- Insecure and unstable loads that move so their weight is not evenly distributed.

- Driving with the load elevated.

- Driving too fast, cornering at excessive speed.

- Sudden braking or acceleration.

- Hitting obstructions, including kerbs, buildings, structures or other vehicles and plant.

- Driving across slopes.

- Driving too close to the edges of slopes, embankments or excavations.

- Driving over debris.

- Driving over soft ground or holes in the ground, such as drains.

- Poorly maintained or uneven road surfaces.

- Mechanical defects that occur because of lack of maintenance.

- Inappropriate or unequal tyre pressures, causing the weight of the load to be poorly distributed or movement of the load.

Figure 6-14: Plant overturned.
Source: Lincsafe.

Failure of vehicle components

There is a risk of injury to people due to the mechanical failure of a part of a vehicle or plant, in particular load-bearing parts. This may be caused by hydraulic failure of a hoist mechanism or a failure of a chain mechanism that is used to lift a load. The sudden collapse of the load may cause injury to pedestrians or others working in relation to the vehicle (loading/unloading), or ejected broken parts may strike the vehicle driver or those nearby. In addition, a vehicle or plant may be overloaded to the point that its structure cannot hold the weight and may collapse. Heavy goods vehicles may be overloaded to the point that the trailer's load carrying capacity is exceeded and it fails. Where articulated vehicles are used there is a risk of the trailer collapsing as it is coupled or de-coupled from the tractor unit, particularly if the trailer supports are not set correctly or when the tractor does not make a good connection with the trailer and leaves the trailer unsupported as it drives away.

Struck by load

The insecurity of a load being moved by a vehicle or plant can lead to the risk of materials falling or a person being struck by the load. If the load has not been secured to the vehicle or plant properly, for example by straps, sheets or nets and this is combined with cornering or driving over an uneven surface/object the load may fall. In addition, if the load extends beyond the vehicle or plant it may hit people, equipment or structures as it moves around the site.

Coming into contact with a vehicle's or plant's structure while travelling in it

As a vehicle or plant moves there is a risk that the driver or a passenger may come into contact with its structure. The people in the vehicle or plant are particularly vulnerable to this, for example, when their head is 'bumped' against the door frame. The risk is increased the rougher the surface is that they travel over and where the person is not using driver/passenger restraint systems. If the vehicle or plant is subject to excessive braking or comes to a sudden stop (when it strikes another object/vehicle) the people retain the vehicle/plant speed and move within the cab, unless they are able to restrain themselves or are wearing seat restraints. The movement of people within the cab can prove fatal as they strike the vehicle cab interior or if they come into contact with other vehicle occupants.

Falling from a moving vehicle or plant

There is a hazard of a person falling from a vehicle or plant, particularly where they are in an exposed seating or standing position. This is most likely to be a problem when the vehicle or plant is travelling too fast, braking, cornering or going over rough terrain and the person in the vehicle is dislodged by the sudden movement of the vehicle, particularly where the person is not using a driver/passenger restraint system.

Person striking a fixed object

As the vehicle or plant is travelling along there is a risk that the driver or passenger may strike a fixed or moving item. This is not likely if an enclosed structure completely surrounds both the driver and passengers. However, many workplace vehicles have open frameworks or windows that can allow a person to put a part of their body, typically their head or hand, outside the protective structure. Many injuries have occurred to fork-lift truck drivers who, while driving forward with a load obstructing their view, lean out to look round the load and their head strikes a fixed object.

Sometimes, passengers and drivers trail an arm or a leg outside the structure of the vehicle or plant they are travelling on and there is a risk of it striking a fixed object as it passes, particularly while manoeuvring in areas where movement is restricted.

Typical hazards related to when vehicles and plant are not moving

Height

There is a significant risk of falling from height in many transport activities conducted when the vehicle or plant is not moving. The risk of a person falling can arise when loading or unloading the vehicle/plant as this may require a person to go onto the vehicle/plant or the top of a load and work at height.

Figure 6-15: Hazard of height when exiting plant loaded on trailer.
Source: RMS.

This work at height, along with slippery or uneven surfaces can create a significant risk of a person falling, particularly where there is no organised means of access. In addition, workers loading or unloading vehicles at height may lose awareness of where the edge of the vehicle or load is, which may lead to falls and major injury.

The risk of falling can also arise when securing the a load or fitting a sheet over the a load, particularly as this can involve strenuous work and the person may not have their attention on the hazard of height and the need to avoid falling. The risk of falling when fitting a sheet is greatly increased when it is done at times when there are strong winds.

Risks of falling also occur when the driver/passenger are climbing up to enter or climbing down to exit the vehicle/plant.

Materials falling

There is a hazard of materials falling on a driver or other people assisting during loading and unloading operations, particularly where vehicles are used to load materials at a height or to remove materials from delivery vehicles. In addition, when vehicles are being unloaded by hand, there is a hazard related to the stability of the load, as it may have moved and become unstable during transportation. A stack of material may then become more unstable and collapse when items are removed, causing serious injury or death.

> "(5) A person must not remain, or be required or permitted to remain, on any vehicle during the loading or unloading of any loose material unless a safe place of work is provided and maintained for that person."

Figure 6-16: Protection from falling loose materials.
Source: CDMR 2015 Regulation 28.

Figure 6-17: Driver not on dumper while it is loaded with material.
Source: RMS.

Unplanned movement of load

The load on a vehicle may unexpectedly move as part of the load is removed leaving it in an unstable state or when restraining straps are removed and gravity causes the load to roll or slide, for example if wheeled plant is on a trailer which is parked on a slope.

Crushed by a moving part of a vehicle

Part of a vehicle may be opened in order to carry out maintenance and it could unexpectedly close and crush the person working on the vehicle. In windy conditions, the door of a vehicle/plant might get blown shut as a driver getting into or out of the vehicle/plant causing crush injuries.

Slippery surfaces while on a vehicle

Surfaces of the vehicle or plant used on construction sites will be exposed to weather conditions and may become contaminated by water, mud or construction materials. This could make the metal surfaces of the vehicle or plant slippery. Similarly, the load on vehicles brought to site may become slippery because of ice and snow, which could present a hazard if they have to be walked on to enable them to be unloaded.

CONTROL MEASURES TO MANAGE WORKPLACE TRANSPORT

The Principal Contractor or contractor, through its managers, needs to carry out an assessment of risk with regard to the safe movement of vehicles and plant as part of the development of the construction phase plan. This includes the use of various types of construction plant and vehicles such as dumper trucks, fork-lift trucks and those vehicles used for delivery of materials to site. Control measures should consider the following: safe site, safe vehicles and safe drivers.

Safe site

To ensure that the site is kept safe, the following factors should be addressed:

- Suitability of traffic routes (including site access and egress).

- Spillage control.

- Management of vehicle and plant movements.

- Environmental considerations (visibility, gradients, changes of level, surface conditions).

- Maintenance and checking of traffic routes.

- Segregating of pedestrians and vehicles/plant and measures to be taken when segregation is not practicable.

- Protective measures for people and structures (barriers, marking signs, warnings of vehicle/plant operating, approaching and reversing).

- Site rules (including speed limits).

Suitability of traffic routes

Suitability of traffic routes should consider the needs of the range of vehicles and plant likely to use them and the people that might be affected by them. This will mean consideration of the needs of large vehicles and plant to turn corners and arrangements to avoid or plan for vehicles/plant reversing. It will also mean consideration of the needs of long, low vehicles to ensure they are able to negotiate speed-retarding humps, the site terrain and slopes. The emphasis should not only be on safe access and egress into the site, from the public highway, but the provision of safe access and egress on traffic routes within the site.

> "(1) A construction site must be organised in such a way that, so far as is reasonably practicable, pedestrians and vehicles can move without risks to health or safety.
>
> (2) Traffic routes must be suitable for the persons or vehicles using them, sufficient in number, in suitable positions and of sufficient size."

Figure 6-18: Duty to ensure safe movement of pedestrians and vehicles on site. Source: CDMR 2015 Regulation 27.

Figure 6-19: Tractor and trailer unit on a construction site. Source: Road Transport Media.

When establishing a safe site, consideration should be given to the following general principles to ensure traffic routes are suitable for pedestrians and vehicles:

- There should be enough traffic routes to prevent overcrowding.

- Traffic routes should be wide enough for the safe movement of the largest vehicle/plant allowed to use them, including delivery vehicles and specific plant used to position materials on site.

- They should be made of a suitable material, and should be constructed to safely bear the weight of vehicles, and their loads, that will pass over them. Particular consideration of load carrying capacity must be made where routes take vehicles/plant across bridges or near excavations. Routes over underground areas must also be considered. Site roads may have to be established at an early stage in the project to provide suitable routes.

- Consideration should be given to vehicle weight and height restriction on routes - signs, barriers and weight checks may be necessary.

- Plan traffic routes to give the safest route between places and avoid vehicle/plant routes passing close to:

  - Dangerous items, unless they are well protected, for example, fuel or chemical tanks or pipes.

  - Any unprotected edge from which vehicles/plant could fall, or where they could become unstable, such as unfenced edges of excavations.

  - Any unprotected and vulnerable features, for example, anything that is likely to collapse or be left in a dangerous condition if hit by a vehicle/plant, such as scaffolding.

- Traffic routes should be clearly marked and signed, including directional information and markings to separate vehicles/plant travelling in different directions and to separate pedestrians from vehicles and plant. Markings and signs should incorporate speed limits, one-way systems, priorities and other factors normal to public roads.

- Clear direction signs and marking of storage areas and buildings can help to avoid unnecessary movement, such as reversing. Large vehicles and plant, especially where articulated or with a drawbar trailer, often need to perform complicated manoeuvres to turn safely. This is because the trailers swing out behind the tractor unit and this often involves a larger turning circle than other vehicles.

- Sharp bends and overhead obstructions should be avoided where possible. Hazards that cannot be removed should be clearly marked with black and yellow diagonal stripes, for example, loading bay edges and pits. Where practicable barriers should be installed.

- Separate routes and designated crossing places should be provided for pedestrians and suitable barriers installed at recognised danger spots. As far as is practicable pedestrians should be kept clear of vehicle/plant operating areas and/or a notice displayed warning them that they are entering an operating area. Similarly, pedestrian-only zones must be clearly marked so that they are clear to drivers and preferably protected by physical barriers.

- Entrances and gateways must be wide enough for the traffic. Where possible, there should be enough space to allow two vehicles to pass each other without causing a blockage. Where this is not possible, traffic management systems should be in place to control the movement of vehicles from each direction, such as traffic lights, the use of a worker to control and signal to drivers or signage to indicate who has priority. If gates or barriers are to be left open, they should be secured in position. There should be separate entrances/exits for pedestrians and vehicles, unless movement of vehicles is controlled while pedestrians are using the same entrance/exit.

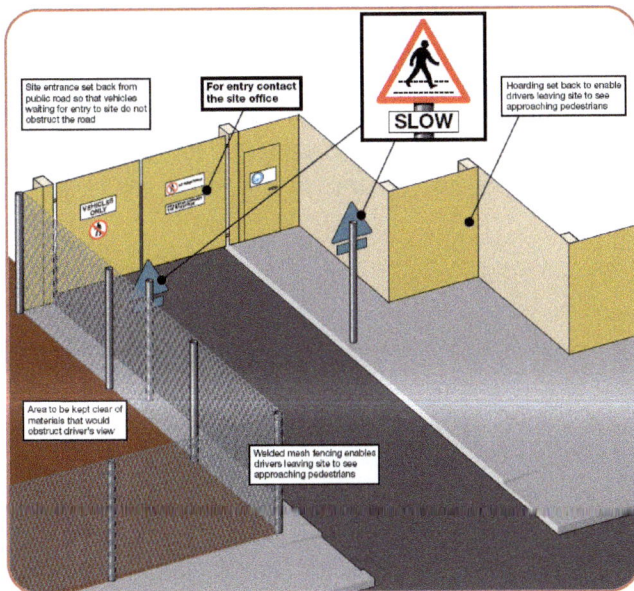

Figure 6-20: Site entrance to construction site.
Source: HSE, HSG150.

Regulation 27 of CDMR 2015 establishes requirements for the organisation and suitability of traffic routes in such a way that:

"(2) Traffic routes must be suitable for the persons or vehicles using them, sufficient in number, in suitable positions and of sufficient size.

(3) A traffic route does not satisfy paragraph (2) unless suitable and sufficient steps are taken to ensure that:

(a) pedestrians or vehicles may use it without causing danger to the health or safety of persons near it;

(b) any door or gate for pedestrians which leads onto a traffic route is sufficiently separated from that traffic route to enable pedestrians to see any approaching vehicle or plant from a place of safety;

(c) there is sufficient separation between vehicles and pedestrians to ensure safety or, where this is not reasonably practicable:

(i) other means for the protection of pedestrians are provided, and

(ii) effective arrangements are used for warning any person liable to be crushed or trapped by any vehicle of its approach;

(d) any loading bay has at least one exit for the exclusive use of pedestrians; and

(e) where it is unsafe for pedestrians to use a gate intended primarily for vehicles, at least one door for pedestrians is provided in the immediate vicinity of the gate, is clearly marked and is kept free from obstruction."

Figure 6-21: Requirements for traffic routes.
Source: CDMR 2015 Regulation 27.

Spillage control

Material and substances spilt onto traffic routes can cause a vehicle to lose control and cause a collision. Arrangements should be in place to control spills. This will include control measures to prevent spills in the form of transporting materials in spill proof containers instead of open containers, the use of containers to hold loose items and binding containers/items together so they do not fall while being moved, for example by using netting, ropes or custom made clamps.

Control measures should include provision of materials to deal with spills, which may need to be located at intervals in the site, for example, sand, special granules or absorbent rolls of material. Where there is a particular risk of large spills it may be necessary to provide drainage. Arrangements should be made to remove obstacles, from traffic routes and place them in a location where they can be returned to use or disposed of, for example items that fall from vehicles.

Management of vehicle and plant movements

Many sites are complex in nature and require the careful management of vehicles and plant in order to ensure that they are brought onto, move around and leave the site safely. This is best considered at the project planning stage as it can cause many problems if a vehicle or plant is brought onto site too early when the site arrangements are not in place to receive it.

It is useful to appoint someone on site to have overall responsibility for vehicle and plant movements. Where a number of contractors share a workplace, it is important that they co-operate with each other and coordinate vehicle movements. Managers should be given formal responsibility for movements in their area or for the work they control.

Where materials are brought to site it may be necessary to manage deliveries to prevent too many vehicles/plant arriving at the site at the same time, causing them to back up into the public highway. Site security arrangements play a significant part in the management of vehicles and plant on site and will assist with controlling vehicles/plant so that they are routed correctly and safely. It is not uncommon for vehicles or plant that are too big or too heavy to access the roadways to be sent to site. Site security staff should be trained to identify these to prevent them accessing the site and causing harm.

Visiting drivers must be carefully managed and made aware of site rules that affect them. They should be clear what their responsibilities are and who is in control of their activity on the site. Arrival of vehicles/plant should be organised so that visiting drivers do not have to enter potentially dangerous areas to move to or from their vehicles to access the site office, toilet or welfare facilities. They must not be allowed to bring unauthorised passengers (for example, children during school holidays) onto the site.

It is important to make sure that visiting drivers are aware of the site layout, the route they need to take, and relevant safe working practices, for example, for parking and unloading. Delivery drivers may never have visited the site before, and may only be on site for a short time.

Vehicle movements, such as bringing a heavy crane onto site, may be organised so that most of the movements take place outside the time that pedestrians are entering/leaving the site when they start and finish work and when the minimum number of workers are on site, enabling it to be moved safely, in a controlled manner. On a smaller scale, pedestrians may be given right of way over vehicles and if pedestrians are in an area, supervisors of vehicle and plant movement may stop all movement until pedestrians have left the area. This could be essential if vehicles and pedestrians share an entrance/exit.

The management of vehicle and plant movements will include ensuring that those who move vehicles/plant are trained and competent to move the specific class of vehicle/plant. This will usually involve ensuring unauthorised people are not allowed to drive, including the use of secure control of keys or electronic access to vehicles and plant.

The need for vehicles and plant to reverse should be avoided where possible. A one-way system can help reduce the need to reverse and the risk of collision. On sites where one-way systems are not practical, it may be appropriate to use a turning point to allow vehicles and plant to turn and drive forwards through the site. Turning arrangements should ideally be by means of a roundabout. If other ways of making reversing safe are not effective enough, the principal contractor or contractor may need to consider providing a competent and authorised signaller/banksman/reversing assistant in order to manage vehicle and plant movements. They should have appropriate high-visibility equipment and use agreed hand signals.

"(2) Where a person may be endangered by the movement of a vehicle, suitable and sufficient steps to give warning to any person who is liable to be at risk from the movement of the vehicle must be taken by either or both:

(a) the driver or operator of the vehicle, or

(b) where another person is directing the driver or operator because, due to the nature of the vehicle or task, the driver or operator does not have full visibility, the person providing directions."

Figure 6-22: Controlling and warning of movement.
Source: CDMR 2015 Regulation 28.

Incidents, including accidents, can be caused where vehicles are unsafely parked as they can be an obstruction and restrict visibility.

There should be clear, designated parking areas that allow the safe checking of loads and sheeting of outgoing vehicles to ensure they are safe before leaving the site.

Environmental considerations

Where vehicles and plant operate, lighting and adverse weather will make a significant impact on their safe operation. **Good lighting**, whether natural or artificial, is vital in promoting health and safety at work. It also has operational benefits, for example, making it easier to read labels. In all working and access areas, sufficient lighting should be provided to enable work activities to be carried out safely at all times during working hours. Where practicable, a suitable standard of lighting must be maintained so that operators of vehicles and plant can see to operate their vehicle equipment and can be seen by others. It is important to avoid areas of glare or shadow that could mask the presence of a person or vehicle/plant.

The relationship between the lighting in the work area and adjacent areas is important; large differences in luminance may cause visual discomfort or affect safety in places where there are frequent movements. For example, if vehicles/plant travel from within buildings to the outside it is important that the light level is maintained at a roughly even level in order to give the driver's eyes time to adjust to the change in light.

Fixed structure hazards should be made as **visible** as possible with additional lighting and/or reflective strips. It is important to ensure good visibility in low levels of light by the provision of adequate lighting, particularly at crossing points and similar danger points where vehicles and pedestrians meet each other. Lighting should be located to provide good illumination of the area without causing glare to pedestrians or drivers. Care should be taken when positioning lighting structures near to roads to ensure they do not present a hazard to vehicles/plant using the roads, particularly when the vehicles/plant have wide loads or are turning. Where the failure of artificial lighting might expose workers to danger, emergency lighting should be provided that is automatically triggered by a failure of the normal lighting system.

Gradients and changes in ground level, such as ramps, represent a specific hazard to plant and vehicle operation. Vehicles and plant have a limit of stability dependent on loading and their 'wheelbase'. These conditions could put them at risk of overturning or cause damage to the coupling of an articulated vehicle. Steep gradients should be avoided as they can make driving vehicles and plant difficult, especially if the surface is made slippery by poor weather. Slopes steeper than 1 in 10 should be avoided for most types of vehicle.

Where *changes in level* are at an edge that a vehicle might approach, and there is risk of falling, a robust barrier or similar means to demarcate the edge must be provided, for example at the edge of an excavation. Particular care must be taken at points where loading and unloading is conducted. Changes in floor level may also cause 'grounding' of part of the vehicle or plant, which may cause them and any load being carried to become unstable.

> "(6) Suitable and sufficient measures must be taken to prevent a vehicle from falling into any excavation or pit, or into water, or overrunning the edge of any embankment or earthwork."

Figure 6-23: Preventing vehicles falling.
Source: CDMR 2015 Regulation 28.

Figure 6-24: Vehicle fallen in to excavation.
Source: RMS.

Figure 6-25: Fencing and barriers protect edge of excavation.
Source: RMS.

It is important to consider the adequacy of *road surfaces* for the adverse weather conditions that may be encountered. It may be necessary to select a rougher surface to provide a better grip for vehicle/plant tyres. Drainage features will need to be large enough and spaced so that they can deal with the greatest expected demands on them. Provisions should be made to clear substances from the surface that might occur as a result of weather conditions, for example, mud, water, ice or snow. These substances may make it difficult for vehicles/plant to grip the surface or present obstructions. There should be a programme in place that anticipates this with regular cleaning of the surface as well as means of dealing with obstructions and material that has fallen on traffic routes.

Excessive ambient noise levels can mask the sound of vehicles working in the area. Therefore additional visual warning of the presence of vehicles should be used, for example, flashing lights. Ventilation should be considered when diesel-powered vehicles are to be used inside a building. Sufficient and suitable parking areas should be provided away from the main work area and located where the risk of unauthorised use of the vehicles and plant will be reduced.

Maintenance and checking of traffic routes

Improvised routes and short cuts can develop if correct routes get obstructed or are poorly marked and demarcated by barriers. This may lead to a reduction in the separation of pedestrians from vehicles/plant and allow them to approach areas where pedestrians exit or need to walk. If traffic routes are not maintained they may become unusable for some vehicles and plant and force them to find alternative routes that are not authorised or increase the risk of collision with pedestrians or structures. Routes may become unusable because rain and high levels of use have turned compacted soil into deep mud or when frozen unyielding ruts. Materials could be placed in traffic routes and cause obstructions to people and vehicles. Debris and waste may also be allowed to build up on traffic routes, causing obstructions that present trip hazards for pedestrians or objects that could destabilise vehicles or plant.

> "(5) No vehicle is to be driven on a traffic route unless, so far as is reasonably practicable, that traffic route is free from obstruction and permits sufficient clearance."

Figure 6-26: Traffic routes kept free from obstruction.
Source: CDMR 2015 Regulation 27.

It is important that traffic routes remain appropriate for the current site layout and are maintained so they are effective, in particular reflecting changes that have occurred as the site develops. Systems of inspection and maintenance of roads, traffic routes and equipment that enables the safe movement of vehicles should take place on a planned, regular basis.

This will include consideration of signs to ensure they provide the correct information. Ensuring signs are removed when they are no longer required will help to ensure full attention is paid to those signs that are current, which reflect the hazards present and direct traffic correctly.

Regulation 27 of CDMR 2015 establishes requirements that seek to ensure that traffic routes for people and vehicles/plant remain useable and effective, which is very important for construction sites that have frequently changing conditions and layouts.

> "(4) Each traffic route must be:
>
> (a) indicated by suitable signs where necessary for reasons of health or safety;
>
> (b) regularly checked; and
>
> (c) properly maintained."

Figure 6-27: Duty to ensure traffic routes are maintained and effective. Source: CDMR 2015 Regulation 27.

Segregation

Segregating pedestrians and vehicles/plant

Where practicable, pedestrians should be segregated from vehicles/plant. There are degrees of segregation that are used; segregation may be total in that pedestrians are removed from the workplace where vehicles/plant operate. This can be difficult to do unless the workplace is very controlled, for example, in some warehouse situations. Some organisations establish segregation of general pedestrians from vehicle/plant operation in this way, but accept that drivers may have to leave their vehicle occasionally, for example, to check loads, to make adjustments to their vehicle or if it breaks down. In this sense, the drivers are authorised pedestrians and all unauthorised pedestrians are segregated from the area.

Where possible, there should be physical segregation of pedestrians from vehicles/plant by the means of separate doors, barriers or fencing. This is a very effective method of ensuring pedestrian safety. Barriers can be used to channel pedestrians into prescribed pedestrian routes, ensuring they are kept sufficiently clear of vehicle entrances and preventing them entering a traffic route directly from buildings.

Clearly defined and marked routes should be provided for the general movement of pedestrians at work. These should be provided for access and egress points to the workplace, temporary accommodation units, car parks, and vehicle delivery routes. Safe crossing places should be provided where pedestrians have to cross main traffic routes. In buildings where vehicles operate, separate doors and walkways should be provided for pedestrian segregation to enable them to get from building to building.

Figure 6-28: Segregating pedestrians and site vehicles/plant. Source: RMS.

Figure 6-29: Segregating pedestrians and vehicles. Source: RMS.

In some situations where physical segregation by barriers is not possible, segregation is based on the provision of a raised walkway for pedestrians. This allows vehicles/plant to pass pedestrians and establishes a safe distance by provision of a sufficiently wide walkway and the raised edge of the walkway; this may be adequate for many workplaces.

Where it is not possible to have a pedestrian route with physical segregation by barriers or raised walkways segregation may be established by means of markings and signs.

In this situation, segregation is established by markings on the ground that indicate safe walk routes that pedestrians can use on which they have right of way over vehicles. Though they do not provide physical segregation, they provide a degree of segregation and improved safety for pedestrians.

> "(3) A traffic route does not satisfy paragraph (2) unless suitable and sufficient steps are taken to ensure that:
>
> (c) there is sufficient separation between vehicles and pedestrians to ensure safety or, where this is not reasonably practicable:
>
> (i) other means for the protection of pedestrians are provided, and
>
> (ii) effective arrangements are used for warning any person liable to be crushed or trapped by any vehicle of its approach;

Figure 6-30: Requirements for seperation of vehicles and pedestrians.
Source: CDMR 2015 Regulation 27.

Measures to be taken when segregation is not practicable

Where segregation is not practicable and vehicles share the same workplace as pedestrians it is important to mark the work areas and indicate main routes followed by vehicles and plant. This will warn drivers to adjust their driving and be more aware of pedestrians when they are leaving a main route. Any traffic route that is used by both pedestrians and vehicles/plant should be wide enough to enable any vehicle/plant likely to use the route to pass pedestrians safely. Where it is not practical to make the route wide enough, passing places or traffic management systems should be provided as necessary. Similarly, pedestrians should be provided with a safe zone at intervals along the traffic route so they can safely get out of the way of vehicles. Audible and visual warnings of the presence of the vehicle/plant would also assist.

Figure 6-31: No segregation.
Source: RMS.

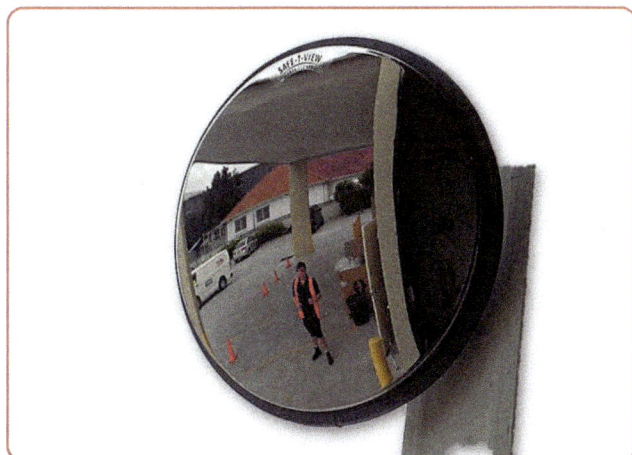

Figure 6-32: Pedestrian's view in mirror fitted to improve vision.
Source: HSE, HSG136.

Locations where the vision of the driver or pedestrian is restricted, at corners and similar 'blind spots', should be dealt with by the careful positioning of mirrors on walls, plant or storage or reducing the height of racking at the end of the aisle.

Where pedestrians and vehicle/plant routes cross, appropriate crossing points should be provided and used. Where necessary, barriers could be provided to guide pedestrians to designated crossing points. Crossing points may be raised to slow vehicles and marked in a suitable way to clearly identify them. At crossing points there should be adequate visibility and open space to enable the pedestrian to see vehicles/plant approaching. Where volumes of vehicle traffic are particularly heavy the provision of suitable bridges or subways to enable pedestrians to cross should be considered.

Figure 6-33: Pedestrian crossing point.
Source: HSE, HSG150.

Where vehicles/plant are dominant, but pedestrians need to access the same part of the workplace, similar means may be used, but for the opposite reason. In this situation the added use of high-visibility personal protective equipment that increases the ability of the pedestrian to be seen, and safety footwear including steel toe protectors, is usually needed. In these situations, it is essential that the pedestrian ensures good positioning and distance to ensure that they are not struck by moving parts of plant or by the plant when it moves to re-position itself. This will involve establishing a safe distance, outside the arc of slewing plant, and positioning so that the driver/operator of the plant can see the pedestrian.

Figure 6-34: Area where vehicles are dominant.
Source: RMS.

Figure 6-35: Safety by distance and positioning.
Source: RMS.

Wearing high-visibility clothing is often mandatory on major construction projects and its use should be considered anywhere where there are significant risks from vehicle movement and visibility of pedestrians is necessary.

The reversing of large vehicles that have a restricted view should be controlled by the use of a signaller/banksman/reversing assistant to guide them. Where it is unavoidable that pedestrians will come into proximity with vehicles, pedestrians should be reminded of the hazards by briefings, site induction and signs, so they are aware at all times.

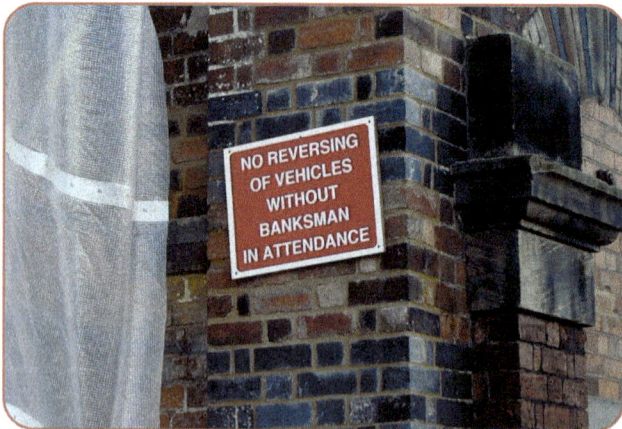

Figure 6-36: Control of vehicle movement.
Source: RMS.

Factors to be considered to reduce the level of risk when pedestrians work in vehicle manoeuvring areas include:

- Defined traffic routes.

- One-way systems.

- Segregated systems for vehicular and pedestrian traffic where possible, including barriers and separate doors.

- Maintaining good visibility by the use of mirrors, transparent doors, provision of lighting and vehicle reversing cameras.

- Signs indicating where vehicles operate in the area.

- Audible warnings and flashing lights on vehicles.

- Established and enforced site rules, including speed limits.

- Traffic control, for example, identification of "no go" or restriction on reversing areas.

- The provision of pedestrian safe zones in vehicle manoeuvring areas, so that workers can go to them and avoid moving vehicles. This is particularly important when vehicles are reversing and where workers might get trapped against fixed objects.

- The wearing of high-visibility clothing, sturdy (suitable) footwear.

- A good standard of housekeeping, to ensure a view of the area and prevent the need to avoid obstructions.

- Training for, and supervision of, all concerned, competency certificates, refresher training, trained signaller/banksman/reversing assistant to direct vehicle.

Figure 6-37: Use of reversing assistant.
Source: RMS.

Protective measures for people and structures
Barriers

It is important to anticipate that drivers of vehicles might misjudge a situation and collide with structures or people whilst operating. Moving vehicles, and in particular large plant, have high impact energy when they are in contact with structures or people. It is therefore essential that vehicles be separated from people and structures.

Plant and parts of the building that are likely to be struck by vehicles/plant should be provided with barriers that continuously surround the plant or alternatively posts can be provided at key positions. It is important to identify locations that warrant protection, for example, steelwork, scaffolds, hoists, falsework, formwork and storage vessels. Clearly, important plant and structures that are at the same level as the vehicle should be protected, but care should also be taken of structures at a height, such as a pipe bridge, roof truss or door lintel. All of these could be damaged by tall vehicles or those that have tipping mechanisms that raise their effective height. Because of the possible high energy involved in collisions the barrier used must be capable of resisting any likely impact. If it is only people that might contact the barrier a simple portable barrier may be adequate, but if it is heavy plant then robust barriers may need to be considered, for example, concrete structures.

Figure 6-38: Barriers and signs - structure.
Source: RMS.

Figure 6-39: Barriers and signs - pedestrians.
Source: RMS.

Markings and signs

Though it may not always be possible to protect all structures, such as a doorway, it is important to apply **markings** to make them more visible or to install height limitation/protection barriers. In addition, **signs** warning of overhead structures or the presence of vehicles in the area will help to increase perception of hazards, raise awareness of the need to take care and avoid collisions.

Figure 6-40: Marking of height restriction.
Source: RMS.

Figure 6-41: Marking of scaffold.
Source: RMS.

Any structure that represents a height or width restriction should be readily identified for people and vehicle drivers. This will include low beams or doorways, pipe bridges and protruding scaffolds, and edges where a risk of falling exists. These markings may be by means of attaching hazard tape or painting the structure to highlight the hazard. Signs should be used to provide information, such as height restrictions, and to warn of hazards on site. Signs may also be used to direct vehicles around workers at a safe distance.

Warnings of vehicle approach and reversing

There is a need to alert people to the hazard when working in or near a vehicle operating area; signs may be used to indicate the general operation of vehicles in the area. These can be supplemented by **visual and audible warning systems** that confirm the presence of a vehicle, such as a flashing light on the top of a dumper truck, see **Figure 6-42**. Warning systems may be operated by the driver, such as a horn on a dumper truck or automatically, such as an audible **reversing** signal on a large road vehicle. These are designed to alert people in the area in order that they can place themselves in a position of safety. They do not provide the driver with authority to reverse the vehicle or to proceed in a work area without caution.

Figure 6-42: Visual warning on a dumper truck.
Source: RMS.

Site Rules

It is important to establish clear and well understood site rules regarding vehicle operations. Drivers that are visiting the site should be made aware of site rules and conditions. These may have to be communicated to visiting drivers by security workers at the time they visit the site.

Site rules would often stipulate site speed limits, where the driver should be whilst vehicles are being loaded, where the keys to the vehicle should be and rules about not reversing without permission. They may also set out what access drivers have to areas of the site for their welfare, for example, availability of refreshment or toilet facilities. If there is a designated smoking or vaping area on site, this should also be clearly signed and drivers informed.

It is important that site workers and visiting drivers are able to communicate effectively; therefore arrangements should be made to communicate with visiting drivers who do not speak or have a limited vocabulary/understanding of the language used on site. Many organisations have site rules prepared to cover different languages that are illustrated with pictograms.

Site rules should also be established for workplace transport that remains on site. This may mean establishing additional rules related to the types of vehicle that operate on the site.

Pedestrians should also know what the site rules for workplace transport are in order to keep themselves safe. Site rules for the safe movement of pedestrians might include such things as using pedestrian exits/entrances or crossing points, not entering hazardous areas or the need to wear personal protective equipment in hazardous areas, and not walking behind a reversing vehicle.

Rules and procedures relating to vehicle/plant movement, for example, speed limits, direction of travel and use of loading/unloading areas must be enforced to ensure compliance with requirements. It is important to have speed limits that are practicable and effective.

Figure 6-43: Suitable clothing and footwear.
Source: RMS.

Speed limit signs should be posted and traffic slowing measures such as speed bumps and ramps may be necessary in certain situations. Monitoring speed limit compliance is necessary, along with action against persistent offenders. Speed limits of 10-15 miles/h (15-25 km/h) are usually considered appropriate for sites with established road routes, although 5 miles/h (8 km/h) may be necessary in certain situations and near workers a walking pace would be appropriate.

"(1) Suitable and sufficient steps must be taken to prevent or control the unintended movement of any vehicle.

(2) Where a person may be endangered by the movement of a vehicle, suitable and sufficient steps to give warning to any person who is liable to be at risk from the movement of the vehicle must be taken by either or both:

(a) The driver or operator of the vehicle.

(b) Where another person is directing the driver or operator because, due to the nature of the vehicle or task, the driver or operator does not have full visibility, the person providing directions.

(3) A vehicle being used for the purposes of construction work must, when being driven, operated or towed, be:

(a) Driven, operated or towed in such a manner as is safe in the circumstances.

(b) Loaded in such a way that it can be driven, operated or towed safely.

(4) A person must not ride, or be required or permitted to ride, on any vehicle being used for the purposes of construction work otherwise than in a safe place in that vehicle provided for that purpose."

Figure 6-44: Vehicles - safe operation and rules.
Source: CDM 2015 Regulation 28.

If rules on vehicle/plant movements into the site are difficult to enforce, physical measures such as gates, barriers, flow plates or control spikes (sprung flaps/spikes that only allow vehicles to cross in one direction) can be very effective.

Figure 6-45: Barriers to ensure direction of travel of vehicles. Source: RMS.

Where construction sites are made up of or join a highway, such as carriageway repairs on a motorway, it is important to identify who has priority - the vehicle or the worker. Site rules may clarify that once the site boundary is crossed, the worker has priority - this has to be clear. The site rules may be reinforced by the provision of additional signs, for example, to clarify a speed limit and the nature of the priority of workers.

CASE STUDY

An organisation was prosecuted and fined after a reversing vehicle at its site killed a delivery driver. The driver was delivering materials to the site and was standing by the side of his vehicle after overseeing the removal of its load when he was struck by a reversing fork-lift truck and died instantly.

The organisation had failed to carry out a suitable risk assessment for the movement of loads at the site. This would have shown the need for fork-lift trucks to avoid reversing for long distances and that delivery drivers should be removed from the danger area. They should have installed suitable barriers to prevent pedestrians gaining access to areas where vehicles were working, and established a formal system for supervising visiting drivers to the site.

Safe vehicles

When undertaking an assessment of vehicle safety, the following factors need to be addressed:

- Vehicle selection.

- Vehicle inspection and maintenance.

Vehicle selection
Suitability

Vehicles and plant should be safe by design, but it is also important to choose the correct vehicle/plant for the task. Consideration will need to be given to the size and weight of what needs to be moved; therefore a suitable vehicle with a sufficient safe working load should be selected. When considering the safe working load, this will include the vehicle capacity when transporting bulk liquids or slurries such as cement. Vehicles should be suitable for any loads carried and it may be important that the vehicle has adequate anchor points to make sure that loads can be carried securely.

Figure 6-46: Rough terrain fork-lift truck, driver protection and lights. Source: RMS.

The vehicle will also need to be suitable for the environment in which it is to be used. This will include consideration of the space available for it to be used and whether it is to be used inside buildings, in which case battery-powered vehicles may be most suitable. The assessment of suitability must consider factors like weather conditions and provide protection from the cold, shade from heat and protection/visibility in the rain. In addition, vehicles to be used in flammable atmospheres should be suitable for work in that environment.

Vehicles/plant have to be suitable for the terrain on which they are going to be used. For example, a standard counterbalance fork-lift truck is suitable for use in a building that has smooth floors, but would be unsuitable for use on a construction site with rough-terrain. A suitable rough terrain lift truck would be required in this environment.

There should be provision of a safe way to get into and out of the driving/passenger position and any other parts of the vehicle that need to be accessed regularly. Access features on vehicles, such as ladders, steps or walkways, should have the same basic features as site-based access systems.

Vehicles should have seats that are safe and comfortable, where they are necessary. Guards should be fitted round dangerous parts of the vehicle/plant, for example, power take-offs, chain drives and exposed hot exhaust pipes.

Figure 6-47: Safe operating platform.
Source: SRTS Limited.

Consider fitting a horn, vehicle lights, reflectors, reversing lights and possibly other warning devices, for example, rotating beacons or reversing alarms. Painting and marking vehicles in distinct colours may be necessary to make the vehicle stand out against the workplace background.

Visibility from vehicles/reversing aids

It is important that drivers are able to see clearly around their vehicle, to allow them to identify hazards and avoid them. Driver visibility is often restricted to the side or rear of the vehicle. Side visibility can be improved greatly with the addition of extra mirrors to the sides of the vehicle. These mirrors provide a good low-level view, assisting with parking and the identification of hazards close to the vehicle sides. Vehicles that have limited driver visibility behind the vehicle may be provided with an extra supplementary mirror that improves vision. For example, a mobile crane may have limited visibility from the driving cab and a mirror in the centre/rear of the vehicle to aid the driver's vision.

A variety of other reversing aids are now in common use for vehicles that travel on ordinary roads. These include simple proximity detectors to the front and rear of the vehicle that trigger an audible alarm when the vehicle approaches an object. The audible alarm tone usually varies as the vehicle gets closer to the object. More sophisticated devices will distinguish between front and rear detection by the use of different sounds.

Figure 6-48: Audible proximity detectors.
Source: Transport support.

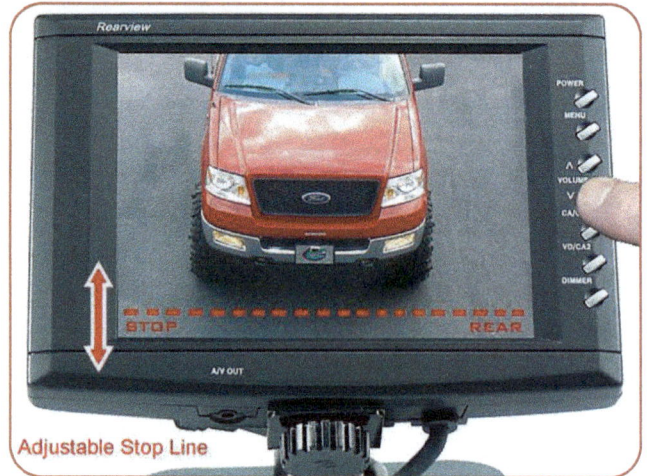

Figure 6-49: Reversing camera systems.
Source: Transport support.

In addition, the device may also display a schematic on a screen in the vehicle cab, indicating the precise direction of the obstruction to the driver.

Further advances in detection methods for vehicles include the use of cameras to the rear or sides of the vehicle with a clear display available to the driver in the cab.

CASE STUDY

A construction site worker was injured by a 360° excavator, which was operating in a poorly organised construction site.

The site worker was carrying out construction activities when the reversing excavator hit him and the track went over his right leg. The excavator was not fitted with devices to improve visibility from the cab, such as rear-mounted convex mirrors or closed-circuit television (CCTV), and the driver had not received formal excavator training. The excavator had been working within 3-4 metres of the injured worker on a daily basis, had knocked him once before and would often lift material over the worker's head.

Roll-over protection and restraint systems

In many work vehicle accidents/incidents drivers are injured because the vehicles do not offer protection when they roll over, or they do not restrain the drivers to prevent them falling out of the vehicles and being injured by the fall or the vehicles falling on them.

Vehicles with open cabs, such as dumper trucks, road rollers and fork-lift trucks, are examples of equipment that may present this risk.

The Provision and Use of Work Equipment (PUWER) Regulations 1998 Part 3 recognises the importance of driver protection and recommends that equipment should be adapted, where practicable, to provide this protection. New equipment purchases should include the provision of these types of protection and restraint systems.

Roll-over protection can take the form of a complete, enclosed cab or of a system of bars that prevent the vehicle driver/passengers from being crushed by the vehicle in the event that it rolls over.

Figure 6-50: Seat belt status alarm beacon.
Source: Transport support.

Restraint systems typically involve some sort of adjustable seat belt that is worn to prevent the driver/passenger from being thrown from the vehicle when travelling or falling from the vehicle in the event that it rolls over. Other devices include seat belt status alarm detectors. This system is widely used on construction equipment to indicate that the operator is wearing the seat belt.

It comprises a retractable lap belt, the buckle of which has a switch that is activated by the belt when fastened. This is connected to a green beacon on the roof of the vehicle to show that the operator is wearing the seat belt.

Figure 6-51: Roll bar and seat restraint.
Source: RMS.

Protection from falling materials

There is a risk of material falling on a vehicle driver where the vehicle is used to load and unload materials at a height or to remove materials from delivery vehicles. In these situations it is important that suitable protection of the driver be provided by the structural design of the vehicle, for example, by providing an enclosed cab. There is also a risk from falling materials where loose materials are loaded into them, for example using an excavator to load loose materials into a dumper truck. The driver of the dumper truck should not remain on the vehicle if the vehicle is not designed to provide protection from loose materials. Protection should also be provided where vehicles work in close proximity to areas where there is a risk from falling materials. Therefore it is advisable to provide suitable protection, where practicable. This will usually also provide protection for roll-over situations.

Vehicle inspection and maintenance

Inspection

Vehicles to be used on site should be inspected when first received and put to use on site and then inspected at regular intervals, the intervals may vary depending on the type of vehicle, amount of use and activities it is used for. Vehicles used on site are subject to Regulation 6 of Provision and Use of Work Equipment Regulations (PUWER) 1998, which lays down requirements for inspecting work equipment to ensure that health and safety conditions are maintained and that any deterioration can be detected and remedied in good time.

"Every employer shall ensure that work equipment exposed to conditions causing deterioration which is liable to result in dangerous situations is inspected:

(a)   At suitable intervals.

(b)   Each time that exceptional circumstances which are liable to jeopardise the safety of the work equipment have occurred, to ensure that health and safety conditions are maintained and that any deterioration can be detected and remedied in good time.

Every employer shall ensure that the result of an inspection made under this Regulation is recorded and kept until the next inspection under this Regulation is recorded."

Figure 6-52: Inspection of work equipment.
Source: PUWER 1998 Regulation 6.

A record of inspections of vehicles subject to PUWER 1998, regulation 6, should be made.

It is also important to conduct a pre-use check of the vehicle. This is usually done by the driver as part of their taking it over for a period of use such as a work shift or day. This check would note the condition of critical items and provide a formal system to identify and consider problems that may affect the safety of the vehicle. Typical checks will include checking the pressure and condition of tyres, testing brakes, lights, horn, windscreen wipers, and checking the adequacy of screen washer fluid.

If the vehicle is to use attachable accessories the driver should ensure they are correctly fitted and secured. For example, for vehicles that pull trailers, the driver should check the adequacy of the trailer coupling and any hydraulic hose/electrical connections between the vehicle and trailer. If there is no designated driver and the vehicle is for general use someone should be nominated to make these checks. A record book or card would usually be used to record the checks and the findings, together with a means of reporting defects.

Maintenance

Employers should ensure that work equipment, including vehicles, is maintained throughout its working life, so that it is in efficient working order and in good repair. The manufacturer's instructions should be taken into account when maintenance work is planned and carried out.

"Every employer shall ensure that work equipment is maintained in an efficient state, in efficient working order and in good repair."

Figure 6-53 Maintenance of work equipment.
Source: PUWER 1998 Regulation 5.

Vehicle maintenance should be planned for at regular intervals and vehicles taken out of use if critical items are not at an acceptable standard.

Site vehicles should be maintained to a suitable standard at a frequency that reflects the demands placed on the vehicle through use on site. Critical items such as tyres and brakes on site vehicles should be kept to the same high standard as one that is used on public roads. Vehicles used on public roads must be maintained to meet the minimum legal requirements of roadworthiness and to remain effective.

Repairs should be carried out promptly to ensure safe operation of the vehicle. Where damage has been caused to a vehicle's exterior body shell this should be assessed and if it is a risk to pedestrians or other road users, the vehicle should be taken out of service until such repairs have been made.

REVIEW

What practices and procedures can be used for reducing risk when vehicles are reversing?

What hazards should be considered during loading and unloading of a lorry in a loading bay?

Safe drivers

All drivers must be safe; the following factors will address driver safety:

- Selection and training of drivers.
- Reversing assistants.

Selection and training of drivers

People are only permitted to operate plant or vehicles after they have been selected, trained and authorised to do so, or are undergoing properly organised formal training under competent supervision.

Selection

The safe usage of plant and vehicles calls for a reasonable degree of both physical and mental fitness. The selection procedure should be devised to identify people who have been shown to be reliable and mature enough to perform their work responsibly and carefully. To avoid wasteful training for workers who lack coordination and the ability to learn, selection tests should be used.

Consideration must be given to any legal age restrictions that apply to vehicles that operate on public roads. A similar approach may be adopted for similar vehicles used on site, though the law is not specific about age limitations in such cases.

Potential drivers should be medically examined prior to employment/training in order to assess their physical ability to cope with this type of work. Drivers should also be medically examined every five years in middle age and after sickness or accident. Points to be considered are:

General: Mobility, having full movement of trunk, neck and limbs.

Vision: Good eyesight, which may be achieved by wearing corrective lenses (spectacles), is important as operators are required to have good judgement of space and distance. In many countries, driving with vision in only one eye is legal. The question of ability to drive usually relates to field of vision. Since removing an eye reduces the visual field by only about 40%, monocular vision alone is not a barrier to driving when the other eye has a very good visual field.

Hearing: The ability to hear instructions and warning signals with each ear (to enable the direction of the source to be located) is important.

Training

It is essential that the immediate supervisors of drivers receive training in the safe operation of vehicles, and that senior management of the organisation appreciate the risks resulting from the interaction of vehicles and the workplace.

For the driver, safety must constitute an integral part of the skill-training programme and not be treated as a separate subject. The driver should be trained to a level consistent with efficient operation and care for the safety of themselves and other persons.

A comprehensive training programme has to be delivered and must consider the following topics:

- Parking and any other restrictions that may be in place, such as speed limits.

- The need to be courteous and give right of way to pedestrians and other vehicles.

- The vehicle's controls and how to drive safely both forwards and in reverse.

- Hazards that are specific to the workplace, such as height restrictions and hazardous materials.

- Safety issues associated with refueling.

- How to carry out pre-use inspections including the procedures for reporting defects.

- Maintenance requirements.

- The issues of substance/alcohol misuse.

- How to secure the vehicle after use and the procedures for key security.

- Any legal requirements that might apply to the driver, such as limitations on driving hours.

On completion of training the worker should be issued with an authority to drive from the organisation. This will often involve a further test related to the site conditions, if the driver has been trained elsewhere. A record of all basic training, refresher training and tests should be maintained in the driver's personal documents file.

"(3) A vehicle being used for the purposes of construction work must, when being driven, operated or towed be—

(a) driven, operated or towed in such a manner as is safe in the circumstances; and

(b) loaded in such a way that it can be driven, operated or towed safely.

(4) A person must not ride, or be required or permitted to ride, on any vehicle being used for the purposes of construction work otherwise than in a safe place in that vehicle provided for that purpose."

Figure 6-54: Safe use of vehicles.
Source: CDMR 2015 Regulation 28.

It should not be assumed that workers who join the workforce as trained drivers have received adequate training to operate safely in their new organisations. Certification of training by other organisations must be checked for adequacy before the driver is given authorisation to drive. Site management must ensure that they have the basic skills and receive training in the organisation's methods (local practices) and procedures for the type of work they are to undertake. They should be examined and tested before issue of an authority to drive.

If high standards are to be maintained, periodic refresher training and testing should be considered.

Many organisations operate local codes of practice and take great care to confirm the authority they have given to drivers by the provision of a licence for that vehicle and sometimes a visible badge to confirm this. Access to vehicles must be supervised and authority checked carefully to confirm that the actual class of vehicle to be driven is within the authority given. It is essential that access to vehicles is restricted to those persons who are competent to drive them, for example, by the control of keys.

"(1) Suitable and sufficient steps must be taken to prevent or control the unintended movement of any vehicle."

Figure 6-55: Preventing unintended movement of vehicles.
Source: CDMR 2015 Regulation 28.

Reversing assistants

The role of the reversing assistant is to assist the driver by extending their field of vision and directing movement of the vehicle. This is achieved by the reversing assistant standing in a position where they can see the area behind the vehicle and signaling to the driver, using recognised signals.

The use of reversing assistants to control reversing operations is not a preferred option and should be subject to a full risk assessment, as it can involve putting the reversing assistant in the potential danger area of a reversing vehicle.

If a reversing assistant has to be used, this is best achieved by providing the correct training in the use of international hand signals to direct the driver. The driver should be instructed to stop the vehicle immediately if the reversing assistant disappears from view.

"(2) Where a person may be endangered by the movement of a vehicle, suitable and sufficient steps to give warning to any person who is liable to be at risk from the movement of the vehicle must be taken by either or both:

(a) the driver or operator of the vehicle, or

(b) where another person is directing the driver or operator because, due to the nature of the vehicle or task, the driver or operator does not have full visibility, the person providing directions."

Figure 6-56: Reversing assisstant to warn of vehicle movement.
Source: CDMR 2015 Regulation 28.

The reversing assistant should wear suitable high-visibility clothing and be taught to place themselves in a safe position that ensures the driver can see their signals and does not put them in danger from the reversing vehicle. They should be aware of the limited view a driver might have from their mirrors, particularly when turning the vehicle as well as reversing, and the effect of driver position when the vehicle is left- or right-hand drive. They should take these factors into account when selecting a position to be seen and be safe.

CONSIDER

What is workplace transport?

How might transport safety be improved in your own workplace?

6.3 Work-related driving

The reversing assistant should take control of the area that the driver is reversing into and warn people who might be at risk of harm from the movement of the vehicle because they are in or are entering the area.

C.J.L Murray and A.D. Lopez conducted a study that analysed global harm due to a range of risk factors. In 1996 they reported in the publication 'Summary, the Global Burden of Disease' that they estimated the number of annual road traffic fatalities in the developed regions of the world to be 222,000. The HSE estimates that up to a third of all road traffic accidents involved somebody who was at work at the time. This may account for over 20 fatalities and 250 serious injuries per week.

"It has been estimated that between 800 and 1,000 road deaths a year are in some way work-related. Many bosses have ignored this problem in the past, but the Health and Safety Executive has now made it clear that employers have duties under health and safety law to manage the risks faced by their workers on the road."

Figure 6-57: Size of the road risk problem.
Source: RoSPA.

For the majority of people, the most dangerous thing they do while at work is drive on the public highway.

Figure 6-58: Road risk.
Source: HSE 1996.

The true cost of work-related driving accidents/incidents to any organisation is nearly always higher than just the costs of repairs and insurance claims. It can include personal costs, such as ill-health, time in hospital, stress on family members and possible sanctions being imposed on a driver's licence following an accident/incident. There is also the potential for drivers to have their licence revoked by the national driver licensing authority that issued them. Therefore the management of driving activities through the introduction of policies, risk assessments and developing safe systems for work-related driving is essential. This will ensure compliance with the Health and Safety at Work etc. Act (HASAWA) 1974 and the Management of Health and Safety at Work Regulations (MHSWR) 1999. The Health and Safety Executive (HSE) provide guidance to assist management in managing driving risk, in the form of the document 'Driving at work - Managing work-related road safety'.

MANAGING WORK-RELATED DRIVING

Plan

Assess the risks

As with other work activities, work-related driving should be risk assessed. This will guide the employer and managers, enabling them to establish effective arrangements for controlling the driving risks. The risk assessments should be used to guide action to plan, do, check and act to manage the risks.

Risk assessments for work-related driving activities should follow the same principles as risk assessments for any other work activity, for example, by following the '5 steps' approach. They should be carried out by a competent person with practical knowledge, relating to the work-related driving activity being assessed.

The range of hazards is likely to be wide and the main areas to consider are the driver, the vehicle and the journey. Also consider the factors that affect these hazards and increase the risk, these factors are described later in this element.

When assessing the risks it is necessary to evaluate the current control measures, consider the control measures described later in this element. Talk to the drivers involved and those that manage them in order to try to determine what happens in practice. Consider different categories of driver, for example, inexperienced drivers, infrequent drivers, those that drive long distances and those that drive specific or unusual vehicles.

Step 1 - Identify the hazards

Hazards will fall into the following categories:

The driver: Competence, fitness and health, alcohol and drug use, behaviour etc.

The vehicle: Suitability, condition, safety equipment, ergonomic considerations, the load,security, etc.

The journey: Route planning, scheduling, time, distance, driving hours, weather conditions, stress, volume of traffic, passengers, etc.

Step 2 - Decide who might be harmed

Obviously the driver, but this might include any passengers, other road users and/or pedestrians. Consideration should also be given to other groups who may be particularly at risk, such as young or newly qualified drivers and those driving long distances.

Step 3 - Evaluate the risk and decide on precautions

Risks may vary depending on whether driving is done at night or in the day, the type of vehicle or driving conditions. Decide the likelihood of the harm and severity (consequence) of any outcome. Consider risk factors such as the load being carried, distance travelled, driving hours, work schedules, traffic levels and weather conditions. Decide on appropriate precautions and once these have been established, decide whether the residual risk is acceptable.

Step 4 - Record the findings and implement them

Significant risk assessment findings need to be recorded and should be made available to the drivers affected.

Step 5 - Review the assessment and update it if necessary

Monitor and review the assessments to ensure the risks are suitably controlled. Systems should be put in place to gather, monitor, record and analyse accidents/incidents that might affect these risk assessments. The vehicle and driver's history should also be recorded. Any changes in the route, new equipment and changes in the vehicle specifications should be reviewed and recorded. This will ensure the effectiveness of controlling the risks.

Policy

All employers who require workers to drive as part of their work should have a policy to address the risks from these activities, whether the employer provides vehicles or expects workers to drive their own vehicles for work purposes. Where it is practical, a commitment to managing work-related driving risks should be included in the employer's general health and safety policy. Where this is not practical employers may establish a separate policy to manage work-related driving risks. Without a specific policy commitment to manage these risks managers may not be given the resources to manage them effectively. It is therefore important that these risks are recognised at senior management level in the health and safety policy and a commitment made to manage them.

Top management taking into account work-related driving

In order to manage the risks of work-related driving it is important there is a top management commitment that ensures it is taken into account when business decisions are being made. This will help to ensure that those with responsibilities for managing the risk put the necessary effort and resources in place to avoid the risks where possible and minimise those that cannot be avoided. Top management should ensure that those managing the risk have the resources and authority to do this effectively.

Role and responsibilities

In a large organisation it is likely that various departments within the organisation will have different responsibilities for the management of work-related driving risk. For example, the dispatch department would often be responsible for planning journeys, the training department may be responsible for driver competence, the human resources department could be responsible for driver selection and checking the validity of driving licences, the maintenance department would usually be responsible for the condition of vehicles and ensuring their roadworthiness and the occupational health department may be responsible for carrying out routine health surveillance of drivers.

In order to establish an organised and coordinated approach the responsibilities of different parts of the organisation and those that manage them should be defined, including the responsibility to co-operate with each other.

Legal responsibilities of individuals on public roads

The specific legal responsibilities of individuals while driving on public roads will vary from country to country. Typical responsibilities of individuals could relate to their fitness to drive, driver competence, the condition of the vehicle and driving behaviour.

Fitness to drive

Legal responsibilities related to the driver's fitness to drive may include:

- The driver having adequate vision for driving. The requirements for adequate vision may be defined in legislation, for example, being able to read (with the aid of vision correcting lenses) a vehicle number plate, in good daylight, from a distance of approximately 20 metres.

- The driver reporting to a competent authority or doctor any health condition likely to affect their driving.

- The driver not driving while affected by fatigue. This may be linked to specific limitations on driving hours and rest periods.

- The driver not driving whilst affected by drugs or alcohol. Legal limits may be set to define when a driver is considered to be affected by drugs or alcohol.

Driver competence

Legal responsibilities related to driver competence may include:

- The driver being of a minimum age to drive the type of vehicle they are driving.

- The driver holding a valid driving licence that confirms their ability to drive the type of vehicle they are driving.

Vehicle condition

Legal responsibilities related to the vehicle condition may include:

- The driver ensuring the vehicle complies with requirements for roadworthiness. This may be a shared responsibility with the employer where the vehicle is provided by the employer. However, the driver will often be held initially responsible for the vehicle they are driving.

- The driver not overloading the vehicle.

- The driver securing any load on a vehicle correctly and ensuring it does not protrude dangerously.

Driving behaviour

Legal responsibilities related to driving behaviour may include:

- The driver driving with due care and attention for others.

- The driver not exceeding speed limits.

- The driver and passengers wearing a seat belt or suitable restraining device.

Do

Co-operation between departments

Where a number of departments are involved in the management of work-related driving it is important to ensure communication between the departments and their co-operation in providing effort to managing driving risks. If arrangements are not put in place this could seriously affect the effectiveness of the management systems. For example, failure of the maintenance department to communicate to the dispatch department that a vehicle needed to be withdrawn from service for maintenance work might lead to the vehicle being driven in an unsafe condition. It is therefore essential that the structure of the organisation enables the co-operation of departments to allow for the easy and reliable interchange of work-related driving risk management information, for example related to licences to drive classes of vehicle, medical conditions, any damage to vehicles and other problems that could increase risk.

Systems to manage work-related driving

In support of their policy, employers should establish management systems to manage their work-related driving risks and managers should be trained to ensure they are competent to manage and implement the work-related driving systems.

It is important that systems are put in place to reduce work-related driving risk by ensuring that driving distances are kept to a minimum, driving hours are controlled and work schedules are organised to avoid pressure on drivers to complete their journeys in an unreasonable time. Specific systems should be established to manage fatigue and the taking of breaks from driving. This should include consideration of workers who might be expected to drive long distances as part of their work after completing a normal work day. Organisations should have systems in place to respond to inclement weather and enable drivers to take safe decisions about the effects of weather on their driving or journey. Systems should also be established to manage vehicle breakdowns, emergencies, problems experienced while travelling and delays that could affect the journey.

Where relevant, systems should be established to manage circumstances where drivers provide their own vehicles for work activities, this should include:

- A requirement that if workers use their own vehicle for work activities they must maintain it in a roadworthy condition.

- A requirement that where the age of the vehicle is significant the vehicle has been confirmed to be roadworthy by an independent organisation. The requirements for initial and regular checks of roadworthiness are sometimes specified by national laws.

For example, national laws may require that vehicles undergo an initial test of roadworthiness when the vehicle is three years old and annual roadworthiness tests thereafter.

- A requirement that workers who drive their own vehicle on behalf of their employer have a current driving licence that allows them to drive the vehicle on public roads.

- A requirement that appropriate insurance for the vehicle is held by the worker and it covers the worker to drive it. The worker should present copies of certificates of insurance that cover work-related driving annually for inspection by the employer.

- A requirement that workers inform their line manager of any changes in circumstances that could affect their ability to drive on the employer's behalf. For example, changes to the vehicle the worker provides or insurance arrangements affecting it, penalty points/citations/endorsements issues for driving offences, the use of any prescription medication or changes to the worker's health that could affect ability to drive safely.

Systems of work to manage work-related driving risk should be formalised, for example, as written procedures. This would help to avoid ambiguity and ensure more consistent management of the risks.

Communication and consultation within the workforce

The arrangements put in place to manage work-related driving risk should include communication and consultation within the workforce, including those that carry out the driving. This will establish a clear understanding of the importance of managing the risk and the role that the workforce plays, this is more likely to lead to them following the systems introduced. They are also more likely to feel that they can report problems, difficulties and incidents related to work-related driving.

Adequate instruction and training

Driver competence is an essential part of managing work-related driving. It is therefore important that systems are established that ensure adequate driver instruction, training and competence for the vehicles they drive. It is important that the drivers hold a valid driving licence for the type of vehicle they drive for work activities. If drivers are expected to drive large goods vehicles (LGVs) or large passenger-carrying vehicles (PCVs) or other specialist vehicles, employers may require confirmation that the driver holds a specific licence or qualifications as proof of the driver's ability to competently drive these vehicles.

Some employers carry out internal assessments of the driving skills of everyone that drives on the employer's

behalf, in addition to confirming that the minimum legal requirements for driving licences are met. These driving skill assessments can be carried out in-house or by an external assessor. Some employers also offer specific training in safe driving techniques for their workers.

Systems should be established that ensure drivers are fit to carry out the work-related driving activities they do. This may involve consideration of medical conditions that could affect driving during a formal medical examination or a driver declaration process. Where drivers drive specialist vehicles, for example, a large goods vehicle, an initial medical examination is often carried out before allowing new drivers to drive the vehicles and a similar medical examination may be carried out periodically. In addition, employers may carry out tests to determine if drivers are driving or attempt to drive while influenced by alcohol or drugs.

Check
Monitoring performance

It is an essential safe working practice and may be a specific legal requirement to monitor health and safety systems related to driving, in particular, driving hours. The performance of an organisation in managing work-related driving risks should also be monitored using a range of active and reactive methods.

Information on driving schedules, including driving hours and distances can be monitored actively to determine if they conform to acceptable standards. The licences of drivers and the condition of vehicles can also be checked regularly. Arrangements for breakdowns and problems drivers experience while driving can be monitored actively by examining reports or records of how they were dealt with and any effects on safety. In addition, poor driving behaviour could be monitored by direct observation or by electronic surveillance systems.

Information on accidents/incidents linked to work-related road safety risk should also be monitored. This should include situations where the driver or other people are injured, any events that result in damage to vehicles or property and near misses. Examples of near misses would include situations where drivers fall asleep momentarily while driving, but do not lose control of their vehicle and cause harm. In addition, certain road traffic accidents/incidents may have to be reported to a competent authority under national law. The ILO's 'P155 - Protocol of 2002 to the Occupational Safety and Health Convention, 1981' (defines the term commuting accident in relation to work-related journeys, see *Figure 6-59.*

"The term commuting accident covers an accident resulting in death or personal injury occurring on the direct way between the place of work and:

i) The worker's principal or secondary residence.

ii) The place where the worker usually takes a meal.

iii) The place where the worker usually receives his or her remuneration."

Figure 6-59: Definition of term commuting accident.
Source: ILO, P155 Protocol of 2002 to the Occupational Safety and health Convention, 1981.

Reporting work-related driving incidents

In order to support the organisation's monitoring of the management of work-related driving risks and any necessary reports to competent authorities it is important that all workers report incidents, including near misses. Where possible this should be extended to reporting problems and difficulties that increased risk so that the organisation can take account of these and take action to reduce the risk in future activities.

Drivers must be required to record information about all incidents involving injury, whether minor or serious. A similar reporting procedure should be in place for reporting significant 'near-misses', driver and manager training regarding this should emphasis how to recognise, analyse and learn from such events. The data provided should be analysed and any changes or improvements noted. These should be communicated to those concerned and the work-related driving systems updated.

Act

Review performance and learn from experience

As with all risks, the organisation should review its performance in the management of work-related driving risks periodically and after a major work-related driving incident. This review should consider the range of factors that influence the risk, the control measures in place to minimise this risk and the effectiveness of these measures. This will inform the management of the organisation and provide an opportunity to learn from experience and determine what measures have been effective and any opportunities for improvement.

REVIEW

What factors should be included in a training programme for drivers to reduce the risk of accidents/incidents to themselves and other workers?

Regularly update the policy

The review of performance should lead the employer to consider any impacts on the current work-related driving policy and make any changes to the policy that would strengthen it and provide an improvement in the management of these risks.

FACTORS THAT INCREASE RISK WHEN DRIVING AT WORK

There are many factors that increase the risks of an accident/incident while driving at work and it is the responsibility of the employer to assess these based on the nature and circumstances of the work. Important considerations include distance travelled, driving hours, work schedules and stressors, such as traffic congestion and inclement weather conditions.

Distance

Road conditions have improved in most countries over the years, which has allowed greater travel distances in shorter periods of time. However the fatigue encountered by drivers often increases due to other drivers' poor performance, creating the need for a particularly high degree of concentration. LGV drivers' travel distance is often regulated by limitations on their driving hours. However, there are often no such limits on drivers of smaller vehicles and private cars. When assessing risk of any kind, an important consideration is the frequency and duration of exposure to hazards. It is logical then that the greater the distance of the journey (i.e. the duration of exposure to the hazard) the greater the risk. It is possible that scheduling of routes may extend or reduce the distance that has to be travelled and therefore affect the level of risk significantly.

Driving hours

If driving hours are excessive the driver is likely to become fatigued, their attention and reaction levels will fall and they are at increased risk of making errors. The driving hours may be excessive because the driver has been driving too long without a break or their cumulative hours in a day have become too much. In some cases their rest period between days may become too short and prevent sufficient recovery. For some drivers the driving hours are in addition to other work hours, for example, for someone travelling to a meeting. These cumulative hours can have a significant effect on fatigue, therefore it is important that cumulative hours are monitored and controlled and that breaks are taken at intervals on longer journeys.

The Highway Code recommends that drivers should take a 15 minute break every two hours. The responsibility for monitoring driving hours and taking breaks is a shared responsibility of both the driver and the employer.

Many commercial vehicles will be fitted with tachometers (digital or analogue), see *Figure 6-60*, which record the time spent driving, or drivers will be required to keep a record/log of driving hours. These driving hours are often limited by national legislation or international agreement.

Driving hours for goods vehicles over 3.5 tonnes and passenger vehicles designed to carry nine or more passengers are regulated by European Agreement Concerning the Work of Crews of Vehicles Engaged in International Road Transport (AETR) rules, EU rules or GB domestic rules depending on where you are driving.

However company-car and van drivers do not have specific legal requirements imposed on them to limit and monitor their driving hours. Therefore safe systems of work and appropriate training should be given to all drivers concerning the hazards and risks associated with excessive work hours, including driving.

Work schedules

Work schedules that are badly organised can put increased pressure on drivers to be in a place by a given time. This can lead to them being tempted to increase speed and take abrupt action to change lanes to improve their progress. This risky action can lead to higher risk of collision and reduced stopping distances.

Work schedules should take account of periods when drivers are most likely to feel sleepy. The high-risk times are 2am to 6am and 2pm to 4pm. Employers should provide drivers with the means to stop and take a break if they feel sleepy, without the fear of recrimination.

Where possible, schedules should be organised so that breaks can naturally and easily be taken, with agreed break points if travel is going or not going as planned. Driving to and from the place of employment does not usually form part of the working day when deciding the working hours of drivers, but may be a factor in the cause of accidents/incidents if travelling home follows a long working day.

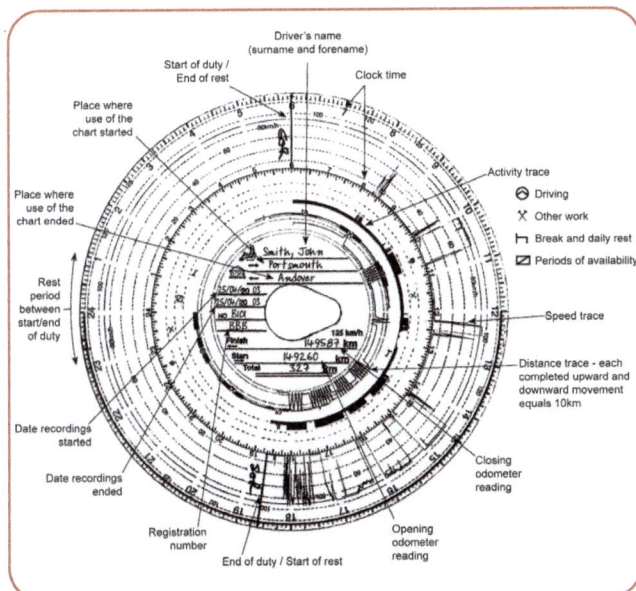

Figure 6-60: Analogue tachograph.
Source: UK Dept. of Transport.

Stress due to traffic

Driving-related stress is likely to be experienced when the demands of the road/traffic environment exceed the driver's ability to cope with or control that environment. Sometimes driving stress can show itself as a 'road rage'. This is an irrational human mechanism that is activated to protect the sufferer from what they perceive to be actual or potential danger.

Some drivers may perceive the circumstances which aggravate the 'road rage' as being either insignificant or trivial, but to the individual experiencing 'road rage' they feel 'very real' and 'very relevant'.

Weather conditions

Weather conditions have a significant impact on the risks of driving. Sudden rainfall, sand storm or snow and fog can lead to poor visibility and sudden braking. Snow and surface water conditions can increase stopping distances. Even good weather can have a negative effect as glare from the sun can limit visibility. This is particularly the case in early morning or evening and is most significant during the winter when the sun is lower in the sky for longer.

EVALUATING THE RISKS

The driver

The level of risk is particularly affected by the driver's **competence.** If someone is new to driving, their skill may be adequate to provide them with a national driving licence but their experience of driving will be low.

Drivers that only drive intermittently also present a high risk as any competence they may have may decay over the time between driving.

The driver's competence has to be appropriate to the driving expected of them. An inexperienced driver driving in heavy traffic in complicated driving settings at night in the winter will present a particularly high risk. Similarly a person that is reasonably experienced in driving their small car may have difficulty when first driving a larger or faster-accelerating vehicle. Some drivers will need to adjust to driving slower vehicles and ones with a different centre of gravity.

Driver **fitness** may influence their ability to see well when driving at night, their ability to travel distances without breaks and may put them in a high-risk category for heart attack or other type of seizure. Pre-existing health conditions such as back injuries and late-term pregnancy could influence a driver's ability to concentrate on road conditions.

The level of training that the driver has received may affect risk in that if no training is given a driver may have developed bad driving habits and not be aware of them. Refresher driver training may help to reduce risk.

The vehicle

The size, weight, centre of gravity and power of a vehicle will all influence its functioning and therefore the risk that may arise from its use. It is important that the vehicle is suitable for the task and the driver.

A large, powerful, fast-acceleration car may be suitable for a specific task but very unsuitable for an inexperienced driver. The condition of the vehicle will have a significant effect on the level of risk. Vehicles that have poor brakes or lights, do not steer well or have poor suspension will represent a high risk in any driving situation. Vehicles with broken or missing mirrors will mean that the driver will not be able to see other road users adequately and will increase the risk when changing lanes.

Much of the *safety equipment* to prevent accidents/incidents, such as anti-lock braking systems (ABS) should be built into most modern vehicles. Other items may be optional, such as 'run-flat' tyres, which may reduce certain road risks. Other safety equipment is designed to reduce the consequences of accidents/incidents and can therefore reduce risk, such as airbags, head restraints, escape kits, warning triangles, fire extinguishers and high-visibility vests. The absence of safety-critical information, like the height of the vehicle, can have an immediate and significant effect on the level of risk. Consideration has to be given to *safety-critical information* related to the vehicle, including its height, width, length, weight and load carrying/towing capacity.

Ergonomic considerations have an effect on both comfort and ability to control the vehicle effectively. Comfort issues can increase fatigue, and ergonomic considerations like seat height adjustment can significantly affect the ability of the driver to see out of the vehicle properly.

The journey

When evaluating the risk related to journeys it is important to take into account such factors as:

- The *route* being taken - motorways are safer than smaller roads; routes using motorways will be lower in risk.

- *Scheduling* - if the journey is to be made early in the morning there might be an increased level of risk due to tiredness, but this may be offset by the reduced level of traffic.

Vehicle Number _____	Date _____	
	Yes	No
Are all departmental vehicles subject to State licensing requirements equipped with the following items in good operating condition:		
Adequate rearview mirrors?	❑	❑
Safety belts?	❑	❑
Windshield wiper blades and fluid?	❑	❑
Horn?	❑	❑
Correctly adjusted headlights?	❑	❑
Brakes with adequate stopping power?	❑	❑
Emergency brake?	❑	❑
Turn/directional signals?	❑	❑
Good tires with adequate tread and correct pressure?	❑	❑
Oil and coolant levels?	❑	❑
Brake lights?	❑	❑
Taillights?	❑	❑
License plate light?	❑	❑
Tight muffler system?	❑	❑
Properly serviced fire extinguisher?	❑	❑
Intact windshield, with no cracks?	❑	❑
Is all seating in the vehicle secured to the frame?	❑	❑
Is there an Automobile Liability ID Card located in the glove compartment or elsewhere in the vehicle?	❑	❑
Are appropriate notices posted in each vehicle as a reminder that all employees and their passengers are required to wear seat belts?	❑	❑
Have all employees been instructed on safe backing practices?	❑	❑
Have employees been informed of what actions to take in the event they are involved in a vehicle accident?	❑	❑
Have employees been informed of appropriate safety guidelines when hauling loads?	❑	❑

Employee Signature _____

Supervisor's Signature _____

Figure 6-61: Typical vehicle pre-use safety checklist. Source: www.tdi.texas.goc.

- *Sufficient time* allowed for travel - if not enough time is allowed for the journey, including an allowance for expected delays, the risk will be higher due to the increased likelihood of the driver speeding or taking unauthorised routes to make up time.

- *Weather* conditions can rapidly increase risks, conditions such as ice and snow and strong winds will have a significant effect.

WORK-RELATED DRIVING CONTROL MEASURES

Avoiding work-related driving activities should be the first consideration, for example:

- The journey may not be necessary; communication may be sufficiently effective by telephone or video conference instead of driving.

- It may be possible to send goods by rail or air freight.

- Some of the driver's journey may be able to be done by rail.

Consider whether your organisation's policy on the allocation of vehicle to workers actively encourages them to drive rather than use other means of transport.

Safe driver

It is the employer's duty to ensure that drivers are *competent*, fit, in good health and capable of doing their work in a way that is safe.

Employers must insist that all drivers produce evidence that they have a current *licence* to drive their class of vehicle. Without a current licence the vehicle insurance may be invalid. Regular *assessments* of driver competence and monitoring the validity of driving documentation, such as a driving licence, should be carried out. LGV drivers and PCV drivers may have to maintain a certificate of professional competence. In the UK, drivers are required to undertake 35 hours of training every 5 years.

Where relevant, the employer must ensure that appropriate *instruction* is given to drivers regarding rules and behaviour with regard to fitness to drive, driving, rests, parking, effects of weather and emergencies.

The driver must make sure that at all times they are driving they remain *fit to drive*. This will include presenting themselves fit at the beginning of a driving period and maintaining themselves fit during the period of driving. A driver's fitness to drive can be affected by such things as drugs (prescription or illegal), alcohol, fatigue or physical and mental health. If drivers are unfit to drive they must inform their employer before they attempt to start driving and if they become affected during driving they must take action to cease driving and inform their employer. If they are affected by tiredness, rather than fatigue, their usual breaks from driving may be adequate to provide recovery without the need to report this to their employer.

Safe vehicle

It is the employer's responsibility to ensure that the vehicle is appropriate for its use and remains *fit for purpose.* It is important that the vehicle is *maintained in a good condition*, safe and suitable for the task to be carried out, including any safety equipment that is required being properly fitted and maintained. This will require the vehicle being *regularly serviced* and *regularly tested* to confirm it meets legal requirements to allow it to continue to being used on public roads. Any safety-critical information should be displayed within the cab, for example, the height or width of the vehicle.

Loads carried by vehicles must be carried securely and account taken of the effects of braking, acceleration and manoeuvring. Where appropriate, adequate *load restraining equipment* should be used for securing goods.

Ergonomic factors related to the use of the vehicle should also be considered. For example, the driver's seat may require additional lumbar cushions. Whole-body vibration caused by the vehicle should be considered, for example, air-suspension seats may be required to reduce the amount of whole-body vibration received by the driver.

The employer should ensure appropriate *safety equipment* is available and used, for example:

- Seat belts and air bags should be installed, maintained and used correctly.

- Two-wheeled vehicle users should use appropriate safety helmets and protective clothing.

- Vehicles could be fitted with speed-limiting devices and electronic trackers to monitor and control the speed of vehicles.

- Many countries require first-aid equipment to be carried in the vehicle for use in the event of an accident/incident. This helps to minimise potential consequences of driving-related injuries.

- Additional equipment may also be required, such as high-visibility clothing, a warning triangle to place in the road for if the vehicle breaks down, warm clothing or blanket for cold weather, shovel and equipment to provide grip under tyres for snow, sand or muddy conditions, portable lighting and welfare facilities.

Safe Journey

Journey planning and scheduling are essential in ensuring the health and safety of workers who drive for work. Investing time in ensuring that journey planning is implemented as a component of the policy, will ensure that where possible, routes are planned thoroughly, *schedules are realistic*, and sufficient time is allocated to complete journeys safely and without fatigue. This will require allowing sufficient time to take *driving breaks* and consideration of *weather conditions*. It may be necessary to plan overnight stops if the journey time extends due to bad weather or traffic conditions. Consideration of the effects of *legal driving hours* will need to be made, where relevant.

Delivery schedules should be adjusted so that unrealistic targets are not set. This will reduce the fatigue and stress of drivers and will not encourage them to drive too fast for the conditions or exceed speed limits.

> **CONSIDER**
>
> Do your management systems ensure that work-related driving is managed effectively?
>
> For example, are you confident that your vehicles are regularly inspected and serviced in accordance with manufacturers recommendations?

Sources of reference

Reference information provided, in particular web links, was correct at time of publication, but may have changed.

A guide to workplace transport safety, HSG136, HSE Books, http://www.hse.gov.uk/pubns/priced/hsg136.pdf

Driving at work, Managing work-related road safety, INDG382, HSE Books, https://www.btpolfed.org.uk/ref/hands4.pdf

Health and Safety in Construction HSG150, 3rd Edition, HSE Books ISBN 978-0-7176-6182-2 www.hse.gov.uk/pubns/priced/hsg150.pdf

Lighting at Work, HSG38, second edition 1997, HSE Books, ISBN 978-0-7176-1232-1 www.hse.gov.uk/pubns/priced/hsg38.pdf

Managing Health and Safety in Construction, Construction (Design and Management) Regulations 2015, Guidance on regulations, L153, HSE Books, ISBN: 978-0-7176-6623-3, http://www.hse.gov.uk/pubns/priced/L153.pdf

Safe Use of Vehicles on Construction Sites, HSG144, HSE Books, ISBN 978-0-7176-6291-3 http://www.hse.gov.uk/pubns/priced/hsg144.pdf

Safety at Street Works and Road Works: A Code of Practice 2013, HMSO ISBN 978-0-115531-45-3 http://www.gov.uk/government/publications/safety-at-street-works-and-road-works

Safe Use of Work Equipment, ACOP and guidance, L22, third edition 2008, HSE Books ISBN 978-0-7176-6295-1 www.hse.gov.uk/pubns/priced/l22.pdf

The Traffic Signs Manual (Chapter 8: The Traffic Safety Measures and Signs for Road Works and Temporary Situations) Department for Transport, http://www.gov.uk/government/publications/traffic-signs-manual:

Part 1: Design. The current edition is dated 2009 - ISBN 978-0-115530-51-7 http://www.gov.uk/government/uploads/system/uploads/attachment_data/file/203669/traffic-signs-manual-chapter-08-part-01.pdf

Part 2: Operations. The current edition is dated 2009 - ISBN 978-0-115530-52-4 http://www.gov.uk/government/uploads/system/uploads/attachment_data/file/203670/traffic-signs-manual-chapter-08-part-02.pdf

Vehicles at work, HSE, http://www.hse.gov.uk/workplacetransport/index.htm

Workplace Transport Safety - Guidance for Employers, HSG136, HSE Books, ISBN 978-0-7176-6154-1 www.hse.gov.uk/pubns/priced/hsg136.pdf

Web links to these references are provided on the RMS Publishing website for ease of use – www.rmspublishing.co.uk

Statutory Provisions

Construction (Design and Management) Regulations (CDMR) 2015 / Construction (Design and Management) Regulations (Northern Ireland) 2016

Health and Safety (Safety Signs and Signals) Regulations (SSSR) 1996 / / Health and Safety (Safety Signs and Signals) Regulations (Northern Ireland) 1996

Provision and Use of Work Equipment Regulations (PUWER) 1998 / Provision and Use of Work Equipment Regulations (Northern Ireland) 1999

Road Traffic Act (RTA) 1988 and 1991

Work at Height Regulations (WAHR) 2005, as amended 2007 / Work at Height Regulations (Northern Ireland) 2005

STUDY QUESTIONS

1) What possible control measures for plant, machinery and vehicles could be used on construction sites?

2) (a) Give FOUR reasons why a dumper truck used on a construction site may overturn?

   (b) What practical measures could be taken to minimise the risk of a dumper truck overturning when used on construction sites?

3) Reversing vehicles within a construction site can be hazardous. How can the risk of accidents from reversing vehicles be reduced?

4) Many construction workers are involved in work-related driving on public roads as part of their work activities. What factors associated with this driving increase the risk of an incident occurring?

5) What practical measures could be used to protect pedestrians on a construction site where segregation of people and vehicles is not possible?

For guidance on how to answer these questions please refer to the 'study question answer guidance' section located at the back of this guide.

Element 7

Working at height

Contents

THE RISKS OF WORKING AT HEIGHT

Vertical distance

Where work is carried out above 2 metres (vertical distance) the risks of major injury or death are considered to be very significant, for example, work on a roof or scaffold. However, major injuries can still occur if a fall results when carrying out tasks at a height of less than 2 metres, for example, while fitting false ceilings or installing utilities inside buildings.

The Work at Height Regulations (WAHR) 2005 state:

> "Work at height" means -
>
> (a) work in any place, including a place at or below ground level;
>
> (b) obtaining access to or egress from such place while at work, except by a staircase in a permanent workplace, where, if measures required by these Regulations were not taken, a person could fall a distance liable to cause personal injury;"

Figure 7-1: Interpretation of "work at height".
Source: Work at Height Regulations 2005.

The Work at Height Regulations 2005 reflect this risk and require controls to be in place to manage the risk of falling, whatever the height. This legislation applies to workplaces in general, and therefore includes construction activities. Under these regulations the interpretation of 'work at height' includes any place of work at ground level, above or below ground level that a person could fall a distance liable to cause personal injury and includes places for obtaining access or egress, except by staircase in a permanent workplace.

WAHR 2005 Regulation 6 states that work at height must only be carried out when it is not reasonably practicable to carry out the work otherwise. If work at height does take place, suitable and sufficient measures must be taken to prevent a fall of any distance, to minimise the distance and the consequences of any fall liable to cause injury. Employers must also make a risk assessment, as required by Regulation 3 of the Management of Health and Safety at Work Regulations (MHSWR) 1999.

Roofs

Working at height and on roofs carries a high risk of accidents, unless proper procedures and precautions are taken. The danger of people or materials falling affects the safety of those working at height and those working beneath. Particular danger arises from two types of roof - fragile roofs and sloping roofs.

Fragile roofs

Any roof that has not been specifically designed to carry a load, other than that which may be imposed on the roof by weather, should be considered a fragile roof. Sheets used in roofing material may deteriorate over time leaving the remaining material in a particularly fragile state.

Materials such as asbestos cement, glass, corrugated metal, slates, tiles or plastic are likely to be unable to bear the weight of a person. A sheet of plastic or metal roofing material may look intact, but it might only have a small fraction of its original strength. In a similar way, plastic roof material used in roof lights will be affected by exposure to sunlight, leaving it brittle and weak.

It should not be assumed that it is safe to walk on newly installed roof material, even though it will have its original strength this may not be enough to bear the weight of a worker. If a worker puts their weight on a fragile roof material it is likely to fail suddenly and cause the worker to fall through the roof. All fragile roofs and/or access routes to them should be marked with an appropriate warning sign.

The WAHR 2005 Regulation 9 states that every employer shall ensure that suitable and sufficient steps are taken to prevent any person at work falling through any fragile surface; and that no person at work may pass across or near, or work on, from or near, fragile surfaces when it is reasonably practicable to carry out work without doing so. If work has to be from a fragile roof then suitable and sufficient means of support must be provided that can sustain foreseeable loads. No person at work should be allowed to pass or work near a fragile surface unless suitable and sufficient guardrails and other means of fall protection is in place. Signs must be situated at a prominent place at or near to works involving fragile surfaces, or persons are made aware of the fragile roof by other means.

Sloping roofs

Sloping roofs are those with a pitch greater than 10 degrees. The slope of the roof may cause workers working on top of the roof to slide off the edge of the roof. Falls from the edge of sloping roofs can cause serious injury even when the eaves are relatively low. The hazard of sloping roofs is less obvious when the pitch is small, causing people to underestimate the possibility of workers sliding off the edge.

The material and therefore the surface of the roof have a significant influence on the hazard, for example, a smooth sheet metal surface can present a significant hazard even when the pitch is small.

The chances of a worker falling are increased when working on roofs that are wet or covered in moss growth and in extreme weather conditions such as high winds.

A build-up of dry particles or grit on a roof can also present a surface that leads to a high risk of slipping as the particles become free to move and form a mobile layer between the roof and the worker's foot. The worker's footwear can also significantly influence the risk, smooth flat soled footwear may seem suitable in dry conditions, but may not provide sufficient grip to deal with surface water in wet conditions.

Figure 7-2: Falls from a height - worker on a roof.
Source: Pixabay

Deterioration of materials

The materials of a structure can deteriorate over time, which can reduce the strength or properties of the material. Materials that have deteriorated particularly present a hazard when people walk on them causing them to break and the workers to fall or when the material cannot hold its own weight and pieces of the material fall on workers below.

The rate of deterioration of materials will accelerate if the structure is exposed to adverse weather conditions (including extremes of temperature) or attack by chemicals, animals, insects etc. Plastic materials may become brittle when exposed to sunlight and metal materials can become reduced in thickness and strength due to rusting. It may not always be obvious that deterioration has occurred and this should be a factor considered at the pre-work assessment.

Unprotected edges

Roofs, scaffolds, unfinished steel work and access platforms may sometimes have open sides. This increases the likelihood of someone or something falling, particularly if people have to approach them, work at them or pass by them repeatedly. It is very easy in these circumstances to lose perception of the hazard and forget that it is there. Errors, such as stepping back over an edge, overreaching, or being pushed over the edge whilst manoeuvring materials can easily lead to fatal falls.

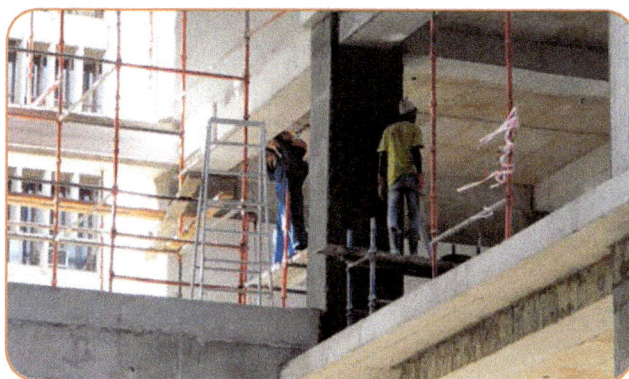

Figure 7-3: Working at unprotected edge.
Source: Richard Neale, ILO Theme summary 14 - Working at height.

Unstable/poorly maintained access equipment

If access equipment like mobile elevating work platforms (MEWPs), ladders and scaffolds is not stable it could move suddenly or fall over causing those using it to fall. The instability is often caused by unstable or uneven ground conditions where the equipment is positioned. The higher the equipment operates the risk of it becoming unstable usually increases, as the ratio of the height compared with the width of the base increases. High winds can also increase the instability of the equipment.

Poorly maintained access equipment can lead to sudden failure of the equipment and workers falling from height. The effective working of the hydraulics of a MEWP is critical and failure to maintain this could lead to a sudden and catastrophic failure of the MEWP whilst it is extended. *See also - **Mobile elevating work platforms - later in this Element.***

Scaffolds rely on the strength and security of its load-bearing parts, failure of these components can lead to the collapse or overturning of the scaffold. Scaffolds need periodic inspection and maintenance to ensure load-bearing parts are still secured and critical items, such as brakes on mobile scaffolds, are in place and effective.

Cracks may occur in the sides of wooden ladders and loose rungs can lead to failure. They may warp or rot if left outside unprotected from the sun and adverse weather. Defects in ladders may be hidden if they are painted or covered in construction materials.

Figure 7-4: Working above ground level.
Source: RMS.

Weather

Adverse weather can have a significant effect on the safety of those working at height. Rain, snow and ice increase the risk of slips and falling from a height. When handling large objects, such as roof panels, high winds can be a serious problem and may cause a worker to be blown off the roof. Extremely cold temperatures can increase the likelihood of brittle failure of materials and therefore increase the likelihood of failure of roof supports, scaffold components and plastic roof lights.

In addition, moisture can freeze; increasing the slipperiness of surfaces and on many occasions the presence of ice is not easily visible. Workers exposed to the cold can lose their dexterity and when hot, sweat may cause them to lose their grip.

Falling materials

The risk of falling materials causing injury is increased where workplaces at height are not kept clear of loose materials and no methods are provided to prevent materials rolling or being kicked off the edges. Workers and people nearby may also be at risk of being injured when materials, such as old roofing material, are thrown from a workplace at height.

Figure 7-5: Risk from falling materials.
Source: RMS.

The risk of materials falling is increased by:

- Poor housekeeping of people working at height.
- Absence of toe boards, edge protection or nets.
- Incorrect hooking and slinging when using a crane.
- Incorrect assembly of gin wheels for raising materials.
- Surplus materials incorrectly stacked.
- Loose materials.
- Absence of safe means to get debris and other materials to the ground.
- Open, unprotected edges.
- Deterioration of materials, causing pieces of the material to fall.
- Gaps in platform surfaces.

APPROACH TO WORKING SAFELY AT HEIGHT

In summary, the approach to working safely at height is:

- Avoid working at height.
- Prevent a fall from occurring by using an existing workplace that is known to be safe or use suitable equipment.
- Minimise the distance and/or consequence of a fall.

Avoiding working at height

Where possible, work at height should be avoided. This could be achieved by using different equipment or methods of work to conduct work at ground level, conducting work from ground level or using other methods. For example, instead of manually filling equipment hoppers at height, bulk delivery or automatic feed systems could be used. Where workers have to lubricate or adjust equipment set at height by hand, options to automate or route the lubrication/adjustment mechanisms to ground level should be considered. Instead of assembling materials or equipment at a height pre-assembly before delivery or on the ground on site should be considered.

Figure 7-6: Using a polo to avoid work at height.
Source: HSE, HSG150.

> "(2) Every employer shall ensure that work is not carried out at height where it is reasonably practicable to carry out the work safely otherwise than at height."

Figure 7-7: Duty to avoid work at height.
Source: WAHR 2005, Regulation 6.

If holes required in materials are drilled at ground level, for example, holes to enable assembly of roof frames, this can greatly reduce the work needed to be done at height. In a similar way, materials can be treated or painted on the ground before assembly or fitting. Long reach or extendable tools can be used to allow cleaning or other tasks to be conducted from the ground. Where equipment, such as light units, requires maintenance an option may be to lower it sufficiently to enable bulbs to be changed and cleaning to be conducted from the ground.

Preventing a fall by using existing workplaces or suitable equipment

Use an existing safe workplace

Do not work at height unless it is absolutely unavoidable. If work must be performed at height, use an existing safe workplace that is:

- Stable and of sufficient strength and rigidity for the intended use - for example a solid roof or platform that is an integral part of a fixed structure.

- Sufficient dimensions to provide a safe working area, having regard to the work to be done – sufficient to permit the safe movement of workers and the safe use of any equipment or materials.

- Suitable and sufficient means for preventing a fall - for example, fixed guardrails.

- A surface which has no gaps that a person, materials or objects could fall through.

- Constructed, used and maintained in a safe condition - to prevent slipping or tripping, and, where it has moving parts, has appropriate devices to prevent its inadvertent movement during work at height or a worker being caught between it and any adjacent structure.

It is best to use an existing workplace that is known to be safe where possible. Erecting a temporary workplace at height is a less desirable option as it has its own added risks. By using more permanent workplaces, design features that provide collective protection of those at risk can be built in to the structure more easily, making the need to use of personal protective measures less likely.

"(3) Where work is carried out at height, every employer shall take suitable and sufficient measures to prevent, so far as is reasonably practicable, any person falling a distance liable to cause personal injury.

4) The measures required by paragraph (3) shall include:

a) His ensuring that the work is carried out:

i) From an existing place of work.

ii) (In the case of obtaining access or egress) using an existing means; which complies with Schedule 1, where

it is reasonably practicable to carry it out safely and under appropriate ergonomic conditions.

b) Where it is not reasonably practicable for the work to be carried out in accordance with sub-paragraph (a), his providing sufficient work equipment for preventing, so far as is reasonably practicable, a fall occurring."

Figure 7-8: Requirement to use existing place of work or provide work equipment. Source: WAHR 2005, Regulation 6.

Provide work equipment to prevent falls

Where no existing, safe workplace is available to enable work at height it will be necessary to provide work equipment that enables work at height and prevents workers falling. This could be in the form of scaffolding, barriers that provide edge protection (temporary or fixed), provision of a mobile elevating work platform (MEWP) or work restraint systems.

Figure 7-9: Use of a work restraint system to prevent worker going near the unprotected edge.
Source: HSE, HSG150.

When planning the work at height it is important to select the most appropriate equipment for the work to be carried out, considering the activity and the work environment. Collective measures must be given priority over personal measures, for example, barriers or scaffolding would provide collective measures and would be preferable to using work restraint systems. Similarly, to maintain safe access and egress, temporary staircases should be considered before the use of ladders.

When using a work restraint system in a mobile elevating work platform (MEWP), it should always be anchored to the inside of the cradle and not the outside (*see Figure 7-65)*.

Minimise the distance and/or consequence of a fall

> "(5) Where the measures taken under paragraph (4) do not eliminate the risk of a fall occurring, every employer shall:
>
> a) So far as is reasonably practicable, provide sufficient work equipment to minimise:
>
>   i) The distance and consequences.
>
>   ii) Where it is not reasonably practicable to minimise the distance, the consequences, of a fall."

Figure 7-10: Duty to minimise the distance/consequences of a fall.
Source: WAHR 2005, Regulation 6.

Collective measures

Collective fall-arrest measures are preferred to personal protective measures because they provide protection for all workers at risk of falling rather than protection for individuals. Collective measures include equipment designed to minimise the distance and consequence of falls by providing an energy absorbing 'soft landing', such as nets and air bags. Collective measures can simplify systems of work and can protect not only workers, but others such as supervisors.

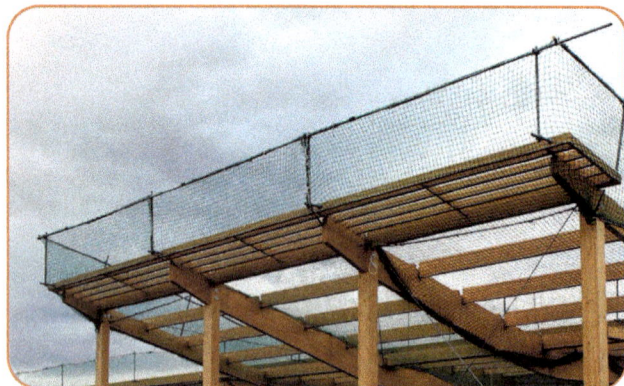

Figure 7-11: Soft landing system - safety net.
Source: ProNet Safety Services.

Safety nets are installed underneath the structure being worked on so that there is an amount of free movement of the net to enable the energy of the person falling to be absorbed. It is essential that there are no obstructions underneath the net that might injure a person falling on the net, such as materials and equipment. Materials that have fallen onto the net should be removed promptly to ensure they do not injure someone who might fall onto them. It is important that measures are in place to rescue anyone who falls onto the net.

Other energy absorbing systems use a number of bags that are filled with an energy absorbing material and are connected together to fill the area where protection is required. They may be filled with air or small pieces of polystyrene or another energy absorbing material. The air filled bags are kept inflated by an air compressor which replaces any air that leaks from the bag. The bags must have sufficient depth and energy absorbent material to absorb the energy of the person falling. Other energy absorbing systems include thick, solid energy absorbing materials formed into mats.

Figure 7-12: Soft landing system - air bags.
Source: thxuk.com.

Where energy absorbing systems, such as safety nets, are used they must be installed as close as possible beneath the surface that people can fall from, securely attached and able to withstand a person falling onto them. They must be installed and maintained by competent personnel. Although they do not prevent people falling, if they are installed correctly they will minimise the distance of the fall and minimise the consequences of the fall by absorbing the energy of the person falling. If the work height increases as work progresses, the energy absorbing system must be repositioned at each higher level to minimise the distance that a worker can fall.

Energy absorbing systems should:

- Provide sufficient clearance from solid surfaces when arresting a fall.
- Have anchors and other fixings that have sufficient strength and stability for the purpose of supporting the foreseeable loading in the event of a fall and during any subsequent rescue.
- Be securely attached to all the required anchor points.
- In the case of energy absorbing bags or mats, be stable.

Personal protective measures

If fall prevention measures (for example, working platforms, barriers or guard rails) or collective fall-arrest measures are not practical, alternative personal protective measures must be used. This could include work positioning systems, rope access systems or personal fall-arrest systems. Personal fall-arrest systems are widely used, and comprise a line, means of anchoring the line, means of absorbing the energy of the fall and a harness worn by the user. It should be remembered that they only provide protection for the user and the harness itself may cause injury when the worker's fall comes to a sudden stop.

Figure 7-13: Personal fall-arrest system.
Source: High Access Solutions.

A fall-arrest system must incorporate means of absorbing the energy of a fall and limiting the forces applied to the user's body. Care should be taken to ensure that when used the fall arrest system it is not at risk of the line being cut and that it is free to operate as intended. Consideration should be made to ensuring at all times it is in use that there is a 'clear zone' around where the worker might fall and that this should take account of any pendulum effect. Arrangements must be made for the rescue of the worker who has fallen.

A personal fall-arrest system should only be used if it is:

- Suitable and of sufficient strength for the purposes for which it is being used - having regard to the work being carried out and any foreseeable loading.

- Correctly fitted.

- Designed to minimise injury to the user and adjusted to prevent the user slipping out of it if they fall - has a means of absorbing the energy of the fall.

- Securely attached to all the required anchor points - anchors must be capable of supporting the foreseeable loading when arresting the fall and during any subsequent rescue.

- Secured in a position that enables the worker to fall safely and for their fall to be arrested.

MAIN PRECAUTIONS NECESSARY TO PREVENT FALLS AND FALLING MATERIAL

Proper planning

Planning requirements
Planning must include the selection of suitable equipment, take account of emergencies and give consideration to weather conditions impacting on safety.

Regulation 4 of the Work at Height Regulations 2005 states:

"Every employer shall ensure that work at height is –

(a) properly planned;

(b) appropriately supervised; and

(c) carried out in a manner which is so far as is reasonably practicable safe, and that its planning includes the selection of work equipment."

Figure 7-14: Organisation and planning.
Source: Work at Height Regulations 2005.

Avoid, prevent, minimise
Work at height must only be carried out when it is not reasonably practicable to carry out the work at ground level or from ground level. If work at height does take place, suitable and sufficient measures must be taken to prevent a fall of any distance and, if this is not possible, to minimise the distance and the consequences of any fall liable to cause injury.

Selection of equipment
Careful consideration during the risk assessment phase of work planning should establish which equipment is best suited for the working environment and the work to be done. Consideration should be given to the working conditions, risks to people and the frequency and duration of use. Give collective measures priority over personal protection measures.

Fall of materials and objects
WAHR 2005 Regulation 10 states that every employer shall take reasonably practicable steps to prevent injury to any person from the fall of any material or object. Where possible this must be by the prevention of materials or objects falling. Where this is not reasonably practicable, measures must be taken to prevent any person being struck by falling material or object, for example, by provision of nets or other structures to catch materials.

In addition, employers should ensure that nothing is thrown or tipped from height in circumstances where it is liable to cause injury to any person. WAHR 2005 Regulation 11 states that where an area presents a risk of a person being struck by an item falling from height it must be equipped with measures that prevent unauthorised persons from entering the area and the area must be clearly indicated. This could involve fencing around the area and provision of warning signs.

Emergencies and rescue
Planning should consider what will need to be done when a worker falls from height and the nature of this emergency. For example, they may have fallen into a net or be attached to a personal fall-arrest system, which could involve injuries and the person being suspended for a period of time before they can be rescued.

Carried out safely

Supervision

It is essential that work at height is carried out safely, which will mean ensuring prevention and protection measures are put into place and used correctly. Although those workers that use the measures have a responsibility to ensure they use them correctly, adequate supervision must be provided to ensure they do.

Avoiding working in adverse weather conditions

Adverse weather can include wind, sand/dust storms, rain, sun, cold, snow and ice. Each of these conditions, particularly in extreme cases, can present a significant hazard to work at height. When long-term projects are planned methods are often adjusted to minimise the effects of these conditions. Work areas can be covered over at an early stage to enable work to be conducted in relative safety. In some cases, adverse weather must be considered formally and work may have to cease until conditions improve, for example, work on a roof in icy conditions or high winds. Similar approaches may have to be taken for operating a mobile elevating work platform (MEWP) in windy conditions. Wind speeds that may cause instability are given by the Prefabricated Access Suppliers and Manufactures Association (PASMA) as winds averaging a speed of 17mph (27Kph) or Beaufort scale force 4. In these type of conditions work from tower scaffolds should cease.

Inspection

Inspection requirements for work equipment specified for use for work at height are set out in Regulation 12 of the WAHR 2005. It is good practice, and a legal requirement, to inspect access equipment routinely and to ensure that the person carrying out the inspection is competent to do so.

Inspection requirements for equipment used for working at height are:

- Where safety depends on how it is installed or assembled - before it is used in that position.

- Temporary work platforms, for example, a scaffold, or mobile work platform used for work in which a person could fall 2 metres or more - regularly whilst in position, typically at intervals not exceeding 7 days.

- Other work equipment for work at height - at suitable intervals.

- Each time it has been affected by exceptional circumstances that are liable to affect the safety of the work equipment, for example, extremes of weather (for example, high winds or flood) or seismic conditions.

Results of an inspection should be recorded and kept until the next inspection. An inspection report containing the particulars set out in the following list should be prepared before the end of the working period within which the inspection is completed and, within 24 hours of completing the inspection, and be provided to the person it was carried out for.

The report should be kept at the site where the inspection was carried out until the work is completed and afterwards at an office of the person on whose behalf it was carried out, for typically 3 months. Reports on inspections should include the following particulars:

1) Name and address of person for whom the inspection is carried out.

2) Location of the access equipment.

3) Description of the access equipment.

4) Date and time of inspection.

5) Details of any matter identified that could give rise to a risk to the health and safety of any person.

6) Details of any action taken as a result of item 5 above.

7) Details of any further action considered necessary.

8) Name and position of the person making the report.

Competent

Those engaged in any activity in relation to work at height must be competent; and, if under training, be supervised by a competent person.

EMERGENCY RESCUE

When selecting equipment for work at height, Regulation 7 of the WAHR 2005 requires the employer to consider the additional risks that may arise from emergencies and the need for evacuation of or rescue from the equipment. This will include the need to establish clear emergency escape routes from major scaffold installations, rescue arrangements for workers where a mobile elevating work platform (MEWP) fails in its raised position and rescue for those on a fall arrest net or in a harnesses. Steps must be taken to minimise the risk of injury due to the fall or contact with the fall-arrest system. Even a short fall onto a net or other fall arresting system could cause minor injuries or fractures. This should be anticipated and workers taught how to minimise the likelihood of injury.

Where a worker has fallen from height, but has been protected by personal fall-arrest equipment, such as a harness, significant health effects, known as suspension fainting, may be experienced if they are not rescued quickly. This is mainly due to blood pooling in the legs, reducing the amount circulating through the rest of the body, which has consequential effects for vital organs, including the brain, heart and kidneys. Unless the individual is rescued quickly the lack of oxygenated blood to vital organs can be fatal. Therefore, a rescue procedure and equipment must be available and rescue practised. The rescue procedure should take into account that the worker's sudden transition from a vertical to a horizontal position, when rescued and laid down, can lead to a massive amount of deoxygenated blood entering the heart, causing cardiac arrest.

INSTRUCTION, TRAINING AND OTHER MEASURES

Instruction and training

Employers should ensure that no person engages in any activity, including organisation, planning and supervision, in relation to work at height unless they are competent to do so. If they are being trained, they must be supervised by a competent person. Workers should receive full training and instruction on the use of work at height equipment to prevent falls and minimise the distance and consequences of a fall. Where personal fall-arrest systems are to be used, this should include how to wear the equipment, how to fit it to anchor points, what is a suitable anchor point and how to attach it at a height that minimises the fall (for example, above the worker's head where possible).

Training should also include the checks that need to be made on collective fall-arrest systems before they are used, this will include checks on the security of nets and the adequacy of airbags. Workers will also need to be trained in how to get off/out of this equipment safely when it has arrested a fall and on emergency arrangements for rescue, where workers cannot assist themselves.

> "Every employer shall ensure that no person engages in any activity, including organisation, planning and supervision, in relation to work at height or work equipment for use in such work unless he is competent to do so or, if being trained, is being supervised by a competent person."

Figure 7-15: Requirement for competence.
Source: WAHR 2005, Regulation 5.

Where a risk of falling remains, after fall protection systems have been provided, additional training and suitable and sufficient other measures should also be taken to prevent injury to any person falling a distance liable to cause injury. This may include training them in techniques that would limit the effect of the fall, controlling what equipment they use that might fall with them, ceasing processes in the area that may cause the worker to be disorientated/overcome leading to a fall, covering items that the person may fall onto and providing means to assist those that have fallen.

Requirements for head protection

Head protection, usually in the form of a hard hat, is required where there is a foreseeable risk of injury to a worker's head from being struck by falling materials.

Actions to be taken include:

- Decide on which areas of the site where hats have to be worn.

- Make site rules and tell everyone in the area.

- Provide workers with hard hats.

- Make sure hard hats are worn and worn correctly.

A wide range of hard hats are available. Let workers try a few and decide which is most suitable for the job and for them. Some hard hats have extra features including a sweatband for the forehead and a soft, or webbing harness. Although these hard hats are slightly more expensive, they are much more comfortable and therefore more likely to be worn.

> **CONSIDER**
>
> How do you decide if someone is 'competent' to work at height?

7.2 Safe working practices for access equipment and roof work

SCAFFOLDING

Design features of scaffolding

Independent tied

This type of scaffold typically uses two sets of standards; one near to the structure and the other set at the width of the work platform. It is erected so that it is independent from the structure and does not rely on it for its primary stability. However, as the name suggests, it is usual to tie the scaffold to the structure in order to prevent the scaffold falling towards or away from the structure.

Putlog

This type of scaffold has a single set of standards erected at the width required for the work platform. The transoms have a flat end that is inserted into the mortar gap in the wall. The structure effectively provides the inner support for the work platform and therefore an inner set of standards is not needed.

Figure 7-16: Independent tied scaffold.
Source: RMS.

Figure 7-17: Independent tied scaffold.
Source: RMS.

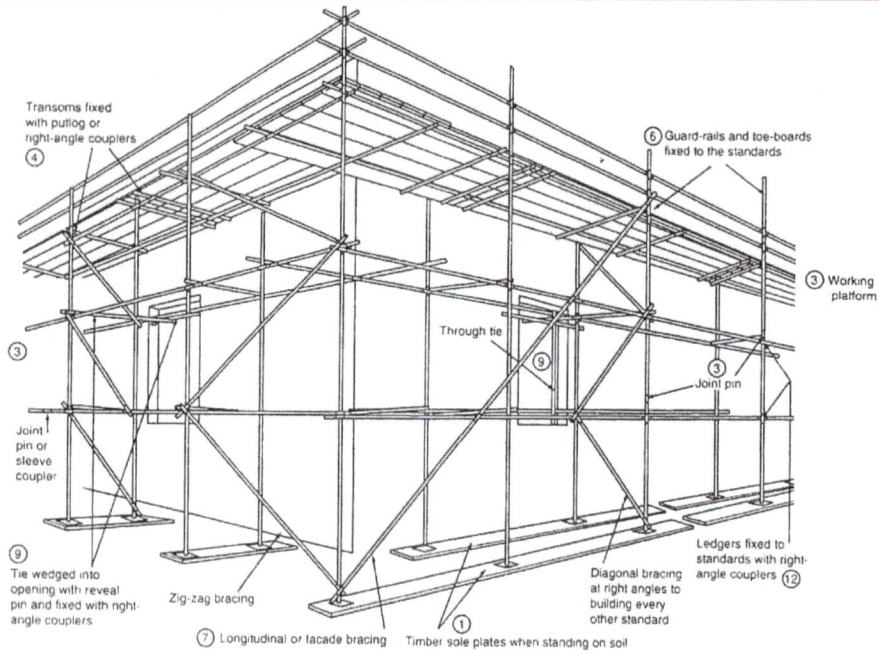

Figure 7-18: Independent tied scaffold.
Source: HSE, HSG150.

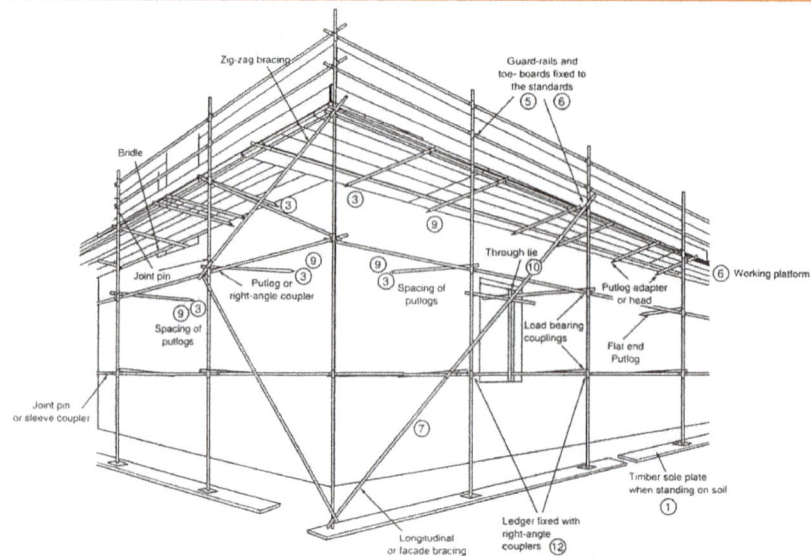

Figure 7-19: Putlog scaffold.
Source: HSE. HSG150.

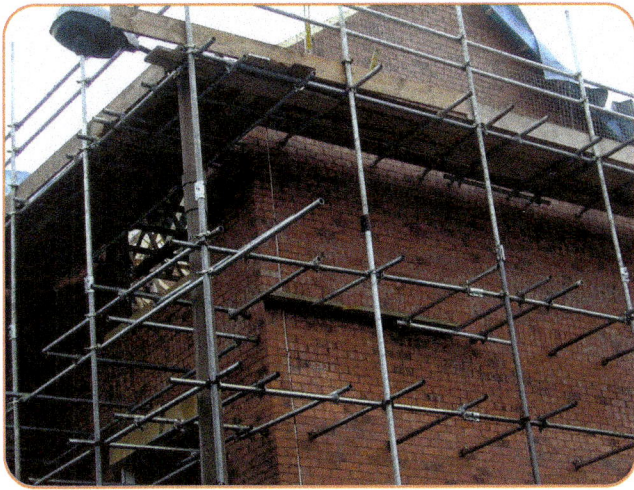

Figure 7-20: Putlog scaffold.
Source: RMS.

Fan

Fans are scaffold boards fixed on scaffold tubes set at an upward angle out from a scaffold in order to catch debris that may fall from the scaffold.

They may be used at the entrances to buildings, to protect people entering and leaving the building that the scaffold is erected against or on the face of the building if it runs alongside a public footpath.

They are also used where a scaffold is erected alongside a pedestrian walkway where there is a need to have an increased confidence that materials which might fall cannot drop onto people below.

In some cases it may be necessary for barriers to be erected to direct pedestrians under the fan.

Figure 7-21: Putlog scaffold.
Source: RMS.

Cantilevered

A cantilever scaffold is usually used in situations where obstructions or limitations of space prevent a conventional independent scaffold with a base on the ground being used. This may be due to features at a lower level, like a balcony, structures that cannot support the weight of the scaffold or work over deep water. Where the cantilever is to be added to an existing independent scaffold it is assembled in the normal way then a section of

scaffold is created that is erected horizontally outwards away from the standards to create a platform. Cantilever scaffolds need to be carefully designed to ensure the main part of the scaffold has sufficient strength and rigidity to counteract the additional forces added by the cantilevered section.

Figure 7-22: Fans.
Source: RMS.

Figure 7-23: Cantilevered scaffold.
Source: Ducker & Young.

Figure 7-24: Cantilevered scaffold.
Source: Rowland Scaffold.

Care must be taken to ensure the structure remains safe and stable. Adding bracing and ties to the main scaffold are normally essential safety precautions. In some situations bracing is provided that angles back from the outer edge of the cantilevered platform to the main scaffold or building.

Mobile tower scaffolds

Mobile scaffold towers are widely used as they are convenient for work which involves frequent access to height over a short period of time in a number of locations that are spaced apart.

However, they are often incorrectly erected or misused and incidents/accidents occur due to people/materials falling or the tower overturning/collapsing.

They must be erected and dismantled by trained, competent workers, strictly in accordance with the supplier's instructions. All parts must be in good condition and from the same manufacturer.

Safe working practices for the use of mobile tower scaffolds include:

- Positioned on firm and level ground.

- The height of an untied, independent tower scaffold must never exceed the manufacturer's recommendations. BS EN 1004 includes a calculation to establish accurate height to base width dimensions. A general guide to acceptable limits are:

  - Outdoor use - 3 times the minimum base width.

  - Indoor use - 3.5 times the minimum base width.

- If the height of the tower scaffold is to exceed these maximum figures then the scaffold must be secured (tied) to the structure or outriggers (rakers) used.

- Working platforms must only be accessed by safe means. Only use internal stairs or fixed ladders and never climb on the outside.

- Before accessing a mobile tower scaffold, the wheels must be turned outwards and the wheel brakes locked 'on'.

- Never move a mobile tower scaffold unless the working platform is clear of people, materials, tools etc.

Figure 7-25: Wheels with brakes.
Source: RMS.

- Mobile tower scaffolds must only be moved by pushing them at base level. They must not be pulled along by workers whilst people are standing on the working platform. Care must be taken to avoid obstructions at base level and overhead.

- Never use a mobile tower scaffold near live overhead power lines or cables.

- Working platforms must always be fully boarded out. Guardrails and toe boards must be fitted if there is a risk of a fall from a height that presents a significant risk, for example, more than two metres. Inspections must be carried out by a competent person - before first use, after substantial alteration and after any event likely to have affected its stability.

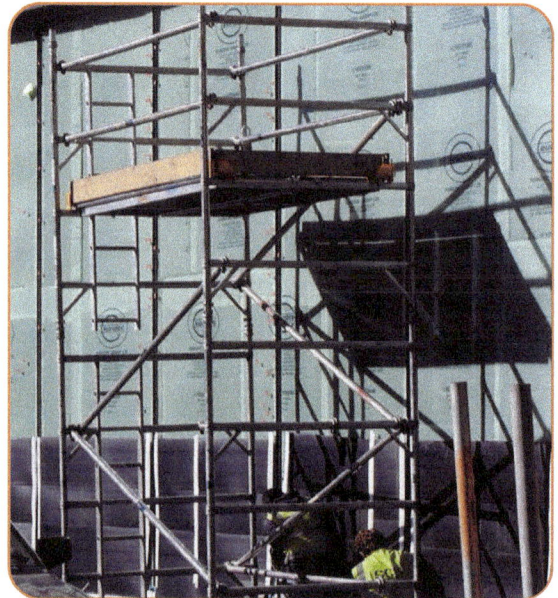

Figure 7-26: Mobile tower scaffold.
Source: RMS.

Figure 7-27: Mobile tower scaffold.
Source: RMS.

Scaffolding terms

Base plate	A small square metal plate that distributes the load from a standard or a raker (scaffold standard used as an outrigger).
Brace	A tube fixed diagonally across two or more members in a scaffold for stability.
Fan	A structure of scaffold boards fixed to standards that is fixed to the outside of a scaffold at an angle inclined upwards and located below a work platform to catch debris that may fall.
Guardrail	A tube fixed horizontally to standards (uprights) to prevent personnel from falling.
Ledger	A tube spanning horizontally and tying the scaffold longitudinally. It may act as a support for put logs or transoms.
Putlog	A tube with a flattened end, spanning from a horizontal member to a bearing in or on a brick wall. It may support scaffold boards.
Raker	Scaffold standard used as an outrigger to prevent the scaffold falling away from a structure.
Reveal pin	A screw jack, fitting in the end of a tube.
Reveal tie	A member joined to a scaffold and a reveal pin that is wedged between two opposite surfaces (for example, window reveals) to make a friction anchorage for tying a scaffold.
Sole board	Large pieces of timber put underneath the base plate to further spread the load over a wide surface area, especially where the ground is soft.
Standard/ column	A vertical or near vertical supporting member.
Tie	A member used for fixing the scaffold to the building or other structure for stability.
Transom	A tube spanning across ledgers to tie a scaffold transversely (normally at 90° to the building face). It may also support a working platform.

Safety features

Base plates and sole boards

Base plates and sole boards are used to spread the weight of the scaffold and to provide a firm surface on which to erect a scaffold. Sole boards are used as well as base plates where the ground is soft. The base plate spreads the load of the standard and helps to keep the standard vertical.

Figure 7-28: Base plates and sole boards.
Source: Limcsafe.

Figure 7-29: Base plate and protection.
Source: RMS.

Safe working practices for the use of base plates and sole boards include:

- A base plate must be used under every standard - it spreads the load and helps to keep the standard vertical.
- Sole boards are used to spread the weight of the scaffold and to provide a firm surface on which to erect a scaffold, particularly on soft ground. Sole boards must be sound and sufficient strength to spread the load, and should run under at least two standards at a time.

Standards

Safe working practices for the use of standards include:

- All standards must be truly vertical (plumb), or leaning only a little towards the structure. Standards out of vertical can 'bow' when put under load and transfer additional loading to other standards, which could lead to the scaffold collapsing.
- Any joints in standards must be staggered.

Ledgers

Safe working practices for the use of ledgers include:

- Ledgers must be truly horizontal, and not more than 2.7 metres above the ground or the ledgers below, on any scaffold.
- Any joints in ledgers must be made with sleeve couplers, and the joints must be staggered.
- Ledger bracings must be fixed to alternate standards on every platform. They must not be fixed to the handrail.

Boards

Boards provide the working platform of a scaffold, walkways and landings for access ladders.

Figure 7-30: Scaffold boards and toe boards - some defective.
Source: RMS.

Safe working practices for the use of boards include:

- Boards supported by transoms or putlogs must be close fitting.

- They must be free from cracks or splits or large knots and must not be damaged in any way that could cause weakness.

Working platforms

Safe working practices for the use of working platforms include:

- Wide enough to allow workers to walk on it safely and to use any equipment or material necessary for their work - at least 600mm wide.

- Working platforms at height must be fitted with guardrails and toe boards. The height at which this requirement applies may be defined under national laws, regulations or codes of practice. A common safe working practice is to fit guardrails and toe boards when the working platform height is two metres or more above the ground or floor level.

- If the platform also carries materials, then the space between guardrails and toe boards must be reduced to a maximum 470mm (UK standard). This can be done with an intermediate rail, mesh or similar material.

- As well as toe boards, the platform itself should be constructed to prevent any object that may be used on the platform from falling through gaps or holes and causing injury to people working below. For scaffolds used on site a close-boarded platform would be enough, however for work over public areas a double-boarded platform sandwiching a polythene sheet may be needed.

- Where a ladder passes through a working platform, the access must be as small as practicable.

- Any access point through a working platform must be covered when it is not being used, and clearly marked to show its purpose.

- If a gap or opening is created in a working platform for work activities, then access to it must immediately be prevented by fitting guard rails etc.

- Trestles must not be put up on a working platform as workers using them would be working above the height of the guard rail.

- If a working platform becomes covered with ice, snow, grease or any other slippery material, then suitable action must be taken to reduce the hazard, by sprinkling sand, salt, sawdust, etc.

- Rubbish or unused materials must not be left on working platforms.

- Working platforms must not be used for "storing" materials - all materials placed on a working platform must be for immediate use only.

- Loading of materials and equipment must be spread evenly over working platforms to the fullest extent possible.

- Where loading of the platform cannot be distributed evenly (as may happen with bricklayers' materials) then the larger weights should be kept nearest to the standards.

- Any working platform near to fragile items, for example, windows, should have sheeting fitted.

- General access must not be allowed to any working platform until the erection of the scaffold has been completed - access must be blocked off to any section of scaffold that is not yet finished.

- All loose materials and equipment must be taken off working platforms before the scaffold and platform is dismantled.

Toe boards

Toe boards are usually scaffold boards placed against the standards at right angles to the surface of the working platform. They help prevent materials from falling from the scaffold and people slipping under guardrails. They must be suitable and sufficient to prevent the fall of any person or any material or object from a place of work.

Safe working practices for the use of toe boards include:

- The toe boards should be fixed to the inside of the standards with toe board clips. Joints must be as near as possible to a standard.

- Toe boards should be continuous around the platform where a guardrail is required.

- Any toe board that is removed temporarily for access or for any other reason must be replaced as soon as possible.

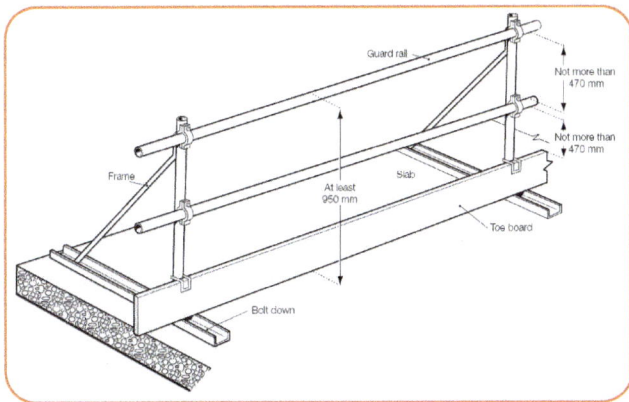

Figure 7-31: Guardrails and toe boards.
Source: HSE, HSG150.

Guardrails

These are horizontal scaffold tubes that help to prevent people falling from a scaffold. Safe working practices for the use of guardrails include:

- They must be fixed to the inside of the standards, at least 950mm (UK standard) from the platform.

- An intermediate guardrail must be positioned such that the gap between it and the top guardrail and the toe boards is no more than 470mm (UK standard).

- Guardrails must always be fitted with load-bearing couplers.

- Guardrails are joined with sleeve couplers.

- Joints in guardrails must be near to a standard.

- Guardrail must go all round the work platform.

- Where there is a gap between the structures then a guardrail must be fitted to the inside of the working platform as well as the outside.

- Guardrails must always be carried around the end of a scaffold, to make a 'stop end'.

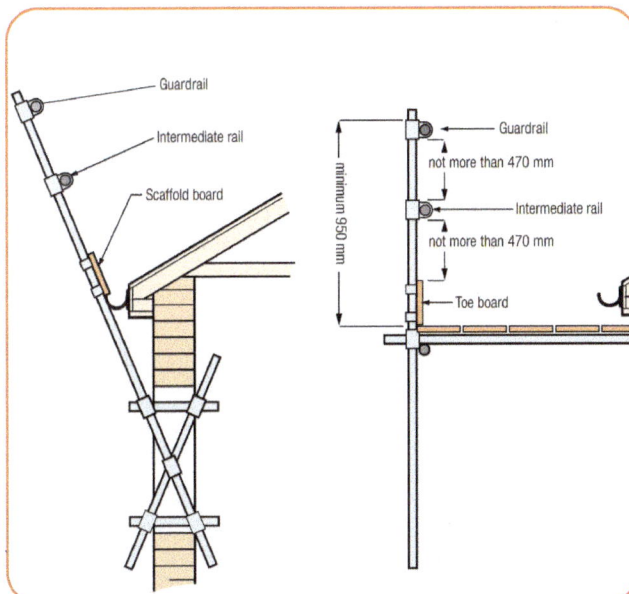

Figure 7-32: Toe boards and sloping roof edge protection.
Source: HSE INDG28.

Mesh (brick) guards

Mesh (brick) guards are fitted to the inside of the guard rail to assist in preventing people and objects from falling between the rails and the toe board. These wire mesh infill panels may be used instead of or as well as an intermediate rail. They sometimes have an integral toe board.

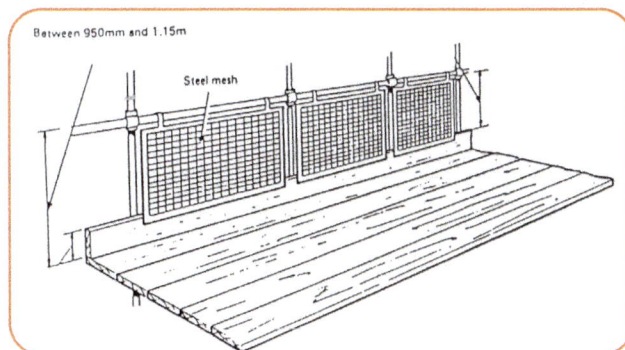

Figure 7-33: Brick guards.
Source: HSE, HSG150.

Debris netting

Debris netting is often fixed to the sides of a scaffold to limit the amount of debris that may come from work being done on it escaping from the scaffold. It provides a tough, durable and inexpensive method of helping to provide protection from the danger of falling debris and windblown waste material. It allows good light transmission and reduces the effects of adverse weather. However, the netting may act as a sail in high winds, so the strength and rigidity of the scaffold needs to take this factor into account at the scaffold design stage.

Figure 7-34: Debris netting.
Source: RMS.

Debris netting may also be slung underneath steelwork or where roof work is being conducted to catch items that may fall. In this situation, it should not be assumed that the debris netting is sufficient to hold the weight of a person who might fall.

Figure 7-35: Debris nets.
Source: RMS.

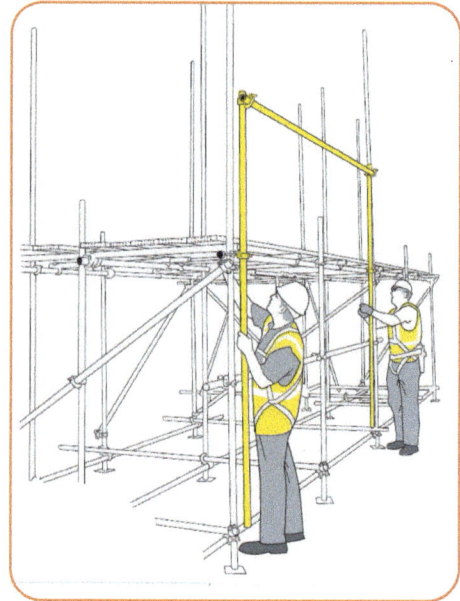

Figure 7-36: Fitting advanced guardrail system.
Source: NASC.

Requirements for scaffold erectors

Scaffold erectors must erect and dismantle scaffold in a way that minimises the risk of falling. Where practicable they should erect intermediate platforms, with guardrails and toe boards, to enable them to build the next scaffold 'lift' safely.

An advanced guardrail may be installed from the level below in order that scaffold erectors can access the platform above and fit guardrails. An advanced guardrail can also be used when scaffolds are being dismantled, in which case the advanced guardrail is fitted to allow removal of the guardrail and toe boards on the level the dismantlers are working on. After which the platform is dismantled from the level below and the advanced guardrail removed and refitted at the lower level, until the process of dismantling is complete.

If a platform is not practicable, falls must be prevented (or their effects minimised) by other means, such as the use of a personal fall arrest system. If a personal fall arrest system is used it is important that the scaffold erector can find and use suitably strong and appropriately positioned anchor points.

The National Access and Scaffolding Confederation (NASC) provide guidance for scaffold workers on the safe erection of scaffolds.

Part 2 of Schedule 3 of the WAHR 2005 requires that scaffolds only be erected by those that have received appropriate and specific training.

"12. Scaffolding may be assembled, dismantled or significantly altered only under the supervision of a competent person and by persons who have received appropriate and specific training in the operations envisaged which addresses specific risks which the operations may entail and precautions to be taken, and more particularly in -

(a) understanding of the plan for the assembly, dismantling or alteration of the scaffolding concerned;

(b) safety during the assembly, dismantling or alteration of the scaffolding concerned;

(c) measures to prevent the risk of persons, materials or objects falling;

(d) safety measures in the event of changing weather conditions which could adversely affect the safety of the scaffolding concerned;

(e) permissible loadings;

(f) any other risks which the assembly, dismantling or alteration of the scaffolding may entail."

Figure 7-37: Competency and training.
Source: WAHR 2005 Schedule 3 Part 2.

Means of access

"So far as is reasonably practicable as regards any place of work under the employer's control, the maintenance of it in a condition that is safe and without risks to health and the provision and maintenance of means of access to and egress from it that are safe and without such risks."

Figure 7-38: Means of access.
Source: HASAWA 1974 s2(2)(d).

General access must not be allowed to any scaffold until its erection has been fully completed.

Access must be prevented to any subsequent sections of scaffold that are not completed and a 'scaffold incomplete' sign displayed. Where practical access to scaffold work platforms should be by means of stairways or lifts in preference to ladders. The stairway will have handrails, be set at an easy tread angle and be provided with a safe means to step off the stairway onto the work platform. System build stairways are available and may be set external to the scaffold. This will ensure easier access and a better route for escape in the event of an emergency. However, ladders are often provide where stairs or other suitable means cannot be provided

Figure 7-39: Stairway access.
Source: RMS.

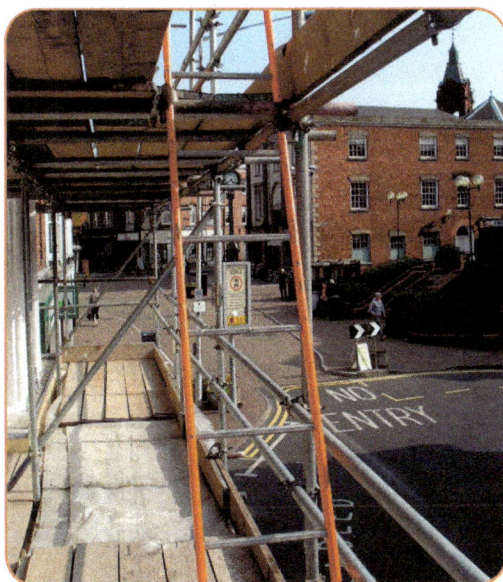

Figure 7-40: Internal ladder access, tied with clips.
Source: RMS.

Safe working practices for the use of ladders for access on a scaffold include:

- They should be free from defect and not painted such that defects would be hidden.

- Placed on a firm footing, with each stile equally supported and the correct way up.

- Positioned so that there is sufficient space at each rung to give an adequate foothold.

- Positioned approximately at an angle of 75° (1 unit horizontally to 4 units vertically).

- Extended to a height of 1 metre above the working platform (unless there is another adequate hand hold).

- Not be so long that it extends excessively past the stepping off point as this can cause the ladder to pivot around the tie/stepping off point and cause the user to fall.

- Positioned so that the vertical height of the ladder running between landings does not exceed 9 metres.

- Securely tied at the top to prevent slipping.

- Be provided with rest platforms if rising more than nine metres.

- Where ladders are external to the scaffold an adequate stepping off point should be provided and the opening protected by a safety gate.

- Where ladders are internal to the scaffold an adequate stepping off point should be provided and the opening protected by a trapdoor or alternative means to prevent falling through the access hole, for example, a guard rail that can be raised to allow access.

Design of loading platforms

Loading platforms are used to provide equipment and materials to the scaffold or work areas at height. They can comprise of single platforms or multi-level platforms. Materials are lifted to the loading platform and placed on it. Loading platforms will need to be specifically designed to take into account the foreseeable concentration of heavy loads that will be placed upon them, for example, equipment, bricks, blocks, mortar or timber.

A loading platform is designed to withstand a weight that would be excessive on the normal working area of a scaffold.

Often the loading platform will be a separate scaffold structure, assembled adjacent to the main working scaffold, tied to both the structure and the main scaffold and will consist of additional braces and sections to provide extra strength and support.

The platform must be correctly signed as a loading area with a safe working load specified.

The loading platform must be provided with a form of fall protection, preferably a safety gate that provides protection when the loading platform is open for loading and when it is closed to enable workers to access the loading platform to retrieve materials and equipment.

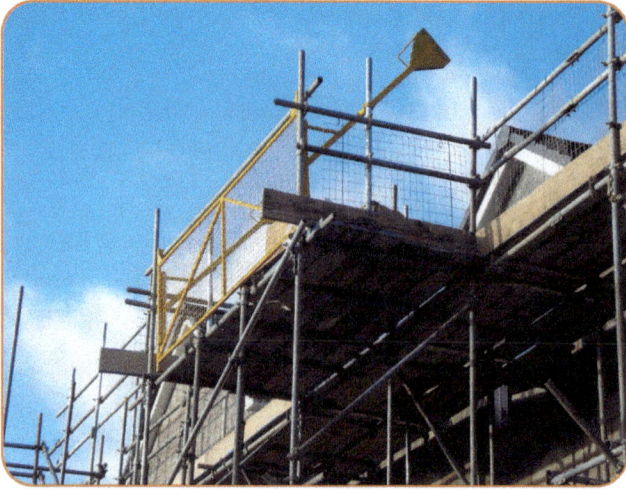

Figure 7-41: Loading platform.
Source: RMS.

Figure 7-42: Multi level loading platforms.
Source: RMS.

Loading platforms used by cranes may not have a safety gate fitted, but will provide fall protection by means of an enclosed system of guardrails and toe boards.

Workers must not be allowed to access the area directly below the platform where the loading and unloading will take place.

Scaffold hoists (persons, materials)

There are various forms of scaffold hoists. In their simplest form they are a **hand operated pulley hoist**, fixed to the scaffold by a scaffold tube that extends out form the scaffold. Alternatively, **electrically powered cable hoists** are available to fit in a similar way or they may have a purpose built hoist beam that is fitted to the scaffold, which rotates to make it easier to retrieve the load.

Inclined hoists are often used to transport materials. Inclined hoists should be erected and used by competent personnel.

Figure 7-43: Powered cable hoist.
Source: BETA.

In addition, scaffold hoists include substantial cage and platform hoist systems for carrying substantial quantities of materials and/or people. The hoist should be protected by a substantial enclosure to prevent anyone from being struck by any moving part of the hoist or material falling down the hoist way. Gates must be provided at all access landings, including at ground level. The gates must be kept shut, except when the platform is at the landing.

The controls should be arranged so that the hoist can be operated from one position only, for example from the ground or by controls inside the hoist cage. All hoist operators must be trained and competent. The hoist's safe working load must be clearly marked.

If the hoist is for materials only there should be a prominent warning notice on the platform or cage to deter people from riding on it.

Figure 7-44: Platform hoist.
Source: RMS.

Hoists should be inspected weekly and in order to comply with requirements of the Lifting Operations and Lifting Equipment Regulations (LOLER) 1998, if used to move people, thoroughly examined at least every six months by a competent person and the results of examination recorded. Hoists used to move materials only must be thoroughly examined at least every 12 months. A thorough examination would be required when it was first installed on site and after exceptional circumstances.

Additionally, there would need to be arrangements for its regular maintenance and for ensuring the guarding of dangerous parts of the hoists machinery and the integrity of any electrical installation.

Protection would need to be provided at the base and top of the hoist to ensure no one enters the area where the cage operates and means provided to ensure the security of the load as it travels to the top of the hoist. Relocation and or dismantling should only be carried out by competent persons.

See also - Element 8.3 - Musculoskeletal health and load handling - 'Lifts and hoists' - for further information.

Ensuring stability

Ties, bracing and rakers

Ensuring stability of a scaffold is critical. In the case of an independent tied scaffold the scaffold is set a small distance away from the structure and ties connect it to the structure and prevent the scaffold falling away from or towards the structure. One of the ways to do this is to use a 'through tie' which is set into place through an opening in the structure such as a window.

Figure 7-45: Tie through window.
Source: RMS.

In addition to the use of ties, it is important that the scaffold is a rigid structure. A scaffold comprising standards, transoms and ledgers alone may not be rigid enough, particularly in the case of tall scaffolds. In order to improve rigidity a system of braces and rakers are used.

Braces (scaffold poles attached across the standards) are used diagonally in opposite directions, to the front and sides of the scaffold, to provide a rigid structure.

Rakers (scaffold poles attached to the ledgers) may be set at a diagonal outwards from the front of the scaffold to the ground to prevent the scaffold falling away from

the structure. They may be used in addition to ties or added to provide stability in a situation where a tie has to be removed in order to do work on the opening where it is fixed.

Effects of materials

Scaffold systems are a means of providing safe access when work at height cannot be avoided. They are not generally designed for storage of materials for long periods.

Materials (bricks, mortar, timber, etc.) when placed on scaffold systems tend to be placed on the outer edge of the scaffold, creating an uneven balance and placing greater forces on the mechanical joints. Similarly, when excessive quantities of materials and equipment are placed on a scaffold the safe working load of the scaffold could be exceeded. These factors can contribute to failure of the boards or joints, or buckling of the tube sections which may ultimately lead to a full or partial collapse of the scaffold.

However, it is acceptable to situate materials on scaffolds in small quantities to reflect the usage rate of the materials by the workers using the scaffold. Provided the safe working load specified for the scaffold is not exceeded, materials may be distributed evenly on the working platform. All loading of scaffolds with materials should be well planned and carried out under supervision to ensure even loading and to limit the load in any one area of the scaffold so that it remains within the safe working load specified for the scaffold. The scaffold should be checked to ensure the safe working load is adhered to at all times.

Where the loading and storage of materials and equipment would exceed the safe working load of the scaffold then a specially strengthened section of scaffold or loading platform should be provided.

Care should be taken to ensure the working platform is not reduced by the addition of materials to a width that compromises access around the scaffold.

Rubbish chutes attached to scaffolds can put additional loading on the structure, which could be sufficient to cause the scaffold to collapse unless it has been considered at the design stage and additional strengthening added to the scaffold in the area of the chute. It is important that the chute is regularly checked for blockages as the additional load of the excess material in the chute could affect the stability of the scaffold.

Materials in the form of debris from the construction processes could build up on the scaffold, including at lower levels than the working platform, this could progressively overload the scaffold and lead to its sudden failure. In addition, material that falls into fans could build up and cause additional loading on the scaffold.

It should be remembered that a scaffold may be able to tolerate individual aspects of loading, but if the various forms of excessive loading acted together it could cause a catastrophic failure of the scaffold.

Weather

Heavy rainfall can lead to soil being washed away from the base of the scaffold or the soil can subside leading to the scaffold becoming unstable. Heavy snowfalls can add additional loading to the scaffold. Freshly fallen snow could become compacted by additional snowfall, leading to significant weight acting on the scaffold, which might exceed the safe working load for the load class of the scaffold.

In addition, strong winds can put additional loading on scaffolds as boarded and sheeted areas may be affected by gusts of wind. This could cause parts of the scaffold to be lifted from the ground momentarily. The forces created by the wind could cause standards to be moved from their stable position on the ground and bend or loosen ties. It should be remembered that, depending on the location and design of the scaffold, strong winds could be amplified as they are funnelled into more confined areas.

Scaffold boards that are only held in place by their own weight could be lifted by high winds, leading to twisting forces on the scaffold and impact damage from the boards if they become airborne.

If a scaffold has been affected by adverse weather conditions it should be inspected prior to work starting.

Adverse weather conditions not only affect the stability of scaffolds, but can also present dangers to workers who are exposed to them, for example cold weather can affect dexterity, awareness and morale. These conditions need to be anticipated and suitable precautions taken. Rain, sleet or snow can make surfaces very slippery and in winter freeze to create ice or frost.

A sudden gust of wind can cause a worker to lose balance. This is usually exaggerated when working at height and handling large sheets of materials. In extreme circumstances, work should be stopped during windy weather as people can easily be thrown off balance while carrying out their work.

When deciding whether to continue or suspend work due to high wind consider:

- Wind speed.

- The measures which have already been taken to prevent falls from the scaffold.

- The position and height of the scaffold and the work being carried out.

- The relationship of the work to other large structures as wind may be tunnelled and amplified at certain points.

Sheeting

Sheeting can be used on the outer part of scaffold systems as a means of preventing materials, drips, splashes, dust and other debris being blown from the working areas of the scaffold onto the construction site or possibly a public area. It can also provide a means of restricting and controlling access to the scaffold and reduce the chilling effects of wind on worker comfort. It is not a means of fall protection.

Figure 7-46: Sheeting.
Source: RMS.

Whilst sheeting is relatively light in weight and causes little stress on the scaffold system, the additional forces created by wind, rain or snow spread over the surface of the sheeting could cause sufficient force to affect the safe stability of the scaffold. It is important that sheeting is securely fixed to the structure and not allowed to flail loosely.

If sheeting is to be added to a scaffold this must be factored in to the load calculations made at the design stage and the effects of weather considered.

Protection from impact of vehicles

Construction sites quite often involve numerous types of vehicles of varying size and weight all presenting hazards to scaffold structures on site. Protection of scaffolds from impact of vehicles is important as an impact at a lower level could cause significant damage to major loadbearing parts of the scaffold and lead to the scaffold collapsing.

Measures that may be implemented in order to maintain a safe environment in relation to vehicles and scaffolds may include:

- Providing signs warning of the presence of scaffolds.

- Marking scaffolds with high-visibility tape or sheath around scaffold standards.

- Closing routes past scaffolds to vehicles.

- Ensuring adequate road width for the size of vehicle accessing the site.

- Provision of one-way systems around sites or turning points (away from scaffold) to minimise the need for reversing.

- Reversing vehicles properly controlled by trained reversing assistants/banksmen/signaller. Competent drivers correctly trained.

- Provision of robust impact protection barriers, for example concrete, wooden or filled plastic blocks strategically placed around scaffold perimeter to limit the proximity of vehicles.

- Providing lighting around the scaffold perimeter, scaffolds located near roads may also be fitted with specific lighting to warn vehicles of its presence.

Inspection requirements

Inspection requirements for scaffolds are set out in Regulation 12 of WAHR 2005 Regulations. All scaffolds used for construction work must be inspected by a competent person before being taken into use for the first time, after any substantial addition, dismantling or other alteration, after any event likely to have affected its strength or stability and where a person could fall 2 metres or more at regular intervals not exceeding 7 days since the last inspection. The result of an inspection must be recorded and kept until the next inspection. A report on the inspection must be completed before the end of the work period and a copy provided to the person on whose behalf it was carried out within 24 hours.

Figure 7-49: Marking of scaffold.
Source: RMS.

Figure 7-47: Impact protection.
Source: RMS.

Figure 7-50: Inspection 'Scaftag'.
Source: RMS.

Reports on inspections to include the following details:

1) Name and address of the person for whom the inspection was carried out.

2) Location of the work equipment inspected.

Figure 7-48: Lighting.
Source: RMS.

3) Description of the work equipment.

4) Date and time of inspection.

5) Details of any matter identified that could give rise to a risk to the health and safety of any person.

6) Details of any action taken as a result of number 5) above.

7) Details of any further action considered necessary.

8) Name and position of the person making the report.

It is useful to indicate to those using the scaffold that the inspection has been conducted by the addition of a tag to the scaffold that shows basic details of the inspection. Employers must keep a copy of the inspection report at the site where it was carried out, until construction work is complete.

After this period they have to be kept at the employer's office for 3 months.

REVIEW

When should scaffolding be inspected?

Outline safe methods of working on a leading edge on a roof.

ACCESS EQUIPMENT

Ladders

Use of ladders

Ladders are primarily a means of vertical access to a workplace. However, they are often used to carry out short-term work at height and this frequently results in injuries. Many injuries involving ladders happen during work lasting 30 minutes or less. Before using a ladder to work from, consider whether it is the right equipment for the job.

Figure 7-51: Ladder as access and workplace.
Source: RMS.

Ladders are only suitable as a workplace for light work of short duration, and for a large majority of activities a mobile scaffold tower or a mobile elevating work platform

(MEWP) is likely to be more suitable and safer. Generally, ladders should be considered as access equipment and use of a ladder as a work platform should be discouraged. There are situations when working from a ladder would be inappropriate, for example:

- Where the equipment or materials used are large or awkward.
- When two hands are needed or the work area is large.
- Excessive height.
- Work of long duration.
- Where the ladder cannot be secured or made stable.
- Where the ladder cannot be protected from vehicles etc.
- Adverse weather conditions.

Safe working practices for the use of ladders include:

- Pre-use selection and inspection. The ladder to be used needs to be long enough to access the point where work will take place without the user overreaching or standing on the top three rungs. The ladder must not be overloaded and needs to be suitable for the worker and the equipment/materials that they may be carrying. Make sure the ladder is in good condition. Check the rungs and stiles for warping, cracking or splintering, the condition of the feet, and for any other defects. Do not use defective or painted ladders.

- Position the ladder properly for safe access and out of the way of vehicles. Do not rest ladders against fragile surfaces.

- Ladders must stand on a firm, level base, and be positioned approximately at an angle of 75° (1 unit horizontally to 4 units vertically), (*see Figure 7-52*; note: a means of securing the ladder is omitted for clarity).

- Ladders must be properly tied near the top, even if only in use for a short time, while being tied a ladder must be footed. If not tied, ladders must be secured near the bottom, footed or weighted.

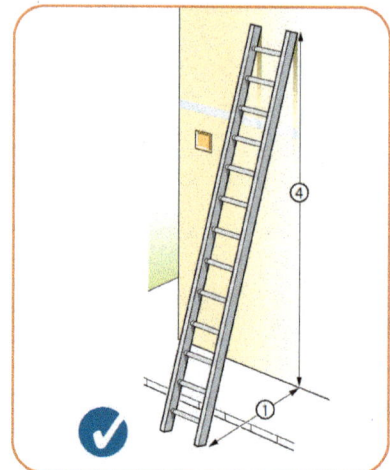
Figure 7-52: Correct 1 in 4 angle.
Source: HSE, INDG402.

Figure 7-53: Improper use of ladder.
Source: RMS.

- If the ladder is being used to gain access to a landing place it should extend about 1 metre above the landing place.

- Both hands should be kept free to grip the ladder when climbing or descending, always maintain three points of contact on the ladder.

- Beware of wet, greasy or icy rungs and make sure soles of footwear are clean.

- Ensure the ladder is not overloaded.

- Only one person to climb at a time.

- Ladders should be used only for work of short duration and not in inclement weather.

- Use a holster or tool bag to carry tools securely.

Figure 7-54: Inappropriate storage.
Source: Lincsafe.

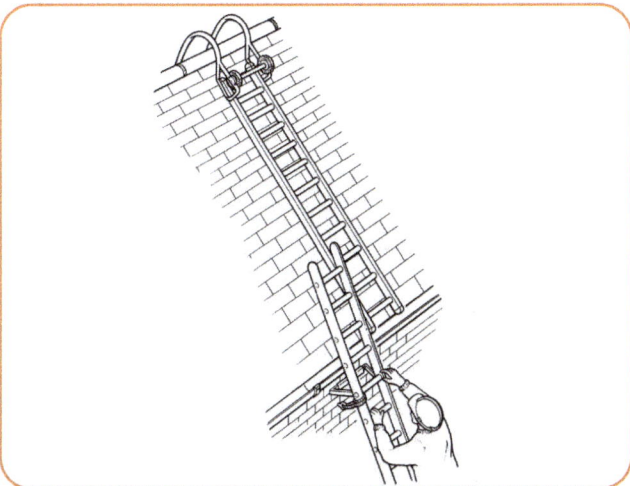

Figure 7-55: Use of roof ladders.
Source: HSE, HSG150.

CONSIDER

When is a ladder right for the job?

Are you using the correct ladders at your workplace?

Step ladders

Step ladders require careful use. They are subject to the same general health and safety rules as ladders. However, in addition, they will not withstand any degree of side loading and overturn very easily. When using a step ladder overreaching should be avoided at all times and care should be taken to avoid side loading.

Figure 7-56: Incorrect use of a stepladder.
Source: HSE, INDG402.

Ensure the step ladder is placed on a firm level surface to minimise the possibility of it overturning sideways.

Step ladder stays must be 'locked out' properly before use. The top step of a stepladder should not be used as a working platform unless it has been specifically designed for that purpose. Typically three clear steps should be left to ensure support and stability, depending on the size and design of the step ladder.

Figure 7-57: Correct use of a step ladder.
Source: HSE, INDG402.

There is a risk of fatigue when climbing long vertical ladders, because the climber needs to physically pull and step their body up the ladder. The ladder hoops are designed to create a 'tunnel' to prevent people falling away from the ladder through fatigue, but do not prevent sliding down the ladder (**see Figure 7-59**). Hoops need to be used in conjunction with rest platforms, at intervals of 9 metres, to allow the climber to rest and recover from fatigue.

Figure 7-59: Ladder hoop.
Source: RMS.

 REVIEW

What are the features of a ladder that would ensure it is suitable for working at height?

Outline a safe method of working on a fragile roof.

Trestles

Trestles are pre-fabricated steel, aluminium or wood supports, of approximately 500mm to 1 metre width, that may be of fixed height or may be height adjustable by means of sliding struts with varying fixing points (pin method) or various cross bars to suit the height required. Scaffold boards are placed on top of them, stretching from one to the other, in order to make a working platform.

These should only be used where work cannot be carried out from the ground, but where a scaffold would be impracticable.

A good example of where they may be used would be where a plasterer is installing and plastering a new ceiling. Typical working heights when using a trestle system range from 300mm to 1 metre. Edge protection should be fitted wherever practical. As with any work carried out above ground level consideration should be given to the application of WAHR 2005 and a risk assessment, as required by the Management of Health and Safety at Work Regulations (MHSWR) 1999, carried out to determine if they are the appropriate equipment for the work to be conducted.

Figure 7-58: Correct use of a step ladder.
Source: HSE, INDG402.

Vertical ladders and ladder hoops

Construction workers may make use of vertical ladders to gain access to a roof or similar areas, they are often used where large process plant is installed over a number of floors and occasional access is required to parts of the plant.

Figure 7-60: A trestle. Source: https://worksafe.govt.nz.

There are many configurations of locking and adjustable trestles. Platforms based on trestles should be fully boarded, adequately supported and provided with edge protection where appropriate. Safe means of access should be provided to trestle platforms, usually by stepladders.

When using trestles the following safe working practices should be used:

Always

- Set up the equipment on a firm, level, non-slip surface.
- On soft ground, stand the equipment on boards to stop it sinking in.
- Place each trestle at 1.5 metre intervals, which allows the scaffold boards to be adequately supported.
- Then open each up to the height required, ensure the locking pins are properly located.

Never

- Do anything that involves applying a lot of side force when using the trestle. The trestle could topple over.
- Exceed the maximum safe working load of a scaffold board (150kg evenly spaced). Remember to include the weight of the user, tools and materials in your calculations.
- Use steps, boxes, etc, placed on top of the trestle boards to gain extra height.

Staging platforms

Staging platforms can be made of metal alloy or wood and are often used for linking trestle systems or tower scaffolds together safely. They also provide a safe work platform for work on fragile roofs. They are produced in various lengths. The same rules apply for edge protection as with other scaffold platforms.

Figure 7-61: Ladder staging platform.
Source: Ladder Safety Devices.

When using staging platforms the following safe working practices should be used.

Ensure the staging platforms are:

- Of sufficient dimensions to allow safe passage and safe use of equipment and materials.
- Free from trip hazards or gaps through which persons or materials could fall.
- Fitted with toe boards and guardrails (if these requirements are not considered necessary for a specific platform, then this should be shown in the risk assessment, i.e. that not installing a toe board and/or a guardrail had been considered and why it was not necessary).
- Kept clean and tidy, for example, mortar and debris should not be allowed to build-up on platforms.
- Not loaded to the extent that there is a risk of collapse or deformation that could affect its safe use. This is particularly relevant in relation to block work loaded on staging platforms.
- Erected on firm level ground to ensure equipment remains stable during use.

Figure 7-62: Staging platform with guard rail edge protection.
Source: RMS.

Figure 7-63: Safety net for leading edge protection.
Source: RMS.

Leading edge protection

Leading edges are created as roof sheets are laid or ones are removed. Falls from a leading edge need to be prevented; work at the leading edge requires careful planning to develop a safe system of work.

Staging platforms, fitted with guardrails or suitable barriers and toe boards, positioned in advance of the leading edge can provide some protection for preventing falls during some leading edge work activities.

However, the staging platforms will need to be used in conjunction with work restraint systems (fall prevention)/ fall-arrest systems (fall protection) attached to a suitable fixing. Close supervision of this system of work will be needed as it is often difficult for the fall prevention/protection equipment to remain safely clipped to the fixing at all times throughout the work activity.

Safety nets are the preferred protection for reducing the distance and consequences of falls at the leading edge, as they provide collective protection to everyone on the roof.

Safety nets should be erected as close to the work surface of the leading edge as possible by trained riggers and be strong enough to take the weight of people who fall onto them.

In some countries, longer term fall protection is added to buildings that have roof sheets to provide protection during construction and at the time roof sheets are removed. This involves the installation of galvanised steel safety mesh on top of the roof steelwork prior to the laying of the roof sheets). The safety mesh is rolled over the structure from work platforms set in position at the sides. It remains in position throughout the life of the building and provides leading edge protection at the time of laying or removal of the roof sheets.

Mobile elevating work platforms

A mobile elevating work platform (MEWP) is, as the name suggests, a means of providing a work platform at height.

Figure 7-64: Mobile elevating work platform (MEWP).
Source: RMS.

The equipment is designed to be movable, under its own power or by being towed, so that it can easily be set up in a location where it is needed. Various mechanical and hydraulic means are used to elevate the work platform to the desired height, including telescopic arms and scissor lifts. The versatility of this equipment, which enables the easy placement of a platform at height, makes it a popular piece of access equipment. Often, to do similar work by other means would take a lot of time or be very difficult. They are now widely available and there is a tendency for people to oversimplify their use and allow people to operate them without prior training and experience. This places users and others at high risk of serious injury.

Some MEWPs can be used on rough terrain. This usually means that they are safe to use on uneven ground. The MEWP's limitations should always be checked in the manufacturer's handbook before moving on to uneven or sloping ground. They should only be operated within their defined stability working area.

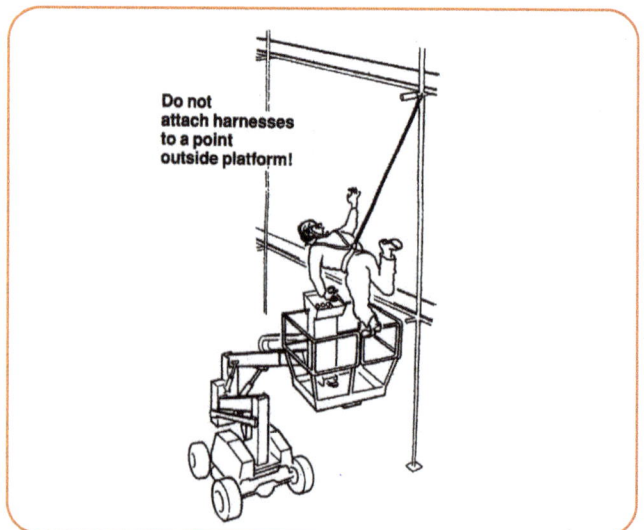

Figure 7-65: Use of harness with a MEWP.
Source: HSE, HSG150.

A harness with a lanyard attached to the platform acts as a work restraint and provides extra protection against falls especially when the platform is being raised or lowered. MEWPs and similar equipment used in poor lighting conditions on or near roads and walkways must use standard vehicle lighting.

Use of mobile elevating work platforms

Mobile elevating work platforms can provide excellent safe access to high level work. When using a MEWP the following safe working practices should be used:

Figure 7-66: Scissor lift MEWP.
Source: HSE, HSG150

Always ensure

- Whoever is operating it is fully trained and competent.
- Any statutory inspections or testing has been carried out.
- The work platform is fitted with guardrails and toe boards.
- It is used on suitable firm and level ground. The ground may have to be prepared in advance.
- Tyres are properly inflated.
- The work area is cordoned off to prevent access below the work platform
- That it is well lit if being used on a public highway or in poor lighting.
- Outriggers are extended and chocked as necessary before raising the platform.
- Harnesses are used by workers.
- All involved know what to do if the machine fails with the platform in the raised position.

Figure 7-67: Scissor lift MEWP.
Source: RMS.

Never

- Operate MEWPs close to overhead cables or dangerous machinery.
- Allow a part of the hand or arm to protrude into a traffic route when working near vehicles.
- Move the equipment with the platform in the raised position unless the equipment is especially designed to allow this to be done safely (the manufacturer's instructions should be checked).
- Overload or overreach from the platform.

Mast climbing work platforms

Mast climbing work platforms (MCWPs) are often used when carrying out repairs or refurbishment to high-rise buildings. They provide excellent access to work positions at a height and enable work to be carried out at different heights as the work progresses.

They are not designed as materials hoists though they have a capacity for some work material and people to be on the platform to carry out the work required.

A particular advantage of using MCWPs is that workers on the platform and their tools/materials can be protected from adverse weather conditions by the provision of an enclosed work platform.

Figure 7-68: Mast climbing work platforms.
Source: HSE, HSG150.

When using MCWPs it is important to ensure that:

- It is erected/installed by a competent specialist who is aware of the purpose of use, likely loadings and weather conditions it will be exposed to.

- Masts are rigidly connected to the structures against which they are operating and outriggers are used when necessary.

- The area below the platform is guarded to prevent people being struck by the platform or by objects that may fall from the platform.

- Working platforms are provided with suitable guard-rails and toe boards.

- The controls only operate from the working platform.

- A handover document is provided by the erector/installer. The document should state its safe working load; cover how to operate the equipment, what checks and maintenance is required and how to deal with emergencies.

- There is a current report of thorough examination for the equipment.

OTHER ACCESS EQUIPMENT

Boatswain's chair

Boatswains' chairs (commonly known as Bosun's chair) and seats can be used as a work positioning system for light, short-term work. The chair is distinguished from a climbing harness by the inclusion of a rigid seat, providing more comfort than padded straps for long-term use. The boatswains chair should be fitted with a harness to prevent the user falling from the seat and may have a foot rest system. This system provides an amount of manoeuvrability that a suspension cradle may not provide, with the ease of use of a rope system. Systems are available with electrically operated rope hoists to enable easy raising and lowering of the seat. They should only be used where it is not practicable to provide a working platform. When using a boatswain's chair it is important to ensure the following points, before, during and after use.

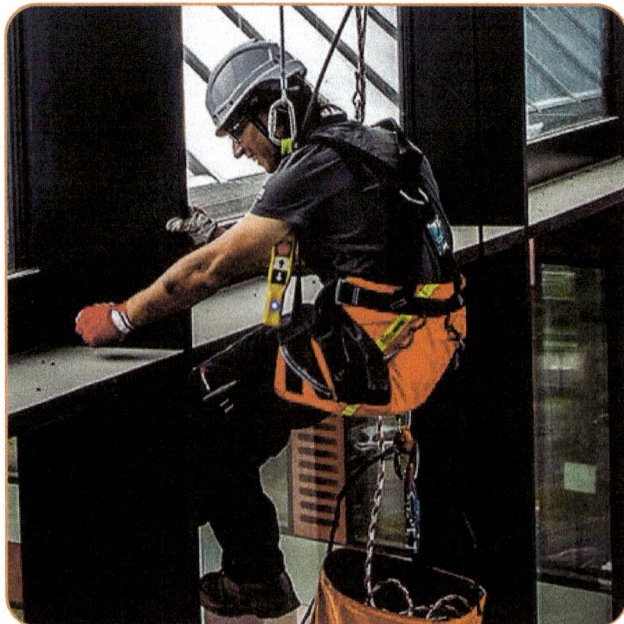

Figure 7-69: Boatswains chair.
Source: High Access Solutions.

Before use

Before use ensure that:

- Installation and use of boatswain's chair to be supervised by trained, experienced and competent person.

- Chair and associated equipment carefully examined for defects.

- Confirm test/examination certificates are valid. Establish safe working load.

- Check that user is both trained and competent in the use of the chair.

- User to wear harness.

- Warning notice displayed and notification of intention to carry out work given.

- Prohibit access to the area below the chair in case materials fall.

In use

When in use ensure that:

- Chairs and ropes are free of material or articles which could interfere with the user's hand-hold.

- The fall rope must be properly tied off in use and always under or around a cleat to act as a brake.

After use

- Chairs and ropes should be left in a safe condition:

- Top rope secured.

- Chair and rope secured to prevent swing.

- Raised when out of use or for overnight storage.

- Inspected for defects.

- Ropes (and chair, if timber) dried before storage.

Cradles (including suspension from cranes)

Suspension cradles comprise of two electric hoists connected to each end of an aluminum cradle. Each hoist is connected to two suspension wires with one acting as a 'fail-safe' wire in case the main suspension wire fails. The suspension wires are hung from a gantry, usually located on the roof of a structure. The suspension cradle is operated from within the cradle to ensure good co-ordination of movement.

Before work starts using a cradle, check that:

- Equipment is installed, modified and dismantled only by competent specialists.

- The suspension cradle should be weight tested and inspected after installation and before use.

- There is a current report of thorough examination for the equipment.

- A handover certificate is provided by the installer. The certificate should cover how to deal with emergencies, operate, check and maintain the equipment, and state its safe working load.

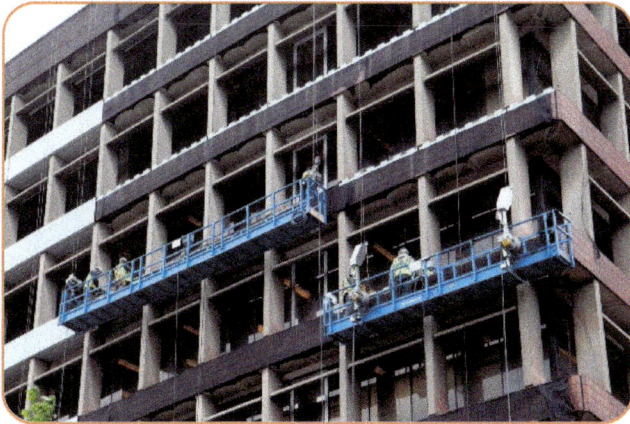
Figure 7-70: Suspension cradle.
Source: Harsco Infrastructure.

Figure 7-71: Crane suspension cradle.
Source: RMS.

- Areas of the site where people may be struck by the cradle or falling materials have been fenced off or similar. Debris fans or covered walkways may also be required.

- Systems are in place to prevent people within the building being struck by the cradle as it rises or descends and prevent the cradle coming into contact with open windows or similar obstructions which could cause it to tip.

- Supports are protected from damage (for example, by being struck by passing vehicles or by interference from vandals).

- The equipment should not be used in adverse weather. High winds will create instability. Establish a maximum safe wind speed for operation. Storms and snow falls can also damage platforms, so they should be inspected before use after severe weather.

- Only trained personnel are to operate.

At the end of each day, check that:

- All power has been switched off and, where appropriate, power cables have been secured and made dead.

- The equipment is secured where it will not be accessible to vandals or trespassers.

- Notices are attached to the equipment warning that it is out of service and must not be used.

- Check the shift report for warnings of malfunction.

Some suspension cradles are designed to be fitted to a crane. Similar precautions apply to those of general suspension cradles. As the crane and cradle carry people they must have been thoroughly examined within a period of 6 months at the time of use. As the crane may have previously been used for lifting loads this is an important additional precaution. The safe working load of the lifting equipment, including the crane, would need to be reduced in order to establish a better factor of safety when lifting people.

The lifting, lowering and positioning of the cradle will mostly be made by the crane operator, therefore it is essential that there is good communication between workers in the cradle and the crane operator. In some cases the cradle will be fitted with its own, additional, electric hoist and duel suspension wire system to assist with making height adjustment to the cradle.

Rope access

Often known as abseiling, this technique is usually used for inspection work rather than construction work.

Like a boatswain's chair, it should only be used when a working platform cannot be provided. If rope access is necessary, then check that:

- A competent person has installed the equipment.

- The user is fully trained.

- There is more than one point securing the equipment.

- Tools and equipment are securely attached or the area below cordoned off.

- The main rope and safety rope are attached to separate points.

Figure 7-72: Personal suspension equipment.
Source: RMS.

FALL ARREST EQUIPMENT

Harnesses

There may be circumstances in which it is not practicable for guardrails etc. to be provided, for example, where guardrails are taken down for short periods to land materials.

In this situation if people approach an open edge from which they would be liable to fall two metres or more, a suitably attached harness and lifeline could allow safe working. When using harnesses and lifelines, ensure:

- An emergency rescue system must be in place before a harness and lanyard is used to protect against a fall.

- A harness will not prevent a fall - it can only minimise the injury if there is a fall. The person who falls may be injured by the impact load to the body when the line goes tight or when they strike against parts of the structure during the fall. An energy absorber fitted to the energy-absorbing lanyard can reduce the risk of injury from impact loads.

- Where possible the energy-absorbing lanyard should be attached above the wearer to reduce the fall distance. Extra free movement can be provided by running temporary horizontal lifelines or inertia reels. Any attachment point must be capable of withstanding the impact load in the event of a fall. Consider how to recover anyone who does fall.

- Anyone who needs to attach themselves should be able to do so from a safe position. They need to be able to attach themselves before they move into a position where they are relying on the protection provided by the harness.

- That there is an adequate fall height to allow the system to operate and arrest the fall.

- A twin lanyard is used where necessary, in situations where the wearer needs to move about. A twin lanyard allows the wearer to clip on one lanyard in a different position before unclipping the other lanyard.

- Installation of equipment to which harnesses will be fixed, for example, a suitable anchor, must be inspected regularly.

- Everyone who uses a harness must be instructed in how to check, wear and adjust it before use and how to connect themselves to the structure or safety line as appropriate.

- The equipment is thoroughly examined at intervals of no more than every six months.

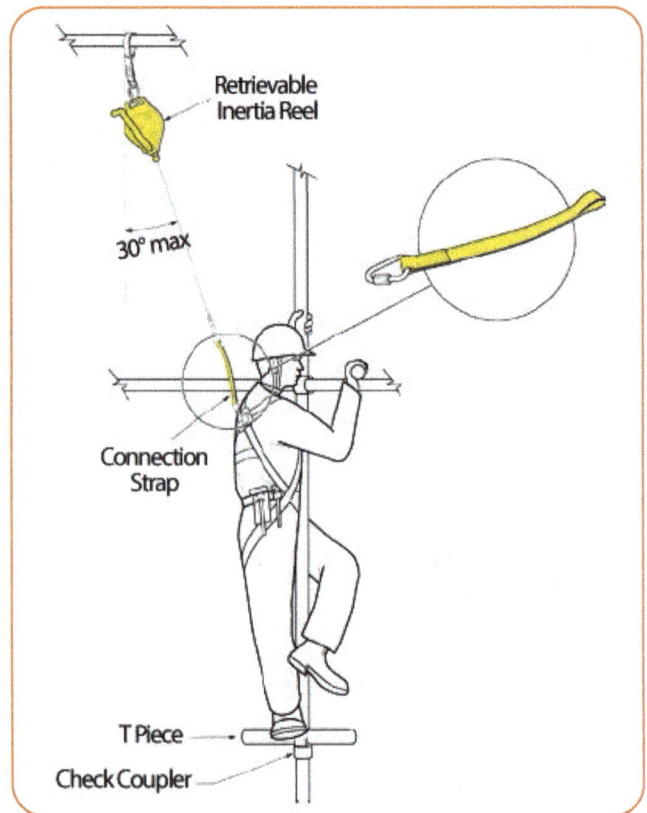

Figure 7-73: Scaffolder using personal fall arrest device with harness.
Source: NASC.

Retractable types of fall arresters tested against EN360 are tested with a 100Kg mass and have to produce an arrest force of less than 6kN. Therefore users greater than 100Kg (15 stone 10 lbs.) in weight should refer to the manufacturer's instruction for use and if not covered within the instruction should contact the manufacturer and obtain information for users weighing greater than 100Kg.

Safety nets

Safety nets are used in a variety of applications where other forms of protection are not reasonably practicable - such as steel erecting and roof work where site personnel are at risk of falling through fragile roofs onto solid surfaces or structures (steel work) below.

Figure 7-74: Safety nets under partially completed roof.
Source: RMS.

Safety nets will arrest the fall of an individual preventing an impact that may cause injury or death. Safety nets must be installed beneath the work area, as a minimum, and consideration should be given to an extension of the protection to allow for people working outside the planned area. Safety nets are only to be installed under supervision by competent installers and are to be inspected weekly and checked daily as the build height progresses.

Relocation of the nets will be necessary if the build height increases to keep the fall distance to a minimum.

It should be remembered that nets are also used to collect debris that may fall. This may be rigged to take a lesser weight than that for protection of people. It is important to identify which type of net is in use in a workplace as reliance on the wrong type may have fatal consequences.

Soft landing systems

Soft landing systems normally consist of inflatable air bags or bean bags that are placed, prior to work commencing, under the work area.

Air bags

Airbags are shock absorption devices that can be employed to protect against the effects of falls from a height. They are made of a toughened nylon, or similar man-made material, and are usually inflated using a powered fan device.

Figure 7-75: Soft fall arrest system.
Source: HSE HSG150.

They are designed so that should a person fall onto one, a volume of air is forced out of the bag creating a cushioning effect that does not bounce or deflect the person in another direction. This equipment must be matched to correspond to the fall height and as such must only be put into use by a competent person.

Soft fall beanbags

Soft fall bean bags use a soft polymer bean to reduce the effect of a fall from up to two metres. The system is used in both traditional building and timber frame constructions.

Crash decks

Collective fall prevention systems (including safety decking) are at the top of the HSE's hierarchy of controls for work at height.

Figure 7-76: Crash deck assembly.
Source: Bondeck Ltd.

Crash decks (passive, collective fall-protection) provide total protection from falls into the building for all trades at all build stages. For example, bricklayers can work directly off the decking for topping out.

Figure 7-77: Passive, collective fall prevention.
Source: Bondeck Ltd.

Because they can be set out at a variety of heights they can remove the need for trestles. The decking can be installed concurrently with, or just before, external scaffold is lifted reducing build time.

Figure 7-78: Crash deck designed for loading.
Source: Bondeck Ld.

Emergency procedures including rescue

It is essential that risk assessments are completed for work at height and they should include planned emergency procedures. Workers should be trained and competent in procedures. These procedures should be rehearsed, so that in the event of an actual incident, workers are competent.

Emergency procedures should take account of the range of emergencies that may particularly affect those working at height, for example, loss of power to powered access equipment, loss of power in areas that need to be illuminated to ensure safety (for example, on a scaffold at night), possibilities of a fire or the occurrence of a fall of a person. In the case of equipment failure, the emergency procedures should include rescue from the work position. In situations where a person has fallen, emergency considerations should ensure that further injury is prevented by prompt rescue, using techniques that do not put the worker in further harm.

Suspension fainting in the use of rope and harnesses

An emergency rescue system must be in place before a harness and lanyard is used by a worker to protect against a fall. The rescue team must be trained to carry out the rescue procedure and in how to deal with a person who is possibly unconscious. The rescue will need to be initiated rapidly (for example, within 15 minutes) to reduce the risk of the person who has fallen and is suspended by the fall arrest equipment experiencing suspension fainting. Suspension fainting is caused by the person being suspended for an extended period resulting in their blood pressure dropping significantly. The fainting is primarily caused by gravity-induced blood pooling in the lower extremities, which in turn reduces return of the blood back to the heart, resulting in decreased cardiac output, lowering of arterial pressure and reduced blood flow to vital organs. Unless rescue is made promptly the resultant lack of flow of oxygenated blood to their vital organs can be fatal. Care has to be taken during rescue when moving the person rescued from a vertical to a horizontal position in order to prevent a surge of deoxygenated blood entering the heart, which could cause cardiac arrest.

ROOF WORK

Flat roofs

Work on a flat roof is high risk. People can fall:

- From the edge of a completed roof.
- From the edge where work is being carried out.
- Through openings or gaps

Edge protection for flat roofs

Unless the roof parapet provides equivalent safety, temporary edge protection will be required during most work on flat roofs. Both the roof edge and any openings in it need to be protected. It will often be more appropriate to securely cover openings rather than put edge protection around them. Any protection should be:

- In place from start to finish of the work.

- Strong enough to withstand people and materials falling against it.

Where possible the edge protection should be supported from ground level, for example, by scaffold standards, so that there is no obstruction on the roof. If the building is too high for this, the roof edge up-stand can support the edge protection provided it is strong enough. Edge protection can also be supported by frames, counterweights or scaffolding on the roof. The protection should be in place at all times. Guarding systems are widely available that enable roof repair work to carry on without removing any guard rails.

Short-duration work on flat roofs

Short-duration means a matter of minutes rather than hours. It includes such jobs as brief inspections or adjusting a television aerial. Work on a flat roof is dangerous even if it only lasts a short time. Appropriate safety measures are essential. It may not be reasonably practicable to provide edge protection during short-duration work. In such cases, anyone working nearer than 2 metres to any unguarded edge should use personal fall prevention equipment (work restraint system) or personal fall protection equipment (fall-arrest system).

Where personal fall prevention/protection equipment is used it needs to be:

Appropriate for the user and in good condition - full harnesses are essential, safety belts are not sufficient.

- Securely attached to an anchorage point of sufficient strength.

- Fitted with as short a lanyard as possible that enables wearers to do their work (significant management discipline is needed to ensure this).

Demarcating safe areas

Full edge protection may not be necessary if limited work on a larger roof involves nobody going any closer than 2 metres to an open edge. In such cases demarcated areas can be set up to limit the area where work can be conducted safely.

Demarcated areas should be:

- Limited to areas from which nobody can fall.

- Indicated by an obvious physical barrier (full edge protection is not necessary but a painted line or bunting is not sufficient).

- Subject to a high level of supervision to make sure that nobody strays outside them (demarcation areas are unacceptable if this standard is not achieved).

Sloping roofs

On traditional pitched roofs most people fall:

- From eaves.

- By slipping down the roof and then over the eaves.

- Through the roof internally, for example, during roof truss erection.

- From gable ends.

Edge protection for sloping roofs

Full edge protection at eaves level will normally be required for work on sloping roofs. The edge protection needs to be strong enough to withstand a person falling against it. The longer the slope and the steeper the pitch the stronger the edge protection needs to be. A properly designed and installed independent scaffold platform at eaves level will usually be enough. Less substantial scaffolding barriers (rather than platforms) may not be strong enough for work on larger or steeper roofs, especially slopes in excess of 30°.

On some larger roofs, the consequences of sliding down the whole roof and hitting the eaves edge protection may be such that intermediate platforms at the work site are needed to prevent this happening.

Figure 7-79: Edge protection and access.
Source: RMS.

Figure 7-80: Edge protection and roof ladder.
Source: RMS.

If the work requires access within 2 metres of gable ends, edge protection will be needed there as well as at the eaves. Powered access platforms can provide good access as an alternative to fixed edge protection. They can be particularly useful in short-duration work.

Means of access

Means of access for roof work is usually by means of a ladder or stair. Though, access may also be provided by using a scissor lift, which can also carry materials. If provided, a ladder should extend at least 1 metre above the edge of the roof, and it should be securely lashed in position, preferably with a wire bond or safety tie. All forms of access should enable workers to easily step onto a safe landing position on the roof or on a landing platform set to the side of the roof, for example, a scaffold tower built for this purpose.

Edge and leading edge protection

When working on a roof it is essential that edge protection be provided, usually in the form of toe boards and guard rails. When doing work on a roof that requires the worker to fit or remove roof materials, such as insulation and cladding, this can cause the worker to be working at a leading edge.

The leading edge is the limit of the roof structure, on one side of the leading edge is an open height hazard and the other is the relatively safer roof. In order to provide protection at the leading edge it is necessary to provide a system of crawling boards that both spread the weight of the worker and provide fall protection in the form of guardrails. The crawling boards must be repeatedly re-set at the new leading edge as the work progresses.

Figure 7-81: Stair access to roof and edge protection.
Source: RMS.

Figure 7-82: Leading edge protection and crawling boards.
Source: RMS.

Crawling boards

On any fragile or angled roof (greater than 30°) or on any roof that is considered hazardous because of its condition or because of the weather; suitable crawling boards must be used. They must be correctly positioned and secure. If it is obvious that the job requires a progression along the roof, extra crawling boards must be provided and no fewer than two such boards shall be taken on any job.

7.3 Protection of others

Demarcation

When working at height is being carried out it is important to protect others from the hazards of use of access equipment and items or objects falling onto them. A simple way of doing this is to identify hazardous zones with hazard tape and barriers. Inform those that may be affected by the work that is being undertaken and warn of the possible dangers that may arise within the demarcated area.

Clear demarcation of areas where construction work is being undertaken and where it is not will assist those not involved in the work to keep out of the area and avoid exposure to hazards. The area should be demarcated at a distance that relates to the hazards, such as materials or tools falling. A safety margin should always be built into the demarcation to ensure risks are controlled.

Barriers

Barriers are a physical means of preventing access to an area that is required to be restricted, for example, where work at height is being carried out. As such, they are particularly appropriate if people, such as the public, might stray past an area marked only with hazard tape.

These can be situated at a proximity that is relative to the hazards associated with the work being carried out. If necessary to ensure a demarcation zone the barriers may be 2 metre high fencing.

Tunnels

Tunnels can be used to 'isolate' others from any surrounding hazards by enclosing an area (i.e. walkway) and thus preventing any contact with likely hazards until through the danger area.

An example of this could be the ground level of a scaffold on a shop front in a busy high street, being enclosed to form a tunnel that allows the public to pass through the works safely whilst allowing the work at height to continue above.

Figure 7-83: Nets, signs and head protection.
Source: RMS.

Signs

Safety signs are used to provide people with information relating to the works being carried out, to control or divert people and most importantly of any dangers or hazards.

Signs may be situated at a location in advance of the work area to give prior warning in addition to the works perimeter and actual work location. Signs may be used to prohibit public access to work areas where work at height is taking place.

Marking

Where equipment used to gain access to height may be collided with it should carry hazard marking. For example, the standards of a scaffold at ground level on a street may be marked with hazard tape. In situations where scaffolds present a hazard to pedestrians, consideration should be given to providing padding in the form of foam sheaths to minimise injury should a pedestrian walk into one of the standards.

Figure 7-84: Marking.
Source: RMS.

Lighting

Suitable and adequate lighting should be provided to allow the works being carried out and any possible hazards to be seen clearly in advance and allow people to take the required actions to avoid interference with the site.

Scaffolds located near roads may be fitted with lighting to warn traffic of its presence. MEWPs and similar equipment used in poor lighting conditions on or near roads and walkways must use standard vehicle lighting.

Sheeting, netting and fans

Sheeting can protect other people at risk by preventing dust, materials or tools being ejected from work activities conducted at height. Sheeting provides a solid barrier to particles and spray that may result from some work and may be preferred for processes where building surfaces are being cleaned.

However, for most construction activities the provision of fine netting is adequate, being lighter and offering less wind resistance it is often preferable to sheeting. Fans offer a robust means of collecting the material that may breach netting and fall onto the public, who would not be wearing head protection. The fans should extend sufficiently to provide protection over walk areas and be set at an angle that would deter material bouncing or rolling off the fan.

Fans add to the loading of the scaffold and consideration of this factor must be made at the time of scaffold design.

See also - 'Sheeting' - earlier in the section - 'Ensuring stability'.

Head protection

There is a need and legal duty to wear head protection on a construction site where there is a risk of injury from falling materials, this includes when people are working around people at height.

This may necessitate the provision of head protection for people other than those involved in the work at height; this could include other workers, visitors to the site or client's workers.

However, it should be remembered that head protection, such as hard hats, provide the user with limited protection from objects falling from height. Therefore it is best, where practicable, to provide other collective means of protection from falling materials.

Figure 7-85: Fans and sheeting.
Source: RMS.

Sources of reference

Reference information provided, in particular web links, was correct at time of publication, but may have changed.

Inspection and reports, HSE Construction Information Sheet no 47 (rev1) Series Code, CIS47REV1 http://www.hse.gov.uk/pubns/cis47.pdf

Health and Safety in Construction, HSG150, HSE Books, ISBN 978-0-7176-6182-2 http://www.hse.gov.uk/pubns/priced/hsg150.pdf

Health and safety in roof work, HSG33, HSE Books, ISBN 978-0-7176-6527-3 http://www.hse.gov.uk/pubns/priced/hsg33.pdf

Managing Health and Safety in Construction, Construction (Design and Management) Regulations 2015, Guidance on Regulations, L153, HSE Books, ISBN: 978-0-7176-6626-3 http://www.hse.gov.uk/pubns/priced/L153.pdf

Personal Protective Equipment at Work (Guidance), L25, HSE Books, ISBN: 978-0-7176-6139-3 http://www.hse.gov.uk/pubns/priced/l25.pdf

Protecting the public: Your next move, HSG151, HSE Books, ISBN 978-0-7176-6294-4 http://www.hse.gov.uk/pubns/priced/hsg151.pdf

Safe use of work equipment, ACOP L22, HSE Books, ISBN 978-0-7176-6295-1 http://www.hse.gov.uk/pubns/priced/l22.pdf

Working at Height: A brief guide, INDG401, HSE Books, http://www.hse.gov.uk/pubns/indg401.pdf

Web links to these references are provided on the RMS Publishing website for ease of use – www.rmspublishing.co.uk

Statutory provisions

Construction (Design and Management) Regulations (CDMR) 2015 / Construction (Design and Management) Regulations (Northern Ireland) 2016

Personal Protective Equipment at Work Regulations (PPER) 1992 (as amended) / Personal Protective Equipment at Work Regulations (Northern Ireland) 1993

Provision and Use of Work Equipment Regulations (PUWER) 1998 / Provision and Use of Work Equipment Regulations (Northern Ireland) 1999

Work at Height Regulations (WAHR) 2005, as amended 2007 / Work at Height Regulations (Northern Ireland) 2005

STUDY QUESTIONS

1) Scaffolding that incorporates a powered hoist has been erected to the outside of an office block in order to undertake repairs to the external cladding of the building. What could affect the stability of the scaffold?

2) A flat roof of a shop in a busy town centre is to be repaired while the shop remains open. What measures could be taken to reduce the risk to workers involved in the roof repair work and others who may be affected by the work?

3) What should be considered when conducting a working at height risk assessment?

4) What are safe working practices associated with the use of a mobile elevating working platform?

5) What are safe working practices associated with work over water?

For guidance on how to answer these questions please refer to the 'study question answer guidance' section located at the back of this guide.

Element 8

Musculoskeletal health and load handling

Contents

Musculoskeletal disorders and work-related upper-limb disorders

MEANING OF MUSCULOSKELETAL DISORDERS AND WORK-RELATED UPPER LIMB DISORDERS

Work-related musculoskeletal disorders (MSDs) are disorders of various parts of the body. These are caused by work and working conditions. Disorders can affect the muscles, joints, tendons, ligaments, nerves, bones and the localised blood circulation system. Most work-related MSDs are cumulative, resulting from repeated exposure to high or low intensity loads over a long period of time.

The main areas of the body affected are the back, neck, shoulders and upper limbs, but the lower limbs may also be affected. Some MSDs, such as carpal tunnel syndrome in the wrist, are specific because of their well-defined signs and symptoms. Others are classed as non-specific because only pain or discomfort exists without clear evidence of a specific disorder or occupational disease.

Work-related upper limb disorder (WRULD) is a generic term for that group of disorders that affect the neck or any part of the arm from the fingers to the shoulders. Recognised WRULDs include carpal tunnel syndrome and tenosynovitis.

The latest statistics from the HSE show that in the UK:

- The total number of work-related musculoskeletal disorder (MSD) cases in all industries in 2020/21 was 470,000 and an estimated 40,000 were in construction.

- Approximately 1.8% of workers in the construction sector reported suffering from a musculoskeletal disorder which they believed was work-related and this rate is significantly higher than that across all industries.

- The number of new cases of work-related MSDs in all industries was 162,000.

- In terms of the affected area for work-related MSDs, upper limbs or neck was 212,000 (44%), back 182,000 (39%) and lower limbs 76,000 (16%).

Prior to the coronavirus pandemic, the rate of self-reported work-related MSDs in all industries showed a gener-ally downward trend. However, to date, work-related MSDs in Great Britain remains an ill-health related condi-tion that places significant burdens on employers and workers accounting for 28% of all work-related ill-health. In the construction industry it represents 54% of all work-related ill-health.

Manual handling, awkward or tiring positions and keyboard work or repetitive action are estimated to be the main causes of work-related MSDs across all industries. Construction trades frequently named as high risk are plasterers, bricklayers and joiners.

Tasks carried out within construction trades require the use of hand and power tools, involving the use of several body regions, constant movement in awkward positions, and repetitive, forceful, use of the back and upper and lower extremities.

Construction along with agriculture, health and social care, transportation, and storage industries all showed elevated rates of musculoskeletal disorders.

Public Health England (PHE) and the Health and Safety Executive (HSE) recognise musculoskeletal disorders as a significant problem:

> "Musculoskeletal conditions are the leading cause of pain and disability in England and account for one of the highest causes of sickness absence and productivity loss." - PHE

"An estimated 6.6 million working days were lost due to work-related MSDs, an average of 14 days lost for each case. Work-related MSDs represent 24% of all days lost due to work-related ill health in Great Britain in 2017/18." - HSE

Figure 8-1: Prevalence of musculoskeletal disorders.
Source: Public Health England and Health and Safety Executive.

EXAMPLES OF REPETITIVE CONSTRUCTION ACTIVITIES THAT CAN CAUSE BY MSDS AND WRULDS

Anyone who over-uses their back, arms and hands repeatedly may develop a musculoskeletal disorder. Those that are involved in repetitive construction activities are particularly at risk of experiencing MSDs and WRULDs. There are a great deal of construction activities that involve repetitive actions, some of these are considered in the following outlines of how the risk presents itself.

Digging

Hand digging becomes progressively more difficult and potentially more injurious the deeper below foot level the work is carried out. This is particularly the case if the spoil being removed has to be lifted above waist height for disposal.

The use of narrow trenching techniques that can restrict posture significantly will also increase risk of musculoskeletal injuries and work-related upper limb disorder.

Kerb laying

Traditionally, kerbs (in one form or another) have been specified on the majority of roads. The standard components used are principally pre-cast concrete and weigh approximately 67kg.

Feature kerbs, stone kerbs or other associated products may be considerably heavier. More recently a number of manufacturers have developed lighter kerbs.

The main hazards associated with the manual handling of kerbs are the weigh, the repetitive nature of the work and poor posture during the work. These hazards create excessive stress and strain on the body, which can cause damage to muscles and tendons, and in the longer term may lead progressively to more serious musculoskeletal injury.

Movement and fixing of plasterboard

Plasterboard is widely used in construction to line internal walls and ceilings. Workers often have to handle the sheets manually and may need to do so in a restricted space, for example, a stairwell. The sheets, which are typically 2.5 metres x 1.25 metres and weigh 32.5kg, are difficult to grip and unwieldy.

The work is physically demanding and the main hazards involve extended reach, and posture flexibility to enable work in a variety of awkward postures, including on raised platforms.

Placement and finishing of concrete slabs

The placement and finishing of concrete slabs will often involve heavy repetitive work and poor posture due to having to work at the level of the worker's feet. The placement and levelling of concrete labs, may involve lifting and repositioning of each slab more than once, putting strain on the fingers, hands and lower back.

Bricklaying

Bricklayers are often self-employed and therefore cannot afford to lose time from work. They will carry on working and ignore the warning signs, such as aches and fatigue, which are often a precursor to more serious musculoskeletal injuries. The size of the bricks being laid, the number of bricks they are expected to lay (often several hundred per day) and the relative position of a wall or structure are significant risk factors.

Erecting/dismantling scaffolds

The erecting and dismantling of scaffolds, by the nature of the work, will involve lifting and placing of materials progressively higher (when erecting) or lower (dismantling). The work involves the need to maintain balance whilst carrying or placing materials, often involving long reaching to position and secure components.

The risk of overreach related injuries to the hands, arms and spine is high when working with materials of extended length that are held at one end, such as scaffold boards and poles.

Use of display screen equipment

Display screen equipment (DSE) is a device or equipment that has a display screen for graphics, words or numbers. DSE includes both conventional display screens and those used in the latest technologies such as laptops, touch-screens and other similar devices.

Traditional drafting of drawings and plans for construction activities has now been replaced by computer aided design programmes (CAD). CAD systems enable architects to modify and amend drawings as often as is necessary. The designs can be manipulated from any number of elevations and many systems allow the construction of three dimensional models (3D). The designs can be sent electronically to the client for approval, speedily enabling early completion of the work.

Display Screen Equipment is also a fundamental part of Building Information Modelling (BIM). BIM is a process for creating and managing information on a construction project across the project lifecycle. One of the key outputs of this process is the digital description of every aspect of the built asset. This model draws on information assembled collaboratively and updated at key stages of a project.

Display screen equipment users working in design and other similar roles may make many thousands of movements an hour, which will create repeated movement of the fingers. Where the DSE user is not fully proficient in using all fingers to operate the keyboard, many of the movements are centred on a small number of fingers. Where the use of the display screen equipment requires the use of a mouse this can also accentuate the use of a small number of fingers centred on the user's dominant hand. Because of the amount of repeated movements, sometimes without taking breaks, the risks of WRULD are high and the effects are cumulative.

POSSIBLE ILL-HEALTH CONDITIONS FROM POORLY DESIGNED TASKS AND WORKSTATIONS

A worker's body will be affected to a varying degree by poorly designed tasks and workstations that involve bending, reaching, twisting, repetitive movements and poor posture. Frequent repetition of work movements and high force demands on the hand are significant risk factors, especially when they occur together. Bent postures of the wrist at work and low environmental temperature have also been identified as risk factors. These poorly designed tasks and workstations can lead to workers developing ill-health conditions in the form of work-related upper limb disorders.

Figure 8-2: WRULDs caused bricklaying work (repetitive, twist & bend).
Source: RAF Mildenhall.

WRULDs were first defined in medical literature as long ago as the 19th century as conditions caused by *forceful, frequent, twisting and repetitive movements.*

WRULDs are musculoskeletal disorders caused by exposure in the workplace affecting the soft tissues, muscles, tendons, ligaments, nerves, blood vessels, joints and bursae (a small fluid-filled sac which provides a cushion from injury between bones and tendons and/or muscles around a joint) of the fingers, hand, wrist, arm, shoulder and neck.

Common WRULDs include carpal tunnel syndrome, tenosynovitis, tendinitis, peritendinitis, epicondylitis and trigger finger (stenosing tenosynovitis). The conditions are usually caused by repetitive movements and aggravated by excessive workloads, inadequate rest periods and sustained or constrained postures.

This can result in pain, soreness or inflammatory conditions of soft tissues, such as muscles and the synovial lining of the tendon sheath. Clinical signs and symptoms are stiffness or pain in joints and inability to straighten or bend the joints or crepitus (a grating sensation in the joint). In addition there can be aches, pain, tenderness, stiffness, weakness, tingling, numbness, cramp, swelling in the muscles of the arms and neck.

Carpal tunnel syndrome

Carpal tunnel syndrome is a condition where the median nerve that passes through the carpal tunnel in the wrist becomes trapped when tendons or ligaments in the wrist become enlarged, often from inflammation. The narrowed tunnel of bones, ligaments and tendons in the wrist pinch the nerves that reach the fingers and the muscles at the base of the thumb.

The first symptoms usually appear at night due to the wrists being flexed during sleep. Symptoms range from a burning, tingling, numbness in the fingers, especially the thumb and the index and middle fingers, to difficulty gripping or making a fist, to dropping things. It is a common condition caused by repetitive bending of the wrist or movement of fingers or where force is exerted.

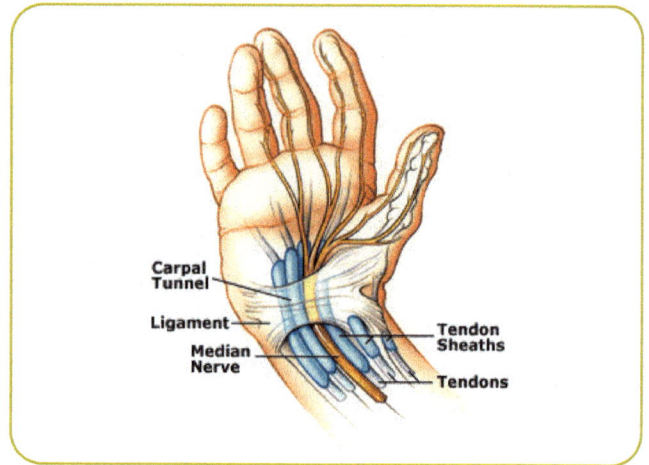

Figure 8-3: Carpal tunnel syndrome.
Source: www.sodahead.com.

Construction activities using hand tools present a risk of this type of harm, such as painting, carpentry or brick laying.

Tenosynovitis

Tenosynovitis is inflammation of the tendon sheath at the wrist. The tendon sheath makes a tiny amount of oily fluid that lies between the tendon and the tendon sheath, which provides lubrication and allowing the tendon to move easily through the sheath. Tenosynovitis occurs when prolonged and repetitive use of the tendon becomes excessive and the tendon sheath can no longer provide sufficient lubricant to the tendon. As a result, the tendon sheath thickens and becomes inflamed. Symptoms include pain and swelling, usually near where the tendon attaches to the bone. It is a common condition often caused by repetitive finger movements, for example, when typing while using display screen equipment.

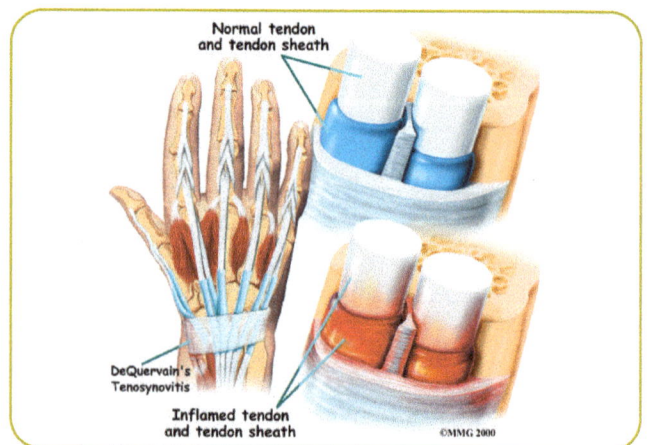

Figure 8-4: Tenosynovitis.
Source: rayur.com.

Tendinitis

Tendinitis involves inflammation of a tendon, the fibrous cord that attaches muscle to bone. It usually affects only one part of the body at a time and lasts a short time. However, in some cases it involves tissues that are continuously irritated. It may result from an injury, activity or exercise that repeats the same movement.

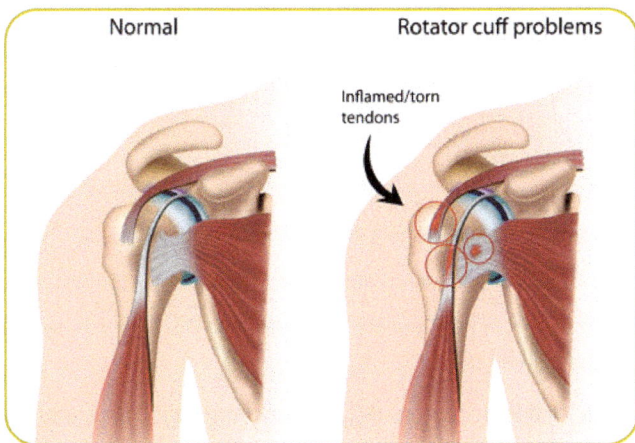

Figure 8-5: Tendinitis.
Source: Beth Israel Deaconess Medical Center.

Rotator cuff tendinitis is one of the most common causes of shoulder pain. Plastering, laying floor screed, or painting may cause rotator cuff tendonitis.

Peritendinitis

Peritendinitis is an inflammation of the sheath that surrounds a tendon. It can be associated with tendinitis, an inflammation of the tendon itself. This condition is most commonly seen in overuse injuries, such as those involving continually repeated tasks (for example, painting, applying sealant using a tube applicator gun and data input into a computer) or for those people who do not get enough rest between tasks or following injury. The primary treatment is resting in order to allow the inflamed tissue to heal.

Epicondylitis

Also referred to as 'tennis elbow' or lateral epicondylitis, this is a condition that occurs when the outer part of the elbow becomes painful and tender, usually because of a specific strain, overuse, or a direct impact (for example, when playing sports, such as tennis). Sometimes no specific cause is found, but it can be caused in the workplace by excessive lifting or twisting at the wrist, for example by using a screwdriver, applying plaster, sawing and painting. Tennis elbow is similar to golfer's elbow (medial epicondylitis), which affects the other side of the elbow.

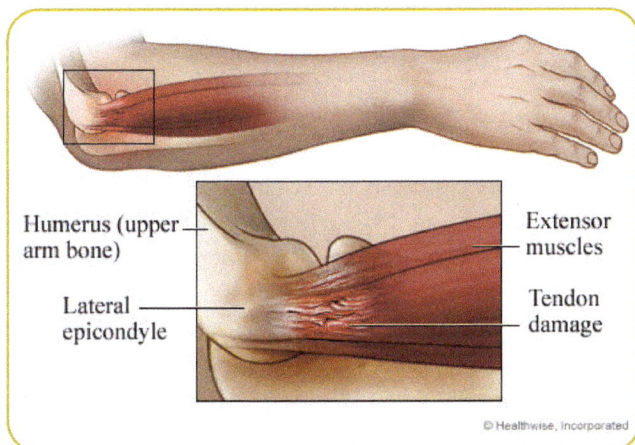

Figure 8-6: Epicondylitis.
Source: Myhealth.alberta.ca.

Trigger finger (stenosing tenosynovitis)

A painful condition that affects the tendons in the hand, which may involve one or more fingers, usually in the dominant hand. It may occur through repetitive movements at work. Trigger finger occurs if there is a problem with the tendon or sheath, such as swelling, which means the tendon can no longer slide easily through the sheath and it can become bunched up to form the nodule. This makes it harder to bend the affected finger or thumb. If the tendon gets caught in the opening of the sheath, the finger can click painfully as it is straightened.

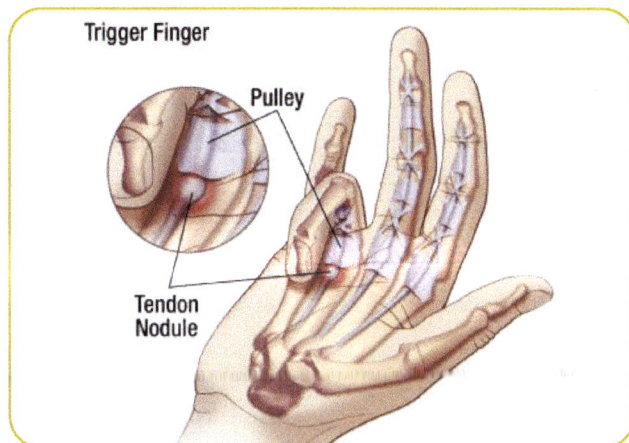

Figure 8-7: Trigger finger.
Source: 3pointproducts.

The occurrence of WRLUDs, varies widely according to the type of work done by workers. High incidences of WRLUD, such as tenosynovitis or peritendinitis, have been reported among construction workers, such as carpenters and roofers.

Industry	Rate per 100,000 workers
Construction	1,830
Human health and social work activities	1,500
All industries	1,130

Figure 8-8: Work related musculoskeletal disorders in Great Britain (WRMSDs), 2020.
Source: Health and Safety Executive Annual Statistics.

Workers who are unaccustomed to hand-intensive work, either as a new worker or after an absence from work, have been found to be at an increased risk. The UK National Health Service reports in their guidance leaflet for employers on upper limb disorders that:

- 1 in 20 adults gets carpal tunnel syndrome.
- 1 in 100 men and 1 in 50 women get tenosynovitis.
- 1 in 75 men and 1 in 90 women get lateral epicondylitis (tennis elbow).

An estimated minimum of 6.6 million working days were lost in Britain due to WRULDs in 2018, with each affected worker taking, on average, 14 days off work. Costs to employers of WRULDs were estimated to be at least £15.0 billion.

The aches, pains and fatigue suffered doing tasks that cause these conditions will eventually impair the worker's ability to do tasks and lead to poor performance. Although present approaches to treatment are largely effective, provided the condition is treated in its early stages, it is essential to consider control measures to avoid or minimise the risk of WRLUDs occurring.

AVOID OR MINIMISING RISKS FROM POORLY DESIGNED TASKS AND WORKSTATIONS

Avoiding or minimising risks from poorly designed tasks and workstations should consider:

- Task - including repetitive or strenuous tasks, the duration of tasks and those that create poor posture.
- Environment - including lighting level or glare and temperature.
- Equipment - including user requirements, adjustability, matching the workplace to individual needs of workers.

Tasks

Anyone who repeatedly uses their hands and arms to perform tasks may be at risk of developing a WRULD. Keyboard operators, workers on factory assembly lines, agricultural workers, musicians, dressmakers, packing workers, bricklayers, supermarket checkout operators, and cleaners are examples of workers at particular risk.

Repetitive

Tasks should be assessed to determine the WRULD risk factors. For example, a task is repetitive in nature if the same series of operations are repeated in a short period of time, such as ten or more times per minute. Repetitive tasks require the same movements to be made using the same muscle groups. If this repetitive movement is carried out over a prolonged period injury may occur to the muscles and ligaments affected.

Many tasks have repetitive movements, for example, stitching pieces of cloth, manufacturing individual parts, packaging individual items, checkout tasks in a super-market, sorting parcels and trimming vegetables. This means that many workers may be exposed to the risks from repetitive movements.

Examples of repetitive tasks found in construction activities include:

- Hammering
- Drilling

- Driving screws
- Sawing
- Laying bricks
- Painting with brushes
- Plastering
- Digging
- Loading and unloading small items – for example, tiles or bricks.

Control measures to avoid or minimise risks of WRULDs from *repetitive* tasks include:

- Breaking up work periods involving a lot of repetition with several short breaks instead of one break in the middle of the work period.

- Allow for short, frequent pauses for very intensive work.

- Use specialist tools to reduce or eliminate part of the repetitive action in the task, for example using an automatic tool to eliminate the manual action and bending needed to tie rebar together with wires.

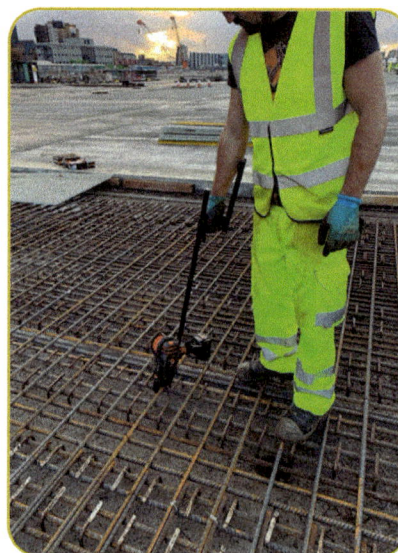

Figure 8-9: Rebar tying gun & extension handle. Source: BAM Nuttall/CCS.

Strenuous

Similarly, tasks that are of a strenuous nature and require a worker to use significant force to carry them out should be considered. For example, building blocks (typically weighing around 10kg) are often used in the construction of walls. The blocks are picked up from a storage stack and moved to their position on the wall, before placing them in position in a controlled manner. In order to do this the worker uses a power grip to hold the block between their thumb, palm and fingers. Tasks such as this require high muscle forces to be applied and can lead to fatigue or injury if there is insufficient time for recovery. The risk is increased if the wrist cannot be held straight when the force is applied, for example resulting from inappropriate working height for the task.

Reduce the amount of **force** required to do the task by such control measures as:

- Reduce the weight of items, or the distance they are moved or slide the item instead of lifting it.

- Provide some means of increasing mechanical advantage, such as levers, or other means of mechanical assistance.

- Distribute force requirements over several fingers rather than one.

- Allow workers to use alternate hands to carry out tasks.

- Enable the stronger muscle groups to be used where possible, for example can a foot operated pedal be used to provide force.

- Ensure any handles and/or controls that are used are well maintained and easy to manipulate without requiring the application of unnecessary force.

- Provide lightweight tools where possible or provide a support, jig or counterbalance to help reduce the force need to use the tool.

- Ensure all tools are well maintained so they operate efficiently and require as little force as is possible to perform the task.

- Ensure the right tools are used for the job, which can reduce the amount of force required to perform tasks.

Duration

The risk of experiencing a WRULD increases with the duration that a task is carried out. Duration refers to the length of time in each shift and the number of working days the task is performed. Duration is an important factor that affects the risk of developing a WRULD. Whilst injury may occur conducting repetitive and strenuous tasks over a short duration if the task requires a lot of effort, work requiring less effort for a long duration is very likely to lead to WRULDs.

Tasks conducted over a long duration could include, holding a polishing tool for a long time while cleaning, drivers keeping their head in a set posture while driving a vehicle for long distances, grasping small tools to carry out medical or dental tasks or holding construction material above a worker's head while another worker fixes it in place.

It is generally accepted that many types of WRULD are cumulative in nature, in that they develop progressively as tasks are carried out over time in situations where the parts of the body doing the work have insufficient time for recovery. Therefore, when the duration of a task is increased the risk of developing a WRULD is increased.

Tasks conducted for a short duration are unlikely to create a significant risk of developing WRULDs, except where the task is exceptionally demanding and/or workers have not been allowed to get accustomed to the demands of the task. This can occur when a worker returns to work after a holiday or if there is a significant increase in the speed of doing the task.

Reduce the effects of the duration of the task to help to avoid or minimise WRULDs by such things as:

- Develop a work/rest regime that provides sufficient time for recovery.

- Share a high-risk task among a team by rotating workers between tasks (each task needs to be sufficiently different to benefit the worker).

- Allow workers to carry out more than one different step of a process, this could reduce the duration of exposure to a WRULD provided the steps are sufficiently different from each other.

- Introduce short frequent breaks for the task, but not necessarily a rest, for example a chance to change posture for a short period.

- Allow workers a short period to get accustomed to the duration of work when starting work for the first time and after return from a holiday or illness.

- Monitor and manage overtime working.

Posture

Tasks that require workers to adopt a particularly awkward posture or to hold a **fixed posture** for a period of time present a high risk of causing WRULDs.

An **awkward posture** is where a part of the body, for example, an arm joint, is used significantly beyond its neutral position. For example, when a worker's arm is hanging straight down with the elbow by the side of the body, the shoulder is in a neutral position. However, when a worker is performing overhead work, such as repairing equipment or accessing objects from a high shelf, their shoulders are moved significantly from the neutral position. When awkward postures are adopted, because muscles are less efficient at the extremes of the joint range, additional muscle effort is needed to do the task. The awkward postures can also lead to compression of soft tissue structures and muscle damage.

Awkward postures include:

- Posture that is unbalanced or asymmetrical.

- Postures that require extreme joint angles or bending and twisting.

- Squatting while servicing plant or a vehicle.

- Working with arms overhead.

- Bending over a desk, table or workbench.

- Using a hand tool that causes the wrist to be bent to the side.

- Kneeling while working concrete to a smooth finish or laying carpet.

- Bending the neck or back to the side to see around bulky items pushed on a trolley.

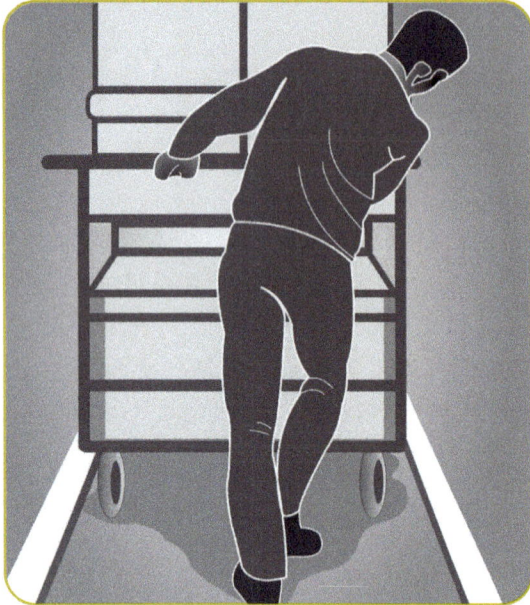

Figure 8-10: Awkward posture.
Source: Australia, Comcare.

Fixed postures occur when a part of a worker's body is held in a fixed position for extended periods without the muscles and other soft tissues being allowed to relax. Fixed postures require the continuous contraction of the same muscles to maintain the posture of a part of the body or to hold the amount of force being exerted constant. For example, when holding and carrying a container or box it is likely that the hands and arms are in a fixed posture. A fixed posture restricts blood flow to the muscles and tendons resulting in less opportunity for recovery and metabolic waste removal. Therefore, muscles held in fixed postures fatigue very quickly. When tasks involve fixed postures and high muscle effort, fatigue will usually force the worker to take a rest. With lower muscle effort the level of fatigue is not so evident, which can lead the worker to spend too long in the same posture without taking a rest, resulting in the development of WRULDs.

Working with the arms extended above the shoulder or extended above the head is common in maintenance and construction work, for example when painting a ceiling or installing light fittings. When working with the arms stretched upwards the small shoulder muscles have to do more in order to hold the weight of the arms.

Figure 8-11: Fixed posture.
Source: Australia, Comcare.

The force on the shoulder muscles is extremely high if the worker also holds a tool or materials in their hand at a distance from the shoulder. In some situations, workers also have to bend their neck backwards to see the work being done, which also strains the neck. These situations significantly increase the risk of shoulder and neck disorders.

Figure 8-12: Poor posture.
Source: Speedy Hire Plc.

Reduce the *awkward and fixed postures* of tasks to help to avoid or minimise WRULDs by such things as:

- Introduce short frequent pauses in the task to provide a chance to change posture for a short period.

- Reduce the size/weight of items that need to be held in a fixed position.

- Provide mechanical devices to hold items raised at height ready for fixing.

- Provide power tools to enable the fixing task to be done quickly.

- Revise the task to enable variation of work to avoid long periods in awkward or fixed postures.

- Reduce the distance items are held away from the body by better layout of workstations.

- Provide equipment, for example a platform, to enable workers to work at the same height as equipment they are maintaining.

- Enable equipment to be easily removed from where it is located to enable maintenance to be conducted without awkward or fixed postures.

- Use specialist tools to avoid the need for awkward postures, for example powered screeding equipment to prevent the need to kneel and bend when screeding concrete.

Figure 8-13: Power Screed.
Source: Northrock

Environment

Working in extremes of **temperature** will make doing simple tasks more fatiguing. In cold environments the circulation of blood to muscles is impaired and maximum strength capacity is diminished, which makes it harder to carry out tasks that are normally relatively easy to do. Cold hands can also result in reduced sensation and more muscle force being needed to hold tools and materials, especially if the worker has to wear gloves. This can lead to the more rapid onset of fatigue and to the development of disorders.

When conducting tasks in high temperatures it can cause workers to sweat, which may cause their hands become slippery with sweat and make it difficult for them to handle things. This can lead to poor grip that requires more muscle force or could cause a sudden movement when their hands slip and lose control of what they are holding, resulting in damage to muscles or other soft tissue.

The **lighting** requirement of tasks are important to consider as the posture a worker adopts to do a task can be influence by their need to see the work. Their visual requirements and posture may be affected by inadequate level of light, shadow, glare, reflections or flickering light.

They can cause workers to adopt a bent neck and poor shoulder postures to enable them to see their work, leading to an increased risk of developing WRULDs.

The work environment may include workstations that have sharp edges or hard surfaces. If the work requires the worker to be in contact with the workstation it can put local pressure on the area of the workers body in contact with the edges and surfaces. If the contact causes significant pressure and/or lasts too long, it can restrict circulation and causes compression of underlying nerves and soft tissue.

Reduce the **environmental** issues affecting tasks to help to avoid or minimise WRULDs by such things as:

- Make sure that the temperature is comfortable.

- Avoid positioning workstations too near to air vents.

- Ensure task illumination is at a level that allows the worker to comfortably view the work without altering their posture. Avoid extreme levels of lighting – low levels (typically below 100-200 lux) or high levels (typically greater than 800 lux).

- Make sure general lighting is good or provide a specific light source for the task.

- Avoid reflected light, for example sunlight on computer monitors.

- Avoid glare by moving light sources, providing blinds on windows or moving workstations.

- Avoid shadows or glare.

- Organise the layout and positioning of items on workstations to minimise the worker having to reach, twist, bend or stoop, for example when collecting components and assembling them.

- Avoid temporary arrangements where materials and equipment are placed on the floor requiring a worker to kneel and bend to conduct tasks, for example where a welder might weld material during part of a maintenance task.

- Avoid tasks that require a worker to be contact with sharp edges and hard surfaces.

- Improve the layout of the workplace to avoid tasks requiring high muscle force in awkward or fixed postures.

Equipment

Equipment may be unduly heavy or not designed to take account of the different size, shape or strength of workers. Objects and attachments may act as obstacles causing the worker to reach over or round them.

Where a workstation has confined space for the worker's legs this can cause poor posture. Seats provided may not be adjustable or where they are adjustable there may not be enough space to make effective use of the adjustable features. Equipment and materials may be positioned so they are not in easy reach of the worker.

Matching the workplace to the individual needs of workers

In managing risks from WRULDs a number of changes may need to be made and ergonomic solutions should be given first consideration. This means matching the workplace, work equipment and the work to the individual needs of the worker, rather than making the worker adapt to fit the workplace, work equipment and work undertaken. Equipment design should therefore take into account the *ergonomic requirements of the worker* and, where possible, allow the worker to *adjust* any settings to suit their needs, for example, setting a workbench height, positioning storage trays for materials used to assemble parts and adjusting the height of their computer screen.

Ergonomics (derived from the Greek *ergon*, meaning work, and *nomos*, meaning natural law) is the scientific study of human work. It is a broad area of study that includes the disciplines of psychology, physiology, anatomy and design engineering. Ergonomics places the human being at the centre of the study, where individual capabilities and the human potential for making mistakes are considered. The fundamental principle of ergonomics is that good job design, by reducing worker error, fatigue and discomfort, is likely to lead to maximum effectiveness.

When matching the workplace to the individual needs of workers it is necessary to take into account that individuals have different physical capabilities due to gender, height, weight, age and level of fitness. This will usually meant making adjustments to limit the risk of developing WRULDs.

New workers, particularly young workers, and those returning to work from a holiday, sickness or injury, may need to be introduced to a slower rate of work than other workers. This enables them to get their body used doing tasks that use their upper limbs, assimilate training more effectively and develop good work practices before having to concentrate on working fast.

Reduce the *equipment* issues affecting tasks to help to avoid or minimise WRULDs by such things as:

- Design workplaces and equipment (including tools) for workers of different sizes, shape, strength and for left-handed workers.

- Tools should have comfortable handles and make

the work easier – avoid rigid hard surfaced handles, sharp edges or narrow handles that put pressure on the hand when they are gripped.

- Tool handles should enable a straight wrist posture (like when a handshake is given) and avoid awkward hand and wrist postures.

- Provide tools with a suitable size of grip, consider the effect of wearing gloves.

- Tools like pliers should not require a wide hand span to use them.

- Use powered tools to reduce the force a worker has to apply to do a task.

- Arrange the position, height and layout of the workstation so that it is appropriate for the workers to do the work - enable work to be done with the joints at about the mid-points of their range of motion.

- Use work fixtures and jigs to hold the work in more accessible positions and to make it easier to see the work to be done.

- Improve the location of objects or attachments to prevent them acting as obstacles and causing poor posture.

- Place equipment and materials that are need more frequently nearest to the worker.

- Ensure seats are adjustable.

- Ensure that there is sufficient space to enable workers to stretch, make changes in leg and foot posture and make effective use of the adjustable features of their chairs.

- Provide platforms, adjustable chairs and footrests for smaller worker to achieve optimal working height to do the work.

- Standing workstations should be used for jobs that require a lot of body movement and greater force.

Figure 8-14: Frequent excessive bending to put mortar on trowel. Source: HSE, COH01.

Figure 8-15: Raised spot board avoids excesive bending.
Source: HSE, COH01.

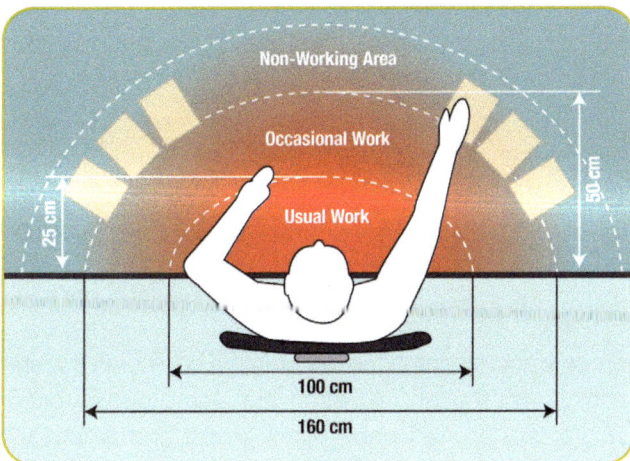

Figure 8-16: Ergonomic workstation assessment.
Source: RMS.

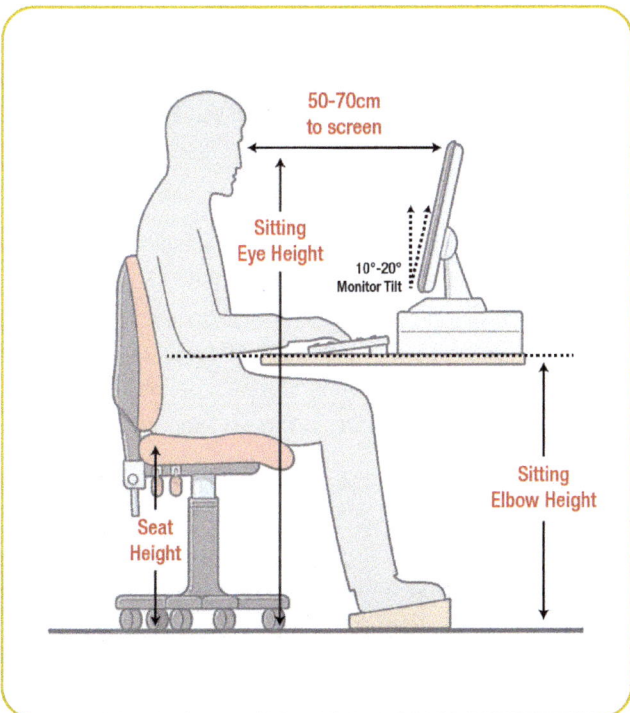

Figure 8-17: DSE user/workstation assessment.
Source: RMS.

CASE STUDY

A call centre company identified the need to display larger characters on a call centre operator's screen. The company acquired and installed suitable software to enable this. Following the change, the operator achieved the same efficiency levels as the rest of the other centre operators and commented that "it enabled me to feel I was making a real contribution to the overall team effort."

REVIEW

What typical work tasks may result in poor posture being adopted by the worker?

What is meant by the term ergonomics?

CONSIDER

In a workplace of your choice, review how an understanding and application of ergonomic principles might improve a work task you have identified.

8.2 Manual handling hazards and control measures

COMMON TYPES OF MANUAL HANDLING INJURIES

In the UK, around 18% of all injuries reported to the Health and Safety Executive, have been attributed to manual handling operations, including the manual lifting and handling of loads. Back injuries generally constitute 39% of musculoskeletal disorders.

Manual handling injuries arise from **hazardous events** such as stooping while lifting, holding the load away from the body, and undertaking twisting movements of the trunk of the body or frequent or prolonged effort. Injuries can also arise from manually handling heavy/bulky/unwieldy/unstable loads, or loads which have sharp/hot/slippery surfaces. Other injuries can be caused by workplace space constraints and the lack of capability of the individual. Manual handling operations can cause many types of **injury.**

The most common injuries are:

- Prolapsed (herniated) spinal disc.
- Muscle strain and sprain.
- Torn or overstretched tendons and ligaments.
- Rupture of a section of the abdominal wall can cause a hernia.
- Loads with sharp edges can cause cuts.
- Dropped loads can result in bruises, fractures and crushing injuries.

Figure 8-18: Manual handling.
Source: Shutterstock.

Damage to intervertebral disc

The spine is made up of individual bones (vertebrae) separated by intervertebral discs. Strain on the disc can cause it to bulge and protrude out between the vertebrae. This is known as a prolapsed or herniated disc. The disc can then press on nearby nerves or the spinal cord. If the fibrous case of the disc collapses the disc can lose its shape and compress – this can cause bones in the spinal column to touch or it can increase pressure on the spinal cord.

The above conditions can cause pain and inflammation in the area of the spinal cord where pressure occurs. Pain may also be felt in the area of the body that the part of the spinal cord affected relates to, such as in the lumbar region of the spine which affects the sciatic nerve (this condition is called sciatica). The damage to the disc is usually caused by too much pressure being applied to it during lifting and handling operations that involve 'top-heavy' bending. This is where the knees are not bent sufficiently and the head and upper body are bent over, causing the spine to bend in a curved manner (sometimes called stooping). This bending creates a high degree of leverage force on the base of the spine, which leads to the extreme pressure exerted on the disc. The force and therefore pressure on the disc is accentuated by the carrying of heavy loads.

Though top-heavy bending is a particular cause of this damage, poor posture due to leaning over for a sustained period could lead to similar damage.

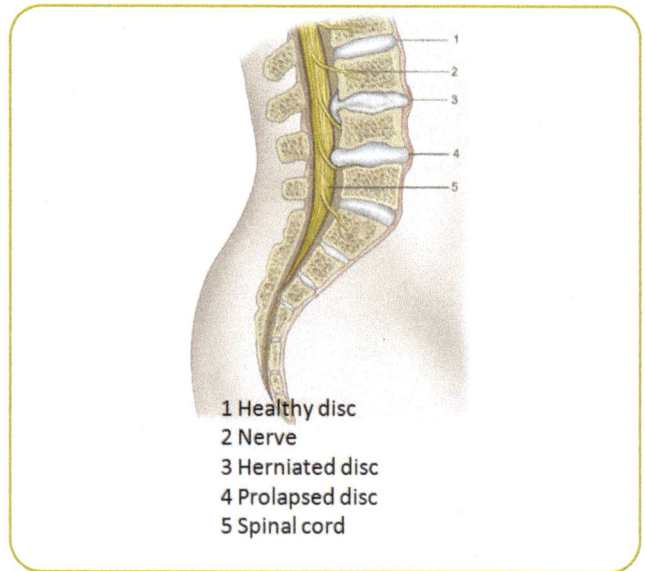

1 Healthy disc
2 Nerve
3 Herniated disc
4 Prolapsed disc
5 Spinal cord

Figure 8-19: Lumber spine illustration.
Source: NHS.

Sprains/strains, fractures and lacerations

The most common injuries that arise from manual handling operations are sprains, strains, fractures and lacerations.

Sprains and strains

Sprains and strains often occur in the back or in the arm and wrists, though injuries in the legs can also occur where the leg has been hyper flexed and where the soft tissues (ligament, tendons and muscles) are overstretched. This can occur through kneeling or handling loads beyond the worker's limit.

- A *sprain* is an injury to a ligament. The ligament is tough fibrous tissue that connects a bone to another bone. Ligament injuries typically involve overstretching or a tearing of this fibrous tissue, i.e. a 'torn ligament'.

- A *strain* is an injury to either a muscle or a tendon that connects muscles to bones. Depending on the severity of the injury, a strain may be a simple overstretch of the muscle or tendon or it may result in a partial or complete tear. Rupture of the muscles in a section of the abdominal wall can cause a hernia.

Sprains and strains tend to occur when a person's body has been overloaded due to a large steady load being applied, a sudden smaller load being applied without the opportunity of the ligament or tendon to stretch, or a part of the body being forced to move in an unusual way. This may be because a person is reaching over to an extreme level (sometimes even without picking up a load, as their own body weight may be sufficient to cause damage) or two people picking up a load together in an uncoordinated way. It could also be due to a load slipping that causes someone to move into an awkward position to prevent it falling completely.

Fractures and lacerations

A fracture is a break in a bone, which is usually the result of trauma where the physical force exerted on the bone is stronger than the bone itself. Fractures of the hands and feet are the most likely type of fractures to arise from manual handling work and may be due to the load that is being lifted accidentally being dropped. Lacerations are often caused by the unprotected handling of loads with sharp corners or edges, or a person's grip slipping when trying to prevent dropping a load.

GOOD HANDLING TECHNIQUE FOR MANUALLY LIFTING LOADS

Lifting techniques using kinetic handling principles

Kinetic lifting is a method of lifting that takes into account the body's characteristics and how it works most effectively. It ensures that lifting is carried out without placing unnecessary strain and pressure on vulnerable body parts. The principles of good lifting and moving techniques have been established for some considerable time. The basic principles are:

- **Think before lifting/handling.** Plan the lift. Can handling aids be used? Where is the load going to be placed? Will help be needed with the load? Remove obstructions such as discarded wrapping materials. When a load is to be lifted or lowered a significant distance, consider resting the load midway on a table or bench. This will provide an opportunity for the worker's body to recover and to renew their grip of the load

- **Keep the load close to the waist.** Keep the load close to the body for as long as possible while lifting. Keep the heaviest side of the load next to the body. If a close approach to the load is not possible before starting to lift it, try to slide the load towards the body before attempting to lift it.

- **Adopt a stable position.** The feet should be apart with one leg slightly forward to maintain balance (alongside the load, if it is on the ground). The worker should be prepared to move their feet during the lift to maintain their stability. Avoid tight clothing or unsuitable footwear, which may make this difficult.

- **Get a good hold of the load.** Where possible, the load should be hugged as close as possible to the body. This may be better than gripping it tightly with hands only.

- **Start in a good posture.** At the start of the lift, slight bending of the back, hips and knees is preferable to fully flexing the back (stooping) or fully flexing the hips and knees (squatting).

- **Do not flex the back any further while lifting.** This can happen if the legs begin to straighten before starting to raise the load.

- **Avoid twisting the back or leaning sideways,** especially while the back is bent. Shoulders should be kept level and facing in the same direction as the hips. Turning by moving the feet is better than twisting and lifting at the same time.

- **Keep the head up when handling.** Look ahead once the load has been held securely, not down at the load.

- **Move smoothly.** The load should not be jerked or snatched as this can make it harder to keep control and may increase the risk of injury.

- **Do not lift or handle more than can be easily managed.** There is a difference between what people can lift and what they can safely lift. If in doubt, seek advice or get help.

- **Put the load down, then adjust its position.** If precise positioning of the load is necessary, put it down first, then slide it into the desired position.

AVOIDING OR MINIMISING MANUAL HANDLING RISKS

Avoiding manual handling risks

Wherever reasonably practicable, manual handling should be avoided, for example, by redesigning the task to avoid moving the load, by automating the process or mechanising it by using mechanical handling aids. Decisions on whether to use automation or mechanisation are best taken at the design stage, or when processes are being updated, as they can involve complex arrange-

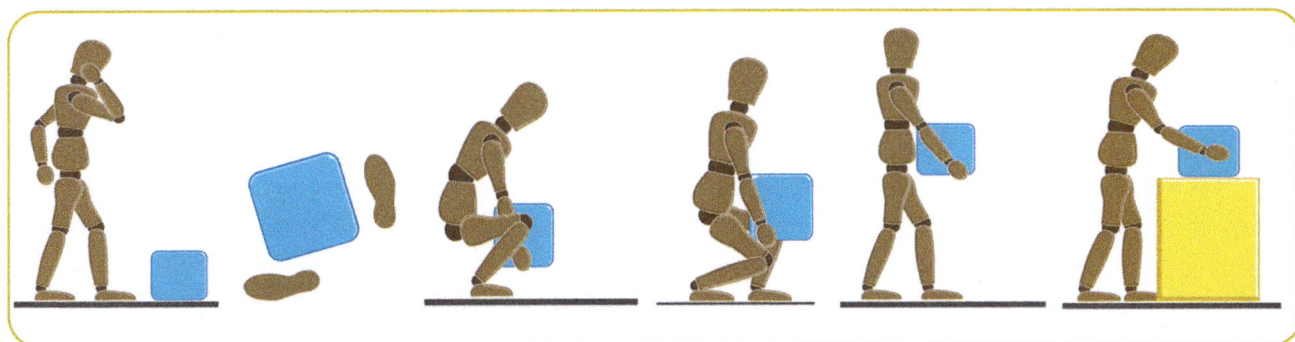

Figure 8-20: Basic lifting principles.
Source: HSE, Guidance L23.

ments and capital expenditure, for example, where the process changes from handling raw materials in sacks or bags to bulk delivery, or where storage changes to using gravity or pneumatic transfer systems.

Avoiding moving the load

In some situations, manual handling can be avoided by not manually moving the load at all. For example, if a heavy load needs to be worked on it might be possible to bring the process to the load, for example, bringing non-destructive testing equipment to where the load is. Prefabrication of construction structures, for example, timber frames, eliminates the need to manually handle timber and assemble frames on site, allowing the prefabricated timber frame to be brought to site and located in position using mechanical aids.

Automation

Whilst automation is a relatively new aspect of construction activities, developments are being made to provide automated equipment that can carry out a range of construction tasks on site, for example bricklaying and concrete surface finishing. In addition, automation of the off-site manufacture of prefabricated house building components, for example timber frames for walls, is well established.

Figure 8-21: Automatic bricklayer.
Source: The Constructor.

Mechanisation

This involves the use of handling aids. Although mechanisation may retain some elements of manual handling, mechanisation means bodily forces are applied more efficiently. Examples are:

- Hoists – Can raise and support the weight of a load, allowing the load to be moved to a new position.

- Trolley, sack truck, roller truck or hoist – Removes the need to carry load and reduces effort to move loads horizontally.

- Chutes and roller conveyors – A way of using gravity to move loads from one place to another.

- Mechanical positioning equipment – Supports the weight of a load and allows a worker to locate and fix it in position, for example window positioning equipment and vacuum kerb positioning equipment.

- The use of an exoskeleton to assist with manual handling tasks.

Figure 8-22 shows a method of moving roofing tiles from ground level to roof height using a hoist system, which eliminates the need to carry tiles up a ladder. *Figure 8-23* shows vacuum kerb moving and positioning equipment, which eliminates the need to lift kerbs, the trolley design reduces the effort to move and position the kerb.

Figure 8-22: Brick/tile hoist.
Source: Mace Industries Ltd.

Figure 8-23: Vacuum lift for heavy concrete kerbs.
Source: The Vacuum Lifting Company.

Minimising manual handling risks

Where risks cannot be avoided, control measures should be put in place to minimise the risk of manual handling injuries to the lowest level reasonably practicable. Practical measures that may be taken to reduce the risk of injury can be decided by considering the task, individual, load, and environment (TILE). This should be carried out with a view to following ergonomic principles and matching the manual handling operation to the individual needs of workers rather than the other way round.

Task

Many tasks have repetitive movements, for example, stacking materials, digging an excavation or laying bricks. This means that workers carrying out these tasks may be exposed to the risk of harm from *repetitive manual handling movements*. Repetitive movements take place when manual handling tasks are made up of a sequence of actions, of fairly short duration, which are repeated a significant number of times and are almost always the same. This requires the same muscle groups to be used repeatedly. Rapid or prolonged repetitive movements may not allow sufficient time for recovery and can cause muscle fatigue or inflammation of soft tissues.

Awkward manual handling movements are where a part of the body is moved well beyond its neutral position, for example, bending at the waist and twisting. Awkward bending movements often takes place where workers do not adopt good lifting techniques, particularly where features of the workplace encourage the worker to use a top heavy bending action, which involves bending at the waist to move a load, for example removing a load from a low storage place. This type of bending action puts severe strain on the lower back and should be avoided. A twisting motion of the spine of 45 degrees will result in a 10% reduction in lifting capability, while a twisting motion of 90 degrees will result in a 20% reduction. Twisting motions can also put severe strain on the lower back and workers should not be expected to be able to lift loads of the same weight as they might be capable of if lifted directly in front of them.

Figure 8-24: Awkward movements.
Source: HSE, HSG60.

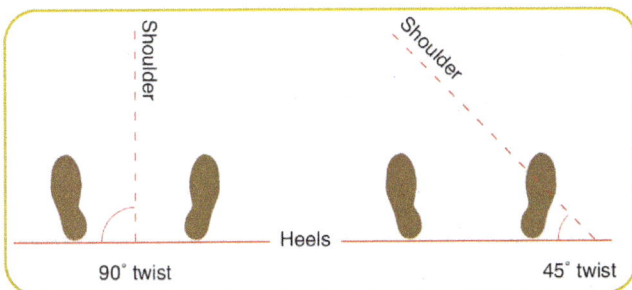

Figure 8-25: Twisting.
Source: HSE, Guidance L23.

When team handling is used for tasks that are beyond the capability of one person it should be remembered that the proportion of the load carried by each member of the team will vary.

Therefore, the load that can be handled by a team will be less than the sum of the loads which each individual could cope with. This should be taken into account when planning to lift loads.

It is important that where it is reasonably practicable risks arising from manual handling tasks are minimised. Consideration should be given to control measures that:

- Avoid large vertical movements - provide somewhere for the load to be placed/rested at about the mid-point in the vertical movement, this will enable recovery of muscles and a change of grip before completing the vertical movement.

- Hold loads close to the body - the use of protective clothing can reduce the risk of manual handling injury as it helps workers to hold loads close to their body without affecting their own clothes.

- Avoid awkward postures like twisting, stooping or reaching upwards – improve the workplace layout and how it is organised, for example avoid storing items at a low level and ensure workers have room to move their feet to avoid twisting.

- Avoid lifting from floor level or above shoulder height, especially heavy loads - organise where loads are placed, put them on a shelf or place on a device that allows the load to be lowered, if necessary, and then raised before lifting it again.

- Reduce carrying distances – restrict carrying distance to less than 10 metres before a rest is taken, use a manual handling aid when moving a load long distances, for example a trolley or wheeled cage.

- Avoid repetitive handling or minimise the effects - vary the work to enable one set of muscles to rest while others are used, reduce the speed required for the repetitive movements to reduce the effort required over a period of time and allow pauses in the work to enable recovery.

- Allow sufficient rest for demanding tasks – strenuous tasks that are carried out for too long or too frequently need rest time to enable muscles affected to recover.

- Allow the work rate to be controlled by the worker – to enable them to identify the need for and take sufficient rest.

- Ensure loads moved while seated remain within the weight capacity of the worker and regard is taken of good practice guidance - for example, limiting loads to 5kg for men and 3kg for women when moved close to their body.

- Push or pull loads instead of lifting and carrying them - generally pushing and pulling loads imposes less loading on the body.

- Avoid strenuous pushing or pulling - push rather than pull and make sure trollies and containers are not overloaded.

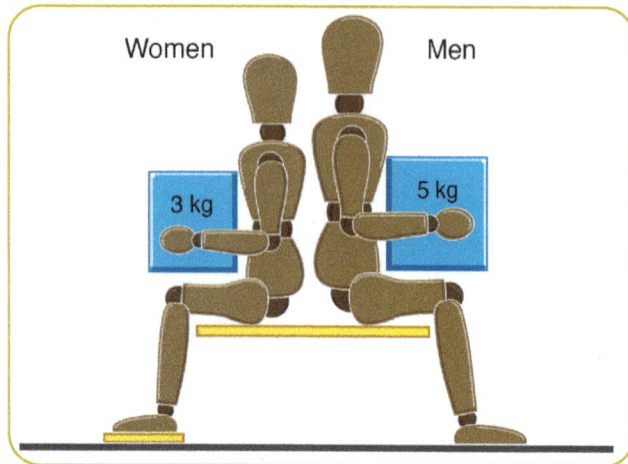

Figure 8-26: Handling while seated.
Source: HSE, Guidance L23.

Figure 8-27: Pushing load.
Source: RMS.

Individual

An individual's physical capability to manually handle loads could significantly affect their ability to perform a manual handling task safely. Their capability can be influenced by a number of factors, including their gender, age, medical condition (general health, previous injuries, pregnancy). In addition, the training and information they are provided with will influence how they carry out tasks. Also, the type of clothing workers wear may inhibit good movement or, if the worker does not have suitable clothing to protect their own clothes, they may not want to hold the load close to their body because it might make their clothes dirty or damage them. Holding a load away from the body greatly influence the forces on the body, for example on the lower back.

To minimise the manual handling risk factors related to the individual consideration should be given to control measures that:

- Ensure workers' clothing and footwear is suitable for their work - provide protective clothing or PPE that does not restrict movement and allows the worker to hold loads close to their body.

- Consider the worker's capability before assigning manual handling tasks to them - a worker's state of health can significantly affect their capability to perform manual handling tasks safely, also, some age groups tend to have reduced physical capability (particularly those under 20 years of age and those above 45 years of age).

- Prevent workers with a health problem or physical disability carrying out manual handling work that is likely to endanger them and assign them to other work that suits their capability.

- Prevent pregnant workers carrying out manual handling work that is likely to endanger them or their unborn child and assign them to other work that suits their capability.

- Provide workers with relevant manual handling information - information on the load and the task, for example the weight of the load and specific precautions the worker must take before and during manual handling tasks are carried out.

- Provide manual handling training to workers and their managers – provide general training (including risks, injuries and control measures, including good manual handling technique) and specific training for more unusual tasks.

- Consider providing workers with an exoskeleton to assist them in carrying out manual handling tasks. *Figure 8-28* shows a worker wearing an unpowered exoskeleton.

Figure 8-28: Exoskeleton.
Source: SuitX.

Load

The nature of a load can contribute to increased manual handling risks due to its size, weight, centre of gravity, stability, how easy it is to grip and any hazardous surfaces (sharp, hot, substance contamination) or contents it may have. Manual handling risks related to the load can be minimised by considering the design of the load, for example, designing it to be smaller by concentrating the substances contained in it or breaking the load up into suitable, smaller sized containers. It might also mean designing in handles or features that make it easier to grip the load, such as 'sticky grip' areas on plastic sacks.

To minimise manual handling risks related to the load, consideration should therefore be given to control measures that take account of the load's weight, size, grip and whether it is sharp or hot and its stability. Each of these risk factors and control measures to minimise the risks are outlined below.

Weight

Regulation 4 of the Manual Handling Operations Regulations (MHOR) 1992 states that employers shall:

"take appropriate steps to provide any of those employees who are undertaking any such manual handling operations with general indications and, where it is reasonably practicable to do so, precise information on –

(a) the weight of each load, and

(b) the heaviest side of any load whose centre of gravity is not positioned centrally."

Figure 8-29: Requirement to provide an indication of weight of loads for manual handling.
Source: Manual Handling Operations Regulations (MHOR) 1992.

Generally, the heavier the load the higher the risk of injury. Therefore, risks can be minimised by ensuring workers only manually handle loads of a reasonable weight. Many countries have used simple maximum weight limits for the lifting of loads. However, this has been found to be an over-simplistic approach. Research has identified that it is not possible to set a simple, specific weight limit for all manual handling tasks because the individual capability of workers and the task being conducted greatly influence what is an acceptable weight. Many of the weight limits set by countries have tended to be too high for average workers and work practices.

Work in the UK has confirmed that the load weight limit adopted by the USA (23kg) is an effective weight limit to reduce risk and the maximum weight limit approach was adapted in the UK to reflect the fact that workers may lift from different positions and that there are differences in the lifting capacity of women and men. This work has been published as the UK Health and Safety Executive's (HSE) lifting guidelines. In the UK, therefore, a single maximum weight limit has been removed from legislation and replaced by the HSE lifting guidelines. The UK guidelines set out an *approximate* boundary within which manual handling operations are unlikely to create a significant risk of injury.

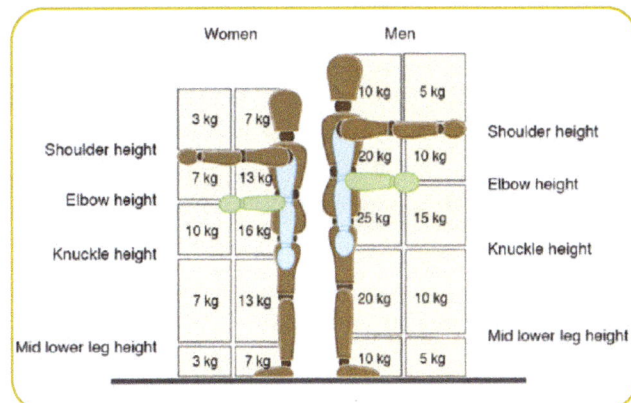

Figure 8-30: Lifting and lowering.
Source: HSE, Guidance L23.

The UK guideline figures are not absolute weight limits. They may be exceeded where a more detailed assessment shows it is safe to do so. However, the guideline figures should not normally be exceeded by more than a factor of about two. The guideline figures for weight will give reasonable protection to nearly all men and between one half and two thirds of women.

It should be remembered that one of the main methods used to reduce the risks from manual handling is to change the load by repackaging it into smaller weights or breaking a bulk load down into smaller batches. This solution avoids the need to lift heavy loads. However, it will increase the frequency of manual handling movements and may lead to a significant repetitive movement risk.

It is often better to find ways of using lightweight materials, for example, replacing pre-cast concrete kerbs and drainage blocks with plastic equivalents.

The HSE explain the benefits of changing the material and design of a load in their case study related to trench blocks used for straight runs of foundations.

Figure 8-31: Lightweight plastic kerbs.
Source: HSE, COH14.

Large heavy trench blocks were used to build foundations. The combination of block weight and poor posture when working in a restricted space exposed the workers to manual handling risks and finger trapping. Lighter trench blocks (weighing less than 20kg) with handholds were designed and made available in place of the traditional heavier units.

Where the weight of a load is beyond the capability of one person the worker should get assistance from another worker or use a manual handling aid, for example vacuum lifters for concrete kerbs and paving.

Size

Large (wide, tall or bulky) loads could make it difficult for workers to see where they are going when moving a load. Also, in order to get a wide or bulky load close to the body, the worker has to open their arms to reach around the load and hold it. Therefore the size of the load may make it hard for a worker to hold the load when it is being moved. Also, the worker's arm muscles cannot produce as much holding force with the arms held open reaching round a load as when the arms are held closer together. In this situation, the muscles will get tired more rapidly. To enable the worker to keep the load as close to the body when lifting and carrying it the size of loads should be limited to enable the worker to hold it comfortably. This could include designing the load to be smaller through concentration of substances contained in it or breaking the load up into suitable-sized containers. Where this is not possible the worker should get assistance or use a manual handling aid.

Grip

Loads that are difficult to grip can result in the load slipping, causing sudden movement of the load. Ordinary gloves usually make gripping a load more difficult than gripping it with bare hands. Special manual handling gloves that have additional surfaces designed to provide grip will help to minimise this risk. Risks can also be minimised by providing the load with handles or using aids for gripping, for example, suction pads when carrying plate material or hand–held hooks for soft material.

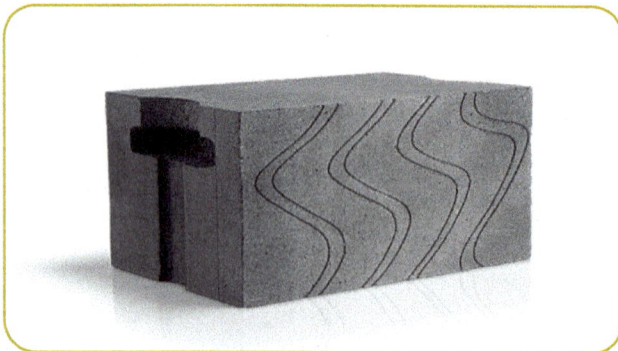

Figure 8-32: Trench block with 'handholds'.
Source: Forterra.

Where the load is contaminated with substances that make griping the load more difficult it may be necessary to clean the surface to remove slippery substances on it before moving it. Where the packaging material makes it difficult to grip the load, it may be possible to change the material or add a slip resistant material to the surface, such as 'sticky grip' areas on plastic sacks.

Sharp, hot, cold, or harmful

Any loads to be handled that have hazardous substances on their surface, sharp corners, jagged edges or rough surfaces increase the risk of workers' injuring their hands when they handle them and may cause the worker to drop the load suddenly. Loads containing hazardous substances or loads that are hot/cold can also injure workers, especially in the event of a collision. Personal protective equipment should be provided to protect the worker or the load should be covered in suitable protective material before moving it. In some cases, it may be appropriate to use handling aids to hold the load or use alternative means to move the load.

Stability

Unstable loads or loads with moving contents, such as a liquid, can cause uneven loading of the muscles and sudden movements of the load can make workers lose their balance and fall. Where possible, the moving parts of a load, for example components in machinery, should be prevented from moving before lifting. Where a load contains a liquid, consider whether the liquid can be emptied out before moving the load or if this is not possible consider the design of the container and whether baffles can be added to restrict movement of the liquid.

Some loads may be unbalanced because the centre of gravity of the load is not where expected and a worker may try to move the load by holding it with the centre of gravity too far away from their body or too much to one side. This can lead to uneven loading of muscles, damage to soft tissues and fatigue. It is important that where the centre of gravity of a load is in a position that is not obvious, sufficient information is given to the worker before they try to lift it so they can take this into account, for example by providing markings on the load. The worker can then reposition the side of the load nearest the centre of gravity close to their body before lifting the load or get assistance.

Working environment

The design or layout of the work environment may cause a worker to take on an awkward posture while moving a load, the temperature of the workplace may lead to fatigue or difficulty in gripping the load and poor lighting level may cause a worker to trip while moving a load. In addition, movement of a load between levels, for example up or down stairs, can present risks. The risks from manual handling may be minimised by the use of good design and layout of the workplace and workstations. This involves placing items where they can be conveniently handled to prevent workers needing to stoop (**see Figures 8-14 and 8-15**), improving work layouts so that travel distances are minimised and arranging that items can be picked up or put down at a suitable height. In general, working levels for workstations should be waist high, with tools, materials and equipment to be handled placed in front of the worker and within easy reach. The layout should suit the worker and there should be adequate room to perform the task safely.

To minimise manual handling risks related to the environment consideration should be given to control measures that:

- Design work layouts to reduce risk – minimise travel distance, reduce reaching (upwards, down and horizontally) and to enable loads to be picked up and put down at a suitable height.

- Remove restrictions on posture - ensure the working environment has sufficient headroom, so that workers do not have to stoop.

- Remove obstructions to free movement - ensure gangways, workspace and working areas are adequate to allow room to manoeuvre during manual handling tasks.

- Provide a suitable floor for good movement – replace uneven floors and ensure good housekeeping, for example spilt liquids are cleaned up. This should help to ensure the worker is well balanced when carrying a load. If manual handling aids are used, for example a sack truck ensure the floor is suitable for it and that the equipment moves easily.

- Avoid variations in floor level, for example stairs, steps, ladders and steep ramps – by taking routes that do not have these variations and using lifts or hoists.

- Ensure workstations are of a uniform height - this would reduce the need for raising or lowering loads when transferring them from one workstation to another.

- Avoid strong air movement when lifting large loads, for example gusts of wind - provide protective barriers or restrict work in strong winds.

- Ensure the general working environment is comfortable and suits manual handling activities - for example, humidity, heating/cooling, ventilation and lighting.

🔍 REVIEW

What factors should be considered before attempting to lift a load manually?

🧠 CONSIDER

Select a manual handling activity that you or other people carry out at work.

What are the risk factors associated with the task, the load, the individual and the environment?

8.3 Load-handling equipment

HAZARDS AND CONTROL MEASURES FOR COMMON TYPES OF LOAD-HANDLING AIDS AND EQUIPMENT

Cranes

Hazards

The principal hazards associated with any crane lifting operation are:

- **Overturning** which can be caused by operating outside the capabilities of the machine, uneven or weak ground (cellars or drains), outriggers not extended, insufficient counterweight, and adverse weather and by striking obstructions.

- **Overloading** by exceeding the operating capacity or operating radii, or by failure of safety devices.

- **Collision** with other cranes, overhead cables or structures.

- **Failure of load-bearing part** from structural components of the crane itself or an accessory fitted to it. This may be due to overloading or degradation of the load-bearing part due to damage, use (wear) or faults (corrosion).

- **Loss of load** from failure of lifting tackle, incorrect hook fittings or slinging procedure.

- **People in and around the crane** may get entangled in or trapped by moving parts.

Factors that will affect all cranes

The factors that can increase the risks from using cranes, which affect all cranes, include:

- Soft or uneven ground conditions.

- Underground voids or cellars that are not capable of bearing the weight of the crane and its load.

- Load-bearing capacity of the crane must be sufficient for the task.

- Adverse weather conditions such as heavy rain, high wind and extremes of temperature.

- Workers or members of the public nearby.

- Insufficient room for the lift.

- Proximity to overhead power lines, buildings or other cranes.

- Tall cranes, for example, construction tower cranes, located near an airport or in a flight path.

Controls

General requirements for cranes

The main control measures associated with any crane and lifting operation include ensuring that:

- Lifting operations are properly planned by a competent person, appropriately supervised and carried out in a safe manner.

- The ground the crane stands on is capable of bearing the load; check for underground services and cellars.

- The ground is level; if not, select a crane with hydraulic level-adjustment stabilisers.

Figure 8-33: Tower crane.
Source: Domson.

- The load-bearing capacity of the crane is sufficient for the task.

- The correct procedure is followed when erecting or dismantling any crane.

- The crane is positioned so that there is enough room for the lift and to avoid collision hazards.

- Non-essential people are kept clear of the work area.

- The crane is not operated in adverse weather conditions.

- The structural integrity of the crane is maintained; check for any signs of corrosion.

Figure 8-34: Ballast at base of tower crane.
Source: Wolffkran.

- A pre-use check by operator is carried out.

- Lifting equipment is of adequate strength and stability for the load. Stresses induced at mounting or fixing points must be taken into account. Similarly, every part of a load, and anything attached to it and used in lifting must be of adequate strength.

Figure 8-35: Lifting operation.
Source: RMS.

- The safe working load (SWL) is clearly marked on lifting machinery, equipment and accessories in order to ensure safe use. Where the SWL depends on the configuration of the machinery, it must be clearly marked for each configuration used and kept with the machinery.

- Passengers are not carried without authorisation, and never on lifting tackle.

- Equipment that is not designed for lifting persons, but which might be used as such, must have appropriate markings to the effect that it is not to be used for passengers.

Figure 8-36: Lifting points on load.
Source: RMS.

- Load indicators are fitted. It is preferable that there are two types. This is a requirement with jib cranes, but beneficial in all cranes.

- Load/radius indicator is fitted. This shows the radius the crane is working at and the safe load for that radi-us. It must be visible to the operator.

- Automatic safe load indicator is fitted. This provides a visible warning when the SWL is approached and audible warning when the SWL is exceeded.

- Controls are clearly identified and are of the 'hold to run' type.

- Over-travel switches are fitted. These are limit switches to prevent the hook or sheave block being wound up to the cable drum.

- Access is provided. Safe access should be provided for the operator and for use during inspection and maintenance/emergency.

- Operating position provides clear visibility of hook and load, with the controls easily reached.

- Lifting tackle, for example, chains, slings, wire ropes, eyebolts and shackles, should be tested before use and thoroughly examined periodically.

Accessories

Lifting accessories include slings, hooks, chains, eyebolts, shackles, lifting beams and cradles. This equipment is designed with the aim of assisting in lifting items without the need for manual force. Because these accessories are in a constantly changing environment and are in and out of use, they need to be protected from damage; a failure of any one item could result in a fatality. For example, lifting eyes need to be correctly fitted, slings have to be used with the correct technique and all equipment must be properly stored when not in use to prevent damage.

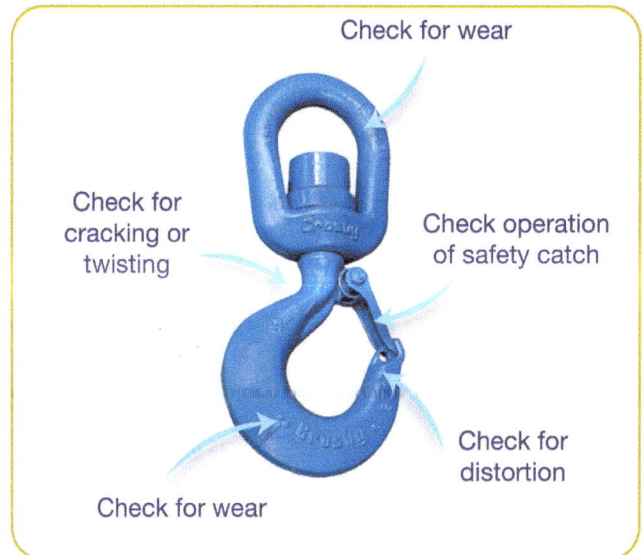

Check for wear

Check for cracking or twisting

Check operation of safety catch

Check for distortion

Check for wear

Figure 8-37: Hook-inspection.
Source: A Noble & Son Ltd.

Figure 8-38: Lifting accessory - sling.
Source: RMS.

Accessories must be attached correctly and safely to the load by a competent person, and then the lifting equipment takes over the task of providing the necessary required power to perform the lift. As with all lifting equipment, accessories must be regularly inspected and certificated and only used by trained and authorised persons.

Operator training and practices

The effective management of lifting operations should involve selecting competent persons including the crane operator and appointed person who will supervise the lifting operation. Crane operators and slingers should be fit and strong enough for the work. A safe system of work, including safety rules, should be developed and communicated to all those involved. Circumstances differ from site to site and additional rules should be added to cover different circumstances, equipment and conditions. Training should be provided for the safe operation of the particular equipment. In particular, crane operators should receive systematic training similar to the approach used to train fork-lift truck operators, discussed earlier in this element. Typically, this would include 3 stages:

Stage 1 - should contain the basic skills and knowledge required to operate the crane safely, to understand the basic mechanics and balance of the machine, and to carry out routine daily checks.

Stage 2 - under strict training conditions in an area which workers not involved in the training and other people that may be harmed are excluded. This stage should include:

- Knowledge of the crane operating principles and controls.

- Use of the crane in likely work conditions and doing the type of work to be undertaken, for example, loading and unloading vehicles.

- On completion of training, the crane operators should be examined and tested to ensure that they have achieved the required standard.

Stage 3 - after successfully completing the first two stages, crane operators should be given further familiarisation and where necessary instruction in the usual place of work for the crane operators.

General rules for safe operation of a crane

Always - Ensure the lift is planned and supervised by a competent person and that the crane operator and slingers are competent.

Always - Select the right appliance and lifting accessories for the job.

Always - Site the crane on solid stable ground away from structures or overhead cables.

Always - Check the crane has been maintained and certified in accordance with local laws.

Always - Ensure the appliance is stable when lifting, for example, not outside lifting radius, and that outriggers are correctly positioned.

Always - Check the weather conditions and follow manufacturer's instructions about maximum wind speeds.

Always - Use correct slinging methods.

Always - Protect sling from sharp edges – pack out and lower onto spacers.

Always - Ensure the sling is securely attached to the hook.

Always - Use standard signals that are understood by those involved in the crane operation.

Always - Ensure the load is lifted to correct height and moved at an appropriate speed.

Figure 8-39: Crane operation.
Source: RMS.

Never - Use equipment if damaged (check before use) – for example, stretched or not free movement, worn or corroded, outside inspection date.

Never - Exceed the safe working load.

Never - Lift with sling angles greater than 120 degrees.

Never - Lift a load over people.

Never - Drag a load or allow sudden shock loading.

Rules should be established for when a crane operator takes over a crane and before use of the crane. For example, before taking over a mobile crane, the driver must always check around it, and check the pressure of tyres, the engine for fuel, lubrication oil, water and the compressed air system. All controls, such as lifting equipment clutches and brakes, should be tested to see that all lifting ropes run smoothly and safe load indicators function correctly.

The following are examples of specific safe operating rules for a *mobile crane.* The operator must:

- Always lower the crane jib onto its rest (if fitted) or to the lowest operating position before travelling unladen, and point in the direction of travel, taking care of steep slopes.

- Always understand the signalling system and observe the signals of the appointed signaller/'banksman'.

- Never permit unauthorised persons to travel on the crane.

- Never use the crane to replace normal means of transport, or as a towing tractor.

- Always check that the crane is on firm and level ground before lifting and that spring locks/outriggers are properly in position.

- Never overload the crane. Always keep a constant watch on the load radius indicator.

- Always ensure movements are made with caution. Violent handling produces excess loading on the crane structure and machinery.

- Always make allowances for adverse weather conditions.

- Never cause loads to swing. Always position the crane so that the pull on the hoist rope is vertical.

- Always ensure that the load is properly slung. A load considered unsafe should not be lifted.

- Always ensure that all persons are in a safe position before any movement is carried out.

- Always make certain before hoisting that the hook is not attached to any anchored load or fixed object.

- Never drag slings when travelling.

- Always ensure that the jib, hook or load is in a position to clear any obstruction when the crane is slewing (swinging in a sideways or circular motion), but the load must not be lifted unnecessarily high.

- Always drive smoothly - drive safely. Remember that cranes are safe only when they are used as recommended by the makers. This applies in particular to speciality cranes.

- Always be on the lookout for overhead obstructions, particularly electric cables.

- Never tamper with or disconnect safe load indicators.

- Always stop the crane if the hoist or jib ropes become slack or out of their grooves, and report the condition.

- Always report all defects to the supervisor and never attempt to use a crane with a suspected serious defect until it has been rectified and certified by a competent person that it is not dangerous.

- Always make sure the crane is safe when leaving it unattended: ensure that the power is off, the engine stopped, the ignition key removed, the load unhooked, and the hook is raised up to a safe position.

- Always ensure special devices (for example, magnets and grabs) are used only for the purpose intended and in accordance with the instruction given.

- Always keep the crane clean and tidy.

- Always park the crane safely after use, remember to apply all brakes, slew locks, and secure rail clamps when fitted. Some cranes, however, particularly tower cranes, must be left to weather vane and the manufacturer's instructions must be clearly adhered to. Park the crane where the weather vaning jib will not strike any object. Lock the cabin before leaving the crane.

Figure 8-40: Poorly planned lifting operation - the rigger is at risk of falling due to lack of fall protection; inappropriate ladder used for access.
Source: RMS.

Lifts and hoists

Hazards

Lifts and hoists are used for transporting people and goods between different levels. They can be found in a variety of buildings and structures, including temporary structures on constructions sites. Small, mobile devices are also used to lift and hoist materials. In general, the hazards associated with lifts and hoists are the same as with any other lifting equipment.

- The lift/hoist may overturn or collapse.

- The lift/hoist may strike people who are near or under the platform or cage.

- The lift/hoist may fail to stop in a safe position at the top, bottom or landing level.

- As the platform or cage moves, the load, including people, may come into contact with fixed or moving structures and objects, for example, landing level structures.

- The supporting ropes may fail and the platform/cage fall to the ground.

- The load, including people, or part of the load may fall.

- The lift/hoist may fail in a high position.

- People in and around the lift/hoist machinery may get trapped or entangled in moving parts.

Figure 8-41: Material host.
Source: HSS.

Controls

All lifts and hoists

If lifts and hoist are properly designed, installed and maintained there is relatively little risk from their operation. All lifts and hoists require:

- Adequate design, including safety devices that are required by national legislation.

- Robust construction.

- Correct selection and installation.

- Holdback mechanisms that operate where the lifting mechanism fails, for example rope failure.

- Overrun trip devices to prevent the platform/cage being lifted or lowered beyond a safe limit. For example, overrun trip devices that prevent the platform/cage contacting the lifting gear at the top or being winched over the top of a drum.

- Guards on lift/hoist machinery, for example, to prevent entanglement with the moving parts of the machinery.

- Guards to prevent contact with the lift cage or platform when it is moving.

- Landing gates (securely closed during operation).

- Suitable lighting on all lift/hoist landings to reduce the risk of people tripping or falling.

- The safe working load of the lift/hoist clearly marked and visible to the user/operator.

- Operation by competent people.

- If the lift/hoist is for materials only, a prominent warning notice on the platform or cage to stop people riding on it.

- Regular inspection to ensure the lift/hoist is in working order. This may be as a daily, pre-use operator check and/or weekly inspection by an engineer.

- Regular servicing by a reputable maintenance company (approximately every three months). The maintenance contract should include removing rubbish as it may cause obstructions and contribute to the risk of a fire. The service report provided should relate to the efficient working of the lift/hoist and is not a substitute for a thorough examination.

- Periodic thorough examination. For example, passenger lifts/hoists should be thoroughly examined every six months by a competent person and goods lifts every 12 months. Alternatively, they should be examined at intervals detailed in an examination scheme drawn up by a competent person based on an assessment of risks.

- The results of inspections and thorough examinations to be recorded. Any remedial work identified should receive prompt attention.

- Unauthorised access to lift/hoist machinery to be prevented by keeping access under key control - which is kept in a secure position, controlled by a responsible person and available at all times to authorised people.

Passenger lifts and hoists

In addition, passenger lifts/hoists require more sophisticated controls measures:

- Operating controls inside the cage.

- Electromagnetic interlocks on the cage doors.

- If within a building, the enclosing shaft must be of fireproof construction.

- The lift/hoist should be protected by a substantial enclosure to prevent anyone from being struck by any moving part of the hoist or material falling down the hoist way.

- Gates must be provided at all access landings, including at ground level. The gates must be kept shut, except when the platform is at the landing. The controls should be arranged so that the lift/hoist can be operated from one position only, which may be from within the lift/hoist.

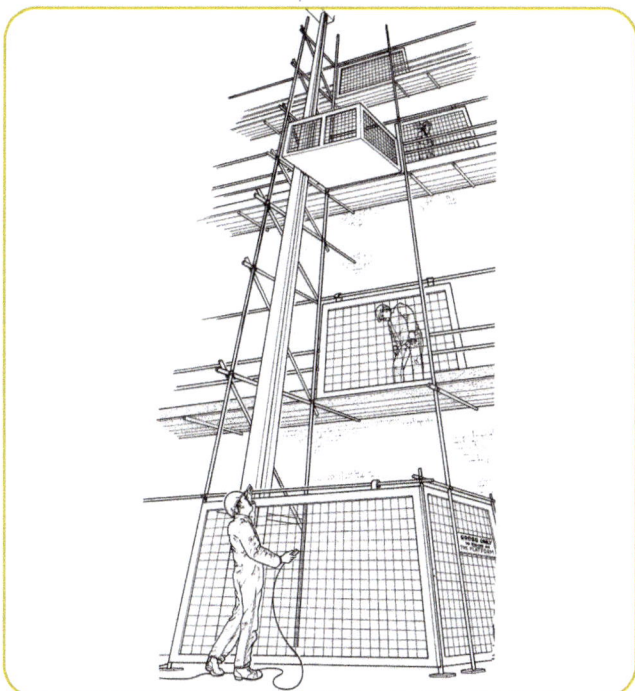

Figure 8-42: Platform material hoist.
Source: HSE, HSG150.

Figure 8-43: Passenger hoist.
Source: HSE, HSG150.

- The safe working load of the lift/hoist must be clearly marked and visible to the operator, and cagecontrolled lifts/hoists must be equipped with effective overload warning devices. Passenger lifts must be fitted with an alarm that can be activated by users from within the cage, if a fault condition arises, trained workers should be available to effect any rescue which may be required.

Procedures should be in place to ensure that any lift/hoist malfunction is reported to the supervisor immediately. A system should be developed for rescuing people trapped in passenger lifts/hoists and where this is to be carried out by workers they should be provide with adequate training on this procedure.

Written rescue procedures should be displayed at appropriate locations and it should be ensured that the alarm bell used by passengers to warn that they are trapped can be activated.

Figure 8-44: Passenger hoist cage & SWl
Source: HSE, HSG150.

REVIEW

Give examples of manual methods of transporting a load that would eliminate or reduce manual handling.

CONSIDER

Are appropriate manual handling aids available to workers in your workplace?

Fork-lift trucks

Hazards

Although the fork-lift truck is a very useful machine for moving materials in many industries, it features prominently in workplace accidents/incidents.

In the UK alone, there are about 20 deaths and 5,000 injuries each year that can be attributed to fork-lift trucks.

These can be analysed as follows:

- Injuries to driver - 40%

- Injuries to assistant - 20%

- Injuries to pedestrians - 40% of which 80% were fractures with some 60% resulting in injuries to ankles and feet

Unless preventive action is taken, these accidents/incidents are likely to increase as fork-lift trucks are increasingly used in the workplace.

From accident investigations it can be seen that about 45% of accidents/incidents can be wholly or partly attributed to operator error, thus proper operator training is essential. There are, however, many other causes, including inadequate premises and gangways, poor truck maintenance and lighting. The hazards related to the use of fork-lift trucks include:

- Overloading – exceeding maximum rated capacity.
- Driving too fast.
- Sudden braking.
- Driving on slopes.
- Driving with the load elevated.
- Driving over debris.
- Under-inflated (pneumatic) tyres or badly cut (solid) tyres.
- Driving with the load incorrectly positioned on the forks.
- Overturning.
- Poor floor surface, for example, holes, such as drains or potholes.
- Failure of load-bearing parts (for example, lifting chain).
- Collision with vehicles, buildings, or pedestrians
- Loss of load.
- Insecure load.

How some of these hazards can occur when using a fork-lift truck is explained below. It should be noted that fork-lift trucks, unless specifically designed to do so, are not intended for carrying passengers. Many of the hazards outlined for fork-lift trucks are equally applicable to other equipment used for load handling and may be seen as generic hazards of mobile powered load handling equipment.

Overturning

The stability of fork-lift trucks is particularly affected by the forces generated when turning, especially at speed, or if the equipment is tilted sideways, for example, by travelling across an incline or by the wheels running into a pothole or over an obstruction. The danger of a fork-lift truck being turned on its side is greater with the load in the raised position (*see Figure 8-45a*), than in the lowered position (*see Figure 8-45b*).

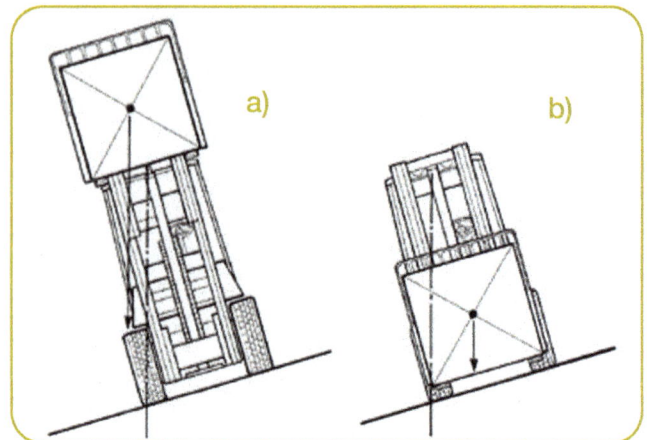

Figure 8-45: Overturning of lift truck.
Source: HSE, HSG6 (no longer available).

Overbalancing

The mass of a counterbalance fork-lift truck acts as a counterweight so that the load can be lifted and moved without the fork-lift truck overbalancing and tipping forward (*see Figure 8-46a*). However, the fork-lift truck can be overbalanced and tip forward if the fork-lift truck is overloaded. Overloading may be caused if the load is too heavy (*see Figure 8-46b*), if the load is incorrectly placed on the forks so that it is too far forward (*see Figure 8-46c*) or if the fork-lift truck accelerates or brakes sharply while carrying a heavy load.

This may not cause the equipment to overturn, but the overbalancing can injure the operator and lead to loss of control of steering. Control of steering may be lost because counterbalance fork-lift trucks usually have rear-wheel steering and overbalancing lifts the rear of a counterbalance fork-lift truck from the ground, preventing the steering wheels from contacting the ground properly.

Collisions with other vehicles, pedestrians or fixed objects

People may appear unexpectedly from a part of a building structure or workers intent on their work may step away from where they are working to collect materials or tools.

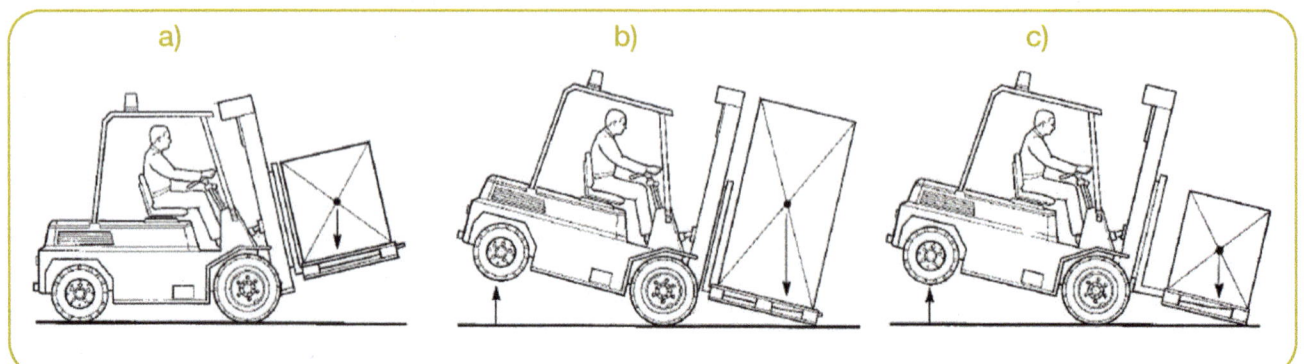

Figure 8-46: Overbalancing of fork-lift truck.
Source: HSE, HSG6 (no longer available).

Often the space in workplaces such as construction sites is restricted. Material may be stored at a height because it is large or to maximise the available space. This in turn leads to restricted visibility, especially at busy junctions where vehicles come together. Loads transported by fork-lift trucks may obstruct the driver's vision and cause them to look to alternate sides of the load in order to see past it. Any of these may lead to collisions with other vehicles and pedestrians or, in the avoidance of these, fixed objects.

Figure 8-47: Poor visibility due to load.
Source: RMS.

When travelling with a load a fork-lift truck should have the forks tilted to cradle the load. If this is not done there is a risk that the load may slide from the forks and hit someone or something.

The shock of impact with ruts and small bumps can be transmitted to the forks and the load, or part of it, may be dislodged and caused to fall.

A fork-lift truck may have an open framework that can allow a person to put a part of their body, typically their head or hand, outside the protective structure. Many injuries have occurred to fork-lift truck drivers who, while driving forward with a load obstructing their view, lean out to look round the load and their head strikes a fixed object.

Materials falling

There is a hazard of materials falling on a driver during loading and unloading operations, particularly where the fork-lift truck is used to load materials at a height or to remove materials from delivery vehicles.

Controls

The following list should only be used as a guide to good practice in the safe operation of fork-lift trucks; it also has some applicability to all mobile work equipment:

- Speeds must be consistent with workplace conditions, special care should be taken on muddy or wet roads.

- Operators should always face the direction of travel.

- When travelling with a load forks should be close to

the ground and tilted back enough to prevent the load sliding off the forks.

- The load must be seated as close to the heel of the forks as possible.

- When travelling up or down a slope, the forks should point up the slope.

- Trucks should not be driven over obstacles or significant holes.

- Trucks should not be driven close to excavations.

- Care should be taken not to drive trucks across a slope.

- Travel slowly and with due care when approaching a road junction, door opening or where pedestrians could be present. Sound the horn as a warning of the presence of the fork-lift truck.

- No passengers, unless the fork-lift truck is specifically designed to accommodate them.

- All parts of the operator's body must be kept within the limits of the fork lift truck and no part of the body placed between the uprights of the mast.

- The fork-lift truck should be stationary with the handbrake applied when the forks are raised or lowered, whether with a load or not.

- Fork-lift trucks should not be used to pull or push loads.

Traffic routes

The precautions related to traffic routes include:

- Where possible, separate routes, designated crossing places and suitable barriers at recognised danger spots.

- Roads, gangways and aisles should be wide enough and have sufficient overhead clearance for the largest fork-lift truck.

- Clear direction signs.

- Sharp bends and overhead obstructions should be avoided.

- Gradients in a fork-lift truck operating area should be kept as small as possible. Trucks should reverse down gradients and not drive across them.

Sufficient and suitable parking areas should be provided away from the main work area. Designated parking areas must be clearly identified. They should be on firm, level ground and located such that they do not obstruct fire points or other traffic routes, especially those for emergency access and exit.

When the fork-lift truck is parked or left unattended the driver should ensure that the mast is tilted slightly forward with the forks resting on the floor, the power is switched off, the brake is applied, and the key removed and returned to a responsible person to prevent unauthorised use.

Protection of people

There is a need to alert people to the hazard when working in or near a vehicle operating area. Signs may be used to indicate the general operation of vehicles in the area. These can be supplemented by **visual and audible warning systems** that confirm the presence of a vehicle, such as a flashing light on the top of a fork-lift truck. The fork-lift truck should be fitted with a horn which can be sounded when approaching blind bends.

Driver protection

Roll over protection must be provided, for example a closed cab, and a restraint system should be fitted, typically an adjustable seat belt.

Selection of equipment

There are many types of fork-lift truck available for a range of activities. Some situations require the use of specialist trucks such as rough terrain and side loading trucks. Many accidents/incidents happen due to the incorrect selection and/or use of fork-lift trucks.

Rough terrain counterbalance fork-lift trucks are similar in design to the standard counterbalanced fork-lift truck, but is equipped with larger wheels and pneumatic tyres, giving it greater ground clearance.

It has greater ability to operate on uneven and soft ground and is mainly used in the construction industry and in agriculture. It may be used with a range of attachments.

Figure 8-48: Rough terrain counterbalance fork-lift truck.
Source: RMS.

When using a side-loading fork-lift truck the driver is positioned at the front and to one side of the loadbearing deck. The load is carried on the deck of the truck, the mast being traversed out sideways to pick up or set down the load.

This type of fork-lift truck is used for stacking and moving long loads such as timber and pipes, and may be fitted with stabilisers for use when picking up or setting down loads.

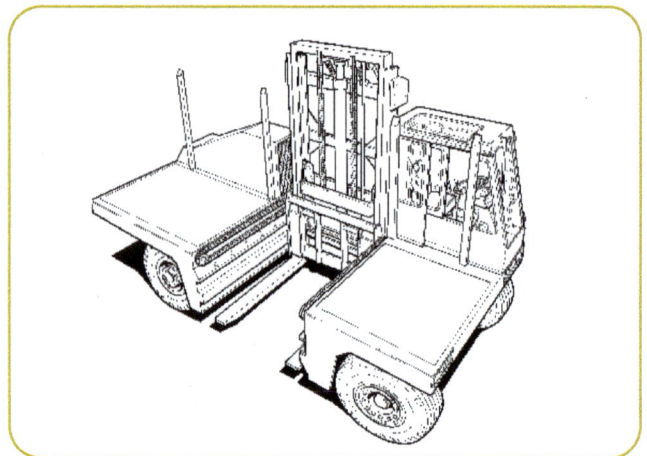

Figure 8-49: Side-loading fork-lift truck.
Source: HSE, HSG6.

When choosing the right truck for the job the following factors should be taken into account:

- Power source – the choice of battery, liquified petroleum gas (LPG) or diesel will depend on whether the truck is to be used indoors or outdoors. Factors to consider include, combustion fumes from LPG and diesel and manual handling with battery and LPG.

- Tyres – solid or pneumatic, depending on the terrain.

- Size and capacity – dependent on the size and nature of the loads to be moved.

- Height of the mast.

- Audible and/or visual warning systems fitted according to the proximity of pedestrians.

- Protection provided for the driver against overturning or the possibility of falling objects.

- Training given to operators must be related specifically to the type of truck.

- Provision of a suitable mechanism to prevent unauthorised use, for example, key or electronic pad.

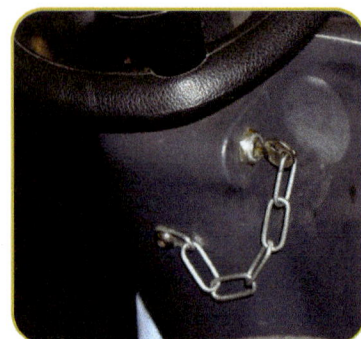

Figure 8-50: Keys - unauthorised use not controlled.
Source: RMS.

Element 8

Figure 8-51: PIN pad to prevent unauthorised use.
Source: RMS.

Provision of information

Drivers and supervisors of fork-lift trucks should be familiar with the following information, which should be shown on the truck:

- Name of the manufacturer (or authorised supplier) of the lift truck.
- Model designation.
- Serial number.
- Unladen weight.
- Rated weight carrying capacity.
- Load centre distance.
- Maximum lift height.
- Inflation pressures of pneumatic tyres.

In addition, the functions of all the controls should be clearly marked so that they can be seen from the driver's position.

Drivers

No person should be permitted to drive a fork-lift truck unless they have been selected, trained and authorised to do so, or are undergoing properly organised formal training.

Daily checks/pre-use inspection

A checklist for daily checks should include items such as the condition and pressure of tyres, the integrity and proper functioning of lights, horns, brakes, steering and mirrors, the absence of oil leaks and that the seat is securely fixed (with properly functioning and intact restraints where fitted). The fork-lift truck should also be checked for obvious signs of damage to its bodywork and lifting forks mechanism and for the security of any equipment fitted such as liquid petroleum gas (LPG) cylinders.

If the fork-lift truck is to use attachment accessories, for example, forks or drum grabs, the driver should ensure they are correctly fitted and secured.

Thorough examination of load-bearing parts

The load-bearing parts of fork-lift trucks, for example, the lifting chain, should be thoroughly examined periodically to determine any wear or damage that could affect the integrity of the lifting equipment that forms part of the fork-lift truck. The Lifting Operations and Lifting Equipment Regulations (LOLER) 1998 specifies that load-bearing parts of fork-lift trucks used only to lift goods must be thorough examinations every 12 months.

Where fork-lift trucks are used to lift people the load-bearing parts must be thoroughly examined every 6 months.

Pallet truck

A pallet truck has two elevating forks for insertion below the top deck of a pallet. When the forks are raised the load is moved clear of the ground to allow movement.

This truck may be designed for pedestrian or rider control. They usually have no mast and cannot be used for stacking. Pallet trucks may be powered or non-powered.

Figure 8-52: Rough terrain pallet truck.
Source: LGH.

Hazards and controls

Pallet trucks can be driven both manually or by a quiet-running electric motor. Hazards include crushing from moving loads or from the momentum of the equipment when stopping. Other hazards include trapping in the forks of the equipment, manual handling when operated manually and electrical hazards from battery power points.

Sack truck

A sack truck is a simple device fitted with two wheels on which the load is pivoted and supported when the truck is manually tilted back and pushed.

Figure 8-53: Sack truck.
Source: RMS.

Hazards and controls

Sack trucks are typically manually powered and the hazards arising from their use are generally of an ergonomic nature relating to posture and over-exertion through manual handling. If large heavy loads are placed on the sack truck this can exert strong downward forces when the truck is tipped back, due to the cantilever effect, the worker then has to apply additional force to overcome that of the load and enable it to be tipped back. The worker may find this strenuous and have difficulty holding it tipped back while moving. This could lead to the truck tipping forward or backwards suddenly.

Mechanical hazards are restricted to the wheels of the truck, which only move when the truck is pushed by the user. Other associated hazards are tripping and falling while using the equipment and manual handling back and strain injuries.

Control measures include indicating a safe working load for the equipment and the provision of information/ instruction in the safe loading and use of the equipment for the operator. A manual handling assessment may be required when using this equipment.

Telescopic handler

A telescopic handler (commonly abbreviated to 'telehandler') consists of a heavy duty chassis, body and lifting gear on large diameter wheels and deep tread pneumatic traction tyres. This equipment is fitted with a boom that is pivoted at the rear of the machine. The boom is raised and lowered by hydraulic rams. In addition, the boom can be extended or retracted to give extra reach or height. These machines may be two or four-wheel drive, and have two-wheel, four-wheel or crab steering.

Telehandlers are all-purpose machines that can be used for construction site preparation, material handling, scaffold erection, elevated work platforms and for final site clean-up. Although a range of attachments may be used with them, they are commonly fitted with forks.

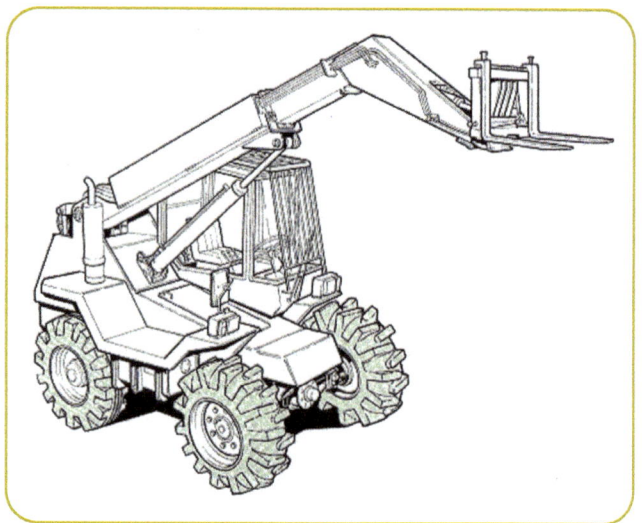

Figure 8-54: Telescopic materials handler.
Source: HSE, HSG6.

Hazards

Hazards include the generic hazards of mobile load handling equipment and are similar to that of a rough terrain fork-lift truck. In the case of a telehandler, hazards include moving large hydraulic rams that actuate the front forks or bucket with hydraulic hoses attached that are under high pressure. There are therefore various mechanical hazards involved, such as high-pressure fluid, impact, crush, trap, and shear. Overturning can result when the boom is extended and the weight of a load moves away from the center of mass of the vehicle and outside the wheelbase of the telehandler.

This can occur when a suspended load swings. Challenging ground conditions (cambers, slopes/rough terrain) can cause further instability. Overturning can also result if load radius charts are used incorrectly and the machine is overloaded because the boom is extended too far for the weight of load being carried. In addition to these there are non-mechanical hazards such as heat, fumes, chemicals and noise.

Controls

Control measures should include using competent operators and seat restraints for the operator, who should be enclosed in a protective cage that can also act as a guard against contact with falling materials and the effects of overturning. Barriers should be provided to segregate pedestrians to a safe distance and good visibility by the operator should be maintained, including with the assistance of mirrors. High-visibility clothing should be worn by those working nearby. The telehandler should also be fitted with reversing alarms and amber flashing lights to warn of its presence. Lateral instability warning technology can help prevent dangerous overturning situations before a critical threshold of instability is reached. Load radius charts (or markings on the boom) should be strictly followed to avoid overloading of the machine.

REQUIREMENTS FOR LIFTING OPERATIONS

Strong, stable and suitable equipment

Strength

Regulation 4 of LOLER 1998 requires every employer to ensure that:

- Lifting equipment is of adequate strength and stability for each load, having regard in particular to the stress induced at its mounting or fixing point.

- Every part of a load and anything attached to it and used in lifting it is of adequate strength.

When assessing the strength of the lifting equipment for its proposed use, the combined weight of the load and lifting accessories should be taken into account. It is important to consider the load, task and environment in order to match the strength of the lifting equipment to the circumstances of use. For example, if the environment is hot or cold this can affect the lifting capacity of the equipment. In order to counteract this effect, equipment with a higher rated safe working load may be needed.

If the load to be lifted is a person, equipment with a generous capacity above the person's weight should be selected in order to provide an increased factor for safety. If the load is likely to move unexpectedly, because of the movement of an animal or liquids in a container, this sudden movement can put additional forces on the equipment and may necessitate equipment with higher strength to be selected.

When lifting a load that is submerged in water, the initial lifting weight will be misleading because the load will be supported by the water. When the load emerges from the water its support will no longer be available and this sudden increase in weight can put additional stress on the crane and its lifting accessories.

When conducting the lifting task, the lifting accessories may be used in such a way that its lifting capacity may be reduced below its stated safe working load; sharp corners on a load and 'back hooking' can have this effect. In these circumstances, accessories with a higher rated safe working load may be required. It is essential to remember that in a lifting operation the equipment only has an overall lifting capacity equivalent to the item with the lowest strength.

For example, in a situation where a crane with a lifting capacity of 50 tonnes is used with a hook of 10 tonnes capacity and a wire rope sling of 5 tonnes capacity, this would give an overall maximum lifting strength/capacity of 5 tonnes. Any alterations or repairs to the crane must be in accordance with the manufacturer's instructions to maintain the strength of the crane.

Stability

A number of factors can affect the stability of lifting equipment, for example, wind conditions, slopes/cambers, stability of ground conditions and how the load is to be lifted. Lifting equipment must be positioned and installed so that it does not tip over when in use.

Anchoring can be achieved by securing with guy ropes, bolting the structure to a foundation, using ballast as counterweights or using outriggers to bring the centre of gravity down to the base area.

Mobile lifting equipment should be sited on firm ground with the wheels or outrigger feet having their weight distributed over a large surface area. Care should be taken that the equipment is not positioned over cellars, drains or underground cavities, or positioned near excavations.

Figure 8-55: Lifting operations.
Source: RMS.

Sloping ground should be avoided as this can shift the load radius out or in, away from the safe working position. In the uphill position, the greatest danger occurs when the load is set down. This can cause the mobile lifting equipment to tip over. In the downhill position, the load moves out of the radius and may cause the equipment to tip forward. If it is necessary to use a crane to lift a load on a slope it is important to select a crane with sufficient load lifting capability and stability to counter-act the effects of the slope on the load and the crane.

Figure 8-56: Siting and stability.
Source: RMS.

Suitability

The employer should ensure that work equipment is used only for operations for which it has been designed, including consideration of the working conditions for each lifting operation. In the UK, Regulation 4 of the Provision and Use of Work Equipment Regulations 1998 states:

> "Every employer shall ensure that work equipment is used only for operations for which, and under conditions for which, it is suitable in order to avoid any reasonably foreseeable risk to the health and safety of any person."

Figure 8-57: Suitability of work equipment.
Source: Provision and Use of Work Equipment Regulations (PUWER) 1998.

In order for lifting equipment to be suitable it must be of the correct type for the task, have a safe working load limit in excess of the load being lifted, and have the correct type and combination of lifting accessories attached.

Lifting equipment used within industry varies and includes mobile cranes, static tower cranes and overhead travelling cranes. The type of lifting equipment selected will depend on a number of factors, including the weight of the load to be lifted, the radius of operation, the height of the lift, the time available, and the frequency of the lifting activities.

This equipment is often very heavy, which means its weight can cause the ground underneath to sink or collapse.

Other factors like height and size may have to be considered as there may be limitations on site roads between structures or there may be overhead restrictions. Careful consideration of these factors must be made when selecting the correct crane. Selecting lifting equipment to carry out a lifting activity should be done at the planning stage, where the most suitable equipment can be identified that is able to meet all of the lifting requirements and the limitations of the location.

Figure 8-58: Danger zone - crane and fixed item.
Source: RMS.

Positioned and installed correctly

Lifting equipment must be positioned or installed so that the risk of the equipment striking a person is as low as is reasonably practicable. Similarly, the risk of a load drifting, falling freely or being unintentionally released must also be considered and equipment positioned to take account of this.

All nearby hazards, including overhead cables and uninsulated power supply conductors, should be identified and removed or covered by safe working procedures such as locking-off and permit systems. The possibility of striking other lifting equipment or structures should also be examined.

Detailed consideration must be given to the location of any heavy piece of lifting equipment due to the fact that additional weight is distributed to the ground through the loading of the equipment when performing a lift. Surveys must be carried out to determine the nature of the ground, whether soft or firm, and what underground hazards are present, such as buried services or hollow voids. If the ground proves to be soft, it can be covered using timber, digger mats or hard core to prevent the equipment or its outriggers sinking when under load.

The surrounding environment must also be taken into consideration and factors may include highways, railways, electricity cables, areas of public interest. The area around where lifting equipment is sited should be securely fenced, including the extremes of the lift radius, with an additional factor of safety to allow for emergency arrangements such as emergency vehicle access or safety in the event of a collapse or fall.

Where practicable, lifting equipment should be positioned and installed such that loads are not carried or suspended over areas occupied by people. Where this is necessary, appropriate systems of work should be used to ensure it is done safely.

If the operator cannot observe the full path of the load, an appointed person (and assistants as appropriate) should be used to communicate the position of the load and provide directions to avoid striking anything or anyone.

Visibly marked with safe working load

The safe working load (SWL) must be clearly marked on lifting machinery, equipment and accessories in order to ensure safe use. Where the SWL depends on the configuration of the machinery, it must be clearly marked for each configuration used and kept with the machinery. Accessories must be marked with supplementary information that indicates the characteristics for its safe use, for example, safe angles of lift.

Equipment designed for lifting people must be clearly marked as such and equipment which is not designed for lifting persons, but which might be used as such, must have appropriate markings to the effect that it is not to be used for lifting people.

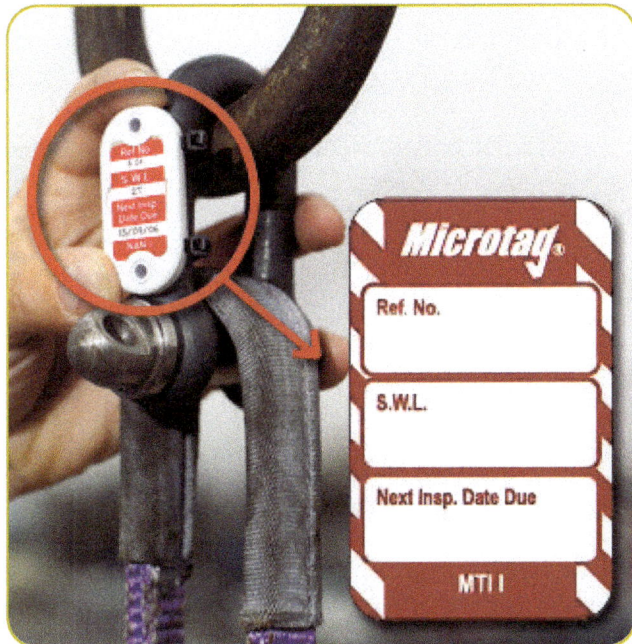

Figure 8-59: Marking of accessories.
Source: Scafftag.

Planned, supervised and carried out in a safe manner by competent people

Regulation 8 of LOLER 1998 requires that every employer shall ensure that every lifting operation involving lifting equipment is:

- Properly planned by a competent person.
- Appropriately supervised.
- Carried out in a safe manner.

The person planning the operation should have adequate practical and theoretical knowledge and experience of planning lifting operations.

The plan should address the risks identified by the risk assessment and identify the resources required, the procedures and the responsibilities so that any lifting operation is carried out safely. The plan should ensure that the lifting equipment remains safe for the range of lifting operations for which the equipment might be used.

The type of lifting equipment that is to be used and the complexity of the lifting operations will dictate the degree of planning required for the lifting operation. Planning combines two parts:

- Initial planning to ensure that lifting equipment is provided which is suitable for the range of tasks that it will have to carry out.

- Planning of individual lifting operations so that they can be carried out safely with the lifting equipment provided.

- Factors that should be considered when formulating a plan include:

- The load that is being lifted - weight, shape, centre of gravity, surface condition, lifting points.

- The equipment and accessories being used for the operation and suitability - certification validity.

- The proposed route that the load will take, including the destination and checks for obstructions.

- The team required to carry out the lift - competencies and numbers required.

- Production of a safe system of work, risk assessments, permits to work.

- The environment in which the lift will take place - ground conditions, weather, local population.

- Securing areas below the lift - information, restrictions, demarcation and barriers.

- A suitable trial to determine the reaction of the lifting equipment prior to full lift.

- Completion of the operation and any dismantling required.

It is important that someone takes supervisory control of lifting operations at the time they are being conducted. Although the operator may be skilled in lifting techniques, this may not be enough to ensure safety, because other factors may influence whether the overall operation is conducted safely, for example, people may stray into the area. The level of supervision necessary is influenced by the nature of the work and the competence of those involved in using the equipment and assisting with the lifting operation. The supervisor of the lifting operation must remain in control and stop the operation if it is not carried out satisfactorily.

Lifting equipment and accessories should be subject to a pre-use check in order to determine their condition and suitability. In addition, care should be taken to ensure the lifting accessories used are compatible with the task and that the load is protected or supported such that it does not disintegrate when lifted.

Lifting operations should not be carried out where adverse weather conditions occur, such as fog, sand/dust storm, poor lighting, strong wind or where heavy rainfall makes ground conditions unstable. It is important that measures are taken to prevent lifting equipment overturning and that there is sufficient room for it to operate without contacting other objects.

Lifting equipment should not be used to drag loads and should not be overloaded. Special arrangements need to be in place when lifting equipment not normally used for lifting people is used for that purpose, for example, de-rating the working load limit, ensuring communication is in place between the people being lifted and the operator, and ensuring the operation controls are manned at all times.

The Health and Safety at Work etc. Act (HASAWA) 1974 places a duty on employers, to their workers, for the provision of information, instruction, training and supervision as is necessary to ensure, so far as is reasonably practicable, the health and safety at work of the workers. In addition to this general duty, a further duty exists under the Provision and Use of Work Equipment Regulations (PUWER) 1998.

Employers must ensure that any person who uses a piece of work equipment has received adequate training for purposes of health and safety, including training in the methods that may be adopted when using work equipment, any risks which such use may entail and precautions to be taken. Various appointments, with specified responsibilities, may be made in order to ensure the safety of lifting operations on site.

The various appointments are as follows:

Competent person	Appointed to plan the operation.
Load handler	Attaches and detaches the load.
Authorised person	Ensures the load safely attached.
Operator	Appointed to operate the equipment.
Responsible person	Appointed to communicate the position of the load (banksman).
Assistants	Appointed to relay communications.

Each person given responsibilities must be competent to carry them out, usually this will mean that they must be adequately trained and experienced. The only exception is when they are under the direct supervision of a competent person for training requirements.

Special requirements for lifting equipment for lifting persons

The employer should ensure that lifting equipment for lifting persons:

- Prevents a person using it being crushed, trapped or struck or falling from the carrier.

- Prevent a person carrying out activities from the carrier being crushed, trapped or struck or falling from the carrier.

- Has suitable devices to prevent the risk of a carrier falling.

- Ensures a person trapped in any carrier is not exposed to danger by being trapped and can be freed.

In addition, the employer should ensure that where devices are not provided to prevent the risk of a carrier falling:

- The carrier has an enhanced safety coefficient suspension rope or chain.

- The rope or chain is inspected by a competent person every working day.

Special arrangements need to be in place when lifting equipment not normally used for people is used for that purpose, for example, de-rating the working load limit, ensuring communication is in place between the people and operator, and ensuring that a competent person is in control of the equipment's operation controls at all times. Lifting equipment for lifting people is subject to specific requirements for statutory examination prescribed by legislation.

CONSIDER

Consider work equipment in your workplace and identify any mechanical lifting equipment in use.

PERIODIC INSPECTION AND EXAMINATION/ TESTING OF LIFTING EQUIPMENT

Statutory requirements for the examination of lifting equipment are set out in Regulation 9 of the LOLER 1998. It requires that lifting equipment must be thoroughly examined by a competent person before being put into service for the first time by a new user.

This does not apply to new lifting equipment (unless its safety depends on installation conditions) that has met European Community declaration of conformity requirements within 12 months of being put into first use or used equipment where it has been certified as being thoroughly examined within the previous 12 months.

Suppliers of used lifting equipment are obliged by Regulation 9 of LOLER 1998 to certify that a thorough examination has been carried out. Where the safety of lifting equipment depends on the installation conditions it must be thoroughly examined by a competent person prior to first use, after assembly and on change of location in order to ensure that it has been installed correctly and is safe to operate.

Lifting equipment exposed to conditions causing deterioration that is liable to result in dangerous situations should be thoroughly examined by a competent person:

- At least every 6 months - lifting equipment for lifting persons and lifting accessories.
- At least every 12 months - other lifting equipment.
- In either case, in accordance with an examination scheme.
- On each occurrence of exceptional circumstances liable to jeopardise the safety of the lifting equipment.

The person who carries out the thorough examination should, as soon as is practicable, make a written report of the results. This should be signed by the competent person who carried out the task and should be kept available for inspection for the period of validity of the report.

Where appropriate to ensure health and safety, inspections should be carried out at suitable intervals between thorough examinations. Examinations and inspections should ensure that the good condition of equipment is maintained and that any deterioration can be detected and remedied in good time.

The term 'competent' is generally taken to mean someone who is qualified and experienced in carrying out such examinations. The Approved Code of Practice that accompanies the Lifting Operations and Lifting Equipment Regulations (LOLER) 1998 identifies a competent person as:

"The person carrying out a thorough examination has such appropriate practical and theoretical knowledge and experience of the lifting equipment to be thoroughly examined as will enable them to detect defects or weaknesses and to assess their importance in relation to the safety and continued use of the lifting equipment."

Figure 8-60: Competent person definition.

Source: Approved Code of Practice to the Lifting Operations and Lifting Equipment Regulations (LOLER) 1998, L113.

REVIEW

What factors should be considered during the planning stage to reduce the risk of injury to workers and others before using lifting equipment?

Sources of reference

Reference information provided, in particular web links, was correct at time of publication, but may have changed.

Ergonomics and human factors at work, INDG90, HSE Books, http://www.hse.gov.uk/pubns/indg90.pdf

Manual Handling, Manual Handling Operations Regulations 1992 (as amended), Guidance on Regulations, L23, HSE Books, ISBN: 978-0-7176-6653-9 http://www.hse.gov.uk/pubns/priced/l23.pdf

Lift-truck training, INDG462, HSE, http://www.hse.gov.uk/pubns/indg462.pdf

Lifting equipment at work – a brief guide, INDG290(rev1), HSE, http://www.hse.gov.uk/pubns/indg290.pdf

Rider-operated lift trucks, L117, 2013, HSE Books, ISBN: 978-0-717664-41-2, http://www.hse.gov.uk/pUbns/priced/l117.pdf

Safe use of lifting equipment, Lifting Operations and Lifting Equipment Regulations 1998, ACoP and Guidance, L113 (amendments made 2018), HSE Books, ISBN: 978-0-7176-6588-0, http://www.hse.gov.uk/pubns/priced/l113.pdf

Safe Use of Work Equipment, Approved Code of Practice and guidance, L22, HSE Books, ISBN: 978-0-7176-6619-5 http://www.hse.gov.uk/pubns/priced/l22.pdf

Seating at Work, HSG57, HSE Books, ISBN: 978-0-7176-1231-4, http://www.hse.gov.uk/pubns/priced/hsg57.pdf

The health and safety toolbox, How to control risks at work, HSG268, HSE Books, ISBN: 978-0-7176-6587-7, http://www.hse.gov.uk/pUbns/priced/hsg268.pdf

Understanding ergonomics at work, INDG90 (rev2), HSE Books, https://www.hse.gov.uk/pubns/indg90.pdf

Upper Limb Disorders in the Workplace, HSG60, HSE Books, ISBN: 978-0-7176-1978-8, http://www.hse.gov.uk/pubns/priced/hsg60.pdf

Working with display screen equipment (DSE): A brief guide, INDG36 (rev4), HSE Books, ISBN: 978-0-717664-72-6 http://www.hse.gov.uk/pubns/indg36.htm

Workplace health, safety and welfare, Workplace (Health, Safety and Welfare) Regulations 1992, ACOP, L24, HSE Books, ISBN: 978-0-7176-6583-9, http://www.hse.gov.uk/pubns/priced/l24.pdf

Work with display screen equipment: Health and Safety (Display Screen Equipment) Regulations 1992 as amended by the Health and Safety (Miscellaneous Amendments) Regulations 2002: Guidance on Regulations, L26, HSE Books ISBN: 978-0-7176-2582-6, http://www.hse.gov.uk/pubns/priced/l26.pdf

Web links to these references are provided on the RMS Publishing website for ease of use –www.rmspublishing.co.uk

Statutory provisions

Health and Safety (Display Screen Equipment) Regulations (DSE) 1992 / Health and Safety (Display Screen Equipment) Regulations (Northern Ireland) 1992

Lifting Operations and Lifting Equipment Regulations (LOLER) 1998 / Lifting Operations and Lifting Equipment Regulations (Northern Ireland) 1999

Manual Handling Operations Regulations (MHOR) 1992 / Manual Handling Operations Regulations (Northern Ireland) 1992

Provision and Use of Work Equipment Regulations (PUWER) 1998 / Provision and Use of Work Equipment Regulations (Northern Ireland) 1999

Workplace (Health, Safety and Welfare) Regulations (WHSWR) 1992 / Workplace (Health, Safety and Welfare) Regulations (Northern Ireland) 1993

STUDY QUESTIONS

1) A welfare unit is to be lifted from the vehicle trailer it is being delivered onto its location on the construction site. What would the person appointed to have overall control of this operation need to ensure so that lifting operations are conducted safely?

2) A 100-metre-long section of a footpath in a residential area with a speed limit of 30mph is to have the kerbstones replaced.

   (a) What hazards could affect the health and safety of the workers involved in this process?

   (b) What control measures could be put into place to reduce the risk?

3) What conditions relating to lifting accessories could form part of a pre-use checklist?

4) What are the requirements for thorough examinations of lifting equipment to be carried out under the Lifting Operations and Lifting Equipment Regulations (LOLER)?

5) What measures should be taken to reduce the risk of a rough terrain fork-lift truck overturning?

6) Manual handling operations during construction activities can cause injuries.

   (a) What TWO types of injury could be caused by the incorrect manual handling of loads?

   (b) With reference to the work environment related to construction manual handling activities, what are the means of reducing the risk of injury during manual handling operations?

For guidance on how to answer these questions, please refer to the assessment section located at the back of this guide.

Element 9

Work equipment

9

Contents

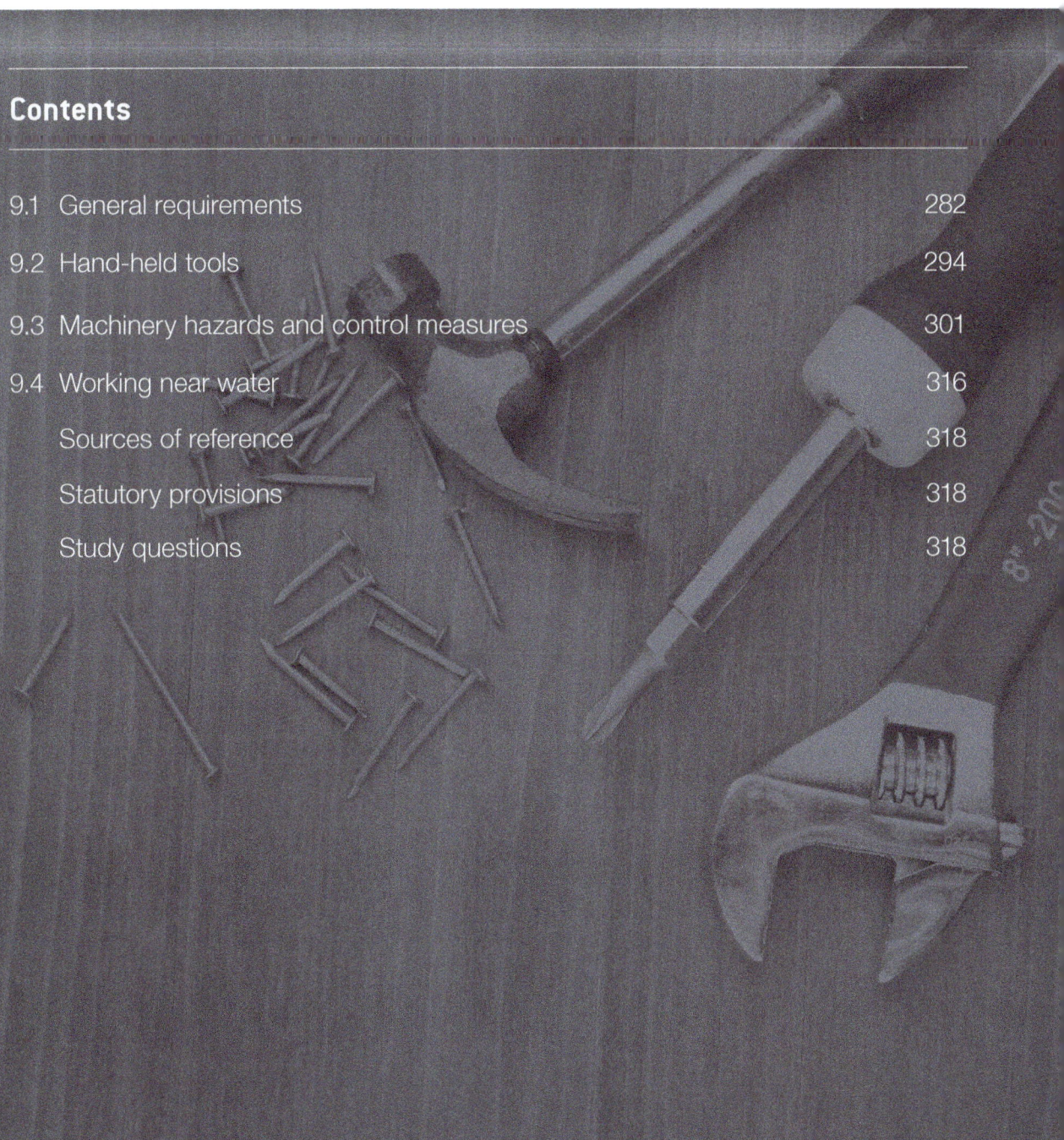

General requirements

TYPES OF WORK EQUIPMENT

The Provision and Use of Work Equipment Regulations (PUWER) 1998 are concerned with most aspects relating to work equipment. Work equipment may be defined as any machinery, appliance, apparatus, tool or assembly of components that are arranged so that they function as a whole. Clearly, the term embraces many hand tools, power tools and types of construction plant.

Examples of work equipment commonly used in the construction activities include:

- Air compressor.
- Cement mixer.
- Circular saw.
- Concrete pump.
- Dumper truck.
- Disc cutter.
- Electricity generator.
- Excavator.
- Hand grinder.
- Hammer.
- Hoist.
- Hop up work platform.
- Laser levelling device.
- Lifting sling.
- Mobile crane.
- Mobile elevating work platform (MEWP).
- Nail gun.
- Portable drill.
- Portable fire alarm call point and sounder.
- Rebar tying device.
- Road laying machines.
- Road roller.
- Sack truck.
- Scaffold.
- Scrabbler.

Not work equipment:

- Livestock.
- Substances.
- Structural items (buildings).
- Private car.

PROVIDING SUITABLE EQUIPMENT

General requirements for suitability of work equipment

The Provision and Use of Work Equipment Regulations (PUWER) 1998 requires duty holders to provide for the safe use of work equipment. The regulation cannot be considered in isolation from other health and safety legislation.

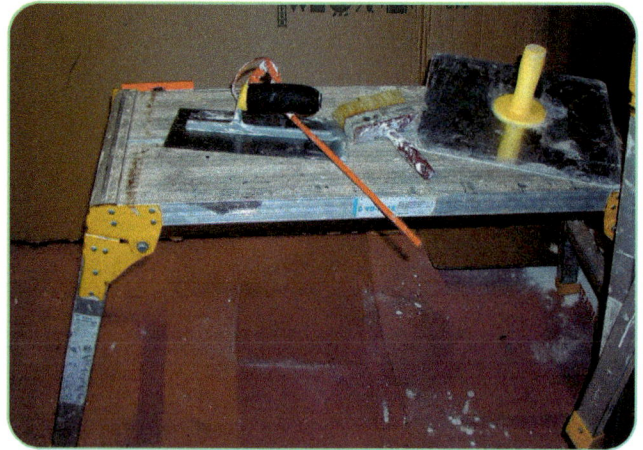
Figure 9-1: Hop up work platform and tools. Source: RMS.

Figure 9-2: Road roller. Source: RMS.

Figure 9-3: Disc cutter. Source: RMS.

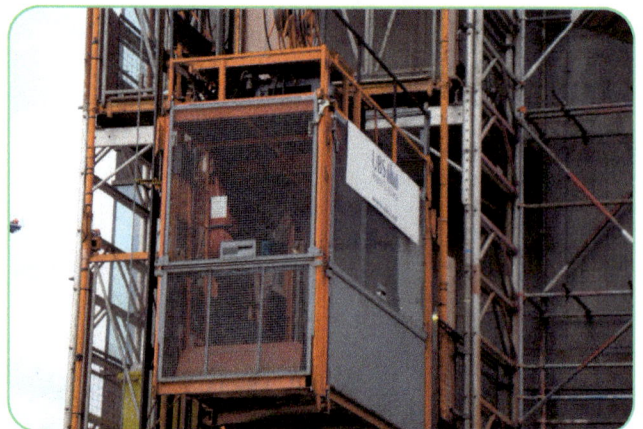
Figure 9-4: Hoist. Source: RMS.

In particular, it needs to be considered with the requirements of the Health and Safety at Work etc. Act 1974. When selecting work equipment, the employer should ensure that it is suitable. Suitability should consider:

- Its initial integrity.
- The place where it will be used.
- The purpose for which it is used.

Integrity - The equipment should be safe through its design, construction and adaptation. This requires the equipment to be made in conformity with relevant health and safety standards. Consideration should be given to manufacturing deficiencies that leave sharp edges on the equipment where it is handled, 'home-made' tools and any equipment adapted from its original design to do a specific task.

Place - The equipment should be suitable for different environments and the risks that relate to them. For example, the environment may have high levels of moisture and humidity or be wet due to the construction activity undertaken, or the atmosphere might be explosive. Account must also be taken of the possibility of the equipment causing a synergistic (combination) hazard in the workplace. For example, a petrol generator used in a confined space could emit toxic exhaust gases, or a hydraulic access platform may be used in a location with a low roof and present a risk of crushing the user. Similarly, the equipment may be designed for indoor use only and not have any protection from the ingress of water, which could present an electrical risk if used outdoors.

Use - The equipment should be suitable for the specific task for which it is used. Equipment used for any activity must be suitable to fulfil the exact requirements of the task. This means considering its strength, durability, power source, portability, protection against the environment, the range of tasks to be carried out, the frequency and duration of use and risks posed due to ergonomic constraints. Whenever equipment is required, the full capacity and limitation requirements should be identified and it should be confirmed that the equipment provided is suitable for the demands/limitations placed upon it. The equipment should be classed as an industrial or commercial type, designed for work activities, and not designed for use in the domestic environment.

For example, a hacksaw may be unsuitable for cutting metal straps used to secure goods to a pallet because the strap would be under tension and not restrained; a purpose-designed tool that cuts and keeps the strap safely would be more suitable than a simple saw. Similarly, a ladder may be unsuitable for work at a height where the worker needs to use both hands to carry out the task, and a scaffold or other access platform may be more suitable. Other examples of unsuitability include

using a crane or fork-lift truck to lift a heavy load that was in excess of the safe working capacity of the equipment or using a heavy electrical tool as a hammer.

In the UK, the main requirements of the Provision and Use of Work Equipment Regulations 1998 is to ensure that the equipment used is suitable for its purpose and maintained to be safe, (*see Figure 9-5*).

> "(1) Every employer shall ensure that work equipment is so constructed or adapted as to be suitable for the purpose for which it is used or provided.
>
> (2) In selecting work equipment, every employer shall have regard to the working conditions and to the risks to the health and safety of persons which exist in the premises or undertaking in which that work equipment is to be used and any additional risk posed by the use of that work equipment.
>
> (3) Every employer shall ensure that work equipment is used only for operations for which, and under conditions for which, it is suitable.
>
> (4) In this regulation "suitable" means suitable in any respect which it is reasonably foreseeable will affect the health or safety of any person."

Figure 9-5: Suitability of work equipment.
Source: Provision and Use of Work Equipment Regulations 1998.

Conformity with basic health and safety standards

PUWER 1998 Regulation 10 requires employers to ensure work equipment conforms to requirements:

> "(1) Every employer shall ensure that an item of work equipment has been designed and constructed in compliance with any essential requirements, that is to say, requirements relating to its design or construction in any of the instruments listed in Schedule 1 (being instruments which give effect to Community directives concerning the safety of products).
>
> (2) Where an essential requirement is applied to the design or construction of an item of work equipment, the requirements of Regulations 11 to 19 and 22 to 29 shall apply in respect of that item only to the extent that the essential requirement did not apply to it.
>
> (3) This Regulation applies to items of work equipment provided for use in the premises or undertaking of the employer for the first time after 31st December 1992."

Figure 9-6: Work equipment conforms to requirements.
Source: Provision and Use of Work Equipment Regulations 1998.

Those involved in the supply (including design and manufacture) of equipment have a responsibility to ensure that it is safe and healthy, so far as is reasonably practicable.

This requires them to take account of all relevant standards, including basic health and safety standards.

The International Organization for Standardization (ISO) produces a number of standards that relate to work equipment. These standards have been established by international agreement and are adopted by countries globally. They set out basic health and safety standards with which designers and manufacturers should comply, for example, ISO 12100: 'Safety of machinery - General principles of design - Risk assessment and risk reduction'. ISO 12100:2010 specifies basic terminology, principles and a methodology for achieving health and safety in the design of machinery. It also specifies principles of risk assessment and risk reduction to help designers in achieving this objective. These principles are based on international knowledge and experience of the design, use, incidents, accidents and risks associated with machinery. Procedures are described for identifying hazards and estimating and evaluating risks during relevant phases of the machine life cycle, and for the elimination of hazards or sufficient risk reduction. Guidance is also given on the documentation and verification of the risk assessment and risk reduction process. ISO 12100:2010 is also intended to be used as a basis for the preparation of type-B (which cover specific safety devices and machine ergonomics) or type-C health and safety standards (which deal with specific classes of machinery and their requirements) by countries. Other ISO standards that relate to work equipment include:

- ISO 13849-1:2006 'Safety of machinery - Safety-related parts of control systems - Part 1: General principles for design'.

- ISO 13850:2006 'Safety of machinery - Emergency stop - Principles for design'.

- ISO 13856-1:2013 'Safety of machinery - Pressure-sensitive protective devices - Part 1: General principles for design and testing of pressure-sensitive mats and pressure-sensitive floors'.

In the European Union, ISO 12100:2010 has been adopted and given an EN prefix to signify this. Equipment made in conformity with harmonised European standards and meeting essential health and safety requirements may carry a CE marking to confirm this. The letters 'CE' are the abbreviation of the French phrase 'Conformité Européene', which literally means 'European Conformity'.

It is a requirement of European Union legislation that employers in any member state ensure that an item of work equipment brought into use in the workplace has been designed and constructed in compliance with the directives/harmonised standards that relate to it and meets any essential health and safety requirements set for it. This includes types of work equipment that are classed as machinery.

Figure 9-7: CE mark.
Source: European Commission.

PREVENTING ACCESS TO DANGEROUS PARTS OF MACHINERY

Dangerous parts of machinery present particularly high risks to anyone that may be exposed to them. Employers should ensure that measures are taken to:

a) Prevent access to any dangerous part of machinery or to any rotating stock-bar.

b) Stop the movement of any dangerous part of machinery or rotating stock-bar before any part of a person enters a danger zone.

See section 9.3 – 'Machinery hazards and control measures', later in this element.

THE NEED TO RESTRICT THE USE AND MAINTENANCE OF EQUIPMENT WITH SPECIFIC RISKS

Where the use of work equipment is likely to involve a specific risk to health or safety, employers should ensure that:

- The use of that work equipment is restricted to those persons given the task of using it.

- Repairs, modifications, maintenance or servicing of that work equipment is restricted to those workers who have been specifically chosen to perform the tasks, and also those who have received adequate training related to the tasks they have been assigned.

For example, in view of the specific risks, it would be appropriate to restrict the use and operation of a telehandler, brush-cutter, abrasive wheel, nail gun, circular saw or mobile elevated work platform to those competent and authorised to use it. In the same way, maintenance should be restricted to persons who have the required competence to undertake the activity, for example, replacement of a grinding wheel, or replacement of load-bearing components of a rough-terrain fork-lift truck. Maintenance workers may not be considered to have the specific expertise or skill necessary to work on the equipment, particularly new equipment, leading to increased risk to themselves and others. It is essential that operation and maintenance of equipment with specific risks be restricted to those who have the appropriate skill and expertise.

PROVIDING INFORMATION, INSTRUCTION AND TRAINING

In the UK, there is a broad-based principle of 'duty of care' that requires workers to be involved in the prevention of health and safety risk and not to rely solely on the actions of their employer. This duty of care establishes the responsibility of workers for the health and safety of themselves and others that may be affected by their acts or omissions.

Users of work equipment, in particular machinery, should therefore use it in accordance with any training or instruction that they have received. They should use safety devices and protective equipment correctly and not render them inoperative.

Furthermore, users should report any faults or defects in their equipment that, if not rectified, could lead to the development of hazardous situations. The expectations placed on workers do not reduce the responsibility of the employer to ensure the use of work equipment is safe and healthy.

> "Every employee shall use any machinery, equipment, dangerous substance, transport equipment, means of production or safety device provided to him by his employer in accordance both with any training in the use of the equipment concerned which has been received by him and the instructions respecting that use which have been provided to him by the said employer in compliance with the requirements and prohibitions imposed upon that employer by or under the relevant statutory provisions."

Figure 9-8: Employee duties.
Source: Regulation 14, Management of Health and Safety at Work Regulations 1999.

Whenever equipment is provided and used in the workplace, employers should ensure that workers who use it are adequately trained and are supplied with adequate health and safety information, and, where appropriate, written instructions relating to its use.

Employers should also ensure that those who supervise or manage the use of work equipment have been adequately trained and are knowledgeable regarding the necessary health and safety information and, where appropriate, have written instructions relating to its use. Training, information and instructions include those concerning the:

- Conditions and methods of use of the work equipment, including the capacities and limitations of the equipment.
- Risks that may arise from use of the equipment.
- Precautions to be taken to avoid and reduce risk, including safe operating procedures provided by the

manufacturer/supplier and those drawn from experience in using the work equipment.

Training may be needed for existing workers, as well as workers using the equipment for the first time (including temporary workers), particularly if they have to use powered machinery. The greater the level of danger, the more substantial the training will need to be. For some high-risk work, such as driving fork-lift trucks, using a chainsaw or operating a crane, training should be carried out by specialist instructors. It should also be remembered that newly trained workers may obtain basic skills through training but may lack experience and judgement and require close supervision for an initial period. The requirement for refresher training at appropriate intervals should also be established. To ensure the effectiveness of training, those participating should be assessed for retention, comprehension and, where appropriate, skill.

Whilst employers should ensure that any written instructions are available to the people directly using the work equipment, they should also ensure such instructions are made available to other appropriate people. For example, maintenance instructions should be made available to the people involved in maintaining the work equipment.

Examples of training required for different workers:

Users - how to carry out pre-use checks, report defects, only to use equipment for the purpose designed. The training of someone to use a grinding machine should cover the proper methods of dressing the abrasive wheels.

Maintenance workers - safe isolation, acceptable replacement parts and adjustments in accordance with manufacturer's manuals. Maintenance workers should be given training on the specific risks related to maintenance, modification or repair work, such as stored energy in electrical systems.

Managers - be aware of the hazards and controls and maintain effective supervision.

In addition, where the use of work equipment is likely to involve a specific risk to health and safety, the employer must provide those authorised to repair, modify, maintain or service the equipment with adequate training.

WHY EQUIPMENT SHOULD BE MAINTAINED AND MAINTENANCE CONDUCTED SAFELY

Equipment to be maintained

Procedures for defective equipment
Under Section 7 of the HASAWA 1974 and Regulation 14 of the Management of Health and Safety at Work Regulations (MHSWR) 1999, workers should co-operate with the employer and notify it of any shortcomings

in the health and safety arrangements, even when no immediate danger exists, so that the employer can take remedial action if needed. This extends to the identification of defective equipment of which any worker becomes aware; they should take reasonable steps to safeguard themselves and others.

Appropriate action includes following procedures to deal with defective equipment and, where appropriate and with the employer's consent, taking it out of use and quarantining it until it can be repaired by a competent person.

Maintenance of work equipment

PUWER 1998, Regulation 5, requires employers to maintain work equipment: "Every employer shall ensure that work equipment is maintained in an efficient state, in efficient working order and in good repair."

Employers should ensure that work equipment is maintained throughout its working life, so that it is in efficient working order and in good repair. The manufacturer's instructions should be taken into account when maintenance work is planned and carried out.

The frequency at which maintenance activities are carried out should also take into account the intensity of use and the operating environment, for example, marine (wet, salt water) or outdoors on a construction site, the variety of operations (whether the machine performs many different tasks or just one task very frequently) and risk to health and safety from malfunction or failure.

The extent and complexity of maintenance may vary from simple checks to integrated programmes for complex plant. Maintenance must, however, be effective and be targeted at the parts of the work equipment where the failure or deterioration could lead to health and safety risks.

Figure 9-9: Inspection and examination of equipment.
Source: RMS.

When carrying out maintenance work on large process machinery, there are many hazards to be aware of. These include contact with dangerous moving parts of the machinery, electricity, stored energy such as heat or pressure; contact with gases, fumes and vapours and exposure to radiation and biological agents; manual handling of heavy machine parts or tools; noise and vibration, working at height or in confined spaces.

Practical measures that should be taken to protect people undertaking this type of work include: where possible, designing the machine to reduce the need to remove guards for routine maintenance and lubrication; the use of a permit-to-work system to assist in administrative controls to ensure that electrical power to the machine is isolated and locked off, and all pipelines leading to the machine are similarly isolated. This may include the need to release stored energy, for example, hydraulic pressure and, where necessary, to allow time for high-temperature equipment to cool down before maintenance starts. Where necessary, means of access such as a scaffold may have to be erected, and barriers and warning signs placed round the machinery to advise workers that maintenance work is in progress.

It is important that only skilled and competent workers undertake maintenance work and that they use appropriate tools, for example, spark-reduced tools in potentially flammable areas, a torque wrench for correct tensioning of securing bolts to pipework flanges. Consideration should be given to ensuring adequate standards of lighting and ventilation in the work area and to arrange for the work to be properly supervised and coordinated with other workers who may be present in the workplace at the time.

Where necessary, workers should be provided with appropriate personal protective equipment (PPE). Typical considerations include, for example, head protection (bump cap or helmet), eye protection (visor for arc welding, goggles for chemical splash) and fall-arrest systems (for example, harness, nets).

Inspections

Regulation 6 of PUWER 1998 lays down requirements for inspecting work equipment to ensure that health and safety conditions are maintained and that any deterioration can be detected and remedied in good time:

"Every employer shall ensure that work equipment exposed to conditions causing deterioration which is liable to result in dangerous situations is inspected:

(a) At suitable intervals.

(b) Each time that exceptional circumstances which are liable to jeopardise the safety of the work equipment have occurred, to ensure that health and safety conditions are maintained and that any deterioration can be detected and remedied in good time.

Every employer shall ensure that the result of an inspection made under this Regulation is recorded and kept until the next inspection under this Regulation is recorded."

Figure 9-10: Inspection of work equipment.
Source: Provision and Use of Work Equipment Regulations 1998.

For new equipment, inspections should be carried out as appropriate to the risks, for example, after installation and often with a commissioning engineer before first use. When existing equipment is modified or relocated, it should be re-inspected, for example when a hoist is erected at a new construction site.

Employers should ensure that work equipment exposed to conditions causing deterioration, for example, corrosion from aggressive construction process materials, or significant cycles in temperature change that may result in dangerous conditions, is inspected at an appropriate frequency. This should be:

- At suitable intervals – this may involve different degrees of inspection, including a thorough inspection by a competent person.

- Each time that exceptional unplanned circumstances occur – this could include natural phenomena (severe weather conditions such as high winds and seismic activity), accidents/incidents or a prolonged inactivity that is liable to result in damage or deterioration.

When equipment is used away from the workplace (for example, off-site) or when equipment is hired to meet a temporary requirement (for example, lifting equipment from a third party), it should be accompanied by relevant documentation indicating that a recent inspection has been carried out (many organisations use a system of labelling for portable electrical appliances, i.e. portable appliance testing (PAT). Equipment must be inspected by the user before use to identify any obvious defects and tested regularly by a competent person to ensure that it remains in a suitable condition. User checks may include guards, cables, casing integrity, cutting or machine parts and safety devices such as cut-outs. Competent persons include a maintenance worker as part of a preventive maintenance programme or it may be necessary to involve a third party, such as an insurance provider inspector, to carry out a periodic inspection of a pressure system or lifting equipment to ensure and validate its safe condition.

Inspections should address a list of identifiable health-and-safety-critical parts and will often use a check sheet as an aid to ensure all parts are considered. The purpose of an inspection check sheet is to record deterioration of specific parts, any abuse or misuse and any corrective action necessary. The results of the inspection will confirm whether or not a piece of equipment is in a safe enough condition to use. The inspection record should be kept for a suitable period and at least until the next inspection. The person carrying out the inspection must be competent. They should be capable of identifying any faults and determining their effects on health and safety. Specific requirements for the inspection, including thorough inspection (examination), of different

work equipment may be specified by national law, for example, equipment used for work at height.

Maintenance can be carried out according to various systems of control. These may include a reactive approach, i.e. breakdown maintenance, or more active approaches such as planned preventive maintenance or condition-based maintenance.

In addition to PUWER 1998, other Regulations, such as the Lifting Operations and Lifting Equipment Regulations (LOLER) 1998 require and set certain statutory inspection requirements. Note, also that Regulation 6(5) of PUWER 1998 has been amended by the Work at Height Regulations (WAH) 2005 to include 'work equipment to which Regulation 12 of the WAH 2005 applies'. *See also 'WAH 2005' in the 'Relevant statutory provisions' section.*

Breakdown maintenance

Breakdown maintenance is concerned with repair when things go wrong, it is reactive and causes delays to the work schedule. If breakdown maintenance is the only approach used, the employer must consider all modes of failure and identify any which may put workers at risk and establish appropriate secondary health and safety controls, for example, spillage containment, respiratory protection, guards and other PPE.

It is widely accepted that a maintenance strategy based solely on repair at the time of component breakdown is neither efficient nor effective and may result in unacceptable unsafe conditions.

Planned preventive maintenance

Planned preventive maintenance (PPM) is an important strategy in maintenance and seeks to maximise the life of the component/equipment. This is done through a number of maintenance methods that are both planned and preventive. The planned method seeks to inspect and replace components on a scheduled basis. Preventive maintenance seeks to keep the condition of the component at its best by carrying out frequent care, for example lubrication, adjustment, cleaning. Although all maintenance is preventive in some respect, the primary aim of PPM is to prevent failures occurring while the equipment is in use.

Benefits of PPM

The main benefits are:

- Extended life of components.
- Assurance of reliability.
- Confirmation of condition of components.
- Reduced risk of loss-producing failure events.
- Ability to carry out work at a suitable time.
- Better utilisation of maintenance staff.
- Less standby facility required.
- Less expensive (last-minute) contracted facility required.

- Cost-effective actions.
- Demonstrates the employer has taken steps to meet the legal duties to maintain safe equipment.

Condition-based maintenance

Condition-based maintenance (CBM) is sometimes referred to as 'predictive' maintenance. Unlike PPM, which uses information provided by manufacturers to set timescales for changing components, CBM is a technique that involves monitoring the condition of the equipment and predicting equipment failure. Typical measurements include:

- Vibration analysis – rotating equipment such as compressors, pumps, motors all exhibit a certain degree of vibration. As they degrade or fall out of alignment, the amount of vibration increases. Vibration sensors can be used to detect when this becomes excessive.
- Infrared – IR cameras can be used to detect high-temperature conditions in energised equipment.
- Ultrasonic – detection of deep sub-surface defects such as boat hull corrosion.
- Acoustic – used to detect gas, liquid or vacuum leaks.
- Oil analysis – to measure the number and size of particles in a sample to determine equipment wear.
- Electrical – motor current readings using clamp-on ammeters.
- Operational performance – sensors throughout a system to measure variables such as pressure, temperature, or flow.

Maintenance action is then determined by this monitoring and prediction information. Many CBM systems are controlled by computers. CBM assumes that all equipment will deteriorate and that partial or complete loss of function will occur at some point. CBM monitors the condition or performance of plant equipment through various technologies. The data is collected, analysed, trended, and used to project equipment failures. Once the timing of equipment failure is known, action (such as replacing an oil filter when a replacement is needed not on a predetermined schedule) can be taken to prevent or delay failure. In this way, the reliability of the equipment can remain high. CBM can rely on visual inspections but often uses technology to gather, store and analyse data that may require a substantial financial investment in measuring equipment and worker up-skilling. The initial costs of implementation for such a system may be high. Also, additional competences are required to interpret and utilise the results from the system analysis correctly.

Maintenance to be conducted safely

No one should be exposed to undue risk during maintenance operations. In order to achieve this, equipment should be stopped and isolated as appropriate before work starts. If it is necessary to keep equipment running then the risks must be justified and the work adequately controlled. Controls may take the form of reducing running speed or range of movement or providing temporary guards.

Maintenance hazards

The principal sources of hazards are associated with maintenance work on:

- Heavy plant.
- Hoists.
- Cranes.
- Concrete pumps.
- Live electrical equipment.
- Storage tanks.
- Crushers.
- Storage tanks.
- Crushers

Figure 9-11: Hoist.
Source: RMS.

Typical hazards associated with maintenance operations include:

Mechanical: Entanglements, machinery traps, contact, shearing traps, in-running nips, ejection, unexpected start-up.

Electrical: Electrocution, shock, burns.

Pressure: Unexpected pressure releases, explosion.

Physical: Extremes of temperature, noise, vibration, dust.

Chemical: Gases, vapours, mists, fumes.

Structural: Obstructions, floor openings, voids.

Access: Work at heights, confined spaces.

Typical accidents/incidents associated with maintenance operations include:

- Crush, by moving machinery.
- Falls.
- Burns.
- Asphyxiation.
- Electrocution.
- Explosions.

One or more of the following factors causes maintenance accidents/incidents:

- Lack of perception of risk by managers/supervisors, often because of lack of necessary training.
- Unsafe or no system of work devised, for example, no permit-to-work system in operation, no facility to lock-off machinery and electricity supply before work starts and until the work has finished.
- No coordination between workers, and lack of communication with other supervisors or managers.
- Lack of perception of risk by workers, including failure to wear protective clothing or equipment.
- Inadequacy of design, installation, siting of plant and equipment.
- Use of sub-contractors with no health and safety control systems, i.e. risk assessments, method statements or who are inadequately briefed on health and safety aspects specific to the site.
- Lack of appreciation of synergistic (combined) hazards in the workplace.

Maintenance control measures

Isolation

This does not simply mean switching off the equipment using the stop button. It includes switching the equipment off at the electrical isolator for the equipment. In new workplaces, individual equipment isolators should be provided, i.e. each piece of equipment has its own isolator near to it. One isolator should not control several items of equipment as it is then impossible to isolate a single piece of equipment.

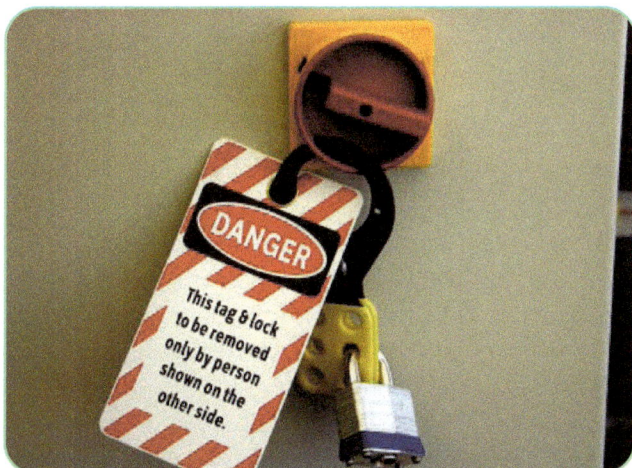

Figure 9-12: Electrical isolator with hole for padlock.
Source: RMS.

Figure 9-13: Physical isolation of valve.
Source: RMS.

Lockout and tagout

Electrical isolation by itself does not afford adequate protection, because there is nothing to prevent the isolator being switched back on or someone replacing removed fuses while the maintenance worker who isolated the equipment is still working on the equipment and is in danger.

To ensure that this does not happen, the isolator needs to be physically locked in the off position, typically using a padlock, and the key hold by the worker that may be in danger if the isolation is removed. Multiple lockout devices are often used where multiple trades are carrying out work on the same equipment or plant. The multiple lockout devices are designed to accommodate a number of isolation padlocks for the different people working on the equipment, and the equipment cannot be energised until all the padlocks are removed, thereby protecting the last worker on the job. It is also a good idea to add a sign or tag – 'Do not switch on...' – on the equipment at the point of isolation. This approach is often called lockout and tagout (LOTO). Furthermore, after lockout and tagout (LOTO), maintenance workers should test the circuit before working on it, to ensure the circuit is no longer live and there is no residual energy in the circuit. Hence LOTO – lockout and tagout becomes LOTOTO – lockout, tagout and testout.

Figure 9-14: Multiple (padlock) lock-off device.
Source: RMS.

The control measures to enable maintenance work to be conducted safely include:

- Plan work in advance – provide safe access, support parts of equipment that could move or fall.
- Use written safe systems of work, method statements or permit-to-work systems as appropriate.
- Plan specific operations using method statements.
- Use physical means of isolating or locking off plant.
- Incorporate two-man working for high-risk operations.
- Integrate safety requirements in the planning of specific high-risk tasks.
- Prevent unauthorised access to the work area by using barriers and signs.
- Ensure the competence of those carrying out the work.
- Ensure the availability and use of appropriate PPE – gloves, eye protection, etc.
- Prevent fire or explosion – thoroughly clean vessels that have contained flammable solids or liquids, gases or dusts and check them thoroughly before hot work is carried out.
- Consider the need for rescue or treatment of workers in the event of an accident/incident.

EMERGENCY OPERATION CONTROLS, STABILITY, LIGHTING, MARKINGS AND WARNINGS, CLEAR WORK SPACE

Operation and emergency controls

Controls for starting or making a significant change in operating conditions

Equipment should be fitted with a specific start-control device. It should only be possible to start equipment by positive, voluntary activation of the control device provided for that purpose. Start controls should be shrouded or otherwise protected to prevent inadvertent operation. Near each start control there should be a stop control. If it is possible to restart or change the operating conditions of the machine without using the start-control device, this should only be done where it does not lead to a hazardous situation (for example, the initiation of certain functions of machinery by the closure of an interlocking guard).

Regulation 14 of PUWER 1998 requires that equipment has appropriate controls for starting and changing conditions:

"(1) Every employer shall ensure that, where appropriate, work equipment is provided with one or more controls for the purposes of:

(a) Starting the work equipment (including re-starting after a stoppage for any reason).

(b) Controlling any change in the speed, pressure or

other operating conditions of the work equipment where such conditions after the change result in risk to health and safety which is greater than or of a different nature from such risks before the change.

Figure 9-15: Control devices for work equipment.
Source: Provision and Use of Work Equipment Regulations 1998.

For equipment that is capable of functioning in automatic mode, the starting of the equipment, restarting after a stoppage, or a change in operating conditions may be possible without intervention provided this does not lead to a hazardous situation.

Where equipment has several start-control devices and users can therefore place one another in danger, additional devices should be fitted to preclude such risks, for example, a system that requires positive start signals from all users before the equipment can start. If starting and stopping must be performed in a specific sequence in order to ensure health and safety, there should be devices that ensure that these operations are performed in the correct order. Any change in the operating conditions should only be possible by the use of a control, unless the change does not increase risk to health or safety. Examples of operating conditions include speed, pressure, temperature and power.

The controls provided should be designed and positioned so as to prevent, as far as possible, inadvertent or accidental operation. Buttons and levers should be of appropriate design, for example, including a shroud or locking facility. It should not be possible for the control to 'operate itself' due to the effects of gravity, vibration, or failure of a spring mechanism, for example.

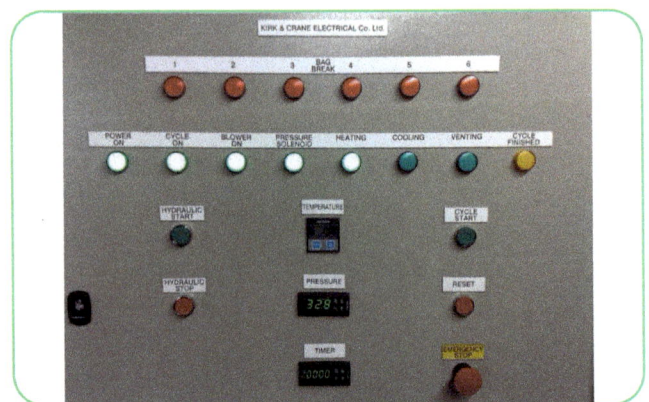

Figure 9-16: Control panel clearly marked; operation status displayed; emergency stop available.
Source: Kirk & Crane Electrical Co. Ltd.

Stop controls

The action of a stop control should bring the equipment to a safe condition in a safe manner. This acknowledges that it is not always desirable to bring all items of work equipment immediately to a complete or instantaneous stop, for example, to prevent the unsafe build-up of heat or pressure or to allow a controlled run-down of large rotating parts. Similarly, stopping the pumping mechanism of a concrete pump during operation could lead to blockages.

The stop control should have priority over the start controls. Once the equipment or its hazardous functions have stopped, the energy supply to the actuators concerned should be cut off, unless this is undesirable for operational reasons; in these circumstances, the stop condition should be monitored and maintained. Accessible dangerous parts must be rendered stationary. However, parts of equipment that do not present a risk, such as suitably guarded cooling fans, do not need to be positively stopped and may be allowed to idle.

"Regulation 15 - Stop controls

(1) Every employer shall ensure that, where appropriate, work equipment is provided with one or more readily accessible controls the operation of which will bring the work equipment to a safe condition in a safe manner.

(2) Any control required by paragraph (1) shall bring the work equipment to a complete stop where necessary for reasons of health and safety.

(3) Any control required by paragraph (1) shall, if necessary for reasons of health and safety, switch off all sources of energy after stopping the functioning of the work equipment.

(4) Any control required by paragraph (1) shall operate in priority to any control, which starts or changes the operating conditions of the work equipment."

Figure 9-17: Stop controls for work equipment.
Source: Provision and Use of Work Equipment Regulations 1998.

Emergency stops are intended to effect a rapid response to potentially dangerous situations and they should not be used as functional stops during normal operation. Emergency stop controls should be:

- Coloured red.
- Positioned in such a way as to be safely operated without hesitation or loss of time and without ambiguity.
- Designed in such a way that the movement of the control device is consistent with its effect.
- Located outside the danger zones, except where necessary for certain control devices such as an emergency stop (for use of maintenance workers) or a teach pendant (hand-held control).
- Positioned in such a way that their operation cannot cause additional risk.
- Designed or protected in such a way that the desired effect can be achieved only by a deliberate action.
- Made in such a way as to withstand any foreseeable forces; particular attention should be paid to emergency stop devices liable to be subjected to considerable force at the time of use.

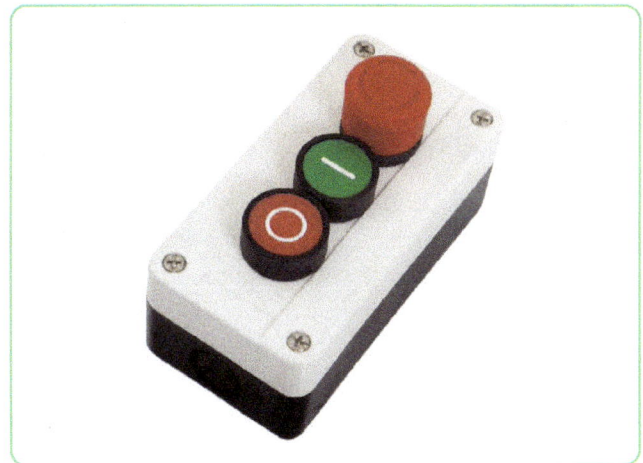

Figure 9-18: Start and stop controls.
Source: Fine Woodworking Tools.

Emergency stop controls should be easily reached and actuated. They should remain available and operational at all times, regardless of the operating mode of the equipment, for example, where special operating modes are applied to the equipment during maintenance work. Common types are mushroom-headed buttons, bars, levers, kick plates or pressure-sensitive cables.

Once the emergency stop control has been activated, the command must remain in place until the control has been specifically overridden by a direct action to disengage the control by resetting it. Disengagement of the emergency stop control must not automatically restart the equipment, but it should require a positive action by the operator to restart it.

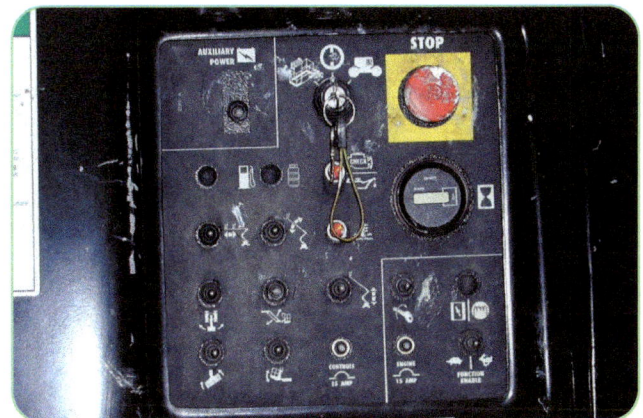

Figure 9-19: Control panel showing emergency stop button.
Source: RMS.

"Regulation 16 - Emergency stop controls

(1) Every employer shall ensure that, where appropriate, work equipment is provided with one or more readily accessible emergency stop controls unless it is not necessary by reason of the nature of the hazards and the time taken for the work equipment to come to a complete stop as a result of the action of any control provided by virtue of regulation 15(1).

(2) Any control required by paragraph (1) shall operate in priority to any control required by regulation 15(1)."

Figure 9-20: Emergency stop controls for work equipment.
Source: Provision and Use of Work Equipment Regulations 1998.

Controls

It should be possible to identify easily what each control does and on which equipment it takes effect.

Controls should be:

- Clearly visible and identifiable.
- Readily distinguishable from one another by their separation, size, shape, colours or feel, and by labelling.
- Labelling should be either with words or unambiguous and easily recognisable symbols to identify the functions of the controls.

Controls should be so arranged that their layout, travel and resistance to operation are compatible with the action to be performed, taking account of ergonomic principles. Equipment should be fitted with such indicators (such as red and green warning lights) as may be required for safe operation. The operator should be able to read them from the control position.

Except where necessary, the employer should ensure that no control for work equipment is in a position where any person operating the control is exposed to a risk to their health. For example, an emergency stop control may be placed inside a danger zone for use of workers during maintenance or setting robot equipment in teach mode.

From each control position, the operator should be able to ensure that nobody is in a danger zone. Alternatively, the control system should be designed and constructed in such a way that starting is prevented while someone is in a danger zone. If neither of these solutions is practicable, an acoustic and visual warning signal should be given before the equipment starts. Any persons exposed should have time to leave the danger zone or prevent the machinery from starting up.

As well as time, suitable means of avoiding the risk should be provided. This may take the form of a device used by workers at risk to prevent start-up or to warn the operator of their presence, for example the provision of a wedge or block to physically prevent the movement of equipment. Otherwise, there must be adequate provision to enable workers at risk to withdraw, for example, sufficient space or exits.

Circumstances will affect the type of warning chosen. If necessary, the employer should ensure that the equipment can only be controlled from control positions located in one or more predetermined zones or locations. Where there is more than one control position, the control system should be designed in such a way that the use of one of them prevents the use of the others, except for stop controls and emergency stops.

Figure 9-21: Controls on top of MEWP.
Source: RMS.

Control systems

Control systems should be designed and constructed in such a way as to ensure that as few hazardous situations as possible arise. They should be designed and constructed taking into account the following aspects:

- It should be able to withstand the intended operating stresses and external influences, taking into account foreseeable abnormal situations. External stresses include humidity, temperature, impurities, vibration and electrical fields.

- Its operation should not create any increased risk to health or safety.

- It does not impede the operation of stop and emergency stop controls.

- The loss of supply of any source of energy used by the work equipment cannot result in additional or increased risk to health or safety.

- A fault in the hardware or software of the control system does not lead to hazardous situations.

- Errors in the control system logic should not give rise to hazardous situations.

- Reasonably foreseeable human error during operation should not give rise to hazardous situations.

There are national, European and international standards (for example, BS EN 60204 series) which provide guidance on design of control systems so as to achieve high levels of performance related to safety. They suggest particular attention should also be paid to:

- Machinery should not start unexpectedly.
- The parameters of the machinery should not change in an uncontrolled way.
- Machinery should not be prevented from stopping if the stop command has already been given.
- No moving part of the machinery or piece held by the machinery should fall or be ejected.
- Automatic or manual stopping of the moving parts, whatever they may be, should be unimpeded.

- Protective devices should remain fully effective or give a stop command.
- Safety-related parts of the control system should apply in a coherent way to the whole of an assembly of machinery and partly completed machinery.
- For cableless control, an automatic stop should be activated when correct control signals are not received, including loss of communication.

Failure of any part of the control system or its power supply should lead to a 'fail-safe' condition (more correctly and realistically called 'minimised failure to danger'), and not impede the operation of the stop or emergency stop controls. The measures that should be taken in the design and application of a control system to mitigate against the effects of failure will need to be balanced against the consequences of any failure, and the greater the risk, the more resistant the control system should be to the effects of failure.

Stability

Work equipment should be designed to be stable enough to avoid overturning, falling or uncontrolled movement during transportation, assembly, installation, dismantling and use. If the shape and weight distribution of the equipment does not provide sufficient stability, appropriate means of stabilisation must be made, including the use of anchorages and additional bracing or support structures. Where equipment has a variable reach or if the weight and centre of gravity may change during use it is important to take this into account.

Most machines used in a fixed position should be bolted or otherwise fastened down so that they do not move or rock during use. This is particularly important where the equipment is tall relative to its base or has a high centre of gravity, for example, a pedestal-mounted abrasive wheel, platform hoists and tower cranes. It has long been recognised that woodworking and other machines (except those specifically designed for portable use) should be bolted to the floor or similarly secured to prevent unexpected movement.

"Regulation 20 - Stability

Every employer shall ensure that work equipment or any part of work equipment is stabilised by clamping or otherwise where necessary for purposes of health or safety."

Figure 9-22: Stability of work equipment.
Source: Provision and Use of Work Equipment Regulations 1998.

Lighting

Any workplace where a person is using work equipment must have adequate and suitable lighting for the operation of the equipment, in particular machinery, so that equipment movements, controls and displays can easily be seen.

Local lighting may be needed to give sufficient view of a dangerous process or to reduce visual fatigue. Equipment should be supplied with integral lighting suitable for the operations concerned where the absence of integral lighting would be likely to cause a risk despite ambient lighting of normal intensity.

Artificial lighting should, where possible, not produce glare, any stroboscopic effects (a visual phenomena that occurs when something that is moving appears to be represented by a series of stationary or short images – this is also known as the 'wagon wheel effect' where wagon wheels on motion pictures can appear to rotate in the opposite direction of travel) or disturbing shadows.

Lighting should be suitable for the environment in which it is used. For example, it should be intrinsically safe in flammable atmospheres. Equipment should be designed and constructed so that there is no area of shadow likely to cause nuisance during its operation and no irritating dazzle. Construction sites may not have sufficient general lighting and it may be necessary to provide equipment with local task lighting to ensure lighting levels are adequate for their safe operation. Internal parts of equipment requiring frequent inspection and adjustment, as well as the maintenance areas, should be provided with appropriate lighting.

"Regulation 21 - Lighting

Every employer shall ensure that suitable and sufficient lighting, which takes account of the operations to be carried out, is provided at any place where a person uses work equipment."

Figure 9-23: Lighting for work equipment.
Source: Provision and Use of Work Equipment Regulations 1998.

Markings and warnings

The employer should ensure that work equipment is marked in a clearly visible, legible and indelible manner with any marking appropriate for reasons of health and safety. Equipment designed and constructed for use in a potentially explosive atmosphere or similar hazard condition should be marked accordingly. Certain markings may also serve as a warning, for example, the maximum working speed, maximum working pressure, maximum working load or the contents of the equipment being of a hazardous nature (for example, the colour coding of gas bottles or service mains).

Where residual risks to health and safety remain following the provision of markings and other physical precautionary measures, warnings and warning devices should be introduced. Warnings are usually in the form of a notice, sign or similar. Examples of warnings are positive instructions (hard hats must be worn); prohibitions (no naked flames) and restrictions (do not heat above 60°C).

Element 9

Warning devices are active units that give out either an audible or visual signal, usually connected to the equipment in order that it operates only when a hazard exists; for example, when equipment such as a conveyor starts to operate and workers need to move to a safe position, or when a fault in the operation of unsupervised equipment may endanger people. Warning devices are also used where the equipment is mobile; a flashing light can warn of the presence of a dumper truck and an audible device may warn of a vehicle reversing. Markings and warnings given by warning devices on work equipment should be unambiguous, easily perceived and easily understood. Text, signs and pictograms should only be used if they are understood in the culture in which the equipment is to be used. Consideration may have to be given to international and national standards concerning colours and safety signs/signals, for example, ISO 7010:2011 'Graphical symbols - Safety colours and safety signs used in the workplace and public areas'.

Clear unobstructed workspace

Workrooms should have enough free space to allow people to get to and from workstations and to move within the room with ease. The workspace that is necessary for the tasks to be carried out should be determined and provided to ensure safe working. Workrooms should be of sufficient height (from floor to ceiling) over most of the room to enable safe access to workstations. In older buildings with obstructions such as low beams, the obstruction should be clearly marked.

Where work equipment such as a circular saw is used in a workplace, care should be taken to ensure that adequate space exists around the equipment to make sure it is not overcrowded, and does not cause risk to users and those passing by the equipment when it is operating. It is important that the workspace allocated for users of work equipment is maintained clear and unobstructed. Obstructions could lead to slips, trips and falls, which could also involve contact with the equipment or moving parts of the equipment. Material being readied to be worked on, material being worked on and any scrap material should not be allowed to build-up in the workspace. Material such as dust, scraps of wood or metal can quickly build-up as the equipment is used. Good control of the input, output and processing of material is essential. Scrap material must be removed from the workspace at suitable intervals, so that it does not build-up and obstruct the workspace.

"(4) A construction site must, so far as is reasonably practicable, have sufficient working space and be arranged so that it is suitable for any person who is working or who is likely to work there, taking account of any necessary work equipment likely to be used there."

Figure 9-24: Workspace on construction sites.
Source: CDMR 2015, Regulation 17.

REVIEW

List two categories of worker that should receive information, instruction and training on the safe use of work equipment.

What hazards are typically associated with maintenance work?

List the requirements for a stop button.

9.2 Hand-held tools

GENERAL CONSIDERATIONS FOR SELECTING HAND-HELD TOOLS (POWERED OR MANUAL)

Requirements for safe use

Workers need to be able to recognise the hazards associated with the different types of tools for their work and know the necessary safety precautions. They should be trained in the proper use of the tools they use. The tools must be suitable for the purpose and conditions (wet, dusty, flammable vapours) of use.

Requirements for safe use of hand-held tools include:

- Keep all tools in good condition with regular maintenance.
- Workers should only use tools they have been trained and authorised to use.
- Use the right tool for the work to be done – for example, use the correct size of spanner and the correct type of saw.
- Examine each tool for damage before use and do not use damaged tools.
- Operate tools according to the manufacturers' instructions.
- Use appropriate personal protective equipment.
- Do not misuse tools, for example using a screwdriver as a chisel or extending the handle of a spanner to improve leverage.
- Frequently used tools are within convenient reach.
- Adequate space is provided in the work area.
- Work benches are of a suitable height to avoid the need for awkward and uncomfortable postures.

Condition and fitness for use

Hand tools should be kept in good condition in order for them to be able to do the work required efficiently and safely, for example, sharp blades enable easy cutting with less effort, which means the worker's hands can be kept away from the blade and the tool is less likely to slip. Examples of condition and fitness for use include:

- Hammers and mallets - avoid split, broken or loose shafts and worn or chipped heads. Make sure the heads are properly secured to the shafts.

- Cutting tools – keep the cutting edge sharp and cover as much of the exposed cutting edge as is possible when in use.

- Files - should be fitted with a proper, secure handle. Never use them as levers.

- Chisels - the cutting edge should be sharpened to the correct angle. Do not allow the head of cold chisels to spread to a mushroom shape – grind off the sides regularly. Use hand guards to prevent impact injuries when using chisels. To prevent damage to wooden handles and the potential for wood splinters, a mallet should be used for striking wood chisels.

Figure 9-25: Damaged chisel.
Source: RMS.

Suitability for purpose

Tools are designed and manufactured for specific tasks, which is why screwdrivers are made with different tip sizes and length. Many injuries happen because workers do not use a tool that is suitable for the work being done. Therefore tools should be suitable for the purpose they are to be used for. The employer should ensure that workers are provided with suitable equipment for the work being done. Consider the demands of the work activity and the suitability of the design of the equipment being used, in terms of size, shape and how appropriate it is for the task. Suitable for purpose will include:

- Quality of materials the tools are made from, manufactured to a recognised standard.

- Insulated tools for electrical work.

- Non-sparking tools for flammable atmospheres.

- The right size, for example the right size of spanner, hammer or screwdriver.

- Matching the features of the tool to the needs of the work being done, for example using pliers with serrated jaws where improved grip of items is required.

- Using low energy power tools where possible, for example battery operated or low pressure air driven tools.

- Not improvising by using a tool not designed for the work, for example, not using a screwdriver as a lever.

Location to be used in (including flammable atmosphere)

It is important to consider the location where hand-held tools are going to be used, for example whether the atmosphere is flammable, wet, dark or at a height.

When working in areas where a flammable atmosphere is present, using non-sparking tools is necessary to prevent the creation of sparks that could cause the flammable atmosphere to ignite. Non-sparking tools are made of materials that do not contain iron (non-ferrous metals) and therefore the risk of a spark being created while the tool is in use is reduced. Common materials used for non-sparking tools include brass, bronze, copper-nickel alloys and copper-aluminum alloys. Non-sparking tools can also be made of wood, leather and plastics. Some common tools that are available in a non-sparking option include hammers, chisels, pliers, screwdrivers, shovels and spanners. Specialist electrically powered hand-held tools are also available for use in flammable atmospheres.

If the location the tools are to be used in is wet it may mean that some powered tools may not be suitable, which could limit the tools that are suitable for this location to those that are manually operated, specialist electrical tools or tools operated by compressed air.

In locations that are dark supplementary lighting will be essential to ensure the worker can operate the tools safely. Where the location is cold this may require the worker to wear gloves and necessitate the use of tools designed to enable them to be gripped easily while wearing gloves. If the location of the work activity is at a height consideration should be made to the possibility of the tools being dropped a significant distance that could harm people and measures put in place to prevent this.

HAZARDS AND CONTROL MEASURES FOR A RANGE OF HAND-HELD TOOLS (MANUALLY OPERATED)

Anyone who uses a hand tool may be at risk of injury, either accidentally or through misuse or equipment failure.

There is a range of injury hazards relating to the use of hand tools. For example, noise-induced hearing loss from cutting or impact tools; respiratory disease from inhalation of dust from hand sanders; eye injury from material thrown off from cutting; punctures and cuts caused by equipment that has features that are sharp such as a nail gun, knives, chisels, saws, planes and screwdrivers. Heat-producing equipment such as blowtorches can cause burns and permanent scarring.

Misuse includes using the wrong tool for the job, for example, using a hammer rather than a screwdriver to drive a screw into material, or using a file as a lever, as it is brittle and may break unexpectedly.

Examples of misuse of hand tools are shown in the following schematics:

Figure 9-26: Hammer shaft split with peg holding on head and very worn hammerhead.
Source: ILO.

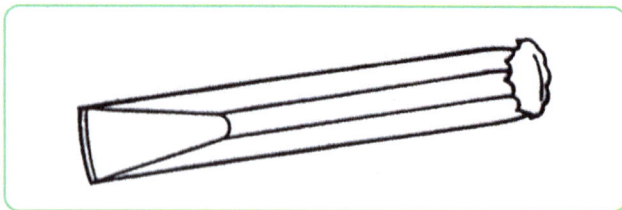

Figure 9-27: Mushroom-shaped head.
Source: ILO.

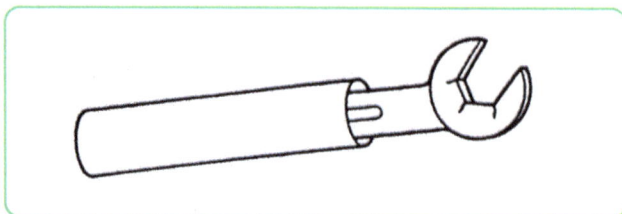

Figure 9-28: Piece of tube used as an extension.
Source: ILO.

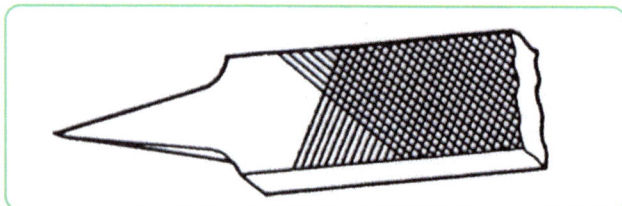

Figure 9-29: Broken file.
Source: ILO.

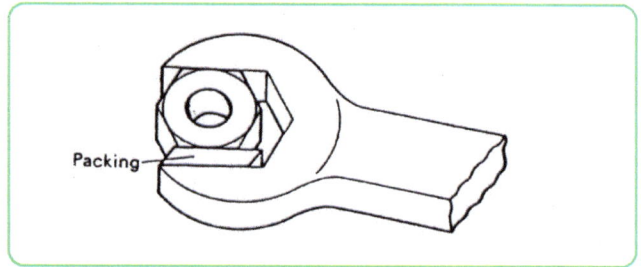

Figure 9-30: Packing material used in spanner head.
Source: ILO.

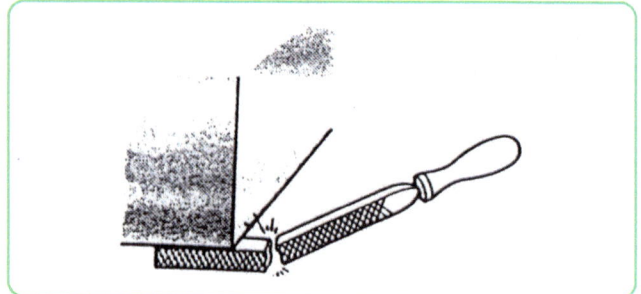

Figure 9-31: File used as a lever.
Source: ILO.

Hammers

Hazards

The main hazard when using a hammer is that the worker may use a hand to hold what is being hit and the hammer may hit the worker's hand or fingers instead of the intended item. Also, the hammer head could hit the intended item, but due to its shape or the angle of the impact the hammer head could slip sideways and impact the worker or twist the handle in the worker's hand, which could damage the wrist. These impacts could break a bone in the wrist, hand or fingers, as well as cause minor scrapes, cuts and bruising to other parts of the body. If the hammer head is not secure on the handle, the head could fly off when in use. The handle could become split and cause splinters or cuts to the worker's hand. If the hammer head becomes chipped, parts of it may fly off when it impacts what is being hit, this could cause eye injury or cuts.

Control measures

The following control measures will help to avoid harm from the use of hammers:

- Check the condition of the hammer – avoid split, broken or loose handles and worn or chipped heads. Heads should be properly secured to the handles.
- Place the work against a hard surface.
- Grip the handle firmly.
- Hold the hammer at the end of the handle.
- Check area is clear around you before swinging the hammer.
- Hit the surface of the work squarely with the hammer.
- Use the whole arm and elbow.
- Work in a natural position.
- Practise good hammering technique.

Files

Hazards

The main hazard when using a file is the tang, which is the pointed tip of the file that is fitted into the handle. A handle must be fitted over the tang before using the file, otherwise there is a good chance of puncturing the hand or other parts of the body with it when using the file. The handle could become split and cause splinters or cuts to the worker's hand.

If the surface of the file becomes blocked with material being filed or the file does not contact the material properly when filing, there is a risk that the file could suddenly slip forward across the surface of the material and cause damage to the worker's hand.

Files are very brittle and can snap easily if too much pressure is applied or if they are used as levers.

Control measures

These should have a proper, well-designed handle. The file should be held firmly in one hand with the fingers of the other hand used only to guide it. The material being filed should be firmly held in clamps or a vice. File strokes should be made away from the user. Take care to avoid the file slipping on the surface of the material causing sudden forward movement of the user. Files must never be used as a lever – they are very brittle and will shatter easily. They should only be cleaned by using a cleaning card and not by striking them against a solid object as this can also cause the file to shatter.

Chisels

Hazards

When a chisel is used there is a risk of injury, particularly to the eyes, from flying particles of the material being chiselled. The power required when using a chisel on metal may cause the workpiece to fly out from its holding device. Chisels have a cutting edge, which for wood chisels is particularly sharp, therefore there is a risk of cuts to the hand when handling them.

If the chisel head is allowed to become "mushroomed" pieces of the head can break off during use and cause eye injury. If a chisel is used as a screwdriver or a lever, the tip of the chisel may break suddenly and fly off, hitting the user or other workers.

Control measures

Choose a chisel large enough for the job, so the blade is used rather than only the point or corner. Never use a chisel with a dull blade – the sharper the tool, the better the performance. Chisels that are bent, cracked, or chipped should be discarded. The cutting edge should be sharpened to the correct angle. Do not allow the head of cold chisels to spread to a mushroom shape – grind off the sides regularly. Use a hand guard on the chisel and hit the chisel squarely. Chisels should not be used as a lever as they may break suddenly.

Screwdrivers

Hazards

If a worker holds a screw in their hand and the screwdriver slips out of the screw head while the worker is applying a powerful twisting motion, the screwdriver could move suddenly and stab or cut the worker's hand. A greasy handle could also cause the screwdriver to slip and lead to injury. The handle could become split and cause splinters or cuts to the worker's hand.

Repetitive use of a screwdriver, involving powerful, twisting motions, may cause health conditions, such as carpal tunnel syndrome. Most screwdrivers have metal shanks, therefore there is a risk of electric shock if the screwdriver is used near a live electrical sources.

Control measures

A screwdriver is one of the most commonly used and abused tools. The practice of using screwdrivers as punches, wedges or levers should be discouraged as this dulls blades and may cause injury if the blade fractures or slips in use. Screwdrivers should be selected so the tip fits the screw. When working on electrical equipment, screwdrivers must be equipped with insulated handles and shanks appropriate to the voltage of the equipment. Screwdrivers with split handles or damaged tips should be taken out of use and discarded safely.

Spanners

Hazards

Hazards of using a spanner may vary depending on the work being done, but could include:

- Powerful, repetitive action to turn the spanner could lead to upper limb disorders, for example damage to soft tissue in the hand, wrist and shoulder.

- The spanner slipping off the nut or other item being turned could cause the worker's hand to impact with things nearby.

- The spanner or item being turned may break with similar effects.

- The workpiece may suddenly break free from the device holding it, causing the worker to lose balance and fall.

Control measures

Avoid spanners with splayed or damaged jaws. Use ring spanners or sockets where possible, as these are less likely to slip. Discard safely any spanners that show signs of slipping. Ensure enough spanners of the correct size are available. Do not improvise by using pipes, etc., as extensions to the handle. Where necessary, use penetrating oil to loosen tight nuts.

Knives

Hazards

The most obvious hazard when using knives is the sharp blade which can easily cut the skin.

Control measures

Knives cause more disabling injuries than any other hand tool. The risks are that the hands may slip from the handle onto the blade or that the knife may strike the body or the free hand. Use knives with handle guards if possible. Make sure that the cutting stroke is always away from the body. Return the knife to a safe place and ensure it is sheathed before conducting other tasks, like moving a sack that the user has opened. Do not hold a knife with an open blade while carrying other objects.

Knives must be kept sharp and in their holders or sheaths when not in use. Dirty or oily knives should be wiped clean so that the user's grip does not slip. To clean the blade, wipe with a towel or cloth with the sharp edge turned away from the wiping hand. Foolish behaviour of any kind involving a knife (throwing, 'fencing', etc.) must not be tolerated. Where the atmosphere may be flammable, use alloy or bronze tools to prevent sparks.

HAZARDS AND CONTROL MEASURES FOR A RANGE OF HAND-HELD TOOLS (POWER OPERATED)

Anyone who uses a hand-held portable power tool may be at risk of injury from the power source, equipment failure or the action of the power tool coming into contact with the worker, either by normal use, accidentally or through misuse. Injury hazards include hand-arm vibration, for example, caused by the use of hand-operated power tools.

Workers who regularly use these as part of their job may be at risk of a permanent injury known as hand-arm vibration syndrome (HAVS). In addition, many portable power tools are powered by electricity and their use in flammable atmospheres could present a risk of fire or explosion.

Electric drill

Electric drills are used for penetrating various materials and in construction are usually of a medium to heavy-duty nature. This equipment involves rotating shafts and tool bits, sharp tools, electricity and flying debris.

Hazards

Obvious hazards include shock and electrocution leading to possible fatalities; however, this potential is reduced by using 110 volt or battery-operated equipment. Other hazards include puncture, entanglement, noise and dust.

Control measures

Control measures include using only equipment that is suitable for the task and ensuring equipment is tested and inspected as safe to use, and the provision of suitable shut-off and isolation measures, goggles and hearing protection. Care should be taken to ensure that drill bits are kept sharp, as injuries can occur when the rotating drill bit gets stuck in material, causing the drill to kick and rotate in the operator's hands.

Sander

Hazards

Sanding equipment is available in a variety of sizes from hand-held equipment to large industrial machines. Sanders are used to provide a smooth finished surface, using a mechanical abrasive action.

Sanding operations are carried out on a wide variety of materials, including wood, minerals such as marble and man-made fibres.

The main hazards associated with the sanding process are vibration and noise. Also, harm can be caused by the inhalation of respirable particles from dust. Where organic materials such as wood are being processed, fire may result from overheated surfaces or explosion from dust by-products.

Associated hazards may include electrocution if supply cables are damaged by hand-held sanders, particularly if the sander is placed on the ground whilst still rotating (orbital sanders). Other risks include trips from trailing cables, cuts from sharp surfaces and strains or sprains from manual handling of process materials.

Figure 9-32: Belt sander.
Source: Clarke international.

Control measures

When using sanding equipment, suitable personal respiratory protective equipment (RPE) is required to protect the user from dust exposure. Where possible, local exhaust ventilation equipment should be used to minimise dust in the atmosphere.

In order to protect against vibration injuries, the operator should be given regular breaks and the equipment maintained at intervals, including the renewal of sanding media to prevent the need for over-exertion by the operator. Where hand-held tools are used, suitable hand protection will also reduce injuries from vibration, cuts and manual handling.

Pre-use inspections by the operator and regular thorough examinations should be carried out to identify potential electrical problems. Care should be taken to ensure that power leads do not create tripping hazards and they are positioned so that the likelihood of mechanical damage is minimised.

Pneumatic drill/chisel

Pneumatic drills or chisels are used commonly where heavy duty tasks are performed such as penetrating tarmac or concrete surfaces. This equipment is usually very heavy and labour intensive, resulting in manual handling hazards and risks. The pneumatic energy is usually delivered through mobile industrial compressors that are capable of supplying adequate power.

Hazards

The compressor is a separate piece of equipment that introduces its own specific hazards. Perhaps the most obvious hazard is the drill or chisel piece which, when operating, presents a risk of impact injury to the feet of users, and others nearby. During normal operation, the drill or chisel tool produces high intensity noise from contact with the surface. Other noise includes the exhausting of the pneumatic pressure from the internal drive of the equipment as the tool operates. The noise sources are located very close to the user's ears, are at a level that is damaging to the ear and require the user to wear hearing protection. The operation of the chisel produces the hazard of flying debris in the form of dust and fragments that pose the risk of abrasion, cuts or eye damage. There are risks associated with high pressure air lines becoming broken or damaged resulting in pipes lashing freely. The energy produced by the compressor is converted into vibration. Vibration introduces the risk of injuries, such as hand arm vibration syndrome (HAVS), with possible long term effects, including damage to the nervous system beginning at the fingertips.

Control measures

In order to control this risk the surface may be damped down and protective goggles worn by the user. Precautions for vibration include using well maintained equipment, using equipment with lower vibration levels, taking frequent breaks or job rotation, exercise to improve circulation and warm the hands and seeking suitable personal protective gloves. Equipment should always be inspected as safe to use, with certification for pressure lines. Screens can be used as a final measure to protect others at the location where the equipment is used.

Disc cutter

Hazards

Disc cutter saws consist of a petrol motor providing power to a circular cutting disc. Disc cutters are used for cutting grooves into surfaces. Hazards associated with disc cutters include cuts from the rotating blade, entanglement with rotating parts, ejection of material, noise, vibration, fire, inhalation of fumes and dust.

Control measures

Goggles must be worn when operating this equipment. Due to the velocity of the flying debris they must be grade 1 impact resistant and totally enclose the eye region of the face. The area being cut into may be damped down to minimise the production of dust clouds into the atmosphere or use water suppression fitted to the cutter.

On-tool extraction systems are available that make their operation almost dust-less when working correctly

Dust masks must also be worn when using this equipment. Sparks arising from cutting operations can result in a fire. In order to control this risk good housekeeping practice is essential, including removing sources of fuel from the work area.

Figure 9-33: Disc cutter.
Source: STIHL.

Misuse of this equipment must be prevented through training and supervision, including using the side of the cutting disc, using over worn discs, using the incorrect disc for the material being cut, leaving the cutter with the disc still spinning and not wearing the appropriate personal protective equipment (PPE).

Cut off saw

Cut off (chop) saws consist of a motor providing power to a circular cutting blade or disc, which has an enclosed guard. Cut off saws are normally powered by electricity, however petrol motors are available. The blade or disc is mounted on a counter sprung or a pneumatic arm which is operated by pulling down onto the materials to be cut. They are usually used to cut brick, stone, timber, steel, aluminium, plastic and wood into manageable lengths.

Hazards

Hazards associated with these saws are cuts, entanglement, electric shock, fire, inhalation of fumes, dusts, noise, vibration, ejection of material, such as metal shavings or small off-cuts.

Figure 9-34: Cut off saw.
Source: Draper Tools.

Control measures

Only trained and competent workers should use this equipment due to the hazards; the saw should be fitted with a method of locking-off the equipment to prevent unauthorised use. The saw should be fitted with an adjustable fixed or self-adjusting guard to enclose the blade. Goggles and protective clothing must be worn to prevent injuries from flying debris or dusts/fumes from harming the human body. In order to control the risk of fire from the sparks ejected when cutting some materials good housekeeping practice is essential, including removing sources of fuel from the work area. The equipment should be isolated from the mains supply prior to changing any cutting discs or replacing saw blades.

Cartridge and pneumatic nail guns

Guns for nails and other fixings are used in many activities in construction including carpentry, steelwork, plastering and surveying. Power sources range from battery power, electrical low voltage, pneumatic and explosive cartridge. They are used for shooting nails or pins into materials in order to secure a section.

Pneumatic guns may be operated by a pressurised changeable cylinder or compressor fed supply. Pneumatic guns usually provide a continuous supply of fixings from a belt or other type of magazine. Explosive cartridge guns are available in single cartridge manual change or fast loading format, the latter often being used to fix roofing and decking to steelwork. Some fixing guns operate using a fuel cell and small linear combustion engine, the controlled combustion/explosion providing the energy to drive the fixing into the material.

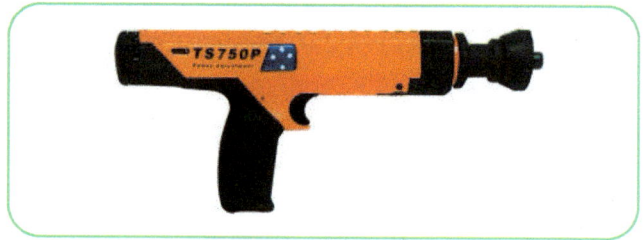

Figure 9-35: Cartridge fixing gun.
Source: ITW.

Figure 9-36: Pneumatic nail gun.
Source: Bostitch.

Hazards

The puncture hazard of fixings fired from the gun is the main hazard and can affect the user and people in the area where the gun is used. There is a significant risk of flying debris from fixing operations and impact/puncture injuries to eyes or hands are likely from the flying debris and fixings released incorrectly from the gun. The noise produced is high intensity impact noise and represents a significant hazard; it may also be amplified by the material being fixed at it vibrates due to the impact.

Vibrations produced by the use of this equipment, which is grasped in the hand, may present a risk of causing hand arm vibration syndrome (HAVS) when in sustained use. The high pressure air used with some fixing guns creates a hazard of injection of air into the body.

The explosive cartridges used with some fixing guns can present an explosive hazard coupled with the hazard of ejection of cartridge materials.

Control measures

It is important that users of these guns are trained to keep their hands away from the firing zone. This may be assisted by using clamps and other devices instead of the user's hand to hold materials in position prior to fixing. Impact resistant goggles or face protection must be worn when using the gun.

Personal hearing protection must be worn at all times due to the risk of hearing damage from the high intensity impact noise. The equipment must be well maintained in order to reduce vibration levels. Operators of the equipment should be instructed to take regular rest breaks and report any prolonged numbness to their fingers or hands to their supervisor.

Procedures should be in place to ensure there is a safe zone around and behind the fixing area. In order to avoid fixings passing through the material being fixed, the lowest power explosive cartridge or air pressure capable of making the fixing should be used. Arrangements should be in place to deal with 'miss-fires' and jams of fixings within the gun. Where explosive cartridges are used, strict control must be placed over the storage and issue of the cartridges.

REVIEW

What factors should be considered to reduce the risk of injury when using a hand-held hammer?

What are the risks from using hand-held power tools?

What are the control measures for the safe use of a hand-held electric drill?

9.3 Machinery hazards and control measures

MAIN MECHANICAL AND OTHER HAZARDS

This section describes the origin of machinery hazards and the potential consequences as a result of contact with mechanical and other hazards identified in ISO 12100:2010 'Safety of machinery - General principles of design - risk assessment and risk reduction', Table B1.

Mechanical hazards

Entanglement

Entanglement is the potential consequence of coming into contact with moving elements of machinery. In particular, the origin of entanglement is contact with rotating elements of machinery.

Figure 9-37: Auger drill – entanglement.
Source: STIHL.

The mere fact that a machine part is revolving can constitute a very real hazard that can lead to entanglement. Loose clothing, jewellery, long hair, etc. increase the risk of entanglement. Examples of entanglement hazards include couplings, drill chucks/bits, flywheels, spindles, shafts (especially those with keys/bolts) and rotating tools like abrasive wheels.

Friction and abrasion

Friction and abrasion are the potential consequences of coming into contact with moving elements of machinery. In particular, the origin of friction and abrasion is the skin coming into contact with moving elements that have a rough surface. Examples of moving elements with rough surfaces include abrasive surfaces of a sanding machine, grinding wheel or conveyor belt.

Figure 9-38: Friction and abrasion - abrasive wheel.
Source: RMS.

Cutting or severing

Cutting or severing are the consequences of coming into contact with sharp edges. The origin of cutting and severing is those sharp elements of a machine that are moving or stationery.

Saw blades, knives and even rough edges, especially when moving at high speed, can result in serious cuts and even amputation injuries. The moving elements of machinery can appear stationary due to the stroboscopic effect under certain lighting conditions, which can increase the likelihood of a worker contacting a sharp edge. Examples of cutting and severing hazards include saws blades, blades on slicing machines, abrasive cutting discs and chainsaw blades.

Figure 9-39: Cutting or severing - circular saw blade.
Source: Speedy Hire plc.

Shearing

Shearing is the consequence of coming in to contact with moving elements of machinery. The origin of shearing is elements of machines that move past each other or stationary objects. It differs from cutting in that the moving parts are not necessarily sharp.

Examples of shearing hazards include scissor lift MEWP and power press tools.

Figure 9-40: Shearing - scissor lift MEWP when lowering.
Source: RMS.

Stabbing and puncture

Examples of stabbing and puncture are the potential consequences of ejection of sharp items from a machine or contact with a sharp operating element of a machine.

Stabbing and puncture hazards include fixing materials deliberately ejected from a machine (such as nails fired from a nail gun), materials ejected during the machine's operating process (waste material ejected when cutting or abrading), part of a machine ejected as it wears out or breaks (part of a cutting blade) and stabbing or puncture by a drill bit.

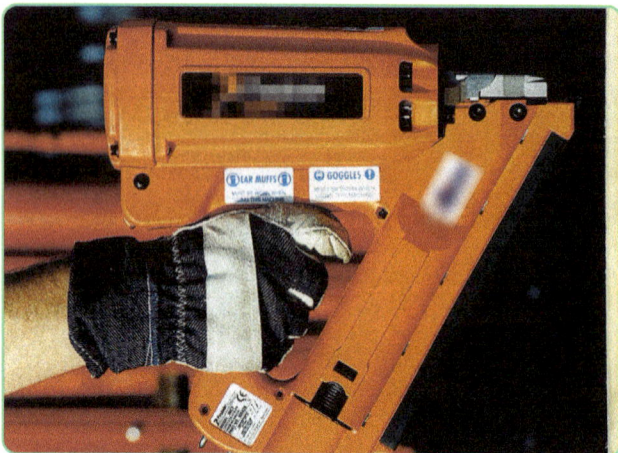

Figure 9-41: Stabbing and puncture - hand-held nail gun.
Source: Speedy Hire plc.

Figure 9-42: Stabbing and puncture - ejected broken disc saw blade.
Source: Water Active.

Impact

Impact is the potential consequence of a moving element of a machine or an object ejected from a machine directly striking a person. The origin of impact could be contact with those elements of a machine that strike the body, but do not penetrate or crush it, or parts of the machine, materials or tools ejected from a machine. The impact may cause the person or part of the person to be moved violently and/or local damage to the body at the site of the impact.

Examples of impact include a person being struck by the jib of a crane/excavator, a robot arm or materials being moved on a hoist or ejection of waste material from a machine or ejection of part of an abrasive wheel when it fractures.

Figure 9-43: Impact - construction work equipment.
Source: RMS.

Crushing

Crushing is the potential consequence of being caught between moving elements of a machine. The origin of crushing is part of the body being squeezed between two moving elements of a machine moving towards each other or a moving element of a machine moving towards a stationery object.

Examples of crushing hazards include the platform of a hoist or MEWP closing together with the ground or parts of the equipment.

Figure 9-44: Crushing - when lowering MEWP arms and platform.
Source: RMS.

Drawing-in or trapping

Drawing-in or trapping is the consequence of a person coming into contact with moving elements of a machine. The origin of drawing-in or trapping is contact with two rotating element that rotate towards each other or a single rotating element that rotates towards a gap in a fixed element. This movement draws in any part of the body presented to it and is therefore described as a drawing-in hazard.

Examples of drawing-in hazards are chain drives of a fork-lift truck lifting mast, V-belts on the drive from a motor to the drum of a cement mixer or the spindle of a drill, rotating rollers of a road roller and conveyor belts as they pass over rollers.

Figure 9-45: Drawing-in – rotating rollers.
Source: RMS.

Figure 9-46: Injection - hydraulic fluid leak from lifting system.
Source: RMS.

Injection

Injection is the potential consequence of contact with high-pressure fluids projected from machinery.

Injection of compressed air or high-pressure fluids through the skin may lead to soft tissue injuries similar to crushing. Air entering the blood stream through the skin may be fatal. Examples of high-pressure fluid injection hazards include diesel injectors, spray painting, compressed air jets for blast cleaning the outside of a building and a high-pressure lance for cutting concrete or a hydraulic fluid leak from lifting systems.

Other (non-mechanical) hazards

Machinery may also present other hazards. The nature of the hazard will determine the measures taken to protect people. The various sources of non-mechanical hazards include the following:

- Electricity – shock and burns.
- Hot surfaces/fire.
- Noise and vibration.
- Biological – viral and bacterial.
- High/low temperatures.
- Manual handling.
- Chemicals that are toxic, irritant, flammable, corrosive, explosive.
- Access - slips, trips and falls; obstructions and projections.
- Ionising and non-ionising radiation.

Mechanical (example of equipment with hazards that can have these consequences)	Non-mechanical
Entanglement (Drilling machine)	Electricity
Friction and Abrasion (Grinding wheel)	Hot surfaces/fire
Cutting (Sharp edges of a circular saw)	Noise
Shear (Scissor lift mechanism)	Vibration
Stabbing and Puncture (Nail gun)	Access
Impact (Moving arm of an excavator)	Chemicals
Crushing (Platform of a hoist, ram of a forge hammer)	Radiation
Drawing-in (Conveyor belt)	Biological
Injection (High pressure hydraulic oil system)	Manual handling
	Extremes of temperature

Figure 9-47: Summary – mechanical and non-mechanical machinery hazards. Source: RMS.

HAZARDS OF A RANGE OF SITE EQUIPMENT

Petrol-driven strimmer/brush-cutter

Hazards are:

- Contact with the moving parts of the strimmer – entanglement and cutting.
- Struck by parts of the strimmer material broken off in the operation - impact.
- Struck by fragments of the material being cut or stones - impact.
- Struck by passing traffic - impact.
- Noise and vibration.
- Fire or explosion associated with flammable fuel and hot petrol engine.
- Burns from contact with the hot casing of petrol engine.
- Exposure to combustion products from the petrol engine.
- Exposure to extreme weather conditions.
- Manual handling.
- Slips, trips and falls.

Chainsaw

The most significant mechanical hazard of using a chainsaw is the cutting or severing hazard of the moving chainsaw blade. There is a particular risk of cutting and severing injuries to the user's legs, feet, head and shoulders as the chainsaw cuts through the timber or slips to the side of what is being cut or if 'kickback' occurs.

Kickback occurs when a chain tooth at the upper quadrant of the guide bar tip cuts into wood without cutting through it. The chain cannot continue moving, and the bar is driven in an upward arc toward the operator, which present a serious risk of injury from the chainsaw blade. Kickback can result in major injuries or death.

Cutting or severing injuries can also result if the chain breaks during operation, for example due to poor maintenance or attempting to cut inappropriate materials.

Figure 9-48: Petrol chainsaw.
Source: Mowdirect Garden Machinery.

In addition to the cutting and severing hazards, operation of a chainsaw presents vibration and noise hazards. The equipment is often powered by a petrol motor, which presents a fire hazard, hazard of hot petrol motor components/exhaust and exposure to combustion products from the exhaust.

Another hazard exists when heavy timber begins to fall or move when a cut is nearly complete – the chainsaw operator may be crushed by the timber or receive impact injuries.

In addition, use of the equipment may cause the operator to take up awkward postures and experience other non-mechanical hazards, include work at height, slips, trips and falls, manual handling hazards, eye injuries from wood chipping/saw dust, cuts and splinters from handling wood.

Cement mixer

A cement mixers is portable construction equipment used for mixing a variety of aggregates and cement. Hazards include:

- Drawing-in hazard of pulleys and drive belts.
- Entanglement hazard of the rotating drum and drive shaft.
- Impact hazard of materials ejected from the drum.
- Shovels and trowels entangled in the mixer blades.
- Risk of fire if fuel source is diesel or petrol.
- Exposure to fumes from petrol/diesel.
- Electricity if this the power source, particularly as water is used in the mixing process.
- Slips and trips from spilt materials.
- Manual handling when loading.
- Exposure to cement dusts and wet cement.

Figure 9-49: Cement mixer.
Source: RMS.

Bench-mounted circular saw

Cutting and severing is the main mechanical hazard, other mechanical hazards include entanglement in power drive shafts, drawing-in to drive mechanisms, such as a pulley and drive belt. Material being cut could be ejected and impact or puncture someone. Non-mechanical hazards include electricity, noise, sawdust, splinters and cuts from handling material and musculoskeletal disorders related to posture and materials handling.

Compressor

Hazards from pneumatic compressors include those from the fuelling of the plant (normally diesel). The hazards of fuelling include the fuel contacting the skin or splashing into eyes, inhalation of fuel vapours and a risk of fire. In operation, exhaust gases are produced, including carbon monoxide.

High levels of noise and vibration could be produced by a compressor, although this may depend on the effectiveness of the soundproofing material incorporated into the compressor and the condition of the equipment. High pressure air is supplied from the compressor to where it is needed through air pressure lines, which present a hazard of injection of air and they are also a trip hazard.

Figure 9-50: Compressor.
Source: Atlas Copco.

Plate compactor

Plate compactors are machines used for consolidating aggregates or loose materials. They are also used for securing block paviors into position by applying a heavy downward force through a flat plate where the force is increased by the vibrating motion.

Hazards arise from the source of power, typically petrol, which has associated fire risks that are increased when refuelling is carried out while the exhaust system is still hot.

This equipment, because of its purpose, has to be heavy, which brings a risk of manual handling strain injuries when loading or unloading from a vehicle. Additional hazards include noise, vibration and crush injury to feet.

Figure 9-51: Plate compactor.
Source: HSS Hire Service Group Ltd.

Ground consolidation equipment

Ground consolidation equipment serves the same purpose as a plate compactor. However, a plate compactor is generally for lighter duty than the large types of ground consolidation equipment, such as ride-on rollers or pedestrian operated rollers.

Ground consolidation equipment moves at a greater speed than lighter duty equipment and its power drives the equipment directly through the roller wheel. The potential harm from drawing in/crush hazards of moving rollers are high, a drawing in/crush sustained from a piece of plant of this type could easily lead to a major injury.

Figure 9-52: Ground consolidation equipment.
Source: RMS.

Road-marking equipment

Road marking usually involves the application of a hot plastic based substance to the surface of a road. The equipment used will often have its own supply of fuel in the form of liquefied petroleum gas (LPG).

This is used to maintain the marking substance at a suitable working temperature in a holding vessel.

Direct hazards from the equipment include the hazard of drawing in/crush involving the wheels of the equipment, contact with the hot marking substance and potential overheating of the marking substance leading to a fire. Other hazards include exhaust gases and falls from road-marking vehicles.

Figure 9-53: Thermoplastic hand marking barrow.
Source: Maxigrip Surfacing Ltd.

Figure 9-54: Road Marking vehicle.
Source: Maxigrip Surfacing Ltd.

Electrical generators

Diesel/petrol powered electrical generators are used for powering portable power tools and other construction equipment. Smaller generators are often used to power road breakers; large versions may be used to provide electricity for site welfare arrangements. They may be equipped with 230V and 110V sockets or provide electricity to a 415 volt distribution system. The main hazards are electrocution, hot surfaces, fire, noise and asphyxiation from carbon monoxide present in the exhaust gases. Other hazards include risks related to handling the fuel and, where they are portable, manual handling hazards related to moving them into the position needed.

Figure 9-55: Petrol 110v generator.
Source: RMS.

Figure 9-56: Large diesel generator.
Source: RMS.

Drones

A drone can refer to any unmanned mobile equipment, which would include aircraft, vessels or other vehicles. The drone is operated remotely from a ground control system (GCS), usually by a human. Perhaps the most common drones in use for work are unmanned aerial vehicles (UAVs). They are often fitted with cameras and used for site surveys, monitoring large areas of land and inspecting equipment that is difficult to gain access to, for example, wind turbines, bridges and large process structures.

One of the main hazards of drones relates to their propulsion system, UAVs used for work usually have multiple rotor blades to enable lift, movement and hovering. These blades rotate at high speed and present a cutting and severing hazard, with some risk of entanglement in the rotating blade mechanism. The other main hazard is that of impact. The drone could impact people when it is moving. The people particularly at risk are those that test the equipment, set it to work/retrieve it and those that the equipment travels near while it is operating. As the drone is operated remotely there is a risk of it moving unexpectedly towards people and striking them, perhaps because of operator error or because of extreme weather or faults in the equipment.

Loss of control of the drone could lead to a high speed impact with people or critical equipment, this could be due to failure of the GCS, failure of the drone guidance equipment or loss of the controlling signal between them.

The drone requires a power supply, which might be a battery or petrol motor. These present specific hazards – the battery is source of high energy electricity and the petrol motor has flammable liquid and an ignition source.

Driver-less vehicles

Driver-less vehicles operate autonomously, without human involvement. They are controlled by computer systems in the vehicle that make them respond to information from various sources, including sensors on the vehicle. In the workplace, at this stage in their development, they are likely to be carrying materials and equipment, rather than workers. Simple autonomous vehicles have been in use in the workplace for some time and have been used for such things as automated storage movement systems. The use of autonomous vehicles to transport goods and people on the roads is a growing area of activity.

The hazards associated with autonomous vehicles include:

- Limitations or defects in the vehicle's sensors could mean they fail to detect objects in their path, for example pedestrians, and could impact with them or crush them.

- As they are powered by electricity they are nearly silent and pose a threat of impact with pedestrians who might walk in front of the moving vehicle because they do not hear it, perhaps because they are wearing hearing protection.

- If there is a failure (error, misinterpretation, loss of signal) in the control system of the vehicle it could fail to turn a corner or stop and result in an impact with other vehicles, pedestrians, structures or other stationery objects.

- Errors in the programming may cause the vehicle to act in an unusual way, for example speed up or stop suddenly, which could cause the occupants to impact with the inside of the vehicle.

- A vehicle may also fail to recognise important information that it needs to take into account when determining safe movement, which could lead to the vehicle not take action to avoid a collision, for example, not recognising the status of traffic lights or the presence of a pedestrian crossings.

- A vehicle being maintained might unexpectedly begin to operate, causing the maintenance worker to be exposed to electrical energy sources, moving elements of the motor/drive mechanism and being struck or crushed by the vehicle.

- As with all battery operated vehicles, there are safety hazards associated with the task of battery removal/connection and battery charging.

- An autonomous vehicle may fail to appropriately respond to road surface conditions, for example changes caused by the weather (snow, ice, rain or mud) or the type of surface or contamination on the surface or objects on the surface. This could lead to lose of control of steering or unsuitable braking, with the likelihood of an uncontrolled impact.

REVIEW

List three mechanical and three non-mechanical hazards associated with a:

a) Cement mixer

b) Chainsaw

c) Compactor

List three different types of drawing-in incidents

THE BASIC REQUIREMENTS FOR GUARDS AND SAFETY DEVICES

Machinery should be designed and constructed to prevent risks. Where risks remain employers should prevent contact with hazards that could cause harm. The hazards of machinery may be mechanical or non-mechanical. In this context, the hazards are referred to as 'dangerous parts' of the machinery. There are various control methods available to prevent harm, including a range of guards and protective devices.

Guards

Guards provide a physical barrier between the hazard and people, preventing people placing themselves in the danger zone where they can contact the hazard. Guards should be used in preference to other control methods where possible.

Fixed guards

General operation of fixed guards

A fixed guard should be used whenever practicable. It should be designed so as to prevent access to the dangerous parts of the machine.

A fixed guard may be designed to enable access to dangerous parts by authorised personnel for maintenance or inspection, but only when those parts have been isolated.

A fixed guard must be fitted such that it cannot be removed other than by the use of specialist tools, which are not available to users of the equipment. ISO 12100:2010 recommends that where fixed guards are held in place by fasteners, the guards should not remain in place without their fasteners.

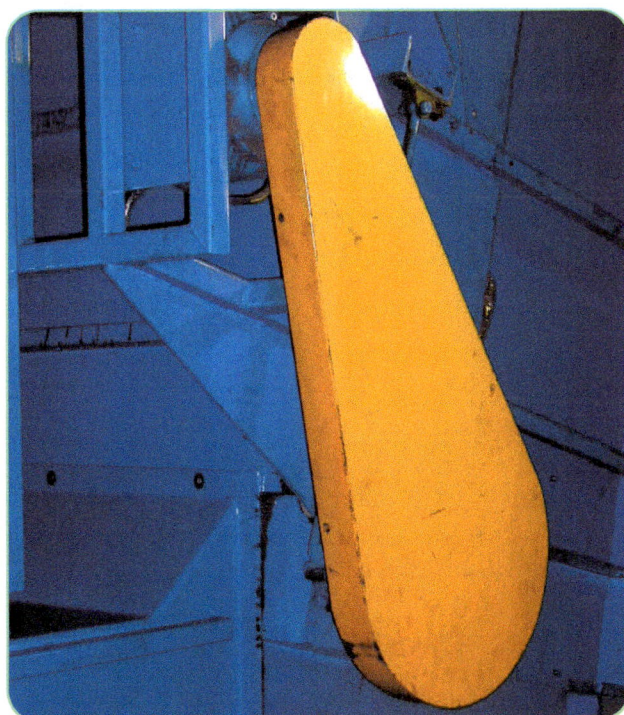

Figure 9-57: Total enclosure fixed guard.
Source: RMS.

In addition, the fasteners should remain attached to the guards when the guards are removed. A common example of a fixed guard is shown in *Figure 9-57.* Not all fixed guards are of solid construction. Some are made of mesh.

The holes in a fixed guard made of mesh should be big enough to allow air circulation to cool the drive belt but positioned far enough away from the drive belt to prevent a finger from penetrating the mesh and sustaining an injury from the belt.

Distance (fixed) guards

Fixed guards do not always completely cover the danger point, the position where dangerous parts can be contacted, but place it out of normal reach. The larger the opening (to feed in material), the greater must be the distance from the opening to the danger point.

For example, a tree-shredding machine uses a fixed distance guard designed to prevent users reaching the dangerous parts of the machine when in use while allowing material to be fed in for shredding.

Figure 9-58: Part of fixed guard removed.
Source: RMS.

Advantages and limitations of fixed guards

The advantages of fixed guards include:

- Create a physical barrier.
- Require a tool to remove.
- No moving parts – therefore they require very little maintenance.

The limitations of fixed guards include:

- Do not disconnect power when not in place, therefore machine can still be operated without guard.
- May cause problems with visibility for inspection.
- If enclosed, may create problems with heat, which, in turn, can increase the risk of explosion.
- May not protect against non-mechanical hazards such as dust/fluids which may be ejected.

Figure 9-59: Fixed guard – mesh and opening too big.
Source: RMS.

Interlocking guards

An interlocking guard is similar to a fixed guard, but it has a movable (usually hinged) part connected to the machine controls in such a way that if the movable part is in the open/lifted position the dangerous parts of the machine at the work point cannot operate. It can also be arranged that the action of closing the guard activates the machine and its working parts (to speed up work efficiently), for example, as is the case with the front panel of a photocopier. Interlocked guards are useful if users of the equipment need regular access to the danger area, for example, to load material in or take material out of a machine.

Everyday examples of interlocking guards are those found on domestic equipment such as dishwashers, microwave cookers and automatic washing machines. Interlocking guards should:

- As far as possible remain attached to the machinery when open.

- Be designed and constructed in such a way that they can only be operated or moved by means of an intentional action.

- Prevent the start of hazardous machinery functions until the guards are closed.

- Give a stop command whenever the guards are opened.

- Ensure that absence or failure of one of their components prevents starting, or stops the hazardous machinery functions.

Figure 9-60: Interlocking guard with viewing panel
Source: RMS.

Where it is possible for an operator to reach the danger zone before the risk due to the hazardous machinery functions has ceased, the interlocking guard should:

- Prevent the start of hazardous machinery functions until the guard is closed and locked.
- Keep the guard closed and locked until the risk of injury from the hazardous machinery functions has ceased.

In the photograph (**see Figure 9-60**), the electrical interlock is positioned halfway down the right-hand side of the panel. The panel is made from transparent material to allow easy visual checks of the products that are manufactured by this equipment.

Advantages and limitations of interlocking guards

The advantages of interlocking guards include:

- Connected to power source, therefore machine cannot be operated with guard open.
- Allows frequent access.

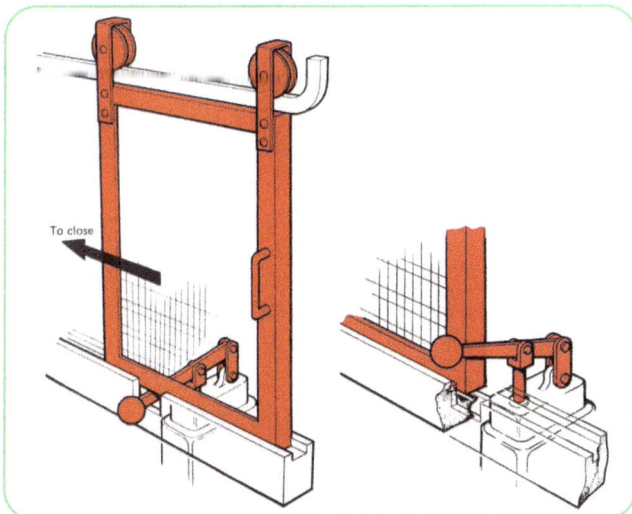

Figure 9-61: Open and closed interlock guard.
Source: BS EN ISO 12100.

The limitations of interlocking guards include:

- Have moving parts, therefore need regular maintenance.
- Can be overridden.
- If the interlock is in the form of a gate, a person can step inside and close the gate behind them (someone else could reactivate machine).

Dangerous parts of machinery may not stop or cease to

be a hazard immediately the guard is opened. It may be necessary to fit a delay timer or brake; for example, the drum on a spin drier does not stop instantly because of the momentum of the drum and contents and therefore a delay may be fitted to prevent the door being opened until the drum is stationary.

Figure 9-62: Interlock at top (right) of hoist gate.
Source: RMS.

Self-adjusting guards

Self-adjusting guards are guards that are fixed to the moving parts of the machine, which close themselves over the dangerous parts as an integral part of the operation of the machine. They prevent accidental access by the operator but allow entry of the material to the machine in such a way that the material forms part of the guarding arrangement itself. For example, a hand-held circular saw.

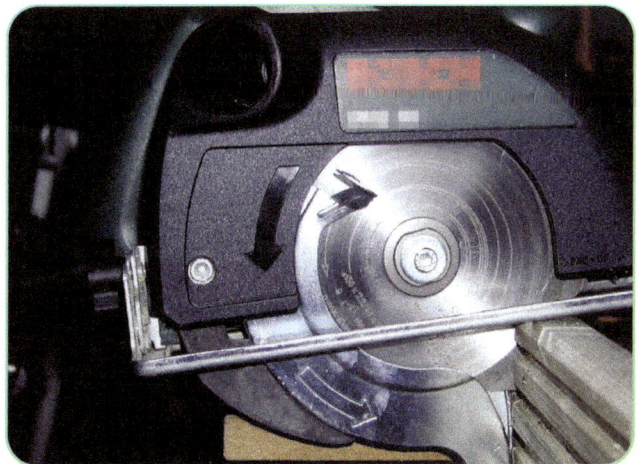

Figure 9-63: Self-adjusting (fixed) guard.
Source: RMS.

Advantages and limitations of self-adjusting guards

The advantages of self-adjusting guards include:

- Close over the dangerous parts to provide protection without the operator needing to do anything.

The limitations of self-adjusting guards include:

- May obscure visibility when in use.
- Are vulnerable to damage in the operation of the equipment.

Adjustable guards

Adjustable guards are fixed guards that incorporate an adjustable element to accommodate a range of positions and size of dangerous parts of the machinery. When adjusted, it remains set in position for the duration of a particular operation.

Advantages and limitations of adjustable guards

The advantages of adjustable guards include:

- Can be adjusted by operator to provide protection.
- Protects operator where size or position of dangerous parts varies.

The limitations of adjustable guards include:

- Are reliant on the operator to adjust to the correct position.
- May obscure visibility when in use.

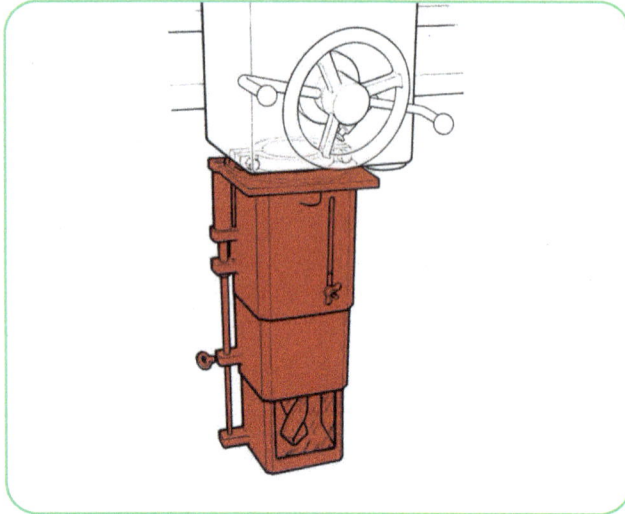

Figure 9-64: Adjustable (fixed) guard.
Source: BS EN ISO 12100.

> **CONSIDER**
>
> Consider a fixed guard in your workplace; give one advantage and one disadvantage.

Protective devices

Because protective devices do not provide a physical barrier to prevent the person's access to the danger zone and rely on a response mechanism they should only be used where a fixed guard is not practical. In addition, some of the protective devices are designed to keep the operator of the machine away from the hazard, do not protect other people.

Sensitive protective equipment

Sensitive protective equipment (sometimes called a trip device) has a sensing mechanism that detects that a person (or part of the person, for example, their hand) is in the danger zone (or approaching the danger zone) and at risk of harm from hazards. The sensing mechanism may be a light beam, pressure mat, rod, cable or other mechanism. The sensing mechanism causes the device to activate a further mechanism that either stops or reverses the machine, preventing further danger to the operator.

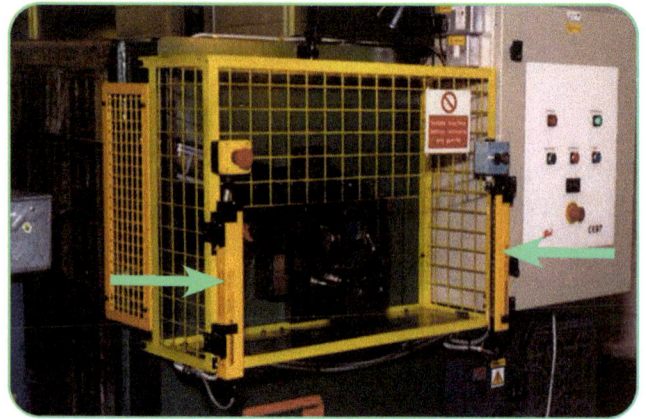

Figure 9-65: Light curtain.
Source: HSE, HSG180.

It is important to note that a sensitive protective device/equipment is not classed as a guard. A guard is something that physically prevents access to the danger zone and the hazard, whereas a sensitive protective device/equipment detects the person in the danger zone and responds to this.

It is essential that the sensing mechanism is positioned so that when a person approaches the danger zone there is enough time for the device/equipment to detect this and cease the action of the machine presenting a hazard before the person is harmed. This makes the location of the sensing mechanism, its sensitivity and the ability of the device/equipment to cease the hazard, critical.

Sensitive protective devices/equipment based on light beam/curtain sensing mechanisms have demonstrated that they can provide a high degree of protection, similar to that provided by an interlocking guard. However, an electro-mechanical telescopic mechanism of the type sometimes used with radial drills does not provide as good protection as an adjustable fixed guard.

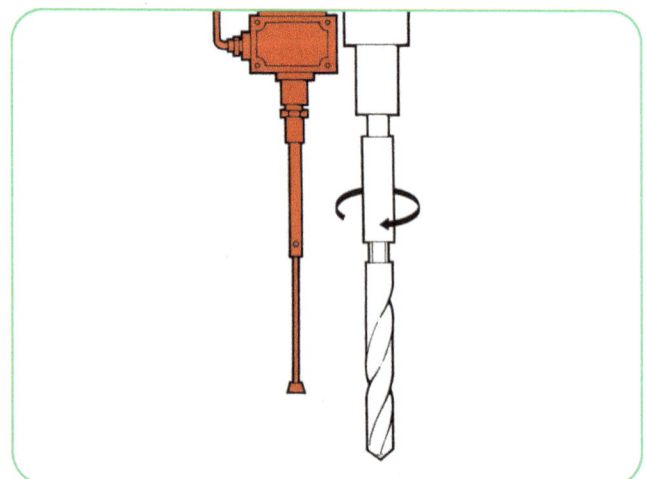

Figure 9-66: Trip device.
Source: BS EN ISO 12100.

Advantages and limitations of sensitive protective devices/equipment

The advantages of sensitive protective devices/equipment include:

- Can be used as an additional risk control measure.
- Can minimise the severity of injury.

The limitations of sensitive protective devices/equipment include:

- Can be overridden.
- May not prevent harm from occurring.
- May cause production delays and increase stress in users with false 'trips'.

Two-hand control device

Two-hand controls (2HC) provide a level of protection where other methods are not practicable, helping to ensure the operator's hands remain outside the danger area.

A two-hand control is a device that requires both hands to operate it. The controls must be operated simultaneously, and this helps to assure that both hands are kept away from the dangerous parts. 2HC devices protect only the operator and then only as long as a colleague does not activate one of the controls. Everyday examples of two-hand controls are a hedge trimmer and a garment press.

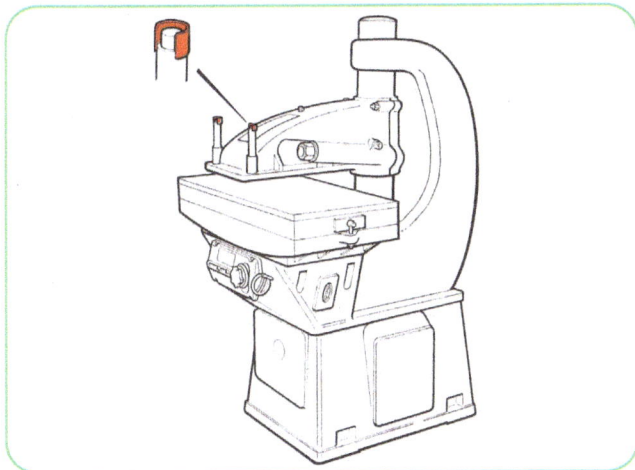

Figure 9-67: Two-hand control device.
Source: BS EN ISO 12100.

Advantages and limitations of two-hand control devices

The advantages of two-hand control devices include:

- Ensures both of the operator's hands are out of the danger area when the machine is operated.

The limitations of two-hand control devices include:

- Only protects the operator from harm.
- May limit speed of operation with delays if controls are not pressed at exactly same time.

Hold-to-run device

The principle of a hold-to-run device is that the operator has to 'hold' a button, stick or foot pedal to 'run' a piece of equipment. A common application of this device is on a domestic lawn mower or hedge trimmer (**see Figure 9-68**).

It is essential that the hold-to-run device, which when released stops the machine, is located far enough away from the danger area to prevent the operator getting access to the moving parts without releasing it.

Advantages and limitations of hold-to-run devices

The advantages of hold-to-run devices include:

- Ensures the operator is out of the danger area when the machine is operated.
- Provides distance between operator and hazard.

The limitations of hold-to-run devices include:

- Only protects the operator from harm.
- There may be residual movement of dangerous parts once the device has been released.

Figure 9-68: Hold-to-run device on hedge trimmer.
Source: RMS.

Emergency stop controls

Emergency stops are intended to effect a rapid response to potentially dangerous situations and they should not be used as functional stops during normal operation. Emergency stop controls should be easily reached and actuated. Common types are mushroom-headed buttons, bars, levers, kick plates, or pressure-sensitive cables. They can be used by the operator or other workers to stop the machine quickly, providing protection from further risk of harm.

They can be useful in quickly bringing a machine to rest in situations where material has got stuck in the machine.

Advantages and limitations of emergency stop controls

The advantages of emergency stop controls include:

- Removes power immediately.
- Equipment has to be reset after use.
- Prevents accidental restarting of the equipment.

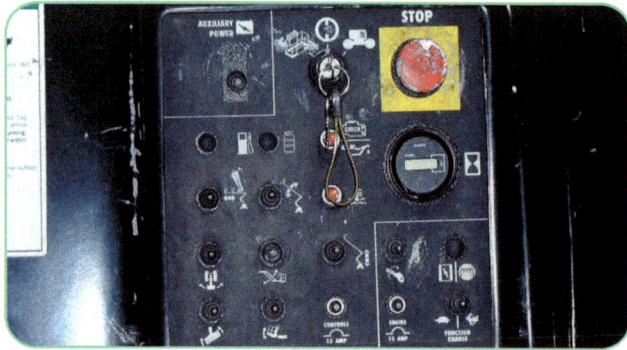
Figure 9-69: Emergency stop control.
Source: RMS.

The limitations of emergency stop controls include:

- Does not prevent access to the danger area.
- May be incorrectly positioned.

Protection appliances

Protection appliances are used to hold or manipulate a work piece at a machine while keeping the operator's hands clear of the danger zone. They are commonly used in conjunction with manually-fed woodworking machines and certain other machines, such as band saws for cutting meat, where it is not possible to fully enclose the cutting tool in a guard. These appliances will normally be used in addition to guards.

Jigs, holders and push-sticks

When the methods of safeguarding explained previously are not practicable, protection appliances such as jigs, holders and push-sticks must be provided. These will help to keep the operator's hands at a safe distance from the danger zone. There is, however, no physical restraint to prevent the operator from placing their hands in danger.

Figure 9-70: Notched push-stick.
Source: Craftsmanspace.

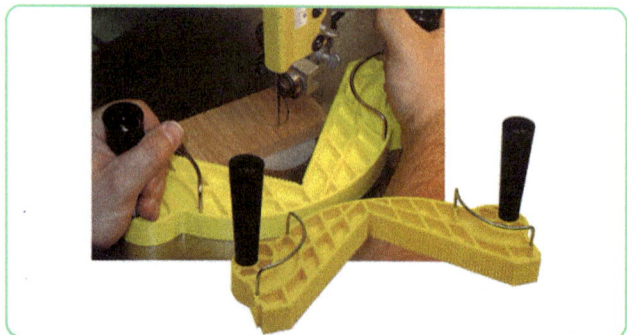
Figure 9-71: Band saw push-stick.
Source: Toolmonger.

Advantages and limitations of protection appliances

The advantages of protection appliances include:

- Provide distance between operator and hazard.
- Inexpensive and easily replaced if damaged.
- May be shaped to suit work being carried out.

The limitations of protection appliances include:

- Harm may still occur from other non-mechanical hazards.
- Some designs can be awkward to use and may result in a lack of control.
- Adjustments have to be made, for example, when using different sizes of wood.
- Failure of the protection appliance (for example, breaking or kickback) may present an additional hazard to the operator.

Figure 9-72: Fixed guard and push-stick used with a circular saw.
Source: Lincsafe.

Information, instruction, training and supervision

The employer should ensure that all persons who use work equipment, and any workers who supervise or manage the use of work equipment, have available to them adequate health and safety information. Where appropriate, written instructions and training pertaining to the use of the work equipment should be made available. This includes information and, where appropriate, written instructions that are comprehensible to those concerned with:

- The conditions in which and the methods by which the work equipment may be used.
- Foreseeable abnormal situations and the action to be taken if such a situation were to occur.
- Any conclusions to be drawn from experience in using the work equipment.

Similarly, employers should ensure that all persons who use work equipment and those who supervise or manage the use of work equipment have received training in any risks involved and precautions to be taken.

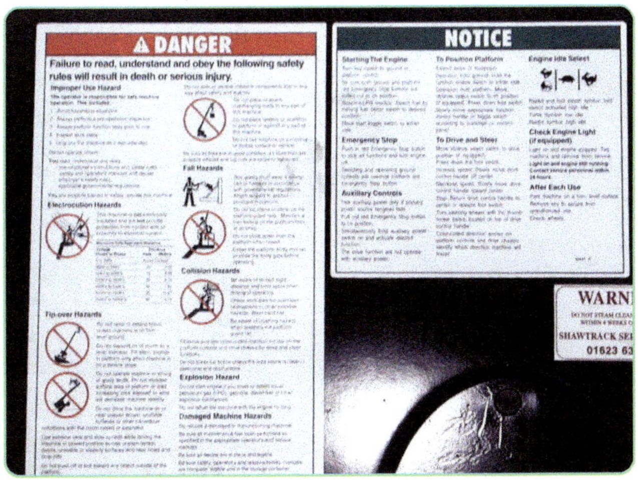

Figure 9-73: Information and Instruction.
Source: RMS.

It is a requirement of both the HASAWA 1974 and PUWER Regulation 11 that employers should provide levels of supervision as necessary. Extra supervision will be required for new, inexperienced or less capable users of equipment.

Advantages and limitations of information, training and supervision

The advantages of information, training and supervision include:

- Easy to reach a wide audience on a variety of subjects.
- Can be applied immediately and adapted to suit specific needs of the user.

The limitations of information, training and supervision include:

- Supervision may not prevent contact with the hazard.
- Relies on the person concerned to follow the instruction.
- May be misunderstood.
- Supervision is needed to a sufficient degree to ensure health and safety; a high degree of supervision may be required for some equipment.

"Regulation 11 -

Dangerous parts of machinery

(1) Every employer shall ensure that measures are taken in accordance with paragraph (2) which are effective:

(a) To prevent access to any dangerous part of machinery or to any rotating stock-bar; or

(b) To stop the movement of any dangerous part of machinery or rotating stock-bar before any part of a person enters a danger zone.

(2)   The measures required by paragraph (1) shall consist of:

(a) The provision of fixed guards enclosing every

dangerous part or rotating stock-bar where and to the extent that it is practicable to do so, but where or to the extent that it is not, then

(b) The provision of other guards or protection devices where and to the extent that it is practicable to do so, but where or to the extent that it is not, then

(c) The provision of jigs, holders, push-sticks or similar protection appliances used in conjunction with the machinery where and to the extent that it is practicable to do so, and

(d) The provision of information, instruction, training and supervision as is necessary.

(5) In this regulation - 'danger zone' means any zone in or around machinery in which a person is exposed to a risk to health or safety from contact with a dangerous part of machinery or a rotating stock-bar; 'stock-bar' means any part of a stock-bar which projects beyond the headstock of a lathe.

Regulation 11(2) gives the measures that an employer should take to fulfil the duty under -Regulation 11(1) a combination of measures may be necessary to satisfy Regulation 11. When deciding on the appropriate level of safeguarding, risk assessment criteria (likelihood of injury, potential severity of injury, numbers at-risk) need to be considered both in relation to the normal operation of the machinery and other operations such as maintenance, repair, setting, tuning, adjustment etc."

Figure 9-74: Dangerous parts of machinery.
Source: Provision and Use of Work Equipment Regulations 1998.

Personal protective equipment

Personal protective equipment (PPE) is a last resort and should only be relied upon when other controls do not adequately control risks. The use of machinery presents a number of mechanical hazards and care has to be taken that PPE is not used in situations where it presents an increased risk of entanglement or drawing-in to machinery, such as might happen with loose overalls and gloves.

The main examples of use of PPE with regard to use of machinery are:

- Eye protection (safety spectacles/glasses, goggles and face shields) – protection for the eyes and the head from flying particles, welding glare, dust, fumes and splashes.

- Head protection (safety helmets or scalp protectors i.e. bump caps) – helmets provide protection from falling objects or the head striking fixed objects, whereas scalp protectors only provide protection from the head striking fixed objects.

- Protective clothing for the body (overalls) – protection from a wide range of hazards, including splashes of coolant used with machinery and specialist clothing for when using chainsaws. Close-fitting overalls designed to avoid entanglement may be required when working with rotating machinery.

- High-visibility clothing – enables those working in or around a dangerous area to be better seen by others working in that area. For example, where there is mobile equipment or where workers have to enter danger areas to remove or place materials.

- Gloves (chain-mail gloves and sleeves) – protection against cuts and abrasions when handling machined components, raw material or machinery cutters.

- Footwear (steel insoles and toecaps) – protection against sharp objects that might be stood on or objects dropped while being handled.

- Hearing protection for noisy machine operations.

Advantages and limitations of PPE

The advantages of PPE include:

- Easy to see if it is being worn.
- Provides protection against a variety of hazards.

The limitations of PPE include:

- Does not substitute for effective guarding of machinery.
- Only protects the user from residual hazards that guards do not deal with.
- May not give adequate protection.
- May pose additional hazards, for example, gloves becoming entangled.

REVIEW

What are merits and limitations associated with the use of PPE in the workplace?

What are the merits of a fixed guard?

Give an example of a 'trip device' and explain how it works

What are the basic requirements for guarding systems?

Figure 9-75: Brush-cutter.
Source: Kawasaki-engines.

CONTROL MEASURES FOR A RANGE OF SITE EQUIPMENT

Petrol-driven strimmer/brush-cutter

Appropriate precautions are:

- Provision of fixed guards around any drive mechanisms and motor/engine.
- Fixed guards to enclose as much of the strimmer material or blade as is practicable.
- PPE (eye and hearing protection).
- Appropriate storage of petrol.
- Regular maintenance by authorised people, including changing of cutters.

Chainsaw

The equipment should have a hold-to-run device and sensitive protection equipment should be fitted in front of the operator's hand that will trip and stop operation if the chainsaw 'kicks-back'. The risks associated with chainsaw use mean that protective clothing, for example, forestry boots, helmet with mesh visor, chainsaw gloves, body/leg protective clothing and hearing protectors should be worn while operating them. In order to limit the effects of vibration, careful selection of equipment with low vibration output is important. Training of users is a critical element of the control strategy for use of this equipment.

Cement mixer

Fixed guards must be provided around drive mechanisms. Motor covers should be closed when in use. Users of the equipment should be warned of the dangers of shovels and trowels becoming caught in the mixer blades when charging. Electrical hazards are reduced by the use of low-voltage power (for example, 110 volts compared with 230 volts mains supply) or a residual current circuit breaker, with suitable heavy-duty protected cable. If the equipment is petrol or diesel driven, hot parts, for example, the exhaust system, should be allowed to cool before refuelling. A no-smoking policy should also be in place.

Other controls include manual handling training, good housekeeping of spilt materials, and avoidance of exposure to dust and fume.

Bench-mounted circular saw

Appropriate precautions are fixed, adjustable and/or self-adjusting guards, jigs, holders and push-sticks. Precautions also include ensuring material being cut is held firm to prevent movement on contact with the saw blade and during the cutting process. Support of the material being cut is advisable to prevent excess side pressure on the blade, particularly on completion of the cut. Care should be taken to ensure the blade has stopped revolving before adjusting the position of material.

Eye and hearing PPE should be worn to reduce the risk of noise induced hearing loss and eye injuries from the waste wood cuttings. Bench-mounted saws should be fitted with suitable local exhaust ventilation equipment. Respiratory protection may also be necessary particularly when cutting hardwoods.

Figure 9-76: Bench cross-cut circular saw.
Source: RMS.

Compressor

Appropriate precautions are:

- Over-pressure relief valve.
- Pressure gauge.
- Hose couplings fitted with pins or chains.
- Hose couplings inspected as part of daily plant inspection.
- Noise control equipment fitted to compressor.

Plate compactor

Appropriate precautions are:

- Use a remote-control rather than a hand-guided compactor to reduce risks from hand-arm vibration.
- Take regular breaks to reduce exposure or rotate use among several people.
- A safe system of work for handling this equipment may be required, which could include a hoist on site vehicles and a trolley to manoeuvre around the site.

Ground consolidation equipment

Control measures include segregation of pedestrians and plant using barriers, warning signs and close supervision.

Road-marking equipment

This equipment requires special training and authorisation for its safe use. Where this equipment is being used, barriers, warning signs and lighting is to be used, in addition to supervision. When this equipment is being used it is normally a requirement to have refuelling facilities available on site. When this is the case suitable isolation from sources of heat should be ensured, with only authorised operatives allowed to carry out refuelling activities.

Electrical generators

Diesel or petrol generators should always be used outdoors to prevent the risk of carbon monoxide accumulation. They should not be sited below scaffolds or near trenches where people are working, because there is a continuous risk of build-up of carbon monoxide whilst the compressor is running. They should never be refuelled when hot and noise limiting covers should remain closed when in use. The generator must be installed and maintained by a competent person. The generator must be of sufficient capacity for the total load required (electrical motors typically require three times their rated wattage on start-up). This should be taken into account before selection and use.

Drones

The safe use of drones for the purposes of work includes the following control measures:

- Establish a clear plan for the flight – considering such things as timing, take off/landing, weather, areas of work activity and possible effects on flight, areas of potential signal interruption, task to be achieved by the drone, restrictions on the flight to protect people/property and limitations due to hazards (flammable areas and obstructions).

- Select a size of drone that suits the task to be done, smaller drones may suit congested areas.

- Obtain a permit to work/fly where necessary.

- Ensure the operator is competent and proficient in relation to the task.

- Ensure the operator has clear, direct, unaided visual line of site (VLOS), use an observer to assist if needed.

- Ensure the operator and any observer has good eyesight.

- Ensure the drone can be clearly seen and controlled safely during night flying operations, additional lighting may be required to enable this.

- Launch/recover the drone from the ground where ever possible - avoid hand launch/recovery and restrict this method to competent people.

- Fly no higher than 120 metres (400 feet) and remain below any surrounding obstacles when possible.

- Remain clear of and do not interfere with manned aircraft operations.

- Keep the drone at least 50 metres (150 feet) away from people and property, including vehicles, keep 150 metres (500ft) away from crowds and built up areas and do not fly over them - unless there is a clear work purpose and risk assessments have been made.

- Do not fly in adverse weather conditions, such as in high winds or reduced visibility.

- Do not fly under the influence of alcohol or drugs.

- Establish an emergency plan for flight propulsion failure.

Figure 9-77: Drone.
Source: Cyberhawk.

Driver-less vehicles

When driver-less vehicles are used in the workplace pedestrians should be excluded from the area, wherever possible. These restricted areas should be clearly identified and suitable warnings provided. Consider interlocking access to the area so that vehicles cannot operate when pedestrians enter the area.

Where pedestrians and driver-less vehicles work in the same area they should be segregated as much as possible, for example, by separate aisles or roadways for the vehicles, separated by physical barriers. Where this is not possible clearly marked, separate routes should be provided and the vehicles must be equipped with suitable sensors that enable the detection of pedestrians in good time to enable the vehicle to stop before contact is made. Because many driver-less vehicles are almost silent in operation, consider adding visual and audible warnings devices to the vehicles to indicate that they are operating.

Figure 9-78: Driver-less work vehicle.
Source: PSA Group.

Also consider reducing the speed of operation of the driver-less vehicles and ensure there is good lighting in the area to provide pedestrians with time to react.

Driver-less vehicles for use in public areas and on public roads require sophisticated sensor and control systems that enable steering and braking in order to ensure they operate safely and avoid collisions with pedestrians and other vehicles.

Further general control measures for the safe operation of driver-less vehicles include:

- Only authorised persons must be permitted to control or maintain a vehicle and its operating system.

- Vehicle safety devices must not be manually overridden when the vehicle is operating in automatic or semi-automatic modes.

- The surfaces over which the vehicle operates must be maintained to ensure that the traction required for travel, steering, and braking performance can be met under the environmental conditions which may be expected on that surface.

- The environment that the vehicle operates in needs to be part of the vehicle design criteria - including temperature, humidity, ambient weather (for example, work on an exposed dock), air quality (for example, aggressive dusts or flammable atmospheres).

- Changes to the driving surface and environment must be evaluated to verify there is no adverse effect on the vehicle's safety systems.

- The vehicle manufacturer's instructions on battery removal/fitting and charging must be followed.

9.4 Working near water

ADDITIONAL CONTROL MEASURES WHEN WORKING NEAR WATER

CDM 2015, Part 4: Duties Relating to Health and Safety on Construction Sites sets out requirements that influence what control measures should be in place when working near water. In particular Regulation 26 sets out requirement to prevent falling into water, minimise the risk of drowning if someone did fall into water and ensure rescue equipment is provided.

Regulation 26 states that:

> "(1) Where, in the course of construction work, a person is at risk of falling into water or other liquid with a risk of drowning, suitable and sufficient steps must be taken to -
>
> (a) prevent, so far as is reasonably practicable, the person falling;
>
> (b) minimise the risk of drowning in the event of a fall; and
>
> (c) ensure that suitable rescue equipment is provided, maintained and, when necessary, used so that a person may be promptly rescued in the event of a fall.
>
> (2) Suitable and sufficient steps must be taken to ensure the safe transport of any person conveyed by water to or from a place of work.
>
> (3) Any vessel used to convey any person by water to or from a place of work must not be overcrowded or overloaded."

Figure 9-79: Prevention of drowning.
Source: CDM 2015 Regulation 26.

Emergency arrangements to minimise the risk of drowning when someone falls into water and rescue arrangements have been explained in **Element 3.4 'Emergency procedures – Suitable emergency arrangements when working near water'**, please refer to this section. The section covers a range of emergency arrangement, including buoyancy aids and safety boats.

In addition, control measures should be established to prevent falling into water. Care should be taken when plant and vehicles are working near water to prevent the collapse of the water edge/bank. This could result in the plant overbalancing and the operator being trapped in the cab in water. This may require the addition of sheeting to stabilise the edge and a safe perimeter should be established by using robust barriers to prevent approach to the water.

Figure 9-80: Safety boat.
Source: Floating Pontoon Solutions.

If this is not practicable because work is to be conducted at the water edge then platforms should be erected in the water to provide a safe distance from the water.

Figure 9-81: Work near water using a pontoon.
Source: Floating Pontoon Solutions.

This will require the provision of suitable barriers, hand rails and toe boards. The platform may be floating or fixed into the ground. Precautions should then include protection for river traffic (advance warnings, lighting of obstructions and consultation with river authorities) and the need for the protection against the possibility of the structure being struck by such traffic.

Where possible, collective protective measures should be used as they provide the best protection for working near water, such as a scaffold system when working over water. Control measures could include the addition of sheeting or nets to scaffolds. There may also be the need for additional guardrails to working platforms or fall arrest equipment such as safety nets or harnesses.

Figure 9-82: Collective protection measures for work near water.
Source: Pontoonworks.

Sources of reference

Reference information provided, in particular web links, was correct at time of publication, but may have changed.

7 ways driverless cars could fail (Forbes), https://www.forbes.com/sites/chunkamui/2016/04/08/7-driverless-doomsdays/#281914d616cf

Buying new machinery, INDG271, HSE Books, http://www.hse.gov.uk/pubns/indg271.pdf

Chainsaws at work, INDG317 (rev2), HSE, http://www.hse.gov.uk/pubns/indg317.pdf

Drone safety risk: an assessment, Civil Aviation Authority, https://publicapps.caa.co.uk/docs/33/CAP1627_Jan2018.pdf

Drones offer risks, underwriting challenges, Risk & Insurance magazine, https://riskandinsurance.com/drones-offer-risks-underwriting-challenges/

Personal Protective Equipment at Work, Guidance on Regulations, L25, HSE Books, ISBN: 978-0-7176-6597-6. http://www.hse.gov.uk/pubns/priced/l25.pdf

Rise of the Drones, Managing the unique risks associated with unmanned aircraft systems, Allianz, https://www.agcs.allianz.com/content/dam/onemarketing/agcs/agcs/reports/AGCS-Riseofthedrones-report.pdf

Safe use of vertical spindle moulding machines - Information Sheet (Woodworking Sheet No. 18), HSE, http://www.hse.gov.uk/pubns/wis18.pdf

Safe use of woodworking machinery - ACOP and guidance, L114 (second edition 2014), HSE Books, ISBN 978-0-7176-6621-8 http://www.hse.gov.uk/pubns/priced/l114.pdf

Safe Use of Work Equipment, Approved Code of Practice and guidance, L22, HSE Books, ISBN: 978-0-7176-6619-5 http://www.hse.gov.uk/pubns/priced/l22.pdf

Safety signs and signals, Guidance on Regulations, L64 HSE Books, ISBN: 978-0-7176-6598-3, http://www.hse.gov.uk/pubns/priced/l64.pdf

The health and safety toolbox, How to control risks at work, HSG268, HSE Books, ISBN: 978-0-7176-6587-7 , http://www.hse.gov.uk/pUbns/priced/hsg268.pdf

Web links to these references are provided on the RMS Publishing website for ease of use - www.rmspublishing.co.uk

Statutory provisions

Electrical Equipment (Safety) Regulations (EESR) 2016

Equipment and Protective Systems Intended for Use in Potentially Explosive Atmospheres Regulations (EPSIUPEA) 2016

Health and Safety (Safety Signs and Signals) Regulations (SSSR) 1996

Personal Protective Equipment at Work Regulations (PPER) 1992 / Personal Protective Equipment at Work Regulations (Northern Ireland) 1993

Provision and Use of Work Equipment Regulations (PUWER) 1998 / Provision and Use of Work Equipment Regulations (Northern Ireland) 1999

Supply of Machinery (Safety) Regulations (SMSR) 2008

STUDY QUESTIONS

1) The Provision and Use of Work Equipment Regulations (PUWER) 1998 require that work equipment used in hostile environments is inspected at suitable intervals. What items on a 360° tracked excavator should be subject to inspection?

2) What are practical control measures for reducing the risk to workers when using a bench mounted electrically powered circular saw on site?

3) What control measures should be used to reduce risks when cutting concrete kerbstones with a petrol disc cutter?

4) In what ways can accidents occur from the use of cartridge-operated nail guns on a construction site?

5) (a) List, and give an example for each, FOUR mechanical hazards of construction machinery?

(b) What are the advantages and limitations of using a fixed guard to protect people from mechanical hazards of construction machinery?

For guidance on how to answer these questions, please refer to the assessment section located at the back of this guide.

Element 10

Electricity

Contents

PRINCIPLES OF ELECTRICITY

Electricity is a facility that we have all come to take for granted, whether for lighting, heating, as a source of motive power or as the driving force behind the computer. Used properly it can be of great benefit to us, but misused it can be very dangerous and often fatal.

Electricity is used in most industries, offices, and homes, and society could now not easily function without it. Despite its convenience to the user, it has a major danger (risk of electric shock, burns, fire or explosion). The normal senses of sight, hearing and smell will not detect electricity. Making contact with exposed conductors at the supply voltage of 230 volts can be lethal.

Unlike many other workplace accidents, the actual number of electrical accidents is small. However, with a reported 10-20 fatalities and about 1000 accidents involving direct contact with electricity each year, the severity is high. Accidents are often caused by complacency, not just by the normally assumed ignorance. It must be recognised by everyone working with electricity that over half of all electrical fatal accidents happen to skilled/competent people.

In order to avoid the causes of electric injury, it is necessary to understand the basic principles of electricity, what it does to the body and what controls are necessary. Two terms appear in common usage with electricity: 'live' and 'dead'. A 'live' system (also known in some countries as a 'hot' system) is one carrying an electrical current. Once the electricity has been disconnected from its power sources, it is described as 'dead'.

Basic circuitry

The flow of electrons through a conductor is known as a current.

Electric current flows due to differences in electrical 'pressure' (or potential difference as it is often known), just as water flows through a pipe because of the pressure behind it.

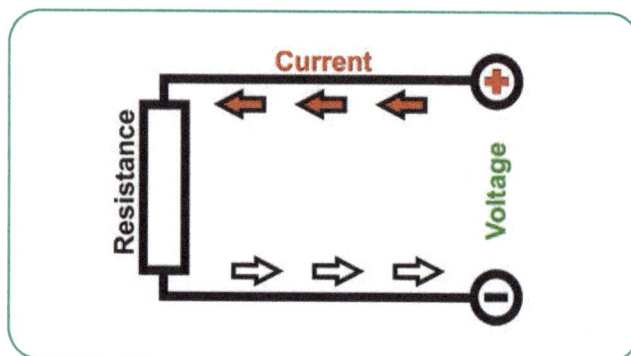
Figure 10-2: A basic electrical circuit.
Source: RMS.

Differences in electrical potential are measured in volts. In some systems the current flows continually in the same direction. This is known as direct current (DC). However, the current may also constantly reverse its direction of flow. This is known as alternating current (AC). Most public electricity supplies are AC.

The UK system reverses its direction 50 times per second and it is said to have a frequency of 50 cycles per second or 50 Hertz (50Hz). DC is little used in standard distribution systems, but is sometimes used in industry for specialist applications. Although there are slight differences in the effects under fault and shock conditions between AC and DC, it is a safe approach to apply the same rules of safety for the treatment and prevention of electric shock.

As a current passes round a circuit under the action of an applied voltage it is impeded in its flow. This may be due to the presence in the circuit of resistance or the opposition to the passage of an electrical current through the circuit conductors. The factors that contribute to conductor resistance are measured in ohms.

Figure 10-3: Electrical hazard warning sign.
Source: Rivington Signs.

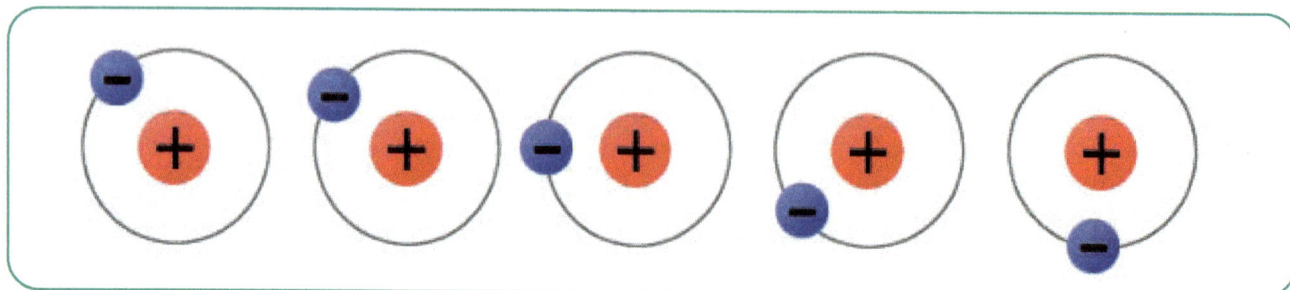
Figure 10-1: Flow of electrons.
Source: RMS.

Relationship between voltage, current and resistance

There is a simple relationship between electrical pressure (volts), current (measured in amperes or milliamperes) and resistance (measured in ohms) represented by Ohm's law:

Voltage (V) = current (I) multiplied by the circuit resistance (R). Therefore, V = I x R or I = V/R.

Hence, given any two values the third can be calculated. Also, if one value changes the other two values will change accordingly. This basic electrical equation can be used to calculate the current that flows in a circuit of a given resistance.

This will need to be done to determine, for example, the fuse or cable rating needed for a particular circuit. Similarly, the current that will flow through a person who touches a live conductor can be calculated.

Resistance in a circuit is dependent on many factors. Most metals, particularly copper and steel, allow current to pass very easily. These have a low resistance and are used as electrical conductors. Other materials such as glass, plastics, rubber and textiles have a high resistance and are used as electrical protective barriers (insulators) with conductors.

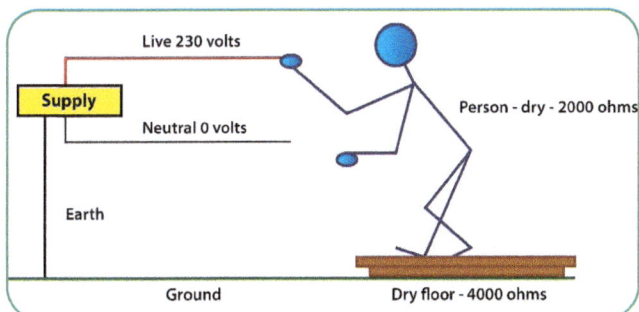

Figure 10-4: An electric circuit under fault conditions showing resistances in the path of a fault current.
Source: RMS.

The schematic in **Figure 10-4** represents a simple electrical circuit in relation to a person. The terms earth and neutral are introduced to represent the two paths (conductors) through which the current returns to the electrical supply generator. This may be through an electrical return conductor (neutral) or literally through the ground.

If the person is on a dry concrete floor, resistance in the body to current flow will be approximately 2,000 ohms and the resistance in the floor approximately 4,000 ohms, therefore, the combined resistance to the current flowing to earth would be 6,000 ohms. Assuming the person is in contact with a live electrical supply at 230 volts, the current flowing through the person in this fault condition can be calculated.

$$\frac{V}{R} = \frac{230 \text{ volts}}{2,000 + 4,000 \text{ ohms}} = 0.038 \text{ amperes}$$

The current flowing through the person to earth will then be about 0.04 amperes or 40 mA (40 milliamperes).

At this voltage (230 volts), there is sufficient potential difference (230 volts to 0 volts) between the supply through the person and the concrete floor to earth, to enable sufficient current to pass through the person and cause a fatal shock.

Electric shock and its effect on the body

The term electric shock is used to describe the unwanted or undesirable exposure to electricity at a detectable level (typically 1mA AC at 50Hz).

When an electric current passes through a material, the resistance to the flow of electrons dissipates energy, usually in the form of heat. If the material is human tissue and the amount of heat generated is sufficient, the tissue may be burnt. The effect is similar to damage caused by an open flame or other high-temperature source of heat, except that electricity has the ability to **burn** tissue well beneath the skin, including internal organs.

Nerve cells communicate by creating electrical signals (at very small voltages and currents) in response to the input of certain chemical compounds (neurotransmitters). If the electric shock current is of sufficient magnitude it will override the electrical impulses normally generated by the neurons, preventing both reflex and volitional (controlled by conscious choice or decision) signals from being able to operate muscles. These effects may be felt as **pain**.

Muscles may also be triggered by a shock current, which will cause them to contract involuntarily. The forearm muscles responsible for bending fingers tend to be better developed than those muscles responsible for extending fingers. If both sets of muscles attempt to **contract**, the 'bending' muscles will be stronger and clench the fingers into a fist.

If the conductor delivering a shock current touches the palm of the hand, the clenching action will force the hand to grasp the conductor firmly, securing contact with the conductor, and it will not be possible for the victim to release their grasp. Even when the current is stopped, the victim may not regain voluntary control over their muscles for a while, as the neurotransmitter chemistry will be in disarray. Involuntary muscle contraction is called tetanus. Shock-induced tetanus can only be interrupted by stopping the current passing through the body.

Electric current is able to affect more than just skeletal muscles; it can also affect breathing and heart function, particularly if the path is across the chest.

The diaphragm muscle controlling the lungs and the heart muscle can also be caused to be in a state of tetanus (involuntary muscle contraction) by a shock current, leading to **respiratory failure** of the lungs and **fibrillation** of the heart or **cardiac arrest**. As discussed previously, fibrillation is a condition where all the heart muscles start moving independently in a disorganised manner, rather than in a coordinated way. It affects the ability of the heart to pump blood, resulting in brain damage and eventual cardiac arrest.

Figure 10-5: Contact with high-voltage buried cable.
Source: www.safetyblog.co.uk.

Factors influencing severity of the effects of electric shock on the body

The factors influencing the severity of the effects of electric shock on the body include the following:

- Voltage.

- Frequency and size of the current involved.

- Duration, the length of contact time (measured in milliseconds).

- The path taken through the body by the current (across the chest/heart/lungs is of significant risk).

- The electrical resistance of the skin in contact with the electrical system (perspiration, sweat) and the internal body resistance as the current flows through the body.

- The resistance of the electrical path of the current flow to earth (factors associated with the footwear/ clothing being worn by the person and the conductivity of the surface they are standing on).

- The general health, gender and age of the person involved (females are more susceptible to electric shock, i.e. at lower current and time of exposure).

The amount of current that flows through the body for a given voltage will depend on the frequency of the supply voltage, on the level of the voltage that is applied and on the state of the point of contact with the body, particularly the moisture condition. The effect of electricity on the body and severity of electric shock results from a combination of the current level and the duration of the passage of that current. The voltage level is relevant mainly in that it causes the passage of the current.

Voltage

Voltage is the driving force behind the flow of electricity, in the same way that pressure in a water pipe influences the amount of water that flows. The correct name for this term is potential difference, as a voltage is the measure of the difference in electrical energy between two points. Any electrical charge that is free to move will move from the higher energy point to the lower one, taking a quantity of that energy difference with it.

Frequency

Alternating current (AC) is preferred over direct current (DC) for power generation/transmission systems and is used domestically in industry and commerce, but AC is 3 to 5 times more dangerous than DC at the same voltage and amperage. AC at frequencies of 50 to 60Hz produces a series of muscle contractions, which will, if the point of contact is the palm of the hand, cause the hand to grasp the source and prolong the exposure. DC is most likely to cause a single convulsive contraction. AC has a greater tendency to cause fibrillation of the heart muscles, whereas DC tends to make the heart muscles stand still. Once the shock current is halted, a still heart has a better chance of regaining a normal beat pattern than one that is in fibrillation. Defibrillating equipment, used by paramedics, utilises a DC shock current to halt fibrillation to enable the heart to recover its normal beat. Though both AC and DC shocks may be fatal, more DC is required to have the same effect as AC, for example, the 'no let go' threshold for DC is reported to be 4-5 times that of AC.

Duration

For an electric shock to have an effect, a person needs to be in contact with the current for sufficient time. At low current levels, the body tolerates the current so the time is not material; however, at higher current levels, for example, 50mA, the person has to remain in contact for sufficient time for the heart to be affected, in the order of milliseconds (ms). In general, the longer a person is in contact with the current, the more harm may be caused.

Resistance

The amount of resistance in a circuit influences the amount of current that is allowed to flow, as explained by Ohm's law. It is possible for a person to be in contact with a circuit and to present sufficiently high resistance that very little current is allowed to flow through their body, see **Figure 10-4**. It should be noted that the level of current flow is also dependent on the voltage; at high voltages an enormous amount of resistance is needed to ensure current flow will remain at a safe level. In an electric shock situation the human body contributes part of the resistance of the circuit, and the amount it contributes depends on the current path taken and other factors such as personal chemical make-up (a large proportion of the body is water), moisture and the thickness of skin, and any clothing that is being worn, such as shoes and gloves.

Current path

The effect of an electric shock on the body is particularly dependent on the current path through the body. Current has to flow through from one point to another as part of a circuit. If the flow was between two points on a finger the effect on the body would be concentrated between the two points. If the current path is between one hand and another, across the chest, this means the flow will pass through major parts of the body, such as the heart, and may cause fibrillation or cardiac arrest.

In a similar way, a contact between hand and foot (feet), causing the current to travel across the chest, can have serious effects on a great many parts of the body, including the heart. These current paths tend to be the ones leading to fatal injuries.

However, a current path from hand to foot down one side of the body may not affect the heart and therefore may not be fatal. Although many people may experience shock from 110 volt or 230 volt supplies, this may not be fatal if they are, for example, standing on or wearing some insulating material. This may be a matter of fortune and, as a rule, this sort of protection should not be relied on.

The amount of current flow through the body has a significant influence on the effect of the electric shock. At low levels of current flow, no effect may be experienced; however, at progressively higher levels of current flow the larger muscles of the body may be affected.

The heart, being made of large muscles, requires a significant current to affect it. At high current levels, burns may occur at the point of current entry and exit from the body, as well as at points along the route the current takes through the body.

Common causes of electric shock

Common causes of electric shock include:

- Work on electrical circuits by unqualified persons.
- Work on live circuits.
- Replacement of fuses and light bulbs on supposedly dead circuits.
- Working on de-energised circuits that accidentally become re-energised.
- Using electrical equipment in a wet environment.
- Faults in electrical systems, which energise parts that are not normally conductors, for example, the casing of electrical equipment.

Direct and indirect contact with an electrical source

Direct

Direct contact relates to when a person makes contact with an electrical source - a charged or energised conductor that is intended to be charged or energised. In these circumstances, the electrical system is operating in its normal or proper condition. This may occur when someone is working on equipment where conductors are exposed and in the live condition.

Indirect

Indirect contact relates to when a person makes contact with electrical conducting material that is normally at a safe potential but has become dangerously live through a fault condition. Conductive parts of equipment that may become live in a fault condition include the conducting casing of equipment and trunking around electrical cables. These are normally safe to touch, but under fault conditions could become dangerously live.

Effects of current flowing in the human body

Current (mA)	Length of time	Likely effects
0–1	Not critical	Threshold of feeling. Undetected by person.
1–15	Not critical	Threshold of cramp. Independent loosening of the hands no longer possible.
15–30	Minutes	Cramp-like pulling together of the arms, breathing difficult. Limit of tolerance.
30–50	Seconds to minutes	Strong cramp-like effects, loss of consciousness due to restricted breathing. Longer time may lead to fibrillation.
50–500	Less than one heart period (750 ms)	No fibrillation. Strong shock effects.
	Greater than one heart period	Fibrillation. Loss of consciousness. Burn marks.
Over 500	Less than one heart period	Fibrillation. Loss of consciousness. Burn marks.

Figure 10-6: Effects of current flowing in the human body.
Source: RMS.

For example, where the casing of equipment has a poor connection to earth, and when a fault occurs on the equipment, the casing may become live.

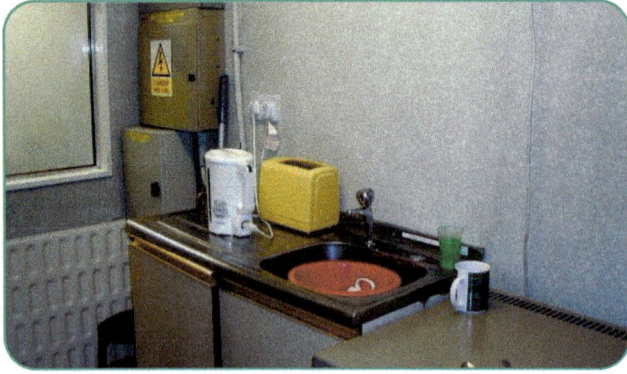

Figure 10-7: Electrical equipment near water supply.
Source: RMS.

Electrical burns

Direct burns

Direct contact with a live electrical source can allow a current to flow through the body, causing a heating effect along the route taken by the electric current as it passes through the body tissue, causing direct burns along its path. Whilst there are likely to be burn marks on the skin at the point of contact (external burns), there may also be a deep-seated burning within the body (internal burns), which is painful and slow to heal.

Indirect burns

Indirect burns can be caused when an electrical discharge (an arc or spark) occurs from a high-voltage system. This discharge contains a lot of energy, and the flash from the discharge can cause burns even without direct contact with the electrical supply.

In addition, a worker may experience indirect burns to external body parts caused by equipment coming into contact with the electrical system. For example, if while working on live equipment, the system is short-circuited by an uninsulated spanner touching live and neutral, this will result in a large and sudden current flow through the spanner. This rapid discharge of energy that follows contact with high voltages not only causes the rapid melting of the spanner, but does so with such violent force that the molten particles of metal are thrown off with huge velocity. When these molten particles contact the parts of a person in the vicinity of the spanner, for example, the hands or face, they can cause serious burns as the molten metal sticks to the skin.

It is not necessary to have high voltages to melt a spanner in this way - it can also occur with batteries that have sufficient stored energy, such as those on a fork-lift truck. Many workers have suffered injury in this way when servicing lift-truck batteries when a spanner has fallen out of the top pocket of their overalls.

External burns

When electrical current makes contact with the skin, it becomes part of the electrical circuit and can cause the point of entry to reach a high temperature, creating immediate tissue damage and charring. The point of entry tends to be limited to small areas and will often appear sunken or hollowed, whereas the exit wound is more extensively damaged and open.

Internal burns

The severity of the damage caused internally by electrical burns will depend upon the pathway along which the current flows. In the human body, the pathways of least resistance are typically, in the first instance, blood vessels, nerves and muscle, then skin, tendon and fat, and, finally, bone. As the outer layer of skin is burnt, the resistance decreases and so the current will increase. The current flowing through the body can cause major injury to internal organs and bone marrow as it passes through them.

Figure 10-8: Used coiled up - risk of overheating.
Source: RMS.

Common causes of electrical fires

Much electrical equipment generates heat or produces sparks and this equipment should not be placed where this could lead to the uncontrolled ignition of any substance. The principal causes of electrical fires are:

- Wiring with defects such as insulation failure due to age or poor maintenance or physical damage.

- Overheating of cables or other electrical equipment through overloading with currents above their design capacity (for example, a coiled extension lead will have a much lower current-carrying capacity than one that is fully uncoiled and will rapidly overheat if this is exceeded).

- Too high a fuse rating for the circuit to be protected (for example, a 13A fuse used for a circuit with a load capacity of 3A).

- Poor connections due to the effects of use/lack of maintenance or unskilled workers (for example, cables not secured by a cable grip inside a drill casing).

Figure 10-9: Max current capacity exceeded.
Source: RMS.

Electrical equipment may itself explode or arc violently and it may also act as a source of ignition of flammable vapours, gases, liquids or dust through electric sparks, arcs or high surface temperatures of equipment. Other causes are heat created by poorly maintained or defective motors, heaters, and lighting.

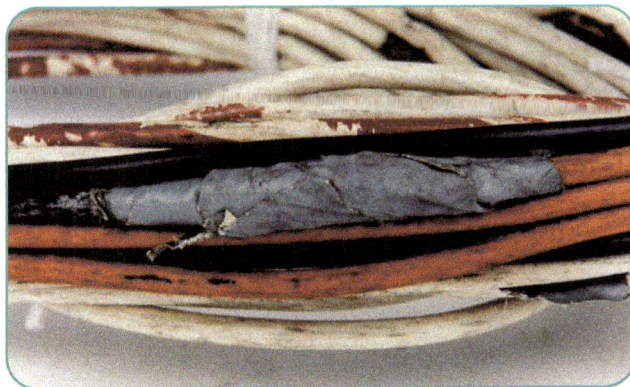

Figure 10-10: Worn cable - risk of electrical fire.
Source: RMS.

Some portable electric equipment, for example, hand held power drills, are powered by a battery that requires to be charged to continue to operate. During the charging process electrical energy passes from the charging device to the battery at a rate determined by the charging device. This transfer of electrical energy creates an amount of heat, but if the correct charging device is used it should remain at an acceptable level. If the incorrect charging device is used the transfer of electric energy may take place too quickly, causing the battery in the device receiving the charge to overheat. This could also occur in a situation where the mechanism that stops the charging process when the battery is at maximum charge fails and charging continues.

Static electricity

Static electricity is different from mains-power and battery-power electricity, as static electricity can be generated naturally. Static electricity is the potential difference (voltage) between surfaces resulting from friction between the surfaces. It is familiar in everyday life as the crackling sound when we remove a woollen sweater and the tiny blue sparks seen in the dark.

Figure 10-11: Evidence of overheating.
Source: RMS.

It is also evident in the clinging together of clothing, paper or sheets of material or the sharp shock we get when we rub and separate from dissimilar surfaces, such as when getting out of a car and then touching the bodywork. Static electricity may be generated in the following situations:

- The flow of liquids and powders through pipes, for example, when re-fuelling with petrol.
- The pouring of powders from insulating plastic bags.
- Spraying
- The unwinding of rolled insulating foils.
- The movement of dust or liquids through air.
- The pouring of liquids, granules or powders from insulated containers.

Static electricity build-up is a potential problem where materials that are not very conductive are in contact with each other, for example, plastics and paper, and where two surfaces are rapidly separated. This is particularly dangerous when a static spark is created in a flammable or explosive atmosphere. Given the right mix of flammable material and oxygen, a static spark with sufficient energy can start a fire, or explosion. The risk to people from direct contact with static electricity is low.

Arcing

Arcing is the flow of electricity through the air from one conductor to another. The arc is a visible plasma discharge between the conductors and is caused by the electrical current ionising gases in the air. Electric arcs occur in nature in the form of lightning. The ability of electricity to arc is used industrially for welding, plasma cutting, and even certain types of lighting, such as fluorescent lighting.

The dangers associated with arcing increase at the higher electrical voltages found in power distribution, where the electricity has the ability to arc distances of 10 metres or more. Being struck by the electric arc will cause an electric shock and probably significant burns. Indirect burns occur from the radiant heat given off by the arc. Damage to the eyes may result from the ultraviolet light (UV) emitted by the arc.

An arc flash is the sudden release of electrical energy through the air when a high-voltage gap exists and there is a breakdown between conductors.

The best way to prevent arc flash or to protect workers in the event of an accident/incident is through effective training. In addition to being 'qualified' to local electrical standards, workers who may be exposed to arc flash hazards need to understand why arc flash occurs and how it can be prevented, and should adopt safe working practices to prevent injury.

Common causes of arc flash include:

- Insulation failure.
- Build-up of dust, impurities, and corrosion on insulating surfaces, which can provide a path for current.
- Equipment failure due to the use of sub-standard parts, improper installation, or even normal wear and tear.
- Birds, bees and rodents snapping leads at connections.
- Human error, including dropped tools, accidental contact with electrical systems, and improper work procedures.

Figure 10-12: Arc discharge.
Source: Solar Power Planet Earth.

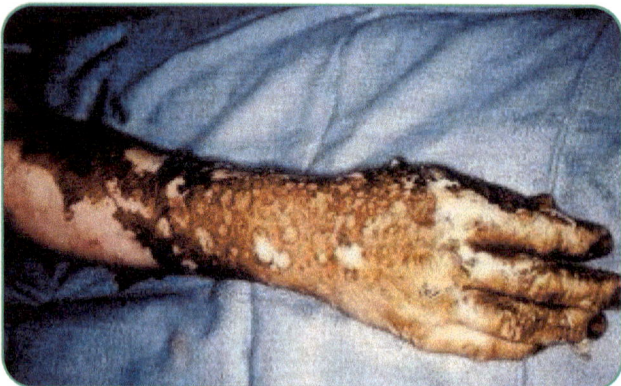

Figure 10-13: Burns from an arc flash of >1000 amps.
Source: SARI/EI.

An arc flash gives off thermal radiation (heat) and bright, intense light that can cause burns. Temperatures have been recorded as high as 20,000°C (36,000°F). High-voltage arcs can also produce considerable pressure waves by rapidly heating the air and creating a blast.

This pressure burst can hit a worker with great force and send molten metal droplets from melted copper and aluminium electrical components great distances at extremely high velocities.

Figure 10-14: Equipment damaged by arc discharge.
Source: GS Engineering Consultants Inc.

Workplace electrical equipment

Use of unsuitable electrical equipment

Regulation 4 of the Electricity at Work Regulations (EWR) 1989 states:

"All systems shall at all times be of such construction as to prevent, so far as reasonably practicable, such danger." "As may be necessary to prevent danger, all systems shall be maintained so as to prevent, so far as reasonably practicable, such danger."

Figure 10-15: Regulation 4 of EWR 1989.
Source: The Electricity at Work Regulations (EWR) 1989.

Many deaths and injuries result from poorly maintained electrical equipment and fires started by faulty electrical appliances. Around 1,000 electrical accidents at work are reported to the Health and safety Executive each year, of these, approximately five people die of their injuries. All electrical equipment should be maintained and checked at appropriate intervals (HSE Guidance 'INDG236 Maintaining portable electrical equipment in low-risk environments' gives some advice on inspection intervals) to ensure it is safe and in good repair.

In particular, managers (and others, such as landlords) responsible for electrical equipment maintenance should ensure:

- Equipment is maintained in a safe condition.
- Information is available to equipment users to ensure safety.
- Safe procedures for inspection and testing are used.
- Records of inspection and testing are maintained.

Accidents/incidents can occur when electrical equipment is used that is unsuitable for the work to be done or for the environment. Electrical equipment used in work conditions where there is a high risk of damage, for example, construction, mining, or quarrying activities, should be of a suitable standard. If equipment is used that is only suitable for low-risk activities, for example,

Element 10

domestic use, cable grips may become loose, casings may get damaged and cable insulation may be cut. All of these conditions could lead to users receiving an electric shock.

Portable electrical equipment that relies on an earth to limit the hazardous effects of its metal casing becoming live could be unsuitable if its arduous use is likely to lead to the earth becoming loose or getting damaged as this would present a serious risk of electric shock for the user. Similarly, if the equipment is not capable of the workload expected of it, it may become overloaded and this may cause it to overheat and cause a fire.

The use of some 230v portable electrical equipment in outdoor or wet work conditions may present a risk of electric shock that could be fatal. In such conditions, reduced-voltage equipment, such as that supplied by a transformer with an output that is centre-tapped to earth, might be more suitable, see *'Element 10.2 - Control measures'* later in this element. Similarly, the use of general electrical equipment in a flammable atmosphere could have catastrophic effects as it might ignite the atmosphere and cause an explosion. The connection of electrical equipment to an electrical supply that does not match the equipment specification, including wrong voltage and current, could overload the conductors, leading to a breakdown of insulation and electric shock or fire. A particular problem can arise through deliberate bad practices by workers. Electrical equipment may be unsuitable because it has been tampered with. For example, a worker may change the fuse for a higher-rated one, and such practices can lead to a fire or electric shock.

Use of defective or poorly maintained electrical equipment

If electrical equipment is poorly maintained or becomes defective in use there is an increased risk of a person making contact with a live conductor that is part of the system and has become exposed. In addition, conductive parts of equipment may become live when a fault occurs. Because such faults may not be obvious, a person may use the equipment unknowingly, exposing them to a high risk of electric shock.

The risk of fatal electric shock is increased if the equipment has a metal casing and the earth connection has become defective. Metal-cased portable electrical equipment poses a particular problem as the flexing of the cable in use can lead to the earth connection failing, often without the knowledge of the user until a fault occurs and they receive an electric shock.

The practice of workers carrying portable electrical equipment by the cable or pulling it along by the cable can lead to conductors becoming loose, which can lead to a fault that could cause a fire or electric shock.

Electrical equipment can become damaged in many ways, for example, plugs or sockets may get broken. This could lead to live conductors being exposed or coming into contact with the outer casing of the equipment, which will significantly increase the risk of electrical shock. Cracks in insulated casings of equipment might allow moisture or liquid to enter the equipment in sufficient quantities for the moisture to conduct electricity from the live conductor to a worker in contact with the casing.

When in use, the insulation of equipment can become less effective than expected as the plastic insulation breaks down over time or when exposed to ultraviolet light/chemicals. This means that where a live conductor comes into contact with the insulation there is a risk of electric shock to anyone touching it.

Figure 10-16: Hazard - fuse wired out.
Source: HSE.

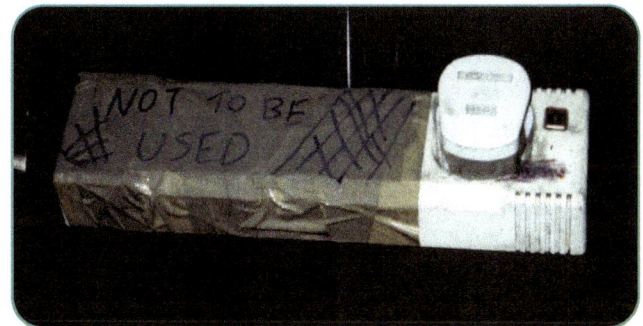

Figure 10-17: Continued use of defective equipment.
Source: HSE.

Figure 10-18: Hazard - damaged cable insulation.
Source: RMS.

Figure 10-19: Hazard - taped joints.
Source: HSE.

Use of electrical equipment in wet environments

Use of electrical equipment in wet conditions increases the risk of harm because the wet conditions increase the conductivity of surrounding surfaces. Where a fault exists on electrical equipment in a wet environment, it may not be necessary to make direct contact with the equipment to receive an electric shock, as the wet substance may act as a conductor, making a circuit between the faulty equipment and the person. These conditions may exist where, for example, a plasterer plasters a wall around a faulty light socket or a pressure water cleaner has a damaged cable lying in the water run-off from the cleaning operation.

Secondary effects

Any injury resulting indirectly from receiving an electric shock is a secondary effect. Secondary effects of electricity occur where the flow of electricity through the body causes muscle spasm or the flow/discharge causes surprise to the worker. Muscular spasm may be severe, particularly if the leg muscles are affected, causing a person to be thrown several metres.

When workers are surprised by an electric shock they receive, they may move their hand away rapidly or step back from where contact was made with the electrical source. Injuries may result from sudden extreme movement, impact with surrounding objects or fall from a height. In addition, a tool may be dropped, causing burns or impact injuries to the user or others nearby. Therefore, a broken bone, cut or bruise may all be injuries caused by 'secondary effects' from receiving an electric shock.

Work near overhead power lines

Contact with live overhead power lines kills a significant number of people and causes serious injuries every year. A high proportion - about one-third - of inadvertent line contacts prove fatal. Overhead power lines may be confused with telephone lines, which can lead people not to identify the risk from contact. Lines may be hard to see at night or against a dark or very bright background.

Power lines typically carry electricity at between 230 volts and 400,000 volts; even contact with an exposed uninsulated 230 volt line can be fatal. Because they are generally not covered in insulation (in rural areas), direct contact can easily be made.

Contact can lead to current passing through a person, and the item that contacted the line, on a path to earth. Rubber-soled shoes would not provide protection from current flow and shock. Actual contact with a power line is not necessary to result in electric shock; a close approach to the line conductors may allow an arc to take place. The risk of arcing increases as the line voltage increases.

Many occupations may require workers to perform their tasks near overhead power lines. Construction workers, truck drivers, tree service workers, mobile equipment operators, agricultural workers, and others may find themselves carrying out their work in the vicinity of live overhead power lines. They may not be trained to recognise the dangers of electrocution if their bodies, equipment, tools, work materials, or vehicles come near to an overhead power line.

See also 'Protection against overhead cables' in 10.2 and 10.3 'Working underneath or near overhead power lines' later in this element.

Contact with underground power cables during excavation work

Underground power cables are present in most locations where excavations are conducted, in shopping areas, on construction or redevelopment sites and even in the countryside. The location of power cables may or may not be known and they may or may not be marked as power cables.

Figure 10-20: Hydraulic breakers strikes a high-voltage electric cable.
Source: PP Construction Safety.

Their presence is not obvious when conducting a visual site survey and so the likelihood of striking a power cable when excavating, drilling or piling is increased. The results of striking an underground power cable can include shock, electrocution, explosion and burns.

As with overhead power lines, any underground service should be treated as live until confirmed dead by a power supply authority. Maps showing the location of underground power cables need to be taken only as an indication of their location, not as a guarantee of accuracy, which means that careful testing needs to take place until indicated power cables are located.

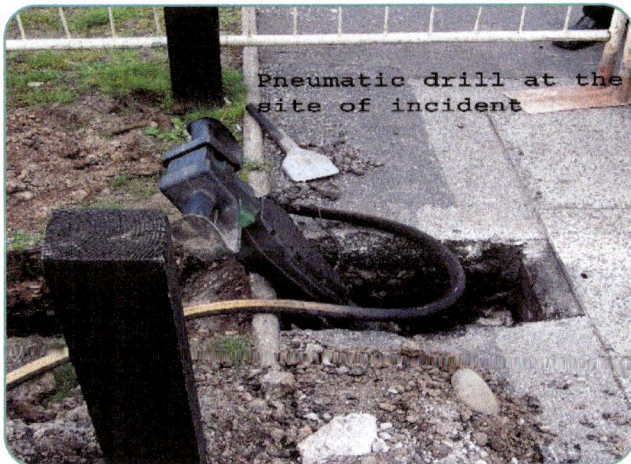

Figure 10-21: 11Kv power cable damaged by pneumatic drill.
Source: PP Construction Safety.

A common problem when installing cables in the ground is that any attempt to lay them in straight lines is often defeated by the presence of other buried services or their natural tendency to coil out of alignment. Caution needs to be exercised as other unmarked power cables in the area may not be easily detected.

See also 'Locating buried services' in 4.2 and 10.4 'Working near underground power cables' later in this element.

Work on mains electrical supplies

Mains electricity supplies, 230 volts, may be treated complacently because these are sometimes viewed as being a common and relatively low voltage. However, a number of deaths occur each year to people working on live mains electricity supplies. It is never absolutely safe to work on live electrical equipment. There are few circumstances where it is necessary to work live, and this must be done only after it has been determined that it is unreasonable for the work to be done dead. In the UK, this is legal requirement under the EWR 1989. Even if working live can be justified, many precautions are needed to make sure that the risk is reduced to an acceptable level.

REVIEW

1) What can affect the severity of injury from contact with electricity?

2) What are common causes of electrical arc flash?

3) Following an electric shock, a worker may experience secondary effects which result in injury. Explain the term secondary effects and outline two common injuries which may occur.

10.2 Control measures

PROTECTION OF CONDUCTORS

Conductors, whether live, neutral, or earthed, require protection from accidental or deliberate contact, interference, misuse or even abuse. This is often in the form of electrical insulation. Insulators are materials that do not readily conduct electricity. Some common insulator materials are glass, plastic, rubber, air, and wood. The insulation must be in good condition.

Insulation appropriate to the environment should be used to give resistance to abrasion, chemicals, heat and impact. Where conductors that are insulated are exposed to a higher risk of damage they may be further protected by a metal casing, which provides reinforcement around the insulation, or ducting, which provides a protective location for the cable.

When conductors are covered with insulation they are often called cables or leads. Flexible cables with multi-strand conductors are required for portable tools and extension leads. Cables must be secured by the outer sheath at their point of entry into the apparatus, including plugs. The individual conductor insulation should not show through the sheath and conductors must not be exposed. Extension leads must be fused.

Temporary wiring should be used in compliance with standards and properly secured, supported and mechanically protected against damage. Taped joints in cables are not allowed and proper line connectors must be used to join the conductors in cables. Connectors should be kept to a minimum to reduce earth path impedance (i.e. the total opposition the AC current faces, so includes resistance and other factors beyond the scope of this course). Many cables are set up on a temporary basis, but these improvised arrangements may get left for a considerable time.

Care should be taken to identify true short-term temporary arrangements and those that warrant full longer-term arrangements, such as placing them in trunking for better protection.

Attention to cables in offices is particularly important to avoid tripping hazards. Cable trips may loosen or damage the conductors' outer insulation sheath or pull the conductors free from terminals within. If this is not noticed, potential shock injuries may occur to workers who subsequently unplug the affected equipment.

Cables routed across roads or pedestrian routes should be covered to protect them from damage. Where cables are to cross a doorway, this is usually best done by taking it around the door frame instead of trailing it across the floor. Overhead cables likely to be hit by vehicles or people carrying ladders, pipes etc. should be highlighted by the use of appropriate signs.

Regular examination should be made for deterioration, cuts (these are best identified by systematic bending of short sections of cable by hand, progressing along the length of the cable - this will open cuts, revealing the conductor), kinks or bend damage (particularly near to the point of entry into the apparatus), exposed conductors, overheat or burn damage, trapping damage, insulation becoming brittle or corrosion.

Figure 10-22: Cover to protect conductors routed across a road.
Source: RMS.

STRENGTH AND CAPABILITY OF EQUIPMENT

It is critical to ensure that all electrical equipment is suitable for its purpose. Electrical equipment must be carefully selected to ensure it is suitable for the electrical system it will become part of, the task it will be expected to perform and the environment in which it will be used. For example, if it is to be used for outdoor work on a construction site in conditions that it might get wet, equipment providing protection from the ingress of water must be selected.

Many tools designed for use in a domestic situation may not be suitable for use in the potentially harsh conditions (for example, wet, high humidity, flammable atmospheres, in constant use) of a workplace. For example, cable-entry grips may be need to be more secure and the outer protection of cables thicker on equipment designed for construction work.

Part of the selection process to ensure the suitability of electrical equipment for work in harsh conditions is to determine situations where low-voltage systems can be used, for example, 110 volts in preference to 230 volts.

The strength and capability of electrical equipment should be considered when determining its suitability. Regulation 5 of the Electricity at Work Regulations (EWR) 1989 requires:

"No electrical equipment shall be put into use where its strength and capability may be exceeded in such a way as may give rise to danger."

Figure 10-23: Regulation 5 of EWR 1989.
Source: The Electricity at Work Regulations (EWR) 1989.

Strength and capability in this context has a wide meaning. Electrical equipment needs to be capable of withstanding the thermal or other effects of the electrical currents that flow through it without failure to danger. This will include normal and fault currents. In order for equipment to remain safe when subjected to sustained fault conditions it may require the inclusion of protective devices that detect the fault and break the circuit, containing the fault. In addition, insulation must be sufficiently effective to enable the equipment to withstand the applied voltage and any transient over-voltages.

When planning construction work consideration should be given to the strength and capability requirements of the electrical system needed for construction activities on site. The plan should include anticipated needs or demand as the site changes, for example, extending provision as a structure is built or orderly withdrawal of provision as a structure is demolished. This may include the location of plant requiring electricity, planning cable runs or ducts, whether high or low voltage is required, control features when in use (switch panels, isolation, and lock off) and protection devices (earthing arrangements, trip switches, residual current devices, panel housings, warnings and instructions).

Planning should also take account of where it is not practicable to provide a permanent mains supplied electrical system and where local power generation is more suitable. In this case the same level of planning is required to ensure the temporary power generation system that will be installed has adequate capability for the demands expected from it. Care should be taken to ensure power is available from supply panels at various points around the construction site, close to where work requiring electricity is carried out. These supply panels must be designed to ensure they have a suitable strength and capability for the harsh conditions related to use on construction sites.

Regulation 25 of the Construction Design and Management Regulations (CDMR) 2015 states:

> "(1) Where necessary to prevent danger, energy distribution installations must be suitably located, periodically checked and clearly indicated."

Figure 10-24: Requirements for energy distribution systems.
Source: CDMR 2015, Regulation 25.

Figure 10-25: Construction site generator.
Source: RMS.

Figure 10-26: Construction site power supply panel.
Source: RMS.

Care should be taken to ensure that power is drawn from agreed points and at agreed, safe, voltages. Where practicable, systems provided for construction work should run at 110V at point of use, or lower, and should be set into place at the earliest safe opportunity; in many cases this can be integrated into ground works for the site.

In some cases ensuring the right amount of power is available where it is needed will require the provision of small, portable electricity generators. These can provide power for construction activities in more remote locations where mains power cannot be provided.

Figure 10-27: 110v generator.
Source: RMS.

Figure 10-28: 110v extension lead.
Source: RMS.

Generating equipment should be installed by a competent person in accordance with BS7375:2010; a code of practice that gives recommendations for the distribution of electricity on construction sites.

If the generator is rated above 55 volts AC, it must be earthed and the effectiveness of the earth checked by a competent person. Generators that operate at or below 55 volts AC do not require an earth. Not all portable generators have a centre tapped earth output of 110 volts, giving 55 volts at point of use, *see 'Reduced and low voltage systems' later in this section.*

Where there is a site distribution system the metal framework of the generator should be connected to a common earth.

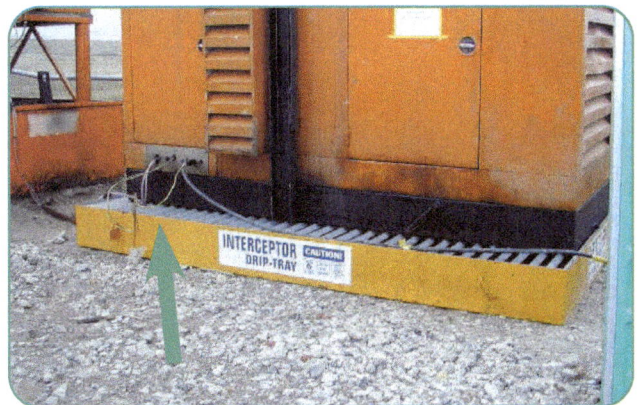

Figure 10-29: Earth provided to generator.
Source: RMS.

Electrical equipment should be used within the manufacturer's rating and the requirements of any instructions supplied. Knowledge of the electrical specification and the tests carried out by the manufacturer, based on the requirements of national standards, may be required. This will assist the employer in identifying the strength and capabilities of the equipment, so that it can be selected and installed appropriately.

Hazardous environments that may affect electrical equipment used in construction activities include:

- Weather - equipment and cables must be capable of withstanding exposure to rain, snow, ice, temperature extremes, etc.

- Natural hazards associated with animals (rats gnawing on cables), solar radiation degradation of insulation, plant growth.
- Extremes of operating temperatures and pressures.
- Corrosive environments (acids/chlorine etc.).
- Flammable substances (gases/dusts/vapours).
- Contamination due to water ingress or dusts.

Additionally, electrical equipment must be sufficiently robust to protect against foreseeable mechanical damage from the environment and from use. If cables are likely to be repeatedly coiled and uncoiled, the movement may result in damage to the cable or connectors.

The cable will, therefore, need to be flexible enough to operate in these conditions.

The British Standard BS 7671:2018 - 'Requirements for electrical installations.

Institution of Engineering and Technology (IET) Wiring Regulations 18th edition', Chapter 13 - 'Fundamental principles for safety', specifies the following needs to ensure electrical installations are appropriate and safe: "Good workmanship and proper materials shall be used. Construction, installation, inspection, testing and maintenance shall be such as to prevent danger. Equipment shall be suitable for the power demanded and the conditions in which it is installed. Additions and alterations to installations shall comply with Regulations. Equipment which requires operation or attention shall be accessible."

ADVANTAGES AND LIMITATIONS OF PROTECTIVE SYSTEMS

Fuses

A fuse is a device designed to automatically cut off the power supply to a circuit within a given time when the current flow in that circuit exceeds a given value. A fuse may be a tinned copper wire in a suitable carrier or a wire or wires in an enclosed cartridge. In effect it is a weak link in the circuit that melts when heat is created by too high a current passing through the thin wire in the fuse case. When this happens the circuit is broken and no more current flows. A fuse usually has a rating in the order of amperes rather than milliamps, which means it has **limited usefulness in protecting people from electric shock.**

The fuse will operate (break the circuit) relatively slowly if the current is just above the fuse rating. Using a fuse with too high a rating means that the circuit will remain intact and the equipment will draw power. This may cause it to overheat, leading to a fire, or if a fault exists, the circuit will remain live and the fault current may pass through the user of the equipment when they touch or operate it.

The following formula should be used to calculate the correct rating for a fuse:

$$\text{Current (amperes (A))} = \frac{\text{Power (watts (W))}}{\text{Voltage (volts (V))}}$$

For example, the correct fuse current rating for a 2 kilowatt kettle on a 230 volt supply would be:

$$\frac{2,000 \text{ W}}{230 \text{ V}} = 8.69 \text{ amperes}$$

Typical fuses for domestic appliances are 3, 5 and 13 ampere ratings. The nearest fuse just above this current level is 13 amperes.

Advantages and limitations of fuses

The limitations of fuses include:

- A weak link in the circuit that melts slowly when heat is created by a fault condition. However, this usually happens too slowly to protect people.
- Easy to replace with wrong rating.
- Needs tools to replace.
- Easy to override by replacing a fuse with one of a higher rating or putting in an improvised 'fuse', such as a nail, that has a high rating.

The advantages of fuses include:

- Offers a good level of protection for the equipment.
- Very cheap and reliable within its limits.

Many countries now use circuit breakers (electro-mechanical devices) instead of fuses in power distribution boards. The advantages of circuit breakers is that they can be readily reset, but only when the fault condition is rectified, unlike a fuse, which might be replaced by one with the wrong rating.

Earthing

A conductor called an earth wire is fitted to the system; it is connected at one end to a plate buried in the ground and the other end connected to the metal casing of the equipment. If for any reason a conductor touches the casing so that the equipment casing becomes 'live', the current will flow to the point of lowest potential, the earth.

	Typical examples of power ratings are:	Suitable fuses at 230 volts:
Computer processor	350 Watts	3 amperes
Electric kettle	1,850–2,200 Watts	13 amperes
Dishwasher	1,380 Watts	13 amperes
Refrigerator	90 Watts	3 amperes

Figure 10-30: Plug - foil fuse, no earth.
Source: RMS.

Figure 10-31: Earthing.
Source: RMS.

The path to this point (earth) is made easier as the wire is designed to have very little resistance. This **may prevent electric shock** provided it is used in association with a correctly rated fuse, or, better still, a residual current device (RCD), and no one is in contact with the equipment at the time the fault occurs.

It should be remembered that earthing is provided where the casing can become live. If the equipment is designed so that this cannot be the case, such as double insulated equipment where the user touches non-conducting surfaces, earthing of the equipment is no advantage. In summary, earthing provides a path of least resistance for 'stray' current and provides protection against indirect shock.

Isolation of supply

Safety devices such as barrier guards or guarding devices are installed on systems to maintain worker safety while these systems are being operated. When non-routine activities such as maintenance, repair, or set-up, or the removal of process material blockages or misaligned feeds are carried out, these safety devices may be removed provided there are alternative methods in place to protect workers from the increased risk of injury from exposure to the unintended or inadvertent release of energy.

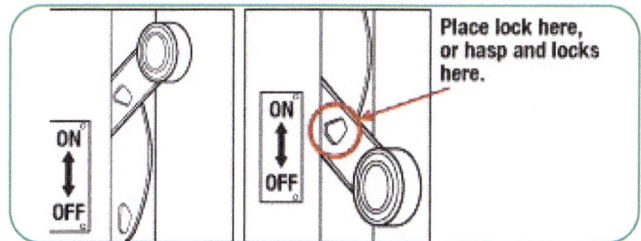

Figure 10-32: Lock out.
Source: Canadian Centre for Occupational H&S.

Isolation of an electrical circuit involves the removal of electrical power from the circuit or system, sometimes referred to as 'making dead' (zero voltage). Isolation of an electrical system is an excellent way of achieving safety for those that need to work on or near the system; for example, isolation of a power supply into a building that is to be refurbished or isolation of plant that is to be maintained. In its simplest form it can mean switching off and unplugging a portable appliance at times it is not in use. Care must be taken to check that the isolation has been adequate and effective before work starts; this can include tests on the system. It is also important to ensure the isolation is secure; 'lock-out' and 'tag-out' (LOTO) systems will assist with this. In practice, lock-out is the isolation of electrical energy from the system (a machine, equipment, or process), which physically locks the system controls in a safe mode.

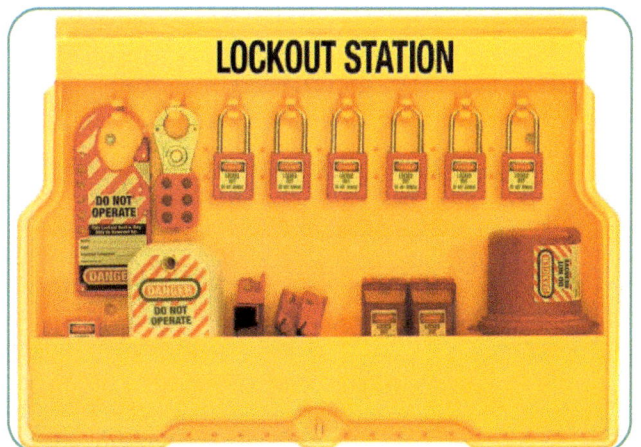

Figure 10-33: Lock-out/tag-out station.
Source: Total Safes.

Figure 10-34: Lock-out/tag-out system in use.
Source: Safety Partners Ltd.

The energy-isolating device can be a manually operated disconnect switch or a circuit breaker (not the user on-off switch). In most cases, these devices will have loops or tabs which can be locked to a stationary item in a de-energised position.

The lock-out device can be any device that has the ability to secure the energy-isolating device in a safe position. Tag-out is a labelling process that is used when lock-out is required. The process of tagging out a system involves attaching a standardised label that includes the following information:

- The reason for lock-out/tag-out (for example, repair or maintenance).
- Time of application of the lock/tag.
- The name of the authorised person who attached the tag and lock to the system.
- To prevent isolated equipment start-up without the authorised individual's knowledge, the LOTO management system should specify that the only person authorised to remove it is the person(s) who placed the lock and tag onto the equipment.

Isolation is a very effective method of protection and ensures people cannot be injured by electrical energy. A limitation of isolation is that certain types of circuit testing and fault finding often have to be carried out with the system live.

Double insulation

This is a common protection device and consists of a layer of insulation around the live electrical parts of the equipment and a second layer of insulated material around this. Commonly, the second layer of insulation is the casing of the equipment (often made from a physical damage-resistant, non-conducting plastic). When the electrical equipment casing is made from an insulator, the equipment is not normally fitted with an earth wire.

Figure 10-35: Double insulated 230V drill.
Source: RMS.

To make sure that the double insulation is not impaired, it must not be pierced by conducting parts, such as metal screws. Neither must insulating screws be used, because there is the possibility that they will be lost and will be replaced by metal screws.

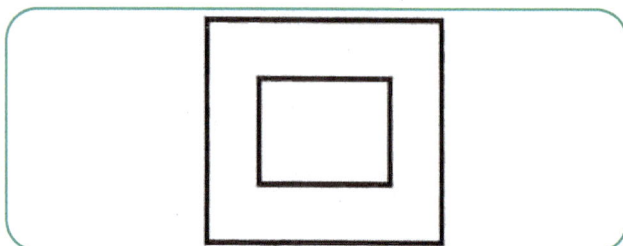

Figure 10-36: Double insulation symbol.
Source: HSG 107.

Any holes in the enclosure of a double insulated appliance, such as those to allow ventilation, must be so small that fingers cannot reach live parts. Each layer of insulation must be sufficient in its own right to give adequate protection against shock. Equipment, that is double insulated has to be identified by being classified at 'Class II' or will carry the symbol shown in **Figure 10-36.**

Figure 10-37: Double-insulation symbol displayed on equipment.
Source: Panasonic.

In summary, double insulated equipment has two layers of insulating material between the live parts of the equipment and the user. If a fault occurs with the live parts and a conductor touches the insulating material surrounding it no current can pass to the user, therefore **no shock occurs.**

Residual current device

A residual current device (RCD) is an electro-mechanical switching device that is used to automatically isolate the supply when there is a difference between the current flowing into a device and the current flowing from the device. Such a difference might result from a fault causing current leakage, with possible fire risks or the risk of a shock when a person touches a system and provides a path to earth for the current. **Figure 10-38**, **schematics 1 to 5**, illustrates the process of disconnection when a fault condition occurs.

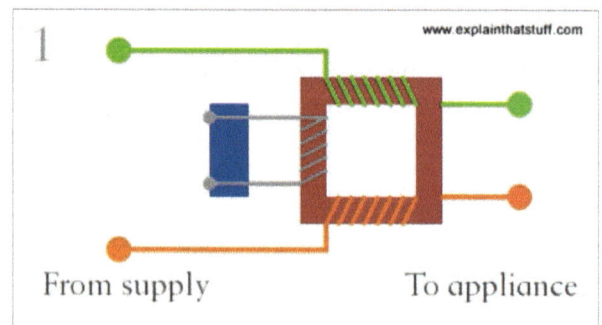

Under normal operation, when power flows from the live wire (green) through the neutral wire (orange) without loss or leakage. The magnetic field is in balance in the core, no electricity flows through the search coil (grey) or relay (blue).

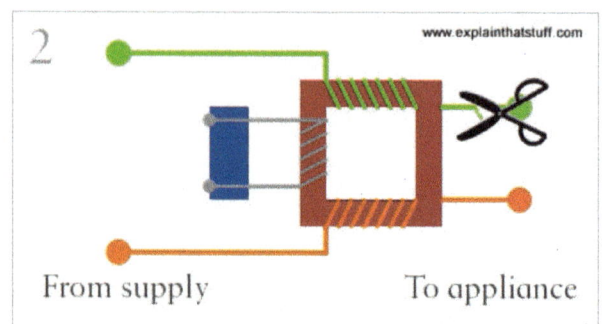

If the live wire (green) is cut (typically by fault or accident).

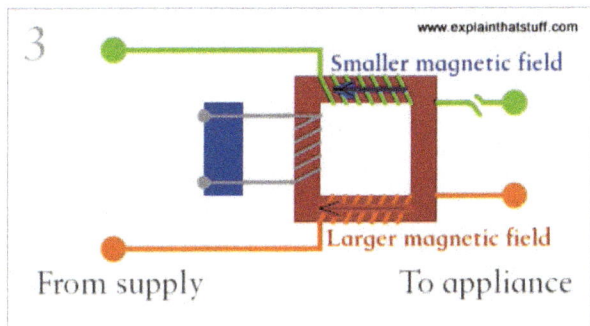

When the live is cut, there is a current imbalance in the core, more current flows through the neutral wire (orange) than through the live wire (green). The two magnetic fields no longer cancel out.

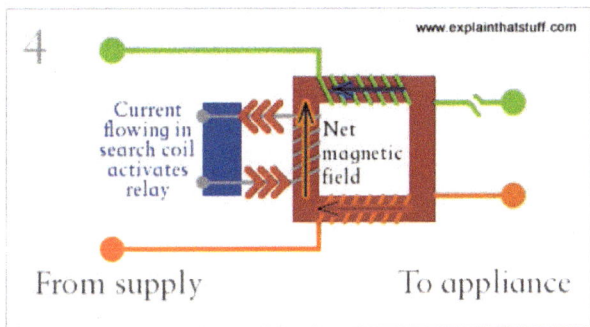

The net magnetic field in the core causes an electric current to flow in the search coil (grey), which activates the relay (blue).

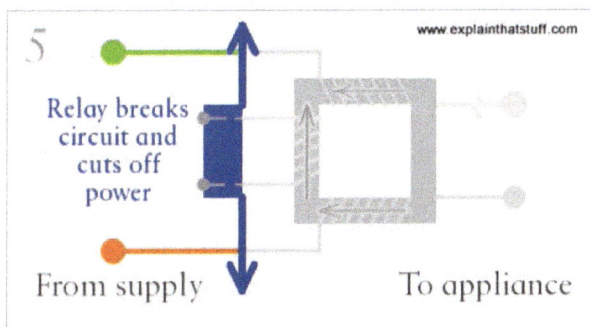

The relay snaps open, breaking the incoming circuit cables and stopping all power from flowing in as little as 30 milliseconds; much faster than the cable can be cut.

Figure 10-38: Operation of an RCD.
Source: www.explainthatstuff.com.

RCDs can be designed to operate at low currents and fast response times (usually 30 mA and 30 milliseconds) and thus they *reduce the effect of an electric shock*. Though they do not prevent the person receiving an electric shock they are very sensitive and operate very quickly and reduce some of the primary effects of the shock, including fibrillation. Despite this, it is still possible for a person to receive injury from the shock, including secondary injuries, such as from a fall caused after receiving the electric shock. The equipment needs to be de-energised from time to time in order to be confident it will work properly when needed. This can easily be done by a simple test routine before use, as equipment is plugged into the RCD.

Figure 10-39: Portable socket with built-in RCD.
Source: Omega.

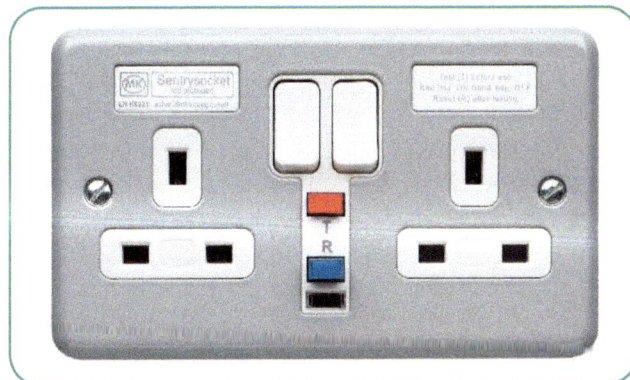

Figure 10-40: Sockets protected by residual current device.
Source: RS Components Ltd.

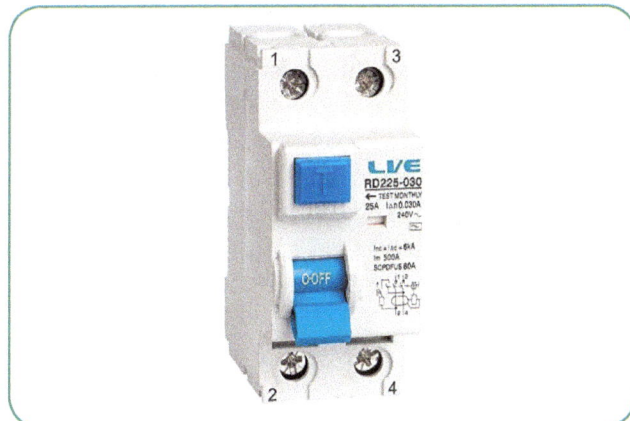

Figure 10-41: RCD/circuit breaker.
Source: RS Components Ltd.

RCD in summary

- Rapid and sensitive.
- Difficult to defeat.
- Easy and safe to test and reset.
- Does not prevent shock, but reduces the effect of a shock.
- They can be prone to 'nuisance' tripping because they are so sensitive to fault conditions.

Reduced and low-voltage systems

One of the best ways to reduce the risk from electricity is to reduce the supply voltage. This is frequently achieved by the use of a transformer.

In simple terms, a transformer consists of two coils of wires around a common iron core.

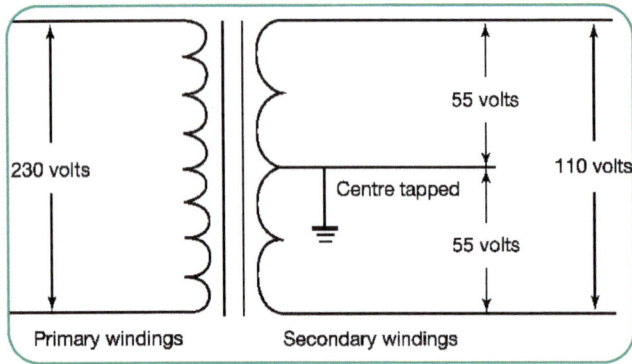

Figure 10-42: Centre-tapped transformer.
Source: RMS.

The first coil of wires is known as the primary winding (and is connected to the supply voltage). The second coil of wires is known as the secondary winding (and generates the output voltage). The ratio between the number of coils in the primary winding and the number of coils in the secondary winding will determine the voltage generated in the secondary winding. If the number of windings in the primary circuit is double that in the secondary, the voltage generated in the secondary circuit will be halved, for example, 230 volts AC to 115 volts AC.

The ratio of winding between the primary and secondary coils can be adjusted (**see Figure 10-42**) to generate any voltage required in the secondary circuit.

In the UK, the primary voltage is 230 volts and the secondary (from a step-down transformer) is 110 volts, which is the voltage the equipment uses to operate. When the transformer is centre-tapped to earth, the generated voltage is 55 volts AC either side of the centre tap, measured relative to the earth. This means that any voltage involved in an electrical shock to earth will be 55 volts and represents a lower risk to a worker if a fault condition occurs.

Using the earlier example of Ohms law, if the voltage is 230 volts then:

$$I = \frac{V}{R} = \frac{230 \text{ volts}}{2000+4000 \text{ ohms}} = 0.038 \text{ amperes or } 38\text{mA}$$

However, if a centre tapped to earth transformer is used, the current is substantially reduced:

$$I = \frac{V}{R} = \frac{55 \text{ volts}}{2000+4000 \text{ ohms}} = 0.009 \text{ amperes or } 9\text{mA}$$

Reference to **Figures 10-4 and 10-7** will clearly show how this **reduces the effects of electric shock** on the body.

An alternative to reduction in voltage by means of a transformer is to provide battery-powered equipment; this will commonly run on 12V-24V but voltages may be higher. The common method is to use a rechargeable battery to power the equipment which eliminates the need for a cable to feed power to the equipment and gives a greater flexibility of use for the user, for example, for drills and power drivers.

Figure 10-43: 110V powered drill.
Source: RMS.

Figure 10-44: 110V centre tapped earth transformer.
Source: RMS.

Figure 10-45: Battery powered drill - 19.2V.
Source: Charles & Hudson.

The appropriate siting of the charger unit (connected to 230 volts) is the main risk associated with the use of battery-powered low-voltage equipment.

USE OF COMPETENT PEOPLE

It is particularly important that anyone who undertakes electrical work is able to satisfy the requirements for competence required by the Health and Safety at Work etc. Act (HASAWA) 1974 and the Electricity at Work Regulations (EWR) 1989. For work on electrical systems below 1,000 volts AC they should be able to work within the guidelines set out in BS7671 'Requirements for electrical installations, IET Wiring Regulations, Eighteenth edition'. Other work should be carried out according to the guidelines set out in the relevant industry standard.

Those who wish to undertake electrical testing work would normally be expected to have more knowledge and to be able to demonstrate competence through the successful completion of a suitable training course. More complex electrical tasks such as motor repair or maintenance of radio frequency heating equipment should only be carried out by someone who has been trained to do them. Work on higher voltage systems must be carried out by specialist electrical engineers trained for this purpose. Work on live electrical systems must be controlled and requires specialist competence and systems of work to ensure safety.

The competent person will need to have:

- Knowledge of electricity.
- Experience of electrical work (similar to that being undertaken).
- An understanding of the hazards presented by electricity and the control measures to reduce the hazard.
- The ability to recognise dangerous situations and whether it is safe for work to continue.

A person who is learning these skills and experience will need to be supervised by a competent person.

CASE STUDY

A person received severe electric shock injuries after incorrectly wiring a machine, which resulted in the machine frame becoming live. The injured person was not competent to undertake such work, yet competent persons were available. The incident occurred in the UK and the employer was fined for their failure to supervise.

USE OF SAFE SYSTEMS OF WORK

Live electrical work

Common terms used when describing live working are given in the UK Guidance note HSG85 'Electricity at work: Safe working practices.' These include:

Charged:	The item has acquired a charge either because it is live or because it has become charged by other means, such as by static or induction charging, or has retained or regained a charge due to capacitance effects, even though it may be disconnected from the rest of the system.

Live:	Equipment that is at a voltage by being connected to a source of electricity. Live parts that are uninsulated and exposed so that they can be touched either directly or indirectly by a conducting object are hazardous if the voltage exceeds 50V AC or 120V DC in dry conditions – see BSI publication PD65193 – and/or if the fault energy level is high.
Live work:	Work on or near conductors that are accessible and 'live' or 'charged'. Live work includes live testing, such as using a test instrument to measure voltage on a live power distribution or control system.

The Electricity at Work Regulations (EWR) 1989, state that:

> "No person shall be engaged in any work activity on or so near any live conductor that danger may arise, unless it is unreasonable in all the circumstances for it to be dead."

Figure 10-46: Work on live conductors, Regulation 14.
Source: The Electricity at Work Regulations (EWR) 1989.

This duty means that if danger could be present, work where possible, should be carried out with the electrical system dead. If it is necessary to work on live conductors, very strict controls must be in place and a safe system of work adhered to. These will include:

- A full justification of why there is no reasonably practicable alternative to live working.
- A live electrical permit-to-work which will be valid for the stated task only.
- At least two competent persons must be present during the work and the more senior of these will sign off the permit when the work is complete and after all equipment has been restored to a fully safe condition.
- Restriction of work area to these competent people.
- Protection of the work area from other hazards, such as vehicles.
- Protection of workers from non-essential live conductors, by isolation or screening.
- Provision of information on the system and task.
- Use of suitable, insulated test equipment and tools.
- Adequate lighting and clear space to work.
- First-aid service immediately available.
- Competent supervision.

Isolation

The Electricity at Work Regulations (EWR) 1989, state that:

> "'Isolation' means the disconnection and separation of the electrical equipment from every source of electrical energy in such a way that this disconnection and separation is secure."

Figure 10-47: Isolation, Regulation 12 of EWR 1989.
Source: The Electricity at Work Regulations (EWR) 1989.

Health and Safety Executive (HSE) booklet HSG85 provides information on isolation procedures when working on both low voltage (LV) and high voltage (HV) systems.

The requirement of the Electricity at Work Regulations (EWR) 1989 means that any isolation method used must be adequate and secure. This means that turning equipment off at a general on-off switch or other mechanism is not adequate. Isolation therefore requires disconnection at the primary isolation mechanism for the equipment or circuit. In order to ensure a high level of security if work on the circuit or equipment takes the person away from the site of the isolator mechanism, a lock or some other means of secure isolation, such as fuse removal, should be considered.

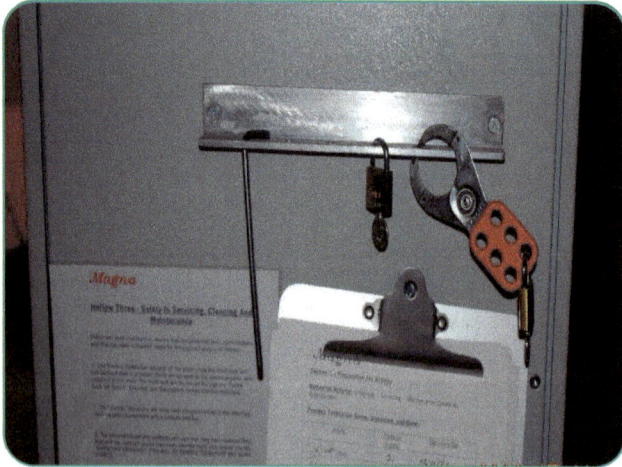

Figure 10-48: Multi-lock system.
Source: RMS.

Isolation procedures, often referred to as 'lock-out/tag-out', in conjunction with a 'permits-to-work' often form part of a 'safe system of work'. Various competences are required to ensure full compliance with these procedures, with specialist training and supervision required. If multi-lock systems are used (**see Figure 10-48**) there must be a contingency plan to deal with the likelihood that someone will lose a key or even leave at the end of a shift without removing their personal padlock.

People who undertake work on electrical equipment or installations should be competent to do so. Those in control of work should instruct workers only to undertake work they are competent to carry out.

CASE STUDY

A worker received a 650 volt AC electric shock when he picked up a cable lying on the ground that was connected to a generator and began to apply insulating tape to exposed wires. On prosecution, the employers were fined £15,000.

Locating buried services

Excavation operations should not begin until all available service location drawings have been identified and thoroughly examined. Record plans and location drawings should not be considered as totally accurate but serve only as an indication of the likelihood of the presence of services, their location and depth.

It is possible for the position of an electricity supply cable to alter if previous works have been carried out in the location due to the flexibility of the cable and movement of surrounding features since original installation. In addition, plans often show a proposed position for the services that does not translate to the ground, such that services are placed in position only approximately where the plan says.

Before groundwork is due to commence it is common and good practice to check for presence of any of the services or hazards by using a detection device. A common service location device used is the cable avoidance tool (CAT) more commonly known as a CAT scanner.

It is important that service location devices such as CAT scanners are used by competent, trained operatives to assist in the identification and marking of the actual location and position of buried services. When identified it is essential that physical markings be placed on the ground to show where these services are located. *See also 10.4 - 'Working near underground power cables' later in this element.*

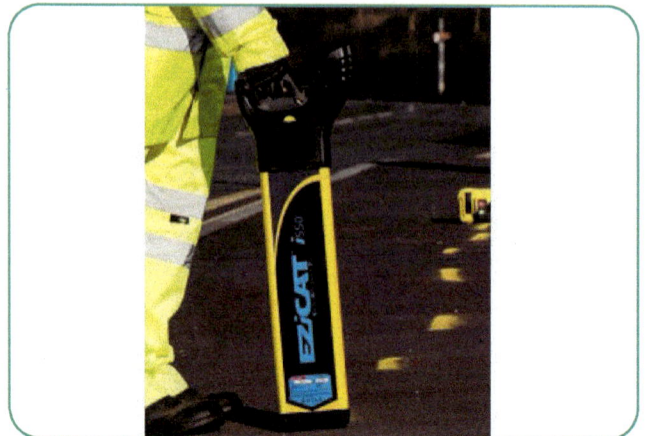

Figure 10-49: Cable Avoidance Tool.
Source: Allpipe Stoppers.

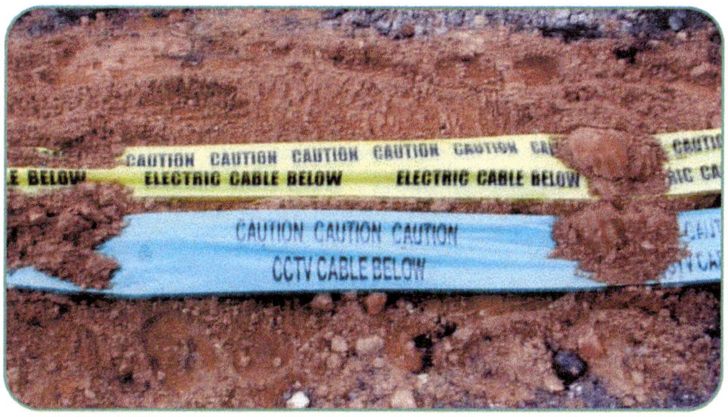

Figure 10-50: Marking of services.
Source: RMS.

Protection against overhead cables

Contact with live overhead lines kills people and causes serious injuries every year. It is important, where possible, to eliminate the danger by:

- Avoidance - carrying out as much work as possible remote from the lines.
- Diversion - diverting all overhead lines clear of the work area.
- Isolation - make lines dead while the work is in progress.
- Insulation - covering overhead lines with suitable insulation.

If the danger cannot be eliminated, the risk must be managed by controlling access to, and work beneath, overhead power lines. For lower-voltage cables it may be possible to temporarily isolate the supply and fit insulation to prevent contact with the live conductor when power is restored.

Figure 10-51: Passing underneath overhead lines, height barriers.
Source: NAL.

The following precautions may also be needed to establish a suitable safe system of work and manage the risk:

- Use barriers to establish a safe zone.
- Clearly mark danger zones with signs and/or bunting.
- Erect 'goal-post' height barriers to define clearance distances for passing underneath overhead power lines.
- Use a signaler/banksman/marshals to move vehicles and plant where appropriate.
- Restrict the use of metal equipment, such as ladders and scaffolds.

See also 10.3 - 'Working underneath or near overhead power cables' later in this element.

EMERGENCY PROCEDURES FOLLOWING AN ELECTRICAL INCIDENT

Anyone working around electrical systems should be aware of what needs to be done for a casualty of electrical shock.

If someone is lying unconscious and in contact with conductors the following actions should be taken in an order depending on the circumstances:

- Assess the situation.
- Summon help, including qualified medical support.
- If possible shut off the power.
- Do not touch the casualty - there may be enough voltage across the body of the casualty to shock the would-be rescuer. The problem with this rule is that the source of power may not be known, or easily found in time to save the casualty from shock.
- Remove the power. If possible prove the system is discharged and dead. If this is not possible take the following action.
- Remove the casualty from the power. It may be possible to dislodge the casualty from the circuit with a dry wooden board or piece of non-metallic material or using a jacket as a loop around the person, holding both sleeves and pulling away.
- Reassess the situation and any remaining danger to yourself and the casualty.
- Once the casualty has been disconnected from the source of electric power, the immediate medical concerns for the casualty should be respiration and circulation (breathing and pulse). If the rescuer is trained in cardio pulmonary resuscitation (CPR), they should follow the appropriate steps of checking breathing (including the airway) and pulse, then applying CPR as necessary to keep the casualty's body from de-oxygenating.
- If the casualty is unconscious, lay them in the recovery position and keep them warm to reduce the chances of physiological shock until qualified medical personnel arrive on the scene.
- Keep the casualty under observation for secondary effects. Cool burns with water.

Figure 10-52: First-aid sign.
Source: RMS.

Further considerations:

- Do not go near the casualty until the electricity supply is proven to be off. This is especially important with overhead high voltage lines: keep yourself and others at least 18 metres away until the electricity supply company personnel advise otherwise.
- Do not delay - after 3 minutes without blood circulation irreversible damage can be done to the casualty.
- Do not wait for an accident to happen - train in emergency procedures and first-aid, plan procedures for an emergency (calling for help, making calls to the emergency services, meeting ambulances and leading them to the casualty) and hold emergency drills.
- Establish if the incident has to be reported under the Reporting of Injuries, Diseases and Dangerous Occurrences Regulations (RIDDOR) 2013.

Figure 10-53: Risk of electric shock due to damage to cable resting on metal checker plat flooring.
Source: RMS.

Figure 10-54: Restriction on work on live circuits.
Source: RMS.

INSPECTION AND MAINTENANCE STRATEGIES

The legal duty for inspection and maintenance

Regulation 4 of the Electricity at Work Regulations (EWR) 1989 states:

"As may be necessary to prevent danger, all systems shall be maintained so as to prevent, so far as reasonably practicable, such danger."

Figure 10-55: Regulation 4 of EWR 1989.
Source: The Electricity at Work Regulations (EWR) 1989.

The Electricity at Work Regulations (EWR) 1989, Regulation 4(2) requires maintenance of electrical systems to prevent danger. Danger is defined as the risk of injury from electric shock, electric burn, fire of electrical origin, electric arcing or explosion initiated or caused by electricity. The Memorandum of Guidance on the EWR 1989 defines a system. In simple terms, it will include any equipment which is, or may be, connected to a common source of electrical energy and includes the source of electricity and the equipment connected to it.

Maintenance is a general term that in practice can include visual inspection, repair, testing and replacement. Maintenance will determine whether equipment is fully serviceable or needing repair.

The HSE guidance HSG107 'Maintaining portable electrical equipment' suggests that cost-effective maintenance can be achieved by a combination of:

- Checks by the user.
- Visual inspections by a person appointed to do this.
- Combined inspection and tests by a competent person or by a contractor.

In order to identify what systems will need maintenance, they should be listed. This same listing can be used as a checklist recording that the appropriate checks have been done. It may also include details of the type of equipment and the checks and tests to be carried out.

User checks

The user of electrical equipment should be encouraged, after basic training, to look critically at apparatus and the source of power. If any defects are found, the apparatus should be marked and not be used again before examination by a competent person.

Obviously, there must be a procedure by which the user brings faults to the attention of a supervisor and/or a competent person who might rectify the fault. Checks by the user are the first line of defence but formal inspection and tests may also be required.

User checks should aim to identify the following:

- Cuts, abrasions and cracks in inner and outer cable insulation.
- Damaged plugs - cracked casing or bent pins, failure of the cord grip.
- Taped or other inadequate cable joints.
- Evidence of bare wires.

- Faulty or ineffective switches.
- Overloaded sockets.
- Lack of formal testing.
- Burn marks or discolouration.
- Damaged casing.
- Loose parts or screws.
- Wet or contaminated equipment.
- Loose or damaged sockets or switches.
- Lack of circuit protection - no RCD.
- Evidence of unauthorised repairs.
- Cables trapped under furniture or in floor boxes
- Vent holes blocked.

Formal inspection and tests

Inspection

The maintenance system should always include *formal visual inspection* of all portable electrical equipment and electrical tests. The factors that might affect the frequency of inspection of, say, a portable drill that is used on a building site, include the nature of the work and the environmental conditions in which the drill is to be used, the frequency and duration of use, the age of the equipment, the intrinsic safety features of the equipment such as double insulation and the use of low voltage, user checks and the number of problems reported, the number and competency of the users, manufacturers' recommendations and best practice guidance, and the results of previous tests and inspections.

The inspection can be done by a person who has been trained in what to look for and has basic electrical knowledge. They should know enough to avoid danger to themselves or others. Visual inspections are likely to need to look for the same types of defects as user checks, but should also include opening accessible parts of equipment/plugs (where this is possible) and inspection of fixed installations.

Opening plugs of portable equipment to check for:

- Use of correctly rated fuse.
- Effective cord grip.
- Secure and correct cable terminations.

Inspection of fixed installations for:

- Damaged or loose cable racks, conduit, or cabling.
- Missing broken or inadequately secured covers.
- Loose or faulty joints.
- Loose earth connections.
- Moisture, corrosion or contamination.
- Burn marks or discolouration.
- Open or inadequately secured panel doors
- Ease of access to switches and isolators.
- Presence of temporary wiring.

User checks and a programme of formal visual inspections are found to pick up some 95% of faults.

Testing

Faults such as loss of earth or broken wires inside an installation or cable cannot be found by visual inspection, so some apparatus needs to have a **combined inspection and test.** This is particularly important for all earthed equipment and leads and plugs connected to hand-held or hand-operated equipment. The electrical system should be tested regularly in accordance with Institution of Engineering and Technology (IET) requirements; tests may include earth continuity and impedance tests and tests of insulation material.

Frequency of inspection and testing

Deciding the frequency

Many approaches to establishing frequency suggest that inspection and testing should be done regularly. The term 'regularly' is not generally specified in terms of a fixed time interval that is applicable for all systems. A risk-based judgement must be made to decide an appropriate frequency.

In effect, the frequency will depend on the conditions in which the system is used; for example, a test of office portable equipment may be sufficient if conducted every three years, whereas equipment used on a construction site may need to be tested every three months. The electrical system as a whole, not just portable equipment, must also be tested periodically. Again, this depends on the conditions of use and may vary from 10 years to 6 months. Factors to be considered when deciding the frequency include:

- Type of equipment.
- Whether it is hand-held.
- Manufacturer's recommendations.
- Its initial integrity and soundness.
- Age.
- Working environment.
- Likelihood of mechanical damage.
- Frequency of use.
- Duration of use.
- Foreseeable use.
- Who uses it?
- Modifications or repairs.
- Past experience.

Records of inspection and testing

In order to identify what systems and equipment will need inspection and testing they should be listed.

This same listing can be used as a checklist recording that the appropriate checks, inspections and tests have been done. It would be usual to include details of the type of the equipment, its location and its age.

It is important that a cumulative record of equipment and its status is available to managers who are responsible for workers using the equipment as well as contractors or suitably qualified workers who are conducting the inspection or test.

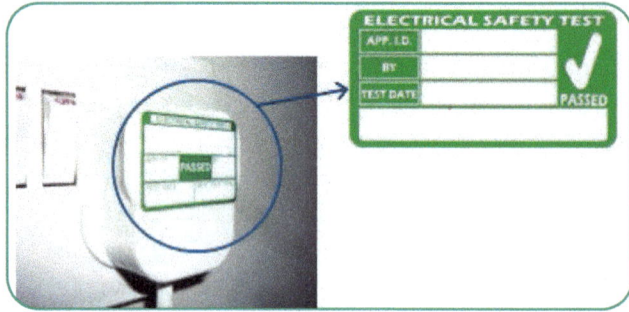

Figure 10-56: PAT labels.
Source: RMS.

In addition, it is common practice to add a label to the system or part of the system (for example, portable appliances) to indicate that an inspection and/or test has taken place and its status following this – portable appliance test (PAT) label.

Some labels show the date that this took place; others prefer to show the date of next inspection or test.

There is a growing trend, especially in offices, for workers to bring to work their own electrically powered equipment including calculators, radios, kettles and coffee makers. The number of electrical accidents has grown accordingly and fires from mobile phone chargers left on overnight are growing in number. All such equipment should be recorded, inspected and tested by a competent person before use and at regular intervals, as if it were the property of the organisation.

> ### CONSIDER
>
> What faults would you look for when carrying out a visual inspection of a portable electrical appliance?

Advantages and limitations of in-service inspection and testing of electrical equipment

Regulation 4(2) of the Electricity at Work Regulations 1989 states that "As may be necessary to prevent danger, all systems shall be maintained so as to prevent, so far as is reasonably practicable, such danger".

All electrical installations deteriorate with age and use. They should therefore be inspected and tested at regular intervals to check whether they are in a satisfactory condition for continued use. Such safety checks are commonly referred to as 'periodic inspection and testing'.

Periodic inspection and testing should be carried out only by electrically competent persons, such as registered electricians.

The Institution of Engineering and Technology (IET) has updated its Code of Practice (COP) for the In-Service Inspection and Testing of Electrical Equipment, (COPISITEE) now in its 5th edition. This update extends to not only portable and transportable electrical equipment but all equipment not covered by other regulations or inspection regimes, for example the British Standard BS 7671:2018 - 'Requirements for electrical installations. Institution of Engineering and Technology (IET) Wiring Regulations 18th edition'.

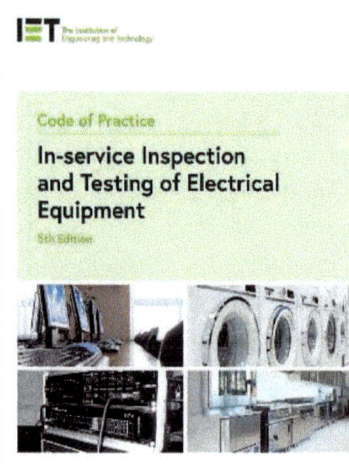

Figure 10.57: COP for In-Service Inspection.
Source: Institution of Engineering and Technology.

This ensures that most, if not all, electrical equipment is subject to some form of inspection and testing. Equipment may not be fitted with a plug and a flex but could be wired to the 'installation', such as an air-conditioning unit or a security access control system.

These types of equipment often fell through the gap of in-service inspection and testing being seen as neither 'portable' nor part of the 'installation'.

The classifications for safety are now described as 'energy source classes', which is subtly different to the previous methods of description. While, to the end-user the changes may be of little consequence in practice, it is important that those performing the inspection and testing of electrical equipment understand the new classifications.

The COP has also removed equipment mobility classifications (i.e. portable, fixed, stationary etc.) with the intent that unless equipment is covered by another inspection and testing regime, the COP will provide the necessary guidance for duty holders regardless of how portable the equipment is.

Portable Appliance Testing or 'PAT' has always been the term commonly used for testing portable electrical equipment and appliances. But with the emphasis on inspection and testing all equipment, whether portable or otherwise, references to 'PAT' have now been removed from the COP.

The 5th Edition promotes greater application of 'User Checks' of electrical equipment. Guidance on the initial frequency of inspection and testing of equipment has been removed in favour of guidance on risk assessments such that intervals between inspections will be more relevant to the environment in which the equipment is being used.

Fixed testing involves testing the electrical installations and systems that conduct electricity around the building. It covers all of the electrical wiring in a building and includes main panels, distribution boards, lighting, socket outlets, air conditioning and other fixed plant.

There is no legal definition of portable equipment. However, it usually means equipment that is intended to be connected to a generator or a fixed installation by means of a flexible cable and either a plug and socket or a spur box, or similar means. This includes equipment that is either hand-held or hand-operated while connected to the supply, intended to be moved while connected to the supply, or likely to be moved while connected to the supply.

In summary, all electrical equipment whether fixed or portable must be inspected periodically y by a suitably qualified person. The testing can be conducted in-house by an employee or outsourced to a third party. Here are some of the issues to consider when doing in-house testing.

Advantages

- Testing is easy to arrange, staff are on hand and an inspection schedule can be developed around their availability.
- The initial cost of training and purchasing equipment can be very high. However, this is usually a one-off investment.
- Employees qualified to carry out inspection and testing will probably have greater ownership and pride in the quality of their work.
- Continuous in-house testing allows for better communication of results between departments.
- Most, if not all, electrical equipment will now be covered by an inspection and testing regime.
- Equipment mobility classifications have been removed.
- Inspections are now based on risk assessments rather than set frequencies.

Disadvantages

- Time limitations and pressures. Employees tasked with the job of inspecting and testing will have their other work to carry out and will be susceptible to time pressure.

- Inspectors may be vulnerable to management pressure to pass equipment if the equipment in question is needed urgently.
- Cost of training and retraining when staff leave or change roles.
- The initial cost of equipment will be high but there are the on-going costs of maintenance and calibration to consider.
- Inspectors need to be familiar with the new classifications of 'energy sources' when carrying out the inspections and testing.

10.3 Control measures for working underneath or near overhead power lines

LEGAL REQUIREMENTS FOR WORKING NEAR POWER LINES

Anyone working near to an overhead power line must manage the risks. Carrying out work in close proximity to live overhead lines should be avoided where possible. If there is no alternative then risks must be properly controlled to an acceptable level. This is in accordance with The Electricity at Work Regulations (EWR) 1989 which require precautions to be taken against the risk of death or personal injury from electricity during work activities and the requirement of The Management of Health and Safety at Work Regulations (MHSWR) 1999 that risks are properly assessed and controlled.

Regulation 25 of the Construction Design and Management Regulations (CDMR) 2015 states:

> "(2) Where there is a risk to construction work from overhead electric power cables:
>
> (a) they must be directed away from the area of risk; or
>
> (b) the power must be isolated and, where necessary, earthed.
>
> (3) If it is not reasonably practicable to comply with paragraph (2)(a) or (b), suitable warning notices must be provided together with one or more of the following:
>
> (a) barriers suitable for excluding work equipment which is not needed;
>
> (b) suspended protections where vehicles need to pass beneath the cables; or
>
> (c) measures providing an equivalent level of safety."

Figure 10-58: Requirements for overhead power cables.
Source: CDMR 2015, Regulation 25.

Owners of overhead power lines have a duty to minimise the risks presented by their lines by regularly inspecting and maintaining their assets, and when asked must advise others on how to control the risks.

The Electricity Safety Quality and Continuity Regulations (ESQCR) 2002 include requirements that the owners of overhead power lines must ensure that they are maintained at the appropriate specified height and meet certain standards.

It is important that site risk assessments are conducted prior to the start of any works, taking into account any site specific circumstances and any advice or information provided by the overhead power lines owner or operator. Adequate control measures must then be put in place including provision of any general health and safety awareness information or training specific to the activity or equipment being used.

HOW TO PREVENT LINE CONTACT ACCIDENTS THROUGH MANAGEMENT, PLANNING AND CONSULTATION WITH RELEVANT PARTIES

Good management, planning, and consultation with interested parties before and during any work close to overhead power lines will reduce the risk of accidents. Planning and consultation with overhead power line owners or operators (i.e. electricity transmission or distribution network operators, Network Rail, private land owners etc.) is essential and forms part of the safe system of work, as it may be necessary to isolate the supply in the event of an incident.

Network owners or operators will also provide information on the route taken by power lines across the site, which will enable accurate identification on site plans. This will establish if work can be carried out away from the lines, if this is not possible the network owner or operator can provide advice on planning access routes to avoid power lines and on the implementation of safe working practices.

Consultation and co-operation should also take place with any other employers or workers who are sharing the site as a workplace, this is particularly important if multiple activities are taking place on site at the same time.

As part of a safe system of work for operating near to overhead power lines, site rules must be put in place. Access routes should be established in consultation and these should be away from overhead power lines. They should be clearly identified, signed and marked. Where this is not possible, then clear passing places with clear goal posts erected at both ends should be established, clearly signed, including clearance height and notice to drivers to lowers jibs. Lighting may be required if work is to take place after sunset.

RISK CONTROL

As with all risk control, the removal of the risk by not carrying out work if there is a risk of contact with or a close approach to the overhead power lines, is the most effective measure. However, if it is not possible to avoid working near or passing closely underneath an overhead power line, the network owner or operator should be contacted to find out if the lines can be permanently diverted away from the work area or replaced with underground cables. If this is not possible or inappropriate as the work is infrequent or short-duration, then find out if the overhead power line can be temporarily switched off or diverted whilst the work is being done. In the event that isolation or temporary diversion of the supply is agreed, it is essential that adequate planning takes place and a safe working procedure has been adopted and approved with the network owner or operator.

If the overhead power line cannot be diverted or isolated and there is no alternative to working near it, a site specific risk assessment must be carried out in order to help identify how to conduct the work safely. The risk assessment should include the following:

- The voltage and height above the ground of the power lines. The height should be measured by a suitably trained person using non-contact measuring devices.

- The nature of the work and if it will be carried out close to or underneath the overhead power line, this should include whether access is needed underneath the power lines.

- The maximum size and reach of any machinery or equipment to be used near the overhead power line.

- The safe clearance distance needed between the power lines and the machinery or equipment and any structures being erected. The overhead power line owner or network operator will be able to provide advice on this.

- The site conditions such as the topography of the land which may reduce safe clearance distances and uneven ground which may affect the stability of plant, vehicles and other equipment.

- The competence, supervision and training of people working at the site.

Figure 10-59: Risks from working near overhead power lines. Source: RMS.

If it is possible that the overhead power line can be isolated for short periods, it may be possible to schedule the movement of large plant and other work activities around the line to take place during these times. This should be planned in agreement with the overhead line owner or network operator.

Care should also be taken not to store or stack items so close to overhead power lines that the safe clearance distance can be infringed by people standing on them.

In situations where work has to be carried out close to or underneath overhead power lines, for example, road works, pipe laying, grass cutting, farming etc. and there is no risk of accidental contact or safe clearance distances being breached, no further precautionary measures would be required. However, risk assessments should take into account any situations that could lead to danger from overhead power lines. Examples of this situation include someone needing to stand on top of a machine or scaffold platform to lift a long item above their head, carrying ladders or poles, or if the combined height of load on a low lorry breaches the safe clearance distance. In these circumstances, additional precautionary measures will need to be taken, for example:

- Ask if the overhead power lines can be isolated for the duration of the work.
- Establish exclusion zones around the overhead power line and any equipment fitted to the pole or pylon.
- Be aware of and allow for unexpected movement of equipment due to weather conditions.
- Carry long objects horizontally and close to the ground.
- Position vehicles so that no part can reach into the exclusion zone.
- Modify machinery such as cranes and excavators by adding physical restraints to prevent them reaching into the exclusion zone;
- Consider fitting plant with insulating guards and proximity warning devices.
- Provide awareness and training for workers regarding the risks and risk prevention measures.
- Provide supervision for the work using someone who is familiar with the risks and required precautions.

Where buildings or structures are to be erected close to or underneath an overhead power line, the risk of contact is increased. This applies to both permanent structures and temporary ones, such as poly-tunnels, tents, marquees, flagpoles, rugby posts, aerials etc. Advice should be sought from the overhead power line owner or network operator in these circumstances to ensure appropriate precautions are in place during the erection of the structure.

Sometimes work needs to be carried out near uninsulated low-voltage overhead power lines, or near uninsulated power lines connected to a building, such as window cleaning, external painting or short-term construction work. If it is not possible to divert or isolate these power lines, then the network owner or operator may be able to fit temporary insulating shrouds to the power lines. People, plant and materials still need to be kept away from these overhead power lines.

USE OF BARRIERS TO ESTABLISH A SAFETY ZONE WHEN WORKING NEAR OVERHEAD LINES

If work is to be carried out near overhead power lines, but no work or passage of equipment underneath the power line it to take place, then the risk of accidental contact can be reduced by erecting ground-level barriers to establish a safety zone to keep people and equipment away from the overhead power lines. In these circumstances barriers should be used in conjunction with warning signs. The recommended distance for the position of barriers is 6 metres (horizontally) from the nearest power line. Some construction vehicles have jib's or telescopic arms that extend long distances, therefore in these circumstances the measurement should be made from the position of the ground barrier to the maximum reach of the jib (measured horizontally).

For very high voltages these distances may be increased significantly. Where plant, such as a crane, is operating in the area additional high-level indication should be erected to warn operators. A line of coloured plastic flags or 'bunting' mounted 3 to 6 metres above ground level over the barriers is often used. Extra care should be taken when erecting bunting and flags to avoid contact or approach near the overhead power lines.

The barriers should be sturdy enough so that they cannot be moved easily and should be colour coded with red and white stripes. Types of barrier commonly used include:

- 200 litre steel drums filled with rubble or concrete, these should be highlighted by painting them with red and white horizontal stripes.
- Railway sleepers.
- A raised earth bank to 1 metre and marked by posts to prevent vehicle entry.
- A tension wire fence with red and white flags attached (this should be earthed in consultation with the electricity network operator).

Ensure that the barriers can be seen at night, this can be done by using white or fluorescent paint or attaching reflective strips.

Figure 10-60: Barriers used to establish safe zone for working near overhead power lines.
Source: HSE, HSG150.

MEANS OF SAFELY PASSING UNDERNEATH OVERHEAD LINES

If equipment, plant or vehicles capable of breaching the safe clearance distance have to pass underneath the overhead power line, it will be necessary to create a safe passageway through the barriers. Access and egress routes beneath overhead power lines should be clearly marked with safe clearance distances using a vertical pole each side of the opening and a height limiting barrier above and between. This work practice is commonly referred to as crossing under 'goal posts'.

The safe clearance distance from the overhead power line depends on the level of voltage of the power line as it is possible for electricity to arc between the line and any conducting material in contact with an earth. This means that the correct clearance may have to be determined by consultation with the electricity network operator responsible for the overhead power line.

The following guidance should be considered when establishing passageways underneath overhead power lines:

- Keep the number of passageways to a minimum.

- Define the route of the passageway using fences and erect goal posts at each end to act as gateways using rigid non-conducting material for example timber or plastic pipe, for the goal posts. Highlight them with red and white stripes.

- If the passageway is too wide to be spanned by a rigid non-conducting goal post, it may be necessary to use tensioned steel wire earthed at each end, or plastic ropes with bunting attached. These should be positioned further away from the overhead power line to prevent them being stretched and the safety clearances being reduced by plant moving towards the line.

- Ensure the surface of the passageway is levelled, formed-up and well maintained to prevent undue tilting or bouncing of the plant and equipment.

- Put warning notices at either side of the passageway, on or near the goal posts and on approaches to the passageway giving the crossbar clearance height and instructing drivers to lower jibs, booms, tipper bodies etc. and to keep below this height while crossing underneath the overhead power lines.

- Illuminate the notices and cross bars at night, or in poor weather conditions, to make sure they are visible.

- Make sure that the barriers and goal posts are maintained.

There may be a requirement to restrict certain equipment, for example, rough terrain fork-lift trucks, telehandlers or cranes from working under overhead power lines due to the possibility of jibs or forks coming into contact with power lines or near enough to draw an arc from the power lines. Similarly the reach of telescopic equipment may have to be restricted by the application of electronic or mechanical restraints.

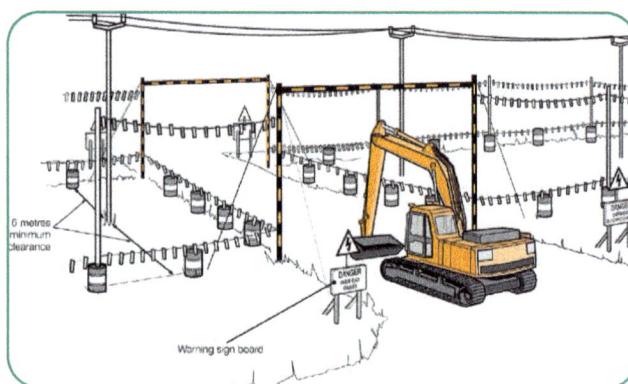

Figure 10-61: Overhead cable protection on traffic route.
Source: HSE, HSG144.

KEY EMERGENCY PROCEDURES IF SOMEONE OR SOMETHING COMES INTO CONTACT WITH AN OVERHEAD LINE

If someone or something comes into contact with an overhead line, it is important that everyone involved knows what action to take to reduce the risk of anyone sustaining an electric shock or burn injuries. Emergency procedures should include the following guidance:

- Never touch the overhead power lines wires.

- Assume that the wires are live, even if they are not arcing or sparking, or if they otherwise appear to be dead.

- Be aware that even if the power lines are dead, they may be switched back on either automatically after a few seconds or remotely after a few minutes or even hours if the network owner or operator is not aware that they have been damaged.

- If possible call the emergency services. Give your location, tell them what has happened and that electricity wires are involved.

- If possible, call the emergency electrical helpline and advise the electricity network operator of the situation.

- If you are in contact with or close to a damaged power line, move away and stand well clear.

- If you are in a vehicle that has touched a power line, remain safe by staying inside the cab of the vehicle. If it is necessary to leave the vehicle, the driver should jump out as far as possible so that no contact is made with the live vehicle. No-one should touch the vehicle whilst standing on the ground.

- Be aware that if a live wire is touching the ground, the area around it may be live also. Keep a safe distance away from the wire or anything else it may be touching and keep others away.

10.4 Control measures for working near underground power lines

PLANNING THE WORK

Damage to underground electrical power cables can cause fatal or severe injury as well as significant disruption, environmental damage, project delays and additional costs. Injuries are usually caused by electrical 'flashover' explosions and associated fire and flames that are emitted when a live cable is penetrated by a sharp object such as the point of a tool.

These explosive events can also occur due to a cable being severely crushed or due to unreported and unrepaired damage to cables, connections and terminations.

Regulation 25 of the Construction Design and Management Regulations (CDMR) 2015 states:

> "(4) Construction work which is liable to create a risk to health or safety from an underground service, or from damage to or disturbance of it, must not be carried out unless suitable and sufficient steps (including any steps required by this regulation) have been taken to prevent the risk, so far as is reasonably practicable."

Figure 10-62: Requirements for underground power cables.
Source: CDMR 2015, Regulation 25.

It is therefore important to effectively plan any excavation work in areas where underground power cables or other services may be present and to use a safe system of work. The extent of the work area should be clearly identified along with the presence of any underground services within the area. The following steps should be carried out to enable the work planning:

- Obtain service drawings from the electricity distribution network operator and other utility companies. Other organisations who hold relevant information about the site may also need to be contacted (particularly for specialist sites i.e. hospitals, airports, Ministry of defence sites, petrochemical sites, nuclear power plants).

- Survey the site to identify the services and other underground structures. Record the location of any services.

- Review and assess the planned work to avoid disturbing services where possible.

- Allow sufficient time and provide sufficient resource to do the work safely.

- Emergency work still needs to be subject to planning and assessment of the risks arising from the work.

Figure 10-63: Risks from working near underground power lines.
Source: RMS.

The provision of pre-construction information

The Construction Design and Management Regulations 2015 allocate specific duties in relation to planning for work near to underground power cables:

- Clients have a duty to make reasonable enquiries about underground services and pass relevant information to the designers and contractors. The most up to date information should be included in the tender information. If clients are unable or unwilling to obtain this information, then the contractor must be allowed sufficient time and resource to do it instead. Clients should also consider how contractors have addressed the risks from underground services.

- Designers have a duty to reduce or 'design out' the risks arising from damage to underground services. Having reduced the risks to a level as low as is reasonably practicable by design, information should be provided to those doing the work about the remaining risks.

- Contractors must prepare safe systems of work for their workers by identifying the hazards they are likely to encounter during the work, making an assessment of the risks presented by those hazards.

Clear information on the type, location, and status of underground services and the tools, equipment and working practices that will be required to avoid damaging the services is essential.

USING CABLE PLANS

Service owners should provide up to date, readable plans which show the recorded line and depth of all known services buried in the proposed work area. This would include underground power cable plans. Information should be made readily available to enquirers and arrangements should be made to respond at short notice and in emergency situations outside of normal office hours.

Obtaining and reviewing plans before any excavation work starts

Plans should always be requested from the service owners or network operators. However, the use of plans alone is not sufficient to identify and locate underground services before starting work. Plans provide an indication of the location, configuration and number of underground services at a particular site, but they do have limitations, such as:

- The position of reference points, such as kerb lines, may have changed since the plans were drawn.

- Re-grading of the surface may mean that the depths shown are no longer correct.

- Services, particularly cables, may have been moved without the knowledge of their owners or operators.

- Service connections may not be marked.

- Services marked as straight lines may, in practice, be looped outside substations or switch rooms.

- Plans may show spare unused ducts.

- The routes of older services may not have been recorded, therefore the absence of a record should not be taken as proof that the area is free of underground services.

Due to these limitations, it is important that further precautions are taken as part of a robust planning stage survey for the proposed work area. Anyone involved in detecting and identifying services must be competent in the proper use of survey tools and detecting devices as well as reading and interpreting plans. There are several different levels of survey:

- Desktop study – service drawings and plans are requested from the owners of underground services and reviewed to determine the likely location of services in the proposed work area. This should

be done for all projects that involve excavation or penetrating the ground.

- Desktop study and site investigation – use the information from the desktop study to assist a physical inspection of the site. Look for physical signs such as inspection hatches, reinstated excavations, street lights and telecoms boxes. Conduct a survey using service location devices.

- Physical identification of the services – in addition to the above, further steps are taken to identify underground service by digging trial holes to verify location, depth and type of underground service. This may also involve passing a tracing device through pipes or tunnels.

The level of survey needed will depend on the nature of the work site and the information available. Once the services have been detected and identified, plans or drawings should be updated as necessary. The status of services should also be confirmed i.e. whether electricity cables are live or gas pipes are pressurised.

What you should do if the information cannot be obtained

If there are no plans or drawings available then caution should be applied and the survey should involve a physical site investigation and activities to identify the presence and location of services.

This will take the form of a site inspection to identify any signs of the presence of services such as lamp posts, illuminated traffic signs, gas service pipes entering buildings, pit covers, pipeline marker posts, valve boxes and housings, or evidence of reinstated trenches. Following this, service location devices should be used by competent operators with sufficient knowledge and experience in the use of survey equipment and techniques. The position of any services identified should be pinpointed as accurately as possible and suitably recorded in a clear and usable format. This should then be shared with those working on the site, the site itself should also be marked with service locations where possible. Where doubt about the location of suspected services remains safe digging methods must be used to remove soil and other material.

Use of equipment for detecting/locating buried services

The selection of detection tools and survey methods must be undertaken by someone who understands the range of methods and tools and their limitations. They need to be aware of the potential for false readings and signals in certain techniques as this could lead to inaccurate information being included in the plan of work, which may lead to serious injury. There are a number of different types of detecting devices or locators which can be classified as follows:

- Hum detectors – are receiving instruments that detect the magnetic field radiated by electricity cables which have current flowing through them. Limitations of this device are that they do not respond to cables where there is little or no current flowing, for example, services to unoccupied premises or street lighting cables in the daytime, direct current cables, some well-balanced high-voltage cables which generate relatively little field, or pot-ended cables.

- Radio frequency detectors – are receiving instruments that respond to low-frequency radio signals, which may be picked up and re-emitted by long metallic pipes and cables. Limitations are that other metallic objects such as abandoned pipes, cables and tram tracks may re-radiate the signal and results can be affected by the locality, length of the underground cable or pipe, the distance from the termination, and geographical orientation.

- Transmitter / receiver instruments – these are small portable transmitters or signal generators which can be connected to a cable or pipe, or placed very close to them so that a signal is introduced into it. Typically the radio frequency detector is used as a receiver for this method. Generally, the location of some part of the cable or pipe needs to be already known in order to correctly position the transmitter. A direct connection is not required, but would improve accuracy if this is possible. This method can be useful if other previous detection methods have not been successful. Use of signal generators will significantly increase the accuracy of the service location.

- Metal detectors – are conventional detectors which will usually locate flat metal covers, joint boxes etc., but may miss round cables or pipes.

- Ground probing radar – a method capable of detecting anomalies in the ground. If the anomalies can be plotted into a continuous line it may indicate a cable, duct or pipe. This technique used alone would not determine the precise nature of the service and should be supported by information available about services present and also by other more conventional forms of detecting device. False readings can also occur when there are boulders and debris in the ground or where the ground has been disturbed. This technique can also be expensive due to equipment costs and specialist operator training required.

- Radio frequency identification (RFID) – a system increasingly used to mark or 'tag' new services. Markers can be programmed with information about the particular service and its depth, which can be

read by detecting devices. The accuracy of the information will depend upon the marker being properly attached to the service. This is a developing system so will only be found on new services and will not assist with older services. RFID marker systems may also require specific detecting tools that may not be compatible with one another.

Some instruments may use more than one of the technologies listed which may also include a depth-measuring facility.

Figure 10-64: Use of service locating device.
Source: SCCS Survey.

USE OF SERVICE LOCATING DEVICES

There are a number of factors which may impact upon the reliability and effectiveness of underground service detection using service location devices:

- The training, skill, hearing and experience of the operator.
- The characteristics of the device being used.
- The calibration and reliability of the detecting device.
- The type, length and depth of the service.
- The magnitude of the current being carried by electricity cables.
- The effects of other nearby services.
- The nature of surface conditions, for example, reinforced concrete.
- The nature of the ground conditions.
- Whether or not a signal generator is being used.

Anyone who uses a service locating device should have received thorough training in its use and limitations. Service locating devices should always be used in accordance with the manufacturer's instructions and should be regularly checked and calibrated to ensure they remain in good working order.

Electricity cables can sometimes be terminated in the ground by means of a seal, which may have external mechanical protection. These 'pot-ended' or 'bottle-ended' cables should be treated as live and should not be assumed to be abandoned or disused. These cables can be difficult to detect with service locating devices even when 'live'.

Other limitations of service locating devices (SLDs) which those operating them need to be aware of are:

- SLDs may not be able to distinguish between cables and pipes running close together and may just present a single signal. The use of signal generators will greatly increase the accuracy of detecting and tracing underground cables.

- SLDs do not detect plastic pipes or other non-metallic services. Exceptions to this are if ground probing radar is used, metallic tracer rods have been laid within the pipe, RFID markers have been utilised, or a small signal transmitter/tracing rod can be pushed along the pipe or ductwork.

- Telecommunications and railway signalling cables cannot always be located by SLDs. Exceptions to this are if metal components are connected to earth, they have been tagged with location markers or they sit in ducts where tracing rods can be used.

If a service recorded on a plan cannot be located, appropriate assistance or advice should be sought from the service owner or operator.

Services should be traced through the full extent of the work area as they may not run in straight lines. Cables will often have kinks or loops and pipes may have joints or bends which may not be shown on service drawings or plans.

It is advisable to use safe digging practices to make trial holes in order to positively identify a service and its depth. Exposing a service safely in this way will allow its status to be checked and may make it easier for a tracing signal to be applied.

Once the service locations and routes have been traced, it should be marked onto the ground with waterproof crayon, chalk/paint on paved surfaces or with wooden pegs in grassed/un-surfaced areas, preferably a short distance to one side of the service.

SAFE DIGGING PRACTICES

Excavation work in areas where underground power cables or other services are believed to be present should follow safe digging practices. Anyone carrying out excavation work to expose services and those supervising them must be competent to do so.

All those involved should receive sufficient information, instruction and training to:

- Understand the risk to safety from damaging services.
- Use service locating devices.
- Use safe digging practices.
- Understand the value of hand digging and the risks from using power tools or mechanical excavators.

Once the service position and route has been determined using a detecting device, excavation can commence with the digging of trial holes using suitable hand tools or vacuum excavation in order to confirm the exact position of any detected services. Extra care should be taken when digging above or close to the assumed line of any services.

Final exposure of the service by horizontal digging is recommended, as the force applied to hand tools can be controlled more effectively. Insulated hand tools should be used when digging near to electricity cables. Spades and shovels with curved edges should be used rather than other tools and they should be eased into the ground with gentle foot pressure. Picks, pins or forks can be used with care to free lumps of stone etc. or to break up hard layers of chalk or sandstone. Picks should not be used in soft clay or soft soils near to underground power cables.

Use of hand held power tools and mechanical excavators is not recommended close to underground power cables.

These tools can be used to break up hard surfaces where the survey has proved that there are no services, or the services are deep enough so as not to be damaged by such tools. Even in these circumstances the use of these tools should be carefully planned and managed. In the event of mechanical excavators being used, a second person should assist the driver from a position where they can see into the excavation and provide warnings of obstacles or services.

It is recommended that hand-held power tools are only used 500mm or more away from the indicated line of a service buried in or below a hard surface. This safety margin may be reduced if congestion of services renders it impractical or where surface obstructions limit the space available, but only if the line of the cable has been positively identified in the plans, confirmed by a locator, and additional precautions taken to prevent damage to the services.

In the event of a buried service being damaged during excavation work, the service owner or operator should be notified immediately, further work should stop and people should be kept clear of the area until it has been made safe by the owner or operator.

THE USE OF APPROPRIATE TOOLS, LOCATING DEVICES, AND ROUTE PLANNING WHEN UNDERTAKING EXCAVATION WORK, INCLUDING VACUUM, AIR AND HYDRO EXCAVATION

Methods and techniques for excavating near underground power cables or other services should be determined before work starts and should take account of:

- The nature and scope of the work.
- The type, position and status of underground services.
- The ground conditions.
- Site constraints.

This information will lead work planners to select appropriate tools and equipment for the specific situation and site conditions, as outlined in previous sections covering service locating devices and safe digging practices.

In addition to the use of hand tools, hand-held power tools and mechanical excavators, safe methods of excavating may also include vacuum excavation. Vacuum excavation equipment may incorporate the use of water jetting and high velocity air jets. These techniques can be very effective in congested excavations where mechanical excavation and the use of hand tools is difficult. However, they do have limitations and will not work on all ground conditions or materials, for example concrete. If jetting tools are utilised it will be necessary to assess:

- Risk of injury to operatives or those nearby from ejected soil and other materials, and implement appropriate controls.

- Use of jetting tools to excavate around cables, due to the potential for old and fragile cables to be damaged by the jet of water or air.

Once the methods and techniques of work are decided upon, those carrying out the work must be provided with a written plan, including information about the location and nature of the underground services. All workers should be competent, provided with appropriate personal protective equipment (PPE) and work equipment, and allowed sufficient time to complete the work.

During any excavation work near underground power cables or other services it is important to make frequent and repeated use of service locating devices as the location of the service is likely to become more accurately detected as covering material is removed.

Once exposed, services may need to be supported, particularly cables with joints as these may be enclosed in earthenware pipes, filled compound or be of cast iron or plastic epoxy filled casings. These will need proper support and should not be treated roughly or manipulated. Never use power cables or other services as handholds or footholds for climbing in or out of excavations.

Sources of reference

Reference the information provided, in particular web links, these are correct at time of publication, but may have changed.

Avoidance of danger from overhead electrical lines, GS6 (fourth edition), HSE Books, http://www.hse.gov.uk/pubns/gs6.pdf

Avoiding danger from underground services, HSG47 (third edition 2014), HSE Books, ISBN 978-0-7176-6584-6 http://www.hse.gov.uk/pubns/priced/hsg47.pdf

Electrical safety and you - A brief guide, INDG231 (Rev1), HSE Books, http://www.hse.gov.uk/pubns/indg231.pdf

Electricity at work, safe working practices, HSG85, HSE Books, ISBN: 978-0-7176-6581-5, http://www.hse.gov.uk/pUbns/priced/hsg85.pdf

Guidance on the management of electrical safety and safe isolation procedures for low voltage installations, Electrical Safety First in partnership with others (including the HSE), https://www.electricalsafetyfirst.org.uk/downloads/

IET Wiring Regulations 18th Edition (first amendment), BS7671:2018

Maintaining Portable and Transportable Electrical Equipment, HSG107, HSE Books, ISBN: 978-0-7176-6606-5, http://www.hse.gov.uk/pubns/priced/hsg107.pdf

Safe Use of Work Equipment, Approved Code of Practice and guidance, L22, HSE Books, ISBN: 978-0-7176-6619-5, http://www.hse.gov.uk/pubns/priced/l22.pdf

The Electricity at Work Regulations 1989, HSR25 HSE Books, ISBN: 978-0-7176-6636-2, http://www.hse.gov.uk/pubns/priced/hsr25.pdf

The health and safety toolbox, How to control risks at work, HSG268, HSE Books, ISBN: 978-0-7176-6587-7 http://www.hse.gov.uk/pUbns/priced/hsg268.pdf

Web links to these references are provided on the RMS Publishing website for ease of use –www.rmspublishing.co.uk

Statutory provisions

Dangerous Substances and Explosive Atmosphere Regulations (DSEAR) 2002 / Dangerous Substances and Explosive Atmosphere Regulations (Northern Ireland) 2003

Electrical Equipment (Safety) Regulations (EESR) 2016

Electricity at Work Regulations (EWR) 1989 / Electricity at Work Regulations (Northern Ireland) 1991

Fire (Scotland) Act (FSA) 2005

Fire and Rescue Services (Northern Ireland) Order 2006

Fire Safety (Scotland) Regulations (FSSR) 2006

Fire Safety Regulations (Northern Ireland) 2010

Health and Safety (First-aid) Regulations (FAR) 1981 (as amended)

Provision and Use of Work Equipment Regulations (PUWER) 1998 / Provision and Use of Work Equipment Regulations (Northern Ireland) 1999

Regulatory Reform (Fire Safety) Order (RRFSO) 2005

STUDY QUESTIONS

1) How can the following two protective measures reduce the risk of electric shock AND, in EACH case, give an example of their use in construction activities?

   (a) Reduced low voltage.

   (b) Double insulation.

2) A fire has started on a construction site immediately adjacent to a bench mounted electric circular saw.

   (a) How might the saw's electrical power system have caused the fire?

   (b) What are the possible effects of an electric shock to the saw operator?

3) A road is being resurfaced in an area that has overhead power lines crossing it. The work involves the use of tipper vehicles, excavators and a road planing machine that has an extended jib. What control measures can reduce the risk from contact with the overhead power lines?

4) (a) What control measures should be used by someone using a hand-held electric drill on a construction site to reduce the risk of electric shock?

   (b) Other than electricity, what FOUR hazards are associated with the use of hand-held electric drills on construction sites?

5) (a) Why does a fuse fitted to hand-held electrical power tools only provide limited electric shock protection to the user of the tool?

   (b) What are the advantages of using a residual current device to protect users of hand held electrical power tools from electric shock?

   (c) What are the advantages of double insulation for hand-held electrical power tools?

For guidance on how to answer these questions please refer to the assessment section located at the back of this guide.

Element 11

Fire

Contents

11.1　Fire principles

BASIC PRINCIPLES OF FIRE
The fire triangle

A simple approach depicts fire as having three essential components: fuel, oxygen and ignition source (heat).

When the three separate parts are brought together, a fire occurs. This is often depicted by the 'fire triangle' as shown in *Figure 11-1*. This traditional approach is useful when considering the 'ingredients' needed to make a fire, and the absence of any one these components will prevent a fire from starting.

Combustion is defined as a chemical reaction during which heat energy and light energy are emitted. If the three components fuel, oxygen and ignition come together in the right proportions, then the chemical reaction of combustion takes place.

For example, when a match is struck, friction (the ignition)

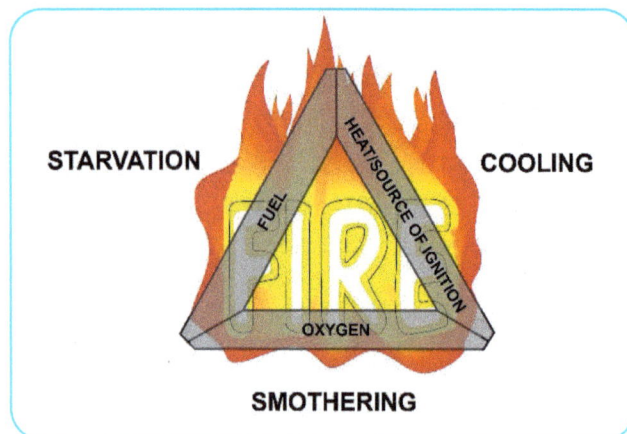

Figure 11-1: Fire triangle.
Source: RMS/Corel.

heats the head to a temperature at which the chemicals react and generate more heat than can escape into the air, therefore burning with a flame. The molecules in the matchstick break down and give off vapour, causing more heat to be released, propagating further chemical reactions. Once a fire has started, a self-sustaining chain reaction begins at the surface of the fuel (solid or liquid), which turns into a vapour, and it is this that burns in the combustion process. A similar process occurs with the molecules of a gas, but because the collective surface area of the molecules is large, the combustion process is very fast. A fire may be defined as a chemical chain reaction in which fuel is reduced by reaction with an oxidiser to produce light, heat and combustion products.

For the combustion process to be maintained, all three of the component parts shown in *Figure 11-1* must remain present. If one or more of these parts of the fire is removed, the fire will be extinguished.

This can be done by:

1) **Cooling** the fire to remove the heat, for example, by use of a water or foam extinguisher.

2) **Starving** the fire of fuel, for example, isolation of a gas supply.

3) **Smothering** the fire by limiting its oxygen supply, for example, by use of a carbon dioxide or foam extinguisher.

4) **Chemical interference** of the flame reactions, for example, wet chemical extinguishing agents.

Sources of ignition

Ignition occurs when a heat source, for example, a spark, contains sufficient heat energy to cause combustion of one or more molecules of a flammable vapour or substance. Any source of heat is a possible ignition source and can be found in many forms.

Examples of sources of ignition could be:

- Discarded smokers' materials (not such a big problem in the UK now that smoking is prohibited in the workplace, but this could remain an issue on some construction sites).

- Naked flames.

- Fixed or portable heaters - particularly those that use liquid fuel.

- Hot processes, for example, welding, cutting and grinding.

- Lighting.

- Deliberate ignition. Many fires in the workplace are started deliberately. The workplace may be targeted by disgruntled workers or unhappy customers or by an act of vandalism.

- Cooking.

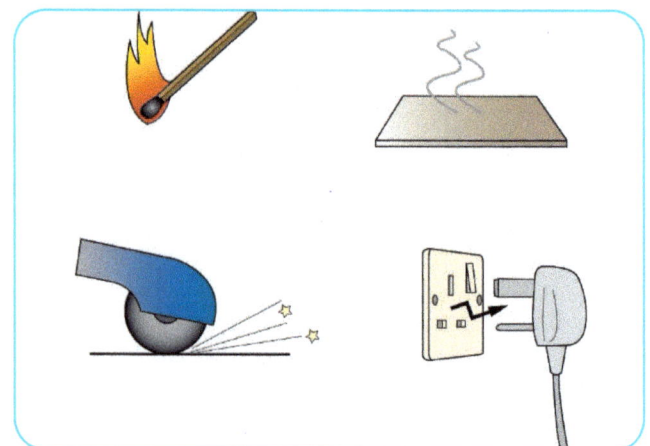

Figure 11-2: Sources of ignition.
Source: RMS/UK, government fire guides.

- Unsafe storage of flammable liquids allowing self-heating as a result of an exothermic reaction which can cause spontaneous combustion, for example, acetylene and chlorine.

- Electrical equipment - overloading electrical circuits.

- Machinery – sparks from electric motors, overheating of drive belts (**see Figure 11-5**) due to over tightening, friction due to lack of lubrication, blunt cutting tools generating heat.

- Static electricity - most commonly from lightning strikes, although sparks from static charges are very dangerous in flammable and explosive atmospheres.

Figure 11-3: Illicit smoking.
Source: FSTC Ltd.

Figure 11-4: Hot lighting bulbs.
Source: HSE, HSG168.

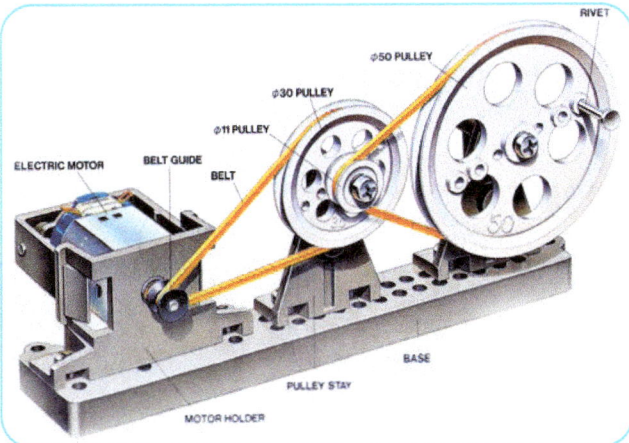

Figure 11-5: Drive belt and pulley.
Source: ServoCity.

Sources of fuel

Anything that burns is a fuel for a fire. Common items that will burn in a typical construction workplace include:

- Flammable liquids, for example, petrol, paints and solvents.

- Flammable gases, for example, acetylene, butane and liquefied petroleum gas (LPG) heater cylinders.

- Wood, for example, pallets, packing material and dust from carpentry processes.

- Plastics, rubber and foam, for example, cladding, sealant, liners and sheeting used as work area dividers.

- Paper and card, for example, packing and office stationery.

- Insulating materials, for example, polystyrene foam in walls and partition components.

- Waste materials, chemicals, waste paper, and general waste.

The structure of the building should also be considered. The building itself could be made from wood or other flammable material or may contain flammable materials as part of the decoration, for example, wallpaper.

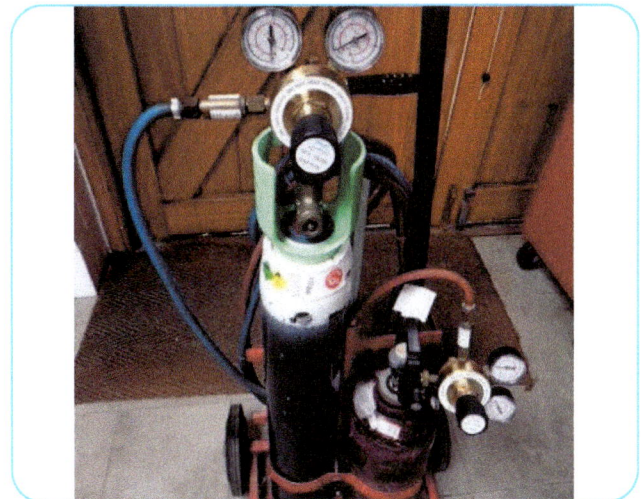

Figure 11-6: Oxy-Acetylene Welding Equipment
Source: Novice Framebuilder.

Figure 11-7: Combustible materials - waste.
Source: Shutterstock.

The words 'flammable' and 'inflammable' mean the same, i.e. that ignition takes place at normal ambient temperatures. The word 'combustible' means likely to catch fire. The terms 'highly flammable' and 'extremely flammable' indicate substances that can be ignited at low ambient temperatures. Petrol exposed to a spark will ignite at temperatures starting from -43°C and so is described as 'extremely flammable'.

Sources of oxygen

The main source of oxygen for a fire is in the air around us. In an enclosed building this is provided by the ventilation system in use. This generally falls into one of two categories: natural airflow through doors, windows and other openings; or mechanical air conditioning systems and air handling systems. In many buildings there will be a combination of systems, which will be capable of introducing/extracting air to and from the building.

Leaks from oxygen supplies, for example, cylinders or piped supply which, combined with poor ventilation can lead to an oxygen enriched atmosphere. Materials that ordinarily will burn only slowly will burn very vigorously in an oxygen-enriched atmosphere. Others such as greases and oils may burst into flames in this kind of atmosphere.

The following precautions should also be considered:

- Never use oxygen instead of compressed air.
- Never use oxygen to improve air quality in a working area or confined space.
- Never use grease or oil on equipment containing oxygen.

Oxidising materials

Combustion involves the oxidation of a combustible (burnable) material. When a combustible material burns, a chemical reaction occurs in which the combustible material combines with oxygen and gives off heat, gases, and generally light.

The usual source of oxygen for combustion is air. However, oxidising materials can supply combustible materials with a source of oxygen and support a fire even when air is not present. Although most oxidising materials do not burn themselves, they can produce very flammable or explosive mixtures when combined with combustible materials.

The involvement of an oxidising material can speed up the development of a fire and make combustion more intense.

Oxidising materials can cause materials that do not readily burn in air to burn rapidly in their presence.

In addition, they can cause combustible materials to burn at ambient temperature without the presence of

an obvious source of ignition, such as a spark or flame. Although oxidising materials may not be widely present on construction sites, construction activities may take place in locations where oxidising materials are present.

What happens when an oxidising material comes into contact with a combustible material mainly depends on the chemical stability of the oxidising material, the less stable an oxidising material is, the greater the chance that it will react in a dangerous way. Oxidising materials come in the form of gases, liquids and solids.

Examples of gases include oxygen and ozone; liquids include nitric acid and chromic acid, and solids include chromate and potassium permanganate.

The United Nations (UN) has established a non-legally binding international agreement called 'Globally Harmonized System of Classification and Labelling of Chemicals (GHS)'. It has been widely accepted globally and is being established in national legislation of the countries that are adopting it.

Within GHS is a classification of materials principally based on the physical hazards they present, including oxidising materials. Materials that come under the class called oxidising gases fall under a single hazard category, whilst those classed as oxidising liquids and oxidising solids have three classifications.

Oxidising gases - Annex 1 of GHS

Category 1

Signal word: Danger

Hazard statement
May cause or intensify fire; oxidiser

Figure 11-8: Oxidising gases.
Source: RMS (Adapted from annex 1 of GHS).

Category 1
Signal word: Danger
Hazard statement
May cause fire or explosion; strong oxidiser

Category 2
Signal word: Danger
Hazard statement
May intensify fire; oxidiser

Category 3
Signal word: Warning
Hazard statement
May cause or intensify fire; oxidiser

Figure 11-9: Oxidising liquids and solids.
Source: RMS (Adapted from annex 1 of GHS).

CONSIDER

Think about your workplace. How many possible sources of ignition, fuel and oxygen can you identify?

CLASSIFICATION OF FIRES

The classification of fires essentially relates to the material that is being combusted and seeks to group similar materials into the same classification.

There is no internationally agreed classification of fires. A common approach is set out in ***Figure 11-10***, which reflects the approach taken in a number of countries, including the UK, and is based on the European Standard 'Classification of fires' - EN 2:1992.

CLASS A
Combustible solids, for example, wood, paper, textiles or plastics (usually material of an organic nature).
CLASS B
Flammable liquids or liquefiable solids – petrol, oil, paint, fat or wax.
CLASS C
Combustible gases, for example, natural gas or liquefied gases, for example, butane or propane.
CLASS D
Combustible metals such as magnesium and lithium. Such specialised fires require a specialised metal powder fire extinguisher to deal with them, and this will be required in scientific labs or where manufacturing processes involve the risk of metal fires. For example, aluminium dust or swarf can catch fire, so any process involving cutting, drilling or milling aluminium poses a potential risk.
CLASS F
Cooking oils and fats usually found in commercial kitchens such as restaurants and fast-food outlets.
ELECTRICAL FIRES
Although this is not a class of fire in the common classification, electricity is often a source of ignition and the presence of electricity will increase the risk of electric shock where water is used as the extinguishing medium.

Figure 11-10: Classification of fires.
Source: RMS.

A basic understanding of the classes of fire needs to be achieved because many fire extinguishers state the classes of fire on which they may be used.

See 'Extinguishing media' in 11.3 later in this element.

PRINCIPLES OF HEAT TRANSMISSION

There are four main methods by which heat may be transmitted: conduction, convection, radiation and direct burning.

a) Convection

This is the movement of hotter gases up through the air (hot air rises and cooler or cold air falls). Convection can quickly move hot gases to another part of a building where they raise the temperature of combustible materials to a point that combustion takes place, for example, hot gases rising up a staircase through an open door.

Control measure: protection of openings by fire doors and the creation of fire-resistant compartments in buildings.

b) Conduction

This follows the principle that heat energy can be transmitted through solid objects. Some materials, such as metal, can absorb heat readily (increasing kinetic energy levels) and transmit it to other rooms by conduction, where it can set fire to combustible items that are in contact with the heated material, for example, metal beams, ducting or pipes transmitting heat through a solid wall.

Control measure: insulation of the surface of a beam or pipe with heat resistant materials.

c) Radiation

This is the transfer of heat as invisible infrared energy waves through the air in a similar way to light (the air or gas is not heated but solids and liquids in contact with the heat are). Any material close to a fire will absorb the heat in the form of energy waves until the material starts to smoulder and then burn. For example, a fire in a waste container stored too near to a building may provide enough radiant heat to transfer the fire to the building.

Control measure: separation distances or fire-resistant barriers.

d) Direct burning

This occurs when combustible materials come into direct contact with a naked flame, for example, process materials affected by a welding flame, or curtains, carpets and other office furnishings that may be consumed by combustion when a fire starts, and enable fire to be transferred along them to other parts of a building.

Control measure: safe systems of work when doing 'hot work', separation distances between stored items and the use of fire retardant materials.

REASONS WHY FIRES SPREAD

Failure of early detection

Detection of fire spread can be delayed by:

- No detection system or fire warden patrols.
- No automatic alarm system in place.
- People not raising the alarm on discovery of a fire.
- Fire starts in unoccupied area, remaining undetected for some time.
- Fire starts out of normal work hours, when workers are not there.
- Building material waste may be being burnt as a normal routine and smoke and other signs of fire may not be seen as unusual and ignored.
- Numerous hot working tasks conducted - therefore smells of burning ignored.
- Frequent occurrence of small, local fires caused by hot work, quickly extinguished and not seen as significant.

Absence of compartments in building structure

Fire compartmentation is the division of the building into discrete fire zones. Fire compartmentation is designed to contain the fire to within the area of the starting point of the fire. The compartmentation approach provides at least some protection for the rest of the building and its occupants even if other fire prevention systems are installed and fail.

Figure 11-11: Absence of fire door.
Source: HSE, HSG168.

Figure 11-12: Suspended ceiling showing voids above.
Source: Borlaug Contracting Inc.

Fire spread within a building can result from an absence of compartmentation by design or modification:

- An open plan layout (reduced compartmentation).
- Suspended ceilings create voids above.
- The structure under construction or alteration is incomplete and has reduced separation between levels and/or sections on a level, for example an absence of fire doors.

Compartments undermined

Fire spread within a building can result from compartments being undermined:

- Fire doors wedged open.
- Poor maintenance of door structure, automatic closing rate or seal ineffective.
- Holes may be designed to pass through compartments and are waiting fitment of services and subsequent sealing.
- Holes cut for ducts or doorways or to provide temporary access to locate/remove equipment.
- Compartmentation may be progressively reduced in buildings under alteration, increasing the risk of fire spread.

Figure 11-13: Fire door wedged open.
Source: RMS.

Figure 11-14: Compartment undermined - holes cut in wall.
Source: RMS.

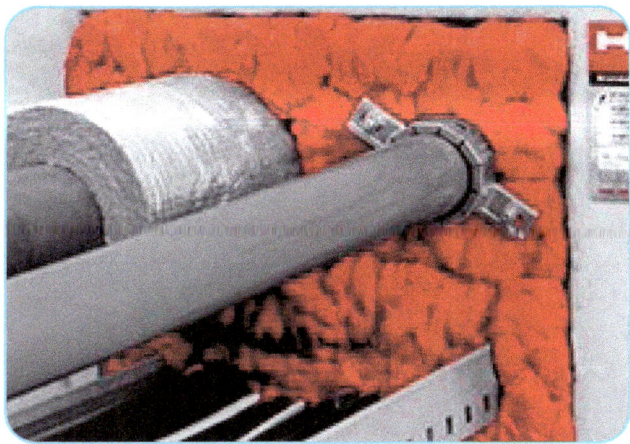
Figure 11-15: Wall gap closed by fire-resistant material.
Source: IndiaMART.

Materials inappropriately stored

Inappropriate storage of materials can cause fire spread:

- Flammable liquids not controlled - too much stored or stored in unsuitable containers.
- Boxes in corridors, under stairways or in access routes, obstructing safe evacuation providing a source of fuel for the fire.
- Off cuts of wood and sawdust left in the areas where work has taken place, increasing the risk of waste ignition.
- Flammable packing materials used in the process, such as pallets, shrink wrap plastic, polystyrene, bubble wrap, allowed to accumulate, increasing the fire (combustible) load on the building.
- Pallets or waste materials inappropriately stored or discarded near to buildings, increasing the risk of arson.
- Flammable waste material removed from a building and discarded outside as part of renovation/ construction work, for example wooden partition or wall covering material.

Figure 11-16: Materials left to accumulate on site.
Source: RMS.

Figure 11-17: Materials inappropriately stored.
Source: RMS.

COMMON CAUSES AND CONSEQUENCES OF FIRE WITHIN THE CONSTRUCTION INDUSTRY

Causes

Causes of fires in the workplace may be split into four main groups: careless actions and accidents, misuse of equipment, defective machinery or equipment and deliberate ignition (arson).

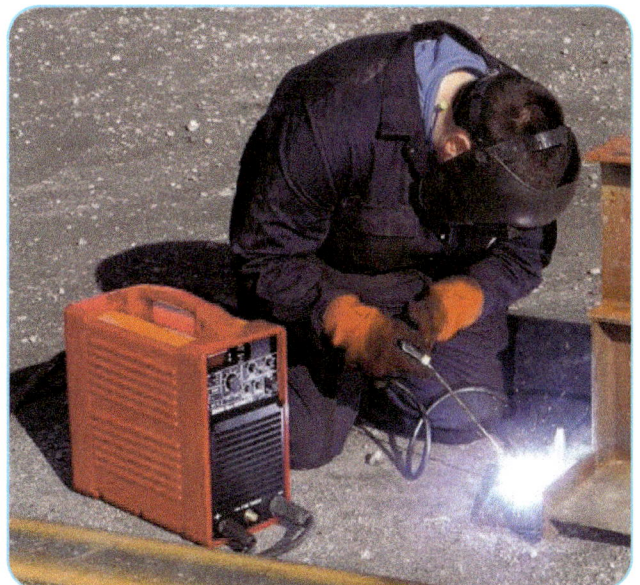
Figure 11-18: Careless action - hot work.
Source: RMS.

Careless actions

Careless actions relate to 'hot works' such as welding, cutting and grinding, discarded lighted cigarettes or matches, smouldering waste, unattended burning of waste materials or inappropriate electrical connections, unsafe use of gases and flammable liquids such as LPG and petrol.

Careless use of chemicals that may react with other substances, for example, oxidising agents (solvents) soaked onto a cloth that give off heat as the cloth dries out.

Misuse of equipment

Misuse of equipment relates to overloading electrical circuits and/or using fuses of too high a rating. Misuse also applies when the servicing instructions are not followed or from failure to repair faulty machinery/equipment promptly. Equipment is misused when it is operated beyond its capacity, for example, a small saw blade cutting through a large section of timber, which can cause overheating.

Misuse of equipment also occurs when it is stored incorrectly, for example, flammable liquids must be stored in a secure compound.

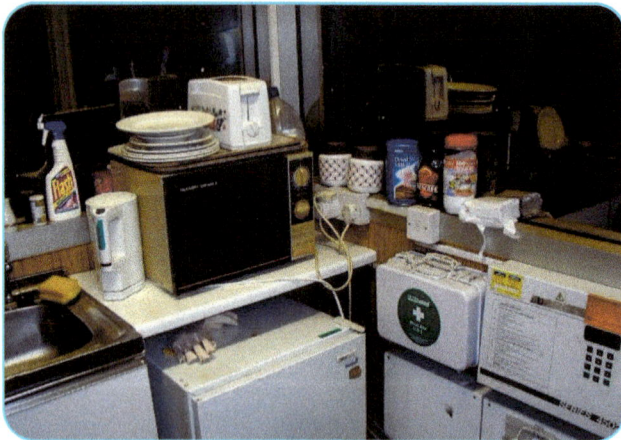

Figure 11-19: Misusing equipment - overloaded electrical sockets.
Source: RMS.

Defective machinery or equipment

Defective machinery or equipment may result in electrical short circuits causing arcs or sparking; an electrical earth fault can cause local overheating.

Figure 11-20: Defective 110v electrical equipment.
Source: Christie.

Electrical insulation failure may occur when affected by heat, damp or chemical corrosion. Mechanical, through generation of heat from friction due to wearing parts, incorrect lubrication or incorrectly tightened fan belts.

Deliberate ignition (arson)

Deliberate ignition is the crime of maliciously and intentionally or recklessly starting a fire or causing an explosion. These acts are often associated with; insurance fraud, aggrieved persons, concealment of another crime, political activists or vandalism.

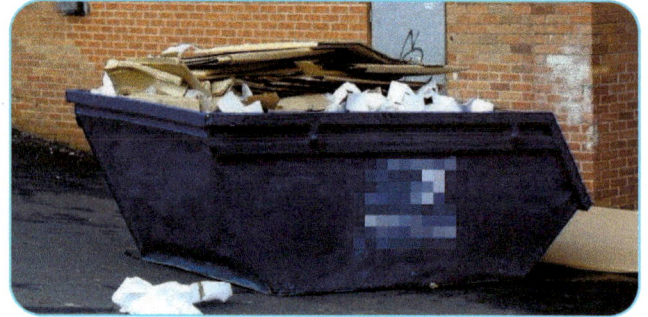

Figure 11-21: Potential for deliberate ignition.
Source: RMS

Consequences

The consequences of fires may be split into four main groups:

1) Human harm.
2) Economic effects.
3) Legal effects.
4) Environmental effects.

Human harm

Although there were 311 fire-related fatalities reported in the UK during 2020/21, the UK government and the Health and Safety Executive (HSE) statistics show that in the UK the loss of life on a construction site due to fire is rare.

The Düsseldorf airport fire in 1996 caused 17 deaths and 62 injuries. The International Labour Organization (ILO) reported that in 1993, a major fire at the Kader Industrial Co. Ltd. factory in Thailand killed 188 workers. A fire in a clothing factory in Dhaka, Bangladesh, killed at least 117 workers in 2012.

Fire has the potential for major loss of life in the workplace due to direct contact with heat and flame or from the effects of smoke and toxic gases. The radiant heat of a fire and contact with flames can give rise to the risk of heat stroke or burns. The degree of exposure to the heat or flames will greatly influence the effect on the body each will have, they could quickly lead to shock, coma or death. Heat is also a respiratory hazard, as superheated gases burn the respiratory tract, which often results in fatalities.

Most fire deaths are not caused by burns but by smoke inhalation. Often smoke incapacitates so quickly that people are overcome and are unable to reach a safe exit.

The use of synthetic materials is now commonplace in the workplace, and when these ignite they produce toxic substances. As a fire grows inside a building, it will often consume most of the available oxygen, which in turn reduces the speed of the burning process. In addition to producing smoke, fire can incapacitate or kill by reducing oxygen levels, either by consuming the oxygen, or by displacing it with other gases.

Smoke is made of components each of which can be lethal in its own way:

- Particles: unburned, partially burned, and completely burned particulates can be so small they penetrate the respiratory system's protective filters and lodge in the lungs. Many products of combustion are toxic or irritating to the eyes and digestive system.

- Vapours: fog-like droplets of poisonous liquid can be inhaled or absorbed through the skin.

- Toxic gases: the most common, carbon monoxide, can cause death, even in small quantities, as it replaces oxygen in the bloodstream. Hydrogen cyanide results from the burning of plastics, such as PVC pipe, and on inhalation interferes with cellular respiration. Phosgene is formed when products such as vinyl materials are burned. At low levels, phosgene can cause itchy eyes and a sore throat; at higher levels it can cause pulmonary oedema (fluid accumulation in the lungs) and death.

Economic effects

Commercial losses from fire are substantial, even though recent workplace injury and death figures have been low. The occurrence of a fire during construction activities can have a disastrous effect on the organisation commissioning the construction work. Business plans and objectives may fail to be achieved and services not provided. The effects may be particularly pronounced in situations where the construction activity is taking place alongside the organisation's normal activities. The Düsseldorf airport fire in 1996 in addition to loss of life and injuries also caused approximately £200 million worth of damage.

The fire at the Buncefield fuel depot in the UK, in December 2005, was the biggest in UK recent history. Explosions and heat from the fire caused severe damage to more than 80 buildings on the industrial estates surrounding the terminal. The cost of the damage is estimated to be between £500 million and £1,000 million.

In recent years there have been a number of fires involving timber frame building sites where the structures have been completely destroyed by fire causing huge economic loses.

When fires do occur in the workplace the business is usually so badly affected it does not resume trading.

Figure 11-22: Fire involving timber frame structure.
Source: Hampshire Fire & Rescue

Legal effects

There is a legal requirement under the Regulatory Reform (Fire Safety) Order (RRFSO) 2005 to prevent fire, protect workers and other relevant persons from the effects of fire and to mitigate the effect of fire on anyone in the vicinity of premises on fire. In addition Regulation 29 of the Construction (Design and Management) Regulations (CDMR) 2015 sets out a general requirement to prevent harm from fires and Regulation 30 establishes requirements for emergency procedures, including fire emergencies. Failure to comply with legislation could result in prosecutions and, if found guilty, fines.

"(1) Suitable and sufficient steps must be taken to prevent, so far as is reasonably practicable, the risk of injury to a person during the carrying out of construction work arising from:

(a) fire or explosion."

Figure 11-23: Fire prevention requirements.
Source: CDMR 2015, Regulation 29.

Environmental

Large uncontrolled fires create pollutants, such as smoke, that enter the atmosphere. The fire itself may cause damage to storage areas with the subsequent leakage of chemicals onto land or into watercourses and the run-off from fire hoses may ultimately enter the water system. The photographs in *Figures 11-24 and 11-25* show some of the damage caused by the Buncefield oil storage depot disaster in December 2005. The plume of smoke was so large it could be seen from space.

Figure 11-24: Buncefield - run-off from fire hoses.
Source: BBC News.

Figure 11-25: Buncefield oil storage depot disaster
Source: Royal Chiltern Air Support Unit.

 REVIEW

Identify the key components of the 'fire triangle'.

Explain how a machine that has not been properly maintained may cause a fire.

Outline TWO common causes of fire in a workplace.

11.2 Preventing fire and fire spread

CONTROL MEASURES TO MINIMISE THE RISK OF FIRE STARTING IN THE WORKPLACE

Elimination and reduction of flammable and combustible materials

Where possible employers should seek to *eliminate* the use of flammable materials in the workplace, for example, replacing adhesives that have a flammable content with those that are water based or to substitute the highly flammable with less flammable substances. Where this is not possible the amount used should be *reduced* and kept to the minimum. Flammable and combustible materials in the workplace must be stored in suitable containers and minimum quantities for immediate work needs. Flammable and combustible materials not in use should be removed to a purpose-designed store in a well-ventilated area, preferably outside the building but in a secure location. Lids should be kept on containers at all times when they are not in use. Any waste containers, contaminated tools or materials in the workplace should be treated in the same way and removed to a store in fresh air until they can be dealt with.

It is important to reduce the presence of flammable and combustible materials by preventing an accumulation of waste. Waste in work areas should be removed to suitable collection points and then removed to

controlled areas ready for recycling or off-site disposal. The efficient handling of waste should ensure that materials are minimised at each stage of handling - in the work area, waste points or controlled waste-handling areas. It is important to remember that containers and contaminated materials also need to be disposed of in a controlled manner so that they do not present a risk of fire. Care has to be taken to control the delivery and therefore the storage of flammable and combustible materials to site. Where possible, deliveries should be staggered to reflect the rate of use in order to minimise the amount stored on site.

STORAGE OF FLAMMABLE LIQUIDS

Terms used with flammable liquids

Flash point

'Flash point' is defined as the lowest temperature at which, in a specific test apparatus, sufficient vapour is produced from a liquid sample for momentary or flash ignition to occur on the application of an ignition source.

The flash point of a substance is the critical characteristic when deciding flammability. It must not be confused with ignition temperature, which may be considerably lower. The ignition temperature of a given substance is the measure of the minimum temperature at which the substance ignites, without the presence of an external spark or flame. Because of the fact that the material auto-ignites at this temperature range, it is also referred to as the substance's auto-ignition temperature.

Flammable

Flammability (or inflammability) is the ease with which the vapours of a substance will ignite, causing fire or combustion. Substances with vapours that will ignite at temperatures commonly encountered are considered flammable. The term flammable has various national definitions that give a temperature requirement relating to flammability.

The UN Globally Harmonised System of Classification and Labelling of Chemicals (GHS) has established an international definition of the term. This has also been reflected in the UN Recommendations on the Transport of Dangerous Goods (UNRTDG) as it relates to packaging for transport.

The GHS has established a definition that 'a flammable liquid is a liquid having a flash point of not more than 93°C'. It further defines that a flammable liquid is classified in one of four categories (*see Figure 11-26*).

These definitions and classifications have been adopted globally and are supported by the internationally agreed labelling under the UN Globally Harmonised System of Classification and Labelling of Chemicals (GHS) - *see Figure 11-27*.

Category	Flash point	Initial boiling point	Symbol	Signal word	Hazard statement
1	<23°C	≤35°C	Flame	Danger	Extremely flammable
2	<23°C	>35°C	Flame	Danger	Highly flammable
3	≥23°C ≤60°C	-	Flame	Warning	Flammable
4	≥60°C ≤93°C	-	No symbol	Warning	Combustible liquid

Figure 11-26: Flammable liquids classification.
Source: UN, Globally Harmonised System of Classification and Labelling of Chemicals.

These definitions, categorisation and labelling are being introduced by many countries progressively. This has often required transitional arrangements to move to the new criteria.

In time, the wider use of the GHS will make uniform the classification and other criteria for all workplace activities involving chemicals.

The flash points of some common solvents are:

- **Ethanol** +12°C.
- **Toluene** +4°C.
- **Methyl ethyl ketone** -9°C.
- **Acetone** -19°C.

General principles for storage and use of flammable liquids

When considering the storage or use of flammable liquids, the HSE Guidance HSG51 'The storage of flammable liquids in containers' advises that the following safety principles should be applied.

Using the acronym 'VICES' can be an aid to remembering these five principles, although there is no order of priority implied by the use of the acronym:

V   Ventilation - plenty of fresh air.

I   Ignition - control of ignition sources.

C   Containment - suitable containers and spillage control.

E   Exchange - try to use a less flammable product to do the task.

S   Separation - keep storage away from process areas, by distance or a physical barrier, for example, a wall or partition.

Figure 11-28: Flammable liquid containers.
Source: HSE, HSG140.

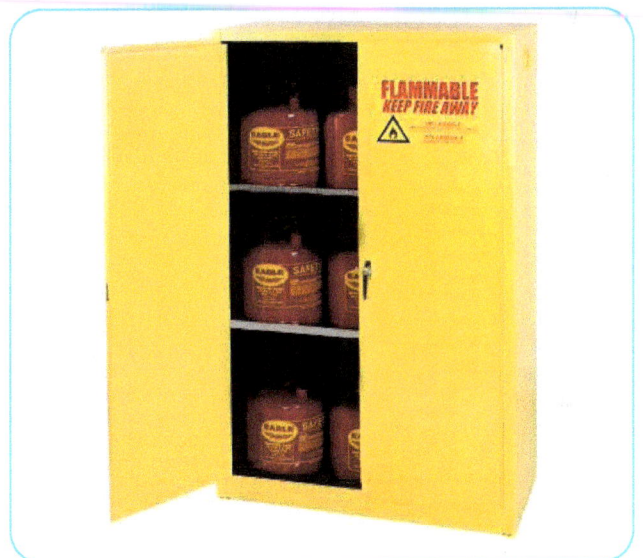

Figure 11-29: Lockable cupboard for flammable liquids.
Source: UniMac.

Label elements for flammable liquids			
Classification	Category 1	Category 2	Category 3
GHS Pictograms			
Signal word	Danger	Danger	Warning
Hazard statement	H224: Extremely flammable liquid and vapour	H225: Highly flammable liquid and vapour	H226: Flammable liquid and vapour

Figure 11-27: Flammable liquids classification and labelling.
Source: EU, Classification, Labelling and Packing (CLP) Regulations.

Storage in the workplace

The objective in controlling the risk from flammable liquids is to remove all unnecessary quantities from the workplace to a recognised storage area outside the building. This may be done as part of a close-down routine at the end of the day.

It is accepted that quantities of flammable liquids may need to be available in a workplace during normal working. In order to keep this to a minimum, the amount kept available in the workplace for immediate use should not exceed that which is required for one work period.

In European countries the maximum recommended quantities that may be stored in suitable cabinets and bins in the workplace are:

- Liquids with a flash point below the maximum ambient temperature of the work area - no more than 50 litres.
- Other flammable liquids with a higher flash point of up to 60°C - no more than 250 litres.

Figure 11-30: Poor storage of flammable liquids.
Source: RMS.

Figure 11-31: Storage of flammable materials.
Source: RMS.

Other control measures for storage in the workplace include:

- In a suitable properly labelled container, to prevent spills and sealed to prevent loss of vapour.

- In a purpose-built cabinet, bin or other storage container that is fire-resistant, with lockable doors, an inbuilt catch tray and clearly signed.
- In a designated, well-ventilated area of the workplace.
- Away from ignition sources, working or process areas.
- Capable of containing any spillage (non-spill caps and bunding) together with an appropriate spill kit.
- In a 30-minute fire-resistant structure.
- Provided with hazard warning signs to illustrate the flammability of the contents.
- Prohibition signs for smoking and naked flame.
- Should not contain other substances or items.
- Suitable emergency procedures.

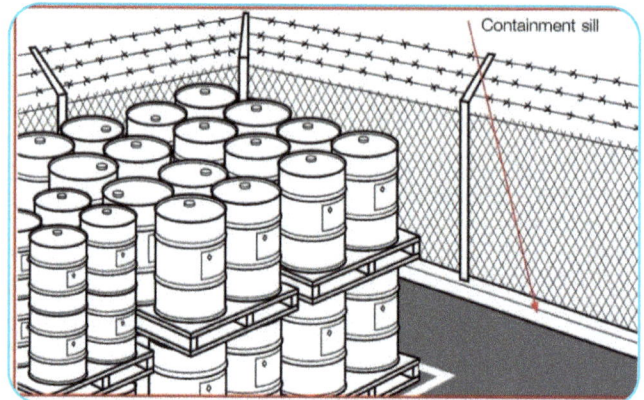

Figure 11-32: Outdoor storage of flammable liquids.
Source: HSE, HSG51.

Storage in the open air

Control measures for storage in the open air include:

- Storage away from potential ignition sources.
- Formal storage area on an impervious concrete pad, with a sump for spills.
- Bunded (impervious compound consisting of a solid floor and kerb or wall to enclose it) all around to take content of largest drum plus an allowance of 10%.
- Located away from other buildings.
- Secure fence and gate 2m high.
- Marked by signs warning of flammability.
- Signs prohibiting smoking or other naked flames.
- Protection from sunlight.
- If lighting is provided within store it must be flameproof.
- Provision for spill-containment materials.
- Fire extinguishers located nearby - consider powder type.

- Full and empty containers separated.

- Clear identification of contents.

- Area kept free of combustible materials.

Liquefied petroleum and other gases in cylinders

Liquefied petroleum gas (LPG) is a term that relates to gas stored in a liquefied state under pressure; common examples are propane and butane. LPG and other gas cylinders should be stored in line with the principles detailed as follows.

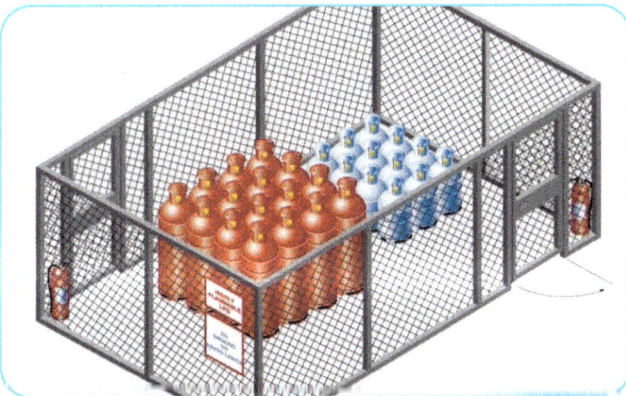

Figure 11-33: Secure LPG storage.
Source: HSE, HSG168.

Storage

Storage requirements for LPG and other gas cylinders include:

- Storage area should preferably be in clear open area and outside (to ensure area is well ventilated).

- Stored in a secure compound - 2m high fence.

- Safe distance from toxic, corrosive, combustible materials, flammable liquids or general waste.

- Located away from any other building.

- If stored inside building, kept away from exit routes; consideration should be given to fire-resisting storage and forced ventilation.

- Well-ventilated area - 2.5% of total floor and wall area as vents, high and low.

- Oxygen cylinders at least 3m away from flammable gas cylinders.

- Acetylene may be stored with LPG if the quantity of LPG is less than 50kg.

- Access to stores should be controlled to prevent LPG etc. being stored in the general workplace when not in use.

- More than one exit (unlocked) may need to be available from any secure storage compound where distance to exit is greater than 12m.

- Storage compound should be locked when not in use.

- Protection from sunlight; take particular care to shield windows from direct sunlight.

- Flameproof lighting.

- Empty containers stored separately from full.

- Fire extinguishers located nearby - consider powder and water types.

Figure 11-34: LPG storage.
Source: Shutterstock.

Transport

Transport requirements for LPG and other gas cylinders include:

- Upright position.

- Secured to prevent falling over.

- Protection in event of accident, for example, position on vehicle.

- Transport in open vehicle preferably.

- Avoid overnight parking while loaded.

- Park in secure areas.

- Driver hazard information and warning signs.

- Driver training.

- Firefighting equipment.

General use

General requirements for use of LPG and other gas cylinders include:

- Cylinder connected for use may be stored in the general workplace; any spare cylinders must be secured in a purpose-built store until required for use.

- Fixed position to prevent falling over, or on wheeled trolley - chained.

- Well-ventilated area.

- Away from combustibles.

- Kept upright unless used on equipment specifically designed for horizontal use, for example, gas-powered lift truck.

- Handled carefully - do not drop.

- Allow to settle after transport and before use.

- Consider manual handling and injury prevention.

- Turn off cylinder before connecting or disconnecting equipment.

- Check equipment before use.

- Any smell of gas during use, turn off cylinder and investigate.

- Use correct gas regulator for equipment/task.

- Use equipment in line with manufacturers' instructions.

Use in temporary accommodation units (TAUs)

Requirements for use in TAUs include:

- Only allow cylinders inside TAU if it is part of a heater (cabinet heater).

- Pipe into site TAUs from a cylinder located outside where possible.

- If a cylinder is outside the TAU use the shortest connecting hose possible.

- TAU to be adequately ventilated, high and low.

- Heaters and cooking equipment should be fitted with flame-failure devices.

- Turn off heater and cooking equipment and cylinder after use and overnight.

- Be aware of the danger of leaks inside TAU, especially overnight, as a severe risk of fire or explosion may occur.

- Keep heaters and cooking equipment away from clothing and other combustibles.

Figure 11-35: Gas cylinders for use with TAU.
Source: RMS.

CONTROL OF IGNITION SOURCES

Hot work

'Hot work' is any process that can be a source of ignition, including welding, cutting, grinding, brazing and soldering processes. Hot work has been responsible for causing many fires.

One of the most tragic fires due to hot work was the Dusseldorf Airport Fire in 1996. The fire was started by welding on an open roadway and resulted in damage in excess of £200 million, 62 injuries and 17 deaths. It is imperative that good safe working practices are utilised. Combustible materials must be removed from the area or covered over.

Consideration must be given to the effects of heat on the surrounding structure, and where sparks, flames, hot residue or heat will travel. Suitable fire extinguishers need to be immediately available and operatives must know how to use them. The work area must be checked thoroughly for some time after the completion of work to ensure there are no smouldering fires. A person should be appointed as a 'fire watcher' to ensure no fires result from hot work while the work is taking place and for some time after. Strong consideration should be given to the use of work permits to control hot work.

Welding and brazing

Welding and brazing activities represent a significant ignition source from the 'naked flame' of an oxy-acetylene torch or electric arc and from the hot materials created by the process, for example, recently welded material that remains hot for some time or from sparks created in the process. In addition, the equipment can represent an explosion risk if it is used incorrectly or not fitted with proper protective devices.

The following would be good practices for welding and would reduce sources of ignition from the process:

- Only use competent trained staff.

- Regulators should be of a recognised standard.

- Colour code hoses:

  - Blue - oxygen.
  - Red - acetylene.
  - Orange - propane.

- Fit non-return valves at blowpipe/torch inlet on both gas lines.

- Fit flashback arrestors incorporating cut-off valves and flame arrestors to the outlet of both gas regulators.

- Use crimped hose connections not jubilee clips.

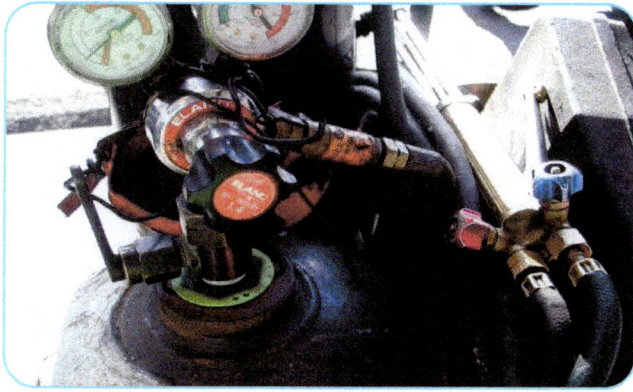

Figure 11-36: Welding equipment.
Source: RMS.

- Do not let oil or grease contaminate an oxygen supply due to explosion hazard.

- Check equipment visually before use, and check new connections with soapy water for leaks.

- Ensure the work area is well ventilated.

- Secure cylinders in an upright position.

- Keep hose lengths to a minimum and check they do not leak.

- Follow a permit-to-work system.

- Do not store standby gases that are not connected to welding apparatus in the workplace.

Smoking

Smoking in public buildings (including the workplace) has been prohibited by UK law for some time. Prohibition of smoking may lead to illicit smoking and extra vigilance may therefore be needed.

Where general national prohibitions do not exist, it may remain necessary for the employer to prohibit smoking where flammable/combustible materials are stored/used or where processes are carried on that release ignitable or explosive dusts or vapours or the production of readily combustible waste.

Smoking should be prohibited in stock rooms and other rooms not under continuous supervision. Any 'no smoking' rule in these areas should be strictly enforced. Where smoking is allowed, provide easily accessible, non-combustible receptacles for cigarette ends and other smoking material and empty these daily.

Smoking should cease half an hour before close-down to enable a check that smoking materials have been extinguished before people leave the workplace.

Arson

Simple, but effective ways to deter arsonists are by giving attention to security, both external and internal, which should encompass the following.

External security includes:

- Control of people having access to the building/site.
- Use of patrol guards.
- Lighting the premises at night - linked to closed circuit television (CCTV).
- Control of keys.
- Structural protection.
- Siting of waste containers/skips at least 8m from buildings.

Figure 11-37: Control arson by external security.
Source: RMS.

Internal security includes:

- Good housekeeping and clear access routes.
- Inspections and audits.
- Visitor supervision.
- Control of delivery and dispatch pick-up areas where third parties may enter the premises.
- Control of sub-contractors.
- Control door access by keypad or electronic locks.

Figure 11-38: Control arson by housekeeping.
Source: RMS.

Further control measures to deter arson include:

- Storing flammable materials in lockable, fire-retardant cabinets.
- Storage of minimum quantities of materials on site.
- No access to ladders or other material that can be used to climb up structures to higher levels out of sight.

Mechanical heat

Mechanical heat, such as friction from drive belts or bearings, can be controlled by routine maintenance in which drive-belt tension is examined and the belt condition checked for signs of overheating. Bearings can be lubricated or greased as well as inspected for wear. Maintenance should also include replacement of wearing parts.

Cooking and heating appliances

Cooking and heating appliances must not be left unattended and their use must be closely supervised. The source of energy for the appliance (electricity or gas) should be maintained, inspected and tested to ensure it functions correctly.

Electricity

Similar to mechanical equipment, electrical equipment must be maintained, inspected and tested to ensure circuits and their insulation has not degraded and that the system is not being overloaded. Portable and fixed electrical appliances should be checked.

Use of suitable electrical equipment in flammable atmospheres

Electrical equipment installed or used in flammable atmospheres must be suitable for use in that environment. There are different classes of flammable atmosphere and types of electrical equipment for use in flammable atmospheres.

The Dangerous Substances and Explosive Atmospheres Regulations (DSEAR) 2002 apply to most workplaces where a flammable atmosphere may occur, including construction sites. DSEAR 2002 requires employers to eliminate or control the risks of explosion from flammable atmospheres

Classification of areas where explosive atmospheres may occur

Employers must classify into zones areas where hazardous flammable atmospheres may occur. The classification given to a particular zone, and its size and location, depends on the likelihood of a flammable atmosphere occurring and its persistence if it does occur. Schedule 2 of DSEAR 2002 contains descriptions of the various classifications of zones for gases and vapours and for dusts.

There are three zones for gases and vapours:

Zone 0	Flammable atmosphere highly likely to be present - may be present for long periods or even continuously.
Zone 1	Flammable atmosphere possible but unlikely to be present for long periods.
Zone 2	Flammable atmosphere unlikely to be present except for short periods of time - typically as a result of a process fault condition.

Similarly, there are three zones for dusts:

Zone 20	Dust cloud likely to be present continuously or for long periods.
Zone 21	Dust cloud likely to be present occasionally in normal operation.
Zone 22	Dust cloud unlikely to occur in normal operation, but if it does, will only exist for a short period.

Selection of equipment and protective systems

Areas classified into zones must be protected from sources of ignition. Electrical equipment for use in hazardous flammable atmospheres needs to be designed and constructed in such a way that it will not provide a source of ignition.

Equipment intended to be used in zoned areas should be selected to meet the requirements of the Equipment and Protective Systems Intended for Use in Potentially Explosive Atmospheres Regulations (EPS) 1996. Zone 0 and zone 20 are the zones with the highest likelihood of a flammable atmosphere occurring and persisting; electrical equipment for this zone needs to be very well protected against providing a source of ignition.

There are a number of ways in which electrical equipment can be designed to prevent ignition of flammable atmospheres, each achieves this in different ways. Different forms of electrical equipment will provide a different equipment protection level (EPL).

The types of protection include:

- 'Intrinsically safe' - cannot produce a spark with sufficient energy to cause ignition - IEC symbol Ex ia (suitable for zone 0, 1, 2) or EX b (suitable for zone 1 or 2).

- 'Flameproof' - ingress of explosive atmosphere is controlled and any ignition is contained in the equipment IEC symbol Ex d (suitable for zone 1 or 2).

- 'Increased safety' equipment - does not produce sparks or hot surfaces - IEC symbol Ex e (suitable for zone 1 or 2).

SAFE SYSTEMS OF WORK

Safe systems of work to minimise fire risks in the workplace combine people, equipment, materials and the environment (the workplace) to produce the safest possible climate in which to work.

When combining these factors to make a safe system of work the following points related to fire risks can be considered.

1) **Safe person**

A safe person element of a system of work begins with raising awareness of potential losses resulting from a fire.

Information can be provided that will identify where to raise the alarm, what the alarm sounds like, how to evacuate and where to muster, responsibility for signing in and out of the site register, fire drill procedures, trained authorised fire appointed persons, use and storage of flammable materials, good housekeeping and use of equipment producing heat or ignition (including hot processes i.e. welding). Safe systems of work must also include consideration of who is at risk, including those persons with special needs such as the young, elderly, infirm or disabled.

2) Safe materials

Safe materials begin with providing information and ensuring safe segregation and storage for materials and sources of ignition/heat. In addition, providing information on the correct way to handle materials and substances, including keeping a substance register that will detail methods of tackling a fire involving hazardous substances.

3) Safe equipment

Safe equipment begins with user information and maintenance to ensure good working and efficient order. Information should also provide the user with a safe method for use and the limitations of and risks from the equipment. Supervision may be necessary to ensure correct use and prevent misuse that may lead to short circuiting or overheating that could result in fire. Where work involves hot processes by nature (welding, grinding, casting, etc.) then permit-to-work procedures may be necessary in order to tightly control the operations.

Other equipment required in relation to fire hazards and control may include smoke or heat-detection equipment, alarm sounders/bells, alarm call-points and appropriate fire-extinguishing apparatus. It should be noted that in the event of a fire alarm, all the passenger lifts should not be used. Under normal circumstances the lift will return to the ground floor and remain in that position with the doors locked in the open position. All equipment should be regularly tested to ensure its conformity and be accompanied with a suitable certificate of validity.

4) A safe environment (workplace)

A safe workplace begins with ensuring that the fabric of the building is designed or planned in a way that will prevent ignition, suppress fire spread and allow for safe, speedy unobstructed evacuation with signs to direct people. Factors to consider will include compartmentation, fire-resistant materials, proper and suitable means of storage, means of detection, means of raising the alarm good housekeeping and regular monitoring and review.

In their simplest form safe systems of work to minimise fire risks might involve replacing lids on containers that contain flammable liquids in order to prevent flammable liquid vapours getting into the atmosphere of the workplace, where they could be ignited.

Other, simple safe systems of work to minimise fire risks include switching off electrical equipment that is not in use to remove it as a potential source of ignition. Systems of work for more complex situations may require written procedures to formally describe them and to help to communicate them. Because hot work presents significant fire hazards it is often controlled by safe systems of work in the form of formal written procedures involving the use of a hot work permit document.

Permit-to-work procedures

A permit-to-work is a formal, documented safe system of work that is used for controlling high risk activities. Implementation is required prior to work beginning to ensure that all precautions have been taken and are securely in place to prevent danger to the workforce.

When managed correctly, a permit-to-work prevents any mistakes or deviations through poor verbal communication by stating the specific health and safety requirements of the work activity. For fire control, a permit-to-work is typically used where there is a requirement to use flammable materials or when hot work or processes are being carried out.

Hot work permits

Hot work permits control and implement a safe system of work whenever work activities utilise heat or flame. If the risk of fire is low, it may not be necessary to implement a hot work permit; however, it should always be considered.

The authorised person issues the hot work permit and will sign the document to declare that all isolations are made and remain in place throughout the duration of the work activity.

Hot work permits should be issued for a specific time, for a specific place, for a specific task, and are issued to a designated competent person. In addition to this, the authorised person will make checks to ensure that all controls to be implemented by the acceptor are in place before work begins.

Figure 11-39: Hot work.
Source: Speedy Hire Plc.

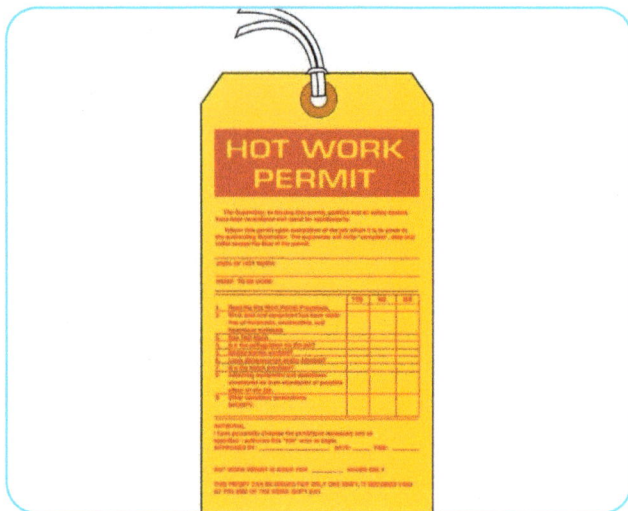

Figure 11-40: Hot work permit tag.
Source: Speedy Hire Plc.

Figure 11-41: Accumulated combustible site rubbish.
Source: HSE, HSG168.

The acceptor of the permit-to-work shall assume responsibility for carrying out the works. The acceptor shall sign the document to declare that the terms and conditions of the permit-to-work are understood and will be complied with fully at all times by the entire work team.

Compliance with a permit-to-work system includes ensuring the required safeguards are implemented and that the work will be restricted to that stated in the permit-to-work document.

Items included in a permit-to-work document for hot work are:

- Permit issue number.
- Authorised person identification.
- Standby fire warden.
- Locations of firefighting equipment.
- Locations of flammable materials.
- Warning information sign locations.
- Emergency muster points.
- Details of the work to be carried out.
- Signature of authoriser.
- Signature of acceptor.
- Signature for works clearance/extension/handover.
- Signature for cancellation.
- Other precautions (risk assessments, method statements, personal protective equipment (PPE)).

See also – 3.3 'Permit-to-work systems' in 'Element 3 – Managing change and procedures'.

GOOD HOUSEKEEPING

Housekeeping and its effect on fire safety

'Housekeeping' means the general tidiness and order of the workplace. It should be remembered that fires need fuel, and a build-up of redundant combustible materials, rubbish and stacks of waste materials can provide that fuel.

Combustible materials cannot be entirely eliminated, but they can be controlled. Any unnecessary build-up of rubbish and waste should be avoided.

If a fire starts in a neatly stacked pile of timber pallets, around which there is a clear space, the fire may be seen and extinguished before it can spread.

However, if the same pile were strewn around in an untidy heap, along with adjacent rubbish, the likelihood is that fire would spread over a larger area and involve other combustible materials before it is seen.

Poor housekeeping can also lead to:

- Blocked fire exits.
- Obstructed escape routes.
- Difficult access to fire alarm call-points/extinguishers/hose reels.
- Obstruction of vital signs and notices.
- A reduction in the effectiveness of automatic fire detectors and sprinklers.

Housekeeping checks

Fire prevention is a matter of good routine, the following checklists are a guide to what to look out for.

List A - Routine checks

Daily at the start of business - including:

- Doors and other access or egress routes, including scaffolding, that may be used for escape purposes - unlocked and escape routes unobstructed.
- Free access to hydrants, extinguishers and fire alarm call-points.
- No deposits on electric motors.

List B - Routine checks

Daily at close-down - including:

- Inspection of whole area of responsibility - to detect any smouldering fires, including smoking materials.
- Fire doors and shutters closed.
- All plant and equipment safely shut down.
- Waste bins emptied (empty bins before they are full).
- No accumulation of combustible process waste, packaging materials or dust deposits.
- Safe disposal of waste, manage waste collection, keep to a minimum and store and secure from arson attack.
- Premises left secure from unauthorised access.

List C - Periodic inspection

During working hours - weekly/monthly/quarterly as decided:

- Goods neatly stored so as not to impede firefighting.
- Clear spaces around stacks of stored materials.
- Gangways kept unobstructed.
- Only minimum quantities of essential flammable and combustible material stored in work areas.
- Materials clear of light fittings.
- Smoking rules known and enforced.

STRUCTURAL MEASURES TO PREVENT SPREAD OF FIRE AND SMOKE

Properties of common building materials

Brickwork

Bricks are resistant to fire because they have already been exposed to high temperatures in the kiln where they were fired. Brickwork will usually perform well in fires. Dependent upon the materials, workmanship, thickness, and the load carried, fire resistance of 30 minutes to 2 hours may be achieved.

Steelwork

Steel and other metals are extensively used in modern building structures. Generally they can be affected by fire at relatively low temperatures unless they are protected from the effects of the fire by some form of fire-retardant materials. This may be done by encasing in concrete, fire retardant boards or spray coatings.

Timber

Timber performs very well in fires as long as it is of sufficient size that, as its outer layers become carbonised (charred), it retains sufficient strength to fulfil its task. Charring slows the combustion process down. Generally timber does not fail rapidly in a fire.

Glass

Glass generally performs poorly in a fire unless it is fire-resistant glass. At high temperatures glass will melt and sag, which is why the traditional fire-resistant glass has wire within it.

Concrete

Concrete is very resistant to fire and, while heat may make the concrete spall (small sections break away), it retains its structural strength for a reasonable time period.

Structural measures to prevent spread

Measures to prevent spread of fire and smoke include:

- Fire-resisting structures.
- Compartmentation to confine the fire to a predetermined size.
- Fire stopping of ducts, flues and holes in fire-resistant structures.
- Fire-resisting self-closing doors.

- Smoke seals and intumescent material (which expands when heat is applied and seals any gap between the door and the door frame) on doors.
- Early and rapid detection of a fire by use of sophisticated fire alarm systems, which may operate response systems automatically.
- Sprinklers in large compartments, in particular 'rapid response' systems to limit the size of the fire.
- Control of smoke and toxic fumes by ventilation systems, so that clear air is maintained at head height level, to enable persons to escape.

Figure 11-42: Magnetic door holder linked to alarm.
Source: RMS.

Fire doors

Fire doors comprise a moving door panel and a frame that should be installed and maintained as an integral set. When the component parts fit together in an effective way they can prevent fire and smoke spread. They should have at least 30 minutes fire-resistance, the required fire resistance will depend on the location/ situation where the door is installed. The door panel will usually have seals around its edges to prevent fire and smoke spread. Generally, an intumescent strip is fitted that swells when exposed to heat from a fire to seal the door. In addition, cold smoke seals may be fitted to prevent spread of smoke in situations where the temperature of the fire has not yet made the intumescent strip swell. The door is usually fitted with a self-closing device to keep the door closed when it is not being used for access.

See **Figures 11-42 and 11-43**, which show mechanical means by which the door can be kept open until the sensors in the devices are activated by sound from the fire alarm system and release the door, which then closes.

Figure 11-43: Door stop with automatic release.
Source: RMS.

Compartmentation

Compartmentation is achieved by use of compartment walls and floors which subdivide the building into smaller areas. The majority of buildings utilise traditional methods for controlling fire spread. This is mainly done by the use of fire-resisting structures and fire-resisting doors to break the building into smaller fire compartments. A fire compartment should withstand a fire for a minimum of 30 minutes, but it may require additional protection, depending upon the purpose of the structure and the use of the building.

If a fire does occur within a compartment it should be confined to that compartment by the nature of the fire-resistant materials. This should have the effect of limiting the damage done to buildings and prevent unchecked fire spread. Stairways, ducts, etc. should also form separate fire compartments to prevent vertical fire spread. In large compartments, sprinkler systems may be fitted in an attempt to limit the size of a fire, and ventilation may be provided to allow heat/smoke to escape. If additional fire safety measures such as sprinklers are installed, then the size of the fire compartments can, in general, be doubled.

Protection of openings and voids

Openings and voids in buildings include lift shafts, service ducts, voids between floors, roof voids, etc. Consideration should be given to the protection of openings and voids by the use of fire barriers such as fire shutters, cavity barriers and fire curtains. It is important that when construction or maintenance work takes place it is managed to minimise the effect on the structure being worked on to keep fire precautions intact as much as possible. This will involve planning for the reinstatement of protection of openings and voids as soon as possible after their breach to do work. The temptation to leave all openings and voids to the end of work and then closing them should be avoided - the longer that openings and voids are left open, the higher the risk from fires.

REVIEW

1) Outline the storage requirements for LPG and other gas cylinders.

2) Identify the safety principles for storage and use of flammable liquids.

3) Outline how the structural features of buildings can help prevent the spread of fire and smoke.

11.3 Fire alarms and firefighting

COMMON FIRE-DETECTION AND ALARM SYSTEMS

Regulation 32 of the Construction (Design and Management) Regulations (CDMR) 2015 establishes specific requirements for fire detection, alarm and fire-fighting equipment. It requires those in control of the site to take account of a number of factors set out in regulation 30 when deciding what would be suitable and sufficient equipment, for example, type of work, number of people, characteristics and size of the site. In addition, the Fire Protection Association sets out advice in the Joint Code of Practice for Fire Prevention on Construction Sites.

"(1) Where necessary in the interests of the health or safety of a person on a construction site, suitable and sufficient fire-fighting equipment and fire detection and alarm systems must be provided and located in suitable places.

(2) The matters in regulation 30(2) must be taken into account when making provision under paragraph (1).

(3) Fire-fighting equipment or fire detection and alarm systems must be examined and tested at suitable intervals and properly maintained."

Figure 11-44: Fire detection, alarm and fire-fighting requirements. Source: CDMR 2015, Regulation 32.

Fire detection

The speed with which a fire in a building is detected is a critical factor in the determination of survival for the occupants of that building. Fires should be detected as soon as they start and building occupants alerted to the presence of the fire by the quickest possible means. It is essential that some form of detection and alarm system is used in the workplace, although the exact type of system will depend on the level of risk in the workplace. Quite often, however, in the early stages of construction work reliance is placed on the vigilance of workers detecting a fire and raising the alarm.

Heat detection

Sensors operate by the melting of a metal (fusion detectors) or expansion of a solid, liquid or gas (thermal expansion detectors).

Radiation detection

Photoelectric cells detect the emission of infra-red/ultra-violet radiation from the fire.

Smoke detection

Smoke may be detected by using ionising radiation, light scatter (smoke scatters beams of light) or obscuration (smoke entering a detector prevents light from reaching a photoelectric cell).

Figure 11-45: Smoke detector.
Source: RMS.

Flammable gas detection

Flammable gas is detected by measuring the amount of flammable gas in the atmosphere and comparing the value with a reference value.

Alarm systems

The purpose of a fire alarm is to give an early warning of a fire in a building for two reasons:

1) To increase the safety of occupants by encouraging them to escape to a place of safety

2) To increase the possibility of early extinction of the fire, thus reducing the loss of or damage to the property.

Types of fire alarms

Alarms must make a distinctive sound, audible in all parts of the workplace it is provided for. Sound levels should be 65 dB(A) or 5 dB(A) above any other noise - whichever is the greater.

Audible alarms may be supplemented by visual or tactile (vibrating) alarms where this would aid hearing-impaired workers and other people.

The meaning of the alarm must be understood by all and readily differentiated from other alarms. Alarms may be manually or automatically operated.

Voice	Simplest and most effective type, but very limited because it is depends on the size of the workplace and background noise levels.
Hand operated	Rotary gong, hand bell or triangle and sounder, but as a single alarm sound it has limited range, suitable for small sites or a small number of temporary accommodation units.
Portable stand-alone call point and sounder	Integrated alarm call point and sounder system, can be portable, if used alone it is also suitable for smaller sites or small number of temporary accommodation units. Additional stand-alone call points and sounders can be located around larger sites.

Call-points with sounders	Standard system, operation of one call-point sounds alarm throughout workplace – may be wired together or portable wireless linked system. Suitable for larger sites.
Automatic system	System as above, with added fire detection to initiate the alarm if it is not raised by a person – may be wired together or portable wireless linked system. Does not rely on a person initiating the alarm.

Figure 11-46: Rotary hand bell.
Source: Besafe Direct.

Alarm call-points should be sited so that no person need travel an excessive distance to sound the alarm. Travelling more than 30 metres (by direct measurement) or 45 metres to sound the alarm, taking into account fixtures, stock and other obstructions (actual), may be considered to be excessive.

There are a number of alarm strategies that may be used depending on circumstances, including single-stage, two-stage and nominated-worker alarms.

Figure 11-47: Stand-alone portable alarm point and sounder.
Source: Howler.

Figure 11-48: Portable wireless-linked system - alarm point, smoke detector and control panel.
Source: Bull Products.

Figure 11-49: Easy operation alarm call-point for fixed installation.
Source: RMS.

Single-stage alarm

The alarm sounds throughout the whole of the building/site and signals a total evacuation.

Two-stage alarm

In certain large/high rise buildings and complex sites it may be better to evacuate the areas of high risk first, usually those closest to the fire or immediately above it. In this case, an evacuation signal is given in the affected area, together with an alert signal in other areas. If this type of system is required, early consultation with the fire and rescue service is essential.

PORTABLE FIRE-FIGHTING EQUIPMENT

Regulation 32 of the Construction (Design and Management) Regulations (CDMR) 2015 establishes specific requirements for fire-fighting equipment, including siting, indication by signs, instruction on use, maintenance and inspection.

"(3) Fire-fighting equipment or fire detection and alarm systems must be examined and tested at suitable intervals and properly maintained.

(4) Fire-fighting equipment which is not designed to come into use automatically must be easily accessible.

(5) Each person at work on a construction site must, so far as is reasonably practicable, be instructed in the correct use of fire-fighting equipment which it may be necessary for the person to use.

(7) Fire-fighting equipment must be indicated by suitable signs."

Figure 11-50: Fire-fighting equipment requirements.
Source: CDMR 2015, Regulation 32.

Siting

Portable fire-fighting equipment in the form of fire extinguishers should always be sited:

- So that it is easily accessible.
- On the line of escape routes.

- Near, but not too near, to danger points.
- Near to room exits inside or outside according to occupancy and/or risk.
- In multi-storey buildings, at the same position on each floor, for example, at the top of stair flights or at corners in corridors.
- Where possible in groups forming fire points.
- So that no person needs to travel an excessive distance to reach an extinguisher. Travelling more than 30 metres (by direct measurement) or 45 metres (actual) taking into account fixtures, stock and other obstructions to reach an extinguisher may be considered to be excessive. In high fire hazard areas 45 metres may be considered excessive and fire extinguishers may need to be located nearer to workers.
- Placed on a purpose designed floor stand or hung on a wall/trolley at a convenient height. If hung on a wall or trolley, this would usually be with the carrying handle about one metre from the floor to facilitate ease of handling/removal from the wall bracket.
- Location indicated by suitable signs.
- Away from excesses of heat, cold, dirt or dust, this may require the equipment to be covered.

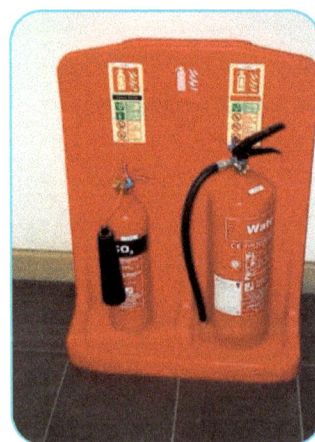

Figure 11-51: Fire extinguishers – in floor stand.
Source: RMS.

Figure 11-52: Fire extinguishers – mobile unit for construction site.
Source: HSE.

Maintenance and inspection

Any fire-fighting equipment provided must be properly maintained and subject to examination and test at suitable intervals such that it remains effective and available for use in an emergency.

Maintenance

This means the servicing of a fire extinguisher by a competent person. It involves thorough examination of the extinguisher (internal/external), refilling and re-pressurisation and is usually done annually.

Inspection

A monthly check (inspection) should be carried out to ensure that extinguishers are in their proper place and that they have not been discharged (the firing pin still tagged and in place), lost pressure or suffered obvious damage.

It may be necessary to increase the frequency of checks made for fire extinguishers located in work environments where there is less control over their use or they may be more likely to suffer damage, for example, on a construction site. Records should be kept of all visual inspections and maintenance checks carried out.

Training requirements

The 'responsible person' must take measures for fire-fighting as necessary. They should, as necessary, nominate competent persons to fight fires, and provide instruction/training and equipment accordingly. This would usually include instruction/training in how to use portable fire extinguishers, including practice in how to use them in a situation that reproduces the circumstances of a fire.

Instruction/training for those who use portable fire extinguishers should include:

- Understanding of the principles of combustion and classification of fires.
- Identification of the various types of fire extinguisher available to them.
- How to identify whether the extinguisher is appropriate to the fire and ready to use.
- Principles of use and limitations of extinguishers.
- Considerations for personal safety and the safety of others.
- How to attack fires with the appropriate extinguisher(s).
- Any specific considerations related to the environment the extinguishers are kept or used in.

Instruction/training should clarify the general and specific rules for use of portable fire extinguishers, for example:

- General - aim at the seat of the fire and direct the extinguishing material across the fire to extinguish it - this is particularly appropriate for Class A fires, involving wood, paper, textiles, etc.

- Specific - if using a foam portable fire extinguisher for a flammable liquid fire the foam is allowed to drop onto the fire by aiming just above it. If this is for a flammable liquid fire contained in an open tank it is possible to get good results by this process or aiming it to the back of the tank and allowing the foam to float over the liquid. For other specific limitations or approaches to the use of individual types of extinguishing media, see the next section.

EXTINGUISHING MEDIA

In the following section reference to a colour code means a code indicating the type of medium used in the fire extinguisher and refers to the European (BS EN 3) system of coding, which has been established for a considerable time and is widely accepted globally.

Water (Portable extinguisher - colour code - red)

Water should only be used on Class A fires - those involving solids like paper and wood. Water works by cooling the burning material to below its ignition temperature, therefore removing the heat part of the fire triangle, and so putting the fire out. Water is the most common form of extinguishing media and can be used on the majority of fires involving solid materials. It must not be used on flammable liquid fires or in the vicinity of live electrical equipment.

Figure 11-53: Water fire point sign. Source: BCW Office Products.

Figure 11-54: Water extinguisher colour coded red above label. Source: Shutterstock.

Foam (Portable extinguisher - colour code - cream)

Foam is especially useful for extinguishing Class B fires - those involving burning liquids and solids which melt and turn to liquids as they burn. Foam works in several ways to extinguish the fire, the main one being to smother the burning liquid, i.e. to stop the oxygen reaching the combustion zone.

Foam can also be used to prevent flammable vapours escaping from spilled volatile liquids and also on Class A fires. It is worth noting that the modern spray foams are more efficient than water on a Class A fire. ***Foam is not be used in the vicinity of live electrical equipment, unless electrically rated.***

Figure 11-55: Foam fire point sign. Source: Warning Signs Direct.

Figure 11-56: Foam extinguisher colour coded cream above label. Source: Low Cost Fire.

Dry powder (Portable extinguisher - colour code - blue)

Powder extinguishers are designed for Class A, B and C (those involving flammable gases) fires, but may only subdue Class A fires for a short while. One of the main ways in which powder works to extinguish a fire is the smothering effect, whereby it forms a thin film of powder on the burning liquid, thus excluding air. Dry powder is also excellent for the rapid knock down (flame suppression) of flammable liquid spills and fires involving flammable gases.

It should be remembered that, if used to extinguish flammable gas fires, the source of the gas must be isolated promptly if re-ignition or explosion is to be prevented.

Figure 11-57: Dry powder fire point sign. Source: Warning Signs Direct.

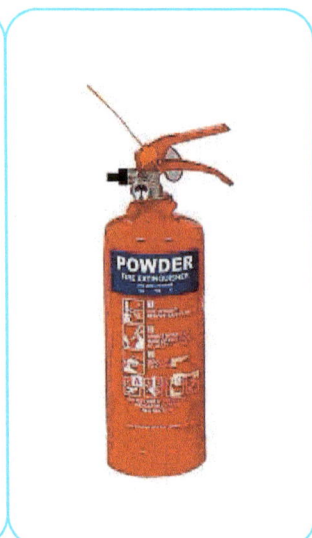

Figure 11-58: Dry powder extinguisher colour coded blue above label. Source: The Sharpedge.

Powders generally provide extinction faster than foam, but there is a greater risk of re-ignition. If used indoors, a powder can cause problems for the operator due to the inhalation of the powder and obscuration of vision. This type of extinguishing media may be used on live electrical equipment.

Carbon dioxide (CO_2) (Portable extinguisher - colour code - black)

Carbon dioxide (CO_2) is safe and excellent for use on live electrical equipment. It may also be used for small Class B fires in their early stages of development, indoors or outdoors with little air movement. CO_2 replaces the oxygen in the atmosphere surrounding the fuel and thus extinguishes the fire.

CO_2 is an asphyxiant and should not be used in confined spaces. As it does not remove the heat there is the possibility of re-ignition. CO_2 extinguishers are very noisy due to the rapid expansion of gas on release; this can surprise people when they operate a portable extinguisher. This expansion causes severe cooling around the discharge horn and can freeze the skin if the operator's hand is in contact with it. As most carbon dioxide portable extinguishers last only a few seconds, only small fires should be tackled with this type of extinguisher. Carbon dioxide extinguishers are not suitable for Class D fires.

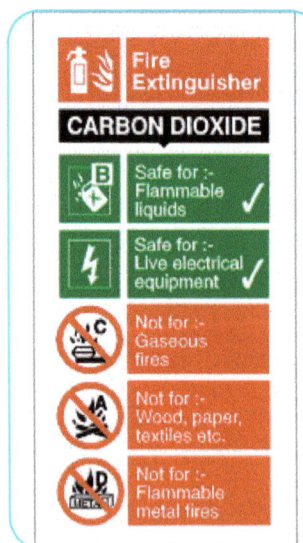

Figure 11-59: CO_2 fire point sign. Source: BCW Office Products.

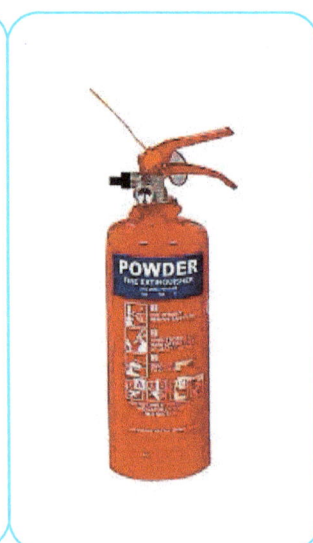

Figure 11-60: CO_2 extinguisher colour coded black above label. Source: Blazetec Fire Protection.

11-61: CO_2 Extinguishing fire. Source: Shutterstock.

Extinguishing media for specific classes of fire

Class C fires

Except in very small occurrences, a Class C fire involving gas should not normally be extinguished. If a gas leak fire is to be extinguished, the gas supply must first be isolated to prevent the gas reigniting and avoid a potential explosion.

Class D fires

Class D fires, those involving combustible metals, are a specialist type of fire and they cannot be extinguished by the use of ordinary extinguishing media. In fact, it may be dangerous to attempt to fight a metal fire with ordinary extinguishing media as the metal may explode or toxic fumes may be produced.

Metal fires can be extinguished by smothering them with dry sand. However, the sand must be absolutely dry or an explosion may occur. Other extinguishing media used may be graphite or salt. All of these extinguishing media essentially operate by the smothering principle.

Figure 11-62: Wet chemical fire point sign.
Source: Midland Fire Ltd.

Figure 11-63: Wet chemical extinguisher colour coded yellow above label.
Source: Midland Fire Ltd.

Class F fires (Wet chemical) (Portable extinguisher - Colour code - Yellow)

Wet chemical extinguishing media have been designed to deal specifically with Class F fires, those involving **cooking oil or fats** where temperatures can exceed 260°C. This type of extinguishing medium congeals on top of the oil and excludes the oxygen. It may also be used on Class A fires depending upon the manufacturer's instructions.

Summary matrix - fire extinguishing media

Figure 11-64 shows a summary of the most common fire-extinguishing media, their method of acting on fires and their effectiveness against different classes of fire. For the purposes of this summary, the classification of fires is that used in the UK and Europe.

ACCESS FOR FIRE AND RESCUE SERVICES AND VEHICLES

As firefighting is generally carried out within buildings it makes good sense to ensure that the fire and rescue services can access a building as quickly as possible in the event of a fire.

The responsible person should ensure that facilities, equipment and devices provided are maintained in an efficient state, in efficient working order and in good repair.

Vehicle access

The fire and rescue services need to get their appliances as close as possible to buildings to prevent time being wasted with running out unnecessary hose. Minimum access requirements for pumping appliances and high-reach appliances will vary dependent upon the height, floor area of the building and whether a fire main is fitted.

Access will be required for a minimum to 15% of the perimeter or within 45m of every point of the footprint of the building, up to a maximum of 100% of the perimeter, dependent upon the factors previously mentioned.

	Method	Class 'A'	Class 'B'	Class 'C'	Class 'D'	Electric	Class 'F'
Water	Cools	Yes	No	No	No	No	No
Spray foam	Smothers	Yes	Yes	No	No	No	No
Dry powder	Smothers & chemical	Yes	Yes	Yes & isolate	Special powders	Yes - low voltage	No
Carbon dioxide	Smothers	No	Yes - small fires	No	No	Yes	No
Wet chemical	Chemical	No	No	No	No	No	Yes
Vapourising liquids	Chemical & smothers	Special uses					

Figure 11-64: Summary matrix - fire extinguishing media.
Source: RMS.

Restrictions or planning consents in force at the time of the construction of the building will cause some variations between similar structures.

Access for firefighting

In low-rise buildings additional access requirements are not normally required. The means of access is more straightforward and ladder access for firefighting is simpler. Access for firefighting appliances or other emergency vehicles must still meet the requirements given in the previous section (vehicle access).

In higher-rise buildings, additional facilities, including firefighting lifts, firefighting stairs and lobbies (usually called a firefighting shaft) are included as good practice. The addition of these measures allows fire and rescue services to quickly reach the floor where the fire is or to access the floor below a fire, from which an operating base can be set up. In general, buildings with floor levels over 18m, or basements more than 10m below a fire, must allow rescue service vehicle access.

It should be noted however that some construction sites could be more remote or inaccessible, such as rural locations, and vehicle access and access for firefighting should be considered at the earliest possible stage in a project.

Figure 11-65: Fire appliance - fully extended turntable ladder.
Source: www.geograph.ie.

Figure 11-66: Firefighter on turntable platform.
Source: www.geograph.ie.

Sources of reference

Reference information provided, in particular web links, was correct at time of publication, but may have changed.

Fire safety, HSE Toolbox, http://www.hse.gov.uk/toolbox/fire.htm

Fire Safety Risk Assessment Series (Home Office publications), https://www.gov.uk/search/all?keywords=Fire+Safety+Risk+Assessment&order=relevance

Fire Safety Risk Assessment - Offices and shops, ISBN-13: 978-1-851128-15-0

Fire Safety Risk Assessment - Factories and warehouses, ISBN-13: 978-1-851128-16-7

Fire Safety Risk Assessment - Sleeping accommodation, ISBN-13: 978-1-851128-17-4

Fire Safety Risk Assessment - Residential care premises, ISBN-13: 978-1-851128-18-1

Fire Safety Risk Assessment - Educational premises, ISBN-13: 978-1-851128-19-8

Fire Safety Risk Assessment - Small and medium places of assembly, ISBN-13: 978-1-851128-20-4

Fire Safety Risk Assessment - Large places of assembly, ISBN-13: 978-1-851128-21-1

Fire Safety Risk Assessment - Theatres, cinemas and similar premises, ISBN-13: 978-1-851128-22-8

Fire Safety Risk Assessment - Open air events and venues, ISBN-13: 978-1-851128-23-5

Fire Safety Risk Assessment - Healthcare premises, ISBN-13: 978-1-851128-24-2

Fire Safety Risk Assessment - Transport premises and facilities, ISBN-13: 978-1-851128-25-9

Regulatory Reform (Fire Safety) Order 2005: a short guide to making your premises safe (Home Office guidance), https://assets.publishing.service.gov.uk/government/uploads/system/uploads/attachment_data/file/14879/making-your-premises-safe-short-guide.pdf

Safety signs and signals, Guidance on Regulations, L64, HSE Books, ISBN: 978-0-7176-6598-3 http://www.hse.gov.uk/pubns/priced/l64.pdf

Safe use and handling of flammable liquids, HSG140, HSE Books http://www.hse.gov.uk/pubns/priced/hsg140.pdf

Storage of flammable liquids in containers, HSG51, HSE Books, https://www.hse.gov.uk/pubns/books/hsg176.htm

Storage of flammable liquids in tanks, HSG176, HSE Books http://www.hse.gov.uk/pubns/priced/hsg176.pdf

The health and safety toolbox, How to control risks at work, HSG268, HSE Books, ISBN: 978-0-7176-6587-7, http://www.hse.gov.uk/pUbns/priced/hsg268.pdf

Web links to these references are provided on the RMS Publishing website for ease of use –www.rmspublishing.co.uk

Statutory provisions

Dangerous Substances and Explosive Atmosphere Regulations (DSEAR) 2002 / Dangerous Substances and Explosive Atmosphere Regulations (Northern Ireland) 2003

Health and Safety (Safety Signs and Signals) Regulations (SSSR) 1996 / Health and Safety (Safety Signs and Signals) Regulations (Northern Ireland) 1996

Fire and Rescue Services (Northern Ireland) Order 2006

Fire Safety Regulations (Northern Ireland) 2010

Fire Safety (Scotland) Regulations (FSSR) 2006

Fire (Scotland) Act (FSA) 2005

Regulatory Reform (Fire Safety) Order (RRFSO) 2005

STUDY QUESTIONS

1)   Arson is the single greatest cause of fire.

(a) What factors make some construction sites vulnerable to arson attacks?

(b) What control measures can be taken to reduce the risk of arson on a construction site?

2)   Liquefied petroleum gas (LPG) is often stored on construction sites. What control measures should be used for the safe storage of LPG on construction sites?

3)   Fire is a significant hazard on refurbishment projects that involve hot works.

(a) Which THREE activities can be considered as 'hot works'?

(b) What precautions may be taken to reduce the risk of a fire occurring during 'hot works'?

4)   What factors would need to be covered in the induction training of workers to enable their response to a fire emergency on a construction site?

5)   What are the factors to be considered when assessing fire risks and the effects on the current fire risk reduction measures of a major refurbishment project on one floor of an occupied multi-floor office block?

For guidance on how to answer these questions, please refer to the assessment section located at the back of this guide.

Element 12

Chemical and biological agents

Contents

FORMS OF CHEMICAL AGENTS

The form taken by a hazardous substance is a contributory factor in its potential for harm. The form affects how easily a substance gains entry to the body, how it is absorbed into the body and how it reaches a susceptible site. Chemical agents take many forms, the most common being the primary forms or states – solids, liquids, gases – and the derivative forms – dusts, fibres, fumes, mists and vapours.

Solids

Solids are materials that are solid at normal temperature and pressure. The atoms, molecules or ions that make up a solid may be arranged in an orderly repeating pattern or irregularly. Materials whose constituents are arranged in a regular pattern are known as crystals. Crystals that are large enough to see and handle are known as crystallites. Other materials are called polycrystalline, which simply means they are composed of many crystallites of varying size and orientation. Almost all common metals, silicon and many ceramics are polycrystalline. Unlike a liquid, a solid object does not flow to take on the shape of its container, nor does it expand to fill the entire volume available to it like a gas. The risk from hazardous solids increases with reduction in particle size, particularly when it becomes a dust that can become airborne.

Liquids

Liquids are substances that are liquid at normal temperature and pressure. Liquids have a definite volume but no fixed shape. Similar to a gas, a liquid is able to flow and take the shape of a container. Some liquids, such as water, resist compression, while others can be compressed. Unlike a gas, a liquid does not disperse to fill every space of a container and maintains a fairly constant density. The density of a liquid is usually close to that of a solid and much higher than that of a gas.

Gases

Gases are formless fluids usually produced by chemical processes involving combustion or by the interaction of chemical substances. A gas will normally seek to fill the space completely into which it is liberated – for example, chlorine gas, carbon monoxide gas and methane gas.

Dusts

Dusts are solid airborne particles, often created by operations such as grinding, crushing, milling, sanding or demolition – for example, silica dust, wood dust and cement dust.

Fibres

Fibres are solid airborne particles that are significantly longer in length than they are in width, for example, cotton fibres, mineral wool fibres and asbestos fibres.

Fumes

Fumes are solid airborne particles formed by condensation from the gaseous state – for example, lead fume, welding fume.

Mists

Mists are finely dispersed liquid droplets suspended in air. Mists are mainly created by spraying, foaming, pickling and electro-plating – for example, mists from a water pressure washer, paint spray, pesticide spray, sprayed oil-based cutting fluids.

Vapour

Vapour is the gaseous form of a material normally encountered in a liquid or solid state at normal room temperature and pressure. Typical examples of vapours are those released from liquid solvents, also known as VOCs (volatile organic compounds), which are used in many construction products such as paints, thinners and glues. These can contain a range of solvents, including toluene, xylene, white spirit, acetone and ethyl acetate, which release vapours when the container holding them is opened.

"The terms 'dust', 'mist' and 'vapours' are formally defined as follows:

> "The terms 'dust', 'mist' and 'vapours' are defined as follows:
>
> **Dust:** Solid particles of a substance or mixture suspended in a gas (usually air).
>
> **Mist:** Liquid droplets of a substance or mixture suspended in a gas (usually air).
>
> **Vapour:** The gaseous form of a substance or mixture released from its liquid or solid state."

Figure 12-1: Definition of dust, mist and vapour.
Source: UN Globally Harmonised System of Classification and Labelling of Chemicals (GHS) - part 3.

FORMS OF BIOLOGICAL AGENTS

Fungi

Fungi are a variety of organisms that act in a parasitic manner, feeding on organic matter. Most are either harmless or positively beneficial to health; however, a number cause harm to humans and may be fatal.

An example of fungi is the aspergillus mould, which can be found in many indoor and outdoor environments, for example heating systems, air conditioning, flood damaged buildings, insulation materials or rotting organic materials, and can therefore be encountered when carrying out a range of construction and demolition work. The spores of the aspergillus mould can cause an allergic reaction in the lungs when they are inhaled, this is called aspergillosis.

Figure 12-2: Mould - Aspergillus.
Source: National Geographic.

Figure 12-2 shows Aspergillus with its characteristic chains of spores emerging from the head. Moulds from the same family can cause ringworm and athlete's foot.

Bacteria

Bacteria are single-cell organisms. Most bacteria are harmless to humans and many are beneficial. The bacteria that can cause disease are called pathogens. Examples of harmful bacteria are leptospira (causing Leptospirosis, known as Weil's disease), bacillus anthracis (causing anthrax), and legionella pneumophila (causing Legionellosis, known as Legionnaires' disease). Construction workers may be exposed to these bacteria when carrying out a variety of construction activities, including refurbishment and demolition. *The risk and controls related to leptospira and legionella are described later in this element in 12.4 'Specific agents'.*

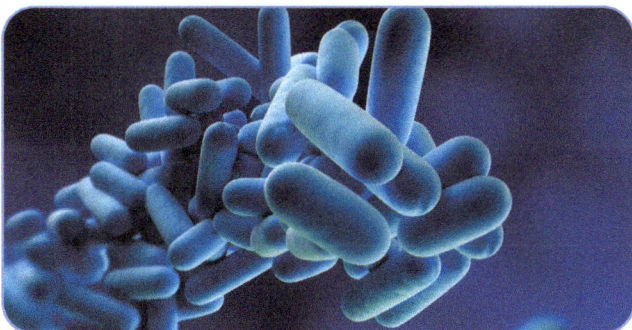

Figure 12-3 Bacteria - legionella pneumophila.
Source: European Hospital.

Viruses

Viruses are the smallest known type of infectious agent. They invade the cells of other organisms, which they take over and where they make copies of themselves, and while not all cause disease many of them do.

Examples of viruses are hepatitis, which can cause liver damage, and the human immunodeficiency virus (HIV), which causes acquired immune deficiency syndrome (AIDS). *The risks and controls related to blood-borne viruses are described later in this element in 12.4 'Specific agents'.*

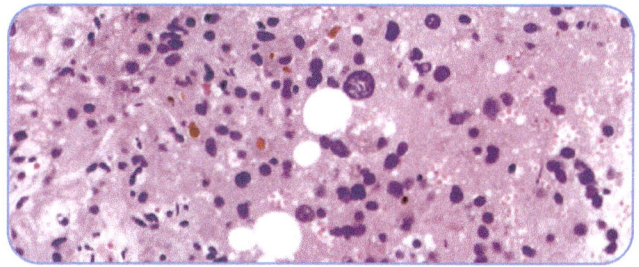

Figure 12-4: Virus - hepatitis C.
Source: National Public Radio.

DIFFERENCE BETWEEN ACUTE AND CHRONIC HEALTH EFFECTS

The effect of a substance on the body depends not only on the substance, but also on the dose and the susceptibility of the individual. No substance can be considered non-toxic; there are only differences in degree of effect.

Acute effect

An acute effect is an immediate or rapidly produced adverse effect following a single or short-term exposure to an offending agent, which is usually reversible (the obvious exception being death). Examples of acute effects are those from exposure to solvents, which affects the central nervous system, causing dizziness and lack of coordination, or carbon monoxide, which affects the level of oxygen in the blood, causing fainting.

Chronic effect

A chronic effect is an adverse health effect produced as a result of prolonged or repeated exposure to an agent. The gradual or latent effect develops over time and is often irreversible. The effect may go unrecognised for a number of years. Examples of chronic effects are lead or mercury poisoning, cancer and asthma.

Other common terms used in the context of occupational health are:

Toxicology

The study of the body's responses to substances. In order to interpret toxicological data and information, the meaning of the following terms should be understood.

Toxicity

The ability of a chemical substance to produce injury once it reaches a susceptible site in or on the body.

A poisonous substance (for example, organic lead) causes harm to biological systems and interferes with the normal functions of the body. The effects may be acute or chronic, local or systemic.

Dose

The level of environmental contamination multiplied by the length of time (duration) of exposure to the contaminant.

Local effect

Usually confined to the initial point of contact. Possible sites affected include the skin, mucous membranes or the eyes, nose or throat.

Examples are burns to the skin by corrosive substances (acids and alkalis), or dermatitis caused by solvents.

Systemic effect

Occurs in parts of the body other than at the point of initial contact. Frequently the circulatory system provides a means to distribute the substance round the body to a target organ/system.

Target organs

An organ of the human body on which a specified toxic material exerts its effects, for example, lungs, liver, brain, skin, bladder or eyes.

Target systems

Central nervous system, circulatory system, respiratory system, and reproductive system.

Examples of substances that have a systemic effect and their target organs/systems are:

- *Alcohol:* central nervous system, liver.
- *Lead:* bone marrow and brain.
- *Mercury:* central nervous system.

It must be noted that many chemicals in use today can have both an acute and chronic effect. A simple example is lead. Symptoms of acute lead toxicity include upset stomach (gastrointestinal problems), dullness, restlessness, irritability, poor attention span, headaches, kidney damage, hypertension and hallucinations. Health effects of chronic exposure to lead are blood disorder effects, such as anaemia, or neurological disturbances, including headache, irritability, lethargy, convulsions, muscle weakness, tremors and paralysis. Chronic lead exposure also causes cardiovascular and renal toxicity. In children, lead exposure may lead to cognitive deficits, such as a decrease in intelligence quotient (IQ). Chronic exposure to lead can cause adverse effects on both male and female reproductive functions.

HEALTH HAZARD CLASSIFICATIONS

UN Globally Harmonized System of Classification and Labelling of Chemicals

The UN has established a non-legally binding international agreement called 'Globally Harmonized System of Classification and Labelling of Chemicals (GHS)'. It has been widely accepted globally and has been established in national legislation of the countries that have adopted it, including the EU and the UK. Within GHS is a classification of chemicals based on their effect on human health. Criteria for classifying chemicals have been developed for the following health hazard classes:

- Acute toxicity.
- Skin corrosion/irritation.
- Serious eye damage/eye irritation.
- Respiratory or skin sensitization.
- Germ cell mutagenicity.
- Carcinogenicity.
- Reproductive toxicity.
- Specific target organ toxicity - single exposure.
- Specific target organ toxicity - repeated exposure.
- Aspiration hazard.

A substance may be classified under more than one category, although those with specific target organ toxicity only apply if other classifications do not.

Acute toxicity

"Acute toxicity refers to those adverse effects occurring following oral or dermal administration of a single dose of a substance, or multiple doses given within 24 hours, or an inhalation exposure of 4 hours." Source: GHS.

The classification is divided into five categories related to the level of toxicity of the chemical. The subdivisions arise from the dose of the substance that would prove lethal to 50% of the population exposed to that substance (hence LD^{50}). The classification of categories and labelling for acute skin and inhalation toxicity are the

Acute oral toxicity - Annex 1 of GHS					
	Category 1	**Category 2**	**Category 3**	**Category 4**	**Category 5**
LD_{50}	≤ 5 mg/kg	>5 <50 mg/kg	>50 <300 mg/kg	>300 <2,000 mg/kg	>2000 <5,000 mg/kg
Pictogram					No symbol
Signal word	Danger	Danger	Danger	Warning	Warning
Hazard statement	Fatal if swallowed	Fatal if swallowed	Toxic if swallowed	Harmful if swallowed	May be harmful if swallowed

Figure 12-5: Acute oral toxicity.
Source: Annex 1 of GHS.

Skin corrosion/irritation - Annex 1 of GHS					
	Category 1A	Category 1B	Category 1C	Category 2	Category 3
Pictogram					No symbol
Signal word	Danger	Danger	Danger	Warning	Warning
Hazard statement	Causes severe skin burns and eye damage	Causes severe skin burns and eye damage	Causes severe skin burns and eye damage	Causes skin irritation	Causes mild skin irritation

Figure 12-6: Skin corrosion/irritation.
Source: Annex 1 of GHS.

same as for oral exposure, except for slightly different hazard statements, (**see Figure 12-5**). An example of a substance with acute oral toxicity is arsenic, which is a systemic poison. Another example is methanol, which is known to cause lethal intoxications in humans (mostly via ingestion) in relatively low doses and is a category 3 substance for acute oral toxicity.

Skin corrosion/irritation

"Skin corrosion is the production of irreversible damage to the skin; namely, visible dead skin cells through the epidermis (outermost skin cells) and into dermis (the middle layer of skin cells), following the application of a test substance for 4 hours. Corrosive reactions are typified by ulcers, bleeding, bloody scabs, and, by the end of observation at 14 days, by discolouration due to blanching of the skin, complete areas of alopecia, and scars." Source: GHS.

The corrosion effect is divided into three categories related to duration of exposure necessary to create an effect, (**see Figure 12-6**).

"Skin irritation is the production of reversible damage to the skin following the application of a test substance for up to 4 hours." Source: GHS.

Sulphuric (battery) acid and sodium hydroxide (caustic soda) are examples of substances classified under skin corrosion/irritation.

Serious eye damage/eye irritation

"Serious eye damage is the production of tissue damage in the eye, or serious physical decay of vision, following application of a test substance to the anterior (outer) surface of the eye, which is not fully reversible within 21 days of application."

"Eye irritation is the production of changes in the eye following the application of a test substance to the anterior surface of the eye, which are fully reversible within 21 days of application." Source: GHS.

This generally relates to effects on the cornea, iris or conjunctiva, (**see Figure 12-7**).

Serious eye injury/eye irritation - Annex 1 of GHS			
	Category 1	Category 2A	Category 2B
Pictogram			No symbol
Signal word	Danger	Warning	Warning
Hazard statement	Causes severe eye damage	Causes severe eye irritation	Causes eye irritation

Figure 12-7: Serious eye injury/eye irritation.
Source: Annex 1 of GHS.

Respiratory sensitisation - Annex 1 of GHS		
	Category 1	Category 1A	Category 1B
Pictogram			
Signal word	Danger	Danger	Danger
Hazard statement	May cause allergy or asthma symptoms or breathing difficulty if inhaled	May cause allergy or asthma symptoms or breathing difficulty if inhaled	May cause allergy or asthma symptoms or breathing difficulty if inhaled

Figure 12-8: Respiratory sensitisation.
Source: Annex 1 of GHS.

The difference between category 1 and 2 for eye injury is whether the harm to the eye is fully reversible within the observation period. A category 2B substance is one where it is considered to be mildly irritating to eyes and fully reversible within 7 days of observation.

Respiratory or skin sensitisation

"A respiratory sensitiser is a substance that will lead to hypersensitivity of the airways following inhalation of the substance'. 'A skin sensitiser is a substance that will lead to an allergic response following skin contact."
Source: GHS.

Sensitisation takes place in two stages. The first is the recognition stage where, following contact with the airways or skin, the substance is recognised by the body as a pathogen (something that can cause harm to health). The second stage is the antibody generation and allergic response to further exposure to the substance.

The three categories of respiratory sensitisers relate to the type and level of evidence that identifies the substance as a sensitiser.

Category 1 has been established by direct evidence that exposure will lead to specific hypersensitivity, for example, asthma, rhinitis/conjunctivitis and alveolitis (inflammation of the alveoli), whereas category 1B substances show a low to moderate frequency of occurrence of sensitisation. Flour dust and isocyanates are examples of respiratory sensitisers.

As with respiratory sensitisers, the three categories for skin sensitisers relate to the type and level of evidence that identifies the substance as a sensitiser.

Category 1 substances are ones where there is strong documented evidence of causing allergic contact dermatitis. Nickel and epoxy resins are examples of skin sensitisers, (**see Figures 12-8 and 12-9**).

Skin sensitisation - Annex 1 of GHS		
	Category 1	Category 1A	Category 1B
Pictogram			
Signal word	Warning	Warning	Warning
Hazard statement	May cause an allergic skin reaction	May cause an allergic skin reaction	May cause an allergic skin reaction

Figure 12-9: Skin sensitisation.
Source: Annex 1 of GHS.

Germ cell mutagenicity - Annex 1 of GHS			
	Category 1A	Category 1B	Category 2
Pictogram			
Signal word	Danger	Danger	Warning
Hazard statement	May cause genetic defects	May cause genetic defects	Suspected of causing genetic defects

Figure 12-10: Germ cell mutagenicity.
Source: Annex 1 of GHS.

Germ cell mutagenicity

"This hazard class is primarily concerned with chemicals that may cause mutations in germ cells of humans that can be transmitted to the descendants of a person."
Source: GHS.

The different categories of mutagenicity reflect the degree of knowledge about the chemical and the indications that it may cause mutagenicity. Category 1 substances are known or are presumed because of related evidence to induce heritable mutations, whereas category 2 substances are ones where there is concern that they may induce heritable mutations. If a specific route of exposure is proven to be the only route causing harm, this route must be stated in the hazard statement, (**see Figure 12-10**).

Carcinogenicity

"The term carcinogen denotes a substance or mixture that induces cancer or increases its incidence.
Substances and mixtures that have induced benign and malignant tumours in well-performed experimental studies on animals are considered also to be presumed or suspected human carcinogens unless there is strong evidence that the mechanism of tumour formation is not relevant to humans." Source: GHS.

As with the different categories of mutagenicity, the categories for carcinogenicity reflect the degree of knowledge about the chemical and the indications that it may cause cancer. Category 1 substances are known or are presumed because of related evidence to induce cancer, whereas category 2 substances are ones where there is concern that they may induce cancer. If a specific route of exposure is proven to be the only route causing harm, this route must be stated in the hazard statement. Benzyl chloride is an example of a carcinogenic substance. Another example is benzene, which affects bone marrow, causing leukaemia, (**see Figure 12-11**).

Carcinogenicity - Annex 1 of GHS			
	Category 1A	Category 1B	Category 2
Pictogram			
Signal word	Danger	Danger	Warning
Hazard statement	May cause cancer	May cause cancer	Suspected of causing cancer

Figure 12-11: Carcinogenicity.
Source: Annex 1 of GHS.

Reproductive toxicity - Annex 1 of GHS			
	Category 1A	Category 1B	Category 2	Additional category on effects on or via lactation
Pictogram				No pictogram
Signal word	Danger	Danger	Warning	No signal word
Hazard statement	May damage fertility or the unborn child	May damage fertility or the unborn child	Suspected of damaging fertility or the unborn child	May cause harm to breast-fed children

Figure 12-12: Reproductive toxicity.
Source: Annex 1 of GHS.

Reproductive toxicity

"Reproductive toxicity includes effects on sexual function and fertility in adult males and females, as well as developmental toxicity in the offspring. The genetically based inheritable effects in offspring come under the classification 'germ cell mutagenicity'." Source: GHS.

The classification covers two main headings:

1) Adverse effects on sexual function and fertility - including alterations to the reproductive system, effects on the onset of puberty or the reproductive cycle, sexual behaviour, fertility and pregnancy outcomes.

2) Adverse effects on development of the offspring - including interference with the development of the foetus or child, before or after birth, resulting from exposure of either parent prior to conception or during development of the offspring.

The classification principally provides a warning for pregnant women and men and women of reproductive capacity. Category 1 substances are known or are presumed because of related evidence to be a human reproductive toxicant, whereas category 2 substances are ones where there is concern that they may be a human reproductive toxicant. If a specific route of exposure is proven to be the only route causing harm, this route must be stated in the hazard statement.

The UN GHS advises that 'for many substances there is no information on their potential to cause adverse effects on the offspring via lactation. However, substances that are absorbed by women and have been shown to interfere with lactation, or which may be present (including metabolites) in breast milk in amounts sufficient to cause concern for the health of a breast-fed child, should be classified to indicate this property hazardous to breast-fed babies'.

An example of a substance classified as reproductive toxic is 2-ethoxyethanol (a solvent used in commercial and industrial cleaning operations), which has been implicated in impairing fertility. In addition, the effect of lead on the development of the brain of an unborn foetus has long been established, *(see Figure 12-12)*.

Specific target organ toxicity - single exposure

This classification is for substances that produce specific, non-lethal target organ toxicity arising from a single exposure. The effects may be reversible or non-reversible, immediate or delayed, but are not covered in other classifications. Specific target organ toxicity can occur by any route and therefore includes oral, dermal and inhalation routes.

Category 1 substances are known to have produced significant toxicity in humans. Category 2 substances are presumed, because of related evidence, to produce significant toxicity in humans, whereas category 3 substances cause transient target organ effects that affect the respiratory tract or have a narcotic effect. Narcotic effects involve depression of the central nervous system, including drowsiness, loss of reflexes, lack of co-ordination, vertigo and reduced alertness. The symptoms may include severe headache, nausea, dizziness, sleepiness, irritability, fatigue, impaired memory function, perception, coordination and reaction time. Some solvents causing narcosis or central nervous system failure may take effect after a single exposure, *(see Figure 12-13 on following page)*.

Specific target organ toxicity - repeated exposure

As with the similar classification for single exposure, this

Specific target organ toxicity - single exposure - Annex 1 of GHS				
	Category 1	Category 2	Category 3	Category 3
Pictogram				
Signal word	Danger	Warning	Warning	Warning
Hazard statement	Causes damage to organs	May cause damage to organs	(Respiratory tract irritation) May cause respiratory irritation	(Narcotic effects) May cause drowsiness or dizziness

Figure 12-13: Specific target organ toxicity - single exposure.
Source: Annex 1 of GHS.

Specific target organ toxicity - repeat exposure - Annex 1 of GHS		
	Category 1	Category 2
Pictogram		
Signal word	Danger	Warning
Hazard statement	Causes damage to organs through prolonged or repeated exposure	May cause damage to organs through prolonged or repeated exposure

Figure 12-14: Specific target organ toxicity - repeat exposure.
Source: Annex 1 of GHS.

classification is for substances that produce specific, non-lethal target organ toxicity, but arising from repeated exposure. The effects may be reversible or non-reversible, immediate or delayed, but are not covered in other classifications. Specific target organ toxicity can occur by any route and therefore includes oral, dermal (via absorption or injection) and inhalation routes.

Category 1 substances are known to have produced significant toxicity in humans. Category 2 substances are presumed, because of related evidence, to produce significant toxicity in humans. If a specific route of exposure is proven to be the only route causing harm, this route must be stated in the hazard statement.

Similarly, if specific organs are known to be affected these organs must be stated, (see Figure 12-14).

Aspiration hazard

"Aspiration means the entry of a liquid or solid chemical directly through the oral or nasal cavity, or indirectly from vomiting, into the trachea and lower respiratory system."
Source: GHS.

Aspiration toxicity includes severe acute effects such as chemical pneumonia, varying degrees of pulmonary injury or death following aspiration.

Category 1 substances are known or are regarded to have caused aspiration toxicity in humans. Category 2 substances are those that cause concern of aspiration toxicity in humans, (see Figure 12-15 on following page).

Aspiration hazard - Annex 1 of GHS		
	Category 1	Category 2
Pictogram		
Signal word	Danger	Warning
Hazard statement	May be fatal if swallowed and enters airways	May be harmful if swallowed and enters airways

Figure 12-15: Aspiration hazard.
Source: Annex 1 of GHS.

REVIEW

What is meant by the terms 'acute' and 'chronic' health effects?

Explain the term 'aspiration hazard'.

Explain the terms 'target organ' and 'target system' and give an example of each.

12.2 Assessment of health risks

ROUTES OF ENTRY OF HAZARDOUS SUBSTANCES INTO THE BODY

Inhalation

The most significant industrial entry route is inhalation. It has been estimated that at least 90% of industrial poisons are absorbed through the lungs.

Harmful substances can directly attack the lung tissue, causing a local effect, or may pass through to the blood system, to be carried round the body and affect target organs such as the liver. The effects of substances that enter the body through inhalation may be local or systemic.

Local effect

A local effect is where the hazardous substance has an effect where it first contacts the body. For example, silicosis, caused by inhalation of silica dust – where dust causes scarring of the lung, causing inelastic fibrous tissue to develop and reducing lung capacity

Systemic effect

A systemic effect is where the hazardous substance has an effect on a different site from where it first contacted the body. For example, anoxia, caused by the inhalation of carbon monoxide - the carbon dioxide replaces oxygen in the bloodstream, affecting the nervous system.

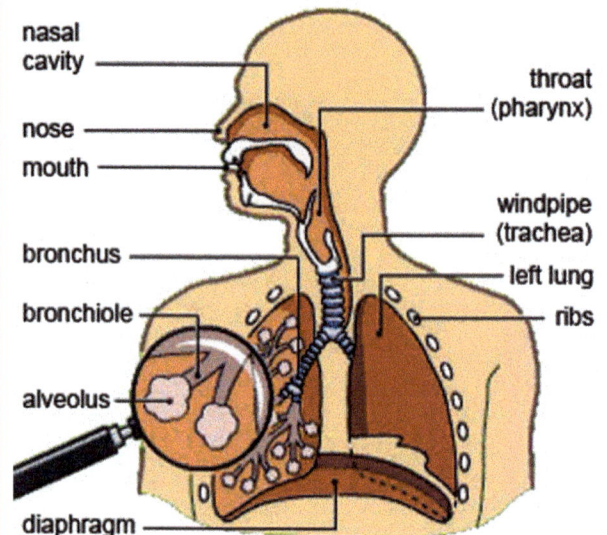

Figure 12-16: Respiratory system.
Source: BBC.

Ingestion

The ingestion route normally presents the least problem as it is unlikely that any significant quantity of harmful liquid or solid will be swallowed without deliberate intent.

However, accidents/incidents will occur where small amounts of contaminant are transferred from the fingers to the mouth if eating, drinking or smoking is allowed in chemical areas or where a substance has been decanted into a container normally used for drinking.

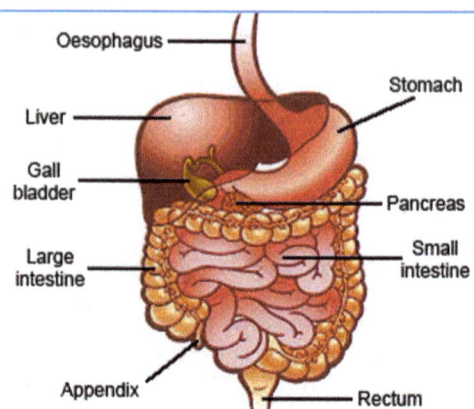

Figure 12-17: Digestive system.
Source: STEM.

The sense of taste will often be a defence if chemicals are taken in through this route, causing the person to spit it out.

If the substance is taken in, vomiting and/or excretion may mean the substance does not cause a systemic problem, though a direct effect, for example, ingestion of an acid, may destroy cells in the mouth, oesophagus or stomach.

Where hazardous substances are ingested, they may pass into the digestive system and be absorbed through the intestine to the blood system and may cause harm in another part of the body.

Absorption (skin contact)

Substances can enter through the skin, via cuts or abrasions and through the conjunctiva of the eye; this is called absorption. Solvents such as organic solvents, for example, toluene and trichloroethylene, can enter due to accidental exposure or if they are used for washing.

The substance may have a local effect, such as de-fatting of the skin, resulting in inflammation and cracking of the outer horny layer of dead skin cells, or pass through into the blood system, causing damage to the brain, bone marrow and liver.

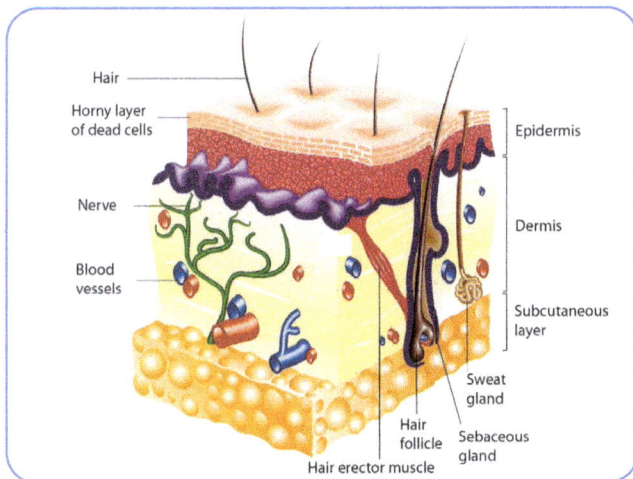

Figure 12-18: Skin layer.
Source: SHP.

Dermatitis

Dermatitis is caused by exposure to substances that interfere with normal skin physiology, leading to inflammation of the skin, usually on the hands, wrists and forearms. The skin turns red and, in some cases, may be itchy. Small blisters may occur and the condition may take the form of dry and cracked skin.

Contact dermatitis (irritant contact dermatitis)

If a person's skin is frequently in contact with some substances or is exposed for a long duration, such persistent contact can lead to irritation and then dermatitis. There are many chemicals used in the workplace that may irritate the skin, leading to this condition, including cement, soaps, detergents, epoxy resins

and hardeners, acrylic sealants, bitumen and solvents used in paints or glues. Removal from contact with the substance usually allows normal cell repair. A similar level of repeat exposure results in the same response. This class of dermatitis is called contact dermatitis.

Sensitisation dermatitis

A second form of dermatitis is called sensitisation dermatitis. In cases of sensitisation dermatitis a person exposed to the substance develops dermatitis in the usual way. When removed from exposure to the substance, the dermatitis usually repairs, but the body gets ready for later exposures by preparing its defence mechanisms. A subsequent small exposure is enough to cause a major response from the immune system.

The person will have become sensitised and will no longer be able to tolerate even small exposures to the substance without a reaction occurring.

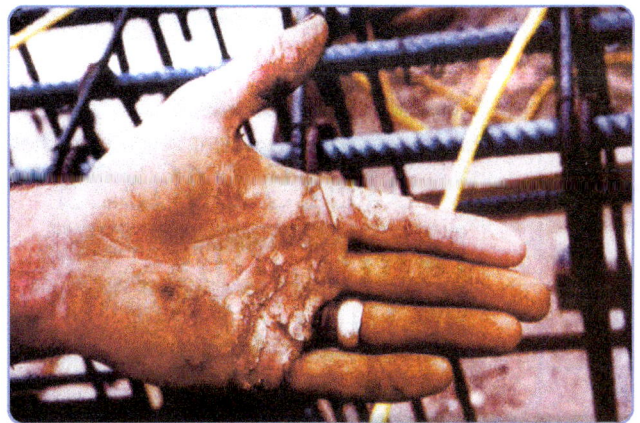

Figure 12-19: Dermatitis.
Source: SHP.

Although the range of substances an individual may become sensitised to varies, some substances have the tendency to cause sensitisation in a large number of people. For example, cement may contain two well-known sensitisers, chromate and cobalt. Epoxy resins also have a tendency to cause sensitisation.

Dermatitis can be prevented by:

- Pre-employment screening for sensitive individuals.

- Careful attention to skin hygiene principles.

- Clean working conditions and properly planned work systems.

- Use of protective equipment.

- Prompt attention to cuts, abrasions and spillages onto the skin.

- Avoid the overuse of hand cleaners containing abrasives such as ground pumice stone.

- Correct selection and use of skin barrier cream, used to repel water, oil or solvents from exposed hands.

Workers at a company premises in Bristol were exposed to hazardous chemicals over a four-year period, leading to the onset of a disease called 'allergic contact dermatitis'. One employee suffered four years of his skin blistering, cracking, splitting and weeping because of this allergic dermatitis.

Two other employees also suffered the symptoms of allergic dermatitis, including fingers and hands becoming so badly swollen and blistered that one could not do up his shirt buttons without his fingers splitting open. All three employees had been working with photographic chemicals.

The company was fined a total of £100,000 and ordered to pay £30,000 costs. It was fined £30,000 for breaching the Health and Safety at Work Act 1974, and £10,000 for 6 separate breaches of the Control of Substances Hazardous to Health (COSHH) Regulations for not making adequate risk assessments, not preventing or controlling exposure of employees to chemicals, and for not providing any 'health surveillance' of employees at risk. It was also fined £10,000 for not reporting a case of allergic contact dermatitis to the HSE.

Injection

Injection is a forceful breach of the skin, perhaps as a result of injury, which can carry harmful substances through the skin barrier. For example, handling broken glass that cuts the skin and transfers a biological or chemical agent. On construction sites there are many items that present a hazard of penetration, such as nails in broken-up timber structures that might be trodden on and penetrate the foot, presenting a risk of infection from tetanus.

In addition, some land or buildings being worked on may have been used by intravenous drug users and their needles may present a risk of injection of a virus such as hepatitis. The forced injection of an agent into the body provides an easy route past the skin, which usually acts as the body's defence mechanism and protects people from the effects of many agents that do not have the ability to penetrate.

When identifying possible routes of entry, it must be remembered that many substances have multiple possibilities. For example, trichloroethylene, which is a known carcinogen, will absorb through the skin. However, when in use, it also gives off very harmful vapours that may be inhaled and, because it is liquid, there is the possibility of accidental ingestion.

THE BODY'S DEFENCE MECHANISMS

The body's response against the invasion of substances likely to cause damage can be divided into external or superficial defences and internal or cellular defences.

Superficial defence mechanisms

Respiratory (inhalation):

NOSE

On inhalation, many substances and minor organisms are successfully trapped by nasal hairs, for example, larger wood and cement dust particles. Gases such as carbon monoxide are inhaled through the nose and easily defeat the defence mechanisms, the most superficial of which is smell. Carbon monoxide is highly poisonous to humans and has no smell. Vehicle and construction plant exhausts are a common source of carbon monoxide.

RESPIRATORY TRACT

The next line of defence against inhalation or substances harmful to health begins here, where a series of reflexes activate the coughing and sneezing mechanisms to forcibly expel the triggering substances. Silica and cement dust generated during construction activities may be trapped in the respiratory tract and ejected by coughing and sneezing.

CILIARY ESCALATOR

The passages of the respiratory system are also lined with mucus and well supplied with fine hair cells which sweep rhythmically towards the outside and pass along large particles.

The respiratory system narrows as it enters the lungs where the ciliary escalator assumes more and more importance as the effective defence. Smaller particles of hazardous substances, such as some silica particles, are dealt with at this stage. The smallest particles, such as organic solvent vapours and asbestos fibres, reach the alveoli and are either deposited or exhaled. Legionella bacteria that are inhaled can also defeat the ciliary escalator and reach the alveoli. Hardwood dusts (oak, mahogany, etc.), which are carcinogenic, if inhaled during cutting or machining operations, are usually trapped by the mucus in the ciliary escalator.

Gastrointestinal (ingestion):

MOUTH

The mouth is used for the ingestion of substances in general. Saliva in the mouth provides a useful defence against hazardous substances that are not excessively acid or alkaline or present in large quantities.

GASTROINTESTINAL TRACT

Acid in the stomach also provides a useful defence similar to saliva. Vomiting and diarrhoea are additional reflex mechanisms which act to remove substances or quantities that the body is not equipped to deal with.

Skin (absorption):

SKIN

The body's largest organ provides a useful barrier against the absorption of many foreign organisms and chemicals (but not against all of them). Its effect is, however, limited by its physical characteristics.

The outer part of the skin is covered in an oily layer and substances have to overcome this before they can damage the skin or enter the body. The outer part of the epidermis is made up of dead skin cells. These are readily sacrificed to substances without harm to the newer cells underneath. Repeated or prolonged exposure could defeat this. When attacked by substances, the skin may blister in order to protect the layers beneath. Openings in the skin such as sweat pores, hair follicles and cuts can allow entry, and the skin itself may be permeable to some chemicals, for example, toluene.

Workers in the construction industry, sewage industry or agriculture (animals) can come into contact with Leptospira bacteria, which can also pass through breaks in the skin. Other blood-borne viruses, such as Hepatitis, may defeat the skin by being injected into the bloodstream. Occupations at risk include those in health-care and construction workers clearing derelict buildings where drug use takes place.

Cellular mechanisms:

The cells of the body possess their own defence systems.

SCAVENGING ACTION

A type of white blood cell called a macrophage attacks invading particles in order to destroy them and remove them from the body. This process is known as phagocytosis.

SECRETION OF DEFENSIVE SUBSTANCES

This is done by some specialised cells.

Adrenaline is a hormone produced by the adrenal glands during high stress or exciting situations. This powerful hormone is part of the human body's acute stress response system, also called the 'fight or flight' response. It works by stimulating the heart rate, contracting blood vessels, and dilating air passages, all of which work to increase blood flow to the muscles and oxygen to the lungs.

Histamine is a chemical that is released when the body is exposed to an allergen. Allergens may include airborne allergens (such as pollen and dust mites), certain foods (such as peanuts and shellfish) or insect venom. Histamine is released in an effort to protect the body from an allergen; however, sometimes an overload of histamine can result in life-threatening symptoms.

PREVENTION OF EXCESSIVE BLOOD LOSS

Reduced circulation through blood clotting and coagulation prevents excessive bleeding and slows or prevents the entry of bacteria.

Heparin is an anticoagulant (blood thinner) that prevents the formation of blood clots and is produced naturally in the lungs and liver.

REPAIR OF DAMAGED TISSUES

This is a necessary defence mechanism that includes removal of dead cells, increased availability of defender cells and replacement of tissue strength, for example, scar tissue caused by silica.

THE LYMPHATIC SYSTEM

Acts as a form of 'drainage system' throughout the body for the removal of foreign substances. Lymphatic glands or nodes at specific points in the system act as selective filters, preventing infection from entering the blood system. In many cases, a localised inflammation occurs in the node at this time.

Other practical measures to complement the body's protection mechanisms

Practical measures include:

- Maintain good personal hygiene.
- Do not apply cosmetics in the workplace.
- Do not allow eating or drinking in the workplace.
- Provide proper containers/storage for food and drink.
- Provide and ensure use of appropriate personal protective equipment.
- Take care when removing contaminated protective clothing.

WHAT NEEDS TO BE TAKEN INTO ACCOUNT WHEN ASSESSING HEALTH RISKS

In order to assess the health risks, it is necessary to consider the following:

- The form of the substance, for example, solid, liquid, dust or gas. The form of the substance directly influences the way that the substance will enter the body. Solid objects, such as concrete or metal, cannot enter the body in their solid state. But if the concrete is cut with a disc cutter, dust is generated, which if not controlled may be inhaled. If the steel contains chromium, when it is machined or welded, particles which become airborne in the form of dust or fume can be inhaled and prove toxic.

- The classification of the hazard, for example, toxic, corrosive or sensitiser. If the substance is toxic, small quantities inhaled, ingested or absorbed through the skin may have a catastrophic effect on the well-being of the individual who is exposed. Sensitisers may cause allergic dermatitis when the worker comes into contact with the substance, or respiratory sensitisers may cause breathing difficulties when inhaled.

- How much of the substance will be present and its concentration. Ammonia occurs naturally in the environment and humans are regularly exposed to low levels of ammonia in air, soil and water. Ammonia exists naturally in the air at levels between 1 and 5 parts in a billion parts of air (ppb). Ingesting liquid ammonia at 35ppm, for example, from the use of a typical household cleaner, may cause burns to the mouth and throat. Exposure to higher concentrations of liquid ammonia in the eyes causes severe chemical eye burns and can lead to blindness.

- The routes of entry into the body, for example, inhalation, ingestion or skin absorption. Some substances cause harm immediately on contact with the skin, for example, acids. Other substances may be present as a gas, mist or fume and be in the air. Humans have to breathe so airborne substances are likely to be inhaled as a route of entry into the body.

- Whether the substance has an acute or chronic effect or both. Acute effects can often warn the person that they have come into contact with a hazardous substance and a reaction occurs almost instantly. This allows the person to take avoidance measures. Lower-level concentrations of substances, however, may not cause significant harm initially, but the damage builds up over time. Substances such as organic solvents can cause skin or eye irritation quickly (acute) and have longer-term exposure consequences (chronic), where damage to the liver and kidneys can occur.

- The extent to which the body's defences will deal with the substance. The body's defence systems try to prevent harm but also help the body to heal itself when it gets injured or sick. However, hazards arising from bacteria, viruses, chemicals, dusts, vapours, extreme temperatures or work processes can break down (weaken) the body's defence systems. In some instances, the body's defence mechanisms can be defeated and illness or death can result.

- The first signs of damage or ill-health. The body has nerve receptors located all over its surface, so often the first sign of damage will be the sensation of pain. Visual signs of damage may be blood from a cut or break in the skin. When a harmful substance has been ingested or inhaled the first signs are often vomiting or diarrhoea or restrictions in breathing.

- The vulnerability of the people involved in the process, for example young persons, pregnant workers, anyone who has existing health problems, such as skin problems or bronchitis. Anyone with a pre-existing medical condition or with underdeveloped defence systems will be at higher risk than workers who are fit and well.

- The effectiveness of existing control measures. Preventing exposure to hazardous substances is the most effective way of controlling health risks. Where exposure cannot be prevented, measures should be taken to eliminate or reduce the risk of harm. The effectiveness with which control measures have been applied will greatly affect the degree of exposure of the worker.

Considering all these issues will help the assessor decide whether the risks to health are tolerable or acceptable or whether further controls are needed.

SOURCES OF INFORMATION

Product labels

Hazardous chemicals have to be labelled so that workers are informed about their effects when they are exposed to them. The label should draw attention to the inherent hazards to those handling or using the chemical and provide information on precautions to prevent harm.

European regulation on classification, labelling and packaging of chemical substances

The European Union (EU) Regulation on Classification, Labelling and Packaging of Chemical Substances and Mixtures (CLP) introduced throughout the EU a system for classifying and labelling chemicals based on the United Nations' Globally Harmonised System (GHS). The CLP regulation is concerned with the hazards of chemical substances and mixtures and how to inform others about them. Its purpose is to protect people and the environment from the effects of those chemicals by requiring suppliers to classify and provide information about the hazards of the chemicals and to package them safely.

It is the responsibility of manufacturers to establish what the hazards of substances and mixtures are before they are placed on the market, and to classify them in line with the identified hazards. When a substance or a mixture is hazardous, it has to be labelled in accordance with CLP so that workers are informed about its effects if it is used. Note that 'mixture' is the same as the term 'preparation', which has been used previously.

All substances available for use in the workplace should be labelled in accordance with CLP, including the use of the appropriate CLP hazard pictogram. Information on the use of symbols to indicate the harm related to various dangerous substance hazard classifications is provided earlier in this element (**see also Section 12.1 'Health hazard classifications'**).

The CLP regulation requires hazardous substances to carry a hazard label in a specified format, which is made up of specific symbols (known as 'pictograms'), warnings and precautions. These pictograms and the wording that supports them are defined in the CLP regulation, which requires chemical suppliers to use them where hazardous properties have been identified. Article 17 of the CLP Regulation requires that hazard labels include the following elements:

- Name, address and telephone number of the supplier(s).
- The nominal quantity of the substance or mixture in the package where this is being made available to the general public, unless this quantity is specified elsewhere on the package.
- Product identifiers.
- Hazard pictograms, where applicable.
- The relevant signal word, where applicable.
- Hazard statements, where applicable.
- Appropriate precautionary statements, where applicable.
- A section for supplemental information, where applicable.

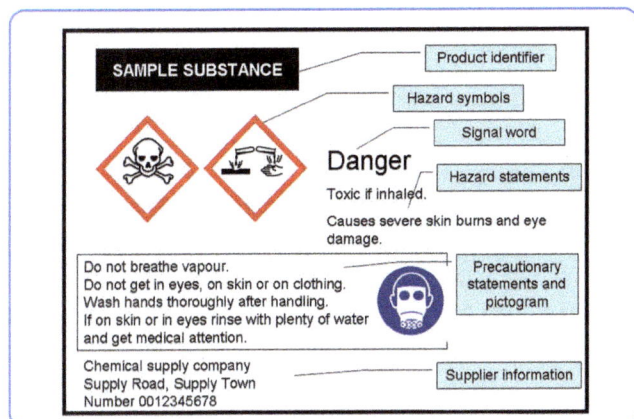

Figure 12-20: Sample GHS chemical labels.
Source: RMS.

A hazard **pictogram** is a graphical composition that includes a symbol plus other graphic elements, such as a border, background pattern or colour that is intended to convey specific information on the hazard concerned. The pictogram forms an integral part of the label and gives an immediate idea of the types of hazards that the substance or mixture may cause.

A **signal word** indicates the relative level of severity of hazardous substances to alert the potential reader of the significance of the hazard. More severe hazards are identified by the signal word 'danger', while less severe hazards are identified by the signal word 'warning'.

A **hazard statement** is a phrase assigned to a hazard class and category that describes the nature and severity of the hazard.

A **precautionary statement** is a phrase and/or pictogram that describe recommended measure(s) to prevent or minimise adverse effects resulting from exposure to a hazardous substance or mixture due to its use. A maximum of six precautionary statements is allowed under CLP requirements. Precautionary statements cover statements for prevention, response to problems or contamination, storage and disposal.

HAZARD STATEMENTS	
H300	FATAL IF SWALLOWED
H320	CAUSES EYE IRRITATION
H336	MAY CAUSE DROWSINESS OR DIZZINESS

PRECAUTIONARY STATEMENTS	
P102	KEEP OUT OF REACH OF CHILDREN
P271	USE ONLY OUTDOORS OR IN WELL VENTILATED AREA
P331	DO **NOT** INDUCE VOMITING

Figure 12-21: Hazard and precautionary statements.
Source: EU, CLP Regulation, Annex III and IV.

Guidance documents

The HSE publishes a range of guidance documents related to hazardous substances, some provide guidance on the hazard presented by substances and information on prevention/control measures, for example guidance for silica, rubber and flour. These are a valuable source of information when assessing risks to health from hazardous substances.

Other guidance documents provide information on acceptable occupational exposure limits for work activities, including concentrations of hazardous substances in workplace air. Occupational exposure limits for hazardous substances represent an important source of information for risk assessment and management of health risks. They provide valuable information on workplace health standards for hazardous substances and carry legal status.

The HSE guidance document EH40 contains lists of occupational exposure limits related to a large number of chemicals, called workplace exposure limits (WELs). Regulation 7 of COSHH 2002 requires employers to prevent or control exposure to hazardous substances, including not exceeding the limits detailed in Guidance Note EH40.

The occupational exposure limits provided in EH40, are updated as new and revised limits are introduced, usually as further knowledge is gained about the health effects of substances. Therefore, it is important to obtain current data on occupational exposure limits (**see Section 12.2 - Occupational exposure limits**).

EH40 provides advice on how to apply and monitor occupational exposure limits, including biological monitoring values for a range of substances. It also contains a description of the limit-setting process, technical definitions and explanatory notes.

Safety data sheets

Safety data sheets are produced by manufacturers and suppliers to provide information on their chemical products to help those that use the chemicals for work to make a risk assessment. They describe the hazards that the chemical presents and give information on its use, handling, storage and measures to be taken in the event of an emergency.

Section 6 of the Health and Safety at Work etc. Act (HASAWA) 1974 requires manufacturers, suppliers and importers to provide information on substances for use at work, this is usually in the form of a SDS.

GHS establishes a requirement to prepare safety data sheets (SDS) for chemicals that constitute a health hazard. This requirement has been adopted globally by many countries, including the EU and the UK. In the EU, the EU regulation known as REACH (Registration, Evaluation, Authorisation and Restriction of Chemicals) is the system for controlling chemicals in EU member states, including the preparation of SDSs.

SDSs established in accordance with the GHS require the following sixteen headings to be included:

1) Identification of the substance/mixture and of the company/undertaking, including the product identifier shown on the label and its recommended use.

2) Hazards identification, including signal words, hazard statements and precautionary statements - the pictogram of hazard symbols may be provided.

3) Composition/information on ingredients.

4) First-aid measures, including the urgency with which it may be needed and symptoms of any effects.

5) Firefighting measures, including suitable media and hazards of combustion.

6) Accidental release measures, including spills containment and recovery.

7) Handling and storage, including any incompatibilities.

8) Exposure controls/personal protection, including exposure limit values and requirements for such things as local exhaust ventilation and specific personal protective equipment (PPE).

9) Physical and chemical properties, such as odour, flash point, vapour pressure.

10) Stability and reactivity, including conditions to avoid, for example, heat or incompatible materials.

11) Toxicological information, including the health hazards, for example, carcinogenicity, route of entry and symptoms.

12) Ecological information, in order to evaluate the environmental impact of dealing with spills and arrangements for waste.

13) Disposal considerations, including methods and properties that may affect landfill or incineration.

14) Transport information, including UN number and transport hazards.

15) Regulatory information, including relevant national information on the regulatory status of the chemical where it is being supplied.

16) Other information, including the date the SDS was prepared and changes from previous versions.

LIMITATIONS OF INFORMATION USED WHEN ASSESSING RISKS TO HEALTH

Although information provided by manufacturers and contained within the HSE guidance document EH40 is very useful when assessing risks to health, it may be very technical and require a specialist to explain its relevance to a specific work activity. Some substances have good toxicological information, usually gained from past experience of harm. However, many others have a limited amount of useful toxicological information available to provide guidance on the likely harm it could cause. This can lead to a reliance on information that is only a 'best understanding at the time' and this may have to be revised as our knowledge of the health effects of the substance changes. This is reflected in the fact that the HSE guidance in EH40 is revised as knowledge on the substances changes and workplace exposure limits are updated. It is therefore important to ensure the information being used when assessing risks to health is up to date.

Individual susceptibility of workers differs by, for example, age, gender or ethnic origin, and this can limit the value of information sourced from manufacturers and guidance documents related to occupational exposure limits.

Exposure history varies over the working life of an individual and current exposure may not indicate that the individual may suffer due to a cumulative effect from earlier exposures. For example, an individual may have been or be engaged in a number of processes within a variety of workplaces or personal pastimes. These limitations are reflected in the use of occupational exposure limits, where the limits are considered to be a maximum and control measures may be applied to control assessed risks to well below this limit.

ROLE AND LIMITATIONS OF HAZARDOUS SUBSTANCE MONITORING

Role of hazardous substance monitoring

The role of hazardous substance monitoring is to determine the level of likely exposure of workers to substances in order to establish the likely effects on the worker. Hazardous substance monitoring can help to identify and assess health risks in the workplace. It can be used to determine compliance with national or other worker exposure limits and determine what controls are required to remain within the limits. Monitoring is required to ensure that exposure has not exceeded a control limit.

Monitoring also helps to identify deterioration in control measures that may result in harm if exposure is not adequately controlled. Monitoring can establish if current controls are adequate to limit exposure and assist in choosing appropriate PPE. Monitoring can provide information on patterns of exposure and levels of risk that managers can use and provide to workers. It can also indicate the need for health surveillance of groups and individual workers.

Limitations of hazardous substance monitoring

As described previously, the health effects of exposure to hazardous substances can be acute or chronic. It is therefore necessary to use appropriate methods of measurement to distinguish these effects. It is also important to understand the limitation of any hazardous substance monitoring method used, for example, the risk of cross-contamination of similar substances and the fact that general workplace monitoring may not represent specific worker exposure. Hazardous substance monitoring of the presence of a chemical in the air may not represent the worker's complete exposure as, for example, there may be additional exposure to the chemical by skin contact or orally through poor hygiene arrangements.

One of the main limitations of hazardous substance monitoring is the competence of the person conducting it. It relies on that person conducting monitoring at the time that represents real exposure of workers and using methods that will give reliable measurements. When carrying out monitoring related to dusts, it is important to discern the amount of dust that can penetrate the airway and cause harm. If the person conducting the monitoring does not understand the difference between 'total inhalable dust' and 'respirable dust' there is a risk that unsuitable measurements may be taken.

'Total inhalable dust' is the amount of airborne material that enters the nose and mouth during breathing and is available for deposition in the body. 'Respirable dust' is the amount of airborne material that penetrates to the gas exchange region of the lung.

General approach to hazardous substance monitoring

When embarking upon a monitoring campaign to assess the hazardous risk to the workforce, several factors need to be considered, the substance(s), who might be affected, the sample period and the method of sampling.

The substance

Considering the substance involves a review of the materials, processes and operating procedures being used within a process, coupled with discussions with management, workers and health and safety practitioners. A brief 'walk-through' survey can also be useful as a guide to the extent of monitoring that may be necessary. Health and safety data sheets are also of use. When the background work has been completed it can then be decided what is to be measured.

Who might be affected

Who might be affected depends on the size and diversity of the group that the hazardous substance monitoring relates to. From the group of workers being considered, the sample of workers to be monitored should be selected, this must be representative of the group and the work undertaken. Selecting the workers with the highest exposure can be a reasonable starting point. If the group is large then random sampling may have to be used, but care has to be exercised to ensure a representative sample is taken. The group should also be informed of the reason for the monitoring to ensure their full co-operation.

Sample period

Where monitoring is not continuous, the sample period selected should reflect the workers' exposure patterns. Factors to consider include the total exposure period and whether exposure is at the same level continuously. If so, a sample taken at any convenient time would be suitable. However, if exposure is occasional, the sample period must reflect this occasional exposure.

The duration of the sample period will also be influenced by whether the hazard presented by the substance is acute or chronic. Substances that present an acute hazard are usually sampled over a shorter sample period that reflects the short-term worker exposure limits set by international or national authorities, whereas substances that represent a chronic hazard may be sampled over a longer period to better reflect the long-term worker exposure limits.

The sample period may also be influenced by the type of equipment available, including whether it provides a continuous, long-term or a short-term measurement.

Measurements to determine	Suitable methods of measurement
Chronic hazard	Continuous personal dose measurement.
	Continuous measurements of average background levels.
	Short-term measurement of contaminant levels at selected positions and times.
Acute hazard	Continuous personal monitoring with rapid response.
	Continuous background monitoring with rapid response.
	Short-term measurement of background contaminant levels at selected positions and times.
Environmental control status	Continuous background monitoring.
	Short-term measurement of background contaminant levels at selected positions and times.
Whether area is safe to enter	Direct-reading instruments that provide an instantaneous measurement.

Figure 12-22: Sampling strategy.
Source: RMS

Method of monitoring

The particular measurement method will be based on the hazard presented by the substance, any limits in detection ability and what resources are available to measure the presence of the substance accurately. A typical approach is outlined in *Figure 12-22*.

Air samples should be collected in the worker's breathing zone. Personal dose measurement devices may conveniently be used for this purpose. Sampling should always be carried out during working hours when the activity involving hazardous substances is in operation. It may be necessary to repeat sampling at different times of the year to take account of seasonal effects on hazardous substances in the workplace.

In order to obtain indications of the distribution of contamination throughout the general atmosphere of the working area, samples should also be taken as close as possible to the sources of the hazardous substances, to indicate the standard of process control at source, and at various places in the working area, to indicate the overall effectiveness of control measures.

The minimum volume of the air sample to be taken for each analysis should be determined according to the sensitivity of the analytical method. The accuracy of the measuring method must be considered and noted with the results of the monitoring, as it can significantly affect results obtained from some monitoring equipment.

The ILO "Occupational Exposure to Airborne Substances Harmful to Health" suggests the time it should take to sample substances in the working environment, *(see Figure 12-23 and refer to Section 7.2 – 'Occupational exposure limits').*

"The sampling of substances for which a ceiling limit exists, or the measurement of the concentration of such substances in the working environment, should normally last 15 minutes."

Figure 12-23: Sampling hazardous substances.
Source: ILO, Occupational exposure to airborne substances harmful to health.

BASIC MONITORING EQUIPMENT

Hazardous substance monitoring is often conducted by a specialist occupational hygienist, therefore detailed knowledge of monitoring equipment may not be required. However information on basic monitoring equipment is provided to give an understanding of how a hazardous monitoring strategy might be applied and outline the role and limitations of basic monitoring equipment.

Short-term samplers

Stain tube detectors (multi-gas/vapour)

A stain tube detector is a simple device for the measurement of contamination on a grab (short-term) sampling basis. It incorporates a glass detector tube filled with inert material. The material is impregnated with a chemical reagent that changes colour ('stains') in proportion to the quantity of contaminant as a known quantity of air is drawn through the tube.

There are several different manufacturers of detector tubes, including Dräger and Gastec. It is important that the literature provided with the pumps and tubes is followed. These provide a quick and easy way to detect the presence of a particular airborne contaminant. However, they possess inherent inaccuracies, and tube manufacturers claim a relative standard deviation of 20% or less (i.e. 1ppm in 5ppm). Types of tube construction include:

- The most common is the simple stain length tube, but it may contain filter layers, drying layers, or oxidation layers.
- Double tube or tube containing separate ampoules, avoids incompatibility or reaction during storage.
- Comparison tube.
- Narrow tube to achieve better resolution at low concentrations.

This list illustrates the main types of tube; however, there are many variations and the manufacturer's operating instructions must be read and fully understood before tubes are used.

There are four types of pump in general use:

1) Bellows pump.

2) Piston pump.

3) Ball pump.

4) Battery-operated pump.

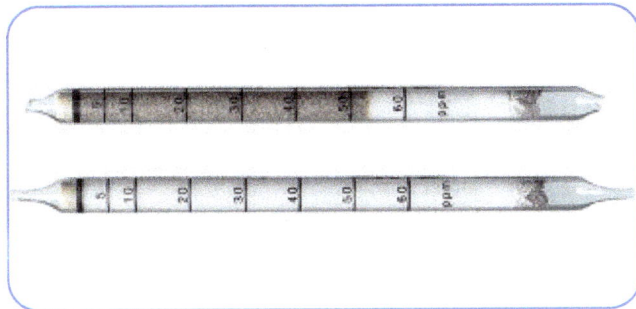

Figure 12-24: Dräger short-term detection tubes (before and after).
Source: Dräger.

Pumps and tubes made by different manufacturers may not be compatible and they should not be mixed when taking measurements.

How to use tubes:

- Choose tube to measure material of interest and expected range.

- Check tubes are in date.

- Check leak tightness of pump.

- Read instructions to ensure there are no limitations due to temperature, pressure, humidity or interfering substances.

- Break off tips of tube, prepare tube if necessary and insert correctly into pump.

- Arrows normally indicate the direction of air flow.

- Draw the requisite number of strokes to cause the given quantity of air to pass through the tube.

- Immediately, unless operating instructions say otherwise, evaluate the amount of contaminant by examining the stain and comparing it against the graduations on the tube.

- Remove tube and discard according to instructions.

- Purge pump to remove any contaminants.

Figure 12-25: Gas detector pump.
Source: Dräger.

Advantages

The advantages of short-term samplers are:

- Quick and easy to use.
- Instant reading without further analysis.
- Does not require much expertise to use.
- Relatively inexpensive.

Disadvantages

The disadvantages of short-term samples are:

- Tubes only measure a specific contaminant.
- Accuracy varies - some are only useful as an indication of the presence of contaminants.
- Is only a grab sample (taken at a single location point and may not represent the workplace as a whole).
- Relies on operator to accurately count pump strokes (manual versions).
- Only suitable for gases and vapours (not dusts).
- The tubes must be used within a short time period to prevent decay of the reagent and loss of reliability.
- The tubes are fragile and break easily.
- Used/expired tubes must be disposed of as chemical waste in accordance with local regulations.

Direct-reading dust sampler

A simple method of measuring dust is by direct observation of the effect of the dust on a strong beam of light, for example, using a dust lamp (Tyndall effect). High levels of small particles of dust show up under this strong beam of light. Other ways are by means of a direct-reading instrument. This establishes the level of dust by, for example, scattering of light. Some also collect the dust sample.

Advantages

The advantages of direct dust samplers are:

- Instant reading.
- Continuous monitoring.
- Can record electronically.
- Can be linked to an alarm.
- Suitable for clean-room environments.

Disadvantages

The disadvantages of direct dust samplers are:

- Some direct-reading instruments can be expensive.
- Do not usually differentiate between dusts of different types.
- Most effective on dusts of a spherical nature.

Long-term samplers

Personal long-term samplers

Passive personal samplers

Passive samplers are so described because they have no mechanism to draw in a sample of the contaminant but instead rely on passive means to sample. As such, they take a time to perform this function, for example, acting as an absorber taking in contaminant vapours over the period of a working day. Some passive samplers, like gas badges, are generally fitted to the lapel and change colour to indicate contamination.

Active personal samplers

Filtration devices are used for dusts, mists and fumes. A known volume of air is pumped through a sampling head and the contaminant filtered out. By comparing the quantity of air with the amount of contaminant a measurement is made. The filter is either weighed or an actual count of particles is done to establish the amount, as with asbestos. The type of dust can be determined by further laboratory analysis. Active samplers are used in two forms, for personal sampling and for static sampling.

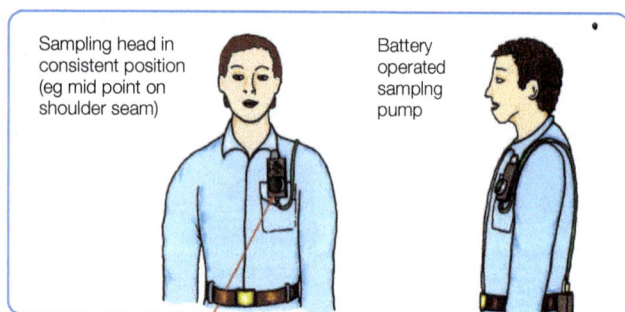

Figure 12-26: Personal sampling equipment.
Source: ROSPA OS&H.

Static long-term samplers

Static long-term sampling devices are placed in the working area. They sample continuously over the length of a shift, or a longer period if necessary. Mains or battery-operated pumps are used. Very small quantities of contaminant may be detected. The techniques employed include absorption, bubblers and filtration; they are similar in principle to personal samplers, but the equipment is designed for static use.

Advantages of static long-term samplers are:

- Will monitor the workplace over a long period of time.
- Will accurately identify 8 hour time weighted average.

Disadvantages of static long-term samplers are:

- Will not generally identify a specific type of contaminant.
- Will not identify multiple exposure i.e. more than one contaminant.
- Does not identify personal exposure.
- Unless very sophisticated, will not read variations in level of contamination during the measurement period.

Smoke tubes

Smoke tubes are simple devices that generate a 'smoke' by means of a chemical reaction. A tube similar in type to those used in stain tube detectors is selected, its ends broken (which starts the chemical reaction) and it is inserted into a small hand bellows. By gently pumping the bellows smoke is emitted. Air flow can be studied by watching the smoke. This can be used to survey extraction and ventilation arrangements to determine their extent of influence.

> **REVIEW**
>
> Explain the term 'dermatitis' and how it may be prevented in the workplace.
>
> What is the function of the ciliary escalator in the respiratory tract?
>
> List eight headings that should be included on a safety data sheet supplied with a hazardous substance.

PURPOSE OF OCCUPATIONAL EXPOSURE LIMITS AND HOW THEY ARE USED

Purpose of occupational exposure limits

The purpose of occupational exposure limits is to assist in the assessment of risk and control of exposure of workers to a variety of substances that can have harmful effects on health. Failure to control exposure to such substances can lead to many forms of ill-health. Therefore, it is important to know in advance what risks may arise from chemicals, what are acceptable occupational exposure limits and how to protect people at work.

Occupational exposure limits are set by the HSE in their document EH40 in the form of workplace exposure limits (WELS). WELs are legally binding occupational exposure limits designed to protect the health of workers by establishing a maximum concentration of an airborne contaminant that a worker may be exposed to, measured over a reference time period. Regulation 7 of COSHH 2002 requires employers to prevent or control exposure to hazardous substances, including not exceeding the limits detailed in EH40.

Occupational exposure limits are maximum concentrations of airborne substances under specified circumstances and are measured in either parts per million or per cubic metre of air. Airborne substances can be solid (dust), liquid (mist/aerosol), gas, vapours or fumes. Solids and liquids can be measured by weight (milligrams – mg), for example, an occupational exposure limit for cement dust may be expressed as 10 mg/m3. Gas, vapours and fumes are weightless.

Therefore, the occupational exposure limit is expressed as a concentration in the atmosphere – for example, an occupational exposure limit for trichloroethylene may be expressed as 100 parts per million (ppm).

The setting and publishing of occupational exposure limits provides employers with information on the substances and on the levels of exposure that they must not exceed. This means that all employers are expected to work to nationally agreed maximum workplace exposure limits.

Occupational exposure limits help to guide action and establish effective control for substances by requiring organisations to put prevention and control measures in place to ensure occupational exposure limits are not exceeded. For example, by limiting the concentration of the chemical in the atmosphere or inhaled by the worker and reducing the time workers are exposed to the chemical.

They do not address safety issues, such as flammable concentrations, and it should be noted that the occupational exposure limits do not generally include exposure to biological substances, as these are usually specified separately.

LONG-TERM AND SHORT-TERM LIMITS

The health effects of exposure to chemicals vary considerably depending on the nature of the chemical and the pattern of exposure. Some effects require prolonged or accumulated exposure, whereas others may arise from very short or a single exposure to a particular chemical. Therefore long-term limits and/or short-term occupational exposure limits may be established for a substance to reflect the different health effects a chemical may have.

Long-term limits

Long-term exposure limits (LTEL) are concerned with the total exposure of the worker to a concentration of the chemical, averaged over a reference period that represents a work period, usually 8 hours. They are therefore a time-weighted average (TWA) worker exposure.

This is appropriate for protecting against the effects of long-term exposure to a chemical that may have chronic health effects. If workers are exposed for longer periods than 8 hours, because they work longer work periods, the long-term exposure limits have to be adjusted to reflect this.

Examples of substances that have long-term occupational exposure limits established by the HSE in their document EH40 are shown in *Figure 12-27*.

Short-term exposure limits

Short-term exposure limits (STEL) are primarily aimed at avoiding the acute health effects of exposure to chemicals, for example eye irritation, or at least reducing the risk of their occurrence. Short-term exposure limits the total exposure of the worker to a concentration of the chemical averaged over a short reference period, usually 15 minutes. For those chemicals that do not have a short-term exposure limit, it is usually recommended that a figure of three times the long-term limit be used as a guideline for controlling short-term peaks in exposure.

Examples of substances that have short-term occupational exposure limits established by the HSE in their document EH40 are shown in *Figure 12-28* on the following page. Note that some of these also have long-term exposure limits.

It can be seen, for example, that trichloroethylene has both a long- and short-term exposure limit to accommodate its acute effect (narcosis) and its chronic effect (possible cancer). It is also expressed as parts per million (ppm) to protect from its harmful vapours and, because processes that use trichloroethylene can create mists, it is also regulated by milligrams per cubic metre.

Substance	8 hour time-weighted average occupational exposure limit
Portland cement	Inhalable dust 10 mg/m³ - Respirable dust 4 mg/m³
Silica, respirable crystalline	0.1 mg/m³ - Respirable fraction
Hardwood dust	3 mg/m³ - Inhalable fraction

Figure 12-27: Examples of substances with long-term occupational exposure limits.
Source: HSE, EH40.

| Substance | Time-weighted average occupational exposure limit | |
	Short-term (15 minutes)	Long-term (8 hours)
Chlorine	0.5ppm or 1.5 mg/m³	None
Phenol	4ppm or 16 mg/m³	2ppm or 7.8 mg/m³
Ethyl acetate	400ppm or 1468 mg/m³	200ppm or 734 mg/m³
Trichloroethylene	150ppm or 820 mg/m³	100ppm or 550 mg/m³

Figure 12-28: Examples of substances with short-term occupational exposure limits.
Source: HSE, EH40.

Ceiling exposure limits

Some workplace activities give rise to frequent short (less than 15 minutes) periods of high exposure that, if averaged over time, do not exceed either an 8-hour LTEL or a 15-minute STEL. These exposures have the potential to cause harm from special risks, such as carcinogenicity, skin absorption or allergic reaction, and may have a 'ceiling' exposure limit set for them. The ceiling exposure limit is a maximum allowable or acceptable concentration (MAC) that should not be exceeded at any time and ensures protection against acute effects, particularly those that may develop from a single exposure.

WHY TIME-WEIGHTED AVERAGES ARE USED

When considering the health effects of a hazardous substance, it is important to know what amount a worker has been exposed to over a period of time. This is known as the dose. The dose will influence the amount of the substance taken into the worker's body and any effects that may result from exposure.

In workplace situations, the concentration of a substance in the air a worker breathes may vary significantly over the reference period, depending on the activity being carried out. For example, the process of mixing powder in a vessel may give a high concentration of the substance in the air the worker breathes while the powder is added, but while the mixing is taking place, the concentration in the air may reduce as water is added to the powder.

If a measurement was taken while filling the vessel with powder this would give the highest concentration, while if a measurement was taken during mixing this would give the lowest concentration. Neither measurement would represent the dose received by the worker over time as one would be too high and the other too low.

For this reason, the high and low concentration measurements taken are averaged over a reference time period.

This is done by determining the amount of time that each of the concentrations is valid and averaging the combined concentrations/times over the reference period. This is the average of the concentrations of substance in the air the worker breathes over the reference period and is called a time weighted average.

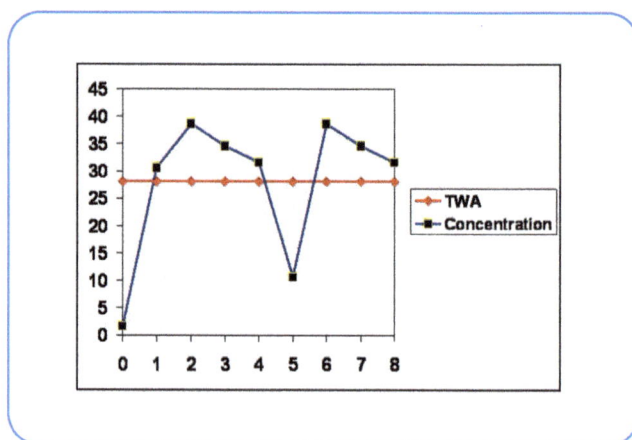

Figure 12-29: Time-weighted average compared with actual concentration.
Source: RMS.

Occupational exposure limits are set in such a way as to control the dose that a worker's body may receive by establishing a time-weighted average concentration of a substance that a worker may be exposed to over a reference period, for example, 8 hours or 15 minutes.

Because occupational exposure limits are set as time-weighted averages, a worker exposed to the occupational exposure limit may be exposed to concentrations of a hazardous substance that are higher than the specified time-weighted average for some of the time.

This is acceptable provided the average concentration is not exceeded. As some work activities conducted in the period may create a concentration that is particularly low, this will offset those when it is high. This effect is particularly important to consider where workers have a variety of tasks and take breaks away from the workplace where exposure is usually negligible.

LIMITATIONS OF EXPOSURE LIMITS

There are many reasons why control of exposure should not be based solely on occupational exposure limits:

- **Compound routes of entry.** Many substances, for example trichloroethylene, have the ability to absorb through the skin. Occupational exposure limits only account for routes of entry by inhalation.

- **Individual susceptibility.** The majority of the work to establish occupational exposure limits has been based on the average male physiology from the countries in which studies were conducted. The occupational exposure limits relate to average males and it is possible for individuals to have a higher than average susceptibility to some substances. In addition, there may be a difference in susceptibility of females to some substances. Some work has been done where specific health-related effects have been noted amongst females, for example, exposure to lead compounds.

- **Origin of exposure limit data.** Work done to date has mainly been based upon exposure to individuals in the developed countries, for example, Europe and the USA.

- **Variations in control.** Local exhaust ventilation systems may not always work consistently because of lack of maintenance, overwhelming levels of contamination, etc. Changes in the environmental conditions, for example, increased humidity or air pressure, may increase the harmful potential of the substance. This may lead to occupational exposure limits being exceeded inadvertently.

- **Errors in monitoring.** Measuring microscopic amounts of contamination requires very accurate and sensitive equipment. Devices have a range of accuracies, and lack of maintenance and misuse can lead to inaccuracies in monitoring due to equipment becoming contaminated.

- **Synergistic effects.** The occupational exposure limits that are available relate to single substances; the effects of multiple substances in the workplace need to be considered.

COMPARISON OF MEASUREMENTS TO RECOGNISED STANDARDS

Comparison of workplace measurements with occupational limits

When identifying hazards and establishing risk controls, it is important to consider relevant occupational exposure limits that set a recognised standard, including workplace exposure limits (WELs) set in the HSE document EH40.

When workplace exposure measurements are made, they should be compared with the limits set by these recognised standards. This will help to establish the presence of a workplace hazard and the level of risk, and will indicate the need to establish prevention and control measures.

The comparison of measurements to occupational exposure limits will also show how effective the new or additional prevention and control measures will need to be if the occupational exposure limits are to be met.

Where occupational exposure limits are exceeded to a significant extent, this indicates the importance of taking immediate measures to reduce the level of occupational exposure to the substance. As measures are put in place, new workplace exposure measurements can be made and compared with the established occupational exposure limits. This will determine the level of progress towards meeting the occupational exposure limits and might indicate success or the need for further work.

Workplace exposure measurements should be made at intervals of time and after changes have been made in the workplace. The results of the measurements should be compared with recognised occupational exposure limits to enable health risks to be monitored and action taken if the performance of control measures declines.

Ensuring the occupational exposure limit is not exceeded

The HSE document EH40 sets workplace exposure limits (WELs) in order to help protect the health of workers. It is a requirement of the Control of Substances Hazardous to Health Regulations (COSHH) 2002 (amended) that WELs must not be exceeded.

WELs are concentrations of hazardous substances in the air, averaged over reference periods and are TWA concentrations. The two reference periods that are used are 8 hours and 15 minutes. The 8-hour WEL is known as a LTEL (long-term exposure limit) and is used to help protect against chronic ill-health effects. 15-minute STELs (short-term exposure limits) are set to protect against acute ill-health effects such as eye irritation, which may happen within minutes of exposure, or where there is the risk of high-level peak exposures over short time intervals during the working day.

Control measures established by employers would need to take into account these two types of limits and that they are time weighted averages. It should be remembered that the concentration of a substance measured in the atmosphere could be higher than the value of the WEL, but only for a short time. When considered over the reference period (8 hours or 15 minutes) the time weighted concentration measured could still be below the value of the WEL.

In order to ensure occupational exposure limits are not exceeded it may be necessary to monitor the concentration of substance in the atmosphere over a period of time, for example using a personal sampler, to reflect the actual exposure of workers on a short-term or long-term basis.

Although occupational exposure limits may be set for substances, employers retain a responsibility to reduce exposure levels to as low as is reasonably practicable. The limits should be considered to be a maximum allowable concentration and it is undesirable to see them as a normal working level if measures can easily be put in place to reduce them significantly below the workplace exposure limit. Work practices and control strategies should be constantly reviewed to ensure the lowest levels of exposure are achieved.

If the levels of exposure are to be maintained with confidence below the workplace exposure limit, it will be necessary to work far enough below them to account for changes in work situation. This is particularly important with those workplace exposure limits that are set for substances that are carcinogens or sensitisers. By working 'at the limit' employers do not allow for sensitive people who may be affected by relatively low exposures. Nor do they account for variations or inaccuracies in monitoring, sudden surges of contaminant or partial failures of control measures.

REVIEW

Explain the significance of 'workplace exposure limits.

Give four reasons why control of exposure should not be based solely on 'workplace exposure' limits.

12.3 Control measures

THE NEED TO PREVENT EXPOSURE OR ADEQUATELY CONTROL IT

Prevention of exposure

Legislation, and good practice establish a clear need to prevent exposure where it is reasonably practicable. If exposure is prevented then no harm to workers will be done. This is the most effective control measure and, although it may involve substantial investment, this is usually the best long-term solution to health risks presented by substances.

Regulation 7(1) of the Control of Substances Hazardous to Health Regulations 2002 (amended) states that:

"Every employer shall ensure that the exposure of his employees to substances hazardous to health is either prevented or, where this is not reasonably practicable, adequately controlled."

Figure 12-30: Regulation 7(1).
Source: UK, Control of Substances Hazardous to Health Regulations (COSHH) 2002 (amended).

Prevention of exposure to hazardous chemicals might be achieved by:

- Changing the method of work so that the operation giving rise to the exposure is no longer present.
- Modifying a process to eliminate the production of a hazardous by-product or waste product.
- Substituting, wherever reasonably practicable, a non-hazardous substance that presents no risk to health where a hazardous substance is currently used.

Adequate control of exposure

In many workplaces, it will not be possible or reasonably practicable to prevent exposure to substances hazardous to health completely. If exposure cannot be prevented Regulation 7(1) of COSHH 2002 (amended) requires employers to adequately control exposure to hazardous substances.

To achieve this employers must consider and apply, where appropriate for the circumstances of the work, the following requirements established by COSHH 2002 (amended):

- The measures set out in Regulation 7(3) in the priority order given.

- The specific measures in Regulation 7(4).

- The principles of good practice for the control of exposure to substances hazardous to health set out in Schedule 2A.

- Any approved WEL for the substance, including the prevention of it being exceeded.

- The need to reduce exposure to as low as is reasonably practicable for substances listed, including carcinogens and sensitisers.

In order to meet the requirements of Regulation 7(3), the employer must apply protection measures appropriate to the activity and consistent with the priority order specified in the Regulation, which involve:

1) Provision of a high level of inherent health and safety by careful design, selection and use of appropriate work processes, systems and engineering controls,

and use of suitable work equipment and materials - where it is necessary to use a hazardous substance, an employer should consider whether it is possible to reduce exposure by using an alternative less hazardous substance, a different form of the same substance, for example, by using a substance in pellet rather than powder form, or a different process. Systems and processes may be used that reduce to the minimum required the amount of hazardous substance used or produced, or equipment may totally enclose the process.

2) Controlling exposure at source - for example, by including adequate ventilation systems and appropriate organisational measures such as reducing to a minimum the number of workers exposed and the level and duration of their exposure.

3) Using personal protective equipment - in addition to the previous measures where those measures alone cannot achieve adequate control.

To meet the requirements of Regulation 7(4) the employer's protection measures must include:

- Arrangements for the safe handling, storage and transport of substances hazardous to health, and of waste containing such substances, at the workplace.

- The adoption of suitable maintenance procedures.

- Reducing, to the minimum required for the work concerned:

  - The number of employees subject to exposure.
  - The level and duration of exposure.
  - The quantity of substances hazardous to health present at the workplace.

- The control of the working environment, including appropriate general ventilation.

- Appropriate hygiene measures including adequate washing facilities.

The selection of measures to control exposure should be determined by the level of the ill-health risk resulting from exposure to the hazardous substance and the scope for reducing the risk to a minimum. It is important to ensure that control measures are put in place to protect maintenance workers as well as those directly affected by the substance.

Where control measures are being established the principles of good practice for the control of exposure to substances hazardous to health should also be followed.

PRINCIPLES OF GOOD PRACTICE

Regulation 7 (7) of the Control of Substances Hazardous to Health Regulations (COSHH) 2002 (amended) requires employers to apply the principles of good practice detailed in schedule 2A of the COSHH 2001 in order to control exposure to hazardous substances in situations where exposure cannot be prevented.

"(7) Without prejudice to the generality of paragraph (1), where there is exposure to a substance hazardous to health, control of that exposure shall only be treated as adequate if:

(a) The principles of good practice for the control of exposure to substances hazardous to health set out in Schedule 2A are applied."

Figure 12-31: Duty to apply principles of good practice.
Source: COSHH 2002 (amended), Regulation 7 (7).

The eight principles of good practice set out in schedule 2A are:

1) Design and operate processes and activities to minimise emission, release and spread of substances hazardous to health.

2) Take into account all relevant routes of exposure - inhalation, skin absorption and ingestion - when developing control measures.

3) Control exposure by measures that are proportionate to the health risk.

4) Choose the most effective and reliable control options, which minimise the escape and spread of substances hazardous to health.

5) Where adequate control of exposure cannot be achieved by other means, provide, in combination with other control measures, suitable personal protective equipment.

6) Check and review regularly all elements of control measures for their continuing effectiveness.

7) Inform and train all employees on the hazards and risks from the substances with which they work and the use of control measures developed to minimise the risks.

8) Ensure that the introduction of control measures does not increase the overall risk to health and safety.

COMMON MEASURES USED TO IMPLEMENT THE PRINCIPLES OF GOOD PRACTICE

Elimination or substitution of hazardous substances

Elimination of hazardous substances represents an extreme form of control and is appropriately used where high risk is present from such things as carcinogens (cancer forming substances). It is usually achieved through the prohibition of use of these substances.

Care must be taken to ensure that all stock is safely disposed of and that controls are in place to prevent the substance being re-introduced to the workplace, even as a sample or for research. In a similar way, an employer may decide to cease activities that use hazardous substances that present a high risk. If this level of control is not achievable then another must be selected.

The substitution of a highly toxic substance with a substance that is less toxic is a very useful control measure, for example, benzene might be replaced with toluene, asbestos might be replaced with glass fibre, or solvent-based adhesives might be replaced with water-based adhesives.

Process changes

An example of process a change is changing from a paint-spraying process, where large volumes of airborne vapours are released, to a process that uses a paint brush. Similarly, the emission of vapours from substances used in a chemical process may be minimised by changing the process to limit the temperature of the substance so that fewer vapours are given off.

It is sometimes possible to change the composition or form of the substance used in a process, for example if a substance, such as cement, requires mixing of its constituent parts it may be ordered pre-mixed, thus avoiding the potential exposure to the substances when particles of the substance become airborne during the mixing process. Similarly, a substance may be used in pellet form, rather than powder form, in order to limit the possible emission of dust into the atmosphere.

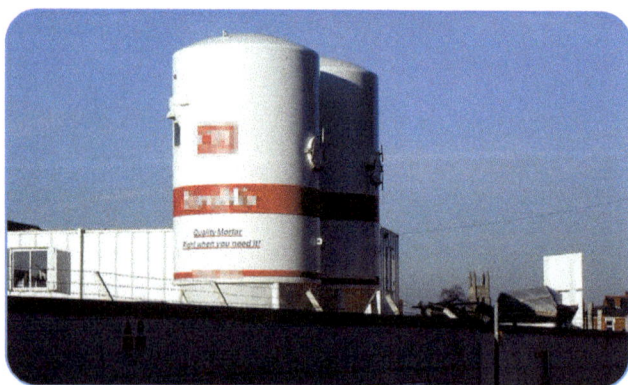

Figure 12-32: Reduced exposure – bulk supply.
Source: RMS.

Processes that usually create airborne dust may also be controlled by changing the process so that water is used to keep the material being worked on wet. For example, material being removed from a building in a demolition process may be watered down so that any small parti-cles become a liquid slurry.

Reduced time exposure

The level of harm to a worker's health is dependent on the concentration of a substance that they are exposed to over time. In situations where the concentration of the substance cannot be reduced easily, reduction in the time workers are exposed to the substance will reduce the harm to their health. Reduction of the exposure time for an individual worker can be achieved by introducing worker rotation during the work period. For example, rotation of four workers over an 8-hour work period so that each worker spends an equal portion of time doing the same task that exposes them to a substance would reduce individual worker exposure by 75%.

Similarly, those workers only needed during a specific part of a process that exposes workers to hazardous substances could move out of the hazardous area to a safe distance when they are not needed and only return when needed.

Enclosure of hazards

Plant and equipment should be designed and installed to contain hazardous substances used at work and prevent their release into the workplace, including vapour and dust from the substances.

Containment of a hazardous substance is best achieved by totally enclosing processes involving the use of hazardous substances. Total enclosure of processes can be more easily achieved where plant and equipment are automated or operated remotely. The opportunity to do this should be considered during the design of plant and equipment and the process to be used.

The strategy of total enclosure of hazardous substances is based on the containment of substance to prevent its free movement in the working environment. It may take a number of forms, for example, enclosure of activities, laboratory 'glove box' for handling toxic substances, pipelines and enclosed conveyors. Total enclosure of the process is used in the removal of asbestos, where the work being done is enclosed in plastic sheeting in order to segregate the work from the surrounding areas. In a similar way, enclosure may be used for building cleaning processes that use impact abrasive spraying technique's, or when spray protection (insulation or paint) is applied to a structure or item during manufacture or maintenance.

Where particularly hazardous substances are used in enclosed plant and equipment, in order to reduce leaks of the substance the plant and equipment should be designed to ensure a slight negative pressure within it, where the process allows. This will ensure that air from the workplace is drawn into any gaps in the enclosure and prevent the release of the substance into the workplace. This will contain the hazardous substance within the plant or equipment, preventing worker exposure to the risk of harm. It will also help to ensure that cleaning of the workplace to remove hazardous substance contamina-tion can be kept to a minimum.

Figure 12-33: Plastic enclosure for asbestos removal.
Source: Merryhill Envirotec

Bulk storage of hazardous substances, with fixed pipework transfer, should be used in preference to small container storage, where appropriate.

In its simplest sense, enclosure of hazards can mean putting lids on substances that have volatile vapours, such as tins of solvent-based products. This action encloses the hazard so that vapours are not released into the atmosphere.

Where total enclosure of a process involving hazardous chemicals is not reasonably practicable, it may be possible to partially enclose the process and supplement this with local exhaust ventilation equipment. This might be the case where it is necessary to place batches of components into a tank of a substance meaning that the hazardous substance cannot be totally enclosed. The tank might be fitted with partial enclosure and local exhaust ventilation over the top of the tank or around the edge of the tank. Similarly, an area in a laboratory used for mixing chemicals might have partial enclosure to three sides and the top, open front and local exhaust ventilation at the back extracting chemical contaminants.

Segregation of process and people

Segregation of process and people is a useful method of controlling the risks from dangerous substances and can take the form of physical separation or segregation by worker characteristics.

Physical segregation

In situations where the hazardous substances cannot be totally enclosed close to their source it may be preferable that the workforce be segregated from the process by providing physical separation in the form of a refuge, for example, a control room of a chemical process. Physical segregation can also be a relatively simple method, such as distancing (segregating) the process from the general workforce.

To further reduce risks from hazardous substances, plant, equipment and storage should be separated from other processes, from incompatible substances and other areas outside the control of the employer.

Segregation by worker characteristics

The protection of the health of young workers in certain work activities is important. A good example is working with lead, where young workers are usually excluded from working in lead processes. There also remains the possibility of gender-linked vulnerability to certain toxic substances, such as lead, and segregation of those particularly at risk from work with these substances affords a high level of control in these circumstances.

Local exhaust ventilation

Where other control measures are not suitable local exhaust ventilation (LEV) equipment should be provided to ensure that workplace exposure limits are controlled so that they do not exceed good practice levels or occupational exposure limits specified in EH40.

Hazards that can be controlled by the effective use of LEV include: dust that could cause coughing, sneezing and various other respiratory diseases; substances that might cause sensitisation or other toxic effects; allergens that could aggravate asthmatic conditions; microorganisms that can cause diseases; asphyxiants that can lead to breathing difficulties, unconsciousness and, ultimately, death.

General applications and principles

Local exhaust ventilation is more frequently being used to control construction dust. Many items of construction equipment that create dust have on-tool extraction to provide local exhaust ventilation.

Figure 12-34: Wall chasing using on-tool extraction.
Source: HSE.

Figure 12-35: Extraction unit.
Source: HSE.

The LEV should be designed, constructed and installed to ensure either the effective removal of contaminated air from the workplace to a suitable place or the filtering/ treatment of the contaminated air and its return to the workplace.

For efficient operation and to prevent exposure of the worker, the LEV intake points should be located as close as possible to the points where the hazardous substances are emitted. The length of ducting and the number of bends in the system should be kept to a minimum to enable efficient operation.

Various LEV systems are in use in the workplace and the main types are called receptors and captors. Receptor systems are designed to use low velocity extraction and are best suited to gases, vapours and lightweight dusts travelling at low velocity from the source. The work process creating the contaminant usually causes the contaminant to move towards the receptor system intake, for example where vapours are given off by a process and rise upwards to a receptor hood. Captor systems are designed to use high velocity extraction and are best suited to remove metal fume from welding operations or waste dust from high speed grinding or cutting equipment. The captor system uses its high velocity of air to capture volatile contaminant particles that may quickly move away from the source of where they are generated and have enough velocity of their own to escape a receptor system.

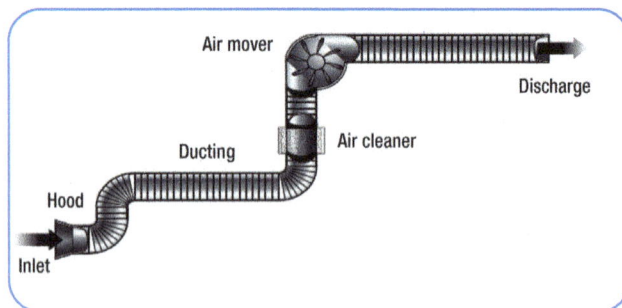

Figure 12-36: Common elements of a simple LEV system.
Source: UK, HSE, INDG408.

The common elements of a simple LEV system are:

- **Hood(s)** to collect airborne contaminants at, or near, where they are created (the source).
- **Ducts** to carry the airborne contaminants away from the process, *(See Figure 12-41)*, which shows the length of ducting, with curves not corners.
- **Air cleaner** to filter and clean the extracted air.
- **Fan** must be the right size and type to deliver sufficient suction at the hood, *(See Figure 12-38)*, which shows the size of fan and motor required for industrial-scale LEV systems.
- **Discharge** the safe release of cleaned, extracted air into the atmosphere.

Figure 12-37: Captor system on circular saw.
Source: RMS.

Figure 12-38: LEV fan and motor.
Source: Custom Dans Australia.

Figure 12-39: Flexible hose and captor hood.
Source: RMS.

Figure 12-40: Portable self-contained unit.
Source: Industrial Air Solutions Inc.

The efficiency of LEV systems can be affected by many factors including the following:

- Damaged ducting.
- Unauthorised alterations.
- Incorrect hood location, *(see Figure 12-39)*, which shows how a captor hood can be repositioned to suit the work activity by the use of a flexible hose.
- Too many bends or sharp bends in ducts.

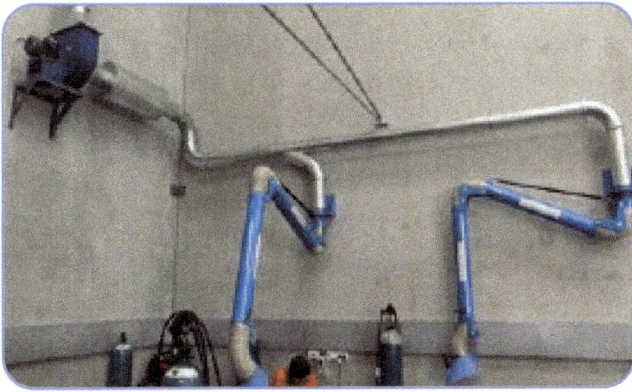

Figure 12-41: Length of ducting with curves.
Source: RMS.

- Blocked or defective filters.
- Leaving too many ports open.
- Process changes leading to overwhelming amounts of contamination.
- Fan strength or incorrect adjustment of fan.
- Excessive amounts of contamination.

The cost of heating/cooling air introduced to the workplace to replace that removed by the LEV system may encourage some employers to reduce extraction rates. When arranging installation of LEV it is vital that the pre- and post-ventilation contamination levels are specified and the required reduction in contamination levels should be part of the commissioning contract.

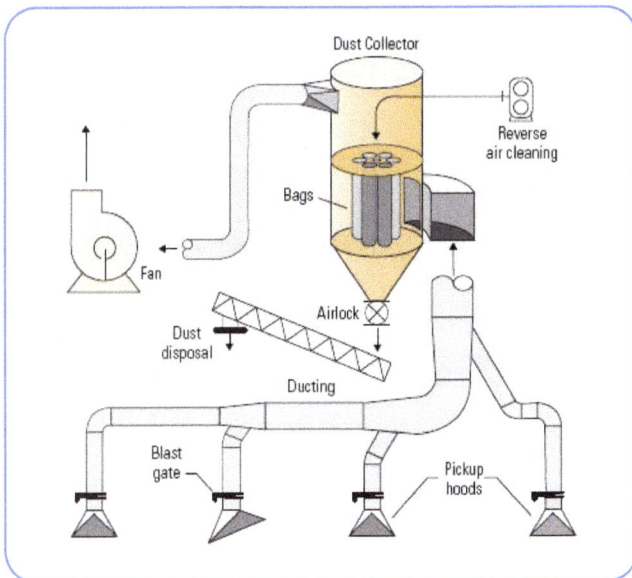

Figure 12-42: Multiple ports on one extraction duct.
Source: Wikipedia.org.

Requirements for inspection and thorough examination

LEV systems should be thoroughly examined at suitable planned intervals to ensure that they are continuing to perform as originally intended. COSHH 2002 (amended) Regulation 9(2) and schedule 4 set out requirements for the thorough examination of LEV systems. A thorough examination and test must take place at least once every 14 months, to ensure effective maintenance is carried out. Thorough examination of LEV systems is more frequent for LEV systems used in those processes listed in schedule 4. For example, LEV systems provided for jute manufacture and blasting of metal castings should be thoroughly examined monthly.

Requirements for records of thorough examination must be kept available for at least 5 years from the date on which the examination was carried out.

Regular inspection of LEV systems should be carried out in addition to the thorough examination. This should identify that no obvious defects exist and that the features for effectiveness are in place, for example, blanking covers for inlets that are not being used.

The majority of ventilation systems, although effective in protecting workers' health from airborne contaminants, can create other hazards. One of the main hazards that needs to be considered when designing a LEV system is the noise the LEV system creates. Even if it has been considered as a design feature when establishing LEV systems, the noise level of the system should be monitored in the workplace on a periodic basis.

Use and limitations of dilution ventilation

Dilution ventilation may be achieved by driving air into a work area, causing the air to flow around it, diluting contaminants in the work area and then allowing the air out of the work area through general leakage or through ventilation ducts to the open air. A variation on this is where air may be forcibly removed from the work area, but not associated with a particular contaminant source, and fresh air is allowed in through ventilation ducts to dilute the contaminated air in the work area. Sometimes a combination of these two approaches is used, for example, general air conditioning provided in an office environment.

Dilution ventilation systems will only affect levels of general contamination in the air in a workplace and will not prevent specific hazardous substances entering a person's breathing zone. Because dilution ventilation does not target any specific source of contaminant and it relies on dispersal and dilution instead of specific removal, it can therefore only be used with nuisance contaminants that are themselves fairly mobile in air. Dilution ventilation may only be used as the only means of control in circumstances where there is:

- Non-toxic contaminant or vapour (not dusts).
- Contaminant that is uniformly produced in small, known quantities.
- No discrete point of release.
- No other practical means of reducing levels.

The limitations of dilution ventilation are:

- It is not suitable for substances of high toxicity.
- Large sudden releases of a contaminant are not coped with well.

- It tends to move dust around rather than remove it.
- Areas in the location may exist where natural airflow does not create air movement.
- Stagnant air or non-moving air is likely to hold high concentrations of the contaminant.

The flow rates of dilution ventilation should be sufficient to change the air of the work area according to any specified requirements of good practices. This will take into account the size of the workplace, the nature and quantity of any contaminant, the working conditions and numbers of workers.

Recirculation of all the air forcibly removed from the workplace should be avoided where possible as it can lead to a decline in the quality of the air being breathed by workers, where recirculation is allowed:

a) Effective methods should be used to decontaminate the air, which should be regularly checked and maintained.

b) Some air should be vented during recirculation and replaced by fresh air to avoid an accumulation of possible contamination.

c) The rate of replacement by fresh air should be designed to ensure that hazardous limits or criteria for the control of the working environment, established by good practice or a competent authority, are not exceeded.

d) Account should be taken in the design of the need to prevent any inadvertent release of hazardous substances into the workplace from causing a hazard and spreading it to other working areas, for example, the provision of emergency shut down of the ventilation system.

Respiratory protective equipment

Regulation 7(3)(c) and principle (e) of the principles of prevention in COSHH 2002 (amended) require the employer to provide workers with suitable personal protective equipment where it is necessary, for example, respiratory protective equipment (RPE). This would be in addition to other control measures where adequate control of exposure cannot be achieved by other means.

Purpose, application and effectiveness

The purpose of RPE is to provide a means of controlling airborne contamination in the air that a worker breathes so that it limits exposure to the contamination to a level that is as low as is reasonably practicable and below the workplace exposure limit set for the substance. RPE does this by either filtering the air breathed or providing clean air to breathe.

It is used in situations where collective measures of protection from airborne contamination are not adequately effective and is used to supplement these measures or on its own. RPE is used to provide protection for a range of contaminants, such as dusts, fumes, vapours, gases or even micro-organisms.

RPE includes a very wide range of devices, from simple respirators offering basic protection against low levels of nuisance dusts to self-contained breathing apparatus.

Types of respiratory protection equipment

There are two main categories of respiratory protection:

1) Respirators.

2) Breathing apparatus.

Respirators

Respirators filter the air breathed but do not provide additional oxygen. There are a number of types of respirator that provide a variety of degrees of protection, from dealing with nuisance dusts to high-efficiency respirators for solvents or asbestos.

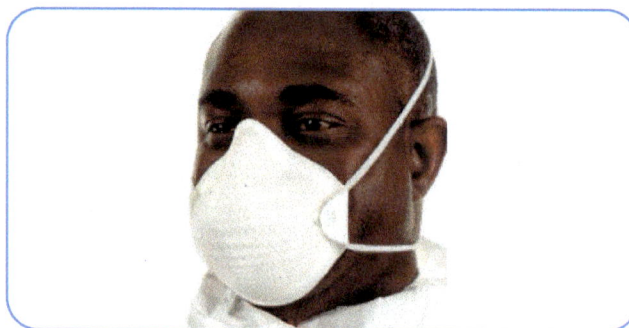

Figure 12-43: Filtering face-piece respirator.
Source: RMS.

Some respirators may be described as providing non-specific protection from contaminants, whereas others are designed to protect the user from a very specific contaminant such as solvent vapours.

There are five main types of respirators:

1) Filtering face-piece.

2) Half-mask respirator.

3) Full-face respirator.

4) Powered air purifying respirator.

5) Powered visor respirator.

Figure 12-44: Half-mask respirator.
Source: Shutterstock.

The advantages of a respirator are:

- Unrestricted movement.
- Often lightweight and comfortable.
- Can be worn for long periods.

The limitations of a respirator are:

- Purifies the air by drawing it through a filter to remove contaminants. Therefore can only be used when there is sufficient oxygen in the atmosphere.
- Requires careful selection and confirmation of correct fit for the user by a competent person.
- Requires regular maintenance.
- Knowing when a filter or cartridge is at the end of its useful life.
- Requires correct storage facilities.
- Can give a 'closed in'/claustrophobic feeling.
- Relies on user for correct use, adjustment etc.
- Incompatible with some other forms of PPE.
- Performance can be affected by beards and long hair.
- Interferes with other senses, for example, sense of smell.

Figure 12-45: Full-face canister respirator.
Source: Shutterstock.

Breathing apparatus

Breathing apparatus provides a separate supply of air (including oxygen) to that which surrounds the person. Because of the self-contained nature of breathing apparatus it may be used to provide a high degree of protection from a variety of toxic contaminants and may be used in situations where the actual contaminant is not known or there is more than one contaminant. There are three types of breathing apparatus:

1) Fresh-air hose apparatus - clean air from uncontaminated source.

2) Compressed air line apparatus from compressed air source.

3) Self-contained breathing apparatus - from cylinder.

The advantages of breathing apparatus are:

- Supplies clean air from an uncontaminated source. Therefore can be worn in oxygen-deficient atmospheres.

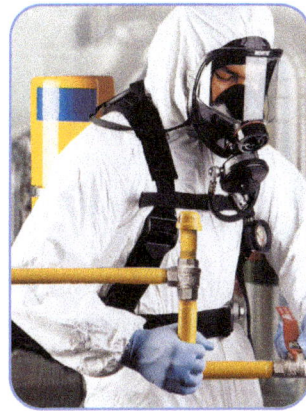

Figure 12-46: Breathing apparatus - air from cylinder.
Source: Haxton Safety.

- Has a high assigned protection factor (APF). Therefore may be used in an atmosphere with high levels of toxic substance.
- Can be worn for long periods if connected to a permanent supply of air.

The limitations of breathing apparatus are:

- Can be heavy and cumbersome, which restricts movement.
- Requires careful selection and confirmation of correct fit for the user by a competent person.
- Requires special training.
- Requires arrangements to monitor/supervise user and for emergencies.
- Can give a 'closed in'/claustrophobic feeling.
- Relies on user for correct use, adjustment etc.
- Incompatible with some other forms of PPE
- Performance can be affected by, for example, long hair.
- Interferes with other senses, for example, sense of smell.
- Requires correct storage facilities.

Selection, use and maintenance

There are a number of issues to consider in the selection of RPE, not least the advantages and limitations shown in the previous list. A general approach must not only take account of the needs derived from the work to be done and the contaminant to be protected from, but must include suitability for the person. This includes issues such as face fit and the ability of the person to use the equipment for a sustained period, if this is required. One of the important factors is to ensure that the equipment will provide the level of protection required.

This is indicated by the assigned protection factor (APF) given to the equipment by the manufacturer – the higher the factor, the more protection provided. With a little knowledge, it is possible to work out what APF is needed using the following formula:

$$APF = \frac{\text{Concentration of contaminent in the workplace}}{\text{Concentration of contaminent in the face-piece}}$$

There are only a small number of APF ratings used, so RPE APFs will be either: 4; 10; 20; 40; 200 or 2000. RPE with an APF of 10 will typically reduce the wearer's exposure by at least a factor of 10 if used properly, or, to put it another way, the wearer will only breathe in one-tenth or less of the amount of substance present in the air.

When selecting RPE always choose RPE with an APF above the calculated value. It is important to understand that this factor is only an indication of what the equipment will provide. Actual protection may be different due to fit and the task being conducted.

Every worker must use any PPE provided in accordance with the training and instructions that they have received. It is important that a worker using RPE has a face fit test to ensure the equipment is providing the protection needed. Where RPE (other than disposable respiratory protective equipment) is provided, the employer must ensure that thorough examination and, where appropriate, testing of that equipment is carried out at suitable intervals.

Other protective equipment and clothing

Hand/arm protection including gloves

There are numerous types of gloves and gauntlets available that offer protection from hazardous substances, such as acids, alkalis and solvents.

Figure 12-47: Gloves.
Source: Speedy Hire plc.

The materials used in the manufacture of these products are an essential feature to consider when making the selection. There are several types of rubber (latex, nitrile, PVC, butyl), all giving different levels of protection against hazardous substances. Care should be taken to ensure that, where necessary, protection is provided for the arms, this may require gloves that extend far enough to provide this protection.

Protective clothing

Protective clothing available that offers protection from hazardous substances might include overalls, aprons, boots and full deluge suits.

Figure 12-48: Protective clothing - gloves.
Source: Haxton Safety.

Eye and face protection

When selecting suitable eye protection, some of the factors to be considered are: type and form of the hazardous substance, for example protection from dusts, liquids, vapours, or gases, amount of protection, comfort and user acceptability, compatibility, maintenance requirements, training requirements and cost.

Figure 12-49: Goggles.
Source: Gempler's.

Figure 12-50: Spectacles.
Source: Pixabay.

Figure 12-51: Eye protection used with ear protection.
Source: Speedy Hire plc.

Footwear - safety boots/shoes

The importance of foot protection is illustrated by the fact that around 21,000 foot and ankle injuries are reported annually.

Inadequate protection and a lack of discipline on the part of the wearer commonly cause these. There are many types of safety footwear on the market, many of them offering different types of protection.

It is vital that the nature of the hazard is considered when selecting appropriate footwear.

Types	Advantages	Limitations
Spectacles	• Lightweight, easy to wear. • Can incorporate prescription lenses. • Do not 'mist up'.	• Do not give all-round protection. • Relies on the wearer for use.
Goggles	• Give all-round protection. • Can be worn over prescription lenses. • Capable of high impact protection. • Can protect against other hazards, for example, dust, molten metal.	• Tendency to 'mist up'. • Uncomfortable when worn for long periods. • Can affect peripheral vision.
Face shields (visors)	• Gives full-face protection against splashes. • Can incorporate a fan which creates air movement for comfort and protection against low-level contaminants. • Can be integrated into other PPE, for example, head protection.	• Require care in use, otherwise can become dirty and scratched. • Can affect peripheral vision. • Unless the visor is provided with extra sealing gusset around the visor, substances may go underneath the visor to the face.

Here are some common examples:

- Falling objects - steel toe-caps.
- Sharp objects - steel in-soles.
- Wet environments - impermeable wellingtons.
- Slippery surfaces- - non-slip soles
- Cold environments - thermally insulated soles.
- Electricity - electrically insulated soles.
- Flammable atmospheres - anti-static footwear.
- Spread of contamination - washable boots.

Figure 12-52: Personal protective equipment.
Source: RMS.

Personal hygiene and protection regimes

Personal hygiene

Personal hygiene has an important role in the protection of the health and safety of workers. Laid-down standards and procedures are necessary for preventing the spread of hazardous substance contamination.

The provision of adequate washing/showering facilities away from areas contaminated with hazardous substances is important to enable workers to remove contamination from their body. These facilities must be located conveniently close to where workers work, get changed, toilet facilities and where they eat to enable them to use them at rest periods and at the end of the day.

The use of barrier creams may also be additional personal protection considerations for hazardous substance risks. Where this is considered appropriate, facilities should be provided.

Facilities for changing and storage of clothing should enable workers to put clean work clothing on at the start of the work period and remove contaminated clothing at the end of the work period. Work clothing needs to be stored separate from non-work clothing.

The provision of laundry facilities for contaminated work clothing reduces workers from repeated exposure to contamination on the clothing. It is important that work clothing contaminated with hazardous substances is not taken home for laundering as this may expose members of the family to unnecessary contamination.

Where personal hygiene is critical, for example, when stripping asbestos, a 'three-room system' is employed. Workers enter the 'clean end' and take non-work clothes off and store them and put work clothes on, leaving by means of the 'dirty end'. When work has been completed they return by means of the 'dirty end', carry out personal hygiene, remove work clothing and put it ready to be cleaned and leave by means of the 'clean end' after dressing in their non-work clothing.

Eating, drinking and smoking

In workplaces that have a risk of contamination by hazardous substances it is important to prohibit eating, drinking and smoking in order to limit the likelihood of workers contacting their face around the mouth and ingesting the substance. In order to encourage workers to follow these prohibitions access to suitable facilities away from the contamination should be provided so they can be carried out. Workers should ensure they have removed contaminated clothing and wash any contamination from their hands before entering the facilities and carrying out any of the above actions.

Vaccination

Certain occupations, such as water treatment/sewage workers, medical profession, have a higher than average risk from some biological hazards. People who work in these occupations may need to be immunised against common high risks, for example, hepatitis B.

Whilst vaccination (if available) can be an effective way of preventing the health effects from exposure to biological agents, it is an intrusive control measure. Some workers may not be willing to receive a vaccination because they fear potential side effects. Employers therefore need to explain the value/limitations of the vaccination and gain the permission of the workers affected by the biological agents before providing vaccinations as a control measure.

HEALTH SURVEILLANCE AND BIOLOGICAL MONITORING

'Health surveillance' is concerned with collecting and using information about a worker's health related to their work and systematically looking for work-related ill-health in workers exposed to certain health risks. Health surveillance of workers may be useful for:

- Assessment of the health of workers in relation to risks caused by exposure to chemicals.
- Early diagnosis of work-related diseases and injuries caused by exposure to hazardous chemicals.
- Assessment of the worker's ability to wear or use required respiratory or other PPE.

The Management of Health and Safety at Work Regulations (MHSWR) 1999, Regulation 6, gives employers a duty to provide health surveillance where it is appropriate. Further details on health surveillance for hazardous substances are contained in Schedule 6 of COSHH 2002 (amended).

Although construction workers are not engaged in the manufacture of these substances they may work in locations where manufacture takes place. Other than the cases stated in Schedule 6 of COSHH 2002 (amended), surveillance may be appropriate where exposure to hazardous substances is such that an identifiable disease or adverse health effect may be linked to the exposure. There must be a reasonable likelihood that the disease or effect may occur under the particular conditions of work prevailing and that valid techniques exist to detect such conditions and effects.

Health surveillance may include medical surveillance conducted by an approved medical practitioner and, where appropriate, examination and investigations that may be necessary to detect exposure levels and early biological effects and responses.

Substances for which health surveillance is appropriate		Processes
Vinyl Chloride Monomer (VCM).		In manufacturing, production, reclamation, storage, discharge, transport, use or polymerization.
Nitro or amino derivatives of phenol and of benzene or its homologues.		In the manufacture of nitro or amino derivatives of phenol and of benzene or its homologues and the making of explosives with the use of any of these substances.
Orthotolidine and its salts.	Dianisidine and its salts. Dichlorbenzidene and its salts.	In manufacture, formation or use of these substances.
Auramine.	Magenta.	In manufacture.
Carbon Disulphide. Disulpher Dichloride. Benzene, including benzol.	Carbon Tetrachloride. Tricholoroethylene.	Process in which these substances are used, or given off as a vapour, in the manufacture of indiarubber or of articles or goods made wholly or partially of indiarubber.
Pitch.		In manufacture of blocks of fuel consisting of coal, coal dust, coke or slurry with pitch as a binding substance.

Figure 12-53: Schedule 6 - medical surveillance.
Source: COSHH 2002 (amended)

Substance	Biological monitoring guidance value	Sampling time
Butan-2-one	70µmol butan-2-one/L in urine	Post shift
Carbon monoxide	30ppm carbon monoxide in end-tidal breath	Post shift
Lindane (Organ chlorine pesticide)	35nmol/L(10µg/L) of Lindane in whole blood (equivalent to 70nmol/Lidane in plasma)	Random
Xylene, o-, m-, p- or mixed isomers	650 mmol methyl hippuric acid/mol creatinine in urine	Post shift

Figure 12-54: Examples of biological monitoring guidance values.
Source: EH40 Workplace exposure limits.

The types of checks that can be included in a health surveillance programme include:

- Clinical examinations.
- Diagnostic tests, such as by X-ray or ultrasonic scan.
- Function measurements, for example, a lung function test.
- Skin checks for signs of rashes.
- Biological monitoring by carrying out tests of blood or urine.

It may also include simple techniques for the early detection of effects on health, for example, a health assessment by questionnaire and self-checks once the symptoms have been explained.

Where a valid and generally accepted method of biological monitoring of workers' health exists for the early detection of the effects on health of exposure to specific occupational risks, it may be used to identify workers who need a detailed medical examination, subject to the individual worker's consent.

Biological monitoring may be particularly useful in circumstances where:

- There is likely to be significant skin absorption and/or gastrointestinal tract uptake following ingestion.
- Control of exposure depends on respiratory protective equipment.
- There is a reasonably well-defined relationship between biological monitoring and effect.
- It gives information on accumulated dose and target organ burden which is related to toxicity.

Biological Monitoring Guidance Values (BMGVs) are set in Table 2 of EH40. These BMGVs are provided for use where they are likely to be of practical value, suitable monitoring methods exist and there are sufficient data available.

BMGVs are non-statutory and any biological monitoring undertaken in association with a guidance value needs to be conducted on a voluntary basis (i.e. with the fully informed consent of all concerned), **see Figure 12-53.**

Where a BMGV is exceeded it does not necessarily mean that any corresponding airborne standard has been exceeded nor that ill health will occur. It is intended that where they are exceeded this will give an indication that investigation into current control measures and work practices is necessary. Similarly, a low BMGV should not suggest that there is no need to reduce workplace exposure further.

It is good practice for the employer to keep records of health surveillance, including biological monitoring, in respect of each worker for an appropriate duration related to the health risk. COSHH 2002 (amended) requires employer to keep records of surveillance in respect of each worker for at least 40 years. This requirement still applies where an organisation ceases to trade, in which case the records must be offered to the HSE.

Figure 12-55: Working with hazardous substances.
Source: Shutterstock.

ADDITIONAL CONTROLS FOR SUBSTANCES WITH SPECIFIC EFFECTS

Carcinogens and genetic damage

Carcinogens are substances that have been identified as having the ability to cause cancer. Examples of these include arsenic, hardwood dusts and used engine oils.

Substances have been identified that cause changes to DNA, increasing the number of genetic mutations above natural background levels. These are known as mutagens. The changes can lead to cancer in the individual affected or be passed to their offspring's genetic material, for example thalidomide and plutonium oxide.

Due to the serious and irreversible nature of cancer and genetic changes, an employer's first objective must be to prevent exposure to carcinogens and mutagens. These substances should not be used or processes carried out with them, if a safer alternative can be used instead. Where this is not feasible, suitable control measures should include:

- Totally enclosed systems.

- Where total enclosure is not possible, exposure to these substances must kept to as low a level as possible through the use of appropriate plant and process control measures, such as handling systems and LEV (these measures should not produce other risks in the workplace).

- Storage of carcinogens/mutagens must be kept to the minimum needed for the process, in closed, labelled containers with warning and hazard signs, including waste products until safe disposal.

- Areas where carcinogens/mutagens are present must be identified and segregated to prevent spread to other areas.

- The number of people exposed and the duration of exposure must be kept to the minimum necessary to do the work.

- PPE is considered a secondary protection measure used in combination with other control measures.

- Measures should be in place for monitoring of workplace exposure and health surveillance for work involving carcinogens and mutagens.

Occupational asthma

Occupational asthma is caused by substances in the workplace that trigger a state of specific airway hyper-responsiveness in an individual, resulting in breathlessness, chest tightness or wheezing. These substances are known as asthmagens or respiratory sensitisers and include such things as flour/grain dust, wood dust, isocyanates and solder. Not everyone who becomes sensitised goes on to get asthma, but once the lungs become hypersensitive, further exposure to the substance, even at quite low levels, may trigger an attack.

Exposure to these substances should be prevented, and where that is not possible, kept as low as reasonably practicable. Control measures used should take account of long-term time-weighted averages and short-term peak exposures to the substance. The additional control measures for carcinogens and mutagens would also be useful for substances that can lead to occupational asthma. It is also important that workers know what emergency action is required if someone has an asthma attack.

If an individual develops occupational asthma, their exposure must be controlled to prevent any further attacks. Because the worker may have become sensitised to the substance they may have to be reassigned to work that does not expose them to breathing the substance causing the asthma, for their own protection. Workers who work with substances that can cause asthma, asthmagens, must have regular health surveillance to detect any changes in respiratory function. The purpose of health surveillance is to monitor and protect the health of individual workers. Collecting simple information may enable early detection of ill health caused by work and identify the need for improved control measures.

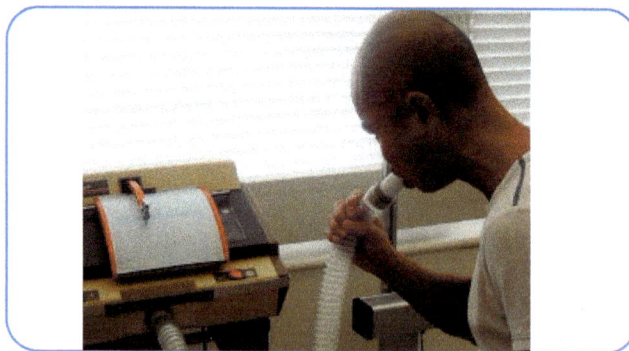

Figure 12-56. Person taking lung function test.
Source: www.elitecycling.co.uk.

CASE STUDY

A worker developed occupational asthma after working for a large multinational company in Gloucester. He was employed between 1995 and 2004 as a solderer and was exposed to rosin-based (colophony) solder fume during his career.

His health was deteriorating from 1999 onwards, and he was taking time off work due to breathing difficulties

The company did not have adequate control measures in place and failed to install fume extraction equipment to remove rosin-based fumes from the workroom air or from the breathing zones of its solderers.

The company did not substitute the rosin-based solder with rosin-free solder until December 2003, despite an assessment having identified the need to in 1999.

Workers, including the asthma sufferer, were not placed under a health surveillance scheme at any time.

As a result of action taken by HSE, the company was fined £100,000 with £30,000 costs. This attracted local and national media attention.

REVIEW

What factors reduce the effectiveness of LEV.?

There are two main categories of respiratory protection, respirators and breathing apparatus.

List two advantages and two disadvantages when using breathing apparatus.

12.4 Specific agents

HEALTH RISKS, CONTROLS AND LIKELY WORKPLACE ACTIVITIES/LOCATIONS

Blood-borne viruses

Blood-borne viruses (BBVs) are viruses that some people carry in their blood that can be passed on to another person. The main BBVs of concern are human immuno-deficiency virus (HIV), hepatitis B and hepatitis C.

BBVs spread when the blood from a person infected with the hepatitis virus gets into the body. This may occur in construction activities through accidental injection with a contaminated needle while carrying out work on derelict buildings or other areas where drug misuse takes place. Infection can also occur where contaminated blood enters the body through cuts or other types of puncture wound. The risk of infection due to needle injection depends on:

- Whether the needle user was infected with a BBV.
- The type of BBV and how infectious it is.
- How much of this infected material enters the bloodstream - a needle attached to a syringe containing blood is likely to be higher risk than a needle on its own - not all exposures result in infection.

The health effects of BBVs will depend on the type of virus:

- HIV - The earliest symptoms of HIV can resemble flu and they generally clear up within a month or two. These symptoms may include fever, headache, fatigue, and swelling in the lymph nodes, particularly those in the neck and groin. However, not everyone who acquires HIV will experience these symptoms. Similarly, for several years, perhaps as long as a decade, a person with HIV might not have any symptoms at all. HIV progresses differently for each person affected. The course of the disease is determined by the specific infections or complications a person with HIV develops. HIV complications can affect different parts of the body. Some are localised to the mouth (white patches in the mouth or pharynx), others in the brain (dementia, difficulty in processing information, brain tumours) and others result in total body changes like losing body weight (loss of muscle as well as fat). Skin conditions in the form of sores or lesions are also common.

- Hepatitis B - Since the virus is extremely infectious - 50 to 100 times more than HIV - even brief, direct contact with infected blood could be enough to cause infection. Once inside the body the virus is transported by the blood to the liver, where it infects the liver cells. Most people infected with hepatitis B remain healthy without any symptoms while they fight off the virus. Some people will not know they have been infected. If there are any symptoms, they will develop on average 60-90 days after exposure to the virus. Common symptoms of hepatitis B include:

- Flu-like symptoms, such as tiredness, general aches and pains, headaches and a high temperature of or above 38°C (100.4°F).
- Loss of appetite and weight loss.
- Feeling sick or being sick.
- Diarrhoea.
- Pain in your upper right-hand side.
- Yellowing of the skin and eyes (jaundice).

Symptoms will usually pass within one to three months. Until the virus has been cleared from those infected, they can pass it onto others.

- Hepatitis C – As with Hepatitis B the virus infects the liver cells. The earliest symptoms of infection by hepatitis C may start about six to seven weeks after exposure to the virus. This is considered the acute phase of the virus and involves mild or no symptoms. Some people are able to completely recover from hepatitis C after the acute phase. If the infection progresses to the chronic phase symptoms may include loss of appetite, fever, nausea, fatigue, vomiting, joint pain, jaundice, dark urine or abdominal pain. Occasionally severe infection with hepatitis C may result in liver failure or cirrhosis (scarring) of the liver. Most people do not know if they are infected and may live for many years without symptoms.

Where a risk of exposure to BBVs has been identified, simple, inexpensive measures to prevent or control risks can be taken:

- Ensure good personal hygiene practices are observed, in particular, hygienic hand-washing.

- Use procedures such as avoiding the use of sharps such as needles, blades, glass, etc.

- Use PPE such as gloves, eye protection, face masks, etc.

- Ensure contaminated waste is disposed of in a safe manner, for example, in a sharps disposal bin.

- Use disposable equipment where there is a risk of BBV contamination, otherwise decontamination procedures must be strictly complied with.

- Ensure workers are aware of immediate steps to be followed upon contamination with blood or other body fluids, for example, needle stick injuries procedure.

- Vaccination where appropriate – immunization is available for hepatitis B, but not other BBVs.

- Decontamination and disinfection procedures.

Carbon monoxide

Carbon monoxide is a colourless and odourless chemical asphyxiant produced as a by-product of incomplete combustion of carbon fuels, for example, gas water heaters, compressors, pumps, dumper trucks or electricity generators. It is a particular risk when operated in poorly ventilated confined areas where workers are forced to inhale it. Carbon monoxide has a great affinity (200 times that of oxygen) with the haemoglobin red blood cells, which means it will inhibit oxygen uptake by red blood cells, resulting in chemical asphyxiation, leading to collapse and death.

Carbon monoxide can have acute and chronic effects:

- **Acute effects:** The earliest symptoms, especially from low-level exposures, are often non-specific and readily confused with other illnesses, typically flu-like viral syndromes, depression, chronic fatigue syndrome, and migraine or other headaches. This often makes the diagnosis of carbon monoxide poisoning difficult. If suspected, the diagnosis can be confirmed by measurement of blood carboxyhaemoglobin. Common problems encountered are difficulty with higher intellectual functions and short-term memory, dementia, irritability, gait disturbance, speech disturbances, Parkinson-like syndromes, cortical blindness, depression and death.

- **Chronic effects:** Long-term, repeat exposures present a greater risk to persons with coronary heart disease and in pregnant women. Chronic exposure may increase the incidence of cardiovascular symptoms in some workers. People exposed to carbon monoxide often complain of persistent headaches, light-headedness, depression, confusion and nausea. When removed from exposure to carbon monoxide the symptoms usually resolve themselves.

Risk control measures for construction activities mainly relate to the careful location of equipment that creates carbon monoxide away from work areas and ensuring workers do not need to work in close proximity to the equipment causing them to breathe in exhaust gases. This is important when workers are doing tasks alongside construction plant or where the plant is located underneath scaffolds or inside buildings.

As acute effects can occur with relatively small exposure levels in a few minutes, care has to be taken when mobile plant that produces carbon monoxide is active near workers for a relatively short time, particularly if they are working in a confined area where the exhaust gases can quickly build up, for example, indoors, in a trench or a confined space.

Carbon monoxide gas is also produced by liquefied petroleum gas (LPG) heaters and cookers therefore it is essential to ensure good ventilation where this equipment is used, for example in temporary accommodation units. The HSE strongly recommends the use of carbon monoxide alarms, but they must not be regarded as a substitute for proper installation and maintenance of gas appliances by a Gas Safe Registered engineer.

Cement

Cement is used extensively in the construction industry as part of the mix for mortar and concrete. Cement can cause harm in the following ways:

- Skin contact - causing contact dermatitis. Dry cement is mildly irritant and contact with cement powder over a period of time can lead to contact dermatitis.

- Skin contact - causing sensitisation dermatitis. Cement contains hexavalent chromium (chromate), which can cause dermatitis due to allergic sensitisation. Once a worker has become sensitised to cement, any future exposure can trigger dermatitis. Manufacturers of cement add an ingredient to lower the hexavalent chromium content and reduce this risk. This ingredient is only effective for a limited period, which is indicated by a date marked on bags. After this period, the level of hexavalent chromium may increase again.

- Skin contact - causing corrosive burns. Wet cement is highly alkaline. If wet cement is trapped inside a worker's boot or glove or their skin is in contact with wet cement in some other way for a period of time then it can cause severe corrosive burns. Burns can develop rapidly and in extreme cases require a skin graft or amputation of a limb.

- Eye contact - causing irritation or inflammation. Wet cement can cause corrosive burns.

- Inhalation - causing irritation of the nose and throat and possible long-term respiratory problems. This could be from exposure to cement powder or dust produced when cutting, drilling etc. dried concrete and mortar.

Figure 12-57 shows severe burns sustained by a worker after kneeling in wet cement for three hours; his right leg had to be amputated.

Avoid exposure to cement powder by using pre-mixed concrete and mortar. Use methods to avoid workers needing to be in contact with substances containing wet cement, for example by using long handled tools. Where possible, exposure to cement in concrete should be eliminated by mechanical batch pouring and using vibrating screeding equipment.

Gloves, waterproof trousers and boots that are resistant to alkalis should be worn when working with wet products containing cement, for example concrete pouring or screeding (particularly if screeding is done kneeling down). Care should be taken to ensure gloves are tight fitting at the cuff and the top of boots are covered by trousers. Any contaminated clothing should be removed and not reused until thoroughly cleaned. Good hygiene standards should be maintained, for example, washing the skin and hands thoroughly after contact with cement.

Where there is a risk of inhalation of dust from cement containing products the dust should be inhibited by wetting down, extracted by on-tool equipment or appropriate respiratory protective equipment used.

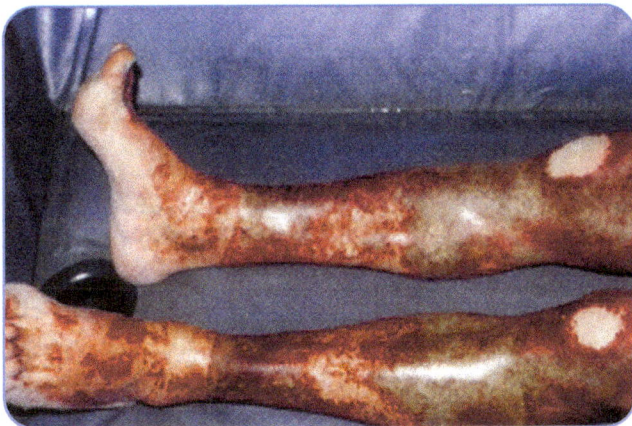
Figure 12-57: Cement burns.
Source: SHP.

Legionnella

Legionella pneumophila bacteria is widely present in natural water sources such as rivers, lakes, reservoirs, but usually in low quantities. It is also frequently present in water-cooling systems and hot water systems. Large workplace buildings are therefore susceptible to contaminated water systems, especially hotels and hospitals.

Legionella bacteria needs certain conditions to multiply, for example, the presence of sludge, scale, algae, rust and organic material that acts as a nutrient, providing food to enable the bacteria to grow. It also needs a suitable temperature to provide enough energy to grow, but not too much that it could not survive - between 20-50°C.

Transmission of the bacteria to people is principally via inhalation of small droplets of water containing the bacteria. These small droplets can be created by processes that create splashes, aeration, water vapours, showers or spray. Greatest risk areas are showers used for washing (for example in temporary accommodation units or asbestos decontamination facilities), high pressure water cleaning equipment, air conditioning sprays, water-cooling towers and recirculating water-cooling systems.

When the legionella bacteria reach the lungs they cause a type of pneumonia called legionnellosis, which is the collective term for diseases caused by the bacteria. This includes Legionnaires' disease, which is the most severe form of this disease and can be fatal. Other forms of the illness that are not so severe are Lochgoilhead fever and Pontiac fever.

Symptoms are aching muscles, headaches and fever followed by a cough. Confusion, emotional disturbance and delirium may follow the acute phase. In some cases, death may occur. The fatality rate for those that contract Legionnaires' disease has ranged between 5% and 30% during outbreaks; for example, those occurring in the UK are reported to be about 12%. The death rate for older people and the infirm who contract the disease is often as high as 50%. Smoking and gender may also affect susceptibility.

To control the risks from legionella bacteria:

- Ensure that the release of water droplets is properly controlled.
- Avoid water temperatures that favour the growth of legionella bacteria, for example heat water to 60°C.
- Ensure water cannot stagnate, for example by keeping pipe lengths as short as possible or removing redundant pipework.
- Keep the water clean and avoid materials that encourage the growth of legionella bacteria.
- Treat water to prevent the growth of legionella or limit its ability to grow - disinfection/chlorination.

Leptospira

Leptospira (spiral-shaped) bacteria are carried by a range of animals. Rodents represent the most common carrier of the bacteria, especially rats (also gerbils, voles and field mice). *(See Figure 12-58: Leptospira Scanning Electron Micrograph (SEM))*.

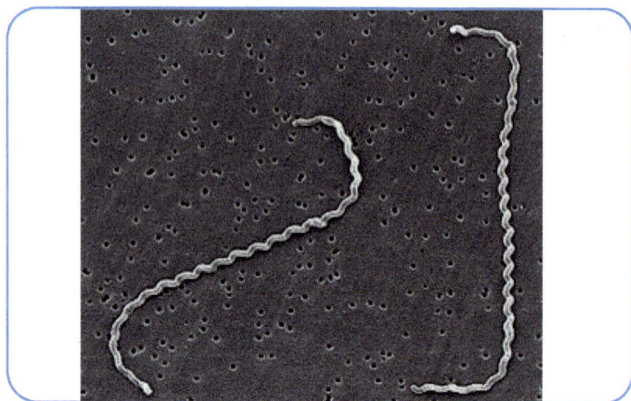

Figure 12-58: Leptospira magnified many times shown on SEM shows spiral shape.
Source: Wikipedia.

Other sources of infection are dogs, hedgehogs, foxes, pigs and cattle. These animals are not necessarily ill, but carry leptospires in their kidneys and excrete them in their urine.

The bacteria can be transmitted to people directly via direct contact with blood, tissues, organs or urine of one of the host animals or indirectly by contact with a contaminated environment, for example soil or water that has been contaminated with urine containing the bacteria.

Those construction workers most at risk from leptospira bacteria are those who work where rats prevail, and include water and sewage work, demolition or refurbishment of old unoccupied buildings, and those working on sites adjoining rivers and other watercourses. The bacteria's survival depends on protection from direct sunlight, so it survives well in watercourses and ditches protected by vegetation.

The leptospira bacteria enter the body through broken skin or mucous membrane of the mouth, nose or eyes - causing leptospirosis (Weil's disease). Symptoms vary but include flu-like illness, conjunctivitis, liver damage (including jaundice), kidney failure and meningitis. If untreated, infection may be fatal.

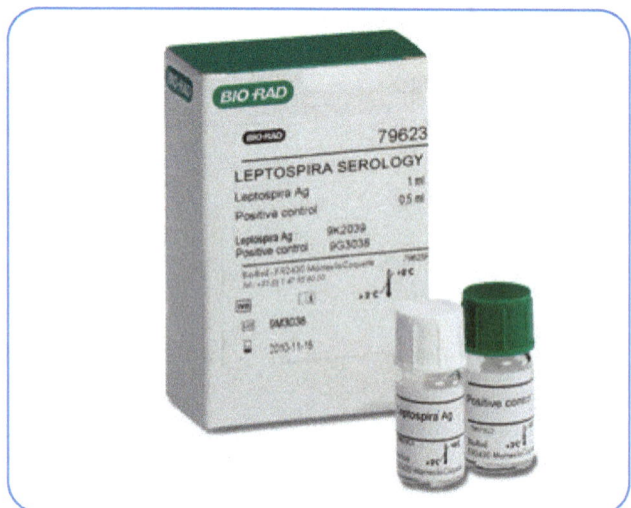

Figure 12-59: Treatment for Leptospira.
Source: Bio-Rad.

Typical control measures include good personal hygiene (for example, hand-washing), PPE (gloves), covering cuts and grazes, preventing animal (rat) infestations and issuing a worker with an 'At Risk' card to show a doctor if they develop flu-like symptoms after being exposed to contaminated water. Anyone with flu-like symptoms who may have been in contact with rat urine should seek medical attention. Antibiotic treatment should be started as soon as possible.

Silica

Silica exists naturally as a crystalline mineral in most rocks, sand and clay. A common variety is quartz (tridymite, cristobalite). In construction activities, it may be encountered in stonework, bricks, concrete or work with quartz-based tiles. The amount of crystalline silica in different materials varies, for example sandstone could be 70-90%, concrete/mortar 25-70% and brick up to 30%.

In construction activities silica dust is created when these materials are cut, sanded, carved etc. Silica dust could become airborne when handling, mixing or sweeping up dry material in powder form. Some of this dust may be fine enough to inhale and travel deep into the lungs. The fine dust is called respirable crystalline silica (RCS) and is too fine to see with normal lighting. Inhalation of crystalline silica dust can result in the following lung diseases:

- Silicosis - a fibrosis of the lung involving inflammation and scarring in the form of nodular lesions in the upper lobes of the lungs. Silicosis makes breathing more difficult and increases the risk of lung infections. Silicosis usually follows exposure to RCS over many years, but extremely high exposures can lead rapidly to ill health. Silicosis (especially the acute form) is characterised by shortness of breath, fever and cyanosis (bluish skin). It may often be misdiagnosed as pulmonary oedema (fluid in the lungs), pneumonia or tuberculosis.

- Chronic obstructive pulmonary disease (COPD) - COPD is a group of lung diseases, including bronchitis and emphysema, resulting in severe breathlessness, prolonged coughing and chronic disability. It can be very disabling and is a leading cause of death in the construction industry.

- Lung cancer - Heavy and prolonged exposure to RCS can cause lung cancer. When someone already has silicosis, there is an increased risk of lung cancer.

Typical control measures include preventing the dust from being inhaled by using alternative methods of processing, using dust suppression (for example by using water) or dust collection (LEV) and a vacuum or wet method of dust removal rather than dry sweeping.

The worker should wear RPE and be subject to health surveillance (lung function testing, etc.).

Wood dust

Exposure to wood dust has long been associated with a variety of adverse health effects, including dermatitis, allergic respiratory effects, mucosal and non-allergic respiratory effects and cancer. Health problems arise from work with both softwoods and hardwoods.

Inhalation of wood dust can result in the following conditions:

- Skin disorders - Contact with the irritant compounds in wood sap contained in the dust can cause dermatitis and other allergic reactions.

- Respiratory disorders - The respiratory effects of exposure to wood dust include asthma, hypersensitivity, pneumonitis (an allergic reaction affecting the alveoli) and chronic bronchitis.

- Cancer - hardwoods are associated with causing nasal cancer.

The quantity and type of wood dust created by construction activities will depend on the wood being cut and the process used. The main activities likely to produce high dust levels are:

- Machining, particularly sawing, routing and turning.

- Sanding by machine and by hand – particularly using a belt sander.

- Cleaning the workplace, especially when compressed air lines are used for blowing dust from walls, ledges and other surfaces.

- Using compressed air lines to blow dust off furniture and other articles before spraying.

- High airborne dust levels can also occur during the bagging of dust from dust-extraction systems.

The biggest risk is from fine dust, as it will spread further and affect more people, not just those involved in the activity. When breathed in, fine dust will go deep into the lungs where it will do the most damage.

Control measures for wood dust include:

- Preventing dust being created on site by ordering pre-cut materials and using pre-finished materials.

- Reducing the effects by using wood types known to be less toxic.

- Reducing the number of workers exposed by using dedicated work areas to minimise the spread of wood dust.

- Using on-tool extraction of the H or M class.

- Wearing RPE – RPE is likely to be needed in most cases when sanding or if cutting wood for periods of more than 15-30 minutes total time over a day – particularly in enclosed spaces.

Tetanus

Tetanus bacteria (Clostridium tetani) are commonly found in soil and in the manure of animals, such as cows and horses. Tetanus bacteria usually enter the body through a cut or puncture wound, which in construction activities may be caused by a nail, splinters, sharp edges of metal or impact damage.

There is a greater risk of developing tetanus if the wound is deep or if it gets dirty with soil or manure, even small wounds can allow enough bacteria to get into the body to cause tetanus.

When the tetanus bacteria enter the body they quickly multiply and release a toxin that affects the nerves, causing symptoms such as muscle stiffness and spasms.

Figure 12-60: Tension of the spine.
Source: Painting by Charles Bell.

Muscles in the jaw first become stiff, and then rigidly fixed (lockjaw). Painful muscle spasms could make breathing and swallowing difficult. This can progress to other parts of the skeletal muscles causing the body to shape into a characteristic form called opisthotonos (behind tension) of the spine.

Tetanus is a serious, but rare condition. It is rare in the UK because children receive a vaccine to prevent its effects, however construction workers born outside the UK may be at increased risk if they have not been vaccinated. Prevention includes skin protection, such as gauntlet gloves and tough material leggings and boots. Prompt medical attention is required for any contaminated deep cuts. Exposed workers can be vaccinated against tetanus; this should be carried out every 10 years or at the time of an injury.

HEALTH RISKS FROM AND CONTROLS FOR WORKING WITH ASBESTOS

Health risks

Asbestos is a general term used to describe a range of mineral fibres (commonly referred to by colour, for example, white, chrysotile; brown, amosite; and blue, crocidolite). Asbestos was mainly used as an insulating and fire-resisting material (and is still used in some countries). Asbestos fibres readily become airborne when disturbed and may enter the lungs, where they cause a number of fatal and serious diseases, the main ones being asbestosis, pleural thickening, mesothelioma and lung cancer.

Asbestosis is a serious scarring condition of the lungs that normally occurs after high levels of exposure to asbestos fibres. The scarring causes the lungs to lose their elasticity and can result in progressive breathlessness, and in severe cases can be fatal.

Asbestosis typically takes more than 10 years to develop. The primary symptom of asbestosis is generally the slow onset of shortness of breath on exertion. In severe, advanced cases, this may lead to respiratory failure. Coughing is not usually a typical symptom, unless the patient has other respiratory tract diseases.

Pleural thickening a problem that generally happens after high levels of exposure to asbestos fibres. Scarring of the lining of the lung (pleura) causes it to thicken and swell. The pleura is a two-layered membrane that surrounds the lungs and lines the inside of the ribcage. If the thickening and swelling covers a large area of the pleura the lung is squeezed, which can cause breathlessness and discomfort in the chest.

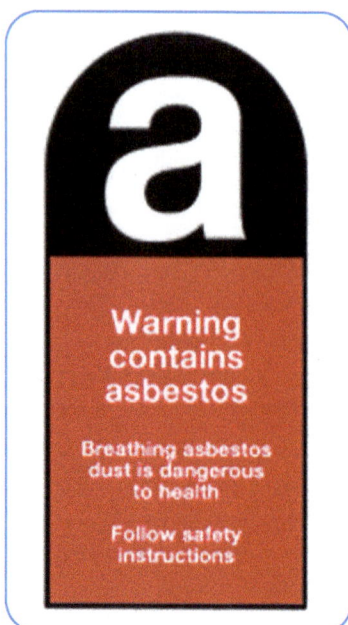

Figure 12-61. Asbestos label.
Source: Scaftag.

Mesothelioma is a cancer which affects the 'mesothelium' (the lining of the lungs (pleura) and the lining surrounding the lower digestive tract (peritoneum)). Even low levels of exposure to asbestos fibres can cause mesothelioma, if exposure is repeated over time. Unfortunately, mesothelioma is incurable and, by the time it is diagnosed, it is almost always fatal. Symptoms of mesothelioma may not appear until 20 to 50 years after exposure to asbestos fibres. Shortness of breath, cough, and pain in the chest due to an accumulation of fluid in the pleural space are often symptoms of pleural mesothelioma.

Asbestos-related lung cancer develops in the tubes that carry air in and out of the lungs, for example, in the bronchus. The cancer can grow within the lung and can spread outside the lung through the rest of the body.

Controls

The Control of Asbestos Regulations (CAR) 2012 place emphasis on assessment of exposure; exposure prevention, exposure reduction and control; adequate information, instruction and training for employees; monitoring and health surveillance. General control measures related to asbestos that are required by CAR 2012 include:

- The presence of asbestos must be identified, recorded and labelled.
- An assessment must be done of work that exposes workers to asbestos.
- Information must be provided on the location and condition of the asbestos to anyone who is liable to work on or disturb it.
- A written plan of work is required for work with asbestos.

Regulation 3 of CAR 2012 requires that specified work with asbestos must be licensed, licensable work is work:

1) Where exposure to asbestos is not sporadic and of low intensity.

2) Where the risk assessment cannot show that the control limit will not be exceeded.

3) The work involves:

  - Work on asbestos coating.
  - Some work on asbestos insulation board or asbestos insulation where exposure is likely to be significant.

Non-licensable work must meet the following conditions:

- Condition 1 – exposure is sporadic and of low intensity.

- Condition 2 – the risk assessment shows that the control limit will not be exceeded.

- Condition 3 – the work falls into one of the following

categories:

- Short, non-continuous maintenance activities in which only non-friable materials are handled.
- Removal of materials in which the asbestos fibres are firmly linked in a matrix.
- Encapsulation or sealing of asbestos-containing materials which are in good condition.
- Air monitoring, collection and analysis of samples.

CAR 2012 introduced a category of work called Notifiable Non-Licensed Work (NNLW). NNLW is work that meets condition 1 and 2 and is not work listed in condition 3. The employer proposing to carry out this work must notify the appropriate enforcing authority.

Other controls required by CAR 2012 related to asbestos include:

- Exposure must be prevented or reduced by controls.
- Controls must be used and maintained.
- The employer is responsible for the cleaning of personal protective clothing.
- Records of work carried out must be kept.
- Workers must have a periodic medical examination.

Awareness training, should be provided for workers that may inadvertently expose themselves to asbestos fibres during their work, such as cable installer's and decorators. This training should enable them to identify possible asbestos containing materials or residues, before or when carrying out cutting or drilling tasks.

DUTY TO MANAGE ASBESTOS

The duty to manage asbestos is covered by Regulation 4 of the CAR 2012. In many cases, the duty holder is the person or organisation that has clear responsibility for the maintenance or repair of non-domestic premises through an explicit agreement such as a tenancy agreement or contract. Where there are domestic premises such as flats the duty holder will be responsible for common areas such as corridors or walkways.

The duty holder has the responsibility to:

- Carrying out an assessment, often involving a survey to inspect accessible parts, to determine if there is asbestos in the premises (or assessing if ACMs are liable to be present and making a presumption that materials contain asbestos, unless there is strong evidence that they do not), its location and its condition.

- Identify and record the location and condition of any ACMs or presumed ACMs in the premises.

- Assess the risks from any material identified.

- Prepare a written plan identifying those parts of the building concerned and setting out in detail how risks identified are to be managed.

- Establish a system for providing information on the location and condition of ACMs for anyone who is liable to work on or disturb it.

- Ensure asbestos is maintained or where necessary safely removed.

- Monitor the condition of the asbestos or materials suspected of containing asbestos.

- Implement the measures specified in the plan and record them.

- Review and revise the plan regularly and when circumstances change.

There is also a requirement on anyone to co-operate as far as is necessary to allow the duty holder to comply with the requirements shown previously. If asbestos containing materials are in good condition they may be left in place and their condition monitored and managed.

Competence for risk assessments

The HSE document 'Managing and work with asbestos', the approved code of practice and guidance to the Control of Asbestos Regulations (CAR) 2012 clarifies competence requirements for those that carry out asbestos risk assessments.

The ACOP to Regulation 6 of CAR 2012, relating to assessment of work which exposes employees to asbestos, states:

"Whoever carries out the assessment should:

Have adequate knowledge, training and expertise in understanding the risks from asbestos and be able to make informed and appropriate decisions about the risks and precautions needed.

Know how the work activity may disturb asbestos.

Be familiar with and understand the requirements of the Regulations ACOP.

Have the ability and the authority to collate all the necessary and relevant information.

Be able to assess other non-asbestos risks on site.

Be able to estimate the expected level of exposure to decide whether or not the control limit is likely to be exceeded."

Figure 12-62: Competence to carry out asbestos risk assessments.
Source: HSE ACOP to CAR 2012.

Asbestos identification

There are three main types of asbestos, which is a naturally occurring mineral. It can be amphibole asbestos which includes crocidolite (blue) and amosite (brown) asbestos, or serpentine asbestos which is chrysotile (white) asbestos. All three of these main types have been used in Great Britain at some time.

Types of survey

Management survey

This is the standard survey used to identify the presence and extent of any asbestos containing materials (ACMs) in the building which could be damaged or disturbed during normal occupancy, including foreseeable maintenance and installation, and to assess their condition.

A management survey might involve minor intrusive work and some disturbance which will vary between premises and depend on what is reasonably practicable for individual properties.

Management surveys can involve a combination of sampling to confirm asbestos is present or presuming asbestos to be present.

Refurbishment and demolition surveys

A refurbishment and demolition survey will be needed before any work of this type is carried out. This survey will be used to locate and describe all ACMs in the area where the work will take place or in the whole building if demolition is planned.

This type of survey is fully intrusive and involves destructive inspection to gain access to all areas, including those that may be difficult to reach.

A refurbishment and demolition survey may also be required in other circumstances, for example, when more intrusive maintenance and repair work will be carried out or for plant removal or dismantling.

Who can undertake them?

An independent expert/specialist organisation must:

- Have adequate training and be experienced in survey work.

- Able to demonstrate independence, impartiality and integrity.
- Have an adequate quality management system.
- Carry out any asbestos survey work in accordance with recommended HSE guidance HSG264 'Asbestos: The Survey Guide'.

HSG264 states that organisations offering an asbestos survey service should be able to demonstrate their competence by holding United Kingdom Accreditation Service (UKAS) accreditation to ISO/IEC 17020 and individuals to ISO/IEC 17024.

Where it can be located

Asbestos has a number of valuable properties; these include physical strength, resistance to chemicals, non-combustibility and good thermal and electrical insulation.

Asbestos is commonly mixed with other materials and is typically used in the following applications:

- As yarn or cloth - for protective clothing, etc.
- Insulation boards - for protection of buildings against fire.
- Asbestos cement products - for construction of buildings and pipes.
- Asbestos resins - for clutch faces and brake linings.
- Various other applications - in gaskets, filters, floor tiles and decorative plasterwork in old buildings.
- Asbestos spraying for thermal and acoustic insulation of buildings and plant, and for fire-resistance in structural steelwork, is no longer carried out, although it is encountered as a problem when being removed.

Asbestos has been used in many parts of buildings, (**see Figure 12-63**) for examples of uses and locations where asbestos can be found.

Asbestos products	What it was used for
Sprayed asbestos (limpet).	Fire protection in ducts and to structural steel work, fire breaks in ceiling voids etc.
Lagging.	Thermal insulation of pipes and boilers.
Asbestos insulating boards (AIB).	Fire protection, thermal insulation, wall partitions, ducts, soffits, ceiling and wall panels.
Asbestos cement products, flat or corrugated sheets.	Roofing and wall cladding, gutters, rainwater pipes, water tanks.
Certain textured coatings.	Decorative plasters, paints.
Bitumen or vinyl materials.	Roofing felt, floor and ceiling tiles.
General uses.	Vehicle brake linings, woven fires, ropes used as high temperature gaskets for furnaces, jet engines, chemical pipelines. Electrical insulation for hotplate wiring, electrical fuse wire holders and in building insulation and sound absorption. Filters for cigarettes. Artificial (chrysotile) snow effects in Hollywood films made in the USA in the 1920's and 1930's.

Figure 12-63: Examples of uses and locations where asbestos can be found.
Source: RMS.

High risk materials

Figure 12-64 Asbestos pipe lagging.
Source: HSE.

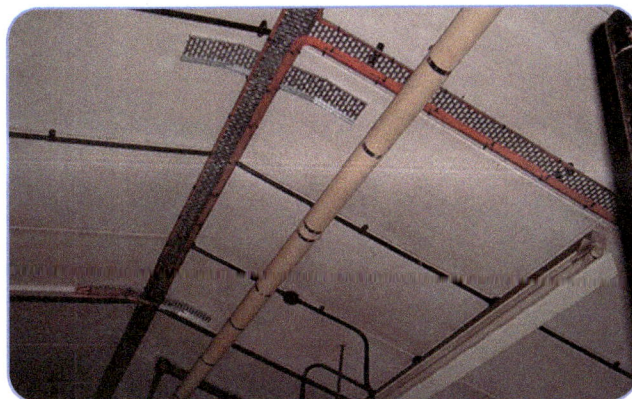

Figure 12-65: Asbestos insulating board (AIB).
Source: HSE.

Figure 12-66: Textured decorative coating.
Source: HSE.

Figure 12-67: Window with AIB panel.
Source: HSE.

Normally lower-risk materials

Figure 12-68: Asbestos cement roof sheeting.
Source: HSE.

Figure 12-69: Asbestos-containing floor tiles.
Source: HSE.

As long as the asbestos-containing material (ACM) is in good condition, and is not being or going to be disturbed or damaged, there is negligible risk. But if it is disturbed or damaged, it can become a danger to health, because people may breathe in any asbestos fibres released into the air.

Workers who may be particularly at risk of being exposed to asbestos when carrying out building maintenance and repair jobs include:

- Construction and demolition contractors, roofers, electricians, painters.

- Decorators, joiners, plumbers, gas fitters, plasterers, shop fitters, heating and ventilation engineers, and surveyors.

- Anyone dealing with electronics, for example, phone and information technology (IT) engineers, and alarm installers.

- General maintenance engineers and others who work on the fabric of a building.

If asbestos is present that can be readily disturbed, is in poor condition and not managed properly, all people in the building could be put at risk.

Procedure for the discovery of asbestos during construction activity

Where unacceptable risks to health and safety are discovered while work is in progress, for example disturbing hidden, missed or incorrectly identified asbestos containing materials, work affecting the asbestos should be stopped immediately, or if necessary after suitable controls in place and prevent further spread.

Where there is extensive damage to asbestos containing materials which causes contamination of the premises, or part of the premises, the area should be immediately evacuated. Work should not restart until a new plan of work is drawn up or until the existing plan is amended. Some measures may need to be carried out by licensed contractors.

Requirements if persons are accidentally exposed to asbestos materials

Regulation 15 of the Control of Asbestos Regulations (CAR) 2012 requires employers to have arrangements in place to deal with accidents, incidents and emergencies. These procedures must include means of raising the alarm and means of evacuation. People not wearing PPE must leave the work area. If people have been contaminated then arrangements must include means of decontaminating. The contaminated area must be cleaned thoroughly by people wearing PPE. Contaminated PPE should be treated as contaminated waste. Supervisors and managers must make sure that the work has been carried out.

Requirements for removal

Non-licensed work

Most asbestos work must be undertaken by a licensed contractor but any decision on whether particular work is licensable is based on the risk. To be classed as non-licensed work it must not be of a type that is defined as licensable work.

Certain non-licensable work is exempt by Regulation 3(2) of CAR 2012 from requirements for notification (Regulation 9), designation of areas (Regulation 18(1)(a)) and health records and medical surveillance (Regulation 22). This work must meet the following conditions:

- Condition 1 - Sporadic and low intensity.
- Condition 2 - Carried out in such a way that the exposure of workers to asbestos will not exceed the legal control limit of 0.1 asbestos fibres per cubic centimetre of air (0.1 f/cm³).
- Condition 3 - the work falls into one of the following categories:

1) A short non-continuous maintenance task.

2) A removal task, where the ACMs are in reasonable condition and are not being deliberately broken up.

3) A task where the ACMs are in good condition and are being sealed or encapsulated.

4) An air monitoring and control task to check fibre concentrations in the air, or it's the collection and analysis of asbestos samples to confirm the presence of asbestos in a material.

Although the work is exempt from the requirements of the three specific regulations the employer will need to comply with other more general requirements set out in CAR 2012.

Notifiable non-licensed work

Some work is not licensable work and not exempt by Regulation 3(2) of CAR 2012. This non-licensed work has a greater risk of fibre release. This work is known as notifiable non-licensed work (NNLW) and is subject to four additional specific requirements:

1) Notification of work.

2) Designation of the area where work is done

3) Medical examinations.

4) Keeping health records.

Licensed work

Certain types of work with ACMs can only be carried out by those who have been issued with a licence by HSE. This is work which meets the definition of 'licensable work with asbestos' in Regulation 2(1).

That is work:

- Where worker exposure to asbestos is not sporadic and of low intensity.

- Where the risk assessment cannot clearly demonstrate that the control limit (0.1 f/cm³ airborne fibres averaged over a four-hour period) will not be exceeded.

- On asbestos coating (surface coatings which contain asbestos for fire protection, heat insulation or sound insulation but not including textured decorative coatings).

- On asbestos insulation or AIB where the risk assessment demonstrates that the work is not sporadic and of low intensity, the control limit will be exceeded and it is not short duration work.

Short duration means the total time spent by all workers working with these materials does not exceed two hours in a seven-day period, including time spent setting up, cleaning and clearing up, and no one person works for more than one hour in a seven-day period.

If licensable work is to be carried out then the appropriate enforcing authority must be notified of details of the proposed work. Employers must carry out a risk assessment of the work. This assessment must be kept at the place where work is being carried out.

Notification and plan of work

If carrying out licensable work, it must be notified to the appropriate enforcing authority in writing 14 days before work commences (the authority may allow a shorter period in an emergency if there is a serious risk to health). The information to be notified is:

a) The particulars specified in Schedule 1 and

b) Any material change, which might affect the particulars notified (including the cessation of the work), in writing and without delay.

For NNLW (work with asbestos which is not licensable work and is not exempted by regulation 3(2)), an employer must notify the appropriate enforcing authority of:

a) The particulars specified in Schedule 1, before work is commenced; and

b) Any material change, which might affect the particulars notified, without delay.

For any work involving asbestos, the employer must draw up a written plan of work. The plan of work must include the following information:

- Nature and duration of the work.
- Number of persons involved.
- Address and location of the work.
- Methods used to prevent or reduce exposure.
- Type of equipment used to protect those carrying out the work and those present or near the worksite.

For licensable work, the plan of work should be site specific and contain the following information:

- The scope of work as identified in the risk assessment.
- Details of hygiene facilities, transit route, vacuum cleaners, air monitoring, protective clothing, respiratory protection equipment (RPE), and communication between the inside and outside of the enclosure.
- Use of barriers and signs, location of enclosures and airlocks, location of skips, negative pressure units, cleaning and clearance certification, emergency procedures.

Work must not take place unless a copy of the plan is readily available on site.

Control measures

Respiratory protection

If, despite other control measures a worker's exposure is likely to exceed the control limit or exceed 0.6f/cm^3 peak level measured over ten minutes, the employer must provide suitable RPE.

RPE must be matched to:

- The exposure concentrations (expected or measured).
- The job.
- The wearer.
- Factors related to the working environment.

Suitable RPE means:

- It provides adequate protection (i.e. reduces the wearer's exposure to asbestos fibres as low as is reasonably practicable, and anyway to below the control limits) during the job in hand and in the specified working environment (for example, confined spaces).
- It provides clean air and the flow rate during the whole wear period at least conforms to the minimum recommended by the manufacturer.
- The face piece fits the wearer correctly.
- It is properly maintained.
- The chosen equipment does not introduce additional hazards that may put the wearer's health and safety at risk.

Protective clothing

Protective clothing must be adequate and suitable. Cuffs, ankles and hoods should be elasticated and provide a tight fit at the face and neck.

Pockets or other attachments which could trap dust should be avoided. Where disposable overalls are used they should be Type 5 (under BS EN ISO 13982-1).

Training

Asbestos awareness training should be given to all workers who could foreseeably be exposed to asbestos. In particular, training should be given to workers whose work is likely to disturb the fabric of the building.

The training should include the following topics:

- The properties of asbestos and its effects on health.
- The types, uses and likely occurrence of asbestos.
- Emergency procedures.
- How to avoid the risks from asbestos.

Air monitoring

Air testing is required for a number of reasons, such as to certify clearance after asbestos has been removed (a regulatory requirement in most cases), to ensure that leaks do not occur during asbestos treatment or removal and to provide reassurance that nobody has been placed at risk.

Personal air monitoring is also necessary to ensure that RPE is providing the appropriate level of protection. Records must be kept for 5 years or, where workers are under health surveillance, for 40 years.

Medical surveillance

Regulation 22 of the Control of Asbestos Regulations (CAR) 2012 requires all employers to ensure that all employees who are exposed to asbestos are under adequate health surveillance.

For licensable work with asbestos every employer must ensure that:

a) A health record is maintained for all employees who are exposed to asbestos.

b) The health record, or a copy of it, is kept available in a suitable form for at least 40 years from the date of the last entry made in it.

c) Each employee who is exposed to asbestos is under adequate medical surveillance by a relevant doctor.

The medical surveillance required by a relevant doctor must include a medical examination not more than 2 years before the beginning of exposure to asbestos and periodic medical examinations at intervals of at least once every 2 years or such shorter time as the relevant doctor may require while exposure to asbestos continues. The medical examinations must include a specific examination of the chest.

For work with asbestos, which is not licensable work, and is not exempted by Regulation 3(2) of CAR 2012, the above requirements in (a) to (c) apply and a medical examination must take place not more than 3 years before the beginning of exposure to asbestos and a periodic medical examination are required at intervals of at least once every 3 years, or such shorter time as the relevant doctor may require while exposure to asbestos continues.

The worker health record to be maintained is not a medical confidential as it contains job exposure information and dates of previous medical examinations.

Requirements for disposal

Licensed carrier

Health and Safety Executive (HSE) requirements on the licensed contractor means that each company must have in place an up to date standard operating procedure manual, which should contain all of the details relating to the safe removal and disposal of asbestos.

This manual must also be sent to the local HSE inspector and constantly updated to reflect changes in guidance and legislation. Waste carriers licence involves a simple application and anyone looking to transport waste asbestos will need to be a Registered Waste Carrier.

Licensed disposal site

Any facility looking to accept waste asbestos has to apply to the Environment Agency for a site licence. This will set out the range and volumes of wastes that can be accepted on the site, the site control and management systems, engineering and infrastructure, manning and qualification requirements and reporting and monitoring regimes.

REVIEW

List two occupations that may expose workers to the risk of asbestosis.

Asbestos has been used in building construction and products for many years, list eight uses of asbestos.

What health effects may occur in workers from high levels of dust in the wood-working industry?

What two chronic health effects are associated with hepatitis B infection?

Sources of reference

Reference information provided, in particular web links, was correct at time of publication, but may have changed.

Asbestos essentials task sheets, HSE, http://www.hse.gov.uk/asbestos/essentials/index.htm

Asbestos: The Survey Guide, HSG264 (second edition), HSE Books, ISBN 978-0-7176-6502-0 http://www.hse.gov.uk/pubns/priced/hsg264.pdf

Control of exposure to silica dust, A guide for employees, INDG463, HSE Books, https://www.hse.gov.uk/pubns/indg463.pdf

Controlling Airborne Contaminants at Work: A Guide to Local Exhaust Ventilation, HSG258, second edition 2011, HSE Books, ISBN: 978-0-717664-15-3 http://www.hse.gov.uk/pubns/books/hsg258.htm

Control of Substances Hazardous to Health (COSHH), HSE, http://www.hse.gov.uk/coshh/index.htm

Control of substances Hazardous to health, Approved Code of Practice and guidance, HSE Books, ISBN: 978-0-7176-6582-2, http://www.hse.gov.uk/pubns/priced/l5.pdf

EC Regulation No 1272/2008 Classification Labelling and Packaging of Substances and Mixtures (CLP) http://www.hse.gov.uk/chemical-classification/legal/clp-regulation.htm

EH40/2005 Workplace exposure limits, third edition, 2018, HSE Books, ISBN: 978-0-7176-6703-1 (relates to Control of Substances Hazardous to Health), http://www.hse.gov.uk/pUbns/priced/eh40.pdf

Globally Harmonized System of Classification and Labelling of Chemicals (GHS) (revision 5 2013), United Nations ISBN 978-92-1-117067-2

Information on the Control of Asbestos Regulations 2012 http://www.hse.gov.uk/asbestos/regulations.htm

Leptospirosis (Weil's disease) – NHS, https://www.nhs.uk/conditions/leptospirosis/

Managing and working with asbestos, Control of Asbestos Regulations 2012. Approved Code of Practice and guidance ISBN: 978-0-717666-18-8 http://www.hse.gov.uk/pubns/books/l143.htm

Managing Asbestos in Buildings: A brief guide, INDG223(rev5) 2012, HSE Books http://www.hse.gov.uk/pubns/indg223.pdf

Managing skin exposure risks at work, HSG262, HSE Books, ISBN: 978-0-7176-6649-2, http://www.hse.gov.uk/pubns/priced/hsg262.pdf

Personal Protective Equipment at Work, Guidance on Regulations, L25, HSE Books, ISBN: 978-0-7176-6597-6, http://www.hse.gov.uk/pubns/priced/l25.pdf

Registration, Evaluation, Authorisation and Restriction of Chemicals (REACH) http://www.hse.gov.uk/reach/

Respiratory protective equipment at work - A practical guide, HSG53 (fourth edition 2013), HSE Books, ISBN 978-0-7176-6454-2 http://www.hse.gov.uk/Pubns/priced/hsg53.pdf

Step by Step Guide to COSHH Assessment, HSG97, second edition 2004, HSE Books, ISBN: 978-0-717627-85-1 http://www.hse.gov.uk/pubns/books/hsg97.htm

The health and safety toolbox, How to control risks at work, HSG268, HSE Books, ISBN: 978-0-7176-6587-7 , http://www.hse.gov.uk/pUbns/priced/hsg268.pdf

Wood dust, Controlling the risks, Woodworking Sheet No 23, HSE, http://www.hse.gov.uk/pubns/wis23.pdf

Wood dust, HSE website, http://www.hse.gov.uk/woodworking/wooddust.htm

Working with substances hazardous to health, a brief guide to COSHH, INDG136, HSE Books, http://www.hse.gov.uk/pubns/indg136.pdf

Web links to these references are provided on the RMS Publishing website for ease of use - www.rmspublishing.co.uk.

Statutory provisions

Control of Substances Hazardous to Health Regulations (COSHH) 2002 / Control of Substances Hazardous to Health Regulations (Northern Ireland 2002

Personal Protective Equipment at Work Regulations (PPER) 1992 / Personal Protective Equipment at Work Regulations (Northern Ireland) 1993

The Control of Asbestos Regulations (CAR) 2012

Workplace (Health, Safety and Welfare) Regulations (WHSWR) 1992 / Workplace (Health, Safety and Welfare) Regulations (Northern Ireland) 1993

STUDY QUESTIONS

1) Insulation board tiles that contain asbestos are to be removed from the ceiling of a storage area located within a high street department store. What measures should be taken into consideration when planning the work to minimise the risks from asbestos?

2) Exposure to cement dust can lead to adverse health effects if breathed in, for example, irritation of the nose and throat.

   (a) What is the meaning of the term 'workplace exposure limit'?

   (b) What control measures could be considered to control levels of cement dust in the workplace?

3) What FOUR hazardous substances are widely encountered in the construction industry AND give an associated health risk for EACH?

4) What information should be included in a safety data sheet (SDS) for hazardous chemicals?

5) (a) What FOUR forms of chemical agents might be found in construction workplaces?

   (b) What are the differences between the acute and chronic effects from exposure to chemical agents?

For guidance on how to answer these questions please refer to the 'study question answer guidance' section located at the back of this guide.

Element 13

Physical and psychological health

Contents

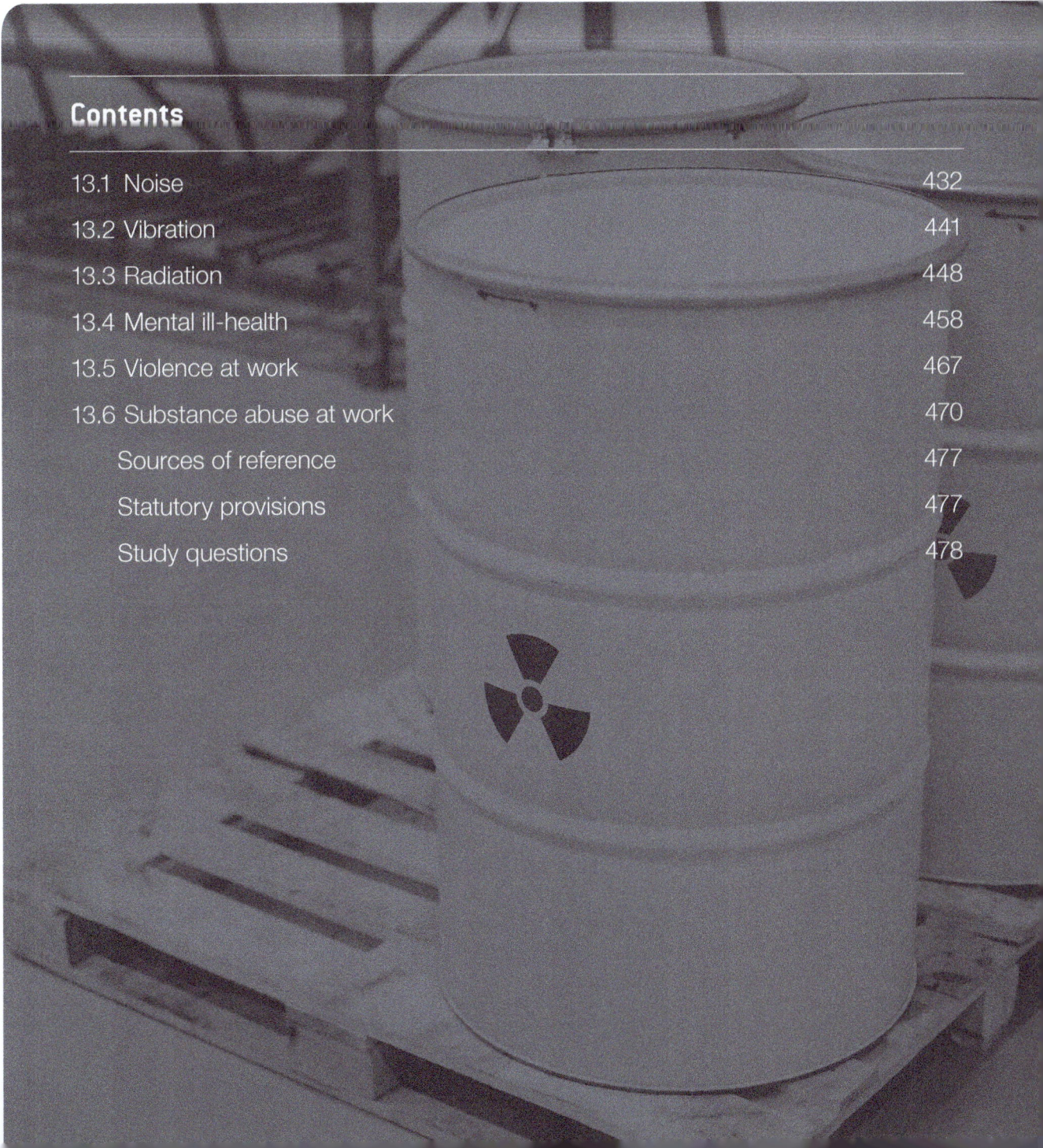

INTRODUCTION

In addition to chemical and biological health hazards there are also physical and psychological hazards which can affect the health of workers. These hazards are covered by regulations that give specific values to exposure and that form the basis for controlling the hazards posed by noise, vibration, ionising and non-ionising radiation, and the effects of work-related stress.

13.1 Noise

PHYSICAL AND PSYCHOLOGICAL EFFECTS ON HEARING OF EXPOSURE TO NOISE

The ear senses **sound**, which is transmitted in the form of pressure waves travelling through a substance, for example, air, water, metals. Any audible sound is noise. The ear has three basic regions **(see Figure 13-1).**

1) The **outer** ear channels the sound pressure waves along the ear canal where they impact on the eardrum, causing it to vibrate.

2) In the **middle** ear, the vibrations of the eardrum are transmitted through the three smallest bones in the body (known as the hammer, anvil and stirrup) to the inner ear.

3) In the **inner ear** the vibrations are transferred to the cochlea (the 'hearing' organ). The cochlea is filled with fluid and contains tiny hair cells (nerves), which respond to the vibrations by bending. Movement of these tiny hair cells causes signals to be sent to the brain via the acoustic/auditory nerve where it is interpreted as recognisable sound.

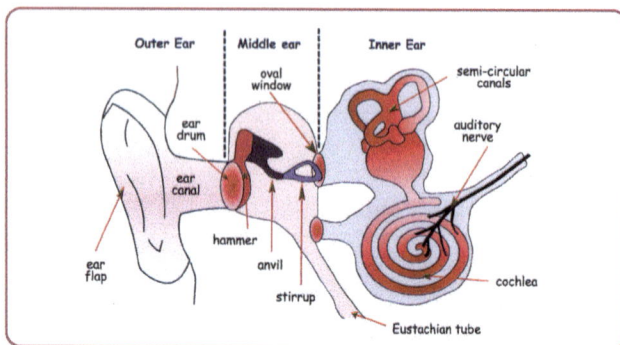

Figure 13-1: Diagram of the ear.
Source: eChalk Ltd.

Physical effects of noise

Exposure to high levels of noise can make the hair cells in the cochlea collapse and flatten. Short-term exposure usually results in the worker experiencing temporary effects, whereas long-term exposure often over a number of years, can result in permanent effects.

Typical hearing effects can include:

- Tinnitus or ringing in the ears, which may be either acute or chronic

- A threshold shift in hearing which can result in a temporary (acute) or permanent (chronic) inability to hear certain sounds. These two conditions are known as Temporary Threshold Shift (TTS) and Permanent Threshold Shift (PTS).

- Noise-induced hearing loss (NIHL), which can be of a temporary or permanent nature and can affect one ear or both. NIHL can be caused by a single exposure to an intense impulse sound, such as an explosion, or by continuous exposure to loud sounds over a period of time. Most NIHL is caused by permanent damage to the hair cells in the cochlea.

The Health and Safety Executive and the Labour Force Survey have both reported statistically on the prevalence of Noise Induced Hearing Loss (NIHL) compared with other diseases:

> "The 2018/2019 Labour Force Survey (LFS) shows an estimated 23,000 individuals who worked in the previous 12 months, who believed their hearing problems were caused or made worse by work."

Figure 13-2: Exposure to noise and noise-induced hearing loss.
Source: Labour Force Survey.

Permanent noise-induced hearing loss is cumulative, occurring gradually over a long period of time, and when established, the worker's hearing cannot recover.

The first sign of NIHL is often indicated by a difficulty in hearing high-pitched sounds, such as consonants (for example, 't', 'd', 's') and the voices of women and children. When more than one person is speaking or there is a background noise, the problem becomes worse.

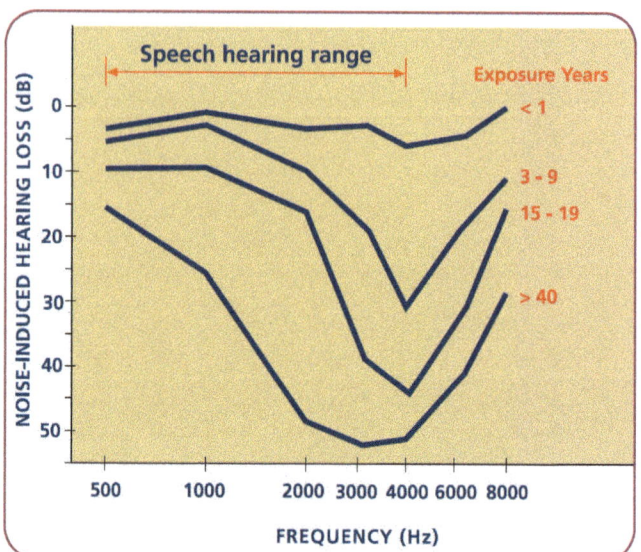

Figure 13-3: Effects of high levels of noise.
Source: Australia, SafeWork SA

In addition to the more constant noise exposure discussed earlier, a single very intense or explosive noise can damage the ear by dislocation of a bone or rupturing the ear drum. This is known as acoustic trauma. Background noise can also cause those with normal hearing ability to fail to hear warnings such as alarms, moving-vehicle warning horns, shouted warnings or instructions and other alarms that may sound.

Psychological effects of noise

Noise is often linked with adverse psychological effects such as stress, sleep disturbance or aggressive behaviour. It is frequently cited as the cause of conflict between workers, particularly in a noisy office environment where some individuals may need to concentrate on complex issues, but they find this difficult or impossible because of background noise levels.

In addition, the loss of hearing in the speech hearing range leads to a feeling of isolation as the person affected cannot contribute so easily to conversations, and pastimes may become affected, for example, listening to music, radio and television.

> ### REVIEW
>
> What are the possible effects on hearing from exposure to noise?
>
> What are the causes of noise-induced hearing loss?

THE MEANING OF COMMONLY USED TERMS

Sound power and pressure

For noise to occur power must be available. It is the sound power of a source (measured in watts) that causes the sound pressure (measured in pascals, Pa) to occur at a specific point.

Intensity and frequency

The *amplitude* of a sound wave represents the intensity of the sound pressure. When measuring the *amplitude* of sound there are two main parameters of *interest (as shown in Figure 13-4).* One is related to the energy in the sound pressure wave and is known as the 'root mean square' (rms) value, and the other is the 'peak' level.

We use the 'rms' sound pressure for the majority of noise measurements, apart from some impulsive types of noise when the peak value is also measured.

A sound can have a *'frequency'* or *'pitch',* which is measured in cycles per second (Hz). Frequency in this context represents the number of times in a given time period that the sound wave repeats itself.

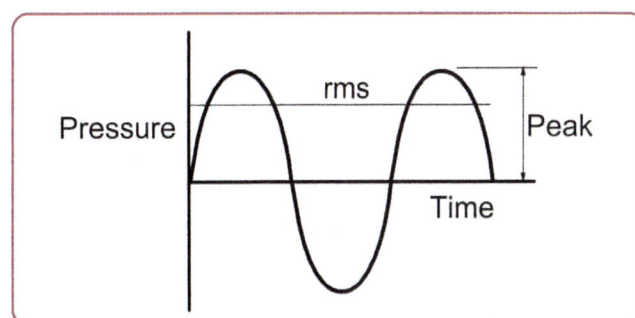

Figure 13-4: Rms and peak levels of a sound wave.
Source: RMS.

The decibel scale

Sound intensity or pressure is measured in a unit known as a pascal (Pa). The ear can detect pressures over a very wide range, from 20 µPa - 20 Pa. To measure in pascals therefore requires a very large range to exist and this is often inconvenient. A more helpful way of measuring sound is to use the decibel range. A decibel (dB) is a unit of sound pressure (intensity) measured on a logarithmic scale from a base level taken to be the threshold of hearing (0 dB). Typical noise levels are listed in *Figure 13-5:*

Source	DD
CHAINSAW	120
SMOKE DETECTOR AT 1 METRE	105
MACHINE SHOP	90
RADIO IN AVERAGE ROOM	70
LIBRARY	30
THRESHOLD OF HEARING	0

Some typical noise levels for equipment used on construction sites are:

DISC CUTTER	99-115 db
HAMMER DRILL	102-111 db
BREAKERS	103-113 db
EARTHMOVER	87-94 db
PILING	100 db
SCRABBLING	100db
CARPENTRY	92 db

Figure 13-5: Typical noise levels in construction.
Source: RMS .

A problem with decibels is that they are based on a logarithmic scale and cannot be added together in the conventional way.

The basic way to interpret decibels is that whenever noise/sound doubles in intensity (loudness) the decibel reading will increase by 3; and when it halves the decibel reading will decrease by 3; for example:

- 2 dB + 2 dB = 5 dB (deciBel arithmetic).
- 85 dB + 85 dB = 88 dB (deciBel arithmetic).
- A 3 dB increase in sound pressure level corresponds to a doubling of the sound energy.
- A 10 dB increase in sound pressure level corresponds to a 10 times increase of the sound energy.
- A 20 dB increase in sound pressure level corresponds to a 100 times increase of the sound energy.

Weighting scales - the terms dB(A) and dB(C)

The human ear can hear sound over a range of frequencies, from 20 Hz up to approximately 20,000 Hz (20 kHz). However, the ear does not hear the same at all frequencies; it naturally reduces (attenuates) low frequencies and very high frequencies, respectively, to the range of speech. To take account of the response of the human ear, sound level meters use weighting scales or filters. The most widely used sound level filter is the A scale. Using this filter, the sound level meter is thus less sensitive to very high and very low frequencies. Measurements made on this scale are expressed as dB(A) or referred to as 'A-weighted'.

The majority of measurements are made in terms of dB(A), although other weightings are used in some circumstances. One of these is the C scale, which is used to assess very high sound pressure levels such as the acoustic emissions of machines. The C scale is used to determine peak sound pressure levels and is particularly useful for impact or explosive noises. It has a broader spectrum than the A-weighted scale and is more accurate at higher levels of noise. Measurements made on this scale are expressed as dB(C).

ASSESSMENT OF EXPOSURE TO NOISE

When exposure should be assessed

Employers have a responsibility under the Control of Noise at Work Regulations (CNWR) 2005 to assess the risks from noise in order to reduce the risk of hearing damage to their workers by controlling exposure to noise. The employer should make a suitable and sufficient assessment of the risk created by work that is liable to expose workers to risk from noise at or above the lower exposure action value of 80 dB.

The UK sets action levels and limits relating to exposure to noise in the workplace, and such levels and limits need to be considered when assessing the risks faced by workers and others.

The assessment should establish which workers are at risk of hearing damage and the level of risk. It will also identify sources of noise that particularly contribute to the noise level workers are exposed to, for example, equipment and specific activities. This will enable analysis of the options available to control the noise at source or by other means. The assessment will also help the employer when providing suitable hearing protection, marking out hearing protection zones and giving information, instruction and training to workers. The results of the risk assessment and controls implemented should be recorded to provide evidence of a structured risk assessment process.

Risk assessment

When conducting the risk assessment, consideration should be given to workers at particular risk. The assessment will need to identify the type, duration, effects of exposure including any additional exposure at the workplace (for example, in rest facilities) to determine whether the exposure limit/action values have been exceeded. Information from health surveillance records and manufacturer's information on noise levels should also be reviewed. The level of noise workers are exposed to should be assessed by:

- Observation.
- Reference to information on expected levels for work conditions and equipment.
- If necessary by measurement of the level of noise to which their workers may be exposed.

It is not always necessary to carry out measurements of noise exposure as part of the assessment; an estimate of noise levels may be enough to decide that controls are required. Estimation by observation may be sufficient to indicate there is a noise problem. One example involves the assessor determining how easy it is for two people to hold a conversation at a distance of one or two metres from each other.

Two metre rule: If conversation is difficult (need to raise the voice or repeat words) at a distance of two metres apart the noise level is likely to be above 85 dB.

One metre rule: If conversation is difficult (need to raise the voice or repeat words) at a distance of one metre apart the noise level is likely to be above 90 dB.

If it is necessary to be more certain whether noise exposure levels exceed acceptable values, measurements of actual noise levels may be required. A detailed assessment should include consideration of:

- Level, type and duration of exposure, including any exposure to peak sound pressure.
- Effects of exposure to noise on workers or groups of workers whose health is at particular risk from such exposure.

Test	Probable noise level	A risk assessment will be needed if the noise is at the level for more than:
The noise is intrusive but normal conversation is possible	80 dB	6 hours
You have to shout to talk to someone 2 m away	85 dB	2 hours
You have to shout to talk to someone 1 m away	90 dB	45 minutes

Figure 13-6: Simple tests to get a rough estimate on whether a risk assessment is required.
Source: RMS.

- Indirect effects on the health and safety of workers resulting from the interaction between noise and audible warning signals.
- Information provided by the manufacturers of work equipment.
- Availability of alternative equipment designed to reduce the emission of noise.
- Any extension of exposure to noise at the workplace beyond normal working hours, including exposure in rest facilities supervised by the employer.
- Appropriate information obtained following health surveillance, including, where possible, published information,
- Availability of personal hearing protectors with adequate noise-reduction (attenuation) characteristics.

Comparison of measurements to exposure limits established by recognised standards

Noise in the workplace is a likely hazard related to machinery and work processes and is an example of a physical agent risk that should be controlled. As people respond differently to noise, the level at which noise will start to cause harm to the hearing of workers varies.

Research has established that long periods of repeated exposure to workplace noise levels between 75dB(A) and 80dB(A) present a small risk of the average worker developing a hearing disability. As noise levels increase, the risk becomes greater. For example, exposure to noise levels of 90dB(A) - 95dB(A) presents a considerably greater risk of a worker developing hearing disability.

Standards concerning acceptable levels of noise, and therefore exposure limits, are usually based on an 8 hour work period, they may also provide exposure limits for shorter and longer working periods.

It should be remembered that exposure to noise below the acceptable noise level does not mean a safe condition exists; it means that an eight-hour exposure to the acceptable noise level is considered to represent an acceptable level of risk to workers hearing in the workplace.

Some workers may be affected by noise exposures below this level and therefore UK legislation will include a requirement to reduce exposure as far as is reasonably practicable and at least below the legally acceptable noise level.

The amount of damage caused by noise depends on the total amount of energy received by the worker's hearing over time. This means, as noise becomes more intense it causes damage to hearing in less time. A 3dB(A) increase in noise level will produce twice the sound energy and cause the same damage in half the time.

Therefore, the acceptable duration of exposure at this noise level is halved. For example, if a worker is exposed to a noise level of 88dB(A) the acceptable duration would be 4 hours, half that of a worker exposed to 85dB(A), 8 hours. Similarly, 15 minutes of working in noise levels of 100dB(A) may cause the same damage as 8 hours working in 85dB(A).

Noise is measured using a sound pressure level meter, which works, in simple terms, by converting pressure variations into an electrical signal.

Figure 13-7: Noise measurement.
Source: Pulsar Instruments Plc.

This is achieved by capturing the sound with a microphone, pre-amplification of the resultant voltage signal and then processing the signal into the information required dependent on the type of meter. For example, 'A' weighting, the A-weighting filter covers the full audio range - 20 Hz to 20,000 Hz and represents the audible response of the human ear at these levels.

The microphone is the most critical component within the meter as its sensitivity and accuracy will determine the accuracy of the final reading. Meters can be set to fast or slow response depending on the characteristics of the noise level. Where levels are rapidly fluctuating, rapid measurements are required and the meter should be set to fast time weighting.

Measurement

In most situations, workers are exposed to variations in noise level over a period of time. When measuring noise we need to determine the average, or 'equivalent continuous level', over a period of time. This is often known as the L_{eq}. Sound pressure level meters are used to measure the:

- Equivalent continuous sound level (L_{eq}) - an average measure of intensity of sound over a reference period, usually the period of time over which the measurement was taken. Measured in dB(A).

- Daily personal exposure level, dB(A), $L_{EP,d}$ - this is equivalent to the L_{eq} over an 8-hour working day. The $L_{EP,d}$ is directly related to the risk of hearing damage.

- Peak pressure level, L_{peak} - this is the peak level of the sound pressure wave with no time constant applied. This is the loudest noise experienced during the measuring process. For noise at work measurements, the peak level should be measured in dB(C).

Personal dosimeters are available for situations where the task of the worker involves movement around the workplace and exposure is likely to vary. Sound pressure level meters should be calibrated using a portable acoustic calibrator and batteries checked before, during and after each measurement session. Laboratory calibration should be carried out annually or according to the manufacturer's instructions.

Workers or their representatives should be consulted, and significant findings, measures taken or planned to control the risks should be recorded. Reassessment should be carried out after action is taken to control exposure, or after a reasonable time, to establish the effectiveness of the controls.

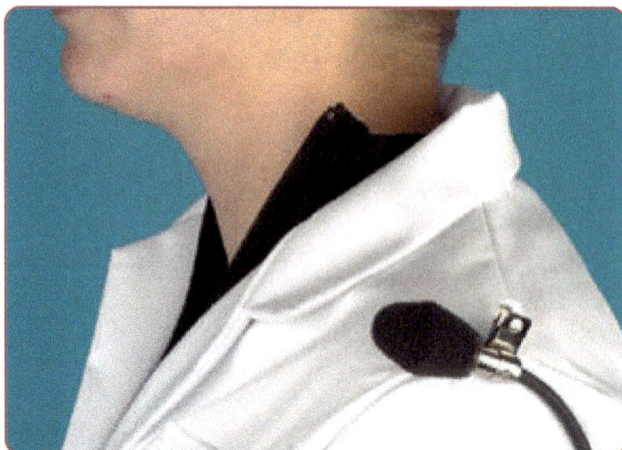

Figure 13-8: Correct position of a personal noise dosimeter.
Source: UK, HSE L108.

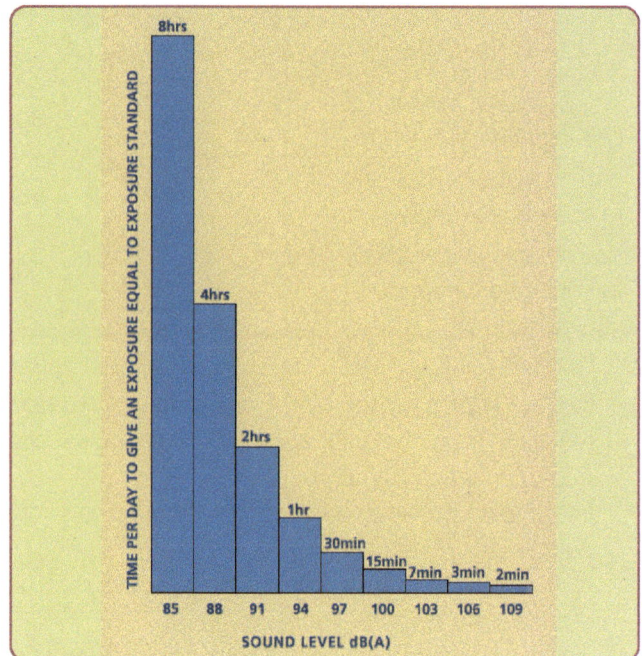

Figure 13-9: Exposure times.
Source: Australia, SafeWork SA.

It should be remembered that exposure to noise below the acceptable noise level does not mean a safe condition exists; it means that an 8-hour exposure to the acceptable noise level is considered to represent an acceptable level of risk to workers' hearing in the workplace. Some workers may be affected by noise exposures below this level, and therefore most national legislation will include a requirement to reduce exposure as far as is reasonably practicable, and at least below the legally acceptable noise level.

The amount of damage caused by noise depends on the total amount received by the worker's hearing over time. This means, as noise becomes more intense it causes damage to hearing in less time. A 3 dB(A) increase in noise level will produce twice the sound energy and cause the same damage in half the time. Therefore, the acceptable duration of exposure at this noise level is halved. For example, if a worker is exposed to a noise level of 88 dB(A) the acceptable duration would be 4 hours, half that of a worker exposed to 85 dB(A), 8 hours.

Similarly, 15 minutes of working in noise levels of 100 dB(A) may cause the same damage as 8 hours working in 85 dB(A).

Action and limit values

The UK's graduated approach to occupational exposure values has established a maximum acceptable noise exposure limit (exposure limit value) and two action levels. The exposure limit value is fixed at 87 dB(A) and the exposure action levels are fixed at 80 dB(A) (known as the lower level); and 85 dB(A) (upper level), taking into account the attenuation provided by any personal hearing protection worn by the workers. These are levels of 'daily average noise exposure', as a time-weighted average over an 8-hour working day. If a worker's exposure to noise in the workplace varies greatly, the employer can choose to calculate weekly instead of daily noise exposure in determining whether levels or limits are exceeded. Weekly noise exposure means the average of daily noise exposures over a week and normalised to 5 working days.

The UK's approach also recognises the potential harm that may be done by particularly high-intensity noise experienced over a short period and sets limit and action levels for this situation in the form of peak sound pressures.

Lower exposure action values

Where a worker is likely to be exposed to noise at or above the lower exposure action values, the daily or weekly exposure of 80dB(A) or a peak sound pressure of 135dB(C), the employer must make hearing protection available upon request and provide the workers and their representatives with suitable and sufficient information, instruction and training. This must include:

- The nature of risks from exposure to noise.

- Organisational and technical measures taken in order to comply.

- Exposure limit values and upper and lower exposure action values.

- Significant findings of the risk assessment, including any measurements taken, with an explanation of those findings.

- Availability and provision of personal hearing protectors and their correct use.

- Why and how to detect and report signs of hearing damage.

- Entitlement to health surveillance.

- Safe working practices to minimise exposure to noise.

- The collective results of any health surveillance in a form calculated to prevent those results from being identified as relating to a particular person.

Figure 13-11: Noise hazard sign.
Source: RMS.

Upper exposure action values

Where the noise exposure of a worker is likely to be at or above the upper exposure action values, the daily or weekly exposure of 85dB(A) or a peak sound pressure level of 137dB(C), the employer must:

- Provide workers with hearing protection, and ensure that it is worn.

- Ensure that the area is designated a hearing protection zone, fitted with mandatory hearing protection signs.

- Ensure access to the area is restricted where practicable.

- So far as reasonably practicable, ensure those workers entering the area wear hearing protection.

Figure 13-12: Mandatory hearing protection sign.
Source: ISO 1710, Graphical symbols: safety colours and safety signs.

	Lower exposure action level values	Upper exposure action level values	Exposure limit values
Daily or weekly personal noise exposure (A-weighted)	80 dB	85 dB	87 dB
Peak sound pressure (C-weighted)	135 dB	137 dB	140 dB

Figure 13-10: Noise exposure values.
Source: UK, Control of Noise at Work Regulations.

Exposure limit values

The employer must ensure that workers are not exposed to noise above an exposure limit value. If an exposure limit value is exceeded the employer must immediately:

- Reduce exposure to below the limit level.
- Identify the reason for the limit value being exceeded.
- Modify the organisational and technical measures to prevent a recurrence.

BASIC NOISE CONTROL MEASURES

The requirements for noise control measures set out in Regulation 6 of Control of Noise at Work Regulations (CNWR) 2005 follow the general principles of prevention set out in the Management of Health and Safety at Work Regulations (MHSWR) 1999. Employers must ensure that risk from the exposure of their employees to noise is either eliminated at source or, where this is not reasonably practicable, reduced to as low as is reasonably practicable. Consideration should be given to reducing noise at source, blocking the transmission path and preventing worker exposure.

Reducing noise at the source

A company policy should be established which purchases only the quietest equipment and replaces outdated, noisy machinery for existing workplaces. This can be done in a number of ways, *see Figure 13-13.*

Blocking the noise transmission path

This can be done by carrying out the following:

- Relocating noisy machines or processes to remote areas of the workplace.
- Fitting sound-absorbent materials to ceilings and walls.

- Enclosing noisy machinery within sound-absorbent materials.
- Mounting noisy floor-standing machinery on rubber pads to isolate the machine from the floor and reduce vibration.
- Fitting flexible or fixed screens or curtains of sound-absorbent (insulation) material.
- Fitting damping material to panels that vibrate.

Noise can be controlled at different points in the transmission path that lead to exposure, *see Figure 13-14.*

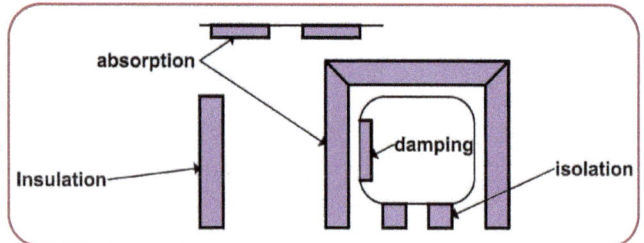

Figure 13-14: Main transmission control methods.
Source: RMS.

Preventing worker exposure to noise (administrative controls)

This can be done by implementing the following:

- Scheduling noisy work for times when as few workers as possible are present.
- Building sound-proof booths (refuges) for operators of noisy equipment to work in and to enable them be insulated from the noise.
- Minimising the amount of time a worker is exposed to noise, for example, by job rotation.

The employer has to take care that noise levels in rest facilities are suitable for their purpose. The employer should adapt measures provided to suit a group of workers or individual worker whose health is likely to be at particular risk from exposure to noise at work. Workers or their representatives should be consulted on the measures used.

Enclosure:	Total enclosure to contain noise at source.
Isolation:	A form of separation between the noise and the worker by distance or use of a barrier of an absorbent character (for example, an acoustic absorbent wall) in the path of noise transmission or relocation of workers into another room remote from the noise source.
Absorption:	When noise passes through porous materials (for example, foam, mineral or wool) some of its energy is absorbed.
Insulation:	Imposing a barrier (for example, a brick wall or lead sheet) between the noise source and the workers.
Lagging:	Insulation of pipework and fluid containers to reduce noise levels.
Damping:	Mechanical vibration can be reduced through conversion into heat by damping materials, for example, use of plastic gears, rubber coating on conveyors or rollers, reinforcement of metal panels.
Silencing:	Pipes/boxes can be designed to reduce air/gas noise (for example, engine exhaust silencers or duct silencers).
Work practice:	Modify material-handling processes to reduce the noise from shock and impact, for example, reducing the distance where objects fall onto hard surfaces or fixing damping material to surfaces or containers. Review frequency of maintenance programmes, for example, equipment lubrication and replacement of worn bearings.

Figure 13-13: Ways of reducing noise at source.
Source: RMS.

REVIEW

What control measures could be used to help reduce noise levels in a workplace?
Explain the 'two metre' rule in relation to noise in the workplace.

PERSONAL HEARING PROTECTION

Purpose

The purpose of personal hearing protection is to protect the user from the adverse effects on hearing caused by exposure to high levels of noise. All hearing protection must be capable of reducing exposure to below the exposure limit values - daily/weekly exposure 87dB(A) averaged over 8 hours per day and peak sound pressure of 140dB(C), established by the CNWR 2005.

The provision of hearing protection should only be considered after all attempts to reduce the exposure to noise by other means have proved ineffective in reducing noise levels satisfactorily, or where exposure above any nationally imposed action levels requiring hearing protection to be provided exists.

Figure 13-16: Ear muff with two-way communication.
Source: Davro Online Safety.

Application and limitations of personal hearing protection

A wide range of custom-fit earplugs is now available for people who work in noisy environments such as working with machinery, maintenance activities or musicians.

Custom-made earplugs ensure a perfect seal to the ear canal. The fit is designed to replicate the exact shape of each individual ear canal and this ensures a comfortable fit, greatly reducing workers' reluctance to wear them.

Custom-fit earplugs require impressions of the ear to be taken first by a trained audiologist so that a proper seal can be created. Normally it takes four to six weeks to make the earplugs after the impressions are taken.

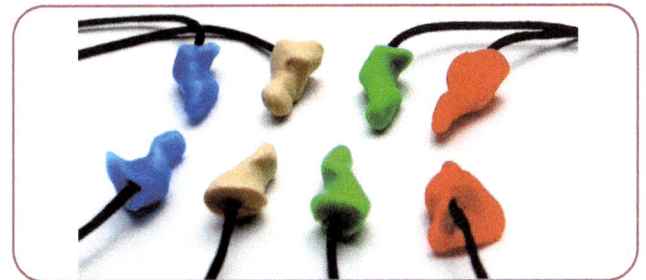

Figure 13-15: Helmet mounted ear muffs.
Source: RPA.

Figure 13-17: Custom-fit earplugs.
Source: Wikimedia Commons.

Earmuffs:	Application:	Limitations:
These completely cover the ear and can be: • Banded. • Helmet mounted. • Communication muffs.	• Worn on the outside of the ear so less chance of infection. • Clearly visible therefore easy to monitor. • Can be integrated into other forms of personal protective equipment (PPE), for example, head protection.	• Can be uncomfortable when worn for long periods. • Incompatibility with other forms of PPE. • Effectiveness may be compromised by, for example, long hair, spectacles. • Requires correct storage facilities and regular maintenance.
Earplugs:	**Application:**	**Limitations:**
These are inserted in the ear canal and can be: • Pre-moulded. • User formable. • Custom moulded. • Banded plugs.	• Easy to use and store - but must be inserted correctly. • Available in many materials and designs, disposable. • Relatively lightweight and comfortable. Can be worn for long periods.	• They are subject to hygiene problems unless care is taken to keep them clean. • Correct size may be required. Should be determined by a competent person. • Interferes with communication. • Worn inside the ear, difficult to monitor.

Figure 13-18: Application and limitations of various types of hearing protection.
Source: RMS.

Selection

When selecting personal hearing protectors employers must take into consideration several factors, including:

- Noise attenuation (reduction) capability.
- Compatibility with other PPE.
- Suitability for the work environment.
- Readily available.
- Comfort and personal choice.
- Issue to visitors.
- Provision of information and training.
- Care and maintenance.

Use

All personal hearing protectors should be used in accordance with employer's instructions, which should be based on manufacturer's instructions for use. Personal hearing protectors should only be used after adequate training has been given. Also, adequate supervision must be provided to ensure that training and instructions are being followed. Personal hearing protectors may not always provide adequate protection, due to any of the following reasons:

- Inadequate or lack of training
- Not fitted properly
- Long hair, spectacles or earrings may cause a poor seal to occur with banded ear muffs.
- Not wearing personal hearing protectors all of the time.
- Personal hearing protectors may become damaged, for example, ear muffs cracked.
- Specification of personal hearing protectors does not provide -sufficient attenuation.

Maintenance

Employers should ensure that any personal hearing protection used are maintained in efficient working order. Simple maintenance can be carried out by the trained wearer, but more intricate repairs should only be carried out by specialist personnel. Maintenance in this context includes actions to keep personal hearing protection in use in good order. This will include the timely disposal of personal hearing protection that no longer affords adequate protection due to the effects of use.

Attenuation factors

It is essential to match the attenuation (noise reduction) provided by a personal hearing protector to the noise level and the desired attenuation performance. In addition, the frequency of the noise may need to be taken into account to ensure the personal hearing protector performs effectively at the frequency of the noise to which the worker is exposed. The attenuation data associated with personal hearing protectors should be supplied by the manufacturer with the product. It may come in a number of forms, using rating numbers that are applicable in different parts of the world. The rating numbers can provide a quick guide to the effectiveness of personal hearing protectors.

SNR (Single Number Rating) - this rating number is used by the European Union and affiliated countries. In addition to an overall rating, the SNR further rates personal hearing protectors in terms of the particular noise environments in which they will be used: H for high-frequency noise environments, M for mid-frequency, and L for low-frequency.

The HML designation does not refer to noise level, but to the spectrum of noise. The noise we hear is made up of a wide range of different frequencies. Frequencies that are close together form a 'bandwidth' which, within the field of noise, is termed an 'octave band'. An octave band is defined where the upper frequency is twice the lower frequency thus a frequency range of 22Hz to 44 Hz would form one band width or Octave. The octave band is given by citing the centre frequency (31.5 Hz in the bandwidth given).

Common octave bands are 31.5 Hz, 63Hz, 125Hz, 250Hz, 500Hz, 1kHz, 2kHz, 4kHz and 8KHz. The standard test frequencies are 63Hz to 8kHz, therefore a personal hearing protector might be rated with an SNR of 26, H=32, M=23, and L=14.

The SNR value is the result of a lengthy mathematical calculation; it gives a single-number rating of a personal hearing protector's attenuation for a specified percentage of the population. The SNR is significantly lower than the average attenuation across all of its test frequencies as the calculation contains correction factors to make it applicable to the broader population. While it is not the perfect measure of attenuation, SNR is a very useful standardised method for describing a personal hearing protector's attenuation in a single number.

For example, if an environment has a weighted noise measurement of 100 dB(A) then by using an earplug with a SNR rating of 26 dB it will reduce the noise to 74 dB(A).

To determine the predicted noise attenuation (PNA) it is necessary that a noise reading be taken in both A-weighting and C-weighting mode, at the worker's ear. If the difference between the A-weighted reading and the C-weighted reading is greater than 2 a formula using the medium and low values is used; if it is less than or equal to 2 a formula using the high and medium value is used. This method enables the effects of the frequency of the noise to be taken into account by using a sound level meter with A and C weighting.

If readings have been taken using an octave band sound level meter, the octave band method can be used to produce a more accurate reflection of the effectiveness of hearing protectors. This can help to ensure a close match between the hearing protector and the noise concerned. It involves quite complicated mathematics, so to assist employers, the UK Health and Safety Executive (HSE) has produced a calculator, which can be found at http://www.hse.gov.uk/noise/calculator.htm. Calculators are also available for SNR and HML methods.

Whichever method is used to determine the PNA of hearing protection, in the UK, the HSE recommends reducing the value by 4 dB to reflect what they see as 'real-world factors'. This takes account of variances in fit or other user factors that may limit the effectiveness of the hearing protection.

THE ROLE OF HEALTH SURVEILLANCE

The role of health surveillance is to provide early detection of work-related ill-health. Audiometry is a medical testing procedure that establishes hearing sensitivity across a range of sound frequencies, which can then be monitored over time. It will assist with the identification of noise hazards and the evaluation of noise control measures.

By conducting health surveillance from the start of a worker's assignment to work it is possible to detect early signs of hearing loss and provide early intervention to limit the continuing effects.

Health surveillance can assist with confirming the success of noise controls at source and support the promotion of personal hearing protection. If a risk assessment indicates a risk to the health and safety of workers who are, or are liable to be, exposed to noise they should be put under suitable health surveillance (including testing of their hearing). The employer must keep and maintain a suitable health record. The employer should, providing reasonable notice is given, allow the worker access to their health record.

Where, as a result of health surveillance, a worker is found to have identifiable hearing damage, the employer should ensure that the worker is examined by a doctor. If the doctor, or any specialist to whom the doctor considers it necessary to refer the worker, considers that the damage is likely to be the result of exposure to work-related noise, the employer should:

- Ensure that a suitably qualified person informs the worker accordingly.
- Review the risk assessment.
- Review any measure taken to control the noise risk.
- Consider assigning the worker to alternative work.
- Ensure continued health surveillance.
- Provide for a review of the health of any other worker who has been similarly exposed.

Workers must, when required by the employer and at the cost of the employer, present themselves during working hours for health surveillance procedures.

🔍 REVIEW

List four items of personal hearing protection.
What control measures could be used to ensure a new worker's hearing protection in a noisy work environment?
Explain why it is important to determine noise action levels in some workplaces.

13.2 Vibration

INTRODUCTION

Hand-held vibrating equipment and machinery can produce risks to the health of workers through hand-arm vibration - often known as HAVS (hand-arm vibration syndrome). Standing or sitting on vibrating plant or machinery can result in WBVS (whole-body vibration syndrome).

Employers should seek to eliminate work which exposes workers to vibration risk or, where this is not reasonably practicable, assess the risk and seek to reduce or control the exposure to the lowest practicable level.

Advice for eliminating, limiting and controlling exposure exists in the Control of Vibration at Work Regulations (CVWR) 2005.

THE EFFECTS ON THE BODY OF EXPOSURE TO VIBRATION

Occupational exposure to vibration may arise in a number of ways, often reaching workers at intensity levels disturbing to comfort, efficiency, health and safety. Long-term, regular exposure to vibration is known to lead to permanent and debilitating health effects such as vibration white finger, loss of sensation, pain, and numbness in the hands, arms, spine and joints. These effects are collectively known as hand-arm or whole-body vibration syndrome. In the case of whole-body vibration it is transmitted to the worker through a contacting or supporting structure, which is itself vibrating, for example, a ship's deck, the seat or floor of a vehicle (tractor or dumper truck, - used for transporting loose material on a construction site), or where a whole structure is shaken by machinery (for example, in the processing of coal, iron ore or concrete), where the vibration is intentionally generated for impacting.

By far the most common route of harm to the human body is through the hands, wrists and arms of the subject - so-called segmental vibration, where there is actual contact with the vibrating source.

Measurement

The assessment of vibration exposure (acceleration) is measured in meters per second squared (m/s2). The risk to health from vibration is affected by the frequency content of the vibration. When vibration is measured in accordance with BS EN ISO 5349-1:2001, vibration frequencies between 8 and 16 Hz are most important, and frequencies above and below this range make a smaller contribution to the measured vibration magnitude.

This process is called frequency weighting. Vibration meters intended for HAV and WBV measurement are equipped with a frequency weighting filter, to modify their sensitivity at different frequencies of vibration. In order to establish a complete measurement of vibration exposure it is necessary to measure the vibration on three axes, 'x', 'y' and 'z'; as shown in **Figure 13-19**, for the hand which relates to the axes in which the vibration is entering the hand and **Figure 13-20**, for the whole body. Where 'x' is taken as before and after vibration, 'y' as side-to-side vibration and 'z' as vertical vibration.

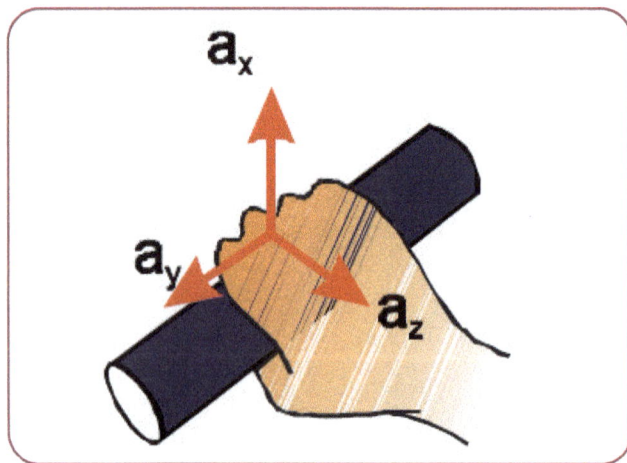

Figure 13-19: Three axes of vibration measurement for the hand.
Source: UK, HSE, RR795.

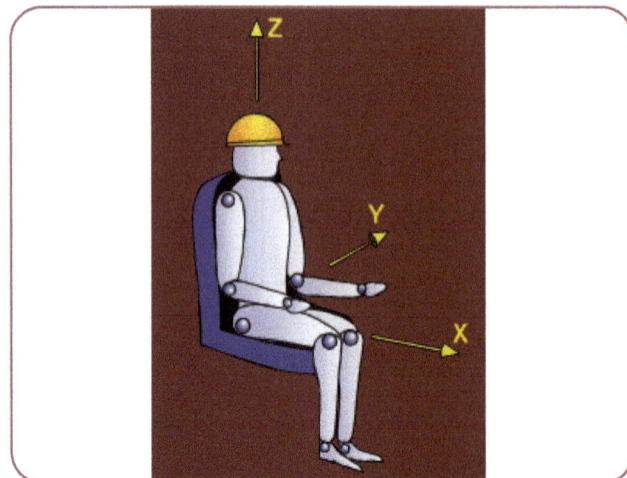

Figure 13-20: Three axes for whole body vibration measurement.
Source: UK, HSE.

The vibration is measured in one axis at a time and once all three axes have been individually measured the combined vibration exposure of the worker can be determined.

The amount of exposure is determined by measuring acceleration in the units of m/s². Several types of instruments are available for measuring acceleration, the rate of change of velocity in speed or direction per unit time (for example, per second). Measuring acceleration can also give information about velocity and amplitude of vibration. The degree of harm is related to the magnitude of acceleration.

A typical vibration measurement system includes a device to sense the vibration (accelerometer), and an instrument to measure the level of vibration. The accelerometer produces an electrical signal and the size of this signal is proportional to the acceleration applied to it. A weighting system provides a single number as a measure of vibration exposure which is expressed as the frequency-weighted vibration exposure in metres per second squared (m/s²), units of acceleration.

The exposure values are expressed in terms of m/s² A(8), which expresses a person's exposure as an average over an eight-hour period. It does not fully represent the risks which are generated by vibration when it includes severe shocks and jolts (for instance, when driving over potholes or large rocks), which are considered to be an important risk factor in back pain. It is possible to make a basic assessment of the severity and frequency of shocks and jolts by observing the working vehicle and the movement of the driver in the seat, or by asking the driver about them.

Hand-arm vibration

The European Physical Agents (Vibration) Directive (2002/44/EC) deals with risks from vibration at work and distinguishes between vibration affecting the hand-arm system and the whole body. **See Figure 13-21** for the Directive's definition of the term 'hand-arm vibration'. This led to the introduction of the CVWR 2005, which sets exposure action and limit values for vibration.

"'Hand-arm vibration': the mechanical vibration that, when transmitted to the human hand-arm system, entails risks to the health and safety of workers, in particular vascular, bone or joint, neurological or muscular disorders."

Figure 13-21: Definition of hand-arm vibration.
Source: EU, Directive 2002/44/EC - on physical agents - vibration.

Effects on the body

Prolonged intense vibration transmitted to the hands and arms by vibrating tools and equipment can lead to a condition known as **hand-arm vibration syndrome (HAVs).** These are a range of conditions relating to long term damage to the circulatory system, nerves, soft tissues, bones and joints. The medical effects of sustained exposure to hand-arm vibration can be serious and permanent and are summarised in the following points:

- Vascular changes in the blood vessels of the fingers.
- Neurological changes in the peripheral nerves.
- Muscle and tendon damage in the fingers, hands, wrists and forearms.
- Suspected bone and joint changes.

Figure 13-22: Use of disc cutter - vibration.
Source: RMS.

Vibration generated by tools and equipment held in the hand can result in a significant reduction in blood flow to the hand. Prolonged exposure can cause the fingers go white and numb.

Leading to probably the best known of the conditions arising from the medical effects of exposure to vibration - vibration white finger (VWF).

This condition is also known as **Raynaud's phenomenon** (it has other causes in addition to exposure to vibration). Other symptoms of the condition are sharp tingling pains in the affected area and possible change in skin colour as blood vessels dilate when exposure to the vibration stops.

Contributory factors

As with all work-related ill-health there are a number of factors that when combined result in the problem -occurring. These include:

- Vibration frequency - frequencies ranging from 2-1,500 Hz are potentially damaging, but the most serious is the 5-20 Hz range.
- Duration of exposure - this is the length of time the individual is exposed to the vibration.
- Contact force - this is the amount of grip or push used to guide or apply the tools or work piece. The tighter the grip, the greater the vibration to the hand.
- Factors affecting circulation - including medication and smoking.
- Individual susceptibility.

Examples of risk activities in construction

- The use of chainsaws.
- The use of hand-held rotary tools in grinding, sanding or cutting.
- The use of hand-held percussive drills for drilling into concrete.
- The use of scrabblers.
- The use of nail and other impact fixing guns.
- The use of disc cutters.
- The use of cut off saws, hand held saws or planners.
- The use of vibrating compactors.
- The use of hand-held powered percussive drills or hand-held powered percussive hammers in demolition, or on roads or footpaths, including road construction.

Whole-body vibration

The European Physical Agents (Vibration) Directive (2002/44/EC) defines the term 'whole-body vibration', *see Figure 13-23.*

"'Whole-body vibration': the mechanical vibration that, when transmitted to the whole body, entails risks to the health and safety of workers, in particular lower-back morbidity and trauma of the spine."

Figure 13-23: Definition of whole-body vibration.
Source: EU, Directive 2002/44/EC - on physical agents - vibration.

Whole-body vibration (WBV) is vibration transmitted to the entire body via the seat or the feet, or both, often through driving or riding in vehicles (including dumper trucks, telehandlers and off-road vehicles). Whole-body vibration can cause lower-back and spine pain, fatigue, insomnia, stomach problems, and headaches shortly after or during exposure.

Studies show that WBV can increase heart rate, respiration and oxygen uptake and can cause changes in blood and urine. Eastern European research has noted workers having an overall ill feeling, which they call vibration sickness.

Prolonged exposure can lead to considerable back pain and time off work and may result in permanent injury and having to give up work.

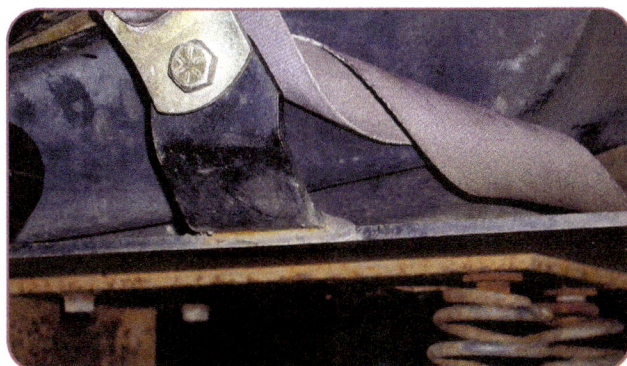

Figure 13-24: Dumper truck seat – anti-vibration mountings.
Source: RMS.

Machine	Daily exposure ranges in the three directions of vibration		
	X (fore and aft vibration)	Y (side-to-side vibration)	Z (vertical vibration)
Backhoe loader - excavating	***0.29 m/s² A(8)***	0.24 m/s² *A*(8)	0.21 m/s² *A*(8)
Backhoe loader - load carrying	0.50 m/s² *A*(8)	***0.71 m/s² A(8)***	0.37 m/s² *A*(8)
Backhoe loader - road travel	***0.52 m/s² A(8)***	0.45 m/s² *A*(8)	0.47 m/s² *A*(8)
3 tonne articulated dumper - soil transport	0.83 m/s² *A*(8)	0.91 m/s² *A*(8)	***0.96 m/s² A(8)***

Figure 13-25: Examples of exposure in construction.
Source: UK, HSE RR400.

High-risk activities include driving rough-terrain vehicles (for example, dumper trucks) and the prolonged use of compactors.

The examples in ***Figure 13-25***, are from a sample of construction equipment. The data was gathered over a sampling period of about half a day.

Most of the exposures were above the UK exposure action value, but all are below the exposure limit value. The highest reading for a machine is shown in bold. Higher levels of WBV exposure could be expected for older equipment and will depend on the actual terrain involved.

ASSESSMENT OF EXPOSURE TO VIBRATION

When exposure should be assessed

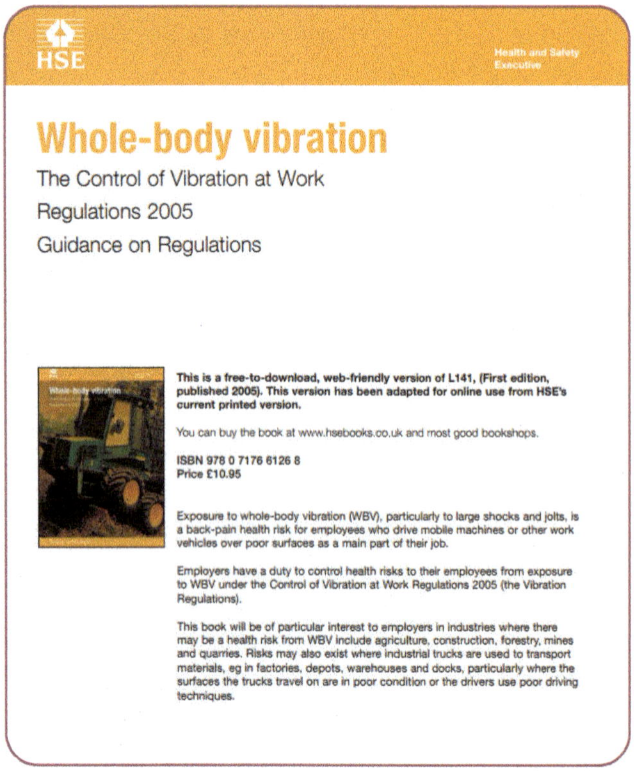
HSE
Health and Safety Executive

Whole-body vibration

The Control of Vibration at Work Regulations 2005

Guidance on Regulations

This is a free-to-download, web-friendly version of L141, (First edition, published 2005). This version has been adapted for online use from HSE's current printed version.

You can buy the book at www.hsebooks.co.uk and most good bookshops.

ISBN 978 0 7176 6126 8
Price £10.95

Exposure to whole-body vibration (WBV), particularly to large shocks and jolts, is a back-pain health risk for employees who drive mobile machines or other work vehicles over poor surfaces as a main part of their job.

Employers have a duty to control health risks to their employees from exposure to WBV under the Control of Vibration at Work Regulations 2005 (the Vibration Regulations).

This book will be of particular interest to employers in industries where there may be a health risk from WBV include agriculture, construction, forestry, mines and quarries. Risks may also exist where industrial trucks are used to transport materials, eg in factories, depots, warehouses and docks, particularly where the surfaces the trucks travel on are in poor condition or the drivers use poor driving techniques.

Figure 13-26: Whole-body vibration.
Source: UK, HSE Guidance L141.

Employers have a responsibility to assess the risks from vibration in order to reduce the risk of damage to their workers' health by controlling exposure to vibration. Regulation 5 of the Control of Vibration at Work Regulations (CVWR) 2005 requires the employer to make a suitable and sufficient assessment of the risk created by work that is liable to expose workers to vibration. Regulation 4 of CVWR 2005 sets action levels and limits relating to the exposure of workers to vibration in the workplace, and such levels and limits need to be considered when assessing the risks faced by workers.

The assessment should establish which workers are at risk of health effects from vibration, and the level of risk. It will also identify sources of vibration that particularly contribute to the vibration level workers are exposed to, for example, equipment and specific activities. This will enable analysis of the options to control the vibration at source or by other means. The assessment will also help the employer when marking equipment or zones that present significant risk, giving information, instruction and training to workers and organising health surveillance for those particularly at risk. If recorded, it will provide evidence of a structured risk assessment process.

Risk assessment

The assessment should observe work practices, make reference to information regarding the magnitude of vibration from equipment and, if necessary, measurement of the magnitude of the vibration.

When conducting the risk assessment, consideration should also be given to the type, duration, effects of exposure, exposure limit/action values, effects on workers at particular risk, the effects of vibration on equipment and the ability to use it, manufacturers' information, availability of replacement equipment, the extension of exposure at the workplace (for example, in rest facilities), temperature and information from health surveillance. The risk assessment should be recorded as soon as is practicable after it is made and be reviewed regularly.

Comparison of vibration exposure levels with recognised exposure limit standards

The International Standardization Organization (ISO) 5349-1:2001 'Mechanical vibration - Measurement and evaluation of human exposure to hand-transmitted vibration - Part 1: General requirements sets out international recommendations for vibration exposure limits. These have been widely adopted globally, sometimes as guidance and at other times as part of a legal framework. For example, a European Union (EU) Directive ratified in 2005 reflected the recommendations in ISO 5349 and gave rise to national legislative vibration limits in European member states. This was introduced into UK legislation as the Control of Vibration at Work (CVWR) Regulations 2005. Regulation 4 states the personal daily exposure action values and daily exposure limit values for vibration, normalised over an 8-hour reference period.

Because the harm from vibration is considered to be related to a combination of the energy from the acceleration and the duration of exposure, it is possible to be exposed to higher levels than the action and limit values provided they are for proportionally shorter periods of time. For example, exposure to a magnitude of 5 m/s² for 4 hours would be equivalent to 2.5 m/s² for 8 hours.

Exposure action and limit values

	Daily exposure action level	Daily exposure limit values
Hand-arm vibration	2.5 m/s²	5 m/s²
Whole-body vibration	0.5 m/s²	1.15 m/s²

Figure 13-27: Vibration action and limit values.
Source: HSE Guidance L141.

In the UK, the HSE have developed a points system, known as the 'Exposure points system and ready reckoner' in the form of a table, **see Figure 13-28**, for calculating daily vibration exposures. All that is needed is the vibration magnitude (level) and exposure time. The ready reckoner covers a range of vibration magnitudes up to 40 m/s² and a range of exposure times up to 10 hours.

The exposures for different combinations of vibration magnitude and exposure time are given in exposure points instead of values in m/s² A(8).

The user may find the exposure points easier to work with than the A(8) values:

- Exposure points change simply with time: twice the exposure time, twice the number of points.
- Exposure points can be added together, for example, where a worker is exposed to two or more different sources of vibration in a day.
- The exposure action value (2.5 m/s² A(8)) is equal to 100 points.
- The exposure limit value (5 m/s² A(8)) is equal to 400 points.

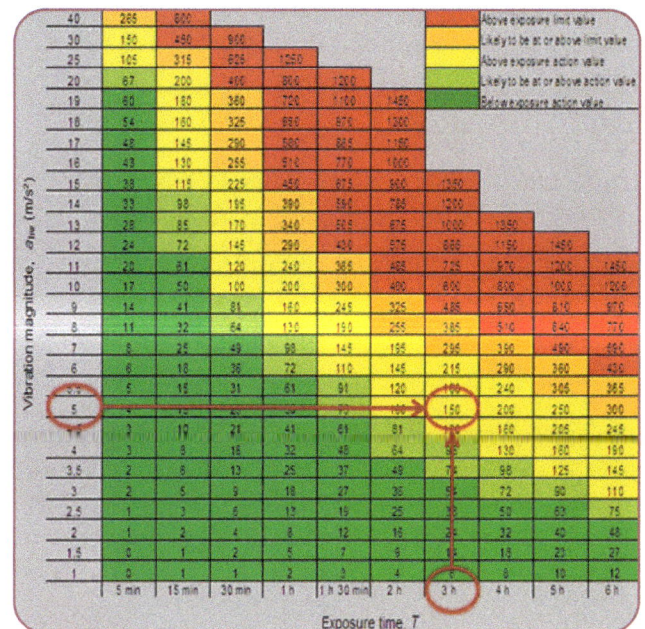

	Above limit value
	Likely to be above limit value
	Above action value
	Likely to be above action value
	Below action value

Figure 13-28: Vibration ready reckoner.
Source: UK, HSE L140.

Using the ready reckoner:

1) Find the vibration magnitude (level) for the tool or process (or the nearest value) on the grey scale on the left of the table.

2) Find the exposure time (or the nearest value) on the grey scale across the bottom of the table.

3) Find the value in the table where the magnitude and time intersect. The illustration shows how it works for a magnitude of 5 m/s2 and an exposure time of 3 hours: in this case, the exposure corresponds to 150 points.

4) Compare the point's value with the exposure action and limit values (100 and 400 points respectively). In this example the score of 150 points lies above the exposure action value.

5) The colour of the square containing the exposure points value tells you whether the exposure exceeds,

or is likely to exceed, the exposure action or limit value:

If a worker is exposed to more than one tool or process during the day, repeat steps 1-3 for each one, add the points, and compare the total with the exposure action value (100) and the exposure limit value (400).

Sourced and adapted from UK, HSE

Exposure action values

Where workers are likely to be exposed to vibration at or above an exposure action level value, the employer should place them under suitable health surveillance and provide them and their representatives with suitable and sufficient information, instruction and training.

Exposure limit values

In general, the employer must ensure that workers are not exposed to vibration above an exposure limit value. If an exposure limit value is exceeded the employer must immediately:

- Reduce exposure to below the limit value.
- Identify the reason for the limit value being exceeded.
- Modify the measures taken to prevent a recurrence.

Where exposure to vibration is usually below the exposure action value, but varies markedly from time to time, the exposure limit value may be occasionally exceeded, providing that:

- Any exposure to vibration averaged over one week is less than the exposure limit value.
- There is evidence to show that the risk from the actual pattern of exposure is less than the corresponding risk from constant exposure at the exposure limit value.
- Risk is reduced to as low a level as is reasonably practicable, taking into account the special circumstances.
- Workers concerned are subject to increased health surveillance.

BASIC VIBRATION CONTROL MEASURES

Preventive and precautionary measures

Regulation 6 of CVWR 2005 states that the employer should seek to eliminate the risk of vibration at source or, if not reasonably practicable, reduce it to as low a level as is reasonably practicable. Where it is not reasonably practicable to take preventive measures that eliminate the risk at source and the personal daily exposure action value is likely to be reached or exceeded, the employer should reduce exposure by implementing a programme of organisational and technical measures. These precautionary measures include the use of alternative methods of work (for example, by *mechanisation*), improved ergonomics, maintenance of equipment, design and layout, rest facilities, information, instruction and training,

limitation by work rotation and breaks and the provision of personal protective equipment to protect from cold and damp. Measures should be adapted to take account of any group or individual worker whose health may be at particular risk from exposure to vibration.

The following precautionary measures should be considered when protecting people who work with vibrating equipment.

Choice of equipment

It is important to consider vibration characteristics when purchasing new equipment or selecting equipment for a task. Some equipment will provide better control of vibration at source; others will have damping measures provided to limit vibration transmission to the user.

Manufacturers should provide details of the 'vibration magnitude' of their equipment to enable employers to determine the extent of any vibration risks in use. Preference should be given to the purchase and use of *low-vibration emission equipment* where this is possible without it impeding work processes so that it takes more time to do the work and leads to an equivalent exposure.

Many manufacturers claim to use composite materials that, when moulded into hand-grips and fitted onto vibrating power tools, reduce vibration by up to 45%. Some equipment may provide beneficial design that can limit the effects of vibration, such as the routing of exhaust gases of portable petrol driven equipment through the operating handles, to keep the user's hands warm in cold weather conditions.

Maintenance

Equipment should be maintained at its optimum performance level, thereby reducing vibration to a minimum. For example, tools such as powered chisels should be kept sharp, and bearings of machinery such as grinders should be *regularly maintained*. Even a relatively small imbalance in set-up can be sufficient to produce high levels of vibration. It is important to maintain the suspension and engine mountings on rough-terrain vehicles in order to minimise the effects of WBV.

Limiting exposure

The work schedule for workers who may conduct work that exposes them to vibration should be examined to identify opportunities to *limit the time workers are exposed* and the magnitude of the vibration to below the action levels set by competent national authorities, for example, by *avoiding long periods of vibration exposure* through *alternating vibration work* with non-vibration work or lower vibration magnitude work or *scheduling rest breaks.*

Where equipment creates a high magnitude of vibration this must be clearly identified and its use by a single worker limited to short periods. Care should be taken to organise work schedules so that rest periods from this high-risk work happen naturally in the process. Where they do not, it may be necessary to use reminders in the form of timed alarms or supervision.

Suitable personal protective equipment

Wearing gloves is recommended for safety of the hands and *protection against the cold* through the retention of heat. Maintaining a good temperature in the hands will help circulation of blood to the fingers. Care needs to be taken to select appropriate gloves, as the absorbent material in some gloves for thermal insulation may introduce a resonance frequency that may increase the total energy input to the hands. Research indicates that anti vibration gloves are unreliable as devices for controlling hand transmitted vibration exposures. Other means of vibration control are far more likely to deliver effective vibration reductions and therefore should be implemented.

Warm clothing can help workers exposed to vibration maintain a good core body temperature, which will assist circulation of blood to the hands. It may be necessary for workers to be provided with a warm location for rest breaks or periods when cold is affecting their circulation. Workers with established HAVS should avoid exposure to cold and thus minimise the number of blanching attacks (*see Figure 13-29*). Workers with advanced HAVS, which *health surveillance* has determined as deteriorating, should be removed from further exposure.

Other precautionary measures

- The vibration characteristics of hand tools should be assessed and reference made to the ISO or national standard organisation guidelines.

- Development of a purchasing policy to include consideration of vibration and, where necessary, vibration isolating devices.

- Training of all exposed workers on the proper use of tools and the minimisation of exposure. The stronger the grip on the tool, the more energy enters the hand. There are working techniques for all tools and the expertise developed over time justifies an initial training period for new starters. Operators of vibrating equipment should be trained to recognise the early symptoms of HAVS and WBV and how to report them.

- A continuous review should be conducted with regard to redesigning tools, rescheduling work methods, or automating the process until such time as the risks associated with vibration are under control.

- As with any management system, the controls in place for vibration should be subject to inspection.

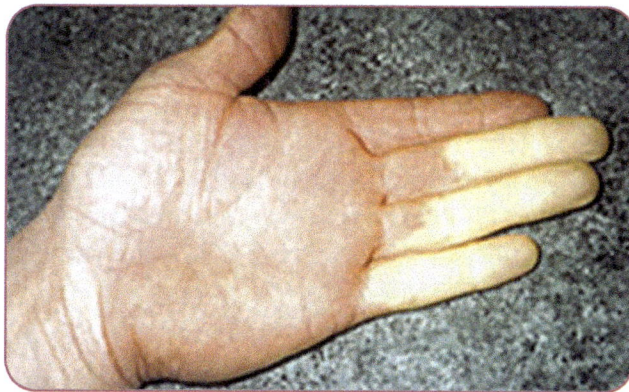

Figure 13-29: Blanched (white) finger typical of hand vibration syndrome (Caucasian race people).
Source: HSE, L140.

Regulation 8 of CVWR 2005 states that employers should provide information, instruction and training to all workers (and their representatives) who are exposed to risk from vibration. This includes any organisational and technical measures taken, exposure limits and values, risk assessment findings, why and how to detect signs of injury, how to report signs of injury, entitlement to health surveillance and any personal or collective results of health surveillance and safe working practices. Information, instruction and training should be updated to take account of changes in the employer's work or methods. The employer should ensure all workers, whether workers or not, who carry out work in connection with the employer's duties have been provided with information, instruction and training.

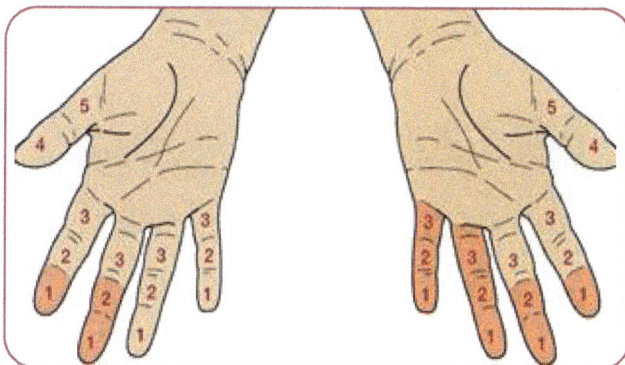

Figure 13-30: Finger zones used to identify degree of hand vibration damage (pin prick test - loss of touch sensitisation).
Source: HSE, L140.

ROLE OF HEALTH SURVEILLANCE

The role of health surveillance is to provide early detection of work-related ill-health, as it will assist with the identification of symptoms of the effects of vibration on health. Health surveillance can therefore prevent or diagnose any health effect linked with exposure to vibration. Health surveillance should be carried out for all workers where there is a risk to their health due to being exposed to vibration.

By conducting health surveillance from the start of a worker's exposure to vibration risks, it is possible to detect early signs of the effects of vibration and provide early intervention to limit the continuing effects.

Health surveillance can also assist with confirming the success of vibration control measures.

Surveillance for HAVS usually involves the worker or an occupational health specialist examining the hands to identify early signs of tingling or blanching, **see Figures 13-29 and 13-30**. For WBV, surveillance may be a simple reporting method or questionnaire relating to experience of lower-back discomfort or pain.

A record of health should be kept of any worker who undergoes health surveillance. The employer should, providing reasonable notice is given, provide the worker with access to their health records and provide copies to a relevant competent authority on request.

If health surveillance identifies a disease or adverse health effect, considered by a doctor or other occupational health professional to be a result of exposure to vibration, the employer should ensure that a qualified person informs the worker and provides information and advice. The employer should ensure they are kept informed of any significant findings from health surveillance, taking into account any medical confidentiality.

In addition, the employer should also:

- Review risk assessments.
- Review the measures taken to comply.
- Consider assigning the worker to other work.
- Review the health of any other worker who has been similarly exposed and consider alternative work.

REVIEW

What control measures can be taken to reduce the risks from WBV?
What are the symptoms of HAVs?

13.3 Radiation

DIFFERENCES BETWEEN NON-IONISING AND IONISING RADIATION

All matter is composed of atoms. Different atomic structures give rise to unique elements. Examples of common elements, which form the basic structure of life, are hydrogen, oxygen and carbon.

Atoms form the building blocks of nature and cannot be further sub-divided by chemical means. The centre of the atom is called the nucleus, which consists of protons and neutrons. Electrons take up orbit around the nucleus.

Protons: Have a unit of mass and carry a positive electrical charge.

Neutrons: These also have mass but no charge.

Electrons: Have a mass about 2,000 times less than that of protons and carry a negative charge.

In an electrically neutral atom the number of electrons equals the number of protons (the positive and negative charges cancel each other out). If the atom loses an electron then a positively charged atom is created. The process of losing or gaining electrons is called **ionisation.**

Figure 13-31: Radiation symbol.
Source: ISO 7010.

If the matter that is ionised is a human cell, the cell chemistry will change and this will lead to functional changes in the body tissue. Some cells can repair radiation damage, others cannot. The cell's sensitivity to radiation is directly proportional to its reproductive function; bone marrow and reproductive organs are the most vulnerable, while muscle and central nervous system tissue are affected to a lesser extent.

Ionising radiation is emitted from radioactive materials either directly or indirectly. It has an energy potential capable of changing the cellular composition of matter by penetrating, ionising and damaging body tissue and organs.

Non-ionising radiation has a relatively long wavelength and does not possess the energy needed to ionise matter. Instead the effect tends to be one of heating up cells rather than changing their composition.

Differences between non-ionising and ionising radiation

Ionising radiation is radiation, typically alpha and beta particles and gamma and X-rays, that has sufficient energy to produce ions by interacting with matter, whereas non-ionising radiation does not possess sufficient energy to cause the ionisation of matter.

TYPES, OCCUPATIONAL SOURCES AND HEALTH EFFECTS OF NON-IONISING RADIATION

Non-ionising radiation exists as optical radiation (ultraviolet, visible light, lasers and infrared) and electromagnetic fields (electrical power transfer line frequencies, microwaves and radio frequencies).

Ultraviolet

Possible occupational sources

There are many possible sources of ultraviolet (UV) radiation to which people may be exposed at work. Sunlight is a natural source of UV radiation. Sunlight UV is filtered by the earth's ozone layer and the shorter high-energy wavelengths, which are more harmful to life forms, are absorbed before they reach ground level.

UV is produced by high temperature sources created by, for example, electrical welding (see **Figure 13-32**) or from fluorescent and tungsten lights and mercury vapour lamps. UV light has many applications; UV will attract insects and is used in food--preparation areas in the design of insect killers. UV is used to sterilise contaminated water in water treatment, to cure adhesives and inks in printing processes, for metal surface inspection and crack detection, forgery detection (paper currency and works of art), for leisure sun beds and tanning lamps. These sources are summarised in the following bullet points:

- The sun.
- Tungsten halogen lamps.
- Mercury vapour lamps.
- Electric arc welding and cutting.
- Insect killers, to attract the insects.
- Crack-detection equipment.
- Water treatment equipment.
- Some lasers.
- Adhesive curing equipment.
- Forged bank note detection equipment.
- Sunbeds and sunlamps.

Potential health effects

Over-exposure to sunlight can cause sunburn and even blindness. Its effect is thermal and photochemical, producing burns and skin thickening, and eventually skin cancer.

Figure 13-32: UV - from welding.
Source: Speedy Hire Plc.

Electric arcs used in welding operations and ultraviolet lamps can produce a harmful effect on the eyes resulting in inflammation (sometimes called 'arc-eye'), and cataract formation.

Natural and artificial light is used in most workplaces and many processes emit radiation in the visible spectrum range.

Visible Light

Possible occupational sources

Visible light can cause hazards in the workplace due to direct or reflected radiation in the range of visible light. Possible sources of high intensity visible light at work include:

- Natural daylight.
- Lasers operating in the visible wavelength, for example, in surveying or level alignment equipment.
- Furnaces or fires.
- Molten metal, ceramics or glass.
- Gas or electrical welding or cutting.
- High-intensity light beams and light bulbs.
- Other high-intensity lights, such as in photocopiers and printers.
- The sun.

Potential health effects

Light in the visible frequency range can cause damage if it is present in sufficiently intense form. The eyes are particularly vulnerable, but skin tissue may also be damaged. Retinal damage may occur if chronic exposure to high levels of light takes place. If the visible light is focused on to the skin it may cause skin burning. Indirect danger may also be created by workers being temporarily dazzled.

Infrared

Possible occupational sources

Any hot source which visibly glows is likely to be a source of infrared radiation, for example:

- Furnaces or fires.
- Molten metal or glass.
- Burning or welding.
- Heat lamps.
- Some lasers.
- The sun.

Potential health effects

Exposure results in a thermal effect such as skin burning and loss of body fluids (heat exhaustion and dehydration). The eyes can be damaged in the cornea and lens, which may become opaque (cataract). Retinal damage may also occur if the radiation is focused on the eye.

Directive 2006/25/EC of the European Parliament and of the Council of 5 April 2006 on the minimum health and safety requirements regarding the exposure of workers to risks arising from physical agents (artificial optical radiation) places a requirement on Member States to assess the risks from artificial optical radiation. This includes ultraviolet, infrared and visible sources (luminaires) in the workplace. The directive does not include sunlight. Legislation was introduced in the UK to manage these risks in the form of the Control of Artificial Optical Radiation at Work Regulations (CAOR) 2010.

Electromagnetic fields

Possible occupational sources

EMFs are a type of non-ionising radiation that is part of the electromagnetic spectrum, the other main component being optical radiation (ultraviolet, visible and infrared light). EMFs are created by electrically charged objects or particles. Electric fields are created by differences in voltage: the higher the voltage, the stronger the field. Magnetic fields, on the other hand, are created when electric current flows: the greater the current, the stronger the field. Electromagnetic fields are present everywhere in our environment but are invisible to the human eye.

EMFs can be divided into two different groups:

1) Low frequency EMF radiation. They have lower frequencies than visible light and are non-ionizing radiation. Examples include EMFs from microwave ovens, computers, visible light, smart meters, WiFi, cell phones, Bluetooth, power lines, and magnetic resonance imaging (MRI).

2) High frequency EMF radiation. These have higher frequencies than visible light, which are ionizing radiation. Examples include ultraviolet (UV) light, X-Rays, and Gamma rays.

Typical activities and equipment that lead to exposure include working with MRI scanners, working near high voltage power lines, working in electricity distribution stations, electric arc welding and working on or near to radar and radio transmitters.

Potential health effects

Short term health effects can include dizziness, nausea and heating up of body tissue. There is no scientific evidence suggesting that EMF has anything other than immediate health consequences. Another issue to bear in mind is that EMF can interfere with implanted electrical devices like pacemakers and defibrillators, or just metal implants.

Radio frequency and microwaves

Possible occupational sources

Radio frequency radiation is produced by radio/television transmitters, intruder detectors and high voltage electricity cables. Microwave radiation is used in commu-

nication systems and in cooking equipment. High-energy radio frequency radiation is used in heat sealers or welders for construction applications in order to fuse together two or more pieces of non-conductive material together, for example PVC plastic. This process is used to join fence screening, weather protection sheets and pipes.

Figure 13-33: Radio mast.
Source: RMS.

Potential health effects

Burns can be caused if workers using this type of equipment allow parts of the body that carry jewellery to enter the radio frequency field. The jewellery will absorb the energy of this form of radiation, get hot and may cause burns.

Intense fields at the source of transmitters will damage the body directly through this absorption/heating process, and particular precautions need to be taken to isolate radio/television transmitters to protect maintenance workers. Microwaves can produce the same deep heating effect in live tissue as they can in cooking.

TYPES, OCCUPATIONAL SOURCES AND HEALTH EFFECTS OF IONISING RADIATION

Types of radiation

Ionising radiation occurs as either electromagnetic rays, for example, gamma rays or X-rays, or in particles, for example, alpha and beta particles.

Radiation is emitted by a wide range of sources and appliances used throughout industry, medicine and research. It is also a naturally occurring part of the environment.

Ionising radiation is found naturally or in the workplace in the form of alpha, beta, gamma, and X-rays. The human body absorbs radiation readily from a wide variety of sources, mostly with adverse effects.

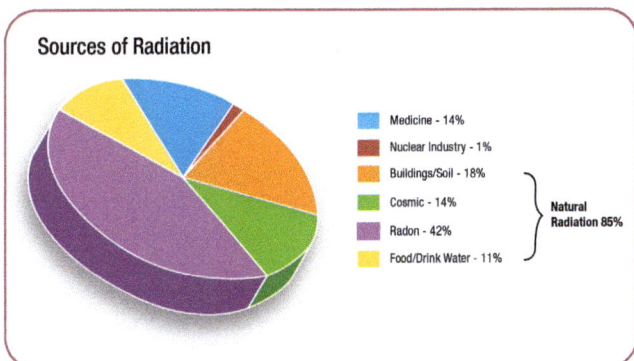
Figure 13-34: Sources of radiation.
Source: World Nuclear Association.

There are a number of different types of ionising -radiation, each with its different powers of penetration and effects on the body. Therefore, the type of radiation will determine the type and level of protection.

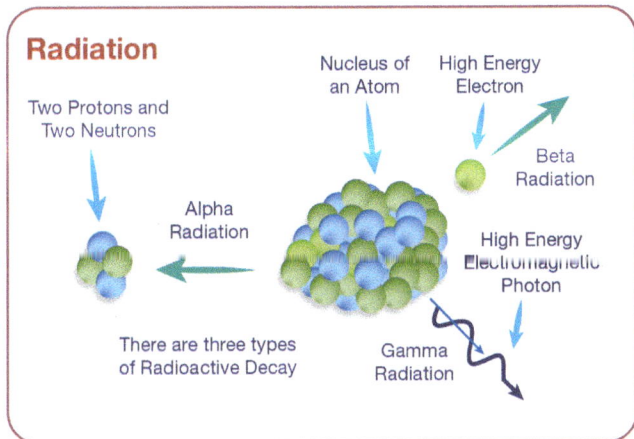
Figure 13-35: Ionising atomic decay.
Source: RMS.

Alpha particles

Alpha particles are comparatively large. Alpha particles travel short distances in dense materials, and are unlikely to penetrate living skin tissue.

The principal risk is through ingestion or inhalation of a source, for example, radon gas emits alpha particles. This might place the material close to vulnerable tissue such as the lungs; when this happens the highly localised energy effect will destroy associated tissue of the organs affected.

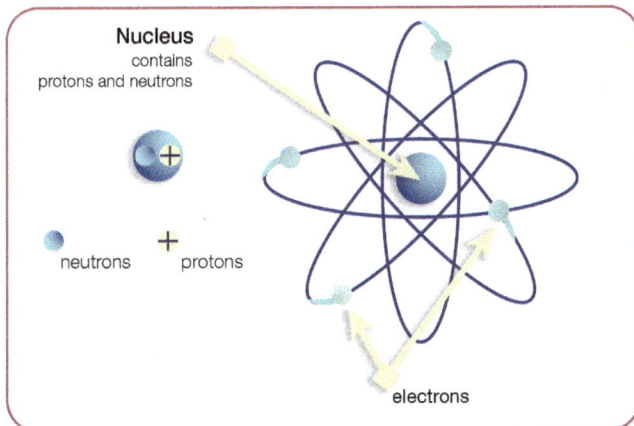
Figure 13-36: Structure of an atom.
Source: RMS.

Beta particles

Beta particles are much smaller and faster moving than alpha particles, and have a longer range, so they can damage and penetrate the skin. While they have greater penetrating power than alpha particles, beta particles are less ionising and take longer to effect the same degree of damage.

Gamma rays

These have great penetrating power. Gamma radiation passing through a normal atom will sometimes force the loss of an electron, leaving the atom positively charged; this is called an ion.

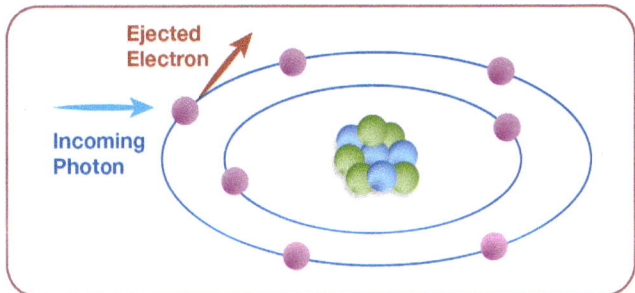
Figure 13-37: Ionisation of a normal atom.
Source: RMS

X-rays

X-rays are very similar in their effects to gamma rays. Both X-rays and gamma rays have **high energy**, and **high penetration power** through fairly dense material. In low-density substances, including air, they may travel long distances.

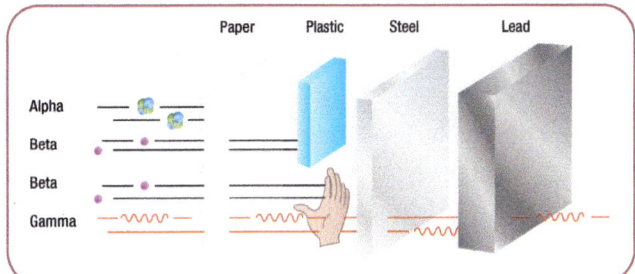
Figure 13-38: Relative penetration power of several types of ionising radiation (note gamma and X-ray have similar penetrating power).
Source: RMS.

Radon

Radon is a naturally occurring colourless, odourless radioactive gas that can seep out of the ground and enter buildings. It is a particularly common source of exposure to radiation and for many people easily exceeds exposure from nuclear power stations or hospital scans and X-rays.

Radon occurs naturally deep within the earth's core from decaying uranium, and it is particularly abundant in regions with granite bedrock. However, the gas disperses outdoors so levels are generally very low. The gas decays readily into other radioactive isotopes of lead, bismuth and polonium, including polonium 210, the dust from which can be inhaled and expose local tissue to intense alpha particle ionisation, resulting in cell death or malignancy. Radon has been identified as a common cause of lung cancer, accounting for more than 1,100

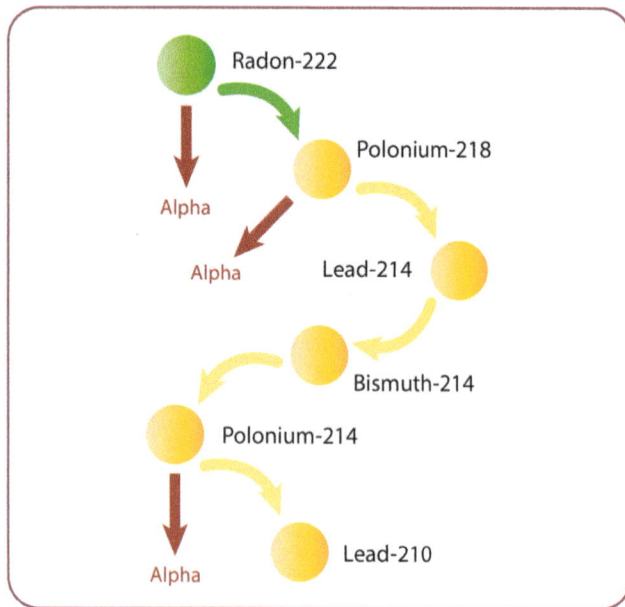

Figure 13-39: Radon decay.
Source: Administration and Business Portal.

deaths in the UK each year.

Occupational sources of ionising radiation

The most familiar examples of ionising radiation in the workplace are in hospitals, dentists' surgeries and veterinary surgeries where X-rays are used extensively. X-ray machines are also used for security purposes at baggage-handling points in airports.

In addition, X-rays and gamma rays are used in non-destructive testing of metals, for example, site radiography of welds in pipelines. In other industries, ionising radiation is used for measurement, for example, in the paper industry for the thickness of paper, and in the food-processing industry for measuring the contents of sealed tins. While all workplaces can be at risk from radon, workplaces at higher risk tend to be those located in specific geological areas. The highest levels of radon are usually found in underground spaces such as basements, cellars, caves and mines, particularly where groundwater is present.

Examples include:

- Underground workers such as miners (uranium miners are exposed to the highest levels).
- Construction workers involved in tunnelling, especially under rivers.
- Utility company workers (when entering service ducts).
- Oil and gas drilling (drilling 'muds' or slurries and the processing of oil).
- Quarry workers.

"Radiation is permanently present throughout the environment, in the air, water, food, soil and in all living organisms. Large proportion of the average annual radiation dose received by people results from natural environmental sources. Each member of the world population is exposed, on average, to 2.4 mSv/yr of ionising radiation from natural sources. In some areas (in different countries of the world) the natural radiation dose may be 5 to 10-times higher to large number of people."

Figure 13-40: Sources of ionising radiation.
Source: World Health Organization.

Potential health effects of ionising radiation

Radiation damage to tissue and/or organs depends on the dose of radiation received. The potential damage from an absorbed dose depends on the type of radiation and the sensitivity of different tissues and organs.

The sievert (Sv) is a unit of radiation weighted dose, also called the effective dose. It is a way to measure ionising radiation in terms of the potential for causing harm. The Sv takes into account the type of radiation and sensitivity of tissues and organs. The Sv is a very large unit so it is more practical to use smaller units such as millisieverts (mSv) or microsieverts (µSv). There are one thousand µSv in one mSv, and one thousand mSv in one Sv. In addition to the amount of radiation (dose), it is often useful to express the rate at which this dose is delivered (dose rate), for example, µSv/hour or mSv/year.

Beyond certain thresholds, radiation can impair the functioning of tissues and/or organs and can produce acute effects such as skin redness, hair loss, radiation burns, or acute radiation syndrome. These effects are more severe at higher doses and higher dose rates. For instance, the dose threshold for acute radiation syndrome is about 1 Sv (1000 mSv). Typical radiation dose limits would be around 20 mSv/year.

If the dose is low or delivered over a long period of time (low dose rate), the damaged cells are more likely to repair themselves successfully. However, long-term effects may still occur if the cell damage is repaired but incorporates errors in the DNA, transforming an irradi-ated cell that still retains its capacity for cell division. This transformation may lead to cancer after years or even decades have passed.

Effects of this type will not always occur, but their likeli-hood is proportional to the radiation dose. This risk is higher for children and adolescents, as they are signifi-cantly more sensitive to radiation exposure than adults. For example, a typical limit for a young person would be around 6 mSv/year.

THE BASIC MEANS OF CONTROLLING EXPOSURE TO IONISING AND NON-IONISING RADIATION

Controls for non-ionising radiation

Employers should conduct a specific risk assessment for risks relating to non-ionising radiation and use it to eliminate or reduce risks. Exposure limit levels should be set for this form of radiation as part of the control measures. Information and training should be provided to those that may be affected, which includes workers and others carrying out work on behalf of the employer. Medical examination and health surveillance should be provided for those workers that receive over-exposure.

Ultraviolet

Protection from natural sources of UV is relatively simple and includes providing outdoor workers with barrier creams, suitable lightweight UV-rated clothing and head protection, and eye protection, as appropriate. In addition, workers should be encouraged to take breaks in the shade where possible and consideration should be given to adjusting work schedules so outside tasks may be conducted at times of day that are less affected by UV. Control of artificial sources of UV includes segregation of UV-emitting processes and the use of warning signs. UV radiation emitted from industrial processes can be isolated by physical shielding such as partitions. Users of UV-emitting equipment, such as welders, can protect themselves by the use of goggles and protective clothing - the latter to avoid 'sunburn'.

Assistants in welding processes often fail to appreciate the extent of their own exposure, and they require similar protection. Workers should check skin exposed to UV frequently to identify possible effects that might lead to skin cancer.

Visible light

The eye detects visible light. It has two protective control mechanisms of its own, the eyelids ('blinking' or closing the eyelids) and the iris (dilates). These are normally sufficient to provide general protection, as the eyelid has a reaction time of 150 milliseconds. However, where this is not adequate because of the intensity of the light or sustained exposure of the eye to it, other precautions should be considered, including confinement of high-intensity sources, matt finishes to nearby paintwork, and provision of protective glasses for outdoor workers in snow, sand or near large bodies of water. Where the high-intensity visible light is artificially created, those not involved in the process must also be protected. Warning signs should be posted and access restricted to the process area.

Infrared

Controls to limit exposure include engineered measures, such as remote controls, screening, interlocks and clamps to hold material to enable the worker to be outside the exposure area. This can be supplemented by forms of PPE where exposure cannot be prevented, for example, face shields, goggles or other protective eyewear, coveralls and gloves, and limiting exposure time through rotation and job-sharing. It is important to protect others not directly involved in the process by using screens, curtains and restricted access.

Radio frequency and microwaves

Radio frequency and microwave radiation can usually be shielded at point of generation to protect the users. If size and function prohibit this, restrictions on entry and working near an energised microwave device will be needed. Metals, tools, flammable and explosive materials should not be left in the electromagnetic field generated by microwave equipment. Appropriate warning devices should be part of the controls for such appliances.

Controls for ionising radiation

The standard approach to controlling ionising radiation involves, time, distance and shielding.

Reduced time

Exposure to ionising radiation can be controlled by reducing the duration of exposure through redesigning work patterns, giving consideration to shift working, job rotation, etc. The dose received will also depend upon the time of the exposure. These factors must be taken into account when devising suitable operator controls.

Increased distance

Radiation intensity reduces the further away a person is from the source. This is subject to the inverse-square law, which means that energy received (dose) is inversely proportional to the square of the distance from the source.

Shielding

The type of shielding required to give adequate protection will depend on the penetration power of the radiation involved. For example, it may vary from thin sheets of silver paper to protect from beta particles through to several centimetres of concrete and lead for protection against gamma or X-rays.

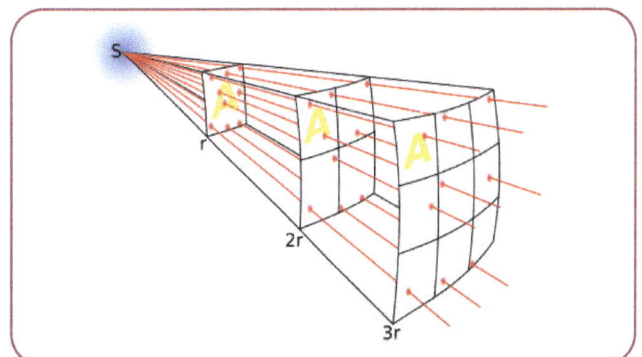

Figure 13-41: The inverse-square law.
Source: Wikipedia.

In addition to the previous specific controls, the following general principles must be observed:

- Radiation should only be introduced to the workplace if there is a positive benefit.

- Safety information must be obtained from suppliers about the type(s) of radiation emitted or likely to be emitted by their equipment.

- Safety procedures must be reviewed regularly.

- Protective equipment provided must be suitable and appropriate, as required by relevant regulations. It must be checked and maintained regularly.

- Emergency plans must cover the potential radiation emergency.

- Written authorisation by permit should be used to account for all purchase/use, storage, transport and disposal of radioactive substances.

The basic means of controlling exposures to radon

Above-ground workplaces

For the vast majority of above-ground workplaces the risk assessment should include radon measurements in appropriate ground-floor rooms where the building is located in a radon-affected area.

Below-ground workplaces

For occupied below-ground workplaces (for example, occupied greater than an average of an hour per week/ 52 hours per year), or those containing an open water source, the risk assessment should include radon measurements.

Control of radon exposure in new buildings can be done by installing a 'radon-proof membrane' within the floor structure. In areas more seriously affected by radon it may be necessary to install a 'radon sump' to vent the gas into the atmosphere. A radon sump has a pipe connecting a space under a solid floor to the outside.

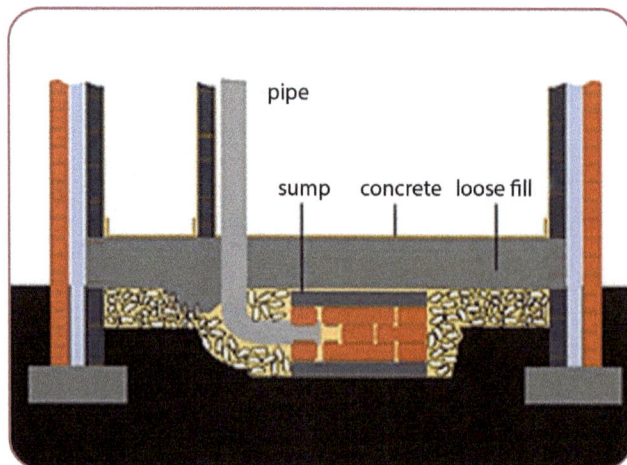

Figure 13-42: Radon sump.
Source: UK, HSE.

A small electric fan in the pipe continually sucks the radon from under the house and expels it harmlessly to the atmosphere. Modern sumps are often constructed from outside the building so there is no disruption inside. In existing buildings, it is not usually possible to provide a radon-proof barrier, so alternative measures are used to control the increasing concentration of radon in the building and subsequent exposure to it.

Such measures include a mixture of active and passive systems such as improved under-floor and indoor ventilation in the area, positive pressure ventilation of occupied areas, installation of radon sumps and extraction pipework and sealing large gaps in floors and walls in contact with the ground.

There are relatively simple tests for radon gas and detectors are safe and simple to use. In some countries these tests are routinely done in areas of known systematic hazards.

Detection involves a collector (a hollow plastic shell containing a piece of clear plastic that records damage caused by radon) that the user hangs in the lowest occupied floor of the premises for 2 to 7 days. The user then sends the collector to a laboratory for analysis. The detectors do not emit anything and do not collect anything dangerous. However, they can be damaged by heat or submersion in water and should not be opened.

Figure 13-43: Radon measurement.
Source: Public Health England

BASIC RADIATION PROTECTION STRATEGIES

Protection strategies for non-ionising radiation

Basic radiation protection strategies relating to non-ionising radiation include:

- Identification of the sources of non-ionising radiation and consideration of the risks related to exposure.

- Identification of classes of worker exposed to non-ionising radiation.

- Where the radiation is emitted from a controlled source, a Laser Protection Supervisor should be appointed to give specialist advice, ensure the energy from the source is reduced as far as reasonably practicable, using engineering measures (i.e. screening), protective eyewear and ensuring information and training for users while monitoring and enforcing control measures.

- Shielding the emission of energy from sources, for example, closing furnace openings when not in use, using glass filters and other forms of enclosure around lighting sources, curtains around welding activities, and enclosure of work areas near radio wave transmitters or provision of UV-protective cream.

- Increasing the distance of workers from the source, for example, the positioning of microwave transmitters so they are high enough above work areas, positioning welding activities away from other workers.

- Limiting the time that a worker is exposed to the source, for example, ensuring workers that work outside in the sun take regular breaks inside or in the shade, or job rotation for those involved in hot processes such as furnace work.

- Provision of information and training on the risks of exposure to non-ionising radiation and the measures to limit exposure.

- Provision of monitoring and health surveillance.

Protection strategies for ionising radiation

Radiation protection strategies for ionising radiation are a little more specific and reflect the principles and recommendations made by the International Commission on Radiological Protection (ICRP) and UK legislation. The three basic principles of radiological protection are:

1) Justification of activities that could cause or affect radiation exposure.

2) Optimisation of protection in order to keep doses as low as reasonably achievable (this includes shielding, minimising time exposed and maximising distance from the radiation source).

3) The use of dose limits.

The recommendations of the ICRP include establishing responsibilities, assessing risk, establishing prevention/control measures and conducting health surveillance. These recommendations are reflected in the Ionising Radiation Regulations (IRR) 2017.

Basic protection strategies for ionising radiation include:

- Assessment of work involving radiation.
- Arrangements for notification, registration and consent by the HSE.
- Appointing a radiation protection advisor(s).
- Appointing a radiation protection supervisor(s).
- Classification of workers.
- Application of prevention and controls measures to limit exposure.
- An information and training programme.

Responsibilities

Every employer should, in relation to any work with ionising radiation, take all necessary steps to restrict so far as is reasonably practicable the extent to which workers are exposed to ionising radiation. Those with responsibilities under the radiation protection programme should be assigned responsibilities in writing. A radiation protection adviser/officer should be appointed to assist the employer with the radiation protection programme.

Radiation protection advisers

With the exception of the operations specified in Schedule 1 of the IRR 2017, at least one radiation protection adviser (RPA) must be appointed in writing by employers using ionising radiation. The numbers of RPA appointed must be appropriate to the risk and the area where advice is needed is to be stated. Employers should consult the RPA on:

- The implementation of controlled and supervised areas.
- The prior examination of plans for installations and the acceptance into service of new or modified sources of ionising radiation in relation to any engineering controls, design features, safety features and warning devices provided to restrict exposure to ionising radiation.
- The regular calibration of equipment provided for monitoring levels of ionising radiation and the regular checking that such equipment is serviceable and correctly used.
- The periodic examination and testing of engineering controls, design features, safety features and warning devices and regular checking of systems of work provided to restrict exposure to ionising radiation.

The employer should provide the RPAs with adequate information and facilities to allow them to fulfil their functions.

Assessment

The employer should make an assessment of work involving exposure to radiation and use the assessment to contribute to the design of a radiation protection programme. The assessment should include consideration of the:

- Limits and technical conditions for the operation of the radiation source.
- Ways in which technical, procedural and behavioural measures related to radiation protection might fail and lead to potential exposures, and the possible consequences of such failures.
- Nature and magnitude of potential exposures and the likelihood of their occurrence.

Arrangements

The employer should follow the procedures for notification, registration or licensing that are laid down by the competent authority.

Prevention and control

Basic radiation protection strategies to prevent and control exposure include the application of the principles of time, distance and shielding. Exposure should be limited to as few people as possible and those individuals should be monitored and exposure levels maintained within limits.

Regulation 9 of the IRR 2017 sets out requirements for the employer to control exposure to ionising radiation:

> "Every employer must, in relation to any work with ionising radiation that he undertakes, take all necessary steps to restrict so far as is reasonably practicable the extent to which his employees and other persons are exposed to ionising radiation."

Figure 13-44: Regulation 9 of IRR 2017.
Source: Ionising Radiation Regulations (IRR) 2017.

In the case of ionising radiation, where possible, only sealed sources should be used and a system developed to minimise dose levels to individuals. This work should be under the control of an RPA (officer) Prevention and control of radiation exposure will include:

- Categorisation/classification of workers.
- Dose limitation, including arrangements for monitoring workers' exposure and radiation levels in the workplace. A system for recording and reporting information related to the control of exposures.
- Designation of controlled or supervised areas.
- Local rules for classified workers and the -supervision of work.
- Information and training programme.

Categorisation/classification of workers

Persons entering a controlled area should either be classified or be entering in accordance with 'written arrangements' designed to ensure that doses will not be exceed, in cases of persons over 18, levels which would otherwise require the persons to be classified. A classified worker is defined in the IRR 2017 as one who is likely to receive an effective dose in excess of 6mSv per year or an equivalent dose that exceeds three-tenths of any relevant dose limit. In the case of other persons, the relevant dose limit is applicable.

Training and information

As part of the radiation protection programme, the employer should establish a training and information programme.

The programme should ensure that all workers receive adequate information on the:

- Health risks due to their occupational exposure to radiation.
- Significance of their actions and how they may affect radiation protection measures.

All radiation workers should receive adequate training on radiation protection. Those assigned responsibilities in the radiation protection programme, for example, the RPA or radiation protection supervisor, should receive appropriate information and training. Appropriate management should receive training on the basic principles of radiological protection, their main responsibility regarding radiation risk management and the principal elements of the radiation protection programme.

Specific information should be provided to female workers who are likely to enter controlled or supervised areas on the risk to an embryo or foetus due to exposure to radiation and on the importance of notifying the employer as soon as she suspects she is pregnant.

THE ROLE OF MONITORING AND HEALTH SURVEILLANCE

The role of radiation monitoring is to ensure that working conditions in workplaces exposed to radiation are kept under review and that the current level of radiation exposure of workers is known. The main functions of monitoring are to:

- Check that areas have been correctly designated for the hazards that exist.
- Identify any changes to radiation exposure levels so that appropriate control measures for restricting exposure can be proposed.
- Detect breakdowns in controls or systems, so as to indicate whether conditions are satisfactory for continuing work in that area.
- Ensure workers use the controls provided and report any defects.
- Ensure workers use personal protection where its use is designated as mandatory.
- Provide information on those who may be at risk and in need of health surveillance.

The role of health surveillance is to provide the identification of symptoms and early detection of ill-health arising from exposure to occupational sources of radiation. Health surveillance can be part of a prevention strategy, allowing interventions to be made that limit further exposure beyond acceptable limits.

The health of the worker should also be established before the individual becomes a classified worker and when the individual ceases to be a classified worker.

Special consideration will be required for classified workers who are new or expectant mothers. The health surveillance undertaken will include examination of the skin and respiratory system, taking into account exposure records and sickness absence records.

Non-ionising radiation

Monitoring of workplace levels of non-ionising radiation is important for both radio frequency and optical forms of radiation as they are not a visible hazard and levels may increase and not be obvious until harm is done to workers. Where workers are exposed to significant levels of non-ionising radiation, employers should arrange for appropriate health surveillance by occupational health specialists, who should assess the possible need for medical examination of workers, including ophthalmic and skin examination.

In particular, employers should arrange for workers using Class 3 or 4 lasers to receive ophthalmic examinations:

a) Pre- and post-assignment to laser work.

b) After an apparent or suspected harmful exposure of the worker's eyes.

The Control of Artificial Optical Radiation at Work Regulations (CAOR) 2010 establishes a specific legal requirement for medical examination and health surveillance of workers that receive over exposure.

Ionising radiation

Where workers are exposed to ionising radiation, Regulation 20 of the Ionising Radiation Regulations (IRR) 2017 requires that the employer who designates an area as a controlled or supervised area should ensure that levels of ionising radiation are adequately monitored and that working conditions in those areas are kept under review.

Figure 13-45: Personal dose monitor.
Source: RMS

Regulation 25(2) of the IRR 2017 requires employers to ensure that classified workers working with ionising radiation are under adequate health surveillance by an appointed doctor or medical specialist to determine their fitness for the work being undertaken.

Figure 13-46: Contamination monitoring.
Source: RMS.

Fitness in this sense is not restricted to possible health effects from exposure to ionising radiation. Those conducting the health surveillance will need to take account of specific features of the work with ionising radiation in relation to the fitness of the individual.

This will include such things as:

- Their ability to wear any personal protective equipment (including respiratory protective equipment) required to restrict exposure.

- Whether they have a skin disease that could affect their ability to undertake work involving unsealed radioactive materials.

- Whether they have serious psychological disorders that could affect their ability to undertake work with radiation sources that involves a special level of responsibility for safety.

REVIEW

Define the terms ionising and non-ionising radiation and identify one industrial source for each.
What are the health risks associated with visible radiation?
What are the monitoring activities which should be undertaken to protect a classified worker?

13.4 Mental ill-health

FREQUENCY AND EXTENT OF MENTAL ILL-HEALTH IN THE CONSTRUCTION INDUSTRY

Common mental ill-health and work-related stress can exist independently – people can experience work-related stress without having anxiety, depression or other mental ill-health problems. They can also have anxiety and depression without experiencing work-related stress.

An existing mental ill-health problem can be aggravated by work-related stress, making it more difficult to control.

If work-related stress reaches a point where it has triggered an existing mental ill-health problem, it becomes hard to separate one from the other. Work-related stress and other mental ill-health problems often go together and the symptoms can be very similar.

Mental ill-health is a significant issue in the general working age population, which is reflected in the statistics compiled by the UK Mental Health Foundation who determined the following global perspective on mental ill-health:

- An index of 301 diseases compiled by the World Health Organisation (WHO) in 2016 found mental health problems to be one of the main causes of the overall disease burden worldwide. They were shown to account for 21.2% of years lived with disability worldwide.

- According to the 2019 Global Burden of Disease study, the predominant mental health conditions worldwide accounting for disability-adjusted life years (DALYS) are depressive disorders followed by anxiety disorders and schizophrenia. It is estimated by the WHO that globally 5% of adults suffer from depression.

- In 2013, depression was the second leading cause of years lived with disability worldwide, behind lower back pain. In 26 countries, depression was the primary driver of disability. Depressive disorders also contribute to the burden of suicide and heart disease on mortality and disability; they have both a direct and an indirect impact on the length and quality of life.

- The World Health Organization (WHO) estimates that between 35% and 50% of people with severe mental health problems in developed countries, and 76 – 85% in developing countries, receive no treatment.

From this information it can be seen that there is a range of mental ill-health that workers could have, but it is not clear their relationship to work, however the main mental ill-health problems were confirmed to be depression and anxiety.

Statistics compiled by the UK Mental Health Foundation, in relation to the UK, determined that according to the UK National Institute for Health and Care Excellence (NICE), common mental health problems include depression, generalised anxiety disorder (GAD), social anxiety disorder, panic disorder, obsessive compulsive disorder (OCD) and post-traumatic stress disorder (PTSD). A study in 2014 confirmed that 17.5% of working-age adults had symptoms of common mental ill-health. Comparing data from 2007 with 2014 the Mental Health Foundation confirmed that the prevalence of depression and general anxiety disorder in adults over 16 years of age had increased.

Mental health problem	2007 (%)	2014 (%)
Depression	2.3	3.3
General anxiety disorder	4.4	5.9

When considering post-traumatic stress disorder (PTSD) a survey conducted in 2014 found that about one third of adults in England reported having at least one traumatic event in their life and 4.4% of adults screened positively for having PTSD in the previous month, whereas a lower number indicated that they believed they had PTSD and only 13% of those screened positive had been diagnosed by a health professional.

The 2014 UK Adult Psychiatric Morbidity Survey concluded that every week approximately 1 in 6 people experience common mental health problems in the workplace and that women in full-time employment are nearly twice as likely to have a common mental health problem as full-time employed men (19.8% vs 10.9%).

Mental ill-health affects the productivity of those in work by impairing their ability to function at full capacity and, in the UK, it is reported that it causes about 40% of all days lost through sickness absence (Sainsbury Centre for Mental Health, 2017). Compared with many common physical conditions, mental health problems are often gradual in onset and long lasting. The Labour Force Survey 2019/2020 identified that on average each mental ill-health sickness absence spell lasts 21.6 days and for this reason, although mental ill-health causes just 25% of absences of less than seven days, they account for 47% of long-term absences. The International Labour Organisation (ILO) reports in their document 'Mental health and work: Impacts, issues and good practices' that the Association of Canadian insurance companies estimates that 230-50% of worker disability allowances are paid in relation to mental ill-health problems.

	Cost per average worker (£)	Total cost to UK employers (£billion)	% of total
Absenteeism	395	10.6	30.4
Presenteeism	790	21.2	60.8
Replacing worker	115	3.1	8.8
Total	1300	34.9	100

Figure 13-47: Estimated annual costs to UK employers of mental ill-health.
Source: Sainsbury Center for Mental Health, 2017

The Sainsbury Centre for Mental Health has estimated (in 2017) that impaired work efficiency (sometimes called 'presenteeism') due to mental ill-health costs the UK £21.2 billion, or £790 for every worker in the UK, which is more than twice the estimated £10.6 billion annual cost of absenteeism. Some US studies have put the cost of presenteeism at four or five times the cost of absenteeism.

The overall cost of depression in England in 2000 was estimated to be £9 billion. More than £8 billion of this cost was due to lost productivity as a result of work days lost, which is 20 times larger than the estimated £0.4 billion costs of the National Health Service (NHS) for dealing with the mental ill-health.

The effect of stress-related conditions can be difficult to calculate, but it has been estimated that the harmful effects of stress cost the UK economy around £4 billion a year, with almost 13 million working days lost and over half a million people reporting that they were suffering from work-related stress.

In the period 2018/19 there were 0.6 million (16,000 in construction) new or longstanding cases of depression, anxiety and stress identified through the self-reporting UK Labour Force Survey. The number of new cases was reported to be 246,000. This had led to an estimated 12.75 million days lost from work by people working in the twelve month period, which equated to an average of 21.2 days lost per case. In 2018/19 stress, depression and anxiety accounted for 44% of all work-related ill health cases and 54% of all working days lost due to ill health. The Institute for Employment Studies Report 2021 found that 42% of construction workers worry about their workload being too high, 37% had difficulties with colleagues and 35% found that they drank more when stressed.

A study in 2005 found that workers in particular jobs are more likely to develop common mental ill-health problems. This includes teachers, nurses, social workers, probation officers and fire fighters, police officers, the armed forces and medical practitioners. Data from the HSE in 2019 showed that 97 per cent of construction workers felt stressed at least once in the last year and 26 percent had considered taking their own lives in 2019.

Mental ill-health problems can cause fatigue, poor memory, impaired attention and difficulty concentrating.

These problems can be made worse by the effects of medication.

The psychological effects of an individual's mental ill-health can result in the following workplace issues:

- Lack of motivation.
- Lack of commitment.
- Poor timekeeping.
- Increase in mistakes.
- Increase in sickness absence.
- Poor decision making.
- Poor planning.
- Tension between colleagues/supervisors.
- Poor service to clients.
- Deterioration in work relations.
- Increase in disciplinary problems.

COMMON SYMPTOMS OF WORKERS WITH MENTAL ILL-HEALTH

Depression

Most people experience short-term feelings of sadness or anxiety during difficult times. This is not usually a sign of depression. Depression is a low mood that lasts for a long time, and affects everyday life.

Depression can affect people in different ways and can cause a wide variety of symptoms. The severity of the symptoms may vary. At its mildest individuals may have a persistently low temperament and at its most severe depression can make those affected feel life is no longer worth living and suicidal.

Common symptoms of depression in individuals might include:

How they feel

- Down, sad, upset or tearful.
- Empty and numb.
- Restless, agitated or irritable.
- Guilty, worthless and negative towards themselves.
- No self-confidence or self-esteem.
- Isolated and unable to relate to other people.
- Finding no pleasure in life or things they used to enjoy.
- A sense of unreality.
- Hopeless and despairing.
- Suicidal.

How they behave

- Difficulty concentrating or remembering things.
- Difficulty speaking, thinking clearly or making decisions.
- Feeling tired all the time.
- Difficulty sleeping, or sleeping too much.
- Moving very slowly, or being restless and agitated.
- Physical aches and pains with no obvious physical cause.
- Avoiding social events and activities they used to enjoy.
- Loss of appetite and weight or eating too much and gaining weight.
- Losing interest in sex.
- Using more tobacco, alcohol or other drugs than usual.
- Self-harming or suicidal behavior.

Depression can lead an individual to feel unable to cope with everyday tasks. Their reduced functional capability means they are no longer able to apply themselves to everyday tasks that they once could do with little difficulty. It may be that they are no longer able to function properly in their workplace or, if they are a student, they may find that they are no longer able to concentrate on studying. In the home this may result in an inability to gather enough energy to perform everyday activities, for example, those related to personal hygiene. This leads the depressed individual to feel incapable of completing basic and important tasks. This then makes the possibility of completing those tasks even more unlikely and this cycle leads to lower self-esteem and further incapability.

Depressed individuals therefore often develop low self-esteem and do not value their own ideas, views and opinions. They tend to constantly worry that what they think and do is not 'good enough'. Eventually, it is not uncommon for depressed individuals to find it impossible to concentrate or gather enough mental energy to hold a conversation.

Treatment for depression involves either medication or talking therapy treatments, or usually a combination of the two. The kind of treatment that a physician recommends will be based on the type and seriousness of the depression diagnosed.

Anxiety/panic attacks

Anxiety and panic disorders are common, but can cause extreme distress to individuals. Anxiety is a feeling of unease, such as worry or fear, which may be mild or severe. Panic attacks are a type of fear response.

They are an exaggeration of your body's normal response to danger, stress or excitement.

Anxiety

Everyone has feelings of anxiety at some point in their life. For example, they may feel worried about sitting an examination or having a medical test or job interview. During times like these, feeling anxious can be perfectly normal. However, some people find it hard to control their anxiety, their feelings of anxiety are more extreme, often constant and affect their ability to cope with everyday tasks. There are a range of anxiety disorders and some of the more common ones are:

Generalised anxiety disorder (GAD) – this means having regular or uncontrollable worries about many different things in everyday life, which could include work.

Social anxiety disorder – means the individual affected experiences extreme anxiety triggered by social situations (such as weddings, workplaces or any situation in which you have to talk to another person). It is also known as social phobia. A phobia is an extreme anxiety triggered by a particular situation (such as social situations or going outside the house) or a particular object (such as spiders).

Obsessive-compulsive disorder (OCD) – this means the individual has anxiety involving repetitive thoughts, behaviours or compulsions.

Anxiety feels different for everyone. An individual might experience some of the symptoms listed below:

How they might feel

- Tense, nervous or unable to relax.
- The world is speeding up or slowing down.
- They cannot stop worrying and have a sense of dread or fear the worst.
- Thinking a lot about bad past experiences, often thinking over a situation again and again.
- Worrying a lot about things that might happen in the future.
- Worrying about anxiety itself, for example worrying about when panic attacks might happen.
- Other people can see they are anxious and are looking at them.
- A need for reassurance from other people or worrying that people are angry or upset with them.
- Worrying that they are losing touch with reality.
- Disconnected from their mind or body, like they are watching someone else or the world around them is not real.

Effects on their body

- An agitated feeling in their stomach.
- Feeling faint, light-headed or dizzy.
- Breathing faster.
- A fast, very strong or irregular heartbeat.
- Sweating, hot flushes or very cold.

- Trembling or shaking.
- Nausea (feeling sick).
- Needing the toilet more or less often.
- Headaches, backache or other aches and pains.
- Feeling restless or unable to sit still.
- Problems sleeping.
- Grinding their teeth, especially at night.
- Changes in their sex drive.
- Having panic attacks.

Symptoms that may particularly show themselves at work may include loss of interest, poor concentration, low mood and irritability.

Panic

Panic disorder is where a person has regular or frequent panic attacks without a clear cause or trigger. Experiencing panic disorder can mean that the person affected feels constantly afraid of having another panic attack, to the point that this anxiety itself can trigger panic attacks.

During a panic attack an individual might feel very afraid that they are losing control, going to faint, having a heart attack or going to die. The physical symptoms of a panic attack can build up very quickly and are similar to the symptoms for extreme anxiety, including:

- A very strong, irregular or fast heartbeat.
- Feeling faint, dizzy or light-headed.
- Feeling very hot, sweating or very cold.
- Nausea (feeling sick).
- Pain in the chest or abdomen.
- Struggling to breathe.
- Legs and hands trembling and shaky.

Work-related stress

Stress is the feeling of being under too much mental or emotional pressure. Pressure turns into stress when an individual feels unable to cope with the pressure. Many of life's pressures can cause stress, particularly work, home life, relationships and financial problems. People have different ways of reacting to pressure, therefore a situation that feels stressful to one person may be motivating to someone else. When an individual feels stressed, it can make it difficult to manage these pressures and can affect everything else in their life. For example, stress caused by work-related pressures could affect a worker's home life.

In the UK, the Health and Safety Executive defines stress as:

"Stress is the adverse reaction people have to excessive pressures or other demands placed on them."

Figure 13-48: Definition of stress.
Source: HSE.

Figure 13-49: Stress words.
Source: Venture Galleries.

Being exposed to situations of excessive pressure can bring about changes in an individual's behaviour, as well as their physical and psychological well-being. In particular, it can lead to worker's developing common mental ill-health problems like depression or anxiety.

Physical effects of stress include:

- Increased heart rate/sweating.
- Headache and dizziness.
- Blurred vision.
- Aching neck and shoulders.
- Skin rashes.
- Lowered resistance to infection.

Behavioural effects of stress include:

- Poor concentration.
- Irritability.
- Erratic sleep patterns.
- Increased smoking.
- Increased alcohol or drug use.

Psychological effects of stress include:

- Increased anxiety.
- Feeling unable to cope with everyday tasks.
- Low self-esteem.
- Depression.

Post-traumatic stress disorder

Post-traumatic Stress Disorder (PTSD) is developed after being involved in or witnessing traumatic events, such as a natural disaster, serious accident, terrorist act, war/combat or other violent assault. The events are therefore usually very frightening, upsetting or distressing. The person may not be exposed to the traumatic event directly, for example and individual may learn about the accidental death of a work colleague or violent death of a close member of their family. It can also occur as a result of repeated exposure to horrible details of trauma, such as what police officers may experience when investigating cases. People working in a range of roles in the emergency services may therefore experience PTSD.

PTSD can affect how an individual feels, thinks, behaves and their body functions. Common symptoms of PTSD include restless sleep, sweating, loss of appetite and difficulty concentrating. This could be because when an individual feels stressed their body release hormones called cortisol and adrenaline. This is the body's automatic way of preparing to respond to a threat (sometimes called the 'fight, flight or freeze' response). Studies have shown that someone with PTSD will continue producing these hormones when they are no longer in danger, which is thought to explain some symptoms, such as extreme alertness and being easily startled.

The symptoms of post-traumatic stress disorder (PTSD) can have a significant impact on an individual's usual life. In most cases, the symptoms develop during the first month after a traumatic event, but in a small number of cases there may be a delay of months or even years before symptoms start to appear. Some people with PTSD experience long periods when their symptoms are less noticeable, followed by periods where they get worse. Other people have constant severe symptoms.

The specific symptoms of PTSD can vary widely between individuals, but generally fall into the following three categories – re-experiencing, avoidance/emotional numbing and hyperarousal.

Re-experiencing

Re-experiencing is the most common symptom of PTSD. This is when an individual involuntarily and vividly relives the traumatic event in the form of:

- Vivid visual recollections or sensations from the traumatic event ('flashbacks'), feeling like the trauma is happening at the time.
- Vivid traumatic dreams ('nightmares').
- Physical sensations, such as pain, sweating, nausea (feeling sick) or trembling.

Some people have constant negative thoughts about their experience, repeatedly asking themselves questions that prevent them coming to terms with the traumatic event. For example, they may wonder why the event happened to them and if they could have done anything to prevent it, which can lead to feelings of self-doubt, blaming themselves, guilt or shame.

Avoidance and emotional numbing

Trying to avoid being reminded of the traumatic event is another common symptom of PTSD. This usually means the individual avoiding certain people or places that remind them of the trauma, or avoiding talking to anyone about their experience. Many people with PTSD try to push memories of the trauma out of their mind, often distracting themselves with work or hobbies. Some people attempt to deal with their feelings by trying not to feel anything at all, known as emotional numbing.

This can lead to the person becoming isolated and withdrawn, and they may also give up activities they used to enjoy. Avoidance and emotional numbing symptoms can include:

- Having to keep busy.
- Avoiding anything that reminds them of the trauma.
- Being unable to remember details of what happened in the traumatic event.
- Feeling emotionally numb or cut off from their feelings.
- Feeling physically numb or detached from their body.
- Being unable to express affection.
- Using alcohol or drugs to avoid memories.

Hyperarousal (feeling 'on edge')

Someone with PTSD may be very anxious and find it difficult to relax. They may be constantly aware of threats, feel nowhere is safe, not trust anyone and be easily startled. This is known as hyperarousal. Hyper-arousal often leads to:

- Irritability.
- Angry outbursts.
- Sleeping problems (insomnia).
- Difficulty concentrating

Because of these, potentially extreme effects on the individual suffering PTSD they sometimes lead to work-related problems and the breakdown of personal relationships.

CAUSES OF MENTAL ILL-HEALTH AND CONTROL MEASURES

Work-related mental ill-health, like other risks, should be subject to a risk assessment in order to identify the causes, evaluate the current controls and decide on actions to manage the risk.

There are six main factors that should be managed to minimise the risk of mental ill-health in the workplace: demands, control, support, working relationships, role and change. These are explained below in the form of causes, standards to be achieved in the workplace that will minimise mental ill-health and the control measures to eliminate or minimise the causes and achieve the standard.

Demand

Causes

Job content issues, for example:

- Monotonous, under-stimulating, meaningless tasks.
- Lack of variety.
- Unpleasant or high risk tasks.
- Environmental aspects of the job, for example, noise, vibration, extremes of temperature and lighting.
- Lone working.

- The job involves making unwelcome interventions, for example, law enforcement activities.
- Involves working with substandard resources, for example equipment that frequently breaks down or fellow workers that absent themselves at short notice.
- Workers are asked to do things they are not capable of doing, for example, they require specific competencies the worker does not have, or are too complex or emotionally demanding.

Workload and work pace, for example:

- Having too much or too little to do.
- Working under time pressures.
- Working to unrealistic deadlines.
- Workload is reactive and depends on external factors, such as demand from customer, national authorities, 'headquarters'.

Working hours, for example:

- Strict and inflexible working schedules.
- Long and unsocial hours.
- Unpredictable working hours.
- Badly designed shift systems.
- Inadequate rest periods.
- Communication equipment is issued to workers that makes them contactable out of normal hours.

Control measures

The standard to be achieved is that the demands of the job are within the capability of the workers and there are systems in place locally to respond to any individual concerns. Control measures to eliminate or minimise the causes and meet this standard include:

- Balancing the demands of the work to the agreed hours of work; consider shift working, the amount of additional hours worked and unsocial hours.
- Provision of regular and suitable breaks from work and rest periods.
- Matching worker skills and abilities to the job demands.
- Designing jobs so they are within the capability of workers.
- Minimising the work environment risks, such as noise and temperature.
- Provision of suitable facilities for work breaks, refreshment and rest.

Control

Causes

Participation and personal control, for example:

- Lack of participation in decision making.
- Lack of control, for example, over work methods, work pace, working hours, rest periods and the work environment.

- Workload and work activities are imposed by external organisations, such as national authorities or 'headquarters'.
- Lack of opportunity for workers to develop their competence.
- Lack of opportunity for workers to use their competence and initiative.

Control measures

The standard required is that workers are able to bring to the attention of the organisation issues to do with their work and that there are systems in place locally to respond to any individual concerns.

Control measures to eliminate or minimise the causes and meet this standard include:

- Providing workers with control over their pace and manner of work.
- Reducing the effects of repetitive and monotonous work by job rotation.
- Encouraging workers to use their skills and initiative to do the work.
- Encouraging workers to develop new skills and so enabling them to do more challenging or new work.
- Providing workers with opportunity to influence when breaks are taken.
- Consulting workers regarding work patterns.
- Encouraging the participation of workers in improving work.
- Providing mechanisms where workplace problems can be discussed.

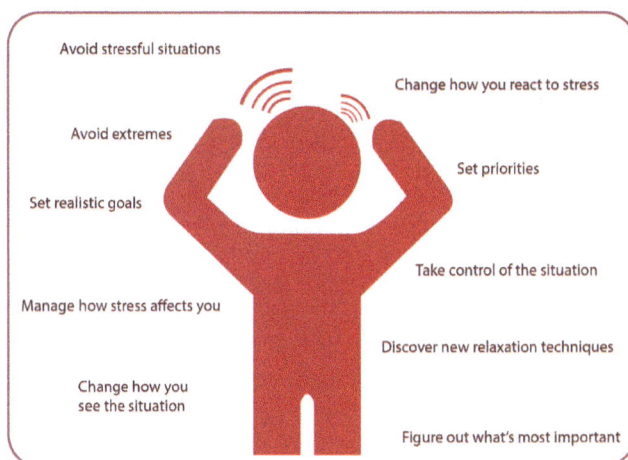

Avoid stressful situations
Change how you react to stress
Avoid extremes
Set priorities
Set realistic goals
Take control of the situation
Manage how stress affects you
Discover new relaxation techniques
Change how you see the situation
Figure out what's most important

Figure 13-50: Managing your mental health.
Source: RMS/Positive Productive.

Support

Causes

Support from the organisation, management and other workers, for example:

- Poor leadership and culture.
- Conflicting priorities, for example, health and safety versus productivity.
- Lack of clarity about organisational objectives and structure.

- Lack of or unclear policies and procedures.
- Poor communication.
- Lack of support from line managers or co-workers or a feeling of isolation.
- Lack of constructive feedback from line managers or co-workers.
- Unclear how to access resources or lack of provision of requested resources to do the job.

Control measures

The standard required is that workers should receive adequate information and support from their colleagues and superiors and there are systems in place locally to respond to any individual concerns.

Control measures to eliminate or minimise the causes and meet this standard include:

- Providing systems to inform workers of important issues and decisions.
- Establishing policies and procedures that provide support, particularly where workers may feel other factors are putting them under pressure.
- Providing systems that enable and encourage managers to identify where workers need support; consider where workers deal with the public in demanding environments, where new workers are introduced and times of high demand.
- Providing systems that enable and encourage managers to provide support to workers; consider particularly those working remotely by virtue of their location or time of working and new workers.
- Encouraging co-workers to support each other.
- Ensuring workers understand what resources and support are available and how they can get it.
- Providing regular constructive feedback to workers.

Work Relationships

Causes

Interpersonal relationships, for example:

- Inadequate, inconsiderate or unsupportive supervision/management.
- Poor relationships with co-workers.
- Bullying, harassment, abuse or violence (**see Figure 13-51**).
- Isolated or solitary work.
- No agreed procedures for dealing with problems or complaints.
- Unacceptable behavior tolerated.
- Workers (at all levels) treated unfairly.
- Poor communication between individuals and management.

> "Persistent, offensive, abusive, intimidating, malicious or insulting behaviour, abuse of power or unfair penal sanctions which make the recipient feel upset, threatened, humiliated, or vulnerable, which undermines their self-confidence and which may cause them to suffer stress. Harassment, in general terms, is unwanted conduct affecting the dignity of men and women in the workplace. It may be related to age, sex, race, disability, religion, sexual orientation, nationality or any personal characteristic of the individual, and may be persistent or an isolated incident. The key is that the actions or comments are viewed as demeaning and unacceptable to the recipient."

Figure 13-51: Definitions of bullying and harassment.
Source: UK, CIPD.

Control measures

The standard required is that workers should feel that they are not subjected to unacceptable behaviours, for example, bullying at work, and there are systems in place locally to respond to any individual concerns. Improving work relationships and attempting to modify people's attitudes and behaviour is a difficult and time-consuming process.

Effective strategies include regular communication with workers, provision of accurate and honest information on the effect of organisational changes on them, adopting partnership approaches to problems and provision of support. Employers should promote a culture that respects the dignity of others. Specific control measures to eliminate or minimise the causes and meet this standard include:

- Promoting positive behaviour that avoids conflict and leads to fairness; consider co-workers, customers and suppliers.

- Establishing policies and procedures that resolve unacceptable behaviour that leads to conflict.

- Encouraging managers to deal with unacceptable behaviour, such as harassment, discrimination or bullying.

- Encouraging workers to report unacceptable behaviour.

- Establishing policies that treat workers equally, for example, male and female, young and old, or disabled and able workers.

- Establishing systems that ensure communication with managers and workers; consider timeliness of communication to those that work isolated by location or time.

- Establishing systems that ensure involvement and consultation, such as regarding the process of conducting risk assessments.

Role

Causes

Role in the organisation, for example:

- Unclear role and responsibilities.
- Conflicting roles within the same job.
- Unclear or conflicting work objectives.
- Responsibility for people.
- Continuously dealing with other people and their problems.
- Organisational change affecting roles.
- Job insecurity.
- Lack of promotion prospects.
- Under-promotion or over-promotion.
- Work of 'low social value'.
- Demanding productivity payment schemes.
- Unclear or unfair performance evaluation systems.
- Being over-skilled or under-skilled for the job.

Control measures

The standard required is that workers must clearly understand their role and responsibilities and there are systems in place locally to respond to any individual concerns.

Control measures to eliminate or minimise the causes and meet this standard include:

- Ensuring workers have the knowledge, skill and experience to conduct their role or are being supported appropriately; consider workers under-taking new or difficult work.

- Ensuring role requirements are compatible, for example, that the need to manage costs does not conflict with health and safety.

- Ensuring role requirements are clear.

- Ensuring workers and their managers understand the roles and responsibilities.

- Providing systems to enable workers to raise concerns about role uncertainty or conflict; particu-larly consider the work-life balance for those workers who provide care for others outside their normal work routine.

Change

Causes

Organisational change, for example:

- Unclear reason for change.
- No prior warning of the change.
- No consultation on change or opportunity to influ-ence the proposed change.
- Extent and impact of the change is unclear.
- Frequency of changes.
- Change is very large.

- Absence of or unclear timetable for changes.
- Absence of support during changes.
- Absence of information and training to enable transi-tion to new circumstances.

Control measures

The standard required is that workers feel the organ-isation engages workers frequently when undergoing organisational change, and there are systems in place locally to respond to any indiviual concerns.

Control measures to eliminate or minimise the causes and meet this standard include:

- Providing workers with timely information to help them understand the change, reasons for it and timing of effects.

- Ensuring worker consultation on proposed changes.

- Providing workers with information on likely impacts of change on their jobs.

- Providing training and support through the period of change.

In addition to the six main factors that are managed in order to minimise the risk of mental ill-health, there are a number of general control measures that should form part of a mental health management strategy. These include:

- Introducing a mental health policy and procedures to demonstrate to managers, workers, worker repre-sentatives and enforcing agencies that the organ-isation recognises mental health as a serious issue worthy of a commitment to manage the risk.

- Providing training and support for workers and all levels of management in the form of mental health awareness and mental health management training, as appropriate.

- Recognising good work performance and the positive contribution of worker and responding to poor performance to ensure equality of treatment.

- Addressing the issue of work-life balance, which may include consideration of job-sharing, part-time work, voluntary reduced hours, home-working, flexible-time, etc.

- Listening to the concerns of workers and managers and providing practical and emotional support.

- Promoting general health and well-being aware-ness initiatives within the organisation, such as diet, exercise and fitness programmes.

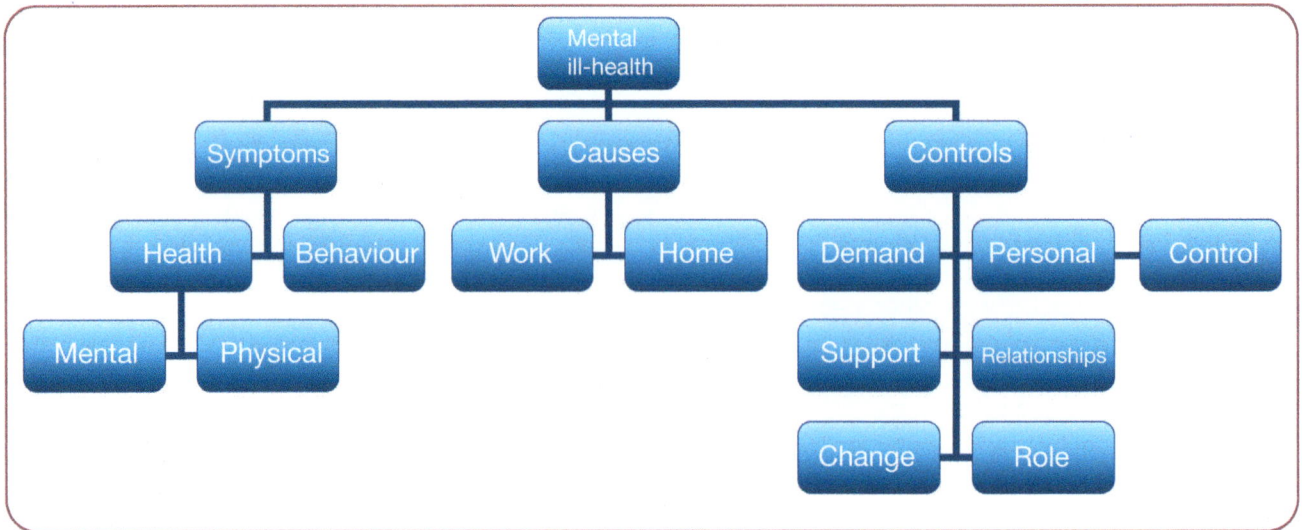

Figure 13-52: Mental ill-health in summary.
Source: RMS.

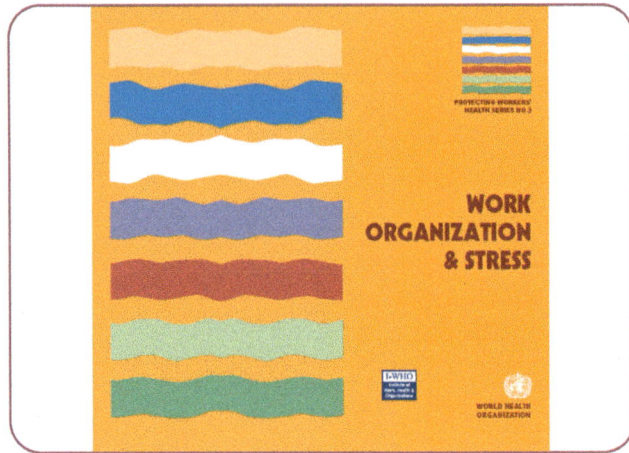

Figure 13-53: Stress prevention at work checkpoints.
Source: ILO.

PEOPLE WITH MENTAL ILL-HEALTH CONTINUING TO WORK

There is an increasing recognition that most people with mental ill-health can continue to work effectively and one of the best things for their wellbeing is for them to be in appropriate work, which takes regard for their mental ill-health.

It is important to remember that studies have shown the longer a worker is off work with mental ill-health the more difficult it becomes for them to return to work and the less likely it is that they will return to work at all. Therefore, where possible, workers should remain in work while receiving treatment or return to work at the earliest opportunity.

For some workers affected by mental ill-health this may mean making some adjustments to their work.

These may be of a temporary nature as they readjust to work after developing the mental ill-health or become unnecessary as their mental ill-health improves and symptoms diminish. It is therefore important to support their return by making appropriate adjustments.

It is also important to remember that for people with

mental ill-health, for example depression, improvement in ability to function well at work may be reduced for a period. In order to improve wellbeing and help recovery, a doctor might confirm a worker to be fit to return to work before they have the full capability to attain previous levels of work performance. This creates the risk that employers interpret poor performance by the worker as lack of effort or motivation or competence and so create the conditions in which it is more likely that the worker believes that they are failing and starting to become ill again. The worker is therefore likely to need an encouraging work environment in order to adjust to work, improve and work well.

There are a range of treatments available for common mental ill-health, including medication and psychological therapies, for example, cognitive behavioural therapy (CBT) and self-help. These can support the individual affected and enable them to continue to work.

With this opportunity and support it is likely that most people with mental ill-health can continue in work.

Good work gives the worker a social identity, purpose, social contacts, a means of structuring/occupying time, activity, involvement and a sense of personal achievement. This can improve the resilience of the worker and reduce the likelihood of them being affected by mental ill-health.

ORGANISATIONS PROVIDING MENTAL WELL-BEING SUPPORT

There are a number of organisations that provide mental well-being support to those in the construction industry, including:

Building Mental Health, Considerate Constructors Scheme, Lighthouse Construction Industry Charity, Mates in Mind, Mental Health at Work including (supported by Mind), Mental Health First Aid England and Scottish Association for Mental Health.

The support they provide is outlined below.

Building Mental Health – Is an organisation run by a group of volunteers from the construction industry. It has a range of resources around the topic of mental health, including links to free downloadable videos and a free editable toolbox talk. For example, one video has been provided by Mace Group called 'Starting the conversation in construction' and one provided by Crossrail focusing on recognising the signs of poor mental health. www.buildingmentalhealth.net

Considerate Constructors Scheme – The scheme is a non-profit making organisation founded to improve the image of the construction industry. It has a specific website page focused on mental health, which provides facts, legal information and a library of examples of good practice from the construction industry.

The good practice illustrated ranges from a 'therapy dog', awareness days, mobile apps and specific interventions. https://ccsbestpractice.org.uk/spotlight-on/mental-health/#Introduction

Lighthouse Construction Industry Charity – Has developed the Construction Industry Helpline to provide confidential mental health support and a free Helpline app to provide advice and guidance to workers. They also support education/training initiatives and provide financial and emotional support to the construction community and their families. www.constructionindustryhelpline.com, www.lighthouseclub.org.

Mates in Mind – Is a charity focusing on the construction industry, promoting awareness of mental health. It has a wide range of 'construction collateral materials', including posters and van stickers. It produces fact sheets, podcasts, advice for employers, information for workers and education/training events. www.matesinmind.org.

Mental Health at Work – Provides information and resources drawn from a range of partner organisations, it is supported by the charity Mind. The resources have a filter option to find those relevant to construction. They make a specific toolkit available for the construction industry called 'Building Mental Health in Construction'. www.mentalhealthatwork.org.uk/toolkit/building-mental-health-in-construction, www.mind.org.uk.

Mental Health First Aid England – Is a social enterprise organisation that provides resources and education, including related to their 'Take 10 Together' campaign to encourage people to talk about mental ill-health. They have developed a toolkit called 'My Whole Self' to encourage openness to mental ill-health and improve how it is accommodated at work.

They also provide training to enable people to identify signs of mental ill-health and offer first aid support. https://mhfaengland.org/.

Scottish Association for Mental Health (SAMH) – SAMH provides case studies, personal stories and direct support in the form of listening, talking and helping people recover from mental ill-health. www.samh.org.uk

CONSIDER

Where can you access valuable advice on how to support workers?

Can you identify specialist organisations that can provide guidance and support?

Does your organisation have a mental health policy?

13.5 Violence at work

TYPES OF VIOLENCE AT WORK

Violence at work is not limited to physical violence, it also includes verbal abuse and threats.

When physical violence is involved, the injuries to those workers affected are obvious. However, those subjected to constant and repeated verbal abuse and threats may suffer less visible psychological effects, such as stress, anxiety and depression.

> "Any incident, in which a person is abused, threatened or assaulted in circumstances relating to their work."

Figure 13-54: Definition of work related violence.
Source: HSE.

Workers may be sworn at, threatened or even attacked. Verbal abuse and threats are the most common types of violence at work, physical attacks are comparatively rare. It is more common in those jobs where workers have face-to-face contact with the public. Physical attacks are obviously dangerous, but serious or persistent verbal abuse or threats can cause psychological damage to a worker's health.

Some forms of behaviour may be considered to be bullying or harassment, rather than violent. In simple terms bullying and harassment is behaviour that makes someone feel intimidated or offended. Some forms of harassment, for example related to age or sex, are unlawful under the Equality Act 2010, but bullying itself is not covered by the Equality Act 2010.

> "Harassment occurs when someone is repeatedly and deliberately abused, threatened and/or humiliated in circumstances relating to work. It may be carried out by one or more manager, worker, service user or member of the public with the purpose or effect of violating a manager's or worker's dignity, affecting his/her health and/or creating a hostile work environment."

Figure 13-55: Definition of harassment at work.
Source: Preventing workplace harassment and violence.

Acas (Advisory, Conciliation and Arbitration Service) characterises bullying as behaviour that is offensive, intimidating, malicious or insulting or an abuse/misuse of power through means that undermine, humiliate, denigrate or injure the recipient.

Examples of bullying or harassing behaviour include:

- Spreading malicious rumours, for example saying a particular site manager with good health and safety standards was the reason why a construction project did not meet target dates and no bonus was given.

- Unfair treatment, for example withholding resources to improve health and safety.

- Picking on or regularly undermining someone, for example for raising health and safety concerns.

- Denying someone training opportunities, for example in relation to managing health and safety in their role.

Often bullying and harassment involves repeated behaviour and may include an accumulation of some of the examples provided above. Bullying and harassment can happen by letter, email, text, social media, phone or face-to-face (including remotely by electronic means). Bullying and harassment can often be hard to recognise as the symptoms may not be obvious to other people that it is not direct toward or it may be hidden in a way that it is only known by the person affected. The person receiving the bullying or harassment may think it is normal behaviour in the organisation, keep it to themselves and try not to be affected by it. They may not reveal that they have been affected because they are anxious that others will consider them weak or not capable of doing their job if they find the behaviour and actions of others intimidating.

CONTROL MEASURES TO REDUCE RISKS FROM VIOLENCE AT WORK

Stage 1 Finding out if you have a problem

Employers should determine the extent to which the job and risk factors described below affect their organisation. In order to determine the scale of the problem employers should:

- Ask workers informally, through managers and health and safety representatives.

- Encourage workers to report all incidents and keep detailed records.

- Classify all incidents according to their actual or potential severity of outcome.

- Try to predict what might happen and how violence may arise.

The most common job and activity risk factors relating to violence are:

- The position a person holds - a worker may hold a position of authority over another person; how they use this authority and the effects on the person can cause disagreement, resentment and could cause the person to be more aggressive.

- The nature of the work - if a worker is working alone this may lead them to be more at risk of violence. The nature of the work may be dealing with people that are emotionally charged, and this may cause them to react to normal things in an unpredictable way.

- The location of the work - if the worker is working isolated from others this may leave them at risk. Some locations may have a tendency for violence, for example, certain parts of a city or housing estates in deprived areas.

- The time of working - working during late evening and early morning may mean there are fewer people around and those people that are around may be more likely to have violent tendencies.

- Alcohol and drugs - can make some people more aggressive. Because their perception and behaviour is more unpredictable, it may lead to misunderstandings and violence.

- Visible appearance - violence may arise simply because someone does not like a worker's visible appearance and what this may represent. This can be accentuated if the appearance is one of privilege and wealth.

- The availability of weapons - if weapons, improvised or normal, are available to be used a sudden outburst that may have been verbal could escalate to involve major personal injury because a weapon was available, for example, knives in a restaurant or home, shotgun on a farm or drinking glass in a bar.

Those jobs where workers have contact with the public and are most at risk are where workers are engaged in caring, education, cash handling, and representing authority. This include police, fire, medical, social workers and those in customer services. Lone workers, those working with people under the influence of drugs and alcohol and those who handle money or valuables are particularly at risk.

In addition, it is necessary to consider possible violence between co-workers. Maintenance and construction activities often need to be completed within a fixed times-cale and can result in a lot of pressure on workers to work quickly and without error. This can lead to a good deal of tension which might result in violent outbursts between co-workers including contractors.

Stage 2 Deciding what action to take

Employers should establish appropriate control measures to eliminate or minimise the effect of the risk factors that affect their organisation.

In order to determine what controls are relevant to the risks, employers should:

- Identify who might be harmed and how.
- Evaluate the risk.
- Record the findings.
- Review and revise the assessment as appropriate.

Deciding who might be harmed and how

Review the jobs and risk factors described earlier to determine what factors affect different workers. Consider the circumstances when they may be affected and the likely outcomes – physical violence, abuse etc. Identify those workers that are most vulnerable, perhaps due to experience or other characteristics. Those workers who have face-to-face contact with the public and work alone are usually most at risk. Those workers whose work brings them into contact with known or potentially violent people should be identified and the risks evaluated, including consideration of current control measures.

Evaluate the risk

Consider the circumstances of the work activities involved and analyse the current arrangements, in particular the:

- Level of training and information provided.
- Work environment involved.
- Design of the job and activities within the job.

Training and information

Consider how the training and information provided enables them to identify the early signs of potential violence and how to avoid or respond to it by using de-escalation techniques (modification of their own behaviour). Determine how good is the provision of critical information that will enable workers to anticipate potentially violent people, for example, by providing case histories. Evaluate the workers' understanding of proto-cols and procedures, including emergency arrange-ments and incident reporting. Assess the measures in place to brief workers on problems that may foresee-ably occur.

Work environment

Evaluate the current arrangements in place in the workplace that deter or de-escalate violence or protect workers, consider:

- The ambience of the environment in terms of colour, lighting, seating, space and even the type of background music.
- Signage to warn violence will not be tolerated.
- Availability/removal of any items that could be used as a weapon. Even pens and pencils can be used as weapons.
- Physical security - for example, closed circuit televi-sion (CCTV) with recording facility, coded locks, wider and higher counters for those who deal with members of the public, panic alarms.
- Place screens between workers and the public.

Design of the job

Evaluate the arrangements in place to minimise risks through the design of the job and specific work activi-ties, consider:

- Cash handling, for example, use credit cards rather than cash, bank money more frequently.
- Check arrangements for lone workers and consider two workers where risks are high.
- Are routines for higher risk activities predictable and therefore at more risk, for example when handling cash and other items that some might consider to be valuable such as drugs and medication.
- What is in place to reduce risks for those that work late at night, for example safe transport or secure car parking.
- Assess those parts of the job that can result in frustra-tion and mean escalation is more likely, for example queues when waiting for attention or malfunction of equipment required to access a service. Consider signing and updating current waiting time, intro-duce pre-appointments whenever possible and ensure sufficient equipment is available to maintain an acceptable service.
- The number of workers available at the location to determine if they are sufficient to avoid lone working.
- Provide a means of raising the alarm and a plan for dealing with violent situations.
- Procedures that are in place for reporting concerns and any occurrence.
- Policy and arrangements for dealing with people involved in alcohol and substance misuse.

Record the findings

Keep a record of the significant findings of your assessment. The record should provide a working document for both managers and workers. The findings should identify what is working and any areas for improvement. This should be used to guide implementation actions for new systems and plans to make improvements.

Review and revise your assessment

Regularly check that your assessment accurately reflection the current work situation, for example, identify changes that have been made to the work environment and work activities. Evaluate if the arrangements put in place to manage risk were effective and if further or different control measures are required. If a violent incident occurs, the assessment should be reviewed to determine if this sort of incident had been considered and what further could be done to prevent such incidents in the future.

Stage 3 Take action

The policy for dealing with violence should be written into the health and safety policy statement so that all workers are aware of it. This will encourage workers to cooperate with the policy and report further incidents. It will also show managers that the organisation has a firm commitment to the prevention of violence, bullying and harassment. This will make them more likely to implement the policies and the arrangements necessary to manage the risks. The actions necessary will depend on the outcome from the assessment and could include the arrangements described above, in stage 2. Clear responsibility should be given for the actions to implement new arrangements or improvements, which should be monitored until their completion. Workers should be involved in the completion of the actions as they will be more likely to support them and might identify early those actions that may not be as effective as was intended.

Stage 4 Check what you have done

Check regularly to see if the arrangements are working by consulting workers and worker health and safety representatives. If arrangements or specific control measures are working well ensure this is communicated to managers and workers. If violence is still a problem, review the circumstances and arrangements – consider work practices and risk controls measures.

Dealing with incidents

If there is a violent incident in the workplace it will be necessary to consider the following:

- Debriefing - victims might need to talk through their experience as soon as possible.
- Providing time off work - individuals may need differing times to recover.

- Support - in some cases victims might need counselling. Consider a phased return to work.
- Legal help - legal assistance may be appropriate in serious cases.
- Other workers - may need guidance or counselling to help them react appropriately.

REVIEW

What control measures could an employer consider to minimise the risk of violence against workers?

List five examples of occupations that are at high risk from violence.

13.6 Substance abuse at work

TYPES OF SUBSTANCES ABUSED AT WORK

Many substances can affect a worker's ability to work safely, including alcohol, drugs and solvents. They do not have to be taken whilst at work to have an effect as substances can have a residual effect even when they are taken outside working hours. This can have a significant effect on the health and safety of the worker and those that may be affected by their work. It is imperative that organisations take substance abuse (misuse) at work seriously in order to protect the health and safety of those who may be affected, especially in high risk activities such as construction, chemical, nuclear, rail and other transport.

Alcohol

Even at low levels of consumption the body is affected by alcohol, reducing reaction time and impeding coordination. It also affects thinking, judgement and mood and can have a significant effect on behaviour when conducting routine work. Whilst large amounts of alcohol in one session can put a strain on the body's functions, in particular the liver, it can also affect muscle function and stamina.

Drinking alcohol raises the drinker's blood pressure and this can increase the risk of coronary heart disease and some kinds of stroke. Regularly drinking smaller amounts also increases the risk of cirrhosis of the liver. The effect of alcohol is not limited to physical effects - people who drink very heavily may develop psychological and emotional problems, including depression.

Brandon was a 25 year old male with alcohol related problems. He had been having an increasing number of absences from work, was often late on duty and with the smell of alcohol on his breath. The manager discussed the issues with Brandon, who agreed that he required help to overcome his excessive drinking. Advice from occupational health resulted in Brandon being referred for alcohol counselling during the working day. With the full support of the manager Brandon responded well, stopped drinking and his attendance improved. Further follow up by occupational health demonstrated that the early intervention and support provided for Brandon prevented his excessive drinking.

Legal/illegal drugs

The effects on the workplace of some drugs can be catastrophic if a worker is a regular user. Even the casual user may not be fully fit for work after taking a drug the night before or the weekend before work.

Legal drugs

Legal drugs include those that are prescribed by a medical practitioner and those that may be obtained without prescription from a pharmacy. Access to drugs with or without prescription varies in different countries. In some countries opium based drugs, such as codeine or morphine, may be prescribed for pain relief. The depressant effects of this drug when used for short-term use under medical supervision would not normally be a concern related to work, however in larger doses they may cause a worker to have reduced mental functions that could affect their use of machinery, tools or driving. Antihistamines, used to treat allergic reactions, may cause drowsiness and impaired thinking that can create risks for workers in a number of jobs. Drugs prescribed to treat depression, antidepressants, can also slow reaction speeds and affect activities that require concentration.

Benzodiazepines are a range of drugs that are in wide use to treat insomnia, anxiety and epilepsy. They have a number of common effects that could affect the health and safety of workers, the main ones being drowsiness, dizziness, confusion, light-headedness and memory loss.

Qualaquin (quinine sulphate) is an antimalarial drug used to treat malaria, a disease caused by parasites. Parasites that cause malaria typically enter the body through the bite of a mosquito. Malaria is common in areas such as Africa, South America, and Southern Asia.

Common side effects of Qualaquin include decreased hearing, dizziness and blurred vision.

The use of legal drugs should be considered where the work being done may be particularly influenced by their effects, for example, work at height, driving or operating mechanical handling equipment or safety critical work like control of a permit-to-work system.

Illegal drugs

Illegal drugs are those that have been classified by the State as requiring control because of the seriousness of their effects. Controls are usually imposed to limit their supply and use, including criminal penalties. There are many types and variants of illegal drugs, some of the common ones include cannabis, depressants and stimulants. These drugs could be taken both in the workplace and away from it. In either case their use will have detrimental effects on the user of them, resulting in higher risks when driving or operating machinery and putting others at risk.

The effects on the individual of using cannabis include:

- Euphoria.
- Slow reactions.
- Poor coordination.
- Short term memory affected.
- Distortion in perception of space and time.
- Drowsiness.
- Inability to think clearly.
- Anxiety and depression.

Depressants include opiates like heroin, opium, morphine and codeine. Opiates begin to affect the central nervous system almost immediately after use. Because the drug suppresses the central nervous system the user experiences 'cloudy' mental function, breathing slows and may reach a point of respiratory failure. They are extremely physically and psychologically addictive drugs.

Stimulants like amphetamines and cocaine and MDMA (3,4-methylenedioxy-N-methylamphetamine) make people feel more lively, awake, energetic, confident and, in the case of MDMA, give a sense of well-being.

The effects on the individual will vary depending on the drug, but generally as the drug wears off the person may become anxious, irritable and restless, however even when they feel desperate for sleep the drug may continue to keep them awake. Finally, exhaustion and often intense mood swings affect the user. The range of adverse health effects include increased heart rate, high blood pressure, high temperature, nausea, chills, sweating, involuntary teeth clenching, muscle cramps and blurred vision.

In the case of MDMA some users report undesired effects like anxiety, agitation and recklessness.

In the hours after taking MDMA it causes significant reduction in mental abilities that can last for a week, particularly those affecting memory.

Solvents

Solvents are readily available in the form of paints, varnishes, inks, glues, aerosols, and degreasing agents, particularly in the workplace. Different solvents can affect an individual in different ways, but in particular the central nervous system is readily affected by breathing in solvent vapours. When an individual is exposed to solvent vapours in low concentrations they tend to cause nausea, headache, dizziness and light-headedness. When higher concentrations of vapours are involved they can cause unconsciousness or death. Long term use can affect the brain, kidneys and liver. The loss of coordination and disorientation from the effects on the central nervous system can lead to workers making errors at work that may affect themselves or others. As with other drugs that affect the central nervous system in a similar way workers that rely on high levels of coordination and concentration, like driving and operating equipment are particularly at risk.

RISK TO HEALTH AND SAFETY FROM SUBSTANCE ABUSE AT WORK

The regular use of substances like alcohol and drugs is becoming increasingly common in some societies. People who start consuming alcohol or drugs usually have a nil or low dependency and do so for recreational reasons. This however can escalate quickly into abuse when larger quantities of alcohol or drugs are consumed more frequently and become habitual. Drugs may be prescribed for a worker with a medical condition as treatment or a worker may use controlled drugs that are illegal, such as cocaine and heroin. In addition, a worker may use solvents that are available through their work.

The effects of substances can vary dependent upon the individual's state of health and fitness, and resilience to the chemicals. Alcohol, drugs and solvents can remain in the body for a considerable time after consumption and continue to affect the individual when at work.

Apart from the general effects on the individual's health (for example, cirrhosis of the liver from alcohol abuse), the risks related to substance misuse and work is derived from the effects the substances have on the individual's biological and psychological systems. These effects create risks to the health and safety of the worker and others.

The ways the effects of substance misuse create health and safety risks include:

- Poor coordination and balance.
- Reduced perception ability.
- Poor concentration and judgement.
- Poor health, including fatigue and stress.
- Poor attitude, lack of adherence to rules.
- Poor memory.
- Increased frustration and aggression.
- Reduction in work rate, which may encourage risk taking when catching up.

The range of sensory impairments that the misuse of substances can have means that a worker's or manager's ability to carry out their work safely may be affected because their coordination, concentration, memory, perception, judgement or balance may be affected. This would make work at height, with equipment, tools, transport and many other activities that rely on care and accuracy to be affected.

It is important to remember that substance abuse affects managers as well as workers and that although middle and senior managers may not work with hazardous equipment they do make decisions that can affect the health and safety of workers and others. These decisions may be seriously influenced by the effects of substances. In addition, abuse of substances may change the behaviour of the worker or manager such that they do not follow normal patterns of behaviour or rules. This could involve taking unnecessary risks, not bothering to use health and safety precautions or communicating important information poorly. In some cases they may become aggressive or violent, particularly if challenged regarding their not following rules, poor behaviour or abuse of substances.

MANAGING SUBSTANCE ABUSE AT WORK

The management of substance abuse at work involves:

- Deciding what the issues are and what to look out for.
- Consulting workers.
- Considering safety-critical work.
- Developing a policy.
- Providing screening.
- Supporting workers with a substance abuse problem.

Deciding what the issues are and what to look out for

Substance abuse is not the same as substance dependence. Substance abuse is the use of illegal drugs and misuse of alcohol, medicines and other substances, for example solvents, in a way that affects the individual at work. The person abusing the substance may or may not have developed a dependency on it.

Potential substance abuse at work may be identified by analysis of absence from work, accidents/incidents and disciplinary records.

This can be supported by observing changes in behaviour, including:

- Sudden mood changes.
- Unusual irritability or aggression.
- Tendency towards confusion.
- Fluctuations in concentration and energy.
- Impaired job performance.
- Poor time keeping.
- Increased sickness.
- Deterioration in relationships - colleagues, managers, customers, family.
- Dishonesty and theft.

In addition, the presence of items related to substance abuse could be identified in the workplace and in an individual's possessions, for example, alcohol containers, syringes and ligatures, scorched tin foils or spoons with their base burnt or small empty packages in which the drugs may have been contained. Although some of these items may have a use not related to substance abuse they can indicate a potential risk.

Risk factors to consider when assessing whether workplace substance abuse exists or may arise include:

- *Availability of substances* - in some workplaces individuals are more likely to be exposed to the presence of alcohol, drugs or solvents, perhaps because they are legitimately involved in their manufacture, storage, transport and distribution, for example, the production and sale of beer, pharmaceuticals or glues. Substance availability may tempt workers to use it.

- *Type of workplace culture* - for example, there may be a culture at work that encourages or accepts consumption of alcohol and/or other drugs at the workplace.

- *Usage of substances in relevant social groups* - if the use of substances socially increases or is a known problem, it may have an impact at the workplace. This includes social groups outside work, but also includes groups that consume drugs and/ or alcohol when they meet before work or at breaks during the working day. In some situations the social group may be a work group, such as those that get involved with sales and promotion and spend time in situations where clients are entertained.

- *Patterns of substance consumption* - different patterns of substance use create different risks. For example, people who use large amounts on infrequent occasions may create different risks compared to people who are regular users in low quantities. Individuals who use alcohol, illegal drugs or solvents in large quantities, but infrequently, may

still be affected by the substance when they return to work because it will take longer for the quantity they have used to leave their body. In a different situation, an individual may use a prescribed drug to treat a medical condition on a single occasion, however the condition may become persistent and the use of the prescription drug and dose levels may increase.

- *Inadequate job design and stress* - unrealistic performance targets and deadlines, excessive responsibility, monotonous work or low job satisfaction may, in some instances, be risk factors. For example, symptoms of stress are sometimes associated with poor health, including substance related problems. Drugs prescribed to moderate the effects of stress may have side effects that create risks to health and safety.

- *Excessive working hours, shift work and the absence of rest* - illegal drugs, such as amphetamines or cocaine, or prescription medication may be taken in order to keep awake or maintain attention.

- *Poor working conditions* - for example, hot or dangerous environments may contribute to substance misuse. Sometimes people drink beer because they feel it is justified by the hot work they do and that they need to replenish lost liquid. In situations where work is perceived to be dangerous or undesirable, a substance may be used to distract the individual from thinking about these aspects of the work, the substance may inhibit thought processes or distort perception.

- *Interpersonal factors* - for example, bullying at work may increase risks.

- *Isolation from family and friends* - workers in isolated areas who are separated from family and friends may be more likely to consume substances due to boredom, loneliness or lack of social activities.

- *Inadequate supervision* - jobs where there is inadequate supervision may increase the risk of substance related problems because there is an absence of guidance and control of behaviour, which over time can reinforce a worker's perception of the acceptability of use of substances.

Consulting workers

It is useful to start communication and consultation with workers early in the process of managing substance abuse, before the policy is agreed.

This will help to:

- Ensure the process is open and transparent, which is important because the emphasis should be on providing support rather than punishing workers.

- Clarify any misunderstandings/concerns and obtain the views of the workforce.

The more that workers are involved in drawing up the policy on substance abuse, the more they will feel they have ownership of it and help it to work successfully. This is particularly important with a policy on substance abuse, because the policy will have an effect on the private lives of some workers and there is a need to establish trust.

Initial communication should aim to raise the general awareness/interest in the issue and encourage a positive approach to colleagues with drug and alcohol-related problems.

To obtain the views of the workforce, consultation can take place using some of the following methods: meetings, worker suggestion schemes, questionnaires, discussions with worker representatives and direct contact with workers.

If this is a new aspect of health and safety that is being managed it may be useful to establishment a working party to consider the issues and develop the policy. This could be drawn from different sectors of the workforce and include representatives from management, occupational health, health and safety, human resources department and people with responsibility for implementing the policy. Where there is an established health and safety committee substance abuse could be included within its remit.

In small organisations it may be sufficient to consult the workers directly. Larger construction organisations may also want to consult the contractors that regularly provide services to them, in addition to their own workers.

The consultation process should allow discussion on things the workers may have concerns about and result in a meaningful policy on substance abuse, with defined procedures and clear lines of accountability.

Consider safety-critical work

Safety-critical work can be considered to be work that requires an individual to have a high level of concentration, perception or co-ordination to carry it out safely and, in this context, where a risk assessment has identified a high level of risk if they are impaired by substances.

Some of the construction work activities that would usually be considered as 'safety-critical' include:

- Operation of mobile plant.
- Work on hi-speed roads.
- Rail trackside work.
- Licensed asbestos removal work.
- Tunnelling work and work in a confined space.
- Work carried out at height where collective preventative measures to control risk are not practicable.

- Operation of lifting equipment.
- Operation machinery, for example high speed machinery.
- Use of devices that explosively or mechanically fire penetrative fixings.

Measures may be put in place for pre-employment and periodic testing for substance abuse by workers that are involved in safety-critical work. Where workers in safety-critical work seek help for substance abuse, it may be necessary to temporarily transfer them to other work.

Develop a policy

All organisations could benefit from an agreed policy on substance abuse.

Developing and implemented an effective policy will help employers to manage substance abuse at work by ensuring:

- A clear understanding within the workplace of the rules relating to substance abuse.
- A greater awareness in the workplace of the effects of substance abuse and early recognition of people affected.
- The necessary procedures and support are in place for those involved in substance abuse.
- Managers are trained to understand the issues involved and have the skills and knowledge to respond appropriately.
- Workers are aware that the organisation will support them if they acknowledge they have developed a dependency problem and need help.

An effective policy should aim to help and support a worker who tells the organisation they have a substance abuse problem rather than lead directly to dismissal. However, it should also highlight when disciplinary or other action will be taken, for example that the organisation will report drug possession or drug dealing at work to the police. If it is decided that screening and testing for substance abuse is to be carried out this should be included in the policy. Substance abuse could be included as part of the organisation's overall health and safety policy.

A substance abuse policy would typically include four main elements:

1) Explanation of why the organisation is concerned about substance abuse, why it would be a problem in the workplace and that everyone is responsible for making the policy work.

2) Explanation of what the organisation will and will not allow in relation to workers' substance use, as it affects their work.

3) Outline of the arrangements the organisation has made to minimise and manage risks related to substance abuse at work.

4) Set out clearly the rules and procedures for managing issues relating to substances, including the supportive and disciplinary elements:

- Supportive procedures should aim to encourage workers to seek help and gain access to support and treatment.

- Performance, capability and disciplinary procedures should explain when and how such procedures would be used. For example the consequences of:

  - Reporting to work unfit due to substance abuse.

  - Impaired performance at work as a result of use of substances outside the workplace.

  - Possessing and/or dealing in drugs at work.

"3.2 - Contents of an alcohol and drug policy

3.2.1 - A policy for the management of alcohol and drugs in the workplace should include information and procedures on:

a) Measures to reduce alcohol and drug-related problems in the workplace through proper personnel management, good employment practices, improved working conditions, proper arrangement of work, and consultation between management and workers and their representatives.

b) Measures to prohibit or restrict the availability of alcohol and drugs in the workplace.

c) Prevention of alcohol and drug-related problems in the workplace through information, education, training and any other relevant programmes.

d) Identification, assessment and referral of those who have alcohol or drug-related problems.

e) Measures relating to intervention and treatment and rehabilitation of individuals with alcohol or drug-related problems.

f) Rules governing conduct in the workplace relating to alcohol and drugs, the violation of which could result in the invoking of disciplinary procedures up to and including dismissal.

g) Equal opportunities for persons who have, or who have previously had, alcohol and drug-related problems, in accordance with national laws and Regulations."

Figure 13-56: Content of an alcohol and drugs policy.
Source: ILO Code of practice on alcohol and drugs in the workplace.

The arrangements to manage and minimise health and safety at work risks from substance abuse could include:

- All workers that come on to a site, such as a hospital, construction or power generation site, might be considered to be in safety-critical work. All workers would then work to the same rules, which would include prohibiting workers being under the influence of substances while at work, offering them opportunities to talk to someone about how this affects them and carrying out random drugs and alcohol tests.

- Reducing the influence of factors that may contribute to stress or fatigue, for example, redesigning jobs, controlling hours of work and providing regular breaks.

- Providing suitable arrangements for workers and managers to take breaks where the environment is conducive to avoiding substance abuse, for example, making non-alcoholic drinks available or providing recreational options where boredom might be a risk factor or where workers are isolated from family and friends.

- Increased supervision of those groups that may be more likely to engage in substance abuse, for example, those with opportunity because of the availability of substances in the workplace or who are involved in dangerous/undesirable work or where their work leads to poor interpersonal relations.

- Disciplinary procedures for those workers who refuse to be tested or who fail a test, refuse to attend rehabilitation or are found to have alcohol or drugs on their person when at work.

The substance abuse policy should be supported by:

- Communication of the policy and procedures on substance abuse and the general expectations for health and safety related to this.

- Encouraging those in management positions to support the policy and procedures.

- Briefing managers and supervisors so they are clear about:

  - How to recognise the signs of substance abuse.

  - The organisation's rules on substance abuse.

  - What to do if they suspect a worker is misusing substances.

  - What to do when a worker tells them about a substance abuse problem

- Providing information, education and training on the risks, symptoms and expectations to everyone that has an influence on substance abuse - managers, workers, contractors, suppliers, neighbouring establishments and families.

- Increasing awareness of new workers by including an explanation of the substance abuse policy in the induction process.

Screening

Some employers adopt screening as part of their substance abuse policy. There is a good argument for carrying out screening related to safety-critical work. In jobs like these the abuse of substances could have disastrous effects on the worker, other workers or members of the public.

The most common form of screening is the 'pre-employment health questionnaire'. This is a self-completed form that usually provides sufficient information to determine a person's fitness to undertake particular work. This could include questions related to use of and dependency on substances. The declaration, in the form of answers to the questionnaire, may be supported by specific medical tests to determine if the potential worker uses substances.

In support of the pre-employment screening periodic or random medical tests for use of substances may be carried out. Where it is proposed these tests are carried out the following should be considered:

- Workers must consent to screening, however if an employer has good grounds for testing and an worker refuses, they may face disciplinary action.

- Screening must be carried out competently to ensure samples are not contaminated or tampered with and testing/analysis are accurate.

- Screening by itself will not solve problems caused by substance abuse, where screening is introduced it should be part of an organisation's overall arrangements to manage substance abuse.

Supporting workers with a substance abuse problem

Organisations may see substance abuse in the workplace as a disciplinary matter or a health concern. Where possible, workers should be given support with a substance abuse problem. Depending on the work they do and the circumstances, if a worker is dismiss because of substance abuse without the employer trying to help them, an employment tribunal may find that they have been dismissed unfairly.

Workers with a substance abuse problem may ask for help at work if they are sure their problem will be dealt with discreetly and confidentially. Someone who is misusing substances has the same rights to confidentiality and support as they would if they had any other medical or psychological condition.

When workers are identified as having a substance abuse problem the organisation should consider the circumstances affecting the person before resorting to disciplinary action. There may be personal or work issues that are influencing their abuse of substances. They would usually need to be identified promptly so that assistance with these underlying issues can be sought, where possible. This assistance may be enough to resolve the substance abuse problem.

Personal issues that may require support include:

- A family member who is dependent on substances.
- Bereavement or breakdown of family relationships.
- Financial difficulties.
- The pressure of being a carer of a dependent person.

Work issues that may influence the substance abuse and require support include:

- Work-related stress.
- Bullying and harassment.

If the worker's normal work is safety-critical it may be necessary to temporarily move them to another job.

The organisation should make arrangements to enable access to rehabilitation advice, counselling and support groups early in the development of a substance abuse problem. Generally, workers should be encouraged to get help from their GP or a specialist substance dependency agency.

REVIEW

List the effects of substance misuse on the safety of those affected at work.

How might substance abuse cause health and safety problems in the workplace?

What measures can employers put in place to reduce such problems?

List three drugs that may be classified as stimulants and three drugs that may be classified as depressants.

Sources of reference

Reference information provided, in particular web links, was correct at time of publication, but may have changed.

Controlling Noise at Work, The Control of Noise at Work Regulations, Guidance on Regulations, second edition 2005, L108, HSE Books, ISBN: 978-0-7176-6164-4, http://www.hse.gov.uk/pubns/priced/l108.pdf

Electromagnetic fields at work, A guide to the Control of Electromagnetic Fields at Work Regulations, HSG281, HSE Books, http://www.hse.gov.uk/pubns/priced/hsg281.pdf

Essentials of Health and Safety at Work, fourth edition 2006, HSE Books, ISBN: 978-0-7176-6179-4

Hand-arm vibration, Control of Vibration at Work Regulations 2005, Guidance on Regulations, L140, HSE Books, ISBN: 978-0-7176-6125-1, http://www.hse.gov.uk/pubns/priced/l140.pdf

How to tackle work-related stress. A guide for employers on making the Management Standards work, INDG430, HSE Books, http://www.hse.gov.uk/pubns/indg430.pdf

HSE Health and Safety Toolbox, online resource, HSE, www.hse.gov.uk/toolbox/index.htm

HSE Stress Management Standards www.hse.gov.uk/stress/standards

Managing drug and alcohol misuse at work, online resource, HSE, https://www.hse.gov.uk/alcoholdrugs/

Personal Protective Equipment at Work, Guidance on Regulations, L25, HSE Books, ISBN: 978-0-7176-6597-6 http://www.hse.gov.uk/pubns/priced/l25.pdf

Radon in the workplace: http://www.hse.gov.uk/radiation/ionising/radon.htm#testingradon

Tackling work-related stress using the Management Standards approach, A step-by-step workbook, WBK1, HSE Books, ISBN 978-0-7176-6715-4, http://www.hse.gov.uk/pubns/wbk01.pdf

The health and safety toolbox, How to control risks at work, HSG268, HSE Books, ISBN: 978-0-7176-6587-7 , http://www.hse.gov.uk/pUbns/priced/hsg268.pdf

Violence at work: A guide for employers, HSE Books, INDG69, www.hse.gov.uk/pubns/indg69.pdf

Whole-body vibration; The Control of Vibration at Work Regulations 2005, Guidance on Regulations, L141, HSE Books, ISBN: 978-0-7176-6126-8, http://www.hse.gov.uk/pubns/priced/l141.pdf

Working with ionising radiation, Approved Code of Practice and guidance, L121 HSE Books, ISBN: 978-0-7176-6662-1, http://www.hse.gov.uk/pubns/priced/l121.pdf

Web links to these references are provided on the RMS Publishing website for ease of use - www.rmspublishing.co.uk

Statutory provisions

Control of Artificial Optical Radiation at Work (CAOR) Regulations 2010

Control of Electromagnetic Fields at Work Regulations (CEMFAW) 2016

Control of Noise at Work Regulations (CNWR) 2005 / Control of Noise at Work Regulations (Northern Ireland) 2006

Control of Vibration at Work Regulations (CVWR) 2005 / Control of Vibration at Work Regulations (Northern Ireland) 2005

Ionising Radiations Regulations (IRR) 2017 / Ionising Radiations Regulations (Northern Ireland) 2017

Personal Protective Equipment at Work Regulations (PPER) 1992 / Personal Protective Equipment at Work Regulations (Northern Ireland) 1993

STUDY QUESTIONS

1) A noise survey carried out on a building site identified that noise exposure action values given in the Control of Noise at Work Regulations 2005 were being exceeded.

   (a) What are the lower AND upper noise exposure action values defined in the Control of Noise at Work Regulations 2005?

   (b) What control measures is an employer required to take by the Control of Noise at Work Regulations 2005 when employees are exposed to noise at or above the upper noise exposure action value?

2) What are the work-related issues that might increase the levels of occupational stress amongst workers on a construction site?

3) As part of major football stadium refurbishment, seat anchors are to be installed by workers using hand-held powered percussive equipment that produce high levels of vibration.

   (a) What are the possible health effects of exposure to vibration?

   (b) What control measures could be used to reduce the risk of health effects due to vibration in this refurbishment work?

4) (a) What are TWO types of ionising radiation that workers in construction activities might encounter?

   (b) What are TWO types of non-ionising radiation that workers in construction activities might encounter?

   (c) What are the range of health effects of exposure to non-ionising radiation?

5) What signs might indicate that a contractor is misusing drugs during working hours?

For guidance on how to answer these questions please refer to the 'study question answer guidance' section located at the back of this guide.

Relevant Statutory Provisions

Contents

Relevant Statutory Provisions

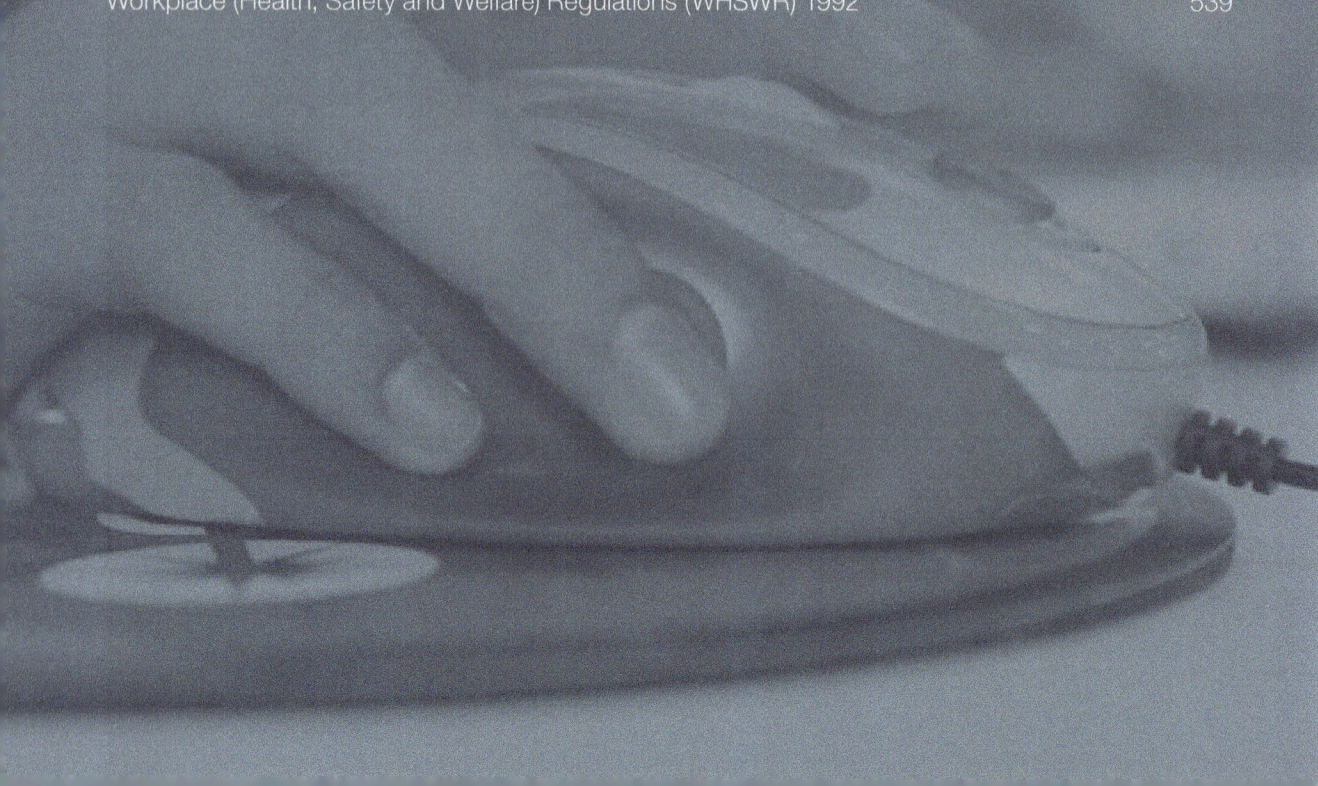

RMS Publishing's technical authors regularly review examiners reports for all NEBOSH qualifications to ensure that the specific publication content is in keeping with the level of study required.

The review considers the core training materials, assessment criteria and relevant legislation. At each stage care is taken to pitch the level of the content to the syllabus requirements.

The syllabus does not require knowledge of all legislation to the same depth, this is reflected in this 'relevant statutory provisions' section of the study book. Relevant statutory provisions abstracts are designed to focus on the specific aspects of the legislation required to meet the syllabus. The study book provides guidance in the form of 'outline of main points' enabling learners to focus on the critical learning points and avoid over studying. In addition, legislation is considered in context in the relevant elements of the study book.

Learners are advised to obtain or gain access to statutory documents, approved codes of practice and guidance related to the relevant statutory provisions as part of their personal development programme, for the purpose of the examination/assessment and future career development. The source documents used in this section may be obtained, free of charge, from www.legislation.gov.uk and www.hse.gov.uk/pubns/books.

NEBOSH do not examine on legislation until it has been in force for 6 months. Learners may show knowledge of new legislation in their answers until that point, learners referring to the former legislation will not lose marks until the 6 month period has passed.

Classification Labelling and Packaging of Substances (CLP) EC Regulation

Considered in context in Element 12

European Regulation (EC) No 1272/2008 on classification, labelling and packaging of substances and mixtures came into force on 20 January 2009 in all EU Member States, including the UK. It is known by its abbreviated form, 'the CLP Regulation' or 'CLP'.

The CLP Regulation adopts the United Nations' Globally Harmonised System (GHS) on the classification and labelling of chemicals across all European Union countries, including the UK. As the GHS is a voluntary agreement rather than a law, it has to be adopted through a suitable national or regional legal mechanism to ensure it becomes legally binding. That is what the CLP Regulation does.

As GHS was heavily influenced by the old EU system, the CLP Regulation is very similar in many ways. The duties on suppliers are broadly the same and cover classification, labelling and packaging. The process suppliers have to follow when they are classifying substances and mixtures has changed and a new set of hazard pictograms are used.

Figure RSP-1: GHS hazard pictograms.
Source: RMS.

ARRANGEMENT OF REGULATION

ANNEX I - CLASSIFICATION AND LABELLING REQUIREMENTS FOR HAZARDOUS SUBSTANCES AND MIXTURES

This annex sets out the criteria for classification in hazard classes and in their differentiations and sets out additional provisions on how the criteria may be met.

PART 1 - GENERAL PRINCIPLES FOR CLASSIFICATION AND LABELLING

1.1. Classification of substances and mixtures.

1.2. Labelling (General rules for the application of labels required by Article 31).

1.3. Derogations from labelling requirements for special cases.

1.4. Request for use of an alternative chemical name.

1.5. Exemptions from labelling and packaging requirements.

PART 2 - PHYSICAL HAZARDS

Including explosive and flammable hazards.

PART 3 - HEALTH HAZARDS

Including acute toxicity and respiratory/skin sensitisation hazards.

PART 4 - ENVIRONMENTAL HAZARDS

Including hazards to the aquatic environment.

PART 5 - ADDITIONAL HAZARDS

Including hazards to the ozone layer.

Annex ii - Special rules for labelling and packaging of certain substances and mixtures.

Annex iii - List of hazard statements, supplemental hazard information and supplemental label elements.

Annex iv - List of precautionary statements.

Annex v - Hazard pictograms.

Annex vi - Harmonised classification and labelling for certain hazardous substances.

Annex vii - Translation table from classification under directive 67/548/EEC to classification under this regulation.

OUTLINE OF MAIN POINTS

The CLP Regulation ensures that the hazards presented by chemicals are clearly communicated to workers and consumers in the European Union, including the UK, through classification and labelling of chemicals. Before placing chemicals on the market, the industry must establish the potential risks to human health and the environment of such substances and mixtures, classifying them in line with the identified hazards. The hazardous chemicals also have to be labelled according to a standardised system so that workers and consumers know about their effects before they handle them. Suppliers must label a substance or mixture contained in packaging according to CLP before placing it on the market. CLP defines the content of the label and the organisation of the various labelling elements. This process ensures the hazards of chemicals are communicated through standard statements and pictograms on labels.

For example, when a supplier identifies a substance as 'acute toxicity category 1 (oral)', the labelling will include the hazard statement 'fatal if swallowed', the word 'Danger' and a pictogram with a skull and crossbones.

CLP provides certain exemptions for substances and mixtures contained in packaging that is small (typically less than 125ml) or is otherwise difficult to label. The exemptions allow the supplier to omit the hazard and/or precautionary statements or the pictograms from the label elements normally required under CLP.

The label includes:

- The name, address and telephone number of the supplier.
- The nominal quantity of a substance or mixture in the packages made available to the general public (unless this quantity is specified elsewhere on the package).
- Product identifiers.
- Where applicable, hazard pictograms, signal words, hazard statements, precautionary statements and supplemental information required by other legislation.

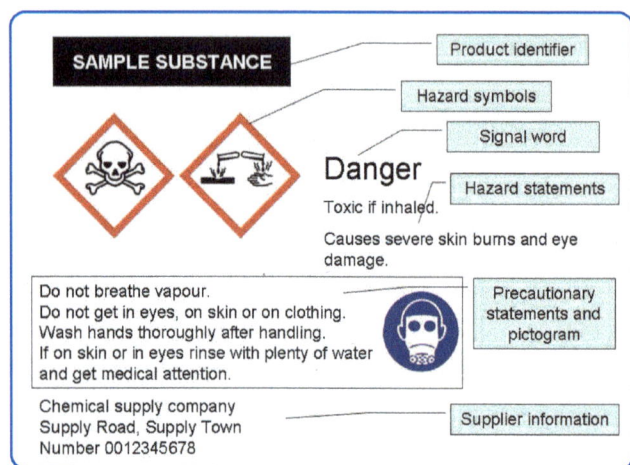

Figure RSP-2: Sample GHS chemical labels.
Source: RMS.

Confined Spaces Regulations (CSR) 1997

Considered in context in Element 4.

ARRANGEMENT OF REGULATIONS

1) Citation, commencement and interpretation.

2) Disapplication of Regulations.

3) Duties.

4) Work in confined spaces.

5) Emergency arrangements.

6) Exemption certificates.

7) Defence in proceedings.

8) Extension outside Great Britain.

9) Repeal and revocations.

OUTLINE OF MAIN POINTS

A failure to appreciate the dangers associated with confined spaces has led not only to the deaths of many workers, but also to the demise of some of those who have attempted to rescue them.

A confined space is not only a space which is small and difficult to enter, exit or work in; it can also be a large space, but with limited/restricted access. It can also be a space which is badly ventilated, for example, a tank or a large tunnel.

The Confined Spaces Regulations (CSR) 1997, define a confined space as any place, including any chamber, tank, vat, silo, pit, pipe, sewer, flue, well, or other similar space, in which, by virtue of its enclosed nature, there arises a reasonably foreseeable specified risk.

The CSR 1997 sets out duties on employers and the self-employed in regard to work carried out in confined spaces. Entry into confined spaces is prohibited unless it is not reasonably practicable to do the work without entering. Work conducted must be carried out in accordance with a safe system of work and arrangements must be in place to deal with emergencies that may arise and to effect a rescue.

Construction (Design and Management) Regulations (CDMR) 2015

Considered in context in Element 1.

The Construction (Design and Management) Regulations (CDMR) 2015 came in to force in April 2015. CDMR 2015 re-emphasises the health, safety and broader business benefits of a well-managed and coordinated approach to the management of health and safety in construction that were introduced in earlier versions of this Regulation.

ARRANGEMENT OF REGULATIONS

PART 1 - COMMENCEMENT, INTERPRETATION AND APPLICATION

1) Citation and commencement.

2) Interpretation.

3) Application.

PART 2 - CLIENT DUTIES

4) Client duties in relation to managing projects.

5) Appointment of the principal designer and the principal contractor.

6) Notification.

7) Application to domestic clients.

PART 3 - HEALTH AND SAFETY DUTIES AND ROLES

8) General duties.

9) Duties of designers.

10) Designs prepared or modified outside Great Britain.

11) Duties of a principal designer in relation to health and safety at the pre-construction phase.

12) Construction phase plan and health and safety file.

13) Duties of a principal contractor in relation to health and safety at the construction phase.

14) Principal contractor's duties to consult and engage with workers.

15) Duties of contractors.

PART 4 - GENERAL REQUIREMENTS FOR ALL CONSTRUCTION SITES

16) Application of Part 4.

17) Safe places of construction work.

18) Good order and site security.

19) Stability of structures.

20) Demolition or dismantling.

21) Explosives.

22) Excavations.

23) Cofferdams and caissons.

24) Reports of inspections.

25) Energy distribution installations.

26) Prevention of drowning.

27) Traffic routes.

28) Vehicles.

29) Prevention of risk from fire, flooding or asphyxiation.

30) Emergency procedures.

31) Emergency routes and exits.

32) Fire detection and firefighting.

33) Fresh air.

34) Temperature and weather protection.

35) Lighting.

PART 5 - GENERAL

36) Enforcement in respect of fire.

37) Transitional and saving provisions.

38) Revocation and consequential amendments.

39) Review.

SCHEDULES

Schedule 1 (Regulation 6) - Particulars to be notified under Regulation 6.

Schedule 2 (Regulation 4 (2)(b), 13(4)(c) and 15(11)) - Minimum welfare facilities required for construction sites.

1) Sanitary conveniences.

2) Washing facilities.

3) Drinking water.

4) Changing rooms and lockers.

5) Facilities for rest.

Schedule 3 (Regulation 12(2)) - Work involving particular risk.

Schedule 4 (Regulation 37) - Transitional and saving provisions.

Schedule 5 (Regulation 38) - Amendments.

Appendix 1 The general principles of prevention.

Appendix 2 Pre-construction information.

Appendix 3 The construction phase plan.

Appendix 4 The health and safety file.

Appendix 5 How different types of information relate to and influence each other in a construction project involving more than one contractor.

Appendix 6 Working for a domestic client.

OUTLINE OF MAIN POINTS

PART 2 - CLIENT DUTIES

Regulation 4 - Client duties for managing projects

(1) A client must make suitable arrangements for managing a project, including the allocation of sufficient time and other resources.

(2) Arrangements are suitable if they ensure that:

(a) The construction work can be carried out, so far as is reasonably practicable, without risks to the health or safety of any person affected by the project.

(b) The facilities required by Schedule 2 are provided in respect of any person carrying out construction work.

(3) A client must ensure that these arrangements are maintained and reviewed throughout the project.

(4) A client must provide pre-construction information as soon as is practicable to every designer and contractor appointed, or being considered for appointment, to the project.

(5) A client must ensure that -

(a) Before the construction phase begins, a construction phase plan is drawn up by the contractor if there is only one contractor, or by the principal contractor.

(b) The principal designer prepares a health and safety file for the project, which:

(i) Complies with the requirements of Regulation 12(5).

(ii) Is revised from time to time as appropriate to incorporate any relevant new information.

(iii) Is kept available for inspection by any person who may need it to comply with the relevant legal requirements.

(6) A client must take reasonable steps to ensure that -

(a) The principal designer complies with any other principal designer duties in Regulations 11 and 12.

(b) The principal contractor complies with any other principal contractor duties in Regulations 12 to 14.

(7) If a client disposes of the client's interest in the structure, the client complies with the duty in paragraph (5)(b)(iii) by providing the health and safety file to the person who acquires the client's interest in the structure and ensuring that that person is aware of the nature and purpose of the file.

(8) Where there is more than one client in relation to a project:

(a) One or more of the clients may agree in writing to be treated for the purposes of these Regulations as the only client or clients.

(b) Except for the duties specified in sub-paragraph (c) only the client or clients agreed in paragraph (a) are subject to the duties owed by a client under these Regulations.

(c) The duties in the following provisions are owed by all clients:

(i) Regulation 8(4).

(ii) Paragraph (4) and Regulation 8(6) to the extent that those duties relate to information in the possession of the client.

Regulation 5 - Appointment of the principal designer and the principal contractor

(1) Where there is more than one contractor, or if it is reasonably foreseeable that more than one contractor will be working on a project at any time, the client must appoint in writing -

(a) A designer with control over the pre-construction phase as principal designer.

(b) A contractor as principal contractor.

(2) The appointments must be made as soon as is practicable, and in any event, before the construction phase begins.

(3) If the client fails to appoint a principal designer, the client must fulfil the duties of the principal designer in Regulations 11 and 12.

(4) If the client fails to appoint a principal contractor, the client must fulfil the duties of the principal contractor in Regulations 12 to 14.

Regulation 6 - Notification

(1) A project is notifiable if the construction work on a construction site is scheduled to -

(a) Last longer than 30 working days and have more than 20 workers working simultaneously at any point in the project.

(b) Exceed 500 person days.

(2) Where a project is notifiable, the client must give notice in writing to the Executive as soon as is practicable before the construction phase begins.

(3) The notice must -

(a) Contain the particulars specified in Schedule 1.

(b) Be clearly displayed in the construction site office in a comprehensible form where it can be read by any worker engaged in the construction work.

(c) If necessary, be periodically updated.

Regulation 7 - Application to domestic clients

(1) Where the client is a domestic client the duties in Regulations 4(1) to (7) and 6 must be carried out by:

(a) The contractor for a project where there is only one contractor.

(b) The principal contractor for a project where there is more than one contractor.

(c) The principal designer where there is a written agreement that the principal designer will fulfil those duties.

(2) If a domestic client fails to make the appointments required by Regulation 5 -

(a) The designer in control of the pre-construction phase of the project is the principal designer.

(b) The contractor in control of the construction phase of the project is the principal contractor.

(3) Regulation 5(3) and (4) does not apply to a domestic client.

PART 3 - HEALTH AND SAFETY DUTIES AND ROLES

Regulation 8 - General duties

(1) A designer (including a principal designer) or contractor (including a principal contractor) appointed to work on a project must have the skills, knowledge and experience and, if they are an organisation, the

organisational capability, necessary to fulfil the role that they are appointed to undertake, in a manner that secures the health and safety of any person affected by the project.

(2) A designer or contractor must not accept an appointment to a project unless they fulfil the conditions in paragraph (1).

(3) A person who is responsible for appointing a designer or contractor to carry out work on a project must take reasonable steps to satisfy themselves that the designer or contractor fulfils the conditions in paragraph (1).

(4) A person with a duty or function under these Regulations must co-operate with any other person working on or in relation to a project at the same or an adjoining construction site to the extent necessary to enable any person with a duty or function to fulfil that duty or function.

(5) A person working on a project under the control of another must report to that person anything they are aware of in relation to the project which is likely to endanger their own health or safety or that of others.

(6) Any person who is required by these Regulations to provide information or instruction must ensure the information or instruction is comprehensible and provided as soon as is practicable.

(7) To the extent that they are applicable to a domestic client, the duties in paragraphs (3), (4) and (6) must be carried out by the person specified in Regulation 7(1).

Regulation 9 - Duties of designers

(1) A designer must not commence work in relation to a project unless satisfied that the client is aware of the duties owed by the client under these Regulations.

(2) When preparing or modifying a design the designer must take into account the general principles of prevention and any pre-construction information to eliminate, so far as is reasonably practicable, foreseeable risks to the health or safety of any person:

(a) Carrying out or liable to be affected by construction work.

(b) Maintaining or cleaning a structure.

(c) Using a structure designed as a workplace.

(3) If it is not possible to eliminate these risks, the designer must, so far as is reasonably practicable:

(a) Take steps to reduce or, if that is not possible, control the risks through the subsequent design process.

(b) Provide information about those risks to the principal designer.

(c) Ensure appropriate information is included in the health and safety file.

(4) A designer must take all reasonable steps to provide, with the design, sufficient information about the design, construction or maintenance of the structure, to adequately assist the client, other designers and contractors to comply with their duties under these Regulations.

Regulation 10 - Designs prepared or modified outside Great Britain

(1) Where a design is prepared or modified outside Great Britain for use in construction work to which these Regulations apply:

(a) The person who commissions it, if established within Great Britain.

(b) If that person is not so established, the client for the project, must ensure that Regulation 9 is complied with.

(2) This Regulation does not apply to a domestic client.

Regulation 11 - Duties of a principal designer in relation to health and safety at the pre-construction phase

(1) The principal designer must plan, manage and monitor the pre-construction phase and coordinate matters relating to health and safety during the pre-construction phase to ensure that, so far as is reasonably practicable, the project is carried out without risks to health or safety.

(2) In fulfilling the duties in paragraph (1), and in particular when:

(a) Design, technical and organisational aspects are being decided in order to plan the various items or stages of work which are to take place simultaneously or in succession.

(b) Estimating the period of time required to complete such work or work stages, the principal designer must take into account the general principles of prevention and, where relevant, the content of any construction phase plan and health and safety file.

(3) In fulfilling the duties in paragraph (1), the principal designer must identify and eliminate or control, so far as is reasonably practicable, foreseeable risks to the health or safety of any person:

(a) Carrying out or liable to be affected by construction work.

(b) Maintaining or cleaning a structure.

(c) Using a structure designed as a workplace.

(4) In fulfilling the duties in paragraph (1), the principal designer must ensure all designers comply with their duties in Regulation 9.

(5) In fulfilling the duty to coordinate health and safety matters in paragraph (1), the principal designer must ensure that all persons working in relation to the pre-construction phase co-operate with the client, the principal designer and each other.

(6) The principal designer must:

(a) Assist the client in the provision of the pre-construction information required by Regulation 4(4).

(b) So far as it is within the principal designer's control, provide pre-construction information, promptly and in a convenient form, to every designer and contractor appointed, or being considered for appointment, to the project.

(7) The principal designer must liaise with the principal contractor for the duration of the principal designer's appointment and share with the principal contractor information relevant to the planning, management and monitoring of the construction phase and the coordination of health and safety matters during the construction phase.

Regulation 12 - Construction phase plan and health and safety file

(1) During the pre-construction phase, and before setting up a construction site, the principal contractor must draw up a construction phase plan or make arrangements for a construction phase plan to be drawn up.

(2) The construction phase plan must set out the health and safety arrangements and site rules taking account, where necessary, of the industrial activities taking place on the construction site and, where applicable, must include specific measures concerning work which falls within one or more of the categories set out in Schedule 3.

(3) The principal designer must assist the principal contractor in preparing the construction phase plan by providing to the principal contractor all information the principal designer holds that is relevant to the construction phase plan including -

(a) Pre-construction information obtained from the client.

(b) Any information obtained from designers under Regulation 9(3)(b).

(4) Throughout the project the principal contractor must ensure that the construction phase plan is appropriately reviewed, updated and revised from time to time so that it continues to be sufficient to ensure that construction work is carried out, so far as is reasonably practicable, without risks to health or safety.

(5) During the pre-construction phase, the principal designer must prepare a health and safety file appropriate to the characteristics of the project which must contain information relating to the project which is likely to be needed during any subsequent project to ensure the health and safety of any person.

(6) The principal designer must ensure that the health and safety file is appropriately reviewed, updated and revised from time to time to take account of the work and any changes that have occurred.

(7) During the project, the principal contractor must provide the principal designer with any information in the principal contractor's possession relevant to the health and safety file, for inclusion in the health and safety file.

(8) If the principal designer's appointment concludes before the end of the project, the principal designer must pass the health and safety file to the principal contractor.

(9) Where the health and safety file is passed to the principal contractor under paragraph (8), the principal contractor must ensure that the health and safety file is appropriately reviewed, updated and revised from time to time to take account of the work and any changes that have occurred.

(10) At the end of the project, the principal designer, or where there is no principal designer the principal contractor, must pass the health and safety file to the client.

Regulation 13 - Duties of a principal contractor in relation to health and safety at the construction phase

(1) The principal contractor must plan, manage and monitor the construction phase and coordinate matters relating to health and safety during the construction phase to ensure that, so far as is reasonably practicable, construction work is carried out without risks to health or safety;

(2) In fulfilling the duties in paragraph (1), and in particular when:

(a) Design, technical and organisational aspects are being decided in order to plan the various items or stages of work which are to take place simultaneously or in succession.

(b) Estimating the period of time required to complete the work or work stages; the principal contractor must take into account the general principles of prevention.

(3) The principal contractor must:

    (a) Organise co-operation between contractors (including successive contractors on the same construction site).

    (b) Coordinate implementation by the contractors of applicable legal requirements for health and safety; (c) ensure that employers and, if necessary for the protection of workers, self-employed persons.

    (i) Apply the general principles of prevention in a consistent manner, and in particular when complying with the provisions of Part 4.

    (ii) Where required, follow the construction phase plan.

(4) The principal contractor must ensure that:

    (a) A suitable site induction is provided.

    (b) The necessary steps are taken to prevent access by unauthorised persons to the construction site.

    (c) Facilities that comply with the requirements of Schedule 2 are provided throughout the construction phase.

(5) The principal contractor must liaise with the principal designer for the duration of the principal designer's appointment and share with the principal designer information relevant to the planning, management and monitoring of the pre-construction phase and the coordination of health and safety matters during the pre-construction phase.

Regulation 14 - Principal contractor's duties to consult and engage with workers

The principal contractor must -

(a) Make and maintain arrangements which will enable the principal contractor and workers engaged in construction work to co-operate effectively in developing, promoting and checking the effectiveness of measures to ensure the health, safety and welfare of the workers.

(b) Consult those workers or their representatives in good time on matters connected with the project which may affect their health, safety or welfare, in so far as they or their representatives have not been similarly consulted by their employer.

(c) Ensure that those workers or their representatives can inspect and take copies of any information which the principal contractor has, or which these Regulations require to be provided to the principal contractor, which relate to the health, safety or welfare of workers at the site, except any information:

    (i) The disclosure of which would be against the interests of national security.

    (ii) Which the principal contractor could not disclose without contravening a prohibition imposed by or under an enactment.

    (iii) Relating specifically to an individual, unless that individual has consented to its being disclosed.

    (iv) The disclosure of which would, for reasons other than its effect on health, safety or welfare at work, cause substantial injury to the principal contractor's undertaking or, where the information was supplied to the principal contractor by another person, to the undertaking of that other person.

    (v) Obtained by the principal contractor for the purpose of bringing, prosecuting or defending any legal proceedings.

Regulation 15 - Duties of contractors

(1) A contractor must not carry out construction work in relation to a project unless satisfied that the client is aware of the duties owed by the client under these Regulations.

(2) A contractor must plan, manage and monitor construction work carried out either by the contractor or by workers under the contractor's control, to ensure that, so far as is reasonably practicable, it is carried out without risks to health and safety.

(3) Where there is more than one contractor working on a project, a contractor must comply with:

    (a) Any directions given by the principal designer or the principal contractor.

    (b) The parts of the construction phase plan that are relevant to that contractor's work on the project.

(4) If there is only one contractor working on the project, the contractor must take account of the general principles of prevention when -

    (a) Design, technical and organisational aspects are being decided in order to plan the various items or stages of work which are to take place simultaneously or in succession.

    (b) Estimating the period of time required to complete the work or work stages.

(5) If there is only one contractor working on the project, the contractor must draw up a construction phase plan, or make arrangements for a construction phase plan to be drawn up, as soon as is practicable prior to setting up a construction site.

(6) The construction phase plan must fulfil the requirements of Regulation 12(2).

(7) A contractor must not employ or appoint a person to work on a construction site unless that person

has, or is in the process of obtaining, the necessary skills, knowledge, training and experience to carry out the tasks allocated to that person in a manner that secures the health and safety of any person working on the construction site.

(8) A contractor must provide each worker under their control with appropriate supervision, instructions and information so that construction work can be carried out, so far as is reasonably practicable, without risks to health and safety.

(9) The information provided must include:

(a) A suitable site induction, where not already provided by the principal contractor.

(b) The procedures to be followed in the event of serious and imminent danger to health and safety.

(c) Information on risks to health and safety.

(i) Identified by the risk assessment under Regulation 3 of the Management Regulations.

(ii) Arising out of the conduct of another contractor's undertaking and of which the contractor in control of the worker ought reasonably to be aware.

(d) Any other information necessary to enable the worker to comply with the relevant statutory provisions.

(10) A contractor must not begin work on a construction site unless reasonable steps have been taken to prevent access by unauthorised persons to that site.

(11) A contractor must ensure, so far as is reasonably practicable, that the requirements of Schedule 2 are complied with so far as they affect the contractor or any worker under that contractor's control.

PART 4 - GENERAL REQUIREMENTS FOR ALL CONSTRUCTION SITES

Part 4 of CDM 2015 sets out general health, safety and welfare requirements that apply to all construction sites involved in construction work.

The Regulations in Part 4 apply to places of work in the ground, at ground level and at height. In essence they require that arrangements are made to ensure risks to health and safety are minimised and appropriate welfare facilities are provided.

Safe place of work, good order and site security (Regulations 17 and 18)

- Safe access to and egress from places of work and other places that workers have access to within a construction site.
- The site must be safe and healthy for those that work there.

- The site must have sufficient working space and be arranged so that it is suitable for those working on it, taking into account any work equipment that is being used.
- The site should be kept in a reasonable state of cleanliness and in good order.
- Depending on the level of risk, the site must have its perimeter identified by signs or be fenced off.
- Removal of material with projecting nails or similar sharp objects that could be a source of danger.

Work on structures (Regulations 19, 20 and 21)

- Prevent accidental collapse of new, existing structures or those under construction.
- Not to load structures unsafely.
- Make sure any dismantling or demolition of any structure is planned and carried out in a safe manner.
- Arrangements for carrying out demolition or dismantling must be recorded in writing.
- Explosive charges must be stored, transported and used safely and securely. Explosive charges can only be fired after steps have been taken to ensure that no one is exposed to risk or injury from the explosion.

Every year there are structural collapses which have the potential to cause serious injury. Demolition or dismantling are recognised as high risk activities. In any cases where this work presents a risk of danger to anyone, it should be planned and carried out under the direct supervision of a competent person.

Excavations, cofferdams and caissons (Regulations 22, 23 and 24)

- Prevent collapse of ground both in and above excavations by the use of supports or battering.
- Prevent people, equipment and materials falling in to excavations.
- Ensure cofferdams and caissons are properly designed, constructed and maintained.
- Inspections of excavations, cofferdams and caissons, and any work equipment/materials that may affect its safety, by a competent person. Prevent work in excavations that have been inspected and found to be unsafe, until they have been made safety.
- Inform the responsible person of the outcome of inspections and provide reports of the inspection, containing the specified details, to the responsible person before the end of the shift. The report must include:

  - The name and address of the person on whose behalf the inspection was carried out.
  - The location of the place of construction work inspected.
  - A description of the place of construction work or part of that place inspected (including any work equipment and materials).

- The date and time of the inspection.
- The details of any matter identified that could give rise to a risk to the safety of any person.
- Details of any action taken as a result of any matter identified above.
- The details of any further action considered necessary.
- The name and position of the person making the report.
- Provide the report or a copy of it, to the person on whose behalf the inspection was carried out, within 24 hours of completing the inspection.
- No further inspection reports are required for subsequent inspection of that place within a 7 day period.

From the outset, and as work progresses, any excavation which has the potential to collapse unless supported should have suitable equipment immediately available to provide such support.

Regular inspection of excavations, cofferdams and caissons is an important aspect of assuring that they are safe to be worked in. When inspections identify problems with them it is essential that they are resolved before people work in them.

Energy distribution installations (Regulation 25)

- Where necessary to prevent danger, energy distribution installations must be suitably located, checked and clearly indicated.
- Where there is a risk from overhead electric power cables, they must be directed away from the area of risk or the power cut off or, if it is not reasonably practicable, provide:
  - Barriers suitable for excluding work equipment which is not needed.
  - Suspended protections where vehicles need to pass beneath the cables.
  - Measures achieving an equivalent level of safety.
- Take suitable and sufficient steps to prevent construction work creating a risk to health or safety from underground services, or from damage to or disturbance of it.

Prevention or avoidance of drowning (Regulation 26)

- Take steps to prevent people from falling into water or other liquid with a risk of drowning and to minimise the risk of drowning in the event of a fall.
- Ensure rescue equipment is immediately available for use and is maintained.
- Make sure safe transport by water is provided and is not overcrowded or overloaded.

Traffic routes and vehicles (Regulations 27 and 28)

- Ensure construction sites are organised so that pedestrians and vehicles can both move safely and without risks to health.
- Make sure traffic routes are suitable and sufficient for the people or vehicles using them.
- Ensure separation between vehicles and pedestrians, including at doors and gate.
- Ensure loading bays have at least one exit for the exclusive use of pedestrians.
- Ensure a separate pedestrian door is provided where gates are used primarily for vehicles.
- Ensure traffic routes are identified, checked and maintained.
- Ensure traffic routes are free from obstruction and wide enough for use.
- Prevent or control the unintended movement of any vehicle.
- Make arrangements for giving a warning of any possible dangerous movement, for example, reversing vehicles.
- Ensure vehicles are loaded and driven safely. Including prohibiting people from remaining on vehicles loading loos materials and riding on vehicles, unless they are in a safe place.
- Ensure measures are taken to prevent vehicles falling into excavations, pits or water.

Fresh air, temperature and lighting (Regulations 33, 34, and 35)

- Ensure sufficient fresh or purified air is available at every workplace. Where plant is used to provide fresh or purified air it must give a visible or audible warning of its failure.
- Make sure a reasonable working temperature is maintained at indoor work places during working hours.
- Provide facilities for protection against adverse weather conditions.
- Make sure suitable and sufficient lighting is available (preferably natural light), including providing secondary lighting where there would be a risk to health or safety if primary or artificial lighting failed.

Prevention and control of emergencies (Regulations 29, 30, 31, 32 and 36)

- Prevent risk from fire, explosion, flooding and asphyxiation.
- Make arrangements for dealing with emergencies, including procedures for evacuating the site and the testing of these procedures.
- Provide emergency routes and exits, which have emergency lighting and are indicated by signs.
- Where necessary, provide firefighting equipment, fire detectors and alarm systems.

These Regulations require the prevention of risk as far as it is reasonably practicable to achieve. However, there are times when emergencies do arise and planning is needed to ensure, for example, that emergency routes are provided and evacuation procedures are in place. These particular Regulations (as well as those on traffic routes, welfare, cleanliness and signing of sites) apply to construction work which is carried out on construction sites. However, the rest of the Regulations apply to all construction work.

The HSE continues to be responsible for inspection of means of escape and firefighting for most sites. However, fire authorities have enforcement responsibility in many premises which remain in normal use during construction work. This continues the sensible arrangement which ensures that the most appropriate enforcement authority regulates fire safety.

PART 5 - GENERAL

Regulation 36 - Enforcement in respect of fire

The enforcing authority for Regulations 30 and 31 (so far as those Regulations relate to fire) and Regulation 32, in respect of a construction site which is contained within or forms part of premises occupied by persons other than those carrying out construction work, or any activity related to this work, is:

- In England and Wales the enforcing authority enforcing the Regulatory Reform (Fire Safety) Order 2005 in the premises.
- In Scotland the enforcing authority enforcing the Fire (Scotland) Act 2005 in the premises.

Regulation 37 - Transitional and saving provisions
Schedule 4, sets out transitional arrangements for projects established before CDM 2015 came into force.

Regulation 38 - Revocation and consequential amendments

1. The 2007 Regulations are revoked.

2. The amendments affecting other legislation are set out in Schedule 5 of CDM 2015.

SCHEDULES

Schedule 1 (Regulation 6) particulars to be notified under regulation 6

Regulation 6 requires the notification of specified categories of project to the enforcing authority, usually the HSE. The particulars to be included in the notification are:

1. The date of forwarding the notice.

2. The address of the construction site or precise description of its location.

3. The name of the Local Authority where the construction site is located.

4. A brief description of the project and the construction work that it entails.

5. The following contact details of the client: name, address, telephone number and (if available) an email address.

6. The following contact details of the principal designer: name, address, telephone number and (if available) an email address.

7. The following contact details of the principal contractor: name, address, telephone number and (if available) an email address.

8. The date planned for the start of the construction phase.

9. The time allocated by the client under Regulation 4(1) for the construction work.

10. The planned duration of the construction phase.

11. The estimated maximum number of people at work on the construction site.

12. The planned number of contractors on the construction site.

13. The name and address of any contractor already appointed.

14. The name and address of any designer already appointed.

15. A declaration signed by or on behalf of the client that the client is aware of the client duties under these Regulations.

Schedule 2 (Regulation 4(2)(b), 13(4)(c)and 15(11)) minimum welfare facilities
Sanitary conveniences

- Provide suitable and sufficient sanitary conveniences or make them available at readily accessible places.
- Provide rooms containing sanitary conveniences that are adequately ventilated and lit.
- Keep the sanitary conveniences and the rooms containing them in a clean and orderly condition.
- Provide separate rooms containing sanitary conveniences for men and women, except where each convenience is in a separate room and the door to the room is capable of being secured from the inside.

Washing facilities

- Provide suitable and sufficient washing facilities, including showers if required by the nature of the work or for health reasons, or make them available at readily accessible places. This includes near sanitary conveniences and changing rooms.

- Provide washing facilities with a supply of clean hot and cold, or warm, running water; soap or other suitable means of cleaning; towels or other suitable means of drying.
- Provide ventilation and lighting for washing facilities and ensure they are kept clean and in an orderly condition.
- Provide separate washing facilities for men and women, except where they are provided in a room and the door to the room can be secured from inside and the facilities in each room are intended to be used by only one person at a time.
- Separate washing facilities are not required where they are only provided for washing hands, forearms and face.

Drinking water

- Provide an adequate supply of wholesome drinking water or make it available at readily accessible and suitable places.
- Mark every supply of drinking water with an appropriate sign where necessary for reasons of health and safety.
- Provide a sufficient number of suitable cups or other drinking vessels, unless the supply of drinking water is in a jet from which people can drink easily.

Changing rooms and lockers

- Provide suitable and sufficient changing rooms or make them available at readily accessible places if a worker has to wear special clothing for the purposes of work and cannot, for reasons of health or propriety, be expected to change elsewhere.
- Where necessary for reasons of propriety, provide separate rooms for men and women.
- Provide changing rooms with seating and, where necessary, facilities to enable the drying of any special clothing and the worker's own clothing and personal effects.
- Provide, where necessary, suitable and sufficient facilities or make them available at readily accessible places to enable persons to lock away any such special clothing which is not taken home, their own clothing which is not worn during working hours, their personal effects.

Facilities for rest

- Provide suitable and sufficient rest rooms or rest areas must be provided or made available at readily accessible places.
- Ensure rest rooms and rest areas:
  - (a) Are equipped with an adequate number of tables and adequate seating with backs for the number of persons at work likely to use them at any one time.
  - (b) Where necessary, include suitable facilities for any person at work who is a pregnant woman or nursing mother to rest lying down.
  - (c) Include suitable arrangements to ensure that meals can be prepared and eaten.
  - (d) Include the means for boiling water.
  - (e) Be maintained at an appropriate temperature.

Schedule 3 (Regulation 12(2)) - work involving particular risks

Regulation 12 requires that special account be taken of work involving risks specified in Schedule 3 when setting out health and safety arrangements in the construction phase plan.

The risks specified in Schedule 3 are:

1. Work which puts workers at risk of burial under earthfalls, engulfment in swampland or falling from a height, where the risk is particularly aggravated by the nature of the work or processes used or by the environment at the place of work or site.

2. Work which puts workers at risk from chemical or biological substances constituting a particular danger to the safety or health of workers or involving a legal requirement for health monitoring.

3. Work with ionizing radiation requiring the designation of controlled or supervised areas under Regulation 16 of the Ionising Radiations Regulations 1999.

4. Work near high voltage power lines.

5. Work exposing workers to the risk of drowning.

6. Work on wells, underground earthworks and tunnels.

7. Work carried out by divers having a system of air supply.

8. Work carried out by workers in caissons with a compressed air atmosphere.

9. Work involving the use of explosives.

10. Work involving the assembly or dismantling of heavy prefabricated components.

Schedule 4 transitional provisions

This schedule established arrangements for projects that were in place at the time the Regulations came into force and provided transitional provisions for this.

They cover situations where a CDM coordinator had been appointed and ones where they had not.

Projects with no existing CDM coordinator or principal contractor

Where no CDM coordinator or principal contractor is appointed, there is or is likely to be more than one contractor and the construction phase has started:

- The client may appoint a principal designer and must appoint a principal contractor and ensure the principal contractor draws up a construction phase plan.

- If the client, other than a domestic client, does not appoint a principal designer the principal must prepare a health and safety file, ensure it is reviewed, updated and revised as the project proceeds.

- Where the client, other than a domestic client, fails to appoint a principal contractor, the client must fulfil the principal contractor duties. If the client is a domestic client the contractor in control of the construction phase is the principal contractor.

Projects with an existing CDM coordinator

Where a CDM coordinator is appointed:

- The appointment of CDM coordinator continues to have effect until a principal designer is appointed or the project comes to an end. Duties in these Regulations to liaise with the principal designer will mean liaising with the CDM coordinator.

- The client may appoint a principal designer for the project before 6th October 2015.

Duties of CDM coordinator during the transitional period

The CDM coordinator must:

- Co-operate with any other person working on or in relation to a project at the same or an adjoining construction site, to the extent necessary to enable any person with a duty or function under these Regulations to fulfil that duty or function.

- Where the CDM coordinator works under the control of another, report to that person anything they are aware of in relation to the project which is likely to endanger their own health or safety or that of others.

- Ensure that suitable arrangements are made and implemented for the coordination of health and safety measures during the planning and preparation for the construction phase, including facilitating:

(i) Co-operation and coordination between all persons working on the pre-construction phase of the project.

(ii) The application of the general principles of prevention.

- Liaise with the principal contractor over:

(i) The content of the health and safety file.

(ii) The information which the principal contractor needs to prepare the construction phase plan.

(iii) Any design development which may affect planning and management of the construction work.

- Where no or partial pre-construction information has been supplied to the CDM coordinator by the client under Regulation 10 of the 2007 Regulations, assist the client to comply with Regulation 4(4) of these Regulations.

- Unless the information has already been provided under Regulation 20(2)(b) of the 2007 Regulations, provide any pre-construction information that is in the possession or control of the CDM coordinator, promptly and in a convenient form, to every designer and contractor appointed, or being considered for appointment, to the project.

- Take all reasonable steps to ensure that designers comply with their duties under Regulation 9 of these Regulations.

- Take all reasonable steps to ensure co-operation between designers and the principal contractor during the construction phase in relation to any design or change to a design.

- If a health and safety file has not been prepared under Regulation 20(2)(e) of the 2007 Regulations, prepare a health and safety file that complies with the requirements of Regulation 12(5) of these Regulations.

- Review, update and revise the health and safety file from time to time to take account of the work and any changes that have occurred.

- If the CDM coordinator's appointment continues to have effect immediately before the project ends, pass the health and safety file to the client at the end of the project.

- If a principal designer is appointed, pass the health and safety file and all other relevant health and safety information in the CDM coordinator's possession to the principal designer, as soon as is practicable after the appointment.

- The CDM coordinator must not arrange for or instruct a worker to carry out or manage design or construction work unless the worker is competent or under the supervision of a competent person.

Projects with one contractor

Where a relevant project has only one contractor and the construction phase has started, the contractor must draw up a construction phase plan, or make arrangements for a construction phase plan to be drawn up as soon as is practicable.

Control of Artificial Optical Radiation at Work Regulations (CAOR) 2010

Considered in context in Element 13.

ARRANGEMENT OF REGULATIONS

1) Citation, commencement and interpretation.

2) Application of these Regulations.

3) Assessment of the risk of adverse health effects to the eyes or skin created by exposure to artificial optical radiation at the workplace.

4) Obligations to eliminate or reduce risks.

5) Information and training.

6) Health surveillance and medical examinations.

7) Extension outside Great Britain.

OUTLINE OF MAIN POINTS

The Regulations came into force on 27th April 2010.

The employer has duties to employees and any other person at work who may be affected by the work carried out.

Assessment of the risk of adverse health effects to the eyes or skin

The employer must make a suitable and sufficient assessment of risk for the purpose of identifying the measures it needs to take to meet the requirements of these Regulations where:

(a) The employer carries out work which could expose any of its employees to levels of artificial optical radiation that could create a reasonably foreseeable risk of adverse health effects to the eyes or skin of the employee.

(b) That employer has not implemented any measures to either eliminate or, where this is not reasonably practicable, reduce to as low a level as is reasonably practicable, that risk based on the general principles of prevention set out in Schedule 1 to the Ionising Radiation Regulations (IRR) 1999.

Obligations to eliminate or reduce risks

An employer must ensure that any risk of adverse health effects to the eyes or skin of employees as a result of exposure to artificial optical radiation which is identified in the risk assessment is eliminated or, where this is not reasonably practicable, reduced to as low a level as is reasonably practicable.

Information and training

If the risk assessment indicates that employees could be exposed to artificial optical radiation which could cause adverse health effects to the eyes or skin of employees, the employer must provide its employees and representatives with suitable and sufficient information and training relating to the outcome of the risk assessment, and this must include the following:

(a) The technical and organisational measures taken in order to comply with the requirements of Regulation 4.

(b) The exposure limit values.

(c) The significant findings of the risk assessment, including any measurements taken, with an explanation of those findings.

(d) Why and how to detect and report adverse health effects to the eyes or skin.

(e) The circumstances in which employees are entitled to appropriate health surveillance.

(f) Safe working practices to minimise the risk of adverse health effects to the eyes or skin from exposure to artificial optical radiation.

(g) The proper use of personal protective equipment.

The employer must ensure that any person, whether or not that person is an employee, who carries out work in connection with the employer's duties under these Regulations has suitable and sufficient information and training.

Health surveillance and medical examinations

If the risk assessment indicates that there is a risk of adverse health effects to the skin of employees, as a result of exposure to artificial optical radiation, the employer must ensure that such employees are placed under suitable health surveillance.

Control of Asbestos Regulations (CAR) 2012

Considered in context in Element 12.

ARRANGEMENT OF REGULATIONS

PART 1 - PRELIMINARY

1) Citation and commencement.

2) Interpretation.

3) Application of these Regulations.

PART 2 - GENERAL REQUIREMENTS

4) Duty to manage asbestos in non-domestic premises.

5) Identification of the presence of asbestos.

6) Assessment of work which exposes employees to asbestos.

7) Plans of work.

8) Licensing of work with asbestos.

9) Notification of work with asbestos.

10) Information, instruction and training.

11) Prevention or reduction of exposure to asbestos.

12) Use of control measures etc.

13) Maintenance of control measures etc.

14) Provision and cleaning of protective clothing.

15) Arrangements to deal with accidents, incidents and emergencies.

16) Duty to prevent or reduce the spread of asbestos.

17) Cleanliness of premises and plant.

18) Designated areas.

19) Air monitoring.

20) Standards for air testing.

21) Standards for analysis.

22) Health records and medical surveillance.

23) Washing and changing facilities.

24) Storage, distribution and labelling of raw asbestos and asbestos waste.

PART 3 - PROHIBITIONS AND RELATED PROVISIONS

25) Interpretation of prohibitions.

26) Prohibitions of exposure to asbestos.

27) Labelling of products containing asbestos.

28) Additional provisions in the case of exceptions and exemptions.

PART 4 - MISCELLANEOUS

29) Exemption certificates.

30) Exemptions relating to the Ministry of Defence.

31) Extension outside Great Britain.

32) Existing licences and exemption certificates.

33) Revocations and savings.

34) Defence.

35) Review.

Schedule 1

Particulars to be included in a notification.

Schedule 2

Appendix 7 to Annex XVII of the REACH Regulation – special provisions on the labelling of articles containing asbestos.

Schedule 3

Amendments.

OUTLINE OF MAIN POINTS

Summary

The **Control of Asbestos Regulations (CAR) 2012** place emphasis on assessment of exposure; exposure prevention, reduction and control; adequate information, instruction and training for employees; monitoring and health surveillance. The Regulations also apply to incidental exposure. The section on prohibitions is now covered by REACH. The amendments in these Regulations have introduced an additional category of work with asbestos. The three categories are: Licensed, Non-Licensed and a new category of Notifiable Non-Licensed (NNLW). A summary of the requirements of each category is detailed in the table below.

Non-licenced work	Notifiable non-licenced work	Licenced work
• Carry out and comply with a risk assessment. • Control exposure. • Provide training and information.	• Notify before work starts. • Provide medical examinations every three years. • Keep health records of employees. • Carry out and comply with a risk assessment. • Control exposure. • Provide training and information.	• Licencing. • Notify fourteen days in advance. • Develop emergency arrangements. • Designate of asbestos areas. • Provide medical examination every two years. • Keep health records of all employees. • Carry out and comply with a risk assessment. • Control exposure. • Provide training and information.

In order to achieve the required changes the Regulations provide a separate definition of licensable work and set out the scope of the work which is exempt from the various requirements as now. Several other amendments have also been necessary and as a result there are changes to the notification requirements and those relating to health records and medical surveillance to distinguish between licensed and non-licensed work and amendments to permit a wider range of medical professionals to carry out the required medical examinations.

The work for which a licence is required is defined as 'Licensable work with asbestos' and is work:

(a) Where the exposure to asbestos of employees is not sporadic and of low intensity.

(b) In relation to which the risk assessment cannot clearly demonstrate that the control limit will not be exceeded.

(c) On asbestos coating.

(d) On asbestos insulating board or asbestos insulation for which the risk assessment:

(i) Demonstrates that the work is not sporadic and of low intensity.

(ii) Cannot clearly demonstrate that the control limit will not be exceeded.

(iii) Demonstrates that the work is not short duration work.

Regulation 3(2) sets out the exemptions for non-licensable work as follows:

Regulations 9 (notification of work with asbestos), 18(1)(a) (designated areas) and 22 (health records and medical surveillance) do not apply where:

(a) The exposure to asbestos of employees is sporadic and of low intensity.

(b) It is clear from the risk assessment that the exposure to asbestos of any employee will not exceed the control limit.

(c) The work involves:

(i) Short, non-continuous maintenance activities in which only non-friable materials are handled.

(ii) Removal without deterioration of non-degraded materials in which the asbestos fibres are firmly linked in a matrix.

(iii) Encapsulation or sealing of asbestos-containing materials which are in good condition.

(iv) Air monitoring and control, and the collection and analysis of samples to ascertain whether a specific material contains asbestos.

Whether a type of asbestos work is either licensable, NNLW or non-licensed work has to be determined in each case and will depend on the type of work being carried out, the type of material being worked on and its condition. The identification of the type of asbestos-containing material (ACM) to be worked on and an assessment of its condition are important parts of your risk assessment, which needs to be completed before work starts. It is the responsibility of the person in charge of the job to assess the ACM to be worked on and decide if the work is NNLW or non-licensed work. This will be a matter of judgement in each case, dependent on consideration of the above factors. A decision flow chart is available from the HSE at www.hse.gov.uk/asbestos/essentials/index.htm.

Duty to manage asbestos in non-domestic premises (Regulation 4)

The Regulations include the 'duty to manage asbestos' in non-domestic premises. Guidance on the duty to manage asbestos can be found in the Approved Code of Practice, Work with Materials Containing Asbestos, L143 (Second Edition), ISBN 9780717662067.

Information, instruction and training (Regulation 10)

The Regulations require mandatory training for anyone liable to be exposed to asbestos fibres at work. This includes maintenance workers and others who may come into contact with or who may disturb asbestos (for example, cable installers) as well as those involved in asbestos removal work.

Prevention or reduction of exposure to asbestos (Regulation 11)

When work with asbestos or which may disturb asbestos is being carried out, the Control of Asbestos Regulations require employers and the self-employed to prevent exposure to asbestos fibres. Where this is not reasonably practicable, they must make sure that exposure is kept as low as reasonably practicable by measures other than the use of respiratory protective equipment. The spread of asbestos must be prevented. The Regulations specify the work methods and controls that should be used to prevent exposure and spread.

Control limits

Worker exposure must be below the airborne exposure limit (Control Limit). The Asbestos Regulations have a single Control Limit for all types of asbestos of 0.1 fibres per cm^3. A Control Limit is a maximum concentration of asbestos fibres in the air (averaged over any continuous 4 hour period) that must not be exceeded. In addition, short term exposures must be strictly controlled and worker exposure should not exceed 0.6 fibres per cm^3 of air averaged over any continuous 10 minute period using respiratory protective equipment if exposure cannot be reduced sufficiently using other means.

Respiratory protective equipment

Respiratory protective equipment is an important part of the control regime but it must not be the sole measure used to reduce exposure and should only be used to supplement other measures. Work methods that control the release of fibres such as those detailed in the **Asbestos Essentials task sheets** (available on the HSE website) for non-licensed work should be used. Respiratory protective equipment must be suitable, must fit properly and must ensure that worker exposure is reduced as low as is reasonably practicable.

Asbestos removal work undertaken by a licensed contractor

Most asbestos removal work must be undertaken by a licensed contractor but any decision on whether particular work is licensable is based on the risk. Work is only exempt from licensing if:

- The exposure of employees to asbestos fibres is sporadic and of low intensity (but exposure cannot be considered to be sporadic and of low intensity if the concentration of asbestos in the air is liable to exceed 0.6 fibres per cm3 measured over 10 minutes).
- It is clear from the risk assessment that the exposure of any employee to asbestos will not exceed the control limit.
- The work involves:

  - Short, non-continuous maintenance activities. Work can only be considered as short, non-continuous maintenance activities if any one person carries out work with these materials for less than one hour in a seven-day period. The total time spent by all workers on the work should not exceed a total of two hours*.

  - Removal of materials in which the asbestos fibres are firmly linked in a matrix. Such materials include: asbestos cement; textured decorative coatings and paints which contain asbestos; articles of bitumen, plastic, resin or rubber which contain asbestos where their thermal or acoustic properties are incidental to their main purpose (for example, vinyl floor tiles, electric cables, roofing felt) and other insulation products which may be used at high temperatures but have no insulation purposes, for example gaskets, washers, ropes and seals.

  - Encapsulation or sealing of asbestos-containing materials which are in good condition.

  - Air monitoring and control, and the collection and analysis of samples to find out if a specific material contains asbestos.

It is important that the amount of time employees spend working with asbestos insulation, asbestos coatings or asbestos insulating board (AIB) is managed to make sure that these time limits are not exceeded. This includes the time for activities such as building enclosures and cleaning.

Under the Asbestos Regulations, anyone carrying out work on asbestos insulation, asbestos coating or AIB needs a licence issued by HSE unless they meet one of the exemptions above. *Although you may not need a licence to carry out a particular job, you still need to comply with the rest of the requirements of the Asbestos Regulations.*

Licensable work - additional duties

If the work is licensable there are a number of additional duties.

The need to:

- Notify the enforcing authority responsible for the site where you are working (for example, HSE or the Local Authority).
- Designate the work area (see Regulation 18 for details).
- Prepare specific asbestos emergency procedures.
- Pay for your employees to undergo medical surveillance.

Non-notifiable licensable work - additional duties

If work is determined to be NNLW, the duties are:

- To notify the enforcing authority responsible for the site where the work is before work starts. (There is no minimum period).
- By 2015 all employees will have to undergo medical examinations which are repeated every three years.
- To have prepared procedures which can be put into effect should an accident, incident or emergency occur.
- To keep a register of all NNLW work for all employees.
- To record the significant findings of and comply with a risk assessment.
- To prevent or reduce exposure so far as is reasonably practicable and to take reasonable steps that all control measures are used.
- To ensure that adequate information, instruction and training is given to employees.

Air monitoring (Regulation 19)

The Asbestos Regulations require any analysis of the concentration of asbestos in the air to be measured in accordance with the 1997 WHO recommended method.

Standards for air testing and site clearance certification (Regulation 20)

From 06 April 2007, a clearance certificate for re-occupation may only be issued by a body accredited to do so. At the moment, such accreditation can only be provided by the United Kingdom Accreditation Service (UKAS). You can find more details of how to undertake work with asbestos containing materials, the type of controls necessary, what training is required and analytical methods in the following HSE publications:

- Approved Code of Practice Work with Materials containing Asbestos, L143, ISBN: 978-0-717662-06-7.

- Asbestos: the Licensed Contractors Guide, HSG 247, ISBN: 978-0-717628-74-2.
- Asbestos: the analysts' guide for sampling, analysis and clearance procedures, HSG 248, ISBN: 978-0-717628-75-9.
- Asbestos Essentials, HSG 210, ISBN: 978-0-717662-63-0. (See also the 'Asbestos Essentials Task Sheets' available on the HSE website).

Other health and safety legislation must be complied with.

Source: HSE Website: www.hse.gov.uk.

Control of Electromagnetic Fields at Work Regulations (CEMFAW) 2016

ARRANGEMENT OF THE REGULATIONS

PART 1 - INTRODUCTION

1) Citation and commencement.

2) Interpretation.

3) Application.

PART 2 - EXPOSURE AND RISK

4) Limitation on exposure to electromagnetic fields.

5) Exposure assessment.

6) Application of regs 7 to 9.

7) Action plan.

8) Risk assessment.

9) Obligation to eliminate or reduce risk.

PART 3 - MISCELLANEOUS

10) Information and training.

11) Health surveillance and medical examinations.

12) Records.

13) Exemptions.

14) Application outside Great Britain.

15) Review.

Schedule.

OUTLINE OF MAIN POINTS

The Control of Electromagnetic Fields at Work Regulations provide the minimum health and safety requirements regarding the exposure of workers to the risks arising from electromagnetic fields. Employers will be required to assess the levels of electromagnetic fields their workers are exposed to against specific sets of exposure levels.

These Regulations implement Directive 2013/35/EU, on the exposure of workers to the risks arising from electromagnetic fields.

Control of Noise at Work Regulations (CNWR) 2005

Considered in context in Element 13.

ARRANGEMENT OF REGULATIONS

1) Citation and commencement.

2) Interpretation.

3) Application and transition.

4) Exposure limit values and action values.

5) Assessment of the risk to health created by exposure to noise at the workplace.

6) Elimination or control of exposure to noise at the workplace.

7) Hearing protection.

8) Maintenance and use of equipment.

9) Health surveillance.

10) Information, instruction and training.

11) Exemption certificates from hearing protection.

12) Exemption certificates for emergency services.

13) Exemption relating to the Ministry of Defence etc.

14) Extension outside Great Britain.

15) Revocations, amendment and savings.

OUTLINE OF MAIN POINTS

Changes to the action levels (Regulation 4)

The Regulations allow the employer to average out the exposure to noise over a one week period in situations where the noise exposure varies on a daily basis. When determining noise levels for the purposes of determining exposure action levels, the noise exposure reducing effects of hearing protection may not be taken in to account. Where exposure is at, or above, the ***lower exposure action value*** (80 dB(A)) the employer has a duty to provide hearing protection to those employees that request it. The employer also has a duty to information, instruction and training on the risks posed by exposure to noise and the control measures to be used.

Where the exposure is at, or above, the ***upper exposure action value*** (85 dB(A)) the employer is also required to introduce a formal programme of control measures. The measures to be taken as part of this programme of control measures will depend on the findings of the noise risk assessment (see following page).

The Control of Noise at Work Regulations 2005 impose a value known as the *exposure limit value*. These are limits set both in terms of daily (or weekly) personal noise exposure (*LEP,d* of 87 dB) and in terms of peak noise (*LCpeak* of 140 dB). The exposure action values, take account of the protection provided by personal hearing protection (unlike the two exposure limit values).

If an employee is exposed to noise at or above the exposure limit value, then the employer must take immediate action to bring the exposure down below this level.

Summary of exposure limit values and action values

The lower exposure action values are:	A daily or weekly personal noise exposure of 80 dB (A-weighted) A peak sound pressure of 135 dB (C-weighted)
The upper exposure action values are:	A daily or weekly personal noise exposure of 85 dB (A-weighted) A peak sound pressure of 137 dB (C-weighted)
The exposure limit values are:	A daily or weekly personal exposure of 87 dB (A-weighted) A peak sound pressure of 140 dB (C-weighted)

Noise risk assessment and control measures (Regulations 5 and 6)

Employers are required (in accordance with the general risk assessment and general principles of prevention contained in Schedule 1 to the Management of Health and Safety at Work Regulation 1999) to ensure that the risks associated with employees' exposure to noise are eliminated where this is reasonably practicable. Where elimination is not reasonably practicable, then the employer must reduce the risks down to as low a level as is reasonably practicable. Regulation 6(2) of the Control of Noise at Work Regulations 2005 introduces the requirement for a formal programme of control measures. If any employee is likely to be exposed to noise at or above an upper exposure action value, the employer shall reduce exposure to a minimum by establishing and implementing a programme of organisational and technical measures, excluding the provision of personal hearing protectors, which is appropriate to the activity and consistent with the risk assessment, and shall include consideration of:

(a) Other working methods which eliminate or reduce exposure to noise.

(b) Choice of appropriate work equipment emitting the least possible noise, taking account of the work to be done.

(c) The design and layout of workplaces, work stations and rest facilities.

(d) Suitable and sufficient information and training for employees, such that work equipment may be used correctly, in order to minimise their exposure to noise.

(e) Reduction of noise by technical means including:

   (i) In the case of airborne noise the use of shields, enclosures, and sound-absorbent coverings.

   (ii) In the case of structure-borne noise by damping and isolation.

(f) Appropriate maintenance programmes for work equipment, the workplace and workplace systems.

(g) Limitation of the duration and intensity of exposure to noise.

(h) Appropriate work schedules with adequate rest periods.

If the risk assessment indicates an employee is likely to be exposed to noise at or above an upper exposure action value, the employer shall ensure that

- The area is designated a Hearing Protection Zone.
- The area is demarcated and identified by means of the sign specified for the purpose of indicating 'ear protection must be worn' (to be consistent with the Health and Safety (Safety Signs and Signals) Regulations 1996).
- The sign shall be accompanied by text that indicates that the area is a Hearing Protection Zone and that employees must wear personal hearing protectors while in that area.
- Access to the area is restricted where this is technically feasible and the risk of exposure justifies it and shall make every effort to ensure that no employee enters that area unless they are wearing personal hearing protectors.

Maintenance and use of equipment (Regulation 8)

There is a duty on the employer to maintain the control introduced to protect employees. This will include maintenance of acoustic enclosures, etc. as well as the maintenance of machinery (as required under the Provision and Use of Work Equipment Regulations 1998) to control noise at source.

Health surveillance (Regulation 9)

Under the Control of Noise at Work Regulations 2005, employees who are regularly exposed to noise levels of 85 dB(A) or higher must be subject to health surveillance, including audiometric testing. Where exposure is between 80 dB and 85 dB, or where employees are only occasionally exposed above the upper exposure action values, health surveillance will only be required if information comes to light that an individual may be particularly sensitive to noise induced hearing loss.

Source: www.lrbconsulting.com and www.hse.gov.uk.

Control of Substances Hazardous to Health Regulations (COSHH) 2002

Considered in context in Element 12.

Amendments to these Regulations were made by the Control of Substances Hazardous to Health (Amendment) Regulations 2004.

ARRANGEMENT OF REGULATIONS

1) Citation and commencement.

2) Interpretation.

3) Duties under these Regulations.

4) Prohibitions on substances.

5) Application of Regulations 6 to 13.

6) Assessment of health risks created by work involving substances hazardous to health.

7) Control of exposure.

8) Use of control measures etc.

9) Maintenance of control measures.

10) Monitoring exposure.

11) Health surveillance.

12) Information etc.

13) Arrangements to deal with accidents, incidents and emergencies.

14) Exemption certificates.

15) Extension outside Great Britain.

16) Defence in proceedings for contravention of these Regulations.

17) Exemptions relating to the Ministry of Defence etc.

18) Revocations, amendments and savings.

19) Extension of meaning of 'work'.

20) Modification of section 3(2) of the Health and Safety at Work etc. Act (HASAWA) 1974.

Schedule 1 - Other substances and processes to which the definition of 'carcinogen' relates.

Schedule 2 - Prohibition of certain substances hazardous to health for certain purposes.

Schedule 3 - Special provisions relating to biological agents.

Schedule 4 - Frequency of thorough examination and test of local exhaust ventilation plant used in certain processes.

Schedule 5 - Specific substances and processes for which monitoring is required.

Schedule 6 - Medical surveillance.

Schedule 7 - Legislation concerned with the labelling of containers and pipes.

Schedule 8 - Fumigations excepted from Regulation 14.

Schedule 9 - Notification of certain fumigations.

Appendix 1 - Control of carcinogenic substances.

Annex 1 - Background note on occupational cancer.

Annex 2 - Special considerations that apply to the control of exposure to vinyl chloride.

Appendix 2 - Additional provisions relating to work with biological agents.

Appendix 3 - Control of substances that cause occupational asthma.

OUTLINE OF MAIN POINTS

Regulations

Reg 2   Interpretation

"Substance hazardous to health" means a substance (including a preparation) –

(a) which is listed in Table 3.2 of part 3 of Annex VI of the CLP Regulation and for which an indication of danger specified for the substance is very toxic, toxic, harmful, corrosive or irritant;

(b) for which the Health and Safety Executive(a) has approved a workplace exposure limit;

(c) which is a biological agent;

(d) which is dust of any kind, except dust which is a substance within paragraph (a) or (b) above, when present at a concentration in air equal to or greater than –

(i) 10 mg/m3, as a time-weighted average over an 8-hour period, of inhalable dust; or

(ii) 4 mg/m3, as a time-weighted average over an 8-hour period, of respirable dust;

(e) which, not being a substance falling within sub-paragraphs (a) to (d), because of its chemical or toxicological properties and the way it is used or is present at the workplace creates a risk to health.

Reg 3   Duties

Are on employer to protect:

- Employees.

Any other person who may be affected, except:

- Duties for health surveillance do not extend to non-employees.
- Duties to give information may extend to non-employees if they work on the premises.

Reg 4   Prohibitions on substances

Certain substances are prohibited from being used in some applications. These are detailed in Schedule 2.

Reg 5   Application of Regulations 6-13

Regulations 6-13 are made to protect a person's health from risks arising from exposure. They do not apply if:

The following Regulations already apply:

- The Control of Lead at Work Regulations (CLAW) 2002.
- The Control of Asbestos at Work Regulations (CAWR) 2002.

The hazard arises from one of the following properties of the substance:

- Radioactivity, explosive, flammable, high or low temperature, high pressure.
- Exposure is for medical treatment.
- Exposure is in a mine.

Reg 6   Assessment

Employers must not carry out work that will expose employees to substances hazardous to health unless they have assessed the risks to health and the steps that need to be taken to meet the requirements of the Regulations. The assessment must be reviewed if there are changes in the work and at least once every five years.

A suitable and sufficient assessment should include:

- An assessment of the risks to health.
- The practicability of preventing exposure.
- Steps needed to achieve adequate control.

An assessment of the risks should involve:

- Types of substance including biological agents.
- Where the substances are present and in what form.
- Effects on the body.
- Who might be affected?
- Existing control measures.

Reg 7   Control of exposure

(1)   Every employer shall ensure that the exposure of his employees to substances hazardous to health is either prevented or, where this is not reasonably practicable, adequately controlled.

(2)   In complying with his duty of prevention under paragraph (1), substitution shall by preference be undertaken, whereby the employer shall avoid, so far as is reasonably practicable, the use of a substance hazardous to health at the workplace by replacing it with a substance or process which, under the conditions of its use, either eliminates or reduces the risk to the health of his employees.

(3)   Where it is not reasonably practicable to prevent exposure to a substance hazardous to health, the employer shall comply with his duty of control under paragraph (1) by applying protection measures appropriate to the activity and consistent with the risk assessment, including, in order of priority—

(a)   the design and use of appropriate work processes, systems and engineering controls and the provision and use of suitable work equipment and materials;

(b)   the control of exposure at source, including adequate ventilation systems and appropriate organisational measures;

(c)   where adequate control of exposure cannot be achieved by other means, the provision of suitable personal protective equipment in addition to the measures required by sub-paragraphs (a) and (b).

(4)   The measures referred to in paragraph (3) shall include—

(a)   arrangements for the safe handling, storage and transport of substances hazardous to health, and of waste containing such substances, at the workplace;

(b)   the adoption of suitable maintenance procedures;

(c)   reducing, to the minimum required for the work concerned—

(i)   the number of employees subject to exposure,

(ii)   the level and duration of exposure, and

(iii)   the quantity of substances hazardous to health present at the workplace;

(d)   the control of the working environment, including appropriate general ventilation; and

(e)   appropriate hygiene measures including adequate washing facilities.

(5)   Without prejudice to the generality of paragraph (1), where it is not reasonably practicable to prevent exposure to a carcinogen, the employer shall apply the following measures in addition to those required by paragraph (3)—

(a)   totally enclosing the process and handling systems, unless this is not reasonably practicable;

(b)   the prohibition of eating, drinking and smoking in areas that may be contaminated by carcinogens;

(c)   cleaning floors, walls and other surfaces at regular intervals and whenever necessary;

(d) designating those areas and installations which may be contaminated by carcinogens and using suitable and sufficient warning signs; and

(e) storing, handling and disposing of carcinogens safely, including using closed and clearly labelled containers.

(6) Without prejudice to the generality of paragraph (1), where it is not reasonably practicable to prevent exposure to a biological agent, the employer shall apply the following measures in addition to those required by paragraph (3)—

(a) displaying suitable and sufficient warning signs, including the biohazard sign shown in Part IV of Schedule 3;

(b) specifying appropriate decontamination and disinfection procedures;

(c) instituting means for the safe collection, storage and disposal of contaminated waste, including the use of secure and identifiable containers, after suitable treatment where appropriate;

(d) testing, where it is necessary and technically possible, for the presence, outside the primary physical confinement, of biological agents used at work;

(e) specifying procedures for working with, and transporting at the workplace, a biological agent or material that may contain such an agent;

(f) where appropriate, making available effective vaccines for those employees who are not already immune to the biological agent to which they are exposed or are liable to be exposed;

(g) instituting hygiene measures compatible with the aim of preventing or reducing the accidental transfer or release of a biological agent from the workplace, including—

(i) the provision of appropriate and adequate washing and toilet facilities, and

(ii) where appropriate, the prohibition of eating, drinking, smoking and the application of cosmetics in working areas where there is a risk of contamination by biological agents; and

(h) where there are human patients or animals which are, or are suspected of being, infected with a Group 3 or 4 biological agent, the employer shall select the most suitable control and containment measures from those listed in Part II of Schedule 3 with a view to controlling adequately the risk of infection.

(7) Without prejudice to the generality of paragraph (1), where there is exposure to a substance for which a maximum exposure limit has been approved, control of exposure shall, so far as the inhalation of that substance is concerned, only be treated as being adequate if the level of exposure is reduced so far as is reasonably practicable and in any case below the maximum exposure limit.

(8) Without prejudice to the generality of paragraph (1), where there is exposure to a substance for which an occupational exposure standard has been approved, control of exposure shall, so far as the inhalation of that substance is concerned, only be treated as being adequate if—

(a) that occupational exposure standard is not exceeded; or

(b) where that occupational exposure standard is exceeded, the employer identifies the reasons for the standard being exceeded and takes appropriate action to remedy the situation as soon as is reasonably practicable.

(9) Personal protective equipment provided by an employer in accordance with this regulation shall be suitable for the purpose and shall—

(a) comply with any provision in the Personal Protective Equipment Regulations 2002(1) which is applicable to that item of personal protective equipment; or

(b) in the case of respiratory protective equipment, where no provision referred to in sub-paragraph (a) applies, be of a type approved or shall conform to a standard approved, in either case, by the Executive.

(10) Without prejudice to the provisions of this regulation, Schedule 3 shall have effect in relation to work with biological agents.

(11) In this regulation, "adequate" means adequate having regard only to the nature of the substance and the nature and degree of exposure to substances hazardous to health and "adequately" shall be construed accordingly.

Reg 8   Employer shall take all reasonable steps to ensure control measures, PPE, etc. are properly used/applied. Employee shall make full and proper use of control measures, PPE etc. and shall report defects to employer.

Reg 9   Maintenance of control measures

Employer providing control measures to comply with Reg 7 shall ensure that it is maintained in an efficient state, in efficient working order and in good repair and in the case of PPE in a clean condition, properly stored in a well-defined place checked at suitable intervals and when discovered to be defective repaired or replaced before further use.

- Contaminated PPE should be kept apart and cleaned, decontaminated or, if necessary destroyed.

- Engineering controls - employer shall ensure thorough examination and tests.
- Local exhaust ventilation (LEV) - once every 14 months unless process specified in Schedule 4.
- Others - at suitable intervals.
- Respiratory protective equipment - employer shall ensure thorough examination and tests at suitable intervals.
- Records of all examinations, tests and repairs kept for 5 years.

Reg 10   Monitoring exposure

Employer shall ensure exposure is monitored if:

- Needed to ensure maintenance of adequate control.
- Otherwise needed to protect health of employees.
- Substance/process specified in Schedule 5.

Records kept if:

- There is an identified exposure of identifiable employee - 40 years.
- Otherwise - 5 years.

Reg 11   Health surveillance

1) Where appropriate for protection of health of employees exposed or liable to be exposed, employer shall ensure suitable health surveillance.

2) Health surveillance is appropriate if:

- Employee exposed to substance/process specified in Schedule 6.
- Exposure to substance is such that an identifiable disease or adverse health effect can result, there is a reasonable likelihood of it occurring and a valid technique exists for detecting the indications of the disease or effect.

3) Health records kept for at least 40 years.

4) If employer ceases business, HSE notified and health records offered to HSE.

5) If employee exposed to substance specified in Schedule 6, then health surveillance shall include medical surveillance, under the supervision of a relevant doctor at 12 monthly intervals - or more frequently if specified by the relevant doctor

6) The relevant doctor can forbid employee to work in process, or specify certain conditions for him to be employed in a process.

7) The relevant doctor can specify that health surveillance is to continue after exposure has ceased. Employer must ensure.

8) Employees to have access to their own health record.

9) Employee must attend for health/medical surveillance and give information to the relevant doctor.

10) The relevant doctor entitled to inspect workplace.

11) Where the relevant doctor suspends employee from work exposing him to substances hazardous to health, employer of employee can apply to HSE in writing within 28 days for that decision to be reviewed.

Reg 12   Information etc.

Employer shall provide suitable and sufficient information, instruction and training for him to know:

- The substances and the risks to health which they present.
- Any relevant workplace exposure limit or similar occupational exposure limit;
- Access to any relevant safety data sheet; and
- Other legislative provisions which concern the hazardous properties of those substances;
- The significant findings of the risk assessment
- Precautions in place.
- Results of monitoring of exposure at workplace.
- Results of collective health surveillance.

If the substances have been assigned a maximum exposure limit, then the employee/safety representative must be notified forthwith if the MEL has been exceeded.

Reg 13   Arrangements to deal with accidents, incidents and emergencies

To protect the health of employees from accidents, incidents and emergencies, the employer shall ensure that:

- Procedures are in place for first-aid and safety drills (tested regularly).
- Information on emergency arrangements is available.
- Warning, communication systems, remedial action and rescue actions are available.
- Information made available to emergency services: external and internal.
- Steps taken to mitigate effects, restore situation to normal and inform employees.
- Only essential persons allowed in area.

These duties do not apply where the risks to health are slight or measures in place Reg 7(1) are sufficient to control the risk. The employee must report any accident or incident which has or may have resulted in the release of a biological agent which could cause severe human disease.

Schedule 3 Additional provisions relating to work with biological agents
Regulation 7(10)

Part I Provision of general application to biological agents

1   Interpretation.

2   Classification of biological agents.

Where no approved classification exists, the employer

shall assign the agent to one of four groups according to the level of risk of infection.

Group 1 - unlikely to cause human disease.

Group 2 - can cause human disease.

Group 3 - can cause severe disease and spread to community.

Group 4 - can cause severe disease, spread to community and there is no effective treatment.

3   Special control measures for laboratories, animal rooms and industrial processes.

Every employer engaged in research, development, teaching or diagnostic work involving group 2, 3 or 4 biological agents; keeping or handling laboratory animals deliberately or naturally infected with those agents, or industrial processes involving those agents, shall control them with the most suitable containment.

4   List of employees exposed to certain biological agents.

The employer shall keep a list of employees exposed to Group 3 or 4 biological agents for at least 10 years. If there is a long latency period then the list should be kept for 40 years.

5   Notification of the use of biological agents.

Employers shall inform the HSE at least 20 days in advance of first time use or storage of group 2, 3 or 4 biological hazards. Consequent substantial changes in procedure or process shall also be reported.

6   Notification of the consignment of biological agents.

The HSE must be informed 30 days before certain biological agents are consigned.

Part II Containment measures for health and veterinary care facilities, laboratories and animal rooms.

Part III Containment measures for industrial processes.

Part IV Biohazard sign.

The biohazard sign required by Regulation 7(6) (a) shall be in the form shown.

Part V Biological agents whose use is to be notified in accordance with paragraph 5(2) of Part I of this Schedule.

- Any Group 3 or 4 agent.
- Certain named Group 2 agents.

Figure RSP-3: Biohazard sign.
Source: COSHH 2002.

Control of Substances Hazardous to Health (Amendment) Regulations (COSHH) 2004

These Regulations make minor amendments to the Control of Substances Hazardous to Health Regulations (COSHH) 2002 and came into force on 17th January 2005 and 6th April 2005.

ARRANGEMENT OF REGULATIONS

1)   Citation and commencement.

2)   Amendments of the Control of Substances Hazardous to Health (Amendment) Regulations 2004.

3)   Amendment of the Chemicals (Hazard Information and Packaging for Supply) Regulations 2002.

4)   Amendments of the Control of Lead at Work Regulations 2002.

OUTLINE OF MAIN POINTS

Reg 2(a)(v) workplace exposure limit for a substance hazardous to health means the exposure limit approved by the Health and Safety Commission for that substance in relation to the specified reference period when calculated by a method approved by the Health and Safety Commission, as contained in HSE publication 'EH/40 Workplace Exposure Limits 2005' as updated from time to time.

Reg 7(7) Schedule 2a - Principles of good practice for the control of exposure to substances hazardous to health.

(a)   Design and operate processes and activities to minimise emission, release and spread of substances hazardous to health.

(b)   Take into account all relevant routes of exposure - inhalation, skin absorption and ingestion - when developing control measures.

(c)   Control exposure by measures that are proportionate to the health risk.

(d)   Choose the most effective and reliable control options which minimise the escape and spread of substances hazardous to health.

(e) Where adequate control of exposure cannot be achieved by other means, provide, in combination with other control measures, suitable personal protective equipment.

(f) Check and review regularly all elements of control measures for their continuing effectiveness.

(g) Inform and train all employees on the hazards and risks from the substances with which they work and the use of control measures developed to minimise the risks.

(h) Ensure that the introduction of control measures does not increase the overall risk to health and safety.

Control of Vibration at Work Regulations (CVWR) 2005

Considered in context in Element 13.

Hand-arm vibration (HAV) and whole body vibration (WBV) are caused by the use of work equipment and work processes that transmit vibration into the hands, arms and bodies of employees in many industries and occupations. Long-term, regular exposure to vibration is known to lead to permanent and debilitating health effects such as vibration white finger, loss of sensation, pain, and numbness in the hands, arms, spine and joints. These effects are collectively known as hand-arm or whole body vibration syndrome.

These Regulations introduce controls, which aim substantially to reduce ill-health caused by exposure to vibration. These Regulations came into force on 6th July 2005.

ARRANGEMENT OF REGULATIONS

1) Citation and commencement.

2) Interpretation.

3) Application and transition.

4) Exposure limit values and action values.

5) Assessment of the risk to health created by vibration at the workplace.

6) Elimination or control of exposure to vibration at the workplace.

7) Health surveillance.

8) Information, instruction and training for persons who may be exposed to risk from vibration.

9) Exemption certificates for emergency services.

10) Exemption certificates for air transport.

11) Exemption relating to the Ministry of Defence etc.

12) Extension outside Great Britain.

13) Amendment.

OUTLINE OF MAIN POINTS

Regulation 3 Duties under most of these Regulations extend not only to employees but to others, whether or not at work that may be affected. The duty does not include Regulation 7 (health surveillance) or Regulation 8 (information instruction and training), these are limited to employees.

Regulation 4 states the personal daily exposure limits and daily exposure action values, normalised over an 8-hour reference period.

	Daily exposure action values	Daily exposure limits
Hand arm vibration	2.5 m/s^2	5 m/s^2
Whole body vibration	0.5 m/s^2	1.15 m/s^2

Regulation 5 requires the employer to make a suitable and sufficient assessment of the risk created by work that is liable to expose employees to risk from vibration. The assessment must observe work practices, make reference to information regarding the magnitude of vibration from equipment and if necessary measurement of the magnitude of the vibration.

Consideration must also be given to the type, duration, effects of exposure, exposures limit/action values, effects on employees at particular risk, the effects of vibration on equipment and the ability to use it, manufacturers' information, availability of replacement equipment, and extension of exposure at the workplace (for example, rest facilities), temperature and information on health surveillance. The risk assessment should be recorded as soon as is practicable after the risk assessment is made and reviewed regularly.

Regulation 6 states that the employer must seek to eliminate the risk of vibration at source or, if not reasonably practicable, reduce it to as low a level as is reasonably practicable. Where the personal daily exposure limit is exceeded the employer must reduce exposure by implementing a programme of organisational and technical measures. Measures include the use of other methods of work, ergonomics, maintenance of equipment, design and layout, rest facilities, information, instruction and training, limitation by schedules and breaks and the provision of personal protective equipment to protect from cold and damp. Measures must be adapted to take account of any group or individual employee whose health may be of particular risk from exposure to vibration.

Regulation 7 states that health surveillance must be carried out if there is a risk to the health of employees liable to be exposed to vibration. This is in order to prevent or diagnose any health effect linked with exposure to vibration. A record of health shall be kept of any employee who undergoes health surveillance. The employer shall, providing reasonable notice is given, provide the employee with access to their health records and provide copies to an enforcing officer on request. If health surveillance identifies a disease or adverse health effect, considered by a doctor or other occupational health professional to be a result of exposure to vibration, the employer shall ensure that a qualified person informs the employee and provides information and advice. The employer must ensure they are kept informed of any significant findings from health surveillance, taking into account any medical confidentiality.

In addition the employer must also:

- Review risk assessments.
- Review the measures taken to comply.
- Consider assigning the employee to other work.
- Review the health of any other employee who has been similarly exposed and consider alternative work.

Regulation 8 states that employers must provide information, instruction and training to all employees who are exposed to risk from vibration and their representatives. This includes any organisational and technical measures taken, exposure limits and values, risk assessment findings, why and how to detect injury, entitlement to and collective results of health surveillance and safe working practices. Information instruction and training shall be updated to take account of changes in the employers work or methods. The employer shall ensure all persons, whether or not an employee, who carries out work in connection with the employer's duties has been provided with information, instruction and training.

Dangerous Substances and Explosive Atmospheres Regulations (DSEAR) 2002

Considered in context in Element 11.

ARRANGEMENT OF REGULATIONS

1) Citation and commencement.
2) Interpretation.
3) Application.
4) Duties under these Regulations.
5) Risk assessment.
6) Elimination or reduction of risks from dangerous substances.
7) Places where explosive atmospheres may occur.
8) Arrangements to deal with accidents, incidents and emergencies.
9) Information, instruction and training.
10) Identification of hazardous contents of containers and pipes.
11) Duty of coordination.
12) Extension outside Great Britain.
13) Exemption certificates.
14) Exemptions for Ministry of Defence etc.
15) Amendments.
16) Repeals and revocations.
17) Transitional provisions.

Schedule 1 - General safety measures.

Schedule 2 - Classification of places where explosive atmospheres may occur.

Schedule 3 - Criteria for the selection of equipment and protective systems.

Schedule 4 - Warning sign for places where explosive atmospheres may occur.

Schedule 5 - Legislation concerned with the marking of containers and pipes.

Schedule 6 - Amendments.

Schedule 7 - Repeal and revocation.

OUTLINE OF MAIN POINTS

These Regulations aim to protect against risks from fire, explosion and similar events arising from dangerous substances that are present in the workplace.

Dangerous substances

These are any substances or preparations that due to their properties or the way in which they are being used could cause harm to people from fires and explosions. They may include petrol, liquefied petroleum gases, paints, varnishes, solvents and dusts.

Application

DSEAR applies in most workplaces where a dangerous substance is present. There are a few exceptions where only certain parts of the Regulations apply, for example:

- Ships.
- Medical treatment areas.
- Explosives/chemically unstable substances.
- Mines.
- Quarries.
- Boreholes.
- Offshore installations.
- Means of transport.

Main requirements

You must:

- Conduct a risk assessment of work activities involving dangerous substances.
- Provide measures to eliminate or reduce risks.
- Provide equipment and procedures to deal with accidents and emergencies.
- Provide information and training for employees.
- Classify places into zones and mark zones where appropriate.

The risk assessment should include:

- The hazardous properties of substance.
- The way they are used or stored.
- Possibility of hazardous explosive atmosphere occurring.
- Potential ignition sources.
- Details of zoned areas
- Coordination between employers.

Safety measures

Where possible eliminate safety risks from dangerous substances or, if not reasonably practicable to do this, control risks and reduce the harmful effects of any fire, explosion or similar event.

Substitution - Replace with totally safe or safer substance (best solution).

Control measures - If risk cannot be eliminated apply the following control measures in the following order:

- Reduce quantity.
- Avoid or minimise releases.
- Control releases at source.
- Prevent formation of explosive atmosphere.
- Collect, contain and remove any release to a safe place, for example, ventilation.
- Avoid ignition sources.
- Avoid adverse conditions, for example, exceeding temperature limits.
- Keep incompatible substances apart.

Mitigation measures - Apply measures to mitigate the effects of any situation.

- Prevent fire and explosions from spreading to other plant, equipment or other parts of the workplace.
- Reduce number of employees exposed.
- Provide process plant that can contain or suppress an explosion, or vent it to a safe place.

Zoned areas

In workplaces where explosive atmospheres may occur, areas should be classified into zones based on the likelihood of an explosive atmosphere occurring. Any equipment in these areas should ideally meet the requirements of the Equipment and Protective Systems Intended for Use in Potentially Explosive Atmospheres Regulations (ATEX) 1996. However equipment in use before July 2003 can continue to be used providing that the risk assessment says that it is safe to do so. Areas may need to be marked with an 'Ex' warning sign at their entry points. Employees may need to be provided with appropriate clothing, for example, anti-static overalls. Before use for the first time, a person competent in the field of explosion protection must confirm hazardous areas as being safe.

Accidents, incidents and emergencies

DSEAR builds on existing requirements for emergency procedures, which are contained in other Regulations. These may need to be supplemented if you assess that a fire, explosion or significant spillage could occur, due to the quantities of dangerous substances present in the workplace. You may need to arrange for:

- Suitable warning systems.
- Escape facilities.
- Emergency procedures.
- Equipment and clothing for essential personnel who may need to deal with the situation.
- Practice drills.
- Make information, instruction and training available to employees and if necessary liaise with the emergency services.

Electricity at Work Regulations (EWR) 1989

Considered in context in Element 10.

ARRANGEMENT OF REGULATIONS

PART I - INTRODUCTION

1) Citation and commencement.

2) Interpretation.

3) Persons on whom duties are imposed by these Regulations.

PART II - GENERAL

4) Systems, work activities and protective equipment.

5) Strength and capability of electrical equipment.

6) Adverse or hazardous environments.

7) Insulation, protection and placing of conductors.

8) Earthing or other suitable precautions.

9) Integrity of referenced conductors.

10) Connections.

11) Means for protecting from excess of current.

12) Means for cutting off the supply and for isolation.

13) Precautions for work on equipment made dead.

14) Work on or near live conductors.

15) Working space, access and lighting.

16) Persons to be competent to prevent danger and injury.

PART III - REGULATIONS APPLYING TO MINES ONLY

17) Provisions applying to mines only.

18) Introduction of electrical equipment.

19) Restriction of equipment in certain zones below ground.

20) Cutting off electricity or making safe where firedamp is found either below ground or at the surface.

21) Approval of certain equipment for use in safety-lamp mines.

22) Means of cutting off electricity to circuits below ground.

23) Oil-filled equipment.

24) Records and information.

25) Electric shock notices.

26) Introduction of battery-powered locomotives and vehicles into safety-lamp mines.

27) Storage, charging and transfer of electrical storage batteries.

28) Disapplication of section 157 of the Mines and Quarries Act 1954.

PART IV - MISCELLANEOUS AND GENERAL

29) Defence.

30) Exemption certificates.

31) Extension outside Great Britain.

32) Disapplication of duties.

33) Revocations and modifications.

Schedule 1 - Provisions applying to mines only and having effect in particular in relation to the use below ground in coal mines of film lighting circuits.

Schedule 2 - Revocations and modifications.

OUTLINE OF MAIN POINTS

Systems, work activities and protective equipment (Regulation 4)

The system and the equipment comprising it must be designed and installed to take account of all reasonably foreseeable conditions of use:

- The system must be maintained so as to prevent danger.

- All work activities must be carried out in such a manner as to not give rise to danger.
- Equipment provided to protect people working on live equipment must be suitable and maintained.

Strength and capability of electrical equipment (Regulation 5)

Strength and capability refers to the equipment's ability to withstand the effects of its load current and any transient overloads or pulses of current.

Adverse or hazardous environments (Regulation 6)

This Regulation requires that electrical equipment is suitable for the environment and conditions that might be reasonably foreseen. In particular, attention should be paid to:

- Mechanical damage caused by, for example, vehicles, people, vibration, etc.
- Weather, natural hazards, temperature or pressure. Ice, snow, lightning, bird droppings, etc.
- Wet, dirty, dusty or corrosive conditions. Conductors, moving parts, insulators and other materials may be affected by the corrosive nature of water, chemicals and solvents. The presence of explosive dusts must be given special consideration.
- Flammable or explosive substances. Electrical equipment may be a source of ignition for liquids, gases, vapours etc.

Insulation, protection and placing of conductors (Regulation 7)

The purpose of this Regulation is to prevent danger from direct contact. Therefore, if none exists, no action is needed. Conductors though will normally need to be insulated and also have some other protection to prevent mechanical damage.

Earthing or other suitable precautions (Regulation 8)

The purpose of this Regulation is to prevent danger from indirect contact. Conductors such as metal casings may become live through fault conditions. The likelihood of danger arising from these circumstances must be prevented by using the techniques described earlier in this section i.e. earthing, double insulation, reduced voltages etc.

Integrity of referenced conductors (Regulation 9)

In many circumstances the reference point is earthed because the majority of power distribution installations are referenced by a deliberate connection to earth at the generators or distribution transformers. The purpose of this Regulation is to ensure that electrical continuity is never broken.

Connections (Regulation 10)

As well as having suitable insulation and conductance, connections must have adequate mechanical protection and strength. Plugs and sockets must conform to recognised standards as must connections between cables. Special attention should be paid to the quality of connections on portable appliances.

Means for protecting from excess current (Regulation 11)

Faults or overloads can occur in electrical systems and protection must be provided against their effects. The type of protection depends on several factors but usually rests between fuses and circuit breakers.

Means for cutting off the supply and for isolation (Regulation 12)

Means must be provided to switch off electrical supplies together with a means of isolation so as to prevent inadvertent reconnection.

Precautions for work on equipment made dead (Regulation 13)

Working dead should be the norm. This Regulation requires that precautions be taken to ensure that the system remains dead and to protect those at work on the system.

Any or all of the following steps should be considered:

- Identify the circuit. Never assume that the labelling is correct.
- Disconnection and isolation. These are the most common methods: isolation switches, fuse removal and plug removal.
- Notices and barriers.
- Proving dead. The test device itself must also be tested before and after testing.
- Earthing.
- Permits to work.

Work on or near live conductors (Regulation 14)

Live work must only be done if it is unreasonable for it to be done dead.

If live work must be carried out then any or all of the following precautions should be taken:

- Competent staff (see Reg 16).
- Adequate information.
- Suitable tools. Insulated tools, protective clothing.
- Barriers or screens.
- Instruments and test probes. To identify what is live and what is dead.
- Accompaniment.
- Designated test areas.

Working space, access and lighting (Regulation 15)

Space - where there are dangerous live exposed conductors, space should be adequate to:

- Allow persons to pull back from the hazard.
- Allow persons to pass each other.

Lighting - the first preference is for natural lighting then for permanent artificial lighting.

Persons to be competent to prevent danger and injury (Regulation 16)

The object of this Regulation is to 'ensure that persons are not placed at risk due to a lack of skills on the part of themselves or others in dealing with electrical equipment'.

In order to meet the requirements of this Regulation a competent person would need:

- An understanding of the concepts of electricity and the risks involved in work associated with it.
- Knowledge of electrical work and some suitable qualification in electrical principles.
- Experience of the type of system to be worked on with an understanding of the hazards and risks involved.
- Knowledge of the systems of work to be employed and the ability to recognise hazards and risks.
- Physical attributes to be able to recognise elements of the system, for example, colour blindness and wiring.

Advice and queries regarding qualifications and training can be directed to the IEE - Institute of Electrical Engineers, London.

Defence (Regulation 29)

In any Regulation where the absolute duty applies, a defence in any criminal proceedings shall exist where a person can show that: 'He took all reasonable steps and exercised due diligence to avoid the commission of the offence'.

Is there a prepared procedure (steps), is the procedure being followed (diligence) and do you have the records or witness to prove it retrospectively?

Fire Safety (Scotland) Regulations (FSSR) 2006

Considered in context in Element 11.

The Fire Safety (Scotland) Regulations (FSSR) 2006 are Regulations that were made by Scottish Ministers under the powers contained in the Fire (Scotland) Act 2005, and further build upon the requirements of that act.

ARRANGEMENT OF REGULATIONS

PART 1 - PRELIMINARY

PART 2 - ASSESSMENTS

PART 3 - FIRE SAFETY

PART 4 - MISCELLANEOUS

Please refer to the Regulatory Reform (Fire Safety) Order (RRFSO) 2005 for the equivalent legislation for England and Wales.

Fire (Scotland) Act (FSA) 2005

Considered in context in Element 11.

ARRANGEMENT OF ACT

PART 1 - FIRE AND RESCUE AUTHORITIES

PART 2 - FIRE AND RESCUE SERVICES

PART 3 - FIRE SAFETY

PART 4 - MISCELLANEOUS

5) Payments in respect of advisory bodies.

6) Abolition of Scottish Central Fire Brigades Advisory Council.

7) False alarms.

8) Disposal of land.

PART 5 - GENERAL

1) Ancillary provision.

2) Orders and Regulations.

3) Minor and consequential amendments and repeals.

4) Commencement.

5) Short title.

Schedule 1 - Joint fire and rescue boards: supplementary provision.

Schedule 2 - Fire safety measures.

Schedule 3 - Minor and consequential amendments.

Schedule 4 - Repeals.

OUTLINE OF MAIN POINTS

The Act is targeted at reducing the number of workplace fires by imposing far reaching responsibilities on all employers, as well as those who have control to any extent of non-domestic premises, to assess and reduce the risks from fire. Fire certificates were abolished under the Act and have been replaced by a fire safety regime based upon the principles of risk assessment and the requirement to take steps to mitigate the detrimental effects of a fire on relevant premises. The regime applies to all employers as well as to anyone who has control of non-domestic premises to any extent including building owners, tenants, occupiers and factors. The overriding duty is to ensure, so far as is reasonably practicable, safety in respect of harm caused by fire in the workplace and is supplemented by a number of prescriptive duties:

- To carry out a fire safety risk assessment of the premises. If you have 5 or more employees, the risk assessment must be recorded in writing.
- Not to employ a young person (a person under 18) unless an assessment of the risks of fire to young persons has been undertaking.
- To appoint a competent person to assist with the discharge of fire safety duties.
- To identify the fire safety measures necessary.
- To put in place arrangements for the planning, organisation, control, monitoring and review of the fire safety measures that are put in place.
- To implement these fire safety measures using risk reduction principles.
- To inform employees of the fire safety risks and provide fire safety training.
- To coordinate and cooperate with other duty holders in the same premises.
- To review the risk assessment.

Health and Safety (Display Screen Equipment) Regulations (DSE) 1992 (as amended)

ARRANGEMENT OF REGULATIONS

2) Every employer shall carry out suitable and sufficient analysis of workstations.

3) Employers shall ensure that equipment provided meets the requirements of the schedule laid down in these Regulations.

4) Employers shall plan activities and provide such breaks or changes in work activity to reduce employees' workload on that equipment.

5) For display screen equipment (DSE) users, the employer shall provide, on request, an eyesight test carried out by a competent person.

6&7) Provision of information and training.

OUTLINE OF MAIN POINTS

Workstation assessments (Reg 2)

Workstation assessments should take account of:

- Screen - positioning, character definition, character stability etc.
- Keyboard - tilt able, character legibility etc.
- Desk - size, matt surface etc.
- Chair - adjustable back and height, footrest available etc.
- Environment - noise, lighting, space etc.
- Software - easy to use, work rate not governed by software.

Information and training (Regs 6&7)

Information and training should include:

- Risks to health.
- Precautions in place (for example, the need for regular breaks).
- How to recognise problems.
- How to report problems.

Health and Safety (First-Aid) Regulations (FAR) 1981 (as amended)

Considered in context in Element 3.

ARRANGEMENT OF REGULATIONS

1) Citation and commencement.

2) Interpretation.

3) Duty of employer to make provision for first-aid.

4) Duty of employer to inform his employees of the arrangements.

5) Duty of self-employed person to provide first-aid equipment.

6) Power to grant exemptions.

7) Cases where these Regulations do not apply.

8) Application to mines.

9) Application offshore.

10) Repeals, revocations and modification.

Schedule 1 - Repeals.

Schedule 1 - Revocations.

OUTLINE OF MAIN POINTS

1) Regulation 2 defines first-aid as: '...treatment for the purpose of preserving life and minimising the consequences of injury or illness until medical (doctor or nurse) help can be obtained. Also, it provides treatment of minor injuries which would otherwise receive no treatment, or which do not need the help of a medical practitioner or nurse'.

2) Requires that every employer must provide equipment and facilities which are adequate and appropriate in the circumstances for administering first-aid to his employees.

3) Employer must inform their employees about the first-aid arrangements, including the location of equipment, facilities and identification of trained personnel.

4) Self-employed people must ensure that adequate and suitable provision is made for administering first-aid while at work.

Health and Safety (Miscellaneous Amendments) Regulations (MAR) 2002

See also, FAR 1981, DSE 1992, MHOR 1992, PPER 1992, WHSWR 1992, PUWER and LOLER.

These Regulations made minor amendments to UK law to come into line with the requirements of the original Directives and came into force on 17th September 2002.

In relation to this publication, the Regulations that are affected by the amendments are:

- Health and Safety (First-Aid) Regulations (FAR) 1981.
- Health and Safety (Display Screen Equipment) Regulations (DSE) 1992.
- Manual Handling Operations Regulations (MHOR) 1992.
- Personal Protective Equipment at Work Regulations (PPER) 1992.
- Workplace (Health, Safety and Welfare) Regulations (WHSWR) 1992.
- Provision and Use of Work Equipment Regulations (PUWER) 1998.
- Lifting Operations and Lifting Equipment Regulations (LOLER) 1998.

ARRANGEMENT OF REGULATIONS

1) Citation and commencement.

2) Amendment of the Health and Safety (First-Aid) Regulations 1981.

3) Amendment of the Health and Safety (Display Screen Equipment) Regulations 1992.

4) Amendment of the Manual Handling Operations Regulations 1992.

5) Amendment of the Personal Protective Equipment at Work Regulations 1992.

6) Amendment of the Workplace (Health, Safety and Welfare) Regulations 1992.

7) Amendment of the Provision and Use of Work Equipment Regulations 1998.

8) Amendment of the Lifting Operations and Lifting Equipment Regulations 1998.

9) Amendment of the Quarries Regulations 1999.

OUTLINE OF MAIN POINTS

Regulation 3 - amendment of the Health and Safety (First-Aid) Regulations 1981

The Health and Safety (First-Aid) Regulations 1981 are amended by adding the additional requirements that any first-aid room provided under requirements of these Regulations must be easily accessible to stretchers and to any other equipment needed to convey patients to and from the room and that the room be sign-posted by use of a sign complying with the Health and Safety (Safety Signs and Signals) Regulations 1996.

Regulation 3 - amendment of the Health and Safety (Display Screen Equipment) Regulations 1992

The Health and Safety (Display Screen Equipment) Regulations 1992 were amended to cover workstations 'used for the purposes of' an employer's undertaking, which includes workstations provided by the employer and others.

The Health and Safety (Display Screen Equipment) Regulations 1992 were amended to provide for people that are users of display screen equipment in an employers undertaking but are not employees of the employer, for example, staff provided through an employment agency. This extension of duty relates to requests for eye site tests from the person.

The employer who carries on the undertaking must ensure an eyesight test is carried out, as soon as is practicable after the request for those currently a user and before they become a user for those who become users. The Health and Safety (Display Screen Equipment) Regulations 1992 were similarly amended with regard to health and safety training for users.

Regulation 4 - amendment of the Manual Handling Operations Regulations 1992

Regulation 4 of the Manual Handling Operations Regulations 1992 were amended by adding the requirement to, when determining whether manual handling operations at work involve a risk of injury and the appropriate steps to reduce that risk, have regard to:

- Physical suitability of the employee to carry out the operations.
- Clothing, footwear or other personal effects they are wearing.
- Knowledge and training.
- Results of any relevant risk assessment conducted for the Management of Health and Safety at Work Regulations.
- Whether the employee is within a group of employees identified by that assessment as being especially at risk.
- Results of any health surveillance provided under the Management of Health and Safety Regulations.

Regulation 5 - amendment of the Personal Protective Equipment at Work Regulations 1992

The Personal Protective Equipment at Work Regulations 1992 were amended so that personal protective equipment (PPE) must also be suitable for the period for which it is worn and account is taken of the characteristics of the workstation of each person.

Provision of personal issue of PPE needs to take place in situations where it is necessary to ensure it is hygienic and free of risk to health. Where an assessment of PPE is made this must consider whether it is compatible with other personal protective equipment that is in use and which an employee would be required to wear simultaneously.

The amendments require that information provided to satisfy Regulation 9 for the provision of information, instruction and training must be kept available to employees. A new, additional duty is created requiring the employer, where appropriate, and at suitable intervals, to organise demonstrations in the wearing of PPE.

Regulation 6 - amendment of the Workplace (Health, Safety and Welfare) Regulations 1992

The Workplace (Health, Safety and Welfare) Regulations 1992 have been amended to improve clarity, include additional Regulations and make provision for the disabled.

An additional Regulation (4A) sets out a requirement where a workplace is in a building, the building shall have a stability and solidity appropriate to the nature of the use of the workplace. The range of things requiring maintenance under these Regulations is extended to equipment and devices intended to prevent or reduce hazards. A new duty requires workplaces to be adequately thermally insulated where it is necessary, having regard to the type of work carried out and the physical activity of the persons carrying out the work. In addition, excessive effects of sunlight on temperature must be avoided.

The Regulations were amended with regard to facilities for changing clothing in that the facilities need to be easily accessible, of sufficient capacity and provided with seating. Requirements were amended such that rest rooms and rest areas must include suitable arrangements to protect non-smokers from discomfort caused by tobacco smoke. They also must be equipped with an adequate number of tables and adequate seating with backs for the number of persons at work likely to use them at any one time and seating which is adequate for the number of disabled persons at work and suitable for them. A new Regulation (25A) was added requiring, where necessary, those parts of the workplace (including in particular doors, passageways, stairs, showers, washbasins, lavatories and workstations) used or occupied directly by disabled persons at work to be organised to take account of such persons.

Amendment of the Provision and Use of Work Equipment Regulations 1998

The Regulations have a small number of amendments affecting 3 main Regulations. Regulation 10, which deals with equipment's conformity with community requirements, is amended such that the requirement to conform to 'essential requirements' is no longer limited to the point at which the equipment was designed and constructed - equipment must now conform at 'all times'. The 'essential requirements' are those that were applicable at the time it was put into first service. Regulation 11 was amended to remove the opportunity of reliance on information, instruction, training and supervision as a separate option in the hierarchy of control of dangerous parts of machinery.

The requirement to provide information, instruction, training and supervision is now amended to apply to each stage of the dangerous parts of machinery control hierarchy. Regulation 18, which deals with control systems, carries a small but important amendment which means the requirement that all control systems of work equipment are 'chosen making due allowance for the failures, faults and constraints to be expected in the planned circumstances of use' is modified from an absolute duty to one of so far as is reasonably practicable.

Amendment of the Lifting Operations and Lifting Equipment Regulations 1998

Minor changes to the definitions in the Lifting Operations and Lifting Equipment Regulations 1998 were made by these Regulations.

(a) In the definition of 'accessory for lifting' in Regulation 2(1), by substituting for the word 'work' the word 'lifting'.

(b) In Regulation 3(4), by substituting for the words '(5)(b)' the words '(3)(b)'.

Health and Safety (Miscellaneous Repeals, Revocations and Amendments) Regulations (MRRA) 2013

See also, PPER 1992 and WHSWR 1992.

These Regulations repeal one Act and revoke twelve instruments (plus a related provision in the Factories Act 1961) and came into force on 6th April 2013.

The Regulations, applicable to this award, that are affected by the amendments are detailed below.

ARRANGEMENT OF REGULATIONS

1) Citation and commencement.

2) Repeals and revocations.

3) Consequential amendments to the Dangerous Substances (Notification and Marking of Sites) Regulations 1990.

4) Consequential amendments to the Workplace (Health, Safety and Welfare) Regulations 1992.

OUTLINE OF MAIN POINTS

Consequential Amendments to the Personal Protective Equipment at Work Regulations (PPE) 1992

The Personal Protective Equipment at Work Regulations

1992 have been amended so that they cover the provision and use of head protection on construction sites thus maintaining the level of legal protection when the Construction (Head Protection) Regulations were revoked as part of the Health and Safety (Miscellaneous Repeals, Revocations and Amendments) Regulations 2013. These measures are being removed because they have either been overtaken by more up to date Regulations, are redundant or do not deliver the intended benefits.

Amendment of the Workplace (Health, Safety and Welfare) Regulations (WHSWR) 1992

The Workplace (Health, Safety and Welfare) Regulations 1992 have been amended to include the requirement for adequate lighting and safe access for workers on ships in a shipyard or harbour undergoing construction, repair or maintenance.

Health and Safety (Safety Signs and Signals) Regulations (SSSR) 1996

Considered in context in Element 2.

ARRANGEMENT OF REGULATIONS

1) Citation and commencement.

2) Interpretation.

3) Application.

4) Provision and maintenance of safety signs.

5) Information, instruction and training.

6) Transitional provisions.

7) Enforcement.

8) Revocations and amendments.

OUTLINE OF MAIN POINTS

The Regulations require employers to provide specific safety signs whenever there is a risk which has not been avoided or controlled by other means, for example, by engineering controls and safe systems of work. Where a safety sign would not help to reduce that risk, or where the sign is not significant, there is no need to provide a sign. They require, where necessary, the use of road traffic signs within workplaces to regulate road traffic.

They also require employers to:

- Maintain the safety signs which are provided by them.
- Explain unfamiliar signs to their employees and tell them what they need to do when they see a safety sign.

The Regulations cover 4 main types of signs:

1) **Prohibition** - circular signs, prime colours red and

white, for example, no pedestrian access.

2) **Warning** - triangular signs, prime colours black on yellow, for example, overhead electrics.

3) **Mandatory** - circular signs, prime colours blue and white, for example, safety helmets must be worn.

4) **Safe condition** - oblong/square signs, prime colours green and white, for example, fire assembly point, first-aid etc.

Supplementary signs provide additional information.

Supplementary signs with yellow/black or red/white diagonal stripes can be used to highlight a hazard, but must not substitute for signs as defined above.

Firefighting, rescue equipment and emergency exit signs have to comply with a separate British Standard.

Ionising Radiations Regulations (IRR) 2017

Considered in context in Element 13.

ARRANGEMENT OF REGULATIONS

PART I - PRELIMINARY

1. Citation and commencement.
2. Interpretation.
3. Application.
4. Duties under the Regulations.

PART II - GENERAL PRINCIPLES AND PROCEDURES

5. Notification of certain work.
6. Registration of certain practices.
7. Consent to carry out specified practices.
8. Radiation risk assessments.
9. Restriction of exposure.
10. Personal protective equipment.
11. Maintenance and examination of engineering controls etc and personal protective equipment.
12. Dose limitation.
13. Contingency plans.

PART III - ARRANGEMENTS FOR THE MANAGEMENT OF RADIATION PROTECTION

14. Radiation protection adviser.
15. Information, instruction and training.
16. Cooperation between employers.

PART IV - DESIGNATED AREAS

17. Designation of controlled or supervised areas.
18. Local rules and radiation protection supervisors.
19. Additional requirements for designated areas.
20. Monitoring of designated areas.

PART V - CLASSIFICATION AND MONITORING OF PERSONS

21. Designation of classified persons.
22. Dose assessment and recording.
23. Estimated doses and special entries.
24. Dosimetry for accidents etc.
25. Medical surveillance.
26. Investigation and notification of overexposure.
27. Dose limitation for overexposed employees.

PART VI - ARRANGEMENTS FOR THE CONTROL OF RADIOACTIVE SUBSTANCES, ARTICLES AND EQUIPMENT

28. Sealed sources and articles containing or embodying radioactive substances.
29. Accounting for radioactive substances.
30. Keeping and moving of radioactive substances.
31. Notification of certain occurrences.
32. Duties of manufacturers etc. of articles for use in work with ionising radiation.
33. Equipment used for medical exposure.
34. Misuse of or interference with sources of ionising radiation.

PART VII - DUTIES OF EMPLOYEES AND MISCELLANEOUS

35. Duties of employees.
36. Approval of dosimetry services.
37. Defence on contravention.
38. Exemption certificates.
39. Extension outside Great Britain.
40. Modifications relating to the ministry of defence etc.
41. Transitional provisions and savings.
42. Modifications and revocation.
43. Review.

Schedule 1 - Work not required to be notified under Regulation 5.

Schedule 2 - Consent to carry out a practice: Indicative list of information.

Schedule 3 - Dose limits.

Schedule 4 - Matters in respect of which a radiation protection adviser must be consulted.

Schedule 5 - Particulars to be entered in the radiation passbook.

Schedule 6 - Particulars to be contained in a health record.

Schedule 7 - Quantities and concentrations of radionuclides.

Schedule 8 - Transitional provisions and savings.

Schedule 9 - Modifications.

OUTLINE OF MAIN POINTS

These regulations supersede and consolidate the Ionising Radiations Regulations 1999. IRR 2017 apply to a large range of workplaces where radioactive substances and electrical equipment emitting ionising radiation are used. They also apply to work with natural radiation, including work in which people are exposed to naturally occurring radon gas and its decay products. Any employer who undertakes work with ionising radiation must comply with IRR 2017. Under IRR 2017 employers who work with ionising radiation are called Radiation Employers.

IRR 2017 requires employers to keep exposure to ionising radiations as low as reasonably practicable. Exposures must not exceed specified dose limits. Restriction of exposure should be achieved first by means of engineering control and design features. Where this is not reasonably practicable employers should introduce safe systems of work and only rely on the provision of personal protective equipment as a last resort.

Lifting Operations and Lifting Equipment Regulations (LOLER) 1998

Considered in context in Elements 7 and 8.

ARRANGEMENTS OF REGULATIONS

1) Citation and commencement.

2) Interpretation.

3) Application.

4) Strength and stability.

5) Lifting equipment for lifting persons.

6) Positioning and installation.

7) Marking of lifting equipment.

8) Organisation of lifting operations.

9) Thorough examination and inspection.

10) Reports and defects.

11) Keeping of information.

12) Exemption for the armed forces.

13) Amendment of the Shipbuilding and Ship-repairing Regulations 1960.

14) Amendment of the Docks Regulation 1988.

15) Repeal of provisions of the Factories Act 1961.

16) Repeal of section 85 of the Mines and Quarries Act 1954.

17) Revocation of instruments.

Schedule 1 - Information to be contained in a report of a thorough examination.

Schedule 2 - Revocation of instruments.

OUTLINE OF MAIN POINTS

The Lifting Operations and Lifting Equipment Regulations (LOLER) 1998 impose health and safety requirements with respect to lifting equipment (as defined in Regulation 2(1)). They are not industry specific and apply to almost all lifting operations.

The Regulations place duties on employers, the self-employed, and certain persons having control of lifting equipment (of persons at work who use or supervise or manage its use, or of the way it is used, to the extent of their control (Regulation 3(3) to (5)). The Regulations make provision with respect to:

- The strength and stability of lifting equipment (Regulation 4).
- The safety of lifting equipment for lifting persons (Regulation 5).
- The way lifting equipment is positioned and installed (Regulation 6).
- The marking of machinery and accessories for lifting, and lifting equipment which is designed for lifting persons or which might so be used in error (Regulation 7).
- The organisation of lifting operations (Regulation 8).
- The thorough examination (defined in (Regulation 2(1)) and inspection of lifting equipment in specified circumstances, (Regulation 9(1) to (3)).
- The evidence of examination to accompany it outside the undertaking (Regulation 9(4)).
- The exception for winding apparatus at mines from Regulation 9 (Regulation 9(5)).
- Transitional arrangements relating to Regulation 9 (Regulation 9(6) and (7)).
- The making of reports of thorough examinations and records of inspections (Regulation 10 and Schedule 1).
- The keeping of information in the reports and records (Regulation 11).

Management of Health and Safety at Work Regulations (MHSWR) 1999

Considered in context in Elements 1,2, and 3.

ARRANGEMENT OF REGULATIONS

1) Citation, commencement and interpretation.

2) Disapplication of these Regulations.

3) Risk assessment.

4) Principles of prevention to be applied.

5) Health and safety arrangements.

6) Health surveillance.

7) Health and safety assistance.

8) Procedures for serious and imminent danger and for danger areas.

9) Contacts with external services.

10) Information for employees.

11) Cooperation and coordination.

12) Persons working in host employers' or self-employed persons' undertakings.

13) Capabilities and training.

14) Employees' duties.

15) Temporary workers.

16) Risk assessment in respect of new or expectant mothers.

17) Certificate from a registered medical practitioner in respect of new or expectant mothers.

18) Notification by new or expectant mothers.

19) Protection of young persons.

20) Exemption certificates.

21) Provisions as to liability.

22) Exclusion of civil liability.

23) Extension outside Great Britain.

24) Amendment of the Health and Safety (First-Aid) Regulations 1981.

25) Amendment of the Offshore Installations and Pipeline Works (First-Aid) Regulations 1989.

26) Amendment of the Mines Miscellaneous Health and Safety Provisions Regulations 1995.

27) Amendment of the Construction (Health, Safety and Welfare) Regulations 1996.

28) Regulations to have effect as health and safety Regulations.

29) Revocations and consequential amendments.

30) Transitional provision.

Schedule 1 - General principles of prevention.

Schedule 2 - Consequential amendments.

OUTLINE OF MAIN POINTS

Management of Health and Safety at Work Regulations (MHSWR) 1999 set out some broad general duties that apply to almost all kinds of work. They are aimed mainly at improving health and safety management. The Regulations work in a similar way to the broad health and safety requirements set out in the Health and Safety at Work etc. Act (HASAWA) 1974, and can be seen as a way of fleshing out what is already in the HASAWA 1974. The principal Regulations are discussed below.

Risk assessment (Regulation 3)

The Regulations require employers (and the self-employed) to assess the risk to the health and safety of their employees and to anyone else who may be affected by their work activity. This is necessary to ensure that the preventive and protective steps can be identified to control hazards in the workplace.

Where an employer is employing or about to employ young persons (under 18 years of age) he must carry out a risk assessment which takes particular account of:

- The inexperience, lack of awareness of risks and immaturity of young persons.
- The layout of the workplace and workstations.
- Exposure to physical, biological and chemical agents.
- Work equipment and the way in which it is handled.
- The extent of health and safety training to be provided.
- Risks from agents, processes and work listed in the Annex to Council Directive 94/33/EC on the protection of young people at work.

Where 5 or more employees are employed, the significant findings of risk assessments must be recorded in writing (the same threshold that is used in respect of having a written safety policy). This record must include details of any employees being identified as being especially at risk.

Principles of prevention to be applied (Regulation 4)

Regulation 4 requires an employer to implement preventive and protective measures on the basis of general principles of prevention specified in Schedule 1 to the Regulations. These are:

1) Avoiding risks.

2) Evaluating the risks which cannot be avoided.

3) Combating the risks at source.

4) Adapting the work to the individual, especially as regards the design of workplaces, the choice of work equipment and the choice of working and production methods, with a view, in particular, to alleviating monotonous work and work at a predetermined work-rate and to reducing their effect on health.

5) Adapting to technical progress.

6) Replacing the dangerous by the non-dangerous or the less dangerous.

7) Developing a coherent overall prevention policy which covers technology, organisation of work, working conditions, social   relationships and the influence of factors relating to the working environment.

8) Giving collective protective measures priority over individual protective measures.

9) Giving appropriate instructions to employees.

Health and safety arrangements (Regulation 5)

Appropriate arrangements must be made for the effective planning, organisation, control, monitoring and review of preventative and protective measures (in other words, for the management of health and safety).

Again, employers with five or more employees must have their arrangements in writing.

Health surveillance (Regulation 6)

In addition to the requirements of other specific Regulations, consideration must be given to carrying out health surveillance of employees, where there is a disease or adverse health condition identified in risk assessments.

Health and safety assistance (Regulation 7)

The employer must appoint one or more competent persons to assist him in complying with the legal obligations imposed on the undertaking. The number of persons appointed should reflect the number of employees and the type of hazards in the workplace.

If more than one competent person is appointed, then arrangements must be made for ensuring adequate cooperation between them. The competent person(s) must be given the necessary time and resources to fulfil their functions. This will depend on the size the undertaking, the risks to which employees are exposed and the distribution of those risks throughout the undertaking.

The employer must ensure that competent person(s) who are not employees are informed of the factors known (or suspected) to affect the health and safety of anyone affected by business activities.

Competent people are defined as those who have sufficient training and experience or knowledge and other qualities to enable them to perform their functions.

Persons may be selected from among existing employees or from outside. Where there is a suitable person in the employer's employment, that person shall be appointed as the 'competent person' in preference to a non-employee.

Procedures for serious and imminent danger and for danger areas (Regulation 8)

Employers are required to set up emergency procedures and appoint **competent persons** to ensure compliance with identified arrangements, to devise control strategies as appropriate and to limit access to areas of risk to ensure that only those persons with adequate health and safety knowledge and instruction are admitted.

The factors to be considered when preparing a procedure to deal with workplace emergencies such as fire, explosion, bomb scare, chemical leakage or other dangerous occurrence should include:

- The identification and training requirements of persons with specific responsibilities.
- The layout of the premises in relation to escape routes etc.
- The number of persons affected.
- Assessment of special needs (disabled persons, children etc.).
- Warning systems.
- Emergency lighting.
- Location of shut off valves, isolation switches, hydrants etc.
- Equipment required to deal with the emergency.
- Location of assembly points.
- Communication with emergency services.
- Training and/or information to be given to employees, visitors, local residents and anyone else who might be affected.

Contacts with external services (Regulation 9)

Employers must ensure that, where necessary, contacts are made with external services. This particularly applies with regard to first-aid, emergency medical care and rescue work.

Information for employees (Regulation 10)

Employees must be provided with relevant information about hazards to their health and safety arising from risks identified by the assessments.

Clear instruction must be provided concerning any preventative or protective control measures including those relating to serious and imminent danger and fire assessments.

Details of any competent persons nominated to discharge specific duties in accordance with the Regulations must also be communicated as should risks arising from contact with other employer's activities (see Regulation 11).

Before employing a child (a person who is not over compulsory school age) the employer must provide those with parental responsibility for the child with information on the risks that have been identified and preventative and protective measures to be taken.

Cooperation and coordination (Regulation 11)

Employers who work together in a common workplace have a duty to cooperate to discharge their duties under relevant statutory provisions. They must also take all reasonable steps to inform their respective employees of risks to their health or safety which may arise out of their work. Specific arrangements must be made to ensure compliance with fire legislation.

Persons working in host employers' or self employed persons' undertakings (Regulation 12)

This Regulation extends the requirements of Regulation 11 to include employees working as sole occupiers of a workplace under the control of another employer. Such employees would include those working under a service of contract and employees in temporary employment businesses under the control of the first employer.

Capabilities and training (Regulation 13)

Employers need to take into account the capabilities of their employees before entrusting tasks. This is necessary to ensure that they have adequate health and safety training and are capable enough at their jobs to avoid risk. To this end, consideration must be given to recruitment including job orientation when transferring between jobs and work departments. Training must also be provided when other factors such as the introduction of new technology and new systems of work or work equipment arise.

Training must:

- Be repeated periodically where appropriate.
- Be adapted to take account of any new or changed risks to the health and safety of the employees concerned.
- Take place during working hours.

Employees' duties (Regulation 14)

Employees are required to follow health and safety instructions by using machinery, substances, transport etc. in accordance with the instructions and training that they have received.

They must also inform their employer (and other employers) of any dangers or shortcoming in the health and safety arrangements, even if there is no risk of imminent danger.

Temporary workers (Regulation 15)

Consideration is given to the special needs of temporary workers. In particular to the provision of particular health and safety information such as qualifications required to perform the task safely or any special arrangements such as the need to provide health screening.

Risks assessment in respect of new or expectant mothers (Regulation 16)

Where the work is of a kind which would involve risk to a new or expectant mother or her baby, then the assessment required by Regulation 3 should take this into account.

If the risk cannot be avoided, then the employer should take reasonable steps to:

- Adjust the hours worked.
- Offer alternative work.
- Give paid leave for as long as is necessary.

Certificate from a registered medical practitioner in respect of new or expectant mothers (Regulation 17)

Where the woman is a night shift worker and has a medical certificate identifying night shift work as a risk then the employer must put her on day shift or give paid leave for as long as is necessary.

Notification by new or expectant mothers (Regulation 18)

The employer need take no action until he is notified in writing by the woman that she is pregnant, has given birth in the last six months, or is breastfeeding.

Protection of young persons (Regulation 19)

Employers of young persons shall ensure that they are not exposed to risk as a consequence of their lack of experience, lack of awareness or lack of maturity.

No employer shall employ young people for work which:

- Is beyond his physical or psychological capacity.
- Involves exposure to agents which chronically affect human health.
- Involves harmful exposure to radiation.
- Involves a risk to health from extremes of temperature, noise or vibration.
- Involves risks which could not be reasonably foreseen by young persons.

This Regulation does not prevent the employment of a young person who is no longer a child for work:

- Where it is necessary for his training.
- Where the young person will be supervised by a competent person.
- Where any risk will be reduced to the lowest level that is reasonably practicable.

(Note: Two HSE publications give guidance on these topics. HSG122 - New and expectant mothers at work: A guide for employers and HSG165 - Young people at work: A guide for employers).

Exemption certificates (Regulation 20)

The Secretary of State for Defence may, in the interests of national security, by a certificate in writing exempt the armed forces, any visiting force or any headquarters from certain obligations imposed by the Regulations.

Provisions as to liability (Regulation 21)

Employers cannot submit a defence in criminal proceedings that contravention was caused by the act or default either of an employee or the competent person appointed under Regulation 7.

Exclusion of civil liability (Regulation 22)

As amended by Health and Safety at Work etc. Act 1974 (Civil Liability) (Exceptions) Regulations 2013):

Regulation 22 specifies:

"(1)   Breach of a duty imposed by Regulation 16, 16A, 17 or 17A shall, so far as it causes damage, be actionable by the new or   expectant mother.

(2)   Any term of an agreement which purports to exclude or restrict any liability for such a breach is void."

Revocations and amendments (Regulations 24-29)

The Regulations:

- Revoke Regulation 6 of the Health and Safety (First-Aid) Regulations (FAR) 1981 which confers power on the Health and Safety Executive to grant exemptions from those Regulations.
- Amend the Offshore Installations and Pipeline Works (First-Aid) Regulations 1989.
- Amend the Mines Miscellaneous Health and Safety Provisions Regulations 2013.

The Regulations also make amendments to the statutory instruments as specified in Schedule 2.

Manual Handling Operations Regulations (MHOR) 1992

Considered in context in Element 8.
See also - Health and Safety (Miscellaneous Amendments) Regulations (MAR) 2002.

ARRANGEMENT OF REGULATIONS

1) Citation and commencement.
2) Interpretation.
3) Disapplication of Regulations.
4) Duties of employers.
5) Duty of employees.
6) Exemption certificates.
7) Extension outside Great Britain.
8) Repeals and revocations.

OUTLINE OF MAIN POINTS

Citation and commencement (1)

Interpretation (2)

'Injury' does not include injury caused by toxic or corrosive substances which:

- Have leaked/spilled from load.
- Are present on the surface but not leaked/spilled from it.
- Are a constituent part of the load.

'Load' includes any person or animal.

'Manual Handling Operations' means transporting or supporting a load including:

- Lifting and putting down.
- Pushing, pulling or moving by hand or bodily force.
- Shall as far as is reasonably practicable.

Disapplication of regulations (3)

Duties of employers (4)

Avoidance of manual handling (4) (1)(a)

The employer's duty is to avoid the need for manual handling operations which involve a risk of their employees being injured - as far as is reasonably practicable.

Assessment of risk (4) (1)(b)(i)

Where not reasonably practicable make a suitable and sufficient assessment of all such manual handling operations.

Reducing the risk of injury (4) (1)(b)(ii)

Take appropriate steps to reduce the risk of injury to the lowest level reasonably practicable.

The load - additional information (4) (1)(b)(iii)

Employers shall provide information on general indications or where reasonably practicable precise information on:

- The weight of each load.
- The heaviest side of any load whose centre of gravity is not central.

Reviewing the assessment (4) (2)

Assessment review:

- Where there is reason to believe the assessment is no longer valid.
- There is sufficient change in manual handling operations.

Duty of employees (5)

Employees shall make full and proper use of any system of work provided for his use by his employer.

Exemption certificates (6)

Extension outside Great Britain (7)

Repeals and revocations (8)

Schedules

Schedule 1 - Factors to which the employer must have regard and questions he must consider when making an assessment of manual handling operations.

Schedule 2 - Repeals and revocations.

Appendix 1 - Numerical guidelines for assessment.

Appendix 2 - Example of an assessment checklist.

Thus the Regulations establish a clear hierarchy of measures:

1. Avoid hazardous manual handling operations so far as is reasonably practicable.

2. Make a suitable and sufficient assessment of any hazardous manual handling operations that cannot be avoided.

3. Reduce the risk of injury so far as is reasonably practicable.

New Roads and Street Works Act (NRSWA) 1991

ARRANGEMENT OF ACT

'Part 1' - New Roads in England and Wales

- Concession agreements.
- Toll orders.
- Further provisions with respect to tolls.
- Annual report.
- Miscellaneous.
- General.

'Part II' - New Roads in Scotland

- Toll Roads
- Further provision with respect to tolls
- Report
- Supplementary provisions

'Part III' - Streetworks in England and Wales

- Introductory provisions.
- The street works register.
- Notice and coordination of works.

- Streets subject to special controls.
- General requirements as to execution of street works.
- Reinstatement.
- Charges, fees and contributions payable by undertakers.
- Duties and liabilities of undertakers with respect to apparatus.
- Apparatus affected by highway, bridge or transport works.
- Provisions with respect to particular authorities and undertakings.
- Power of street authority or district council to undertake street works.
- Supplementary provisions.

Part IV - Road Works in Scotland

- Introductory provisions.
- The road works register.
- Notice and coordination of works.
- Roads subject to special controls.
- General requirements as to execution of road works.
- Reinstatement.
- Charges, fees and contributions payable by undertakers.
- Duties and liabilities of undertakers with respect to apparatus.
- Apparatus affected by road, bridge or transport works.
- Provisions with respect to particular authorities and undertakings.
- Power of road works authority or district council to undertake road works.
- Supplementary provisions.

Part V - General

Schedules

Schedule 1 - Supplementary provisions as to termination of concession.

Schedule 2 - Procedure in connection with toll orders.

Schedule 3 - Street works licences.

Schedule 4 - Streets with special engineering difficulties.

Schedule 5 - Procedure for making certain orders under Part III.

Schedule 6 - Roads with special engineering difficulties.

Schedule 7 - Procedure for making certain orders under Part IV.

Schedule 8 - Minor and consequential amendments.

Schedule 9 - Repeals.

OUTLINE OF MAIN POINTS

One of the most important elements of the street works legislation is the duty on street authorities to coordinate all works in the highway. As important is the parallel duty on undertakers to co-operate in this process. It is essential that both street authorities and undertakers take these responsibilities seriously. The New Roads and Street Works Act 1991 spells out the objectives of the coordination function.

They are to:

- Ensure safety.
- Minimise inconvenience to people using a street, including a specific reference to people with a disability.
- Protect the structure of the street and apparatus in it.

These objectives should be taken into account by everyone responsible for planning and carrying out works in the highway.

The NRSWA 1991 sets out three requirements for a successful coordination framework:

- The notice system - the notices themselves provide vital information to aid the coordination process, while the notice periods provide time within which appropriate steps can be taken. The best method of exchanging notices is electronically.

- Streets subject to special controls - these designation procedures provide a mechanism by which attention can be focused on particularly sensitive streets.

- The coordination tools - the Act provides a range of tools to facilitate the coordination process. These include: the power to direct the timing of street works, the power to restrict street works following substantial road works and the requirement on undertakers to avoid unnecessary delay and obstruction.

There are three important principles to which undertakers and street authorities must adhere if coordination is to be effective. They are:

- The need to balance the potentially conflicting interests of road users and undertakers' customers.

- The importance of close co-operation and liaison between street authorities and undertakers.

- An acknowledgement on all sides of the fact that works programmes and practices may have to be adjusted in order to ensure that the objectives of the coordination provisions are achieved.

In carrying out their responsibilities under the NRSWA 1991 street authorities and undertakers should endeavour to ensure that their works are planned in such a way as to minimise inconvenience to all road users including disability groups.

Authorities and undertakers are expected to work particularly closely in relation to 'urgent work', which is that work that falls short of emergencies as defined by the NRSWA 1991.

These works will need to be executed in circumstances where planning and coordination time is limited. Where work is needed in traffic sensitive streets considerable disruption may result and both parties are expected to take an objective but balanced view on the necessity of the work and arrangements surrounding it.

Personal Protective Equipment at Work Regulations (PPER) 1992 (as amended)

See also - Health and Safety (Miscellaneous Amendments) Regulations (MAR) 2002 and Health and Safety (Miscellaneous Repeals, Revocations and Amendments) Regulations (MRRA) 2002.

Considered in context in Element 2.

1) Citation and commencement.

2) Interpretation.

3) Disapplication of these Regulations.

4) Provision of personal protective equipment.

5) Compatibility of personal protective equipment.

6) Assessment of personal protective equipment.

7) Maintenance and replacement of personal protective equipment.

8) Accommodation for personal protective equipment.

9) Information, instruction and training.

10) Use of personal protective equipment.

11) Reporting loss or defect.

12) Exemption certificates.

13) Extension outside Great Britain.

14) Modifications, repeal and revocations directive.

Schedule 1 - Relevant Community

Schedule 2 - Modifications

Schedule 3 - Revocations

OUTLINE OF MAIN POINTS

2) Personal protective equipment (PPE) means all equipment (including clothing provided for protection against adverse weather) which is intended to be worn or held by a person at work and which protects him against risks to his health or safety.

3) These Regulations do not apply to:
- Ordinary working clothes/uniforms.
- Offensive weapons.
- Portable detectors which signal risk.
- Equipment used whilst playing competitive sports.
- Equipment provided for travelling on a road.

The Regulations do not apply to situations already controlled by other Regulations i.e.
- Control of Lead at Work Regulations 2002.
- Ionising Radiation Regulations 1999.
- Control of Asbestos at Work Regulations 2002.
- COSHH Regulations 2002 (as amended).
- Control of Noise at Work Regulations 2005.

4) Suitable PPE must be provided when risks cannot be adequately controlled by other means. Reg. 4 requires that PPE will not be suitable unless it:

- Is appropriate for the risk and conditions.
- It takes account of ergonomic requirements.
- It takes account of the state of health of users.
- It takes account of the characteristics of the worker's workstation.
- Is capable of fitting the wearer, if required after adjustment.
- Is effective in controlling risks, without increase in overall risk.
- Complies with EU directives.

Where it is necessary to ensure hygiene or prevention of health risk personal issue will be made.

5) Equipment must be compatible with any other PPE which has to be worn.

6) Before issuing PPE, the employer must carry out a risk assessment to ensure that the equipment is suitable.
- Assess risks not avoided by other means.
- Define characteristics of PPE and of the risk of the equipment itself.
- Compare characteristics of PPE to defined requirement.
- Repeat assessment when no longer valid, or significant change has taken place.

7) PPE must be maintained.
- In an efficient state.
- In efficient working order.
- In good repair.

8) Accommodation must be provided for equipment when it is not being used.

9) Information, instruction and training must be given on:

- The risks PPE will eliminate or limit.
- Why the PPE is to be used.
- How the PPE is to be used.
- How to maintain the PPE.

Information and instruction must be comprehensible to the wearer/user and kept available to them.

10) Employers shall take reasonable steps to ensure PPE is worn.
- Every employee shall use PPE that has been provided.
- Every employee shall take reasonable steps to return PPE to storage.

11) Employees must report any loss or defect.

The Guidance on the Regulations points out:

"Whatever PPE is chosen, it should be remembered that, although some types of equipment do provide very high levels of protection, none provides 100%."

From 6 April 2022, existing PPE regulations were amended, and the Personal Protective Equipment at Work (Amendment) Regulations 2022 came into force. They extend employers' and employees' duties regarding PPE to a wider group of workers. These are defined as 'limb (b) workers' to include those who have more casual employment relationships than employees. This means businesses will be required to supply their PPE and the workers themselves will be required to use it in line with the training you provide.

Typically, the limb (b) workers who will now also fall within the PPE regulations will:

- Carry out casual or irregular work for one or more organisations.

- After one month of continuous service, receive holiday pay but no other employment rights such as the minimum period of statutory notice.

- Only carry out work if they choose to.

- Have a contract or other arrangement todo work or services personally for a reward (the contract doesn't have to be written) and only have a limited right to send someone else to do the work, for example swapping shifts with someone on a pre-approved list (subcontracting).

- Not be in business for themselves (they do not advertise services directly to customers who can then also book their services directly).

Note the changes will not affect self-employed individuals in normal contracted for service arrangements such as electricians, handymen and gardeners.

Provision and Use of Work Equipment Regulations (PUWER) 1998

See also - Health and Safety (Miscellaneous Amendments) Regulations (MAR) 2002.
Considered in context in Element 9.

ARRANGEMENT OF REGULATIONS

PART I - INTRODUCTION

1) Citation and commencement.
2) Interpretation.
3) Application.

PART II - GENERAL

4) Suitability of work equipment.
5) Maintenance.
6) Inspection.
7) Specific risks.
8) Information and instructions.
9) Training.
10) Conformity with Community requirements.
11) Dangerous parts of machinery.
12) Protection against specified hazards.
13) High or very low temperature.
14) Controls for starting or making a significant change in operating conditions.
15) Stop controls.
16) Emergency stop controls.
17) Controls.
18) Control systems.
19) Isolation from sources of energy.
20) Stability.
21) Lighting.
22) Maintenance operations.
23) Markings.
24) Warnings.

PART III - MOBILE WORK EQUIPMENT

25) Employees carried on mobile work equipment.
26) Rolling over of mobile work equipment.
27) Overturning of fork-lift trucks.
28) Self-propelled work equipment.
29) Remote-controlled self-propelled work equipment.
30) Drive shafts.

PART IV - POWER PRESSES

31) Power presses to which Part IV does not apply.
32) Thorough examination of power presses, guards and protection devices.
33) Inspection of guards and protection devices.
34) Reports.
35) Keeping of information.

PART V - MISCELLANEOUS

36) Exemption for the armed forces.
37) Transitional provision.
38) Repeal of enactment.
39) Revocation of instruments.

Schedule 1 - Instruments which give effect to Community directives concerning the safety of products.

Schedule 2 - Power presses to which Regulations 32 to 35 do not apply.

Schedule 3 - Information to be contained in a report of a thorough examination of a power press, guard or protection device.

Schedule 4 - Revocation of instruments.

OUTLINE OF MAIN POINTS

General requirements

PUWER impose health and safety requirements with respect to the provision and use of work equipment, which is defined as 'any machinery, appliance, apparatus, tool or installation for use at work (whether exclusively or not)'.

The Regulations:

- Place general duties on employers.
- Certain persons having control of work equipment, of persons at work who use or supervise or manage its use or of the way it is used, to the extent of their control.
- List minimum requirements for work equipment to deal with selected hazards whatever the industry.
- 'Use' includes any activity involving work equipment and includes starting, stopping, programming, setting, transporting, repairing, modifying, maintaining, servicing and cleaning.

The general duties require the employer to:

- Make sure that equipment is suitable for the use that will be made of it.
- Take into account the working conditions and hazards in the workplace when selecting equipment.
- Ensure equipment is used only for operations for which, and under conditions for which, it is suitable.

- Ensure that equipment is maintained in an efficient state, in efficient working order and in good repair.
- Ensure the inspection of work equipment in specified circumstances by a competent person; keep a record of the result for specified periods; and ensure that evidence of the last inspection accompany work equipment used outside the undertaking.
- Give adequate information, instruction and training.
- Provide equipment that conforms to EU product safety directives.

Specific requirements cover

- Guarding of dangerous parts of machinery.
- Protection against specified hazards i.e. articles and substances falling/ejected, rupture/disintegration of work equipment parts, equipment catching fire or overheating, unintended or premature discharge of articles and substances, explosion.
- Work equipment parts and substances at high or very low temperatures.
- Control systems and control devices.
- Isolation of equipment from sources of energy.
- Stability of equipment.
- Lighting.
- Maintenance operations.
- Warnings and markings.

Mobile work equipment

Mobile work equipment must have provision as to:

- Its suitability for carrying persons and its safety features.
- Means to minimise the risk to safety from its rolling over.
- Means to reduce the risk to safety from the rolling over of a fork-lift truck.
- The safety of self-propelled work equipment and remote-controlled self-propelled work equipment.
- The drive shafts of mobile work equipment.

Power presses

The Regulations provide for:

- The thorough examination (defined in Regulation 2(1)) of power presses and their guards and protection devices (Regulation 32).
- Their inspection after setting, re-setting or adjustment of their tools, and every working period (Regulation 33).
- The making (Regulation 34 and Schedule 3) and keeping (Regulation 35) of reports.
- The Regulations implement an EU directive aimed at the protection of workers. There are other directives setting out conditions which new equipment (especially machinery) will have to satisfy before it can be sold in EU member states.

Registration, Evaluation, Authorisation and Restriction of Chemicals (REACH) EC Regulation

Considered in context in Element 12.

REACH (Registration, Evaluation, Authorisation and restriction of CHemicals) is the system for controlling chemicals in the EU. REACH is an EU Regulation that applies throughout the EU, including the UK. It has been adopted to improve the protection of human health and the environment from the risks that can be posed by chemicals. In principle, REACH applies to all chemical substances; not only those used in industrial processes, but also used by the public, for example in cleaning products, paints as well as in articles such as clothes, furniture and electrical appliances. Therefore, the Regulation has an impact on most companies across the EU. It also promotes alternative methods for the hazard assessment of substances in order to reduce the number of tests on animals. REACH provides a common EU system to gather hazard information, assess risks, classify, label, and restrict the marketing and use of individual chemicals and mixtures. This is known as the REACH system:

R egistration of basic information of substances to be submitted by companies, to a central database.

E valuation of the registered information to determine hazards and risks.

A uthorisation and restriction requirements imposed on the use of substances of very high concern.

CH emicals.

REACH establishes procedures for collecting and assessing information on the properties and hazards of substances. Companies need to register their substances and to do this they need to work together with other companies who are registering the same substance. The European Chemicals Agency (the Agency; ECHA) receives and evaluates individual registrations for their compliance, and the EU Member States evaluate selected substances to clarify initial concerns for human health or for the environment. Authorities in member states and ECHA's scientific committees assess whether the risks of substances can be managed.

Authorities can ban hazardous substances if their risks are unmanageable. They can also decide to restrict a use or make it subject to a prior authorisation. If the risks cannot be managed, authorities can restrict the use of substances in different ways. In time, the most hazardous substances should be substituted with less dangerous ones. REACH requires information, in the form of Safety data sheets, to be provided to users by those that supply chemicals.

'Supply' means making a chemical available to another person. Manufacturers, importers, distributors, wholesalers and retailers are all examples of suppliers.

ARRANGEMENT OF REGULATIONS

TITLE I - GENERAL ISSUES

Chapter 1 - Aim, scope and application.

Chapter 2 - Definitions and general provision.

TITLE II - REGISTRATION OF SUBSTANCES

Chapter 1 - General obligation to register and information requirements.

Chapter 2 - Substances regarded as being registered.

Chapter 3 - Obligation to register and information requirements for certain types of isolated intermediates.

Chapter 4 - Common provisions for all registrations.

Chapter 5 - Transitional provisions applicable to phase-in substances and notified substances.

TITLE III - DATA SHARING AND AVOIDANCE OF UNNECESSARY TESTING

Chapter 1 - Objectives and general rules.

Chapter 2 - Rules for non-phase-in substances and registrants of phase-in substances who have not pre-registered.

Chapter 3 - Rules for phase-in-substances.

TITLE IV - INFORMATION IN THE SUPPLY CHAIN

TITLE V - DOWNSTREAM USERS

TITLE VI - EVALUATION

Chapter 1 - Dossier evaluation.

Chapter 2 - Substance evaluation.

Chapter 3 - Evaluation of intermediates.

Chapter 4 - Common provisions.

TITLE VII - AUTHORISATION

Chapter 1 - Authorisation requirement.

Chapter 2 - Granting of authorisations.

Chapter 3 - Authorisations in the supply chain.

TITLE VIII - RESTRICTIONS ON THE MANUFACTURING, PLACING ON THE MARKET AND USE OF CERTAIN DANGEROUS SUBSTANCES AND MIXTURES

Chapter 1 - General issues.

Chapter 2 - Restrictions process.

TITLE IX - FEES AND CHARGES

TITLE X - AGENCY

TITLE XII - INFORMATION

TITLE XIII - COMPETENT AUTHORITIES

TITLE XIV - ENFORCEMENT

TITLE XV - TRANSITIONAL AND FINAL PROVISIONS

Annex I - General provisions for assessing substances and preparing chemical safety reports.

Annex II - Requirements for the compilation of safety data sheets.

Annex III - Criteria for substances registered in quantities between 1 and 10 tonnes.

Annex IV - Exemptions from the obligation to register in accordance with article 2(7)(a).

Annex V - Exemptions from the obligation to register in accordance with article 2(7)(b).

Annex VI - Information requirements referred to in article 10.

Annex VII - Standard information requirements for substances manufactured or imported in quantities of one tonne or more.

Annex VIII - Standard information requirements for substances manufactured or imported in quantities of 10 tonnes or more.

Annex IX - Standard information requirements for substances manufactured or imported in quantities of 100 tonnes or more.

Annex X - Standard information requirements for substances manufactured or imported in quantities of 1,000 tonnes or more.

Annex XI - General rules for adaptation of the standard testing regime set out in annexes VII to X.

Annex XII - General provisions for downstream users to assess substances and prepare chemical safety reports.

Annex XIII - Criteria for the identification of persistent, bioaccumulative and toxic substances, and very persistent and very bioaccumulative substances.

Annex XIV - List of substances subject to authorisation.

Annex XV - Dossiers.

Annex XVI - Socio-economic analysis.

Annex XVII - Restrictions on the manufacture, placing on the market and use of certain dangerous substances, mixtures and articles.

OUTLINE OF MAIN POINTS

Scope and exemptions

REACH applies to substances manufactured or imported into the EU in quantities of 1 tonne or more per year. Generally, it applies to all individual chemical substances on their own, in preparations or in articles (if the substance is intended to be released during normal and reasonably foreseeable conditions of use from an article).

Some substances are specifically excluded:

- Radioactive substances.
- Substances under customs supervision.
- The transport of substances.
- Non-isolated intermediates.
- Waste.
- Some naturally occurring low-hazard substances.

Some substances, covered by more specific legislation, have tailored provisions, including:

- Human and veterinary medicines.
- Food and foodstuff additives.
- Plant protection products and biocides.

Other substances have tailored provisions within the REACH legislation, as long they are used in specified conditions:

- Isolated intermediates.
- Substances used for research and development.

REACH places the burden of proof of compliance on companies. To comply with the Regulation, companies must identify and manage the risks linked to the substances they manufacture and supply in the EU. They have to demonstrate to ECHA how the substance can be safely and healthily used, and they must communicate the risk management measures to the users.

Registration

Companies have the responsibility of collecting information on the properties and the uses of substances that they manufacture or import at or above one tonne per year. They also have to make an assessment of the hazards and potential risks presented by the substance. They have to register this by submitting a dossier to ECHA, based in Helsinki. The dossier contains details of the substance's properties, other relevant information about risks and how these risks can be managed. They will not be able to manufacture or import a substance within the EU, or import an article that intentionally releases a substance, unless the substance has been registered.

Evaluation

ECHA and the Member States evaluate the information submitted by companies to examine the quality of the registration dossiers and the testing proposals and to clarify if a given substance constitutes a risk to human health or the environment. REACH promotes the use of alternative methods for the assessment of the hazardous properties of substances in order to reduce the number of tests on animals, for example, quantitative structure-activity relationships (QSAR) and read across.

Authorisation

The authorisation procedure aims to assure that the risks from Substances of Very High Concern (SVHC) are properly controlled and that these substances are progressively replaced by suitable alternatives, while ensuring the good functioning of the EU internal market. SVHCs will need to be authorised for specific uses if they appear in Annex XIV of REACH. There will not be a 'blanket' authorisation for a substance to be used generally. Instead, applications for authorisation for a specific use may be made by companies that register the substances, or by those that use them. When a substance is placed on Annex XIV of REACH, a 'sunset date' will be set after which its use will be prohibited, unless an authorisation has been granted.

Restriction

Restrictions are set to protect human health and the environment from unacceptable risks posed by chemicals. Restrictions may limit or ban the manufacture, placing on the market or use of a substance. This is a direct means of controlling the risks associated with any given hazardous substance. Restriction will be used when it is felt that action at the European level is needed. Restriction decisions will also be made by the European Commission based on advice from the ECHA and consultation with EU Member States and others. Restrictions in place already from previous legislation are carried over into REACH in Annex XVII of REACH; further restrictions will be added to this Annex as necessary.

Information

The passage of information up and down the supply chain is one of the important features of REACH - users should be able to understand what manufacturers and importers know about the dangers involved in using chemicals, and they can also pass information back up the supply chain. REACH adopts and builds on the existing system for passing information in a structured way down to chemicals users - the Safety Data Sheet (SDS). This should accompany materials down through the supply chain, providing the information that users need to ensure chemicals are safely managed. REACH will also allow for information on uses of chemicals to be passed back up the supply chain, so that these can be reflected in the SDS.

Users of chemicals - Article 37 - 39 particularly

REACH aims to reduce the use of (and risks from) hazardous chemicals. Certain substances will be identified as substances of very high concern (SVHC), and may become subject to 'authorisation'. Substances that are subject to authorisation cannot be used in the EU unless a company (and their identified users) have been authorised to do so. This will mean that these substances are eventually phased out of all non-essential uses. Other substances are subject to a 'restriction'. Restrictions limit or ban the manufacture, placing on the market or use of substances that pose an unacceptable risk to human health and the environment.

REACH may make things better for users of chemicals as it is designed to provide more information on chemicals and increase confidence in their safe and healthy use. In particular, better information on the hazards of chemicals and how to use them safely and healthily will be passed down the supply chain by chemical manufacturers and importers through improved Safety Data Sheets. Users have a duty to apply risk reduction measures identified in Safety Data Sheets provided to them form chemicals (Article 07(5)).

If users of chemicals use a chemical in a novel way that is perhaps not expected, then they will need to consider letting their supplier know. This use will need to be considered for registration by the supplier. If users do not want to let your supplier know about this use (for example, because of commercial concerns) then they do not have to, but it will mean that they will have to let ECHA know about this use and possibly have to submit their own risk assessment. Once the chemical has been registered and they have been provided with a safety data sheet listing the registration number, they have a maximum of 6 months to provide information about their use to ECHA if it is not included in the supplier's registration. The user's supplier should be able to tell them which uses are covered by the registration. Users of hazardous substances should check whether any of the substances they use are subject to a 'restriction'. This information should be communicated to users in the safety data sheet or other communication given to them by their supplier. If users use restricted substances, they will need to ensure their use of the substance meets the conditions of the restriction. Users are able to contribute to the public consultations whenever a new restriction is proposed, or an existing restriction is amended. Users of hazardous substances should check whether any of the substances they use are subject to authorisation, or are likely to become subject to authorisation in the future. If a substance is subject to authorisation, users will only be able to use the substance if they, or someone else further up the supply chain, have been granted an authorisation for that use. Information about whether a substance is subject to authorisation, and whether an authorisation has been granted or refused, should be communicated to users in the safety data sheet or other communication given by the supplier.

Safety data sheets under REACH - Article 31 particularly

Once chemicals are registered, safety data sheets will list registration numbers. Safety data sheets may also include exposure scenarios. An exposure scenario describes the operating conditions and risk management measures that have been identified by the supplier as necessary to use the chemical safely and healthily in the users processes. Under Article 31 of REACH the supplier of a substance or mixture must provide 'the recipient', including end users, with a safety data sheet compiled in accordance with Annex II of REACH. Safety Data Sheets need not be supplied where substances are provided to the general public and they are provided with sufficient information to enable them to take the necessary measures, unless they are requested by a distributor or user. REACH requires users to follow the advice on risk management measures given in the exposure scenario attached to the safety data sheet. If users use different risk management measures to those described in the exposure scenario, then they should be able to justify why their measures offer an equivalent (or better) level of protection for human health and the environment to those described in the exposure scenario. If users need to change the risk management measures used at their workplace in order to comply with the exposure scenario then they have a maximum of 12 months (after they have received a safety data sheet listing a registration number for a chemical) to do this.

REACH, Article 31 and Annex II, specifies the scope of the content of safety data sheets for chemicals:

1. Identification of the substance/mixture and of the company/undertaking.

2. Hazards identification.

3. Composition/information on ingredients.

4. First-aid measures.

5. Firefighting measures.

6. Accidental release measures.

7. Handling and storage.

8. Exposure controls/personal protection.

9. Physical and chemical properties.

10. Stability and reactivity.

11. Toxicological information.

12. Ecological information.

13.   Disposal considerations.

14.   Transport information.

15.   Regulatory information.

16.   Other information.

Regulatory Reform (Fire Safety) Order (RRFSO) 2005

Considered in context in Element 11.

Introduction

The amount of legislation covering the risk of fire has grown considerably over time.

The situation was identified as unwieldy; many different Regulations existed, often with conflicting definitions and requirements.

In order to simply this and remove confusion the Regulatory Reform (Fire Safety) Order 2005 was introduced. There were four principal pieces of legislation that covered fire safety in the workplace that have been affected by the RRFSO:

- Fire Precautions Act (FPA).
- Fire Precautions (Workplace) Regulations (FPWR).
- Management of Health and Safety at Work Regulations (MHSWR).
- Dangerous Substances and Explosive Atmosphere Regulations (DSEAR).

Fire Precautions Act

This legislation has been repealed by the RRFSO 2005.

Fire Precautions Workplace Regulations

These Regulations have been repealed by the implementation of the RRFSO, however their content has been incorporated within the RRFSO 2005.

Management of Health and Safety at Work Regulations

It is this Regulation that makes the legal requirement for risk assessments. In addition, it made various requirements for the management of fire safety within workplaces, which have now been revoked. This Regulation continues as a stand-alone health and safety Regulation as the relevant fire aspects of this Regulation have been incorporated within the RRFSO 2005.

Dangerous Substances and Explosive Atmosphere Regulations

This Regulation outlines the safety and control measures that need to be taken if dangerous or flammable/explosive substances are present. This Regulation will continue as a stand-alone health and safety Regulation. Again the relevant fire aspects of this Regulation have been incorporated within the RRFSO 2005.

Regulatory Reform (Fire Safety) Order 2005

This is an all encompassing fire safety order, which came into force in England and Wales on 1st October 2006. As shown earlier it has aspects of other legislation within it and is compiled in such a way as to present a cohesive structure for fire safety legislation. The order is split into five parts, each part is then subdivided into the individual points, or articles as they are called in the Order:

- Part 1   General.
- Part 2   Fire safety duties.
- Part 3   Enforcement.
- Part 4   Offences and appeals.
- Part 5   Miscellaneous.

OUTLINE OF MAIN POINTS

PART 1 - GENERAL

This part covers various issues such as the interpretation of terminology used, definition of responsible person, definition of general fire precautions, duties under the order, and its application.

PART 2 - FIRE SAFETY DUTIES

This part imposes a duty on the responsible person to carry out a fire risk assessment to identify what the necessary general fire precautions should be. It also outlines the principles of prevention that should be applied and the necessary arrangements for the management of fire safety. The following areas are also covered:

- Firefighting and fire detection.
- Emergency routes and exits.
- Procedures for serious and imminent danger and for danger areas.
- Additional emergency measures re dangerous substances.
- Maintenance.
- Safety assistance.
- Provision of information to employees, employers and self-employed.
- Capabilities and training.
- Cooperation and coordination.
- General duties of employees.

PART 3 - ENFORCEMENT

This part details who the enforcing authority is, (which in the main is the Fire Authority), and it states they must enforce the order. It also details the powers of inspectors. It also details the different types of enforcement that can be taken:

- Alterations notice.
- Enforcement notice.
- Prohibition notice.

PART 4 - OFFENCES AND APPEALS

This part details the 13 offences that may occur and the subsequent punishments and appeals procedure. It also explains that the legal onus for proving that an offence was not committed is on the accused. A disputes procedure is also outlined within this part.

PART 5 - MISCELLANEOUS

Various matters are covered within this part, the principal points being:

- 'Firefighters switches' for luminous tube signs etc.
- Maintenance of measures provided for the protection of firefighters.
- Civil liability.
- Duty to consult employees.
- Special provisions for licensed premises.
- Application to crown premises.

There is then a schedule that covers the risk assessment process.

Reporting of Injuries, Diseases and Dangerous Occurrences Regulations (RIDDOR) 2013

Considered in context in Element 3.

ARRANGEMENT OF REGULATIONS

1) Citation and commencement.
2) Interpretation.
3) Responsible person.
4) Non-fatal injuries to workers.
5) Non-fatal injuries to non-workers.
6) Work-related fatalities.
7) Dangerous occurrences.
8) Occupational diseases.
9) Exposure to carcinogens, mutagens and biological agents.
10) Diseases offshore.
11) Gas-related injuries and hazards.
12) Recording and record-keeping.
13) Mines, quarries and offshore site disturbance.
14) Restrictions on the application of Regulations 4 to 10.
15) Restriction on parallel requirements.
16) Defence.
17) Certificates of exemption.
18) Revocations, amendments and savings.
19) Extension outside Great Britain.
20) Review.

Responsible person

Reportable event	To	Responsible person
Death, specified non-fatal injury, over 7 day injury, disease.	Employee.	Employer.
	Self-employed person working in someone else's premises.	Person in control of the premises: • At the time of the event. • In connection with trade, business or undertaking.
Specified injury, over 7 day injury, disease.	Self-employed in own premises.	Self-employed person or someone acting for them.
Death, being taken to hospital or a specified non-fatal injury on hospital premises.	A person not at work.	Person in control of the premises: • At the time of the event. • In connection with trade, business or undertaking.
Dangerous occurrences - general.		Person in control of the premises where, or in connection with the work going on at which, the dangerous occurrence happened: • At the time of the event. • In connection with trade, business or undertaking.

Schedules

Schedule 1 - Reporting and Recording Procedures.

Schedule 2 - Dangerous Occurrences.

Schedule 3 - Diseases Reportable Offshore.

Schedule 4 - Revocations and Amendments.

OUTLINE OF MAIN POINTS

The Reporting of Injuries, Diseases and Dangerous Occurrences Regulations (RIDDOR) 2013 covers the requirement to report certain categories of injury and disease sustained at work, along with specified dangerous occurrences and gas incidents, to the relevant enforcing authority. These reports are used to compile statistics to show trends and to highlight problem areas, in particular industries or companies.

The main points of RIDDOR

Reporting

1) When a person *dies or suffers any specified injury* listed in Regulation 4 *(non-fatal injuries to workers)* as a result of a work-related accident or dies as a result of occupational exposure to a biological agent or an incident occurs of the type listed as a dangerous occurrences in Schedule 2 the responsible person must notify the relevant enforcing authority by the quickest practicable means (usually by telephone) without delay and must send them a report in an approved manner (online) within 10 days. This therefore includes accidents connected with work where:

   - An employee or a self-employed person at work is killed or suffers a specified injury (including as a result of physical violence).
   - A member of the public is killed or taken to hospital.

A work-related *'accident'* in the context of RIDDOR 2013 includes an act of non-consensual physical violence done to a person at work.

2) In cases of work-related diseases that are listed in Regulation 8 and 10 the responsible person must send a report of the diagnosis in an approved manner (online) to the relevant enforcing authorities without delay. In cases of diseases related to carcinogens, mutagens and biological agents that are listed in Regulation 9 the responsible person must notify the relevant enforcing authority in an approved manner.

3) If personal injury results in *more than 7 days (excluding the day of the accident) incapacity* for routine work, but is not one of the specified non-fatal injuries, the responsible person must send a report to the relevant enforcing authority in an approved manner (online) as soon as is practicable and in any event within *15 days* of the accident. The day of the accident is not counted, but any days which would not have been working days are included.

4) If there is an accident connected with work (including an act of physical violence) and an employee, or a self-employed person at work, suffers an over-three-day injury it must be recorded by the employer.

5) The enforcing authority for most workplaces is either the Health and Safety Executive or the Local Authority, for railway operations it is the Office of Rail Regulation (ORR).

Road traffic accidents

Road traffic accidents only have to be reported if:

- Death or injury results from an accident involving a train.
- Death or injury results from exposure to a substance being conveyed by a vehicle.
- Death or injury results from the person being engaged in work connected with loading or unloading of any article/substance or results from another person engaged in these activities.
- Death or injury results from the person being engaged in work on or alongside a road or results from another person engaged in these activities.

Work on or alongside a road means work concerned with the construction, demolition, alteration, repair or maintenance of:

- The road or the markings or equipment on the road.
- The verges, fences, hedges or other boundaries of the road.
- Pipes or cables on, under, over or adjacent to the road.
- Buildings or structures adjacent to or over the road.

Non-employee

The responsible person must not only report non-employee deaths, but also cases that involve major injury or being taken to hospital if caused by an accident out of or in connection with their work.

Employee death

Where an employee dies within one year of the date of an accident, as a result of a reportable injury, as soon as the employer knows they must inform the enforcing authority in writing of the death.

Gas incidents

Specified gas incidents are notified without delay and reported within 14 days to the Health and Safety Executive.

Injury under medical supervision

Reporting and recording requirements do not apply in situations where the injury or death of a person arises out of the conduct of an operation, examination or other medical treatment of that person whilst under the supervision of a registered medical practitioner or dentist.

Self-employed people

If a self-employed person suffers a specified non-fatal injury while working at premises that are owned or occupied by themselves they do not need to notify the enforcing authority immediately.

However, they or someone acting for them must report the injury within 10 days. Where an injury is not a specified non-fatal injury, but causes a self-employed person to be incapacitated from routine work for more than seven consecutive days, the self-employed person or someone acting for them must report it within 15 days of the accident.

There is no reporting requirement for situations where a self-employed person suffers a fatal accident or fatal exposure on premises controlled by that self-employed person.

Recording

In the case of an accident at work, the following details must be recorded:

- Date and time.
- Name.
- Occupation.
- Nature of injury.
- Place of accident.
- Brief description of the circumstances in which the accident happened.
- In the case of a person not a work, instead of occupation a record of their status should be made (for example, passenger, customer, visitor or bystander).
- The date on which the accident was first notified or reported to the relevant enforcing authority and the method used.

For non-reportable injuries that incapacitate for more than 3 days a record of notification or reporting is not relevant, but a record of the accident must be maintained.

Similar information should be recorded for dangerous occurrences, except that details of injured persons will not be relevant. In the case of a diagnosis of a reportable disease, the following details must be recorded:

- The date of diagnosis of the disease.
- The name of the person affected.
- The occupation of the person affected.
- The name or nature of the disease.
- The date on which the disease was first reported to the relevant enforcing authority.
- The method by which the disease was reported.

Records must be kept for at least 3 years and kept at the place where the work it relates to is carried out or at the usual place of business of the responsible person.

Defences

A person must prove that they were not aware of the event requiring reporting and that they had taken all reasonable steps to be made aware, in sufficient time.

Specified injuries (RIDDOR 2013 - Regulation 4)

The list of specified non-fatal injuries is:

- Any bone fracture diagnosed by a registered medical practitioner, other than to a finger, thumb or toe.
- Amputation of an arm, hand, finger, thumb, leg, foot or toe.
- Any injury diagnosed by a registered medical practitioner as being likely to cause permanent blinding or reduction in sight in one or both eyes.
- Any crush injury to the head or torso causing damage to the brain or internal organs in the chest or abdomen.
- Any burn injury (including scalding) which:

    - Covers more than 10% of the whole body's total surface area.

    - Causes significant damage to the eyes, respiratory system or other vital organs.

- Any degree of scalping requiring hospital treatment.
- Loss of consciousness caused by head injury or asphyxia.
- Any other injury arising from working in an enclosed space which:

    - Leads to hypothermia or heat-induced illness.

    - Requires resuscitation or admittance to hospital for more than 24 hours.

Diseases (RIDDOR 2013 - Regulations 8, 9 and 10)

Regulation 8 - Occupational diseases

- Carpal Tunnel Syndrome (CTS), where the person's work involves regular use of percussive or vibrating tools.
- Cramp in the hand or forearm, where the person's work involves prolonged periods of repetitive movement of the fingers, hand or arm.
- Occupational dermatitis, where the person's work involves significant or regular exposure to a known skin sensitiser or irritant.
- Hand Arm Vibration Syndrome (HAVS), where the person's work involves regular use of percussive or vibrating tools, or the holding of materials which are subject to percussive processes, or processes causing vibration.
- Occupational asthma, where the person's work involves significant or regular exposure to a known respiratory sensitizer.
- Tendonitis or tenosynovitis in the hand or forearm, where the person's work is physically demanding and involves frequent, repetitive movements.

Regulation 9 - Exposure to carcinogens, mutagens and biological agents

- Any cancer attributed to an occupational exposure to a known human carcinogen or mutagen (including ionising radiation).
- Any disease attributed to an occupational exposure to a biological agent.

Regulation 10 and Schedule 3 - Diseases offshore

Examples of diseases listed in this Schedule are:

- Chickenpox.
- Cholera.
- Diphtheria.
- Dysentery (amoebic or bacillary).
- Food poisoning.
- Legionellosis.
- Malaria.
- Measles.
- Meningitis.
- Mumps.

Dangerous occurrences (RIDDOR 2013 - schedule 2)

Dangerous occurrences are events that have the potential to cause death or serious injury and so must be reported whether anyone is injured or not.

Examples of dangerous occurrences that must be reported are:

- The failure of any load-bearing part of any lifting equipment, other than an accessory for lifting.
- The failure of any pressurised closed vessel or any associated pipework.
- Any unintentional incident in which plant or equipment either:
  - Comes into contact with an uninsulated overhead electric line in which the voltage exceeds 200 volts.
  - Causes an electrical discharge from such an electric line by coming into close proximity to it.
- Electrical short-circuit or overload attended by fire or explosion which results in the stoppage of the plant involved for more than 24 hours.

The schedule also identifies Dangerous Occurrences that are specific to mines, quarries, transport systems, and offshore workplaces.

Road Traffic Act (RTA) 1991

Considered in context in Element 6.

ARRANGEMENT OF MAIN PART OF ACT

PART 1 - GENERAL

Driving offences

1. Offences of dangerous driving.
2. Careless, and inconsiderate, driving.

Drink and drugs

3. Causing death by careless driving when under influence of drink or drugs.
4. Driving under influence of drink or drugs.

Motoring events

5. Disapplication of sections 1 to 3 of the Road Traffic Act 1988 for authorised motoring events.

Danger to road-users

6. Causing danger to road-users.

Cycling

7. Cycling offences.

Construction and use

8. Construction and use of vehicles.
9. Vehicle examiners.
10. Testing vehicles on roads.
11. Inspection of vehicles.
12. Power to prohibit driving of unfit vehicles.
13. Power to prohibit driving of overloaded vehicles.
14. Unfit and overloaded vehicles: offences.
15. Removal of prohibitions.
16. Supply of unroadworthy vehicles etc.

Licensing of drivers

17. Requirement of licence.
18. Physical fitness.
19. Effects of disqualification.

Insurance

20. Exception from requirement of third-party insurance.

Information

21. Information as to identity of driver etc.

Trial

22. Amendment of Schedule 1 to the Road Traffic Offenders Act 1988.
23. Speeding offences etc admissibility of certain evidence.
24. Alternative verdicts.
25. Interim disqualification.

OUTLINE OF MAIN POINTS

SECTION 2 - CARELESS, AND INCONSIDERATE, DRIVING

"If a person drives a vehicle on a road or other public place without due care and attention, or without reasonable consideration for other persons using the road or place, he is guilty of an offence."

SECTION 3 - DRINK AND DRUGS

Causing death by careless driving when under influence of drink or drugs - if a person causes the death of another person by driving a vehicle on a road or other public place without due care and attention, or without reasonable consideration for other persons using the road or place, and

- At the time of driving is unfit to drive through drink or drugs.
- Has consumed so much alcohol that the proportion of it in his breath, blood or urine at that time exceeds the prescribed limit.
- Is within 18 hours after that time, required to provide a specimen in pursuance of section 7 of this Act, but without reasonable excuse fails to provide it, he is guilty of an offence.

For the purposes of this section a person shall be taken to be unfit to drive at any time when his ability to drive properly is impaired.

SECTION 6 - DANGER TO ROAD-USERS

Causing danger to road-users - a person is guilty of an offence if he intentionally and without lawful authority or reasonable cause.

- Causes anything to be on or over a road.
- Interferes with a motor vehicle, trailer or cycle.
- Interferes (directly or indirectly) with traffic equipment, in such circumstances that it would be obvious to a reasonable person that to do so would be dangerous.

SECTION 8 - CONSTRUCTION AND USE OF VEHICLES

A person is guilty of an offence if he uses, or causes or permits another to use, a motor vehicle or trailer on a road when:

- The condition of the motor vehicle or trailer.
- Its accessories or equipment.
- The purpose for which it is used.
- The number of passengers carried by it, or the manner in which they are carried.
- The weight, position or distribution of its load, or the manner in which it is secured, is such that the use of the motor vehicle or trailer involves a danger of injury to any person.

Breach of requirement as to weight (goods and passenger vehicles) - a person who:

- Contravenes or fails to comply with a construction and use requirement as to any description of weight applicable to.
- A goods vehicle.
- A motor vehicle or trailer adapted to carry more than eight passengers.
- Uses on a road a vehicle which does not comply with such a requirement, or causes or permits a vehicle to be so used, is guilty of an offence.

SECTION 21 - INFORMATION AS TO THE DRIVER ETC.

Where the driver of a vehicle is alleged to be guilty of an offence to which this section applies:

- The person keeping the vehicle shall give such information as to the identity of the driver as he may be required to give by or on behalf of a chief officer of police.
- Any other person shall if required as stated above give any information which it is in his power to give and may lead to identification of the driver.

Supply of Machinery (Safety) Regulations (SMSR) 2008

Considered in context in Element 9.

OUTLINE OF MAIN POINTS

Previously the Supply of Machinery (Safety) Regulations 1992 as amended by the Supply of Machinery (Safety) (Amendment) Regulations 1994 and the Supply of Machinery (Safety) (Amendment) Regulations 2005.

The SMSR 2008 imposes duties upon those who place machinery and safety components onto the market, or put them into service (this includes second-hand machinery which is 'new' to Europe). They set out the essential requirements which must be met before machinery or safety components may be placed on the market or put into service in the UK. They implement the latest version of the Machinery Directive 2006/42/EC and came into force on 29 December 2009, replacing the previous Supply of Machinery (Safety) Regulations 1992, as amended in 1994 and 2005.

Meeting the requirements

The duty to meet the requirements mainly falls to the **'responsible person'** who is defined as the manufacturer or the manufacturer's representative. If the manufacturer is not established in the EEA, the person who first supplies the machinery in the EEA may be the **responsible person**, which can be a user who manufactures or imports a machine for their own use.

Conformity assessment

The responsible person should ensure that machinery and safety components satisfy the Essential Health and Safety Requirements (EHSRs), see Schedule 2, Part 1 of the Regulations, and that appropriate conformity assessment procedures have been carried out. These requirements are intended to ensure that all machinery throughout the EC is constructed to the same safety standards. In addition, the responsible person must draw up a technical file (see below). For certain classes of dangerous machine and safety component, a more rigorous procedure is required. The SMSR places even greater emphasis on the EHSR requirements. However, still extant are the requirements of primary legislation by way of the Health and Safety at Work Act 1974 (HASAWA), which makes a specific reference to the duties of manufacturers and suppliers of workplace machinery (including second hand machinery).

Section 6 of HASAWA (general duties of manufacturers) states:

"It shall be the duty of any person who designs, manufactures, imports or supplies any article for use at work....

(a) To ensure , so far as is reasonably practicable, that the article is so designed and constructed that it will be safe and without risks to health at all times when it is being set, used, cleaned or maintained by a person at work."

HASAWA is reinforced by the duties under Part 3 of SMSR 2008 (general prohibitions and obligations), which states:

"No responsible person shall place machinery on the market or put it into service unless it is safe; and before machinery is placed on the market or put into service, the responsible person must:

(a) Ensure that the applicable essential health and safety requirements are satisfied in respect of it...."

Declaration procedure

The responsible person must issue one of two forms of declaration.

Declaration of conformity

This declaration should be issued with the finished product so that it is available to the user. It will contain various details such as the manufacturer's address, the machinery type and serial number, and Harmonised European or other Standards used in design.

Declaration of incorporation

Where machinery is intended for incorporation into other machinery, the responsible person can draw up a declaration of incorporation. This should state that the machinery must not be put into service until the machinery into which it is to be incorporated has been given a Declaration of Conformity. A CE mark is not affixed at this intermediate stage.

Marking

When the first two steps have been satisfactorily completed, the responsible person or the person assembling the final product should affix the CE mark.

Enforcement

In the UK the Health and Safety Executive is responsible for enforcing these Regulations in relation to machinery and safety components designed for use at work. Trading Standards Officers are responsible for enforcing these Regulations in relation to consumer goods.

Detailed advice for the designer and manufacturer

Technical file contents

The responsible person (defined above) is required to draw up a technical file for all machinery and safety components covered by these Regulations. The file or documents should comprise:

a) An overall drawing of the product together with the drawings of the control circuits.

b) Full detailed drawings, accompanied by any calculation notes, test results etc. required to check the conformity of the product with the essential health and safety requirements.

c) A list of the essential health and safety requirements, transposed harmonised standards, national standards and other technical specifications which were used when the product was designed.

d) A description of methods adopted to eliminate hazards presented by the machinery or safety component.

e) If the responsible person so desires, any technical report or certificate obtained from a component body or laboratory.

f) If the responsible person declares conformity with a transposed harmonised standard, any technical report giving the results of tests.

g) A copy of the instructions for the product.

For series manufacture, the responsible person must also have available documentation on the necessary administrative measures that the manufacturer will take to ensure that the product meets requirements.

Technical file procedure

The technical file document need not be on a permanent file, but it should be possible to assemble and make them available to an enforcement authority. The technical file documents should be retained and kept available for at least ten years following the date of manufacture of the product or of the last unit produced, in the case of a series manufacture.

If the technical file documents are drawn up in the United Kingdom, they should be in English unless they are to be submitted to an Approved/Notified Body in another Member State, in which case they should be in a language acceptable to that approved Body. In all cases the instructions for the machinery should be in accordance with the language requirements of the EHSRs.

Work at Height Regulations (WAHR) 2005

Considered in context in Elements 7 and 8.
See also, PUWER 1998 and WHSWR 1992.

ARRANGEMENT OF REGULATIONS

1) Citation and commencement.

2) Interpretation.

3) Application.

4) Organisation and planning.

5) Competence.

6) Avoidance of risks from work at height.

7) Selection of work equipment for work at height.

8) Requirements for particular work equipment.

9) Fragile surfaces.

10) Falling objects.

11) Danger areas.

12) Inspection of work equipment.

13) Inspection of places of work at height.

14) Duties of persons at work.

15) Exemption by the Health and Safety Executive.

16) Exemption for the Armed Forces.

17) Amendment to the Provision and Use of Work Equipment Regulations (PUWER) 1998.

18) Repeal of section 24 of the Factories Act 1961.

19) Revocation of instruments.

Schedules

Schedule 1	Requirements for existing places of work and means of access or egress at height.
Schedule 2	Requirements for guard-rails, toe-boards, barriers and similar collective means of protection.
Schedule 3	Requirements for working platforms.
Part 1	Requirements for all working platforms.
Part 2	Additional requirements for scaffolding.
Schedule 4	Requirements for collective safeguards for arresting falls.
Schedule 5	Requirements for personal fall protection systems.
Part 1	Requirements for all personal fall protection systems.
Part 2	Additional requirements for work positioning systems.

Part 3	Additional requirements for rope access and positioning techniques.
Part 4	Additional requirements for fall arrest systems.
Part 5	Additional requirements for work restraint systems.
Schedule 6	Requirements for ladders.
Schedule 7	Particulars to be included in a report of inspection.
Schedule 8	Revocation of instruments.

OUTLINE OF MAIN POINTS

The final version of the Regulations, designated the Work at Height Regulations (WAHR) 2005 came into force 6th April 2005.

Under these Regulations the interpretation of 'work at height' includes any place of work at ground level, above or below ground level that a person could fall a distance liable to cause personal injury and includes places for obtaining access or egress, except by staircase in a permanent workplace.

AMENDMENTS TO OTHER REGULATIONS AS A RESULT OF THE WORK AT HEIGHT REGULATIONS 2005

WAHR 2005 makes an amendment to the Provision and Use of Work Equipment Regulations 1998; they also replace certain Regulations in the Workplace (Health and Safety) Regulations; and amend definitions in the Construction (Health, Safety and Welfare) Regulations:

Regulation 17 - Amendment of the Provision and Use of Work Equipment Regulations (PUWER) 1998. There shall be added to Regulation 6(5) of the Provision and Use of Work Equipment Regulations 1998 the following sub-paragraph:

(f) "Work equipment to which Regulation 12 of the Work at Height Regulations 2005 applies."

Schedule 8 - Revocation of instruments

Workplace (Health and Safety) Regulations 1992 - extent of revocation: Regulation 13(1) to (4).

Construction (Health, Safety and Welfare) Regulations 1996 - extent of revocation: in Regulation 2(1), the definitions of 'fragile material', 'personal suspension equipment' and 'working platform'; Regulations 6 to 8; in Regulation 29(2) the word 'scaffold' in both instances; Regulation 30(5) and (6) (a); Schedules 1 to 5; and the entry first mentioned in columns 1 and 2 of Schedule 7.

Work at Height (Amendment) Regulations 2007

These Regulations amended WAHR 2005 to remove the dis-application of WAHR 2005 to certain work concerning the provision of instruction or leadership to people engaged in caving or climbing by way of sport, recreation, team building or similar activities. They introduce a new duty under Regulation 14A that takes into account the special circumstances of work at height in caving and climbing.

Under these Regulations the interpretation of 'work at height' includes any place of work at ground level, above or below ground level that a person could fall a distance liable to cause personal injury and includes places for obtaining access or egress, except by staircase in a permanent workplace.

Regulation 4 states that all work at height must be properly planned, supervised and be carried out so far as is reasonably practicable safe. Planning must include the selection of suitable equipment, take account of emergencies and give consideration to weather conditions impacting on safety.

Regulation 5 states that those engaged in any activity in relation to work at height must be competent; and, if under training, are supervised by a competent person.

Regulation 6 states that work at height must only be carried out when it is not reasonably practicable to carry out the work otherwise. If work at height does take place, suitable and sufficient measures must be taken to prevent a fall of any distance, to minimise the distance and the consequences of any fall liable to cause injury. Employers must also make a risk assessment, as required by Regulation 3 of the Management of Health and Safety at Work Regulations.

Regulation 7 states that when selecting equipment for use in work at height the employer shall take account of working conditions and any risk to persons in connection with the place where the equipment is to be used. The selection of work equipment must have regard in particular to the purposes specified in Regulation 6.

Regulation 8 sets out requirements for particular equipment to conform to standards expressed in schedules to the Regulations. It includes guard-rails, toe-boards, working platforms, nets, airbags, personal fall arrest equipment rope access and ladders.

Regulation 9 states that every employer shall ensure that suitable and sufficient steps are taken to prevent any person at work falling through any fragile surface; and that no work may pass across or near, or work on, from or near, fragile surfaces when it is reasonably practicable to carry out work without doing so. If work has to be from a fragile roof then suitable and sufficient means of support must be provided that can sustain foreseeable loads. No person at work should be allowed to pass or work near a fragile surface unless suitable and sufficient guard rails and other means of fall protection is in place. Signs must be situated at a prominent place at or near to works involving fragile surfaces, or persons are made aware of the fragile roof by other means.

Regulation 10 states that every employer shall take reasonably practicable steps to prevent injury to any person from the fall of any material or object; and where it is not reasonably practicable to do so, to take similar steps to prevent any person being struck by any falling material or object which is liable to cause personal injury. Also, that no material is thrown or tipped from height in circumstances where it is liable to cause injury to any person. Materials and objects must be stored in such a way as to prevent risk to any person arising from the collapse, overturning or unintended movement of the materials or objects.

Regulation 11 states that every employer shall ensure that where an area presents a risk of falling from height or being struck from an item falling at height that the area is equipped with devices preventing unauthorised persons from entering such areas and the area is clearly indicated.

Regulation 12 states that every employer shall ensure that, where the safety of work equipment depends on how it is installed or assembled, it is not used after installation or assembly in any position unless it has been inspected in that position.

Also, that work equipment is inspected at suitable intervals and each time that exceptional circumstances which are liable to jeopardise the safety of the work equipment occur. Specific requirements exist for periodic (every 7 days) inspection of a working platform where someone could fall 2 metres or more.

Regulation 13 states that every employer shall ensure that fall protection measures of every place of work at height are visually inspected before use.

Regulation 14 states the duties of persons at work to report defects and use equipment in accordance with training/instruction.

Workplace (Health, Safety and Welfare) Regulations (WHSWR) 1992

See also - Health and Safety (Miscellaneous Amendments) Regulations (MAR) 2002 and Health and Safety (Miscellaneous Repeals, Revocations and Amendments) Regulations (MRRA) 2002.

Considered in context in Element 1.

ARRANGEMENT OF REGULATIONS

1) Citation and commencement.

2) Interpretation.

3) Application of these Regulations.

4) Requirements under these Regulations.

5) Maintenance of workplace, and of equipment, devices and systems.

6) Ventilation.

7) Temperature in indoor workplaces.

8) Lighting.

9) Cleanliness and waste materials.

10) Room dimensions and space.

11) Workstations and seating.

12) Condition of floors and traffic routes.

13) Falls or falling objects (*Revoked in part by WAH 2005*).

14) Windows, and transparent or translucent doors, gates and walls.

15) Windows, skylights and ventilators.

16) Ability to clean windows etc. safely.

17) Organisation etc. of traffic routes.

18) Doors and gates.

19) Escalators and moving walkways.

20) Sanitary conveniences.

21) Washing facilities.

22) Drinking water.

23) Accommodation for clothing.

24) Facilities for changing clothing.

25) Facilities for rest and to eat meals.

26) Exemption certificates.

27) Repeals, saving and revocations.

Schedule 1 - Provisions applicable to factories which are not new workplaces, extensions or conversions.

Schedule 2 - Repeals and revocations.

OUTLINE OF MAIN POINTS

Summary

The main requirements of the Workplace (Health, Safety and Welfare) Regulations (WHSWR) 1992 are:

1) **Maintenance** of the workplace and equipment.

2) **Safety** of those carrying out maintenance work and others who might be at risk (for example, segregation of pedestrians and vehicles, provision of handrails etc.).

3) Provision of **welfare** facilities (for example, rest rooms, changing rooms etc.).

4) Provision of a safe **environment** (for example, lighting, ventilation etc.).

Environment

Reg 1 New workplaces, extensions and modifications must comply.

Reg 4 Requires employers, persons in control of premises and occupiers of factories to comply with the Regulations.

Reg 6 Ventilation - enclosed workplaces should be ventilated with a sufficient quantity of fresh or purified air (5 to 8 litres per second per occupant).

Reg 7 Temperature indoors - this needs to be reasonable and the heating device must not cause injurious fumes. Thermometers must be provided. Temperature should be a minimum of 16°C or 13°C if there is physical effort.

Reg 8 Lighting - must be suitable and sufficient. Natural light if possible. Emergency lighting should be provided if danger exists.

Reg 10 Room dimensions and space - every room where persons work shall have sufficient floor area, height and unoccupied space (min 11 cu. m per person).

Reg 11 Workstations and seating have to be suitable for the person and the work being done.

Safety

Reg 12 Floors and traffic routes must be of suitable construction. This includes absence of holes, slope, uneven or slippery surface. Drainage where necessary. Handrails and guards to be provided on slopes and staircases.

Reg 13 Tanks and pits containing dangerous substances to be covered or fenced where people could fall into them and traffic routes fenced.

Reg 14 Windows and transparent doors, where necessary for health and safety, must be of safety material and be marked to make it apparent.

Reg 15 Windows, skylights and ventilators must be capable of opening without putting anyone at risk.

Reg 17 Traffic routes for pedestrians and vehicles must be organised in such a way that they can move safely.

Reg 18 Doors and gates must be suitably constructed and fitted with any necessary safety devices.

Reg 19 Escalators and moving walkways shall function safely, be equipped with any necessary safety devices and be fitted with emergency stop.

Housekeeping

Reg 5 Workplace and equipment, devices and systems must be maintained in efficient working order and good repair.

Reg 9 Cleanliness and waste materials - workplaces must be kept sufficiently clean. Floors, walls and ceilings must be capable of being kept sufficiently clean. Waste materials shall not be allowed to accumulate, except in suitable receptacles.

Reg 16 Windows etc. must be designed so that they can be cleaned safety.

Facilities

Reg 20 Sanitary conveniences must be suitable and sufficient and in readily accessible places. They must be adequately ventilated, kept clean and there must be separate provision for men and women.

Reg 21 Washing facilities must be suitable and sufficient. Showers if required (a table gives minimum numbers of toilets and washing facilities).

Reg 22 Drinking water - an adequate supply of wholesome drinking water must be provided.

Reg 23 Accommodation for clothing must be suitable and sufficient.

Reg 24 Facilities for changing clothes must be suitable and sufficient, where a person has to use special clothing for work.

Reg 25 Facilities for rest and eating meals must be suitable and sufficient.

The WHSWR 1992 were amended by the Health and Safety (Miscellaneous Amendments) Regulations (MAR) 2002 to establish specific requirements that rest rooms be equipped with:

- An adequate number of tables and adequate seating with backs for the number of persons at work likely to use them at any one time.
- Seating which is adequate for the number of disabled persons at work and suitable for them.

In addition, WHSWR 1992 were amended to take account of disability arrangements. Where necessary, those parts of the workplace used or occupied directly by disabled persons at work, including in particular doors, passageways, stairs, showers, washbasins, lavatories and workstations, must be organised to take account of such persons.

Assessment

Contents

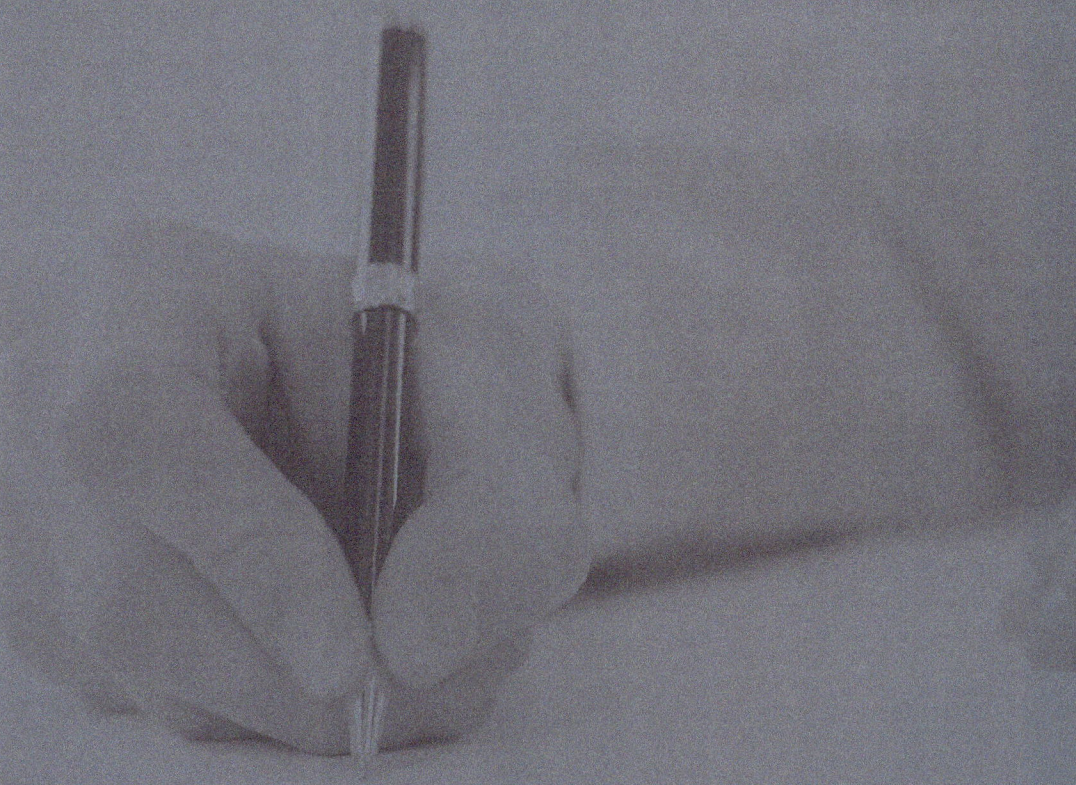

A.1 Assessments of understanding

It is understood that those using this publication may be doing so to broaden their understanding of this important topic, management of health and safety in construction, and others will be studying in order to obtain a specific construction related qualification. The approach taken by those learners of construction health and safety will vary.

This element provides information on how one such qualification is assessed, the NEBOSH Health and Safety Management for Construction (NC) – Unit NC1 Managing construction safely.

UNIT NC1 – OPEN BOOK EXAMINATION (OBE)

The NC1 Managing construction safely open book examination (OBE) will be based on a fictional construction site and will assess the knowledge and understanding across all the elements 1 to 13 and how it can be applied practically. You will be given 48 hours to research and complete the examination but you do not have to complete the examination all in one go.

As part of the open book examination, you are able to use your study or course notes, textbooks, e-learning software and the internet to research during the exam. You are expected to properly plan, draft, amend and submit the online paper within this 48-hour period. This does not necessarily make the online exam any easier. For example, NEBOSH suggest that the online NC1 exam should take around 8 hours to complete.

The open book assessment will begin by giving a realistic scenario to set the scene. This will typically describe a realistic organisation and workplace, with an outline of normal operational activities and worker behaviour. The scenario may go on to outline a developing situation, such as an incident or safety intervention. You may be asked to imagine that you hold a specific role, such as a Site Safety Manager, in the construction workplace described.

You will be asked to carry out a series of tasks relating to this scenario, which may be split into several sub-tasks. NEBOSH will specify the maximum marks available for each task to help manage the time. These tasks will partially or entirely draw on the sign posts and evidence within the scenario. Answers therefore need to be relevant to the scenario and, in some cases, specific evidence will be required to support answers.

Dependent on the specific scenario, an example question might be:

Based on the scenario only, comment on the health and safety culture on site. (10)

The examination will also specify a word count of 6,000 words and the overall word count should be no more than +10%. This recommended word count is there to aid learners in focusing on the subject and context requirements of the OBE and to avoid digression into areas that are not mark worthy.

You should not just copy and paste answer information from another source. All answers must be in your own words, but should material be used straight from textbooks or the web, then it must be referenced on your submission.

Upon completion of a NEBOSH open book exam, learners are expected to complete a closing interview to confirm that the paper that has been submitted is their own work. Closing interviews are a mandatory requirement for the assessment, and failure to show up to your closing interview could mean the submission not being accepted or marked.

The interview will take place via a video link and will be conducted by an interviewer nominated by the Learning Partner. The interviewer is usually a NEBOSH-qualified tutor who will review your submission prior to the scheduled interview. They will then ask questions based on the answers submitted, such as "Why did you come to these conclusions?", or "What element of your answer do you think is the most important?"

At every examination a number of learners - including some good ones - perform less well than they might because of poor examination technique. It is essential that you practice answering questions and learn to manage your time accordingly to answer all the questions set, remembering it will take longer to read and understand the scenario-based questions.

See also the 'Revision and exam guidance' section later in this study book.

RESULTS AND CERTIFICATION

The exam paper will be marked by an examiner appointed by NEBOSH. The pass rate for the NC1 open book exam is 45%. The grading boundaries are as follows:

Distinction	75+
Credit	65-74
Pass	45-64
Refer	0-44

STUDY QUESTIONS

In order that learners may check and reinforce their understanding of the topics covered, a number of questions have been included at the end of each element.

Information regarding the required content of answers to the study questions is provided on the following pages. Any resemblance to NEBOSH examination questions is coincidental.

A.2 Study questions - answer guidance

The following study questions and answer guidance are provided to aid confirmation of understanding, learning and revision.

They are not sample NEBOSH questions but have been developed by RMS, however please note these are not always complete or exhaustive answers and you are required to add appropriate detail where applicable.

ELEMENT 1 - THE FOUNDATIONS OF CONSTRUCTION HEALTH AND SAFETY MANAGEMENT

1) What information needs to be provided when completing an F10 in accordance with the Construction (Design and Management) Regulations 2015?

The answer to this question is taken from Schedule 1 of the Construction (Design and Management) Regulations 2015 and the information which needs to be provided includes:

1) The date of forwarding the notice

2) The address of the construction site or precise description of its location

3) The name of the local authority where the construction site is located

4) A brief description of the project and the construction work that it entails

5) Contact details of the client: name, address, telephone number and (if available) an email address

6) Contact details of the principal designer: name, address, telephone number and (if available) an email address

7) Contact details of the principal contractor: name, address, telephone number and (if available) an email address

8) The date planned for the start of the construction phase

9) The time allocated by the client under Regulation 4(1) for the construction work

10) The planned duration of the construction phase

11) The estimated maximum number of people at work on the construction site

12) The planned number of contractors on the construction site

13) The name and address of any contractor already appointed

14) The name and address of any designer already appointed

15) A declaration signed by or on behalf of the client that the client is aware of the client duties under these Regulations

2) What topics should be included in an induction for site visitors?

In answer to this question the following topics should be included:

- *The sign in and sign out procedures*

- *Escort arrangements for the visit*

- *Information on the site layout*

- *Relevant site rules*

- *Information on hazards that might be present on site*

- *Areas of the site to which visitors would not be allowed*

- *Information on the personal protective equipment which visitors would be expected to wear*

- *Information on emergency procedures in place*

- *Identification of the fire assembly point*

- *The location of the welfare*

- *The location of the first-aid facilities*

- *The arrangements for reporting an accident if one should occur*

3) What should be considered when assessing the adequacy of lighting on a construction site?

In answer to this question the following should be considered:

Lighting is required for:

- *The case of an emergency*

- Task specific activities

- Vehicles

- Road/traffic

- Pedestrians/paths

- Temporary lighting is required when:

- Working in the building prior to lighting being installed

- When natural lighting is poor in:

- Winter (dark mornings/evenings)

- Poor weather

4) What features of a Temporary Accommodation Unit (TAU) should be addressed in order to meet the health, safety, and welfare requirements for workers?

Features which should be addressed include the following:

- *The availability of natural lighting and the adequacy of the artificial lighting provided particularly for specific areas such as stairs and corridors*

- *The provision of effective and sufficient ventilation*

- *An adequate heating system to provide and maintain a reasonable temperature throughout the building*

- *The provision of suitable floor surfaces and stairs*

- *The provision of suitable workstations and seating and the avoidance of space constraints*

- *The provision of doors with transparent panels of safety glass*

- *Ensuring that windows and skylights, designed to open, did not project into an area where people were likely to collide with them and that adequate arrangements were in place for cleaning them in safety*

- *The availability of separate male and female sanitary conveniences, washing facilities provided with hot and cold water and means of drying and accommodation for clothing*

- *The provision of facilities for rest and taking meals including a supply of wholesome drinking water*

ELEMENT 2 - IMPROVING HEALTH AND SAFETY CULTURE AND ASSESSING RISK

1) What are ways in which organisations can positively influence the health and safety behaviours of their workers?

- The preparation of a health and safety policy backed up by an obvious display of management commitment to the observance of laid down standards while demonstrating leadership by example

- The introduction of formal consultation

- The employment of competent personnel backed up by the provision of any additional training that may be required

- The provision of a good working environment

- Involving workers in risk assessments, accident investigations and the development of safe systems and procedures

- Ensuring action is taken quickly to rectify any non-compliance with standard procedures and disciplinary measures taken when required

- Introducing an effective two-way communication system

- Introducing a system of incentives

- Rewards to recognise achievement

2) a) Explain the role of worker participation in health and safety

   b) What are the benefits of participation for the employer?

For (a), the role of worker participation in health and safety is to provide the employer with a wider view of how risks affect workers, their view on the effectiveness of current control measures and on proposed control measures. In addition, the role of worker participation is to show management commitment and motivate workers to work safely and healthily.

For (b), the benefit of worker participation of this type is that it has been identified as one of the most significant factors to influence health and safety behaviour and promote a positive health and safety culture in organisations. It can lead to shared health and safety values and the motivation of those involved to work together to improve health and safety.

3) What should be considered when preparing and presenting a training session on health and safety?

For a training session the following should be considered:

- *Training objectives*

- *Content of the training session (breadth and depth)*

- *Training style and methods (for example, lecture/ video/group work/role play, etc.)*

- *Target audience (for example, existing knowledge and skills/relevance/motivation)*

- *The number of trainees*

- *The time available*

- *The suitability of the training environment (location/ size/layout/lighting, etc.)*

- *The skills required of the trainer(s)*

- *Use of visual aids*

- *How the effectiveness of the training is to be measured, both at the time of presentation (for example, course evaluations) and subsequently improvements in health and safety performance (for example, compliance with procedures, reduction in accidents, etc.)*

4) What is the criteria which must be met for a risk assessment to be 'suitable and sufficient'?

For a risk assessment to be deemed suitable and sufficient, the following criteria must be met:

- *Indicate the competence of the assessor together with any specialist advice that has been sought- including consultation with workers*
- *Identify all significant hazards and risks arising from or connected with the activity to be carried out*
- *Identify all the persons at risk including workers, other workers and members of the public with reference to those who might be particularly at risk*
- *Evaluate the adequacy and effectiveness of existing control measures*
- *Identify other protective measures that may be required*
- *Enable priorities to be set; record the significant findings of the assessment*
- *Identify the period of time for which it is likely to remain valid*

5) What are the sources of information that might be consulted to help determine the severity of the outcome of an incident?

The sources of information include the following:

- *The nature of the hazard, for example, high or low voltage, the height of a fall, etc.*
- *Individual opinions of those who do the job*
- *Internal statistics, for example, accident book*
- *External statistics, for example, HSE website*
- *Insurance companies*
- *ACOPs and guidance*
- *Industry guidance*

ELEMENT 3 - MANAGING CHANGE AND PROCEDURES

1) An organisation is introducing a new work activity that requires a safe system of work.

   a) Why it is important to involve workers in the development of a safe system of work?

   b) Why it is important for safe systems of work to have written procedures?

For part (a), it is important to involve workers in the development of a safe system of work because of their knowledge of the particular working environment involved and what will work in practice. Additionally, their involvement will establish their ownership of the system and will encourage them to use and follow it once it has been finalised and introduced. Finally their involvement will emphasise management's commitment to health and safety and help to raise its profile within the organisation.

For part (b), once a safe system of work is developed, it is imperative that a clear method of communicating its procedures to the workforce is used and this would be better achieved in writing rather than orally. The procedures may contain complex information that will need to be consulted on more than one occasion to ensure the correct sequence of operations is followed. Additionally, different people will need to be aware of the procedures and it is preferable to have them written down rather than pass them on by word of mouth, a method that may not always guarantee consistency in their presentation. A written document will also be needed for audit purposes and could be used as evidence in defending an enforcement action or a civil claim. Finally, the use of written procedures may well be a requirement of the organisation's quality assurance procedures.

2) What are the general details that should be included in a permit-to-work?

The first would be a description and assessment of the work to be performed including the plant involved, its location and the possible hazards associated with the task. This will determine the need for, and nature of, other relevant contents of the permit such as for example, the isolation of sources of energy and product inlets, and the additional precautions required such as atmospheric monitoring, the provision and use of personal protective equipment, the emergency procedures to be followed and the duration of the permit.

An essential element of a permit-to-work system is, of course, the operation of the permit itself. By means of signatures, the permit should be issued by an authorised person, and accepted by the competent person responsible for the work. On completion of the work, the competent person would need to indicate on the permit that the area had been made safe in order for the permit to be cancelled by the authorised person, after which isolations could be removed.

3) What are the factors to consider when making an assessment of first-aid provision in a workplace?

When making the assessment, consideration should be given to a number of factors including:

- *The size of the organisation - in particular, the number and distribution of workers that need to be covered by the first-aid provision*

- *The type of workforce - including the special needs of trainees, younger and older workers, those with known medical conditions and the disabled*

- *The need to provide first-aid at all times when workers are on-site - considering shift patterns and shift change-over periods, any work done outside normal working hours and cover for first-aiders due to their sickness or other absences*

- *Different work activities - first-aid needs will depend on the type of work being done. Some activities, such as office work, have relatively few hazards and low levels of risk and will need less first-aid provisions than other activities that have more hazards or more specific hazards (construction or chemical sites)*

- *Ease of access to medical treatment - where external emergency medical services are not easily available, more first-aid provision may be required, for example, a first-aid room.*

- *Workers working away from the employer's premises - for some work activities first-aid arrangements may have to be made for workers working in locations away from the employer's fixed work places. For example, mobile teams of workers, particularly those working in remote locations*

- *National legal requirements - the assessment should consider legal requirements and ensure minimum requirements are met or exceeded*

4) An engineer was involved in a near-miss incident when he dropped a component whilst working on an overhead crane. The component narrowly missed an employee who was passing below.

What are four reasons why the near-miss incident should be investigated?

The incident should be investigated in the first instance to prevent a recurrence by identifying its causes and any weaknesses in the existing procedures and systems. It would also help to set priorities for any remedial action found to be necessary if for instance risk assessments

and safe systems of work need to be amended.

An investigation would additionally help to demonstrate management commitment to health and safety and assist in maintaining and even improving the health and safety culture of the organisation and the morale of the workforce.

ELEMENT 4 – EXCAVATION

1) What are the main hazards associated with excavation work adjacent to a busy dual carriageway with a speed limit of 40mph?

The hazards associated with the excavation work include:

- *Collapse of the sides of the excavation.*

- *The effect of the excavation work on adjacent structures, including the road, leading to collapse of the structure.*

- *Water ingress into the excavation, following adverse weather conditions or due to a low water table, causing flooding and a risk of drowning.*

- *Ill health effects due to coming into contact with hazardous substances in contaminated land.*

- *Possible presence of an unexploded bomb.*

- *Contact with buried services, for example, gas, electricity and water.*

- *Possible chemical asphyxiation due to a build-up of carbon monoxide in the excavation from vehicles, including work vehicles.*

- *Lack of oxygen due to a build-up in carbon dioxide in the excavation from chalk deposits.*

- *Fire and explosion from flammable substances.*

- *Movement of excavation machinery, including contact people and overhead lines.*

- *The risk of vehicles, people or materials falling into unprotected excavations.*

- *Biological hazards, for example, leptospirosis.*

- *High noise levels from the vehicles on the duel carriageway.*

- *Being struck by vehicles on the duel carriage way.*

2) What are the control measures that could be considered for excavation work on a construction site?

The control measures for excavation work include:

- *Identification/detection and marking of buried services.*

- *Safe digging methods for where buried services are identified.*

- *Support of the sides of the excavation, for example, shoring by using sheets or using support boxes.*

- *Support for structures where their stability may be affected by the excavation.*

- *Means of access to and out of the excavation.*

- *Crossing points to get from one side of the excavation to the other.*

- *Barriers around the excavation area.*

- *Lighting of the excavation area and signs to warn of its presence.*

- *Safe positioning and storage of spoil from the excavation and material to go in it.*

- *Prevention of water ingress and de-watering, for example, by means of pumps.*

- *Positioning and routing of vehicles and other equipment to avoid collapse of the excavation and build-up of carbon monoxide in the trench.*

- *Provision of personal protective equipment.*

- *Provision of welfare facilities to assist with maintaining hygiene, for example, washing hands.*

- *Checking the atmosphere where there is a risk of flammable/toxic substances or asphyxiation.*

- *Controlling the movement of excavation machinery to avoid contact people and obstructions.*

3) (a) Give FOUR examples of a confined space that may be encountered in construction activities?

Examples of confined spaces in construction activities include (four only required):

- *Sewers - brick or concrete structures for carrying liquid waste.*

- *Tunnels - for access purposes or movement of liquids or other materials.*

- *Trench/excavations and pits - areas where earth has been removed.*

- *Tanks - storage tank, for example, a fuel storage tank.*

- *Chambers - an interception chamber for water/waste.*

- *Silos - may be an above ground structure for storing material, for example, cement or mortar.*

- *Cellars - an enclosed low level room or space below ground level usually found in old buildings.*

- *Well - a deep source of water.*

(b) What are FOUR hazards associated with work in a confined space encountered in construction activities?

The hazards associated with confined space work include (four only required):

- *Flammable or explosive gases or vapours.*

- *High temperatures that could cause loss of consciousness.*

- *Gases, fumes and vapours that could cause loss of consciousness.*

- *Lack of oxygen that could cause loss of consciousness.*

- *Ingress of liquid that changes the level of liquid in the confined space leading to a risk of drowning.*

- *Free flowing solids that could move when stood on, causing a person to be trapped in the solids and become asphyxiated, for example, in a silo.*

- *Falls of materials, particularly during tunnelling work*

- *Contact with the structure of the confined space due to restricted space, particularly the head.*

- *Claustrophobic effects.*

- *Biological hazards arising from the presence of vermin.*

4) A leaking underground petrol tank has been emptied so that it can be visually inspected internally prior to repair. What are the features of a safe system of work for the inspection team in order to satisfy the requirements of the Confined Spaces Regulations 1997?

A safe system of work for the internal inspection of the tank would include:

- *Control of the activity by a competent person.*

- *A preliminary remote inspection, followed by the completion of a risk assessment.*

- *Planning and preparation, including establishing a health and safety method statement that requires the use of a permit to work system for confined space entry.*

- *The issue of a confined space entry permit to work by a competent person.*

- *Confirmation of isolation and lock off of fuel supply and other relevant services.*

- *The provision of a safe means of access and egress.*

- The stipulation of a minimum number in the work team, at least three workers, one of them should remain outside the confined space.

- The selection and training of workers for the working in a confined space.

- The provision of monitoring of the atmosphere inside the tank while the work is in progress.

- The provision of adequate and suitable lighting and ventilation.

- Arrangements for the installation of an effective means of communication between those inside and outside the tank.

- The provision of a rescue plan, including raising the alarm, equipment required and getting extra assistance.

- The provision of rescue equipment, for example, breathing apparatus, resuscitation equipment and a tripod hoist to assist in the removal of the workers in the event of an emergency.

- The provision and use of personal protective equipment such as hard hats, goggles and overalls.

- The provision of a harness to assist with emergency escape and consideration of a life line.

5) A trench is to be excavated to a depth of 1.5 metres through a park in order to repair the water mains supply to the on-site café.

(a) What are the possible hazards associated with this work?

Possible hazards include:

- Collapse of the trench sides.
- Lack of oxygen, fumes, etc.
- Water ingress.
- Unprotected trench sides – falls of vehicles/equipment/people into trench.
- Existing services.
- Materials/spoil heaps falling.
- Moving vehicles.
- Noise and/or vibration.
- Potential land contamination.
- Bio-hazards.

(b) What control measures should be taken to minimise the risks to the workers and others who may be affected by the work?

Control measures include:

- Risk assessments, with examples.
- Access into park is safe/managed/monitored.
- Air monitoring.
- Erection of barriers/fencing to work area.
- Trench support for excavation.
- Inspection and recording of excavation/excavators.
- Hand/safe digging near services.
- Detection/support/protection of services.
- Isolate water, dewater, protection against flooding.
- Safe access/egress into/out of trench.
- Keep materials/plant away from excavation (stop blocks, banksmen).
- Edge protection/cover.
- Erect appropriate signs.
- Use of appropriate PPE, for example, ear, hand, head protection, hi-vis jackets, overalls, etc.
- Competent persons.
- Provision of information to nearby residents.
- Security of equipment/materials.
- Precautions for biological hazards, for example, welfare facilities/sharp survey.
- Emergency procedures in place/communicated.

6) A tank measuring 3 metres long, 2 metres wide and 2 metres deep is to be placed into the ground with the soil then back filled around and over it. What factors need to be considered in the risk assessment before excavation starts?

Factors to be considered are:

- Size of excavation – allowance for base/cover/service connections.
- Type of ground – contamination/rock/strength.
- Location – adjacent structures/existing services/access/public.
- Type of support – allow tank to be placed.
- Plant and equipment – wheeled/tracked excavators.

- *Weather – water table/water ingress.*

- *Prevention of falls – barriers/stop blocks.*

- *Access into excavation – secure ladder/within the trench supports.*

7) What control measures are needed to reduce the risks of excavation work on a construction site?

Control measures are:

- *Detection of underground services, gas/electricity, using plans, cable detectors etc.*

- *Identify overhead cables and protect.*

- *Support where there is risk of collapse, for example, shoring, benching.*

- *Keep materials and plant vehicles away from edge, use stop blocks/banksmen.*

- *Use edge protection/cover.*

- *Protect against flooding, for example, provide pumps.*

- *Support adjacent buildings against collapse.*

- *Provide safe access/egress.*

- *Provide adequate lighting.*

- *Test deep excavations for fumes/ventilate or other appropriate action.*

- *Inspection by a competent person.*

- *Precautions for contaminants/biological hazards.*

- *Hand dug close to services/support services.*

ELEMENT 5 – DEMOLITION

1) A multi-storey car park in a city centre is to be demolished.

(a) What are FOUR possible hazards to the environment that could be caused by the demolition?

Possible hazards to the environment include:

- *Airborne dust, fumes and fibres.*

- *Noise.*

- *Spilled fuel.*

- *Silt affecting the drainage systems.*

- *Mud and debris being deposited on adjacent roads.*

- *Congestion of surrounding roads due to site traffic.*

(b) What control measures should be considered to reduce the risk to the environment during the demolition work?

Control measures to consider include:

- *Consideration of the method of demolition.*

- *Damping down the structure and debris to reduce the production of dust.*

- *Sheeting vehicles leaving the demolition site with waste material.*

- *Providing wheel wash facilities and washing down vehicles leaving the demolition site.*

- *Using street cleaners to remove dust, debris and mud from the roads.*

- *Providing noise controls to minimise noise at source, for example, the use of acoustic covers and exhaust silencers on equipment. Orientation of equipment so that noise is directed away from sensitive areas.*

- *Providing noise controls to minimise noise travelling outside the site, for example, sound absorbent barriers, using spoil heaps and stores/welfare facilities to provide an acoustic barrier and distance from the noise to sensitive areas.*

- *Fitting filters or stoppers to site drain gullies.*

- *Bunding fuel tanks and introducing effective controls for waste, including the prohibition of burning rubbish on site.*

2) What are the control measures that should be considered when demolishing a group of ten terraced houses in a series of twenty terraced houses within an occupied street?

Control measures for the demolition activity would include:

Control of the activity by a competent person.

- *A preliminary survey, followed by the completion of a risk assessment. Including the identification of hazardous substances, for example, asbestos and lead.*

- *Planning and preparation, including establishing a health and safety method statement that requires a systematic approach to the demolition and maintains the structure of the other properties.*

- *Perimeter fencing/barriers around the demolition area.*

- *Re-routing pedestrians and vehicles away from the demolition area.*

- Means of access to and out of the demolition area, for equipment and workers.

- Checking properties for vagrants, possible drug users or children and securing access to the buildings not yet demolished to prevent re-entry.

- Identification and isolation of services to the properties to be demolished.

- Provision of shoring of the other properties.

- Liaising with the occupants of the other houses, including agreeing when noisy operations could take place.

- Giving consideration to relocating the people in the houses most affected by the work.

- Safe positioning and storage of material from the demolition.

- If demolition is to be done by hand, safe means of access to height, for example, by the provision of independent scaffolds.

- Prevention of materials falling from height by provision of toe boards, mesh guards, nets and fans.

- Selection and use of suitable demolition machinery that provides protection to the user from falling materials.

- Controlling the movement of demolition machinery to avoid contact with people and obstructions.

- Site security at night.

- Provision of personal protective equipment, including hard hats, goggles, gloves and boots.

- Provision of welfare facilities to assist with maintaining hygiene, for example, washing hands.

- Control of dust caused by the demolition process, for example, by wetting down.

3) What are the main issues to be addressed in a pre-demolition survey of a three storey office block in a city centre?

The main issues to be addressed in a pre-demolition survey would include:

- The location and nature of adjacent properties.

- Methods used in the construction of the office block including any pre-tensioned or post-tensioned structural items.

- Review of drawings available for the building.

- The need for structural calculations to be made.

- The extent of dilapidation of the structure.

- The presence of retaining walls and attendant structural issues.

- Specific risks that may need to be considered in the demolition process, for example, removal of lift machinery or heavy water tanks.

- The presence of underground cellars.

- The potential for asbestos to be present.

- The location and condition of existing services, including gas, electricity and sewers.

- The need for and provision of temporary services to the site based on what is in place.

- Fire risks that would need to be managed during the demolition.

- Restricted access and egress to the site.

- Unauthorised use of the site, including the possibility of vagrants, drug users or children using the building and how to secure access to the building before being demolished, to prevent re-entry.

- Security issues due to its proximity in a city.

- Environmental issues such as noise, vibration and dust caused by the demolition process and the presence of any protected species, for example, bats.

- Evidence of biological hazards such as the presence of vermin, including rats and pigeons.

- The proximity of the building to roads and pathways.

- The proximity of the building to watercourses.

- The existence of a health and safety file.

4) A steel framed warehouse that was part of a large factory is to be demolished. What are the main control measures to be included in a demolition method statement?

Control measures for the demolition method statement would include:

- Control of the activity by a competent person.

- The demolition sequence and the methods of work to be used.

- The isolation of unnecessary existing services.

- The provision of temporary services to the site.

- The control and protection of the workers working on the factory site and the public by means, for example, of creating exclusion zones and erecting perimeter fencing.

- Provision of access and egress to the site for vehicles, other equipment and workers (including separation).

- Soft strip requirements.

- The stability of adjacent buildings.

- The use of temporary shoring to prevent premature collapse of the structure.

- Controlling the movement of demolition machinery to avoid contact with people and obstructions.

- Working at height, including access, for example, by scaffold or MEWP.

- The control of noise and dust.

- The precautions to be taken in dealing with hazardous materials, for example, asbestos, lead, paints and solvents.

- The temporary storage and removal of waste from the site.

- The control, communication and co-ordination of site activities.

- Procedures for reporting hazards that are identified.

- The competencies of the workers to be involved in the demolition operation.

- Provision of personal protective equipment, including hard hat, goggles, gloves and boots.

- Provision of welfare facilities to assist with maintaining hygiene, for example, washing hands.

- Procedures to be followed in the event of an emergency.

5) A single-storey farm building is to be demolished using an excavator. What control measures should the method statement contain for this activity?

The method statement should contain:

- Isolation of existing services/diversion of existing services.

- Requirement for initial soft strip.

- Appropriate removal of hazardous material.

- Means of controlling unauthorised access.

- Access/egress from site.

- Identification of responsible person.

- Emergency procedures.

- Competence of workforce.

- Separation of recyclable and non-recyclable material, on-site segregation.

- Appropriate off-site disposal.

- Welfare facilities/first-aid provision.

- Co-ordination of demolition and waste removal activities.

- Protection of workers/adjacent structures.

- Demolition sequence.

- PPE requirements.

6) What hazards may affect workers or the public during the demolition of a building?

Possible hazards are:

- Work at height.

- Falling debris/masonry.

- Unstable structures/premature collapse.

- Pre-post tension components.

- Build up of demolition materials/debris on floor.

- Services, electricity/gas/water.

- Dust and fumes.

- Hazardous substances/asbestos.

- Biological hazards, pigeon/bat droppings, rat urine.

- Heavy items.

- Use of explosives.

- Use of heavy plant.

- Noise/vibration.

- Hazardous materials from previous use, acids/flammable substances.

- Presence of storage tanks above/below ground – content.

- Presence of cellars/vaults affecting structures/adjoining property.

- Hot work/burning/cutting.

- Environmental issues/silting drains.

ELEMENT 6 – MOBILE PLANT AND VEHICLES

1) What possible control measures for plant, machinery and vehicles could be used on construction sites?

Control measures for the use of the equipment would include:

- *The proper selection of equipment for the work, ensuring that the equipment is capable of moving the maximum foreseeable load.*

- *Ensuring the competency of drivers and operators, by checking licences/certificates, and the experience they have of construction sites.*

- *Where necessary, ensuring the use of competent banksmen.*

- *The need to ensure that dangerous parts of machines are properly guarded.*

- *Providing a regime of planned inspections and maintenance, with procedures for reporting defects.*

- *The need to ensure good all-round visibility for drivers and operators.*

Site control measures to support the use of the equipment include:

- *Planning and maintenance of traffic routes to segregate pedestrians and vehicles.*

- *Imposing and enforcing speed restrictions in respect of vehicles.*

- *Preventing the unauthorised use or movement of plant and vehicles.*

- *Ensuring good standards of lighting and adequate signage.*

- *Avoiding vehicles reversing where possible.*

- *Providing competent supervision to ensure equipment is used properly and for its proper purpose.*

- *Providing awareness training for all workers on site, covering the hazards associated with the use of plant, machinery and vehicles and the precautions that must be observed.*

- *Providing workers with high visibility clothing.*

2) (a) Give FOUR reasons why a dumper truck used on a construction site may overturn?

Overturning can occur if a dumper truck is:

- *Being driven over obstacles, uneven or unstable ground, on or across slopes or too close to an excavation or embankment.*

- *Poorly maintained with tyres that are not inflated to the recommended pressure.*

- *Overloaded or the bucket unevenly loaded.*

- *Driven by people who are not competent or experienced at driving on a construction site, particularly if they drive at excessive speeds round corners or if they fail to avoid the many obstacles that are found on a construction site.*

(b) What practical measures could be taken to minimise the risk of a dumper truck overturning when used on construction sites?

Practical measures that could be taken to minimise the risks of a dumper truck overturning include:

- *Use of trucks with a wide wheel base and a low centre of gravity.*

- *Carrying out regular maintenance with particular attention being given to brakes, steering and tyre pressures and arranging for drivers to carry out pre-use checks.*

- *Providing designated traffic routes, which are kept free from obstructions.*

- *Providing edge protection around excavations.*

- *Setting and enforcing speed limits.*

- *Ensuring that the loads carried by trucks are within their safe working loads.*

- *Ensuring the driver's line of sight is not impaired in any way to avoid sudden steering adjustments.*

- *Avoiding driving across slopes.*

- *Using trained, competent and experienced drivers and monitoring their performance on a regular basis.*

- *Traffic routes positioned not too close to excavations.*

- *Installing smooth roads at the earliest opportunity to avoid rough terrain.*

3) Reversing vehicles within a construction site can be hazardous. How can the risk of accidents from reversing vehicles be reduced?

When reducing the risk of accidents from reversing vehicles consider:

- *Avoiding the need for vehicles to reverse if possible, by provision of one-way systems and good signage.*

- *Where this is not possible then a banksman should be used to ensure the area is clear and to control the reversing activity.*

- *Providing high visibility clothing for those people who need to work in areas where vehicles reverse, for example, a banksman.*

- Providing physical separation of vehicles and pedestrians in areas where vehicles reverse frequently, for example, by means of barriers.

- Providing audible alarms and visual warnings on vehicles to warn of their presence.

- Ensuring vehicles are fitted with mirrors or other reversing aids to assist the driver in manoeuvring, particularly providing a view of the rear of the vehicle.

- Providing mirrors at blind corners.

- Providing a refuge for pedestrians to enable vehicles to pass when space is restricted.

- Providing good lighting for the area where vehicles reverse, well directed lighting so that it avoids glare in the mirrors of the vehicles and shadows in the area being reversed into.

- Providing training for drivers (for example, the use of a banksman) and pedestrians (for example, the need for restricted access to revering areas), signage and enforcement of site rules.

4) Many construction workers are involved in work-related driving on public roads as part of their work activities. What factors associated with this driving increase the risk of an incident occurring?

The factors that increase risks related to work-related driving include:

- Distance - that the greater the distance of the journey (i.e. the duration of exposure to the hazard) the greater the risk.

- Driving hours - if driving hours related to construction activities are excessive the driver is likely to become tired, their attention and reaction levels will fall and they are at an increased risk of making mistakes. The hours may be excessive because the driver has been driving too long without a break, their total hours in a day have become too much or their rest period between days has become too short. For some drivers the driving hours are in addition to other work hours, for example, for someone travelling to a meeting at a site these total hours can have a significant effect on fatigue.

- Work schedules - work schedules that are badly organised can put increased pressure on drivers to be in a place by a given time. This can lead to them being tempted to increase speed and take abrupt action to change lanes to improve their progress. This risky action can lead to higher risk of collision and reduced stopping distances.

- Amount of traffic and delays - driving-related stress is likely to be experienced when the demands of the road/traffic environment exceed the driver's ability to cope with or control that environment. A person in this situation may resort to irrational behaviour in order to avoid certain situations or events that are putting them under pressure.

- Weather conditions - Weather conditions have a significant impact on the risks of driving. Sudden rainfall or snow and fog can lead to poor visibility and sudden breaking. Snow and surface water conditions can increase stopping distances. Even good weather can have a negative effect as glare from the sun can limit visibility. This is particularly the case in early morning or evening and is most significant during the winter when the sun is lower in the sky for longer.

5) What practical measures could be used to protect pedestrians on a construction site where segregation of people and vehicles is not possible?

The practical measures to protect pedestrians on a construction site where segregation of people and vehicles is not possible include:

- Where vehicles share the same workspace as pedestrians it is important to mark the work area as separate from vehicle routes to warn drivers to adjust their approach, be aware of pedestrians and give priority to pedestrians.

- Removable barriers or hi-vis tape and signs can be used to indicate work areas that are shared by vehicles.

- The use of high visibility clothing by pedestrians will assist drivers to see them.

- Drivers should have a clear view at all times, including the use of mirrors or similar devices.

- Audible and visual warnings fitted to vehicles would also assist.

- The reversing of vehicles should be done with the use of a banksman.

- Providing mirrors at blind corners.

- Providing a refuge for pedestrians to enable vehicles to pass when space is restricted.

- Providing good general lighting.

- Workers and drivers should be briefed on the hazards and control measures, including at site inductions.

- Signage, such as speed limit reductions, should also be used.

- A higher level of supervision than usual will be required.

ELEMENT 7 – WORKING AT HEIGHT

1) Scaffolding that incorporates a powered hoist has been erected to the outside of an office block in order to undertake repairs to the external cladding of the building. What could affect the stability of the scaffold?

Stability can be affected by:

- *The design of the scaffold which could have an effect on its stability, for example, whether the scaffold extends round more than one side of the building or the arrangements made for ground floor access points into the office building.*

- *Erecting the scaffold on soft or inadequately consolidated ground or without sound foundations, for example, sole boards and base plates.*

- *Inadequate bracing.*

- *Insufficient ties or an inappropriate mix of ties.*

- *Erecting the scaffold using incompetent workmen, for example, building it with an inadequate number of standards and poorly fitted couplings.*

- *The use of defective materials and incorrect components.*

- *The unauthorised removal of ties by workers.*

- *Adverse weather conditions that might overload the scaffold, for example, high winds and heavy rain or snow.*

- *Overloading of the scaffold with people, tools, equipment and materials.*

- *Inadequate strength of scaffold, bracing and ties in the location where the hoist is fitted.*

- *Overloading the hoist with materials.*

- *The scaffold being struck by vehicles.*

- *The scaffold being undermined by excavation work carried out in close proximity.*

2) A flat roof of a shop in a busy town centre is to be repaired while the shop remains open. What measures could be taken to reduce the risk to workers involved in the roof repair work and others who may be affected by the work?

The measures that could be taken include:

- *Assessing the condition of the roof with regard to its safe working load and possible fragile materials.*

- *The erection of barriers and signs at ground level to prevent access to the working area by unauthorised persons.*

- *The provision of safe means of access to the working area for people and materials, for example, by the use of a scissor lift (mobile elevating work platform).*

- *Segregation of the area where the scissor lift is operating.*

- *Provision of a landing place at the level to the roof so that people can get safe access to the roof.*

- *Provision of fall protection at the roof edge and landing place, for example, double guard rails and toe boards.*

- *Provision of barriers, including guard rails and toe boards, at any openings in the roof where someone could fall.*

- *Provision of fall arrest equipment where necessary.*

- *Precautions to be observed in the use of a LPG heated bitumen boiler, including with regard to the risk of fire and hot boiling liquid spills.*

- *Control of the removal of materials from the roof, for example, using the scissor lift and/or chutes.*

- *Control of placing and storage of vehicles, materials and waste at ground level so that it does not present a risk to pedestrians or other vehicles, including using barriers, signs and lighting.*

- *Control of work at ground level so that it does not obstruct fire exits from the shop or other buildings.*

- *The provision and use of personal protective equipment, for example, head protection, eyes, hands and feet. Special protection of the body will be required for those handling hot bitumen.*

- *The employment of competent workers who are fully aware of the risks associated with the work and the precautions that should be taken.*

- *First aid provision and procedures to be followed in the event of an emergency.*

- *Ensuring the occupiers of the shop understand the risks and precautions to be observed.*

3) What should be considered when conducting a working at height risk assessment?

Considerations for a working at height risk assessment include:

- *Determining whether the work at height could be avoided.*

- *The nature and duration of the task.*

- *Arrangements for planning and supervising the work.*

- *The suitability and state of repair of the access equipment.*

- The level of competence and training of the workers involved.

- The prevailing weather conditions.

- Means of access and egress, including the use of access boards or platforms provided with guard rails and toe boards or fencing.

- The use of fall arrest systems, for example, harnesses.

- The use of personal protection, for example, hard hats.

- The arrangements for emergency rescue.

- Evaluation of any health conditions of the workers involved in the work, for example, vertigo, heart or balance problems.

4) What are safe working practices associated with the use of a mobile elevating working platform?

Safe working practices which should be adopted include:

- The use of competent persons to operate and work from the platform.

- Using a MEWP that has received a thorough examination no more than 6 months before use.

- Using a MEWP that has had periodic inspections and a pre-use check before use.

- Its use only on firm level ground.

- Checking for the presence of buried services, drains and overhead obstructions.

- The use of outriggers for stability.

- The fitting of guard rails and toe boards to the platform, if they are not an integral part of the MEWP.

- The use of harnesses by workers on the platform.

- The erection of barriers around the area where the platform is to be sited and used.

- Ensuring the platform is not overloaded with people, materials and equipment.

- Securing all tools and equipment before the platform is elevated.

- Not moving the location of the MEWP when it is elevated or with people in it.

- Developing and introducing emergency rescue procedures.

5) What are safe working practices associated with work over water?

Safe working practices which should be adopted include:

- The provision of barriers, including guard rails and toe boards where practicable.

- The provision of fall arrest equipment, for example, safety nets and/or harnesses.

- Provision of personal buoyancy aids.

- Provision of a safety boat.

- Placing safety nets or safety lines across fast flowing water for people to hang on to until rescued.

- Close supervision and checks to ensure no one is missing.

- Workers working in pairs or larger groups where practicable.

- Developing and introducing emergency rescue procedures.

- Equipment for raising the alarm.

- Provision of rescue lines.

- Provision of communication equipment between rescue boats and the shore.

ELEMENT 8 – MUSCULOSKELETAL HEALTH AND LOAD HANDLING

1) A welfare unit is to be lifted from the vehicle trailer it is being delivered onto its location on the construction site. What would the person appointed to have overall control of this operation need to ensure so that lifting operations are conducted safely?

The person appointed to control the lifting operation would need to ensure:

- An initial assessment of the operation was carried out, considering access to where the lift was to take place, restriction affecting siting of the crane and movement of the jib when lifting, and ground conditions.

- That the lifting appliance and the associated equipment were suitable for lifting the load, for example, the load carrying capacity and reach of the jib and were marked with their safe working loads. Suitability would include the stability of the crane when lifting, for example, its need for outriggers.

- Documentary evidence to show that the required examinations, inspections and maintenance had been carried out on the crane and accessories.

- *Following consultation with everyone involved, that a plan was in place to enable effective co-ordination of the lifting operation.*

- *That all people involved, including the crane driver, slinger and banksman, were competent to carry out their duties and fully conversant with the hand signals or other means of communication that were to be used.*

- *Systems for control of the lifting area were put in place, including demarcation of it, and supervising the lifting operation.*

- *Lifting points on the welfare unit and lifting accessories were suitable to enable the lift to take place safely and not damage the welfare unit.*

- *A system for reporting defects and incidents was put in place.*

- *Adverse weather conditions were taken into account, especially the presence of high winds which would have an effect on the control of the load.*

2) A 100-metre-long section of a footpath in a residential area with a speed limit of 30mph is to have the kerbstones replaced.

(a) What hazards could affect the health and safety of the workers involved in this process?

The hazards involved in this activity include:

- *Being struck by other road vehicles.*

- *Frequent handling of the heavy kerbstones in an awkward position.*

- *Dropping heavy kerbstones onto fingers or toes.*

- *Silica dust from cutting kerbstones to size.*

- *Noise from road traffic and the range of tools being used.*

- *Vibration from prolonged use of saws and other hand held power tools.*

- *Wet cement used to bed the kerbstones in may cause cement burns and dermatitis effects.*

(b) What control measures could be put into place to reduce the risk?

Control measures will include:

- *The development of a traffic management plan to provide a safe working area.*

- *Ensuring the use of mechanical aids to transport the kerbstones or even using lighter kerbstones (a range is available made from plastic). At worst, using team lifting techniques.*

- *All workers should be trained in manual handling.*

- *Good welfare facilities should be provided so that hygiene standards can be maintained.*

- *Providing tools with in-built suppression methods (water and/or on tool extraction) to prevent/control dust.*

- *Maintaining the tools used so that cutting blade remain sharp.*

- *Safe systems of work may also be considered to reduce vibration exposure, for example, job rotation.*

- *The workers should be issued with a range of PPE, such as safety boots to protect from dropped kerbstones, hearing protection, dust masks and hi-visibility clothing.*

3) What conditions relating to lifting accessories could form part of a pre-use checklist?

The conditions of lifting accessories that could be included in a pre-use check list include:

- *Abrasion or cuts on fibre rope slings.*

- *Stretched fabric of fibre rope slings.*

- *Broken wires or kinks in wire ropes and slings.*

- *Deformed hooks.*

- *Deformed, stretched and cracked links in chains.*

- *Damaged or missing safety clips from hooks and slings.*

- *Evidence of corrosion or misalignment.*

- *Evidence of alterations to accessories, such as shortening or joining items together and adaption of shackles by using bolts to replace shackle pins.*

- *Marking of accessories with their safe working loads.*

4) What are the requirements for thorough examinations of lifting equipment to be carried out under the Lifting Operations and Lifting Equipment Regulations (LOLER)?

The Lifting Operations and Lifting Equipment Regulations require that lifting equipment must be thoroughly examined:

- *Before being put into use for the first time.*

- *After assembly and before being taken into use where the safety of the equipment depends on the installation conditions, for example, a tower crane.*

- *Every six months in the case of accessories.*

- *Every six months in the case of equipment used for lifting persons.*

- *Every twelve months for other lifting equipment.*

- *When exceptional circumstances have occurred which could jeopardise the safety of the equipment.*

5) What measures should be taken to reduce the risk of a rough terrain fork-lift truck overturning?

The measures to take to reduce the risk of a rough terrain fork-lift truck overturning include:

- *Driver training and experience with the use of rough terrain fork-lift trucks.*

- *Control measures to deter excessive speed, such as rules, signs and enforcement of a speed limit through supervision and monitoring.*

- *Pre-use checks by the vehicle driver.*

- *The vehicle tyres should be fully inflated*

- *A maintenance programme to ensure the tyres have no punctures/cuts and are replaced before they become too worn.*

- *Driver training to prevent travelling with the load raised.*

- *Ensuring the vehicle is not overloaded.*

- *Ensuring that the load is stable on the forks.*

- *Traffic routes positioned not too close to excavations.*

- *Avoiding driving across slopes.*

- *Installing smooth roads at the earliest opportunity to avoid rough terrain.*

- *Ensuring the driver's line of sight is not impaired in any way to avoid sudden steering adjustments.*

6) Manual handling operations during construction activities can cause injuries.

(a) What TWO types of injury could be caused by the incorrect manual handling of loads?

Potential injuries include:

- *Spinal disc compression or prolapsed disc.*

- *Strains to tendons or muscles.*

- *Hernia or ruptures.*

- *Cuts and broken bones, for example affecting the feet.*

(b) With reference to the work environment related to construction manual handling activities, what are the means of reducing the risk of injury during manual handling operations?

Means of reducing the risk of injury during manual handling operations include:

- *Ensuring work positions, workstations or work surfaces are at the optimum heights to avoid unnecessary stooping and reaching.*

- *Ensuring the surface of the floor is flat, clear of obstruction and without changes in level so that the worker does not stumble or have to step over objects when carrying loads.*

- *Ensuring the lighting is suitable and sufficient so that workers can see what they are doing and can see how best to grip and move the load.*

- *Controlling the work temperature to avoid excess cold, which reduces dexterity, and excess of heat, which may cause the hands to sweat and lose grip.*

- *Ensuring sufficient space, so that the worker can adopt kinetic handling techniques.*

ELEMENT 9 – WORK EQUIPMENT

1) The Provision and Use of Work Equipment Regulations (PUWER) 1998 require that work equipment used in hostile environments is inspected at suitable intervals. What items on a 360° tracked excavator should be subject to inspection?

Items for inspection include:

- *The provision and condition of roll over protection, for example, the cab.*

- *The provision and condition of driver restraints, for example, seats belts.*

- *The seats for the security of their fittings, function of vibration attenuation and condition for comfort.*

- *The condition of the bodywork, for damage that could affect motion or be a problem to pedestrians.*

- *Windscreens and windscreen cleaning equipment.*

- *The condition of rear view and other mirrors.*

- *The condition of lights and warning devices.*

- *The condition of the tracks and linkages.*

- *The effectiveness of the braking system and steering.*

- *The condition of the hydraulic arm and bucket for signs of overloading damage.*

- *The performance of the bucket release and tilt mechanisms.*

- *The integrity of fuel, oil and hydraulic systems and the legibility of labels and signs.*

2) What are practical control measures for reducing the risk to workers when using a bench mounted electrically powered circular saw on site?

The control measures to take to reduce the risks of using a circular saw include:

- *Ensuring the saw bench is set up on level ground and secured against movement.*

- *Ensuring guards are in place on the circular saw and correctly adjusted for protection against moving parts, particularly the saw blade.*

- *Using a reduced voltage of 110 volts for the circular saw and connecting it to the supply through a residual current device (RCD).*

- *Using heavy duty cables or cable covers to provide protection against moving vehicles or similar damage.*

- *Placing materials close to the saw bench to minimise the amount of manual handling to be carried out.*

- *Providing local exhaust extraction to reduce the dust created by the cutting operation.*

- *The use of appropriate personal protective equipment, including hearing and eye protection and gloves to counter the noise, dust and wood splinters.*

- *The provision of information and training to users, supervisors, those that maintain the equipment and workers in the area it operates on the health and safety risks involved and the precautions to be taken.*

- *Restriction of use of the circular saw to suitable and competent people, trained and experienced in the specific risks of the equipment.*

- *Ensuring that the saw is not overloaded when cutting and that the manufacturer's instructions with regard to use are followed, particularly for feeding wood to the saw blade, a push stick may be required.*

- *Providing a good level of supervision to ensure the precautionary measures are followed.*

3) What control measures should be used to reduce risks when cutting concrete kerbstones with a petrol disc cutter?

The control measures that should be used to reduce the risks when cutting concrete kerbstones include:

- *The selection of suitable equipment and its inspection prior to use.*

- *The provision of information on and training in safe use of the equipment and manual handling for users of the equipment, those that supervise them and other workers that could be affected by the hazards of the cutting activity.*

- *Restriction of use of the petrol disc cutter to competent people, trained and experienced in the specific risks of the equipment.*

- *The use of a competent person to change the disc of the disc cutter when needed.*

- *The provision of an adequate top guard for the disc cutter.*

- *Dust suppression by wet cutting and dust extraction to control dust.*

- *Using the equipment in a ventilated area away from other operations.*

- *Stopping the engine and applying the brake when the cutter is not being used or when it is being carried.*

- *Maintaining the disc cutter so that the cutting blade remains sharp.*

- *Protecting the operator against HAVS by, for example, the provision of insulation for the handle and introducing job rotation.*

- *Safe systems of work may be used to reduce vibration exposure, for example, job rotation.*

- *Ensuring a safe system of work for refuelling and keeping the minimum amount of fuel on site.*

- *The provision and use of appropriate personal protective equipment, including goggles, hearing protection, gloves, respiratory protection and safety boots.*

- *Good welfare facilities should be provided so that hygiene standards can be maintained.*

- *Providing a good level of supervision to ensure the precautionary measures are followed.*

4) In what ways can accidents occur from the use of cartridge-operated nail guns on a construction site?

The ways accidents can occur when using cartridge-operated nail guns includes:

- *The charge strength of the explosive cartridge may be too much for the material the nail is to be fixed to causing penetration.*

- The charge may be correct, but the where the nail is being fixed may be a weak area in the material or there may be a void behind it.

- The material being fixed to may be too thin or thinner than expected.

- The operator of the gun or other workers may hold material being fixed, putting their hand too close to the gun or the nail may penetrate the operator's hand behind the material.

- The nail may ricochet and injure the operator of the gun or other workers near-by.

- The nail may strike a particularly hard part of the material it is being fixed to, for example, a reinforced rod or hard aggregate.

- The operator may attempt to fire a nail into a hole of a previous attempt at fixing, where the material is damaged and fragile.

- Attempting to fix a nail too near an edge of the material, causing the material to break away and deflecting the nail.

- Firing a nail into material that is too hard for the charge used and the nail selected.

- Not firing the nail square to the material causing it to travel at an angle and bounce off the surface of the material.

- A misfire of the cartridge may occur and the cartridge might explode while someone is attempting to remove it forcibly from the gun.

- When a misfire occurs an operator may look at the front of the gun, where the detonation is delayed it could fire the nail into the face of the operator.

- Strain injuries may be caused by the recoil of the gun at time of explosion of the cartridge and operators might lose their footing and fall.

5) (a) List, and give an example for each, FOUR mechanical hazards of construction machinery?

Mechanical hazards of construction machinery include (four only required):

- Entanglement, for example, the rotating chuck of a drilling machine.

- Friction and abrasion, for example, the wheel of a bench grinding machine.

- Cutting, for example, the blade of a circular saw.

- Shear, for example, the scissor mechanism of a scissor lift mobile elevating work platform (MEWP).

- Stabbing and puncture, for example, the nail fired from a nail gun.

- Impact, for example, hit by the moving arm of an excavator.

- Crushing, for example, between the folding arms of a MEWP.

- Drawing in, for example, the rollers of a surface roller.

- Injection, for example, high pressure oil ejected from the hydraulic system of a rough terrain fork-lift truck.

- Ejection, for example, dust particle ejected from a disc cutter.

(b) What are the advantages and limitations of using a fixed guard to protect people from mechanical hazards of construction machinery?

The advantages and limitations of a fixed guard include:

Advantages:

- Create a physical barrier.

- Require a tool to remove.

- No moving parts - therefore they require very little maintenance.

- Can be designed to enable visibility through clear panels or mesh.

- Suitable for adverse environments.

Limitations

- Do not disconnect power when not in place, there-fore the machine can still be operated without the guard being fitted.

- May cause problems with visibility for inspection.

- If enclosed, may create problems with heat which, in turn, can increase the risk of explosion.

- May not protect against non-mechanical hazards such as dust/fluids which may be ejected.

ELEMENT 10 - ELECTRICITY

1) How can the following two protective measures reduce the risk of electric shock AND, in EACH case, give an example of their use in construction activities?

a) Reduced low voltage.

b) Double insulation.

(a) 'Reduced low voltage' involves the reduction of mains voltage by a transformer to a lower, safer voltage - typically 110 volts. Any shock voltage can be restricted to 55 volts by means of a transformer that is centre tapped to earth. Reduced voltage systems are commonly used for portable electrical hand tools on construction sites. Additionally, safety extra low voltage (SELV) - a voltage less than 50 volts - is used in hand lamps.

(b) In 'double insulation' equipment internal live parts of a piece of equipment have two layers of insulation around them, which prevent the live parts contacting exposed conductive parts, such as the outer metal casing of the equipment. Consequently, an internal electrical fault condition cannot make any part of the casing live and pass current to the user.

Double insulation is used on portable electrical hand tools.

2) A fire has started on a construction site immediately adjacent to a bench mounted electric circular saw.

(a) How might the saw's electrical power system have caused the fire?

The saw's electrical power system could cause a fire by:

- Arcing or sparking from the motor inside the saw body may have resulted in wood dust igniting.

- The cables from the power source to the saw may have been left tightly coiled and the electrical resistance generated has built up sufficient heat to ignite.

- The power cables could have become damaged so that there was a short circuit, which caused heat sufficient to ignite the cable insulation and wood dust.

- The connection of the power cable into the saw may have become loose because of vibration and other movement causing a poor connection that created heat and ignition of wood dust.

(b) What are the possible effects of an electric shock to the saw operator?

The possible effects of an electric shock to the saw operator includes:

- The operator may experience muscular contraction in the hand and be unable to break the grip of the equipment.

- The operator could experience respiratory arrest as electricity interferes with the muscles that enable breathing.

- The operator could experience cardiac arrest or interruption of the heart beat rhythm, fibrillation.

- The heat generated by the body's resistance to the flow of electricity can cause burns that may be internal or external to the body, for example, at the point of contact with the electricity.

- The electric shock may also cause secondary injuries, for example, amputation of fingers if the saw blade is still running and the operator comes into contact with it.

3) A road is being resurfaced in an area that has overhead power lines crossing it. The work involves the use of tipper vehicles, excavators and a road planing machine that has an extended jib. What control measures can reduce the risk from contact with the overhead power lines?

The control measures that should be used to reduce the risks from contact with the overhead power lines include:

- Consulting the electricity supply company with a view to making the lines dead and obtaining written confirmation that isolation has been achieved.

- If isolation was not possible, advice should be obtained from the supply company on a safe working distance and precautions introduced to maintain this distance, for example, the erection of goalpost barriers and warning signs.

- Controlling access to the area to ensure only authorised vehicles and other equipment are allowed access and the rules are known by their drivers/ operators.

- Fitting plant with height restrictors, coupled with audible warnings in the cab.

- Designating areas for dropping materials and ensuring that spoil heaps are positioned away from the overhead lines.

- Using competent drivers and banksmen.

- Communicating the safety precautions to the operatives by means of, for example, tool box talks.

4) (a) What control measures should be used by someone using a hand-held electric drill on a construction site to reduce the risk of electric shock?

Control measures to reduce the risk of electric shock include:

- A pre-use check of the equipment by the user, including the general condition of the drill for damage, the condition of both plug and socket, evidence of overheating or burning, the integrity of the cables, the rating of the fuse.

- *The use of reduced/low voltage where possible.*

- *The provision of a residual current device.*

- *The siting and protection of cables from damage by vehicles and people.*

- *Keep drill and cables/connectors out of and protect them from water.*

- *Positioning the cable so that it does not get damaged in use or when the drill is put down.*

- *Check for evidence that a portable appliance test (PAT) has been carried out.*

(b) Other than electricity, what FOUR hazards are associated with the use of hand-held electric drills on construction sites?

Other hazards associated with the use of the drill include:

- *Entanglement with the chuck or bit.*

- *Stabbing or puncture by the drill bit.*

- *Cuts from waste metal from the drilling operation.*

- *Noise from the drill cutting material.*

- *Hot surfaces from the heat generated by the drill's cutting process.*

- *Tripping hazards associated with trailing cables.*

5) (a) Why does a fuse fitted to hand-held electrical power tools only provide limited electric shock protection to the user of the tool?

A fuse protects the equipment rather than the user. This is because it responds at current levels that are often fatal, for example, 3 amperes and above. They are also relatively slow to react to a fault and isolate the flow of current.

The time taken to react may be long enough to cause fibrillation in a person in contact with the fault current.

(b) What are the advantages of using a residual current device to protect users of hand held electrical power tools from electric shock?

The advantages of a residual current device are that it operates at low levels of current, for example, 30 mA and provides rapid isolation from the current, for example, within 30ms. This sensitivity means that it responds to a fault at lower current thresholds and faster than a fuse, limiting the duration of electric shock, making it less likely that the current will cause fibrillation in someone in contact with the fault. In addition, they are difficult to override, can be portable and easy to test.

(c) What are the advantages of double insulation for hand-held electrical power tools?

The advantages of double insulation hand-held electrical power tools are that the internal live parts of a hand-held electrical power tool have two layers of insulation around them, which prevent the live parts contacting exposed conductive parts, such as the outer metal casing of the equipment. Consequently, an internal electrical fault cannot make any part of the casing live and pass current to the user.

ELEMENT 11 – FIRE

1) Arson is the single greatest cause of fire.

(a) What factors make some construction sites vulnerable to arson attacks?

Factors that may make some construction sites might be vulnerable to arson attacks include:

- *A lack of or damaged perimeter fencing.*

- *Inadequate supervision of visitors to the site.*

- *Failure to provide CCTV or security guards during non-working hours.*

- *A poor standard of housekeeping with combustible materials allowed to accumulate and the use of open skips.*

- *A failure to store flammable materials in a secure store when work has finished for the day.*

- *The location of the site, perhaps in socially deprived areas or areas with a history of vandalism.*

- *A failure to engage with the local community, in particular schools, to encourage fire prevention.*

(b) What control measures can be taken to reduce the risk of arson on a construction site?

Control measures that can be taken to reduce the risk of arson include:

- *Erecting and maintaining an effective perimeter fence.*

- *Controlling the entry of persons to the site, such as visitors, sub-contractors and ex-workers.*

- *Providing security lighting and using security guards/ CCTV during non-working hours.*

- *Ensuring combustible waste does not accumulate or is in skips that are secured when not being used.*

- *Ensuring flammable materials are kept in a secure storage area.*

- *Erecting warning signs and liaising closely with local police and neighbouring schools.*

2) Liquefied petroleum gas (LPG) is often stored on construction sites. What control measures should be used for the safe storage of LPG on construction sites?

Factors for the safe storage of LPG on construction sites include:

- *Storage in a clear open space.*

- *Secured compound.*

- *2m high fencing.*

- *Safe distance from toxic, corrosive and combustible substances.*

- *Kept away from exit routes.*

- *Stored in well ventilated areas.*

- *Oxygen cylinders, if stored, is at least 3m away.*

- *Protected from sunlight.*

- *Empty containers stored separate to full ones.*

- *An appropriate fire extinguisher nearby.*

3) Fire is a significant hazard on refurbishment projects that involve hot works.

   (a) Which THREE activities can be considered as 'hot works'?

Activities that are considered to be 'hot works' include (three only required):

- *Cutting operations, for example, during the removal of structural steels.*

- *When welding operations are taking place, for example, steel fabrications.*

- *During plumbing operations, for example, where boilers are connected to water distribution systems.*

- *Activities that could create sparks, for example, grinding activities using portable angle grinders.*

   (b) What precautions may be taken to reduce the risk of a fire occurring during 'hot works'?

Precautions to reduce the risk of fire from 'hot work' include:

- *The use of permits to work for 'hot work'.*

- *Arranging for combustible and flammable materials to be removed from the immediate area where 'hot work' takes place. For example, removing accumulated rubbish and the proper storage of flammable materials away from the 'hot work'.*

- *Organising work that involves flammable materials in the area of 'hot work' does not take place at the same time as 'hot work'.*

- *Ensuring the area where 'hot work' is to take place is well ventilated after work that involves flammable materials.*

- *Covering combustible materials that cannot be removed easily with non-combustible sheets.*

- *Provision of someone to watch for and respond to possible ignition of combustible and flammable materials.*

- *Provision of fire-fighting equipment, for example, portable extinguishers and fire blankets.*

- *Supervision of work activities that follow the 'hot work' to ensure combustible and flammable materials are not placed where the 'hot work' took place until the areas affected are sufficiently cool.*

- *A check of the 'hot work' area for a period of time after the work is completed to check for possible fires.*

4) What factors would need to be covered in the induction training of workers to enable their response to a fire emergency on a construction site?

Factors for fire emergency training on a construction site include:

- *Location and operation of call-points and other means of raising the alarm.*

- *How the fire service is called.*

- *Recognition of fire alarms and the actions to be taken.*

- *Understanding the emergency signs, particularly those that show exit routes and the assembly point.*

- *The importance of maintaining emergency lighting.*

- *Location of fire escape routes and assembly points.*

- *Requirements for safe evacuation (for example, non-use of lifts, do not run etc.).*

- *Consideration of people with special needs.*

- *Location, use and limitations of fire-fighting equipment.*

- *Identity and role of fire marshals.*

- *Arrangements to resume work.*

5) What are the factors to be considered when assessing fire risks and the effects on the current fire risk reduction measures of a major refurbishment project on one floor of an occupied multi-floor office block?

Factors to assess include:

- *Sources of ignition - increase in power tools, extension leads, sources of heating and lighting, grinding and welding activities.*

- *Potential sources of fuel - increase in combustible materials in the form of wood, packaging, flammable liquids, located/stored in large volume and in corridors and on stairs.*

- *Sources of oxygen - potential for oxyacetylene equipment.*

- *Effects on structural integrity of the building - fire doors removed or held open, deconstruction of partitioning, openings created in compartmentalised areas.*

- *Fire detection - false alarms due to dust rather than fire, some detectors covered or taken out of use, arrangements for when doing hot work.*

- *Fire warning - possibility of it being decommissioned on the floor being renovated and specific sections of the floor affected directly by the refurbishment work.*

- *Fire extinguishing - need to isolate sprinkler system on floor being refurbished to do certain work, new partitions etc. affecting effectiveness of system.*

- *Fire escape routes and assembly points - may be blocked by materials and contractors' equipment.*

- *Requirements for safe evacuation (for example, non-use of lifts, do not run etc.) - contractors may change and there would be a need to keep them all informed of requirements.*

- *Consideration of people with special needs - occupants of the building that may be affected by the availability of means of escape.*

- *The need for fire marshals to ensure contractor evacuation.*

ELEMENT 12 - CHEMICAL AND BIOLOGICAL AGENTS

1) Insulation board tiles that contain asbestos are to be removed from the ceiling of a storage area located within a high street department store. What measures should be taken into consideration when planning the work to minimise the risks from asbestos?

Measures to be taken to minimise asbestos risks include:

- *Consideration of the condition of the tiles and the type of asbestos, in part, this would determine whether the work is notifiable to the HSE. A survey may be required to determine this.*

- *The possibility that asbestos might be present in the storage area in other forms than the tiles and that it might become disturbed by the work.*

- *Preventing access to the storage area by isolating the storeroom and locking the door if the condition of the tiles is poor.*

- *Provision of information to the department store staff regarding risks from asbestos and the measures being taken.*

- *The appointment of a licensed contractor for the removal of the asbestos.*

- *Ensuring that, if the work is notifiable, HSE is given fourteen days' notice of the work to be carried out.*

- *The contractor's need to erect sheeting to contain the asbestos during its removal and dependent on the removal method and the condition of the tiles, the need to maintain the sheeted enclosure under negative pressure.*

- *The contractor's need to have a controlled route in and out of the storage area that prevents asbestos escaping.*

- *The contractor's provision and siting of their decontamination and welfare facilities.*

- *The arrangements for independent monitoring of airborne fibre levels during the operation and for carrying out an air clearance test and the issue of a certificate when the work is completed.*

- *The arrangements for the safe disposal of the asbestos waste by means of a licensed carrier.*

2) Exposure to cement dust can lead to adverse health effects if breathed in, for example, irritation of the nose and throat.

(a) What is the meaning of the term 'workplace exposure limit'?

A workplace exposure limit defines the maximum permitted concentration of an airborne hazardous substance averaged over a specified period of time (eight hours or fifteen minutes) and referred to as a time weighted average.

(b) What control measures could be considered to control levels of cement dust in the workplace?

Measures to be taken to minimise cement dust in the workplace include:

- *Having the process that produces the dust carried out off site in a controlled environment.*

- *Providing materials that contain cement in a wet form for distribution round the site.*

- *Providing the cement in hoppers that can more carefully control the amount of cement provided when putting it into containers for mixing, avoiding the use of bags that have residual dust in them.*

- *Suppressing the cement dust as it arises with water sprays.*

- *Enclosing processes that require dry cement to be added for mixing and installing local exhaust ventilation on these mixing processes.*

- *Carrying out processes that create cement dust, for example, drilling, using wet dust suppression or on tool local exhaust ventilation.*

- *Changing processes, for example, using a block splitter instead of a cut-off saw.*

- *Removal of cement (and materials that contain cement) spills/residue before it dries.*

- *Introducing a high standard of housekeeping using a vacuum cleaning system as opposed to hand sweeping.*

3) What FOUR hazardous substances are widely encountered in the construction industry AND give an associated health risk for EACH?

Hazardous substances widely encountered include (four only required):

- *Silica - inhalation of silica may result in silicosis, which is a fibrosis of the lung.*

- *Cement - prolonged or repeated exposure to cement dust can lead to dermatitis. Continuous contact with wet cement may cause burns or skin ulcers. Prolonged or repeated breathing of cement dust can lead to silicosis in the lungs.*

- *Wood dust - inhalation of the dust arising from sanding or cutting of wood may cause allergic or non-allergic respiratory symptoms and cancer (depending on the type of wood).*

- *Solvents - inhalation of vapours from organic solvents, for example, trichloroethylene, can cause drowsiness, it can depress the central nervous system and may lead to liver failure.*

- *Isocyanates - inhalation of vapours, particularly when the substance is sprayed, can lead to allergic respiratory symptoms, for example, occupational asthma.*

- *Lead - lead poisoning results from the inhalation of fumes produced from the heating of lead, for example, when oxyacetylene cutting of metal coated with lead paint.*

- *Asbestos - asbestos fibres readily become airborne when disturbed and can enter the lungs where they can cause asbestosis, mesothelioma (cancer of the pleural lining of the lungs) and thickening of the pleural lining.*

4) What information should be included in a safety data sheet (SDS) for hazardous chemicals?

The safety data sheet for chemicals should include:

- *Identification of the substance/mixture and of the company/undertaking, including the product identifier shown on the label, its recommended use and contact information for the supplier, including for emergencies.*

- *Hazard identification, including signal words, hazard statements and precautionary statements – the pictogram of hazard symbols may be provided.*

- *Composition/information on ingredients.*

- *First-aid measures for initial care must be described in a way that can be understood by an untrained responder. Whether medical attention or special treatment is required, including the urgency with which it may be needed and symptoms of any effects.*

- *Firefighting measures, including suitable media, hazards of combustion and advice to firefighters.*

- *Accidental release measures for non-emergency responders and emergency responders, including spills containment and cleaning up.*

- *Handling and storage, including any incompatibilities and the effects of things like weather.*

- *Exposure controls/personal protection, including workplace exposure limit values and requirements for such things as local exhaust ventilation and specific personal protective equipment (PPE).*

- *Physical and chemical properties, such as odour, flash point, vapour pressure.*

- *Stability and reactivity, including conditions to avoid, for example, heat or incompatible materials.*

- *Toxicological information, including the health hazards, for example, carcinogenicity, route of entry, symptoms and the effects, including delayed and immediate effects from short and long-term exposure.*

- *Ecological information, in order to evaluate the environmental impact of dealing with spills and arrangements for waste.*

- *Disposal considerations, including methods and properties that may affect landfill or incineration.*

- *Transport information, including UN number and transport hazards.*

- *Regulatory information, including relevant national information on the regulatory status of the chemical where it is being supplied.*

- *Other information, including the date the SDS was prepared, changes from previous versions and advice on training appropriate for workers to ensure protection of human health.*

5) (a) What FOUR forms of chemical agents might be found in construction workplaces?

Dusts (silica), fibres (asbestos), fumes (welding), vapours (solvent), gases (carbon monoxide), liquids (sulphuric acid), solids (lead), mists and aerosols (sprayed isocyanates) (four only required).

   (b) What are the differences between the acute and chronic effects from exposure to chemical agents?

The differences between acute and chronic effects are:

- *Acute health effects from exposure to chemical agents include an immediate or rapid adverse effect, following a single or short-term exposure. The effects are usually reversible.*

- *Chronic effects from exposure to chemical agents produce a gradual response and they may not be recognised for a number of years. The effects appear after prolonged or repeated exposure to chemical agents and are often irreversible.*

ELEMENT 13 - PHYSICAL AND PSYCHOLOGICAL HEALTH

1) A noise survey carried out on a building site identified that noise exposure action values given in the Control of Noise at Work Regulations 2005 were being exceeded.

(a) What are the lower AND upper noise exposure action values defined in the Control of Noise at Work Regulations 2005?

The lower noise exposure action value is 80 dB(A) daily average noise exposure measured over an 8hr period and the upper noise exposure action level is 85 dB(A).

(b) What control measures is an employer required to take by the Control of Noise at Work Regulations 2005 when employees are exposed to noise at or above the upper noise exposure action value?

Control measures required include:

- *The elimination of noise at source or its reduction to as low a level as reasonably practicable by implementing a programme of organisational and technical measures.*

- *The control of exposure to noise by the application of the general principles of prevention.*

- *The designation and marking of hearing protection zones.*

- *Provide employees who are exposed to noise above the upper action level with personal hearing protectors.*

- *Ensure, so far as is reasonably practicable, that employees entering hearing protection zones wear hearing protection.*

- *The provision to workers of information, instruction and training on the risks of exposure to noise and the controls that are introduced.*

- *Undertaking health surveillance where a risk assessment indicates that this is necessary.*

2) What are the work-related issues that might increase the levels of occupational stress amongst workers on a construction site?

The issues that could increase the level of stress include:

- *Demands of the work - workload and work patterns. Stress may be due to an individual not being able to cope with the demands of the job, for example, the expectation of working of excessively long hours or to demanding schedules. If the work pattern or job demands are such that the worker does not get adequate breaks this can also lead to stress. In addition, the worker may have to work unsociable working hours, which can increase the risk of stress.*

- *Demands of the work environments - excessive noise levels, poor levels of lighting, inadequate welfare facilities and the poor standard of housekeeping.*

- *Control - if a construction worker has little control over the pace and way they work this can create stress.*

- *Support - if a construction worker perceives a lack of encouragement or resources provided by the organisation, line management or co-workers a feeling of remoteness can be created. Failure to provide adequate resources and properly maintain equipment can increase the risk of stress. Failure to provide good levels of communication, seek the views of employees, involve them in decisions and recognise good performance when it occurs can also lead to stress.*

- *Work relationships - failure to introduce procedures and implement policies to cover harassment, discrimination, violence and to investigate complaints can increase the risk of stress.*

- *Role - if workers do not understand their role in the organisation, or have conflicting roles, this can increase the potential for stress.*

- *Change - if the management of organisational change is not conducted or communicated well, individuals can be confused or feel threatened by the change and this could lead to stress.*

3) As part of major football stadium refurbishment, seat anchors are to be installed by workers using hand-held powered percussive equipment that produce high levels of vibration.

(a) What are the possible health effects of exposure to vibration?

Effects of vibration could include:

- *Numbness, tingling and blanching of the fingers.*

- *Swollen and painful joints.*

- *A reduction in strength, grip, dexterity and in sensory perception.*

- *Involuntary muscular movement.*

(b) What control measures could be used to reduce the risk of health effects due to vibration in this refurbishment work?

Control measures to consider would be:

- *Elimination of the vibration producing task by considering an alternative means of fixing.*

- *Substituting the percussion tools with lower vibration equipment.*

- *Reducing the time of exposure of the workers using the percussion equipment by providing frequent breaks and/or job rotation.*

- *Modifying the equipment to improve the grip on the tools to reduce the vibration transmitted.*

- *Introducing a planned maintenance programme for the tools.*

- *Providing appropriate personal protective equipment, including gloves to keep the hands warm.*

- *Introducing a programme of health surveillance.*

- *Providing the workers with information, instruction and training on the hazards associated with the use of the tools and the control measures that should be taken.*

4) (a) What are TWO types of ionising radiation that workers in construction activities might encounter?

Two types of ionising radiation are:

- *Gamma rays and x-rays used to check welds or the integrity of structures.*

(b) What are TWO types of non-ionising radiation that workers in construction activities might encounter?

Two types of non-ionising radiation are (only two required):

- *Microwaves (for example, from equipment on the roof of office buildings).*

- *Lasers (used for levelling and measurements).*

- *Ultraviolet light (from sunlight and arc welding).*

- *Infrared (from portable lighting).*

(c) What are the range of health effects of exposure to non-ionising radiation?

The range of health effects from exposure to non-ionising radiation include:

- *Heating of the surface of the skin or deeper levels, depending on the energy of the radiation.*

- *Burns to the skin or eyes, for example, sun burn from exposure to ultraviolet light.*

- *Eye damage, for example, the retina from lasers and cataracts from infrared and ultraviolet light.*

- *Skin cancer, from ultraviolet light.*

5) What signs might indicate that a contractor is misusing drugs during working hours?

Signs include:

- *A tendency to be confused.*

- *Abnormal fluctuations in concentration and energy.*

- *Sudden mood changes.*

- *Impaired job performance/taking risks.*

- *Perception adversely affected.*

- *Erratic behaviour/unusual irritability/unusual aggression.*

- *Poor time-keeping.*

- *Increased short-term sickness absence.*

- *Deterioration in relationships with colleagues/customers/management.*

- *Dishonesty/theft to maintain expensive habit.*

- *Unexplained absence during working hours.*

- *Dilated pupils.*

- *Drugs/drug taking equipment found at work.*

This page is intentionally blank

Revision and examination guidance

Contents

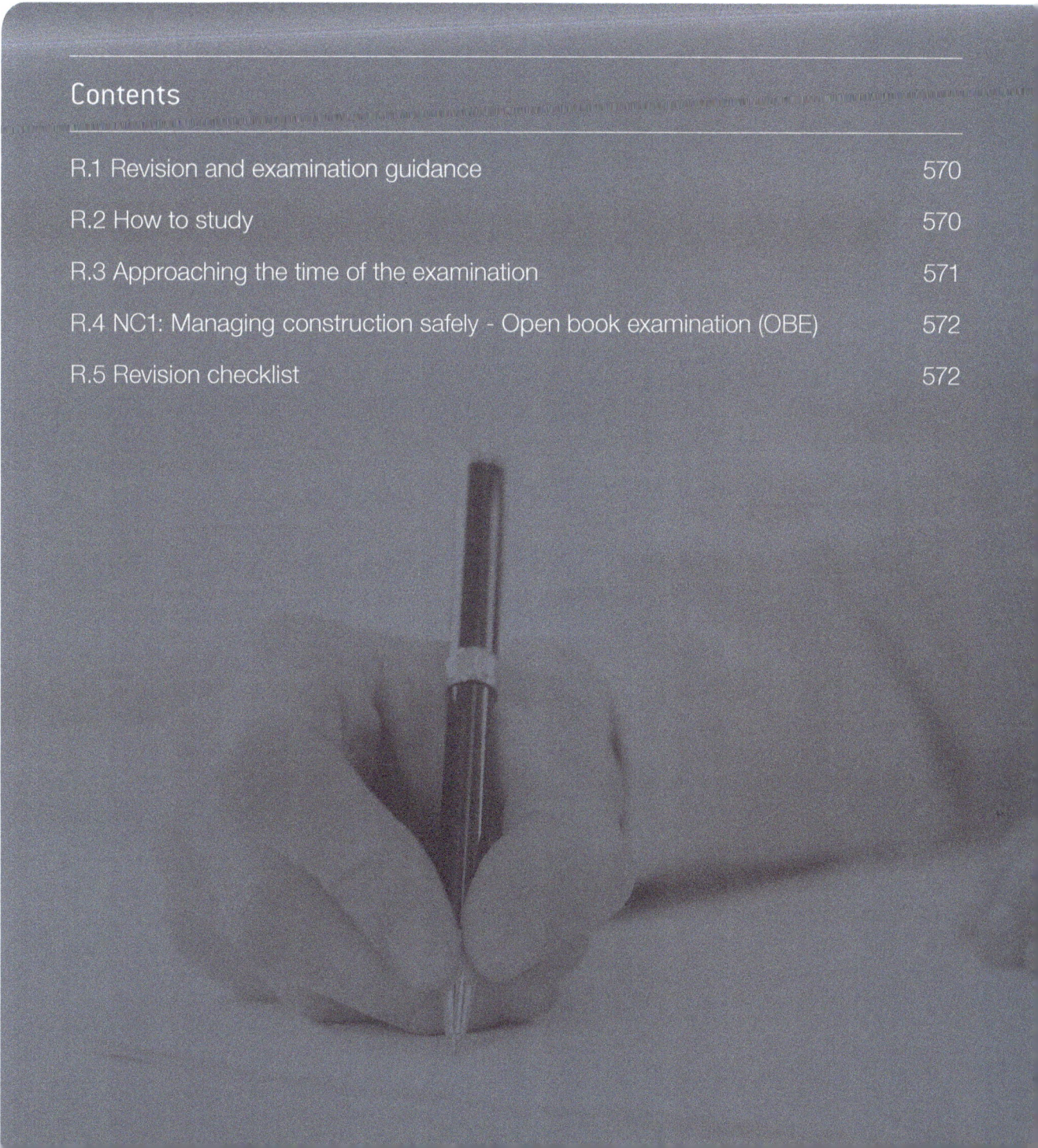

R.1   Revision and examination guidance

Being successful in gaining a NEBOSH qualification means having more than an in-depth knowledge of the subject matter, it means also having the necessary study and revision technique. Studying and revising are activities in their own right and there are tips and techniques that you can adopt which will help you to be more effective and give yourself a greater chance of success.

This section is aimed at providing some useful advice on how to improve your study skills and how to revise. It will also advise you on exam technique and help you to better organise your time.

R.2   How to study

TIME MANAGEMENT

Most people live busy lives and finding time to study and take exams can be difficult. So, do not commit to studying 'all day', make a study plan. Be realistic and disciplined - don't plan a schedule you can't manage.

Remember, too much work can be as unproductive as too little work. A good way to start is to work for 50 minutes, then give yourself a 10-minute break. Also remember to allow breaks in your day for food, relaxation and exercise, but not all at once. Physical exercise will help to increase your concentration. Even a short, brisk walk will improve your attention span.

The best time to do difficult tasks is when you are at your most productive so schedule these topics at the start of your study period. Decide on the time of day when your concentration is at its best; for some this might be first thing in the morning, so consider rising earlier and study before going to work; for others it may be late in the evening, but always remember most people's day jobs are demanding and you may well be too tired!

Another consideration might be to plan to do some work at lunch time, whatever you decide, aim to work to a regular regime that suits you and your family. Try to give each subject equal time; do not concentrate on one subject at the expense of another.

Don't be inflexible to the plan, intrusions will occur that takes priority over your plan - always leave time at the end of each study/revision period for reviewing what you have done and what you must still do.

FINDING A SUITABLE ENVIRONMENT

It is important that you work in comfortable surroundings. Your study environment should be calm and quiet and free from distraction (i.e. e-mails, mobile phone, family, TV, Facebook).

Ensure your room has adequate lighting to prevent eye strain, is at a comfortable temperature and is well ventilated with plenty of fresh air to keep you awake.

You should also ensure that you have a good chair (an upright chair is better than an armchair) and a spacious desk to take all your books and other study aids.

REVISION AND EXAM TECHNIQUE

It may have been some time since you last prepared for and took a professional exam. Remember, success in an exam depends mainly on:

- *Revision* - your ability to remember, recall and apply the information contained in this book.
- *Exam technique* - your ability to understand the questions and write good answers which answer the question, draws on relevant information from the scenario, adheres to the word count and within a reasonable timeframe.

Revision and exam technique are skills that can be learned. We will now deal with both of these skills so that you can prepare yourself for the exam. There is a saying that 'proper planning and preparation prevents a poor performance'. This was never truer than in an exam.

REVISION TIPS

You should recognise the knowledge that you already possess that is relevant to the syllabus. Don't underestimate the importance of this. But, it is never too early to start revising!

Throughout the course or learning you should have been thinking about this and noting the topics that you are finding it difficult to deal with.

- Simply reading this publication repeatedly will not normally help you to remember the information; from the start of your study revisit each day's study and abstract (handwritten preferred) the essence of the subject. This process will ensure greater retention and recall because we remember better that which we do, rather than that which we have just read. This process will identify any ambiguity you might have, early in the study period and enable you to seek clarity from your tutor, if available, before training is complete.
- As the study progresses, condense the information you have abstracted on to revision cards. This is a revision technique in itself. It means that you can carry a lot of information with you and read them in a spare moment.

Other techniques, which you may consider worth trying, include: mind maps, key words and mnemonics.

MIND MAPS/KEY WORDS

Mind maps (often referred to as a spider gram) are a powerful graphic technique which provides a universal key to unlock the potential of the brain. It uses the visualisation part of the brain to retain information in such a way that it can be quickly recovered when required. Tony Buzan is recognised by many as the inventor of mind maps in the 1960's.

Mind maps are best used for a single topic, such as learning a language, the technique is used to identify all the components necessary to achieve this and possibly the order in which they should be carried out. This part is an essential aspect of revision; the fact that you have abstracted that which is required will imbed the information into the brain and enable threads of connectivity to be established simplifying recall in the future.

The mind map should start in the centre of a blank page turned sideways. The blank page, does not restrict the brain to one dimension, as with lined paper, thus enabling the brain to work in a two-dimensional way and to spread out in all directions.

Connect the main branches to the centre start point, avoid straight lines and use lines which are curved as this increases the brains attention. The brain works best by association so that when the lines are connected to the second and third level branches you will understand and remember a lot more easily. Use images or sketches and colours throughout to make this more interesting for you and increase the brains stimulation and retention.

Use one key word per line; single key words give your mind map more power and flexibility. An example of an approach to learning a foreign language is illustrated in **Figure REG-1** below:

MNEMONICS

The use of mnemonics is another brain association technique to aid revision and memory recall. The technique relies on the use of key words to trigger memory recall, often when there are several key headings which require recall to fully answer a question.

Commonly encountered mnemonics are often used for lists and in auditory form, such as short poems, acronyms, or memorable phrases.

- I before E, except after C, and 'weird' is just weird.

A common acronym, based on People, Equipment, Material, Environment (PEME) is used to identify the considerations required to determining a safe system of work.

The acronym PEME is sufficient to trigger the answer, thus:

- A safe system of work involves the combination of People, Equipment, Materials and working Environment in the correct way to ensure a safe outcome.

R.3 Approaching the time of the examination

As the exam gets nearer, you will need to think about organising all the information from your studies in preparation for revision.

NEBOSH have advised that as part of preparations for the new open book exam that the learner completes wider reading so they don't have to do this during the examination. You should instead be able to focus on the scenario and the questions being asked, and be capable of retrieving relevant information fast.

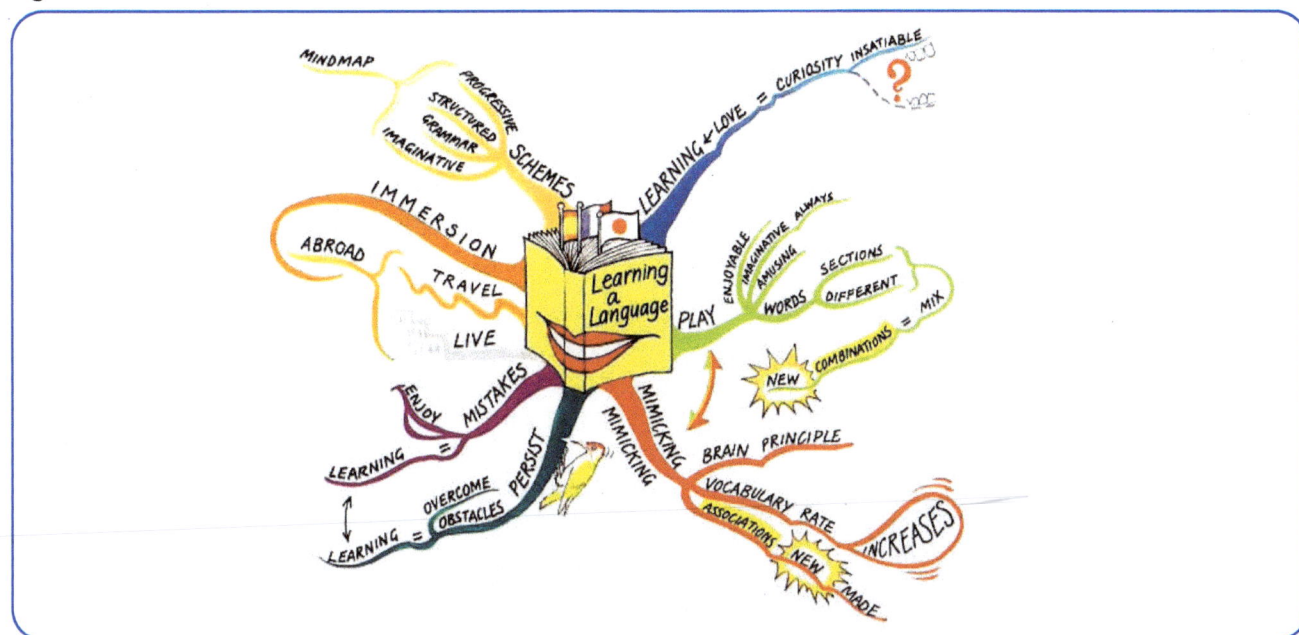

Figure REG-1: Learning a language mind map.
Source: Tony Buzan.

Therefore, it is recommended to download any relevant sources of reference associated with the syllabus prior to the exam, please refer to the NEBOSH website www.nebosh.org.uk/references/

A range of resources have been made available by NEBOSH to assist in the preparation for the examination, including a sample exam paper, video tutorials, and OBE learner guides. Please refer to www.nebosh.org.uk/open-book-examinations/resources/

We recommend that you refer to the revision checklist located at the end of this section, which contains the syllabus content (Elements 1-13). If a topic is in the syllabus then it is possible that there will be an examination question on that topic. You can assess your level of knowledge and recall against the revision checklist. Look at the content listed for each element in the revision checklist. Ask yourself the following question:

> "If there is a question in the exam about that topic. Could I answer it?"

This technique will help you to identify the areas of weakness and then to prepare a plan of work which concentrates on the knowledge that you need to develop. You can rate your current level of knowledge for each topic in each element of the syllabus guide by using the categories 'High', 'Medium' and 'Low'. Use your ranking as an indication of your personal strengths and weaknesses.

For example, if you scored 'High' for the topic, then there is little work to do on that subject to complete your understanding. However, if you scored yourself as low for the subject then you have identified an area of weakness to focus upon.

R.4 — NC1: Managing construction safely - Open book examination (OBE)

As detailed in the earlier 'Assessment' section of this study book, the NC1 assessment is an open book examination, which will be based on a realistic scenario, that will test both what you know and what you can do. You will be asked to carry out a series of tasks using evidence from the scenario where possible and the underpinning knowledge gained through your studies and revision. You will have 48 hours to download, complete and submit your assessment.

It is recommended that you organise the following prior to the commencement of your exam:

- A desk or table that gives space to refer to reading materials and prepare answers.

- A comfortable chair.
- Good lighting.
- Comfortable temperature.
- A clock to monitor the time and meet the submission deadline (the deadline will be UK local time so if you are completing the examination outside of the UK you will need to make sure you consider any time zone differences).
- A supply of refreshments and snacks to maintain energy levels.

Finally, before the exam you should check the following:

- Have you agreed an NG1 OBE date and registered with your Learning Partner?
- Have you revised and prepared for your OBE having also referred to a range of resources provided on the NEBOSH website at www.nebosh.org.uk/obe-resources
- Have you provided an accurate email address to enable OBE registration?
- Have you received a confirmation of registration email approx. 2 weeks before your chosen OBE?
- Have you received your logon email from NEBOSH with a unique username and temporary password? (approx. 8 working days before your OBE)
- Have you logged in, created a new password and familiarised yourself with the platform?
- Have you received details for your closing interview?

Good luck, but always remember:

> "The greatest barrier to success is the fear of failure."

Source: Sven Goran Eriksson.

R.5 — Revision checklist

The revision checklist is designed to help those wishing to take the NEBOSH Health and Safety Management for Construction. By now you should have covered the requirements necessary to fulfil the syllabus requirements and have a good understanding in readiness for the OBE.

The purpose of the following revision checklist is for you to review the topics against your understanding. Column 1 is the element and sub element number. Column 2 is the element and sub element topics. Column 3 and 4 are for you to identify if you are generally comfortable at that level (OK) or not (require more revision – Rev).

	Element topics	Ok	Rev
1.0	**The foundations of construction health and safety management**		
1.1	**Morals and money**		
	• Moral expectations of good standards of health and safety.		
	• The financial costs of incidents (insured and uninsured costs)		
1.2	**The Construction (Design and Management) Regulations 2015**		
	Roles, competence and duties of the following: • Client • Principal designer • Designer • Principal contractor • Contractors • Workers • Domestic clients		
	When the HSE need to be notified		
	Pre-selection and management of contractors, including third-party auditing schemes		
	Effective planning and co-ordination of contracted work, including interaction with existing staff		
	Preparation of pre-construction information, construction phase plan, health and safety file (including the purpose, requirements and an example of a plan)		
1.3	**Type, range and issues relating to construction activities**		
	Types of construction work and range of activities: construction, alteration and maintenance of premises; demolition or dismantling; clearance; excavation; structural work; site movements; service maintenance		
	Why you need to maintain the stability of structures		
1.4	**Site assessment and control measures**		
	Initially assessing the site: historical and current use, likelihood of asbestos, and contaminants		
	Area of site, topography and features of the surrounding area		
	Site control measures: site planning, preparation for specialist activities, security and client/occupier arrangements		
1.5	**Site order and security**		
	The need for safe entry and exit from the site		
	Safe and suitable arrangement of the working space, including housekeeping arrangements		
	The requirement to identify the site perimeter, either with suitable signs or fencing		
	Any out-of-hours security arrangements (if necessary)		
1.6	**Management of temporary works**		
	Management of parts of the works that allow or enable construction of, protect, support or provide access to, the permanent works (which may or may not remain in place at the completion of the works) eg falsework/ formwork, excavations and temporary equipment foundations.		
1.7	**Other construction issues including welfare arrangements**		
	Welfare requirements for: • Toilets and washing facilities • Changing rooms and lockers • Rest and eating facilities • Drinking water		

		Element topics	Ok	Rev
		The types of temporary accommodation units (TAU) required for sites		
		Requirements of location for TAU		
		Particular construction issues relating to: • Use of migrant workers • Temporary nature of construction activities and the constantly changing workplace • Time pressures • Weather conditions • Levels of numeracy and literacy of workers • Non-English speaking workers		©RMS
2.0		**Improving health and safety culture and assessing risk**		
2.1		**Health and safety culture**		
		Meaning of the term 'health and safety culture'		
		Relationship between health and safety culture and health and safety performance		
		Influence of peers on health and safety culture		
2.2		**How human factors influence behavior positively or negatively**		
		Organisational factors, including: culture, leadership, resources, work patterns, communications		
		Job factors, including: task, workload, environment, display and controls, procedures		
		Individual factors, including: competence, skills, personality, attitude and risk perception		
		Link between individual, job and organisational factors		
2.3		**Improving health and safety culture**		
		Gaining management committment		
		Promoting health and safety standards by leadership and example and appropriate use of disciplinary procedures		
		Competent workers		

		Good communication within the organisation
		
		• Benefits and limitations of different methods of communication (verbal, written and graphic)
		• Use and effectiveness of noticeboards and health and safety media
		• Co-operation and consultation with the workforce and contractors, including:
		  • Appointment, functions and entitlements of worker representatives (trade union appointed and elected)
		  • Benefits of worker participations (including worker feedback)		

		When training is needed: • Induction (key health and safety topics to be covered) • Job change • Process change • Introduction of new legislation • Introduction of new technology		
2.4		**Assessing risk**		
		Meaning of hazard, risk and risk assessment		
		Purpose of risk assessment and the 'suitable and sufficient' standard it needs to reach (see HSG65: 'Managing for health and safety')		

	Element topics	Ok	Rev

	A general approach to risk assessment:
	
	 • Identify hazards: - Sources and form of harm; sources of information to consult; use of task analysis, legislation, manufacturers' information, incident data, guidance
	
	 • Identify people at risk: Including workers, operators, maintenance staff, cleaners, contractors, visitors, public
	
	 • Evaluate risk (taking account of what you already do) and decide if you need to do more:
	
	    • Likelihood of harm and probable severity
	    • Possible acute and chronic health effects
	    • Risk rating
	    • Principles to consider when controlling risk (Regulation 4 and Schedule 1 of the Management of Health and Safety at Work Regulations 1999)
	    • Practical application of the principles – applying the general hierarchy of control (clause 8.1.2 of ISO 45001:2018)
	    • Application based on prioritisation of risk
	    • Use of guidance; sources and examples of legislation
	    • Applying controls to specified hazards
	    • Residual risk; acceptable/tolerable risk levels
	    • Distinction between priorities and timescales
	 • Record significant findings
	
	 • Reasons for review		

	Application of risk assessment for specific types of risk and special cases: • Examples of when they are required, including fire, DSE, manual handling, hazardous substances, noise, vibration • Why specific risk assessment methods are used for certain risks		
	Special case applications to young people, expectant and nursing mothers; also consideration of disabled workers and lone workers (see Regulations 16, 18 and 19 of the Management of Health and Safety at Work Regulations 1999*)		
3.0	Managing change and procedures		
3.1	Managing change		
	Typical types of change faced in the workplace and its possible impact, including: construction works, change of process, change of equipment, change in working practices		
	Managing the impact of change: • Communication and co-operation • Risk assessment • Appointment of competent people • Segregation of work areas • Amendment of emergency procedures • Welfare provision		
	Review of change (during and after)		
3.2	Safe systems of work for general work activities		
	Why workers should be involved when developing safe systems of work		
	Why procedures should be recorded/written down		
	The differences between technical, procedural and behavioural controls		

	Element topics	Ok	Rev
	Developing a safe system of work:		
	• Analysing tasks, identifying hazards and assessing risks • Introducing controls and formulating procedures • Instruction and training in how to use the system		
	Monitoring the system		©RMS
3.3	**Permit-to-work systems**		
	Meaning of a permit-to-work system		
	Why permit-to-work systems are used		
	How permit-to-work systems work and are used		
	When to use a permit-to-work system, including: hot work, work on non-live (isolated) electrical systems, machinery maintenance, confined spaces, work at height		
3.4	**Emergency procedures**		
	Why emergency procedures need to be developed		
	What arrangements must be made when planning emergency procedures and first aid provisions (with reference to Regulation 30 of the CDM Regulations)		
	Principles of fire evacuation, including: means of escape, emergency evacuation procedures, role and appointment of fire marshals/wardens, fire drills and provisions for people with disabilities		
	Suitable emergency arrangements when working near water		
	Continual review of emergency procedures as a build continues		
	Inclusion of TAU within the emergency plan		
3.5	**Learning from incidents**		
	The different levels of investigations: minimal, low, medium and high (see HSG245)		
	Basic incident investigation steps: • Step one: gathering the information • Step two: analysing the information • Step three: identifying risk and control measures • Step four: the action plan and its implementation		
	How fatalities, specific injuries, 'over 3- or 7-day injuries', diseases and dangerous occurrences must be recorded and reported		
4.0	**Excavation work hazards and assessment**		
4.1	**Managing change**		
	The hazards of work in and around excavations: buried services, falls of people/equipment/material into excavation, collapse of sides, collapse of adjacent structures, water ingress, use of cofferdams and caissons, contaminated ground, toxic and asphyxiating atmospheres, mechanical hazards		
	Overhead hazards, including power lines (cross-reference to electricity)		
	Risk assessment: factors to consider (depth, soil type, type of work, use of mechanical equipment, proximity of roadways/structures/etc, presence of public, weather, etc)		

	Element topics	Ok	Rev
4.2	**Control measures for excavation work**		
	Controls		
	• Identification/detection and marking of buried services; safe digging methods • Methods of supporting excavations (eg steel sheets, support boxes) • Means of access • Crossing points • Barriers, lighting and signs • Safe storage of spoil • De-watering methods, including well points and sump points • Positioning and routeing of vehicles, plant and equipment • Personal protective equipment		
	Particular requirements for contaminated ground (soil testing, welfare facilities, health surveillance, etc)		
	Inspection requirements for excavations and excavation support systems (see above)		
4.3	**Safe working in confined spaces**		
	Types of confined spaces and why they are dangerous		
	The main hazards associated with working within a confined space		
	What should be considered when assessing risks from a confined space		
	The precautions to be included in a safe system of work for confined spaces		
	When a permit-to-work for confined spaces would not be required		
5.0	**Demolition**		
5.1	**Demolition and dismantling hazards**		
	The meaning of:		
	• Deconstruction • Piecemeal demolition • Deliberate controlled collapse		
	Selection of the appropriate method		
	Hazards and control measures relating to deconstruction and dismantling		
5.2	**Purpose and scope of pre-demolition, deconstruction or refurbishment survey**		
	Duties of the property owner for carrying out a pre-demolition survey for both notifiable and non-notifiable demolition projects		
	Identification of key structural elements including pre and post tensioned components		
	Identification of location and type of services		
	Identification, significance and extent of any dilapidation of the structure		
	Review of drawings, structural calculations and health and safety file related to the structure		
	Review of all structural alterations carried out on the structure in the past		
6.0	**Mobile plant and vehicles**		
6.1	**Safe movement of people**		
	Hazards to pedestrians:		
	• Being struck by moving, flying or falling objects • Collisions with moving vehicles • Striking against fixed or stationary objects		
6.2	**Safe use of vehicles and plant**		

		Element topics	Ok	Rev
		Hazards from workplace transport operations and plant (vehicle movement, non-movement)		
		Control measures to manage workplace transport: • Safe site (design and activity) • Suitability of traffic routes (including site access and egress pedestrian-only zones and crossing points) • Spillage control • Management of vehicle movements • Environmental considerations: visibility/lighting, gradients, changes of level, surface conditions (use of non-slip coatings) • Maintenance and checking of traffic routes • Segregating pedestrians and vehicles and measures to be taken when segregation is not practicable • Protective measures for people and structures (barriers, marking signs, warnings of vehicle approach and reversing) • Site rules (including speed limits) and signage		
		• Safe vehicles • Vehicle selection • Vehicle inspection and maintenance		
		• Safe drivers • Selection and training of drivers • Types of reversing assistant		
6.3		**Work-related driving**		
		Managing work-related driving Plan • Assess the risks • Policy • Top management taking into account work-related driving • Roles and responsibilities Do • Co-operation between departments (where relevant) • Adequate systems in place, including maintenance strategies • Communication and consultation within the workforce • Provision of adequate instruction and training Check • Monitor performance (to ensure the policy is working correctly) • Ensure all workers report work-related road incidents or near misses Act • Review performance and learn from experience • Regularly update the policy		
		Work-related driving control measures: • Safe driver (competence – checks on level of skill/experience, validity of driving licence; provision of instruction; fitness to drive) • Safe vehicle (vehicles fit for purpose for which they are being used; maintained in a safe condition – checks on MOT/service history; adequate restraints for securing goods) • Safe journey (planning of routes; realistic work schedule – enough time to complete the journey safely, allowing for driving breaks; consideration of weather conditions; consideration of legal driving hours where relevant)		

	Element topics	Ok	Rev
7.0	**Working at height**		
7.1	**Working at height hazards and controls**		
	The risks of working at height, including vertical distance, fragile roofs, deterioration of materials, unprotected edges, unstable/poorly maintained access equipment, weather conditions and falling materials		
	Approach to working safely at height: Avoid working at height by, for example, using extendable tools to work from ground level; assembly of components/equipment at ground level Prevent a fall from occurring by using an existing workplace that is known to be safe, such as a solid roof with fixed guardrails; use of suitable equipment such as mobile elevating work platforms (MEWPs), scaffolds; work restraint systems Minimise the distance and/or consequence of a fall, by collective measures such as safety nets and airbags installed close to the level of work, and personal protective measures such as fall-arrest systems		
	Main precautions necessary to prevent falls and falling materials, including proper planning and supervision of work, avoiding working in adverse weather conditions		
	Emergency rescue		
	Provision of training, instruction and other measures		
7.2	**Safe working practices for access equipment and roof work**		
	Scaffolding • Design features of independent tied, putlog, fan, cantilevered and mobile tower scaffolds • Safety features (including sole-boards, base plates, toe-boards, guardrails, boarding, brick guards, debris netting) • Requirements for scaffold erectors • Means of access • Design of loading platforms • Scaffold hoists (people, materials) • Ensuring stability: effects of materials, weather, sheeting, etc; protection from impact of vehicles; inspection requirements		
	Use of ladders, stepladders, trestles, staging platforms, leading edge protection system, MEWPs and mast climbing work platforms (MCWPs)		
	Other techniques: • Boatswain's chair and training required for use • Cradles (including suspension from cranes) • Rope access		
	How fall arrest equipment is used and what its aims are: • Harnesses • Safety nets • Soft landing systems • Crash decks • Emergency procedures (including rescue) • Suspension fainting in the use of rope and harnesses		
	Roof work: • Means of access • Edge and leading edge protection • Crawling boards • Fall arrest equipment (as above)		
7.3	**Protection of others**		

		Element topics	Ok	Rev
		Demarcation, barriers, tunnels, signs		
		Marking, lighting		
		Sheeting, netting and fans		
		Head protection		
8.0		**Musculoskeletal health and load handling**		
8.1		**Musculoskeletal disorders and work-related upper limb disorders**		
		Meaning of musculoskeletal disorders (MSDs) and work-related upper limb disorders (WRULDs)		
		Examples of repetitive construction activities that can cause MSDs and WRULDs		
		Possible ill-health conditions from poorly designed tasks and workstations		
		Avoiding/minimising risks from poorly designed tasks and workstations by considering: • Task (including repetitive, strenuous) • Environment (including lighting, glare) • Equipment (including user requirements, adjustability, matching the workplace to individual needs of workers)		
8.2		**Manual handling hazards and control measures**		
		Common types of manual handling injuries		
		Good handling technique for manually lifting loads		
		Avoiding/minimising manual handling risks by considering the task, the individual, the load and the working environment		
8.3		**Load-handling equipment**		
		Hazards and controls for common types of load-handling aids and equipment: cranes; hoists; fork-lift trucks; pallet trucks; sack trucks; telescopic handlers		
		Requirements for lifting operations: • Strong, stable and suitable equipment • Positioned and installed correctly • Visibly marked with safe working load • Lifting operations are planned, supervised and carried out in a safe manner by competent people • Special requirements for lifting equipment used for lifting people		
		Periodic inspection and examination/testing of lifting equipment		
9.0		**Work Equipment**		
9.1		**General requirements for work equipment**		
		Providing suitable equipment, including the requirement for CE marking within the UK and Europe		
		Preventing access to dangerous parts of machinery		
		When the use and maintenance of equipment with specific risks needs to be restricted		
		Providing information, instruction and training about specific risks to people at risk, including users, maintenance staff and managers		
		Why equipment should be maintained and maintenance conducted safely		
		Emergency operations controls, stability, lighting, marking and warnings, clear workspace		
9.2		**Hand-held tools**		
		General considerations for selecting hand-held tools (whether powered or manual): • Requirements for safe use • Condition and fitness for use • Suitability for purpose • Locations to be used in (including flammable atmosphere)		

	Element topics	Ok	Rev
	Hazards of a range of hand-held tools (whether powered or manual) and how these hazards are controlled		
9.3	**Machinery hazards and control measures**		
	Potential consequences as a result of contact with, or exposure to, mechanical or other hazards as identified in ISO 12100:2010 (Table B.1)		
	Hazards of a range of site equipment, including: strimmer; chainsaw; cement mixer; bench-mounted circular saw; compressor; plate compactor; ground consolidation equipment; road-marking equipment; electrical generators; drones; driver-less vehicles		
	Control measures of the above equipment and the basic requirements for guards and safety devices		
9.4	**Working near water**		
	Additional appropriate control measures when working near water (including buoyancy aids and safety boat)		
10.0	**Electricity**		
10.1	**Hazards and risks**		
	Risks of electricity: • Electric shock and its effect on the body; factors that effect severity (voltage, frequency, duration, resistance, current path); electrical burns (from direct and indirect contact with an electrical source) • Common causes of electrical fires, including portable devices overheating during charging • Workplace electrical equipment including portable items: what is likely to lead to accidents (unsuitable equipment inadequate maintenance; use of defective/poorly maintained electrical equipment; use of electrical equipment in wet environments) • Secondary effects, including falls from height • Work near overhead power lines and contact with underground power cables during excavation work • Work on mains electricity supplies		
10.2	**Control measures**		
	Protection of conductors		
	Strength and capability of equipment		
	Advantages and limitations of protective systems: fuses, earthing, isolation of supply, double insulation, residual current devices, reduced and low voltage systems		
	Use of competent people		
	Use of safe systems of work (no live working unless no other option, isolation, locating buried services, protection against overhead cables)		
	Emergency procedures following an electrical incident		
	Inspection and maintenance strategies: user checks, formal inspection and tests of the electrical installation and equipment; frequency of inspection and testing; records of inspection and testing; advantages and limitations of in-service inspection and testing of electrical equipment		
10.3	**Control measures for working underneath or near overhead power lines**		
	Legal requirements for working near power lines		
	How to prevent line contact accidents through management, planning and consultation with relevant parties		
	Use of barriers to establish a safety zone when working near overhead lines		
	Means of safely passing underneath overhead lines		
	Key emergency procedures if someone or something comes into contact with an overhead line		

	Element topics	Ok	Rev
10.4	**Control measures for working near underground power cables**		
	Planning the work • The provision of pre-construction information		
	Using cable plans • Obtaining and reviewing plans before any excavation work starts • What you should do if the information cannot be obtained • Use of equipment for detecting/locating buried services		
	Use of service locating devices		
	Safe digging practices		
	The use of appropriate tools, locating devices and route planning when undertaking excavation work, including vacuum, air and hydro excavation		
11.0	**Fire**		
11.1	**Fire principles**		
	The fire triangle: sources of ignition, fuel and oxygen in a construction workplace; oxidising materials		
	Classification of fires (A, B, C, D and F) and electrical fires		
	Basic principles of heat transmission and fire spread; convection, conduction, radiation and direct burning		
	Common causes and consequences of fires within the construction industry		
11.2	**Preventing fire and spread**		
	Control measures to minimise the risk of fire starting in a construction workplace: • Eliminate/reduce quantities of flammable and combustible materials used or stored; storage of highly flammable materials • Control ignition sources, including suitable electrical equipment in flammable atmospheres Use safe systems of work • Good housekeeping • Structural measures to prevent the spread of fire and smoke: properties of common building materials (including fire doors); compartmentation; protection of openings and voids		
11.3	**Fire alarms and fire-fighting**		
	Common fire detection and alarm systems		
	Portable fire-fighting equipment: siting, maintenance and training requirements		
	Extinguishing media: water, foam, dry powder, carbon dioxide, wet chemical; advantages and limitations		
	Access for fire and rescue services and vehicles		
12.0	**Chemical and biological agents**		
12.1	**Hazardous substances**		
	Forms of chemical agent: dust, fibres, fumes, gases, mists, vapours and liquids		
	Forms of biological agents: fungi, bacteria, viruses		
	Health hazards classifications: acute toxicity; skin corrosion/irritation; serious eye damage/eye irritation; respiratory or skin sensitisation; germ cell mutagenicity; carcinogenicity; reproductive toxicity; specific target organ toxicity (single and repeated exposure); aspiration hazard		
12.2	**Assessment of health risks**		
	Routes of entry of hazardous substances into the body		
	Factors that need to be taken into account when assessing health risks		

		Element topics	Ok	Rev
		Sources of information: • Product labels • Safety data sheets (who must provide them and information that they must contain)		
		Limitations of information used when assessing risks to health		
		Role and limitations of hazardous substance monitoring		
		Purpose of occupational exposure limits and how they are used		
12.3		**Control measures**		
		The need to prevent exposure or, where this is not reasonable practicable, adequately control it		
		Principles of good practice (see Control of Substances Hazardous to Health Regulations)		
		Common measures used to implement the Principles of Good Practice above		
		Additional controls that are needed for substances that can cause cancer, asthma or genetic damage that can be passed from one generation to another		
12.4		**Specific agents**		
		Health risks, controls and likely workplace activities/locations where the following specific agents can be found • Blood-borne viruses • Carbon monoxide • Cement • Legionella • Leptospira • Silica • Wood dust • Tetanus		
		Health risks from and controls for working with asbestos		
		Duty to manage asbestos: • Asbestos identification (types of survey and who can undertake them, where it can be located) • Procedure for the discovery of asbestos during construction activity • Requirements if people are accidentally exposed to asbestos materials • Requirements for removal (non-licensed, licensed, notification and plan of work) • Respiratory equipment, protective clothing, training, air monitoring and medical surveillance • Requirements for disposal (licenced carrier, notification, licenced disposal site)		
13.0		**Chemical and biological agents**		
13.1		**Noise**		
		The physical and psychological effects of exposure to noise		
		The meaning of commonly used terms in the measurement of sound: sound pressure, intensity, frequency, the decibel scale, dB(A) and dB(C)		
		When exposure should be assessed; comparison of measurements to exposure limits established by recognised standards		
		Basic noise control measures including: isolation, absorption, insulation, damping and silencing; the purpose, application and limitations of personal hearing protection (types, selection, use, maintenance and attenuation factors)		
		Role of health surveillance		
13.2		**Vibration**		
		The effects on the body of exposure to hand-arm vibration and whole body vibration		

	Element topics	Ok	Rev
	When exposure should be assessed; comparison of measurements to exposure limits established by recognised standards		
	Basic vibration control measures including: alternative methods of working (mechanisation where possible); low vibration emission tools; selection of suitable equipment; maintenance programmes; limiting the time workers are exposed to vibration (use of rotas, planning work to avoid long periods of exposure); suitable PPE		
	Role of health surveillance		
13.3	**Radiation**		
	The types of, and differences between, non-ionising and ionising radiation (including radon) and their health effects		
	Typical occupational sources of non-ionising and ionising radiation		
	The basic ways of controlling exposures to non-ionising and ionising radiation		
	Basic radiation protection strategies, including the role of the competent person in the workplace		
	The role of monitoring and health surveillance		
13.4	**Mental ill-health**		
	The frequency and extent of mental ill-health in the construction industry		
	Recognising common symptoms of workers with mental ill-health: depression; anxiety/panic attacks, posttraumatic stress disorder (PTSD)		
	The causes of, and controls for, work-related mental ill-health (see the HSE's Management Standards): • Demands • Control • Support • Relationships • Role • Change		
	Recognition that most people with mental ill-health can continue to work effectively		
	Awareness of organisations which provide mental well-being support to those in the construction industry		
13.5	**Violence at work**		
	Types of violence at work including: physical, psychological, verbal, bullying and harassment		
	Effective management of violence at work, according to HSE Violence at Work, A Guide for Employers		
13.6	**Substance abuse at work**		
	Risks to health and safety from substance abuse at work (alcohol, legal/illegal drugs and solvents)		
	Managing substance abuse at work: • What the issues are and what to look out for • Consult employees • Consider safety-critical work • Develop a policy • Screening • Supporting employees with a substance abuse problem		

INDEX